Environmental Footprints and Eco-design of Products and Processes

Series Editor

Subramanian Senthilkannan Muthu, Head of Sustainability - SgT Group and API, Hong Kong, Kowloon, Hong Kong

Indexed by Scopus

This series aims to broadly cover all the aspects related to environmental assessment of products, development of environmental and ecological indicators and eco-design of various products and processes. Below are the areas fall under the aims and scope of this series, but not limited to: Environmental Life Cycle Assessment; Social Life Cycle Assessment; Organizational and Product Carbon Footprints; Ecological, Energy and Water Footprints; Life cycle costing; Environmental and sustainable indicators; Environmental impact assessment methods and tools; Eco-design (sustainable design) aspects and tools; Biodegradation studies; Recycling; Solid waste management; Environmental and social audits; Green Purchasing and tools; Product environmental footprints; Environmental management standards and regulations; Eco-labels; Green Claims and green washing; Assessment of sustainability aspects.

More information about this series at https://link.springer.com/bookseries/13340

Aldo Alvarez-Risco ·
Subramanian Senthilkannan Muthu ·
Shyla Del-Aguila-Arcentales
Editors

Circular Economy

Impact on Carbon and Water Footprint

Springer

Editors
Aldo Alvarez-Risco
Universidad de Lima
Lima, Peru

Subramanian Senthilkannan Muthu
SgT Group & API
Hong Kong, Kowloon, Hong Kong

Shyla Del-Aguila-Arcentales
Escuela Nacional de Marina Mercante
"Almirante Miguel Grau"
Callao, Peru

ISSN 2345-7651 ISSN 2345-766X (electronic)
Environmental Footprints and Eco-design of Products and Processes
ISBN 978-981-19-0548-3 ISBN 978-981-19-0549-0 (eBook)
https://doi.org/10.1007/978-981-19-0549-0

This Springer imprint is published by the registered company Springer Nature Singapore Pte Ltd.
The registered company address is: 152 Beach Road, #21-01/04 Gateway East, Singapore 189721, Singapore

Preface

Although governments, companies, and citizens have been motivated for years, few efforts have been made to achieve concrete results related to sustainability. Various indicators, programs, laws, initiatives, and other activities have been generated, but little has been achieved. However, the world transformed by the COVID-19 pandemic has increased the interest in environmental issues since it has been possible to understand the need for efficient use of resources due to the profound economic impact caused by the pandemic. Since the signing of the Sustainable Development Goals in 2015, different orientations were created to contribute to environmental content due to the global need, for the eco-efficient use of water, air, land, and other natural resources. With these efforts, the concept of ecological footprint gains much importance.

This book aims to provide the theoretical and practical support needed to implement a circular economy and a footprint linked to water, food, and other sources. These can guide other institutions and policymakers in establishing regulations and implementation and monitoring processes.

Part I presents the circular economy approach, mentioning the strategies, the aspects of deglobalization due to the pandemic, and the necessary leadership needed for the global crisis to be successfully addressed at both the governmental and business levels. Chapter "Measuring Circular Economy" presents various strategies for measuring the circular economy and the progress being made in the world. In Chapter "Better Students, Better Companies, Better Life: Circular Learning", a promising approach to optimizing the education of students in universities is presented so that the improvement of their academic training is accompanied by better results in companies and, ultimately, a better individual and collective life; in other words, the origin of circular learning to build a circular world is proposed. Chapter "Leadership for Sustainability in Crisis Time" shows the global leadership approach needed at different levels, adapted to the current activities of companies, and presents resources for future research in this field and different green certification. The end of this chapter shows the scientific evidence of efforts to comply with the Paris Agreement and recent evidence of the level of compliance in countries. Chapter "Circular Economy for Food Loss Reduction and Water foodprint" discusses that food availability is a critical issue in the world and requires substantial improvement of processes, including the use of new technologies, while at the same time using water efficiently during the food production process. Chapter "3D Print, Circularity and Footprints" presents a very innovative and current topic such as 3D printing, showing the contributions in the saving of resources generated by this technology and the contribution achieved in reducing pollution in the construction sector.

In part II, the applications of footprints in different approaches are presented in more detail. Thus, Chapter "Circular Economy for Packaging and Carbon Footprint" describes the efforts and achievements of the circular economy in product packaging and its relationship with the impact on the carbon footprint through these production processes. Chapter "Circular Economy for Waste Reduction and Carbon Footprint" shows the management of waste in terms of circular economy, detailing the regulation, programs, and impact on carbon footprint. In the case of Chapter "Food Loss Reduction and Carbon Footprint Practices Worldwide: A Benchmarking Approach of Circular Economy", it is described food loss reduction programs because food management is an old and big problem that needs to follow the circular economy guidelines to contribute to global practices and can impact the achievement of better values in the carbon footprint. It is known that the management of clothing manufacturing usually has a great challenge to ensure that it is carried out circularly and that, in addition, waste management is also a necessity that current processes are needed. The description of these contents is seen in Chapter "Fashion and Textile Circularity and Waste Footprint". Chapter "Material Selection for Circularity and Footprints" highlights the need for material selection in companies since more and more new materials have better attributes and are mainly less polluting.

Part III shows substantial efforts and, above all, the innovation of tourist and educational activities. Chapter "Water Footprint in the Textile and Food Supply Chain Management: Trends to Become Circular and Sustainable" describes advances in water management in clothing manufacturing and the food supply chain. These massive products are essential because the ecological management of these production processes has a tremendous global impact. Chapter "Virtual Tourism, Carbon Footprint and Circularity" and Chapter "Virtual Education: Carbon Footprint and Circularity" address how tourism and education have been modified, leading to a global offer via virtual and how a positive impact has been achieved in reducing the ecological footprint on consumers. However, it remains to be discussed whether a more significant benefit can be achieved with these practices and, at the same time, project what else must be planned so that tourist and educational activities in the blended world are successful.

All chapters were reviewed. Some authors assisted in reviewing chapters written by others.

This book shows the experience and knowledge of authors and editors in a highly relevant and timely global change movement: the circular economy and footprint. We expect the book to guide governments, universities, schools, multinational and local companies, and citizens. Finally, the circular economy and footprint are concrete ways to contribute to the UN Sustainable Development Goals, which allow circular international business and contribute with footprint goals.

Lima, Peru — Dr. Aldo Alvarez-Risco
Hong Kong, Hong Kong — Subramanian Senthilkannan Muthu
Lima, Peru — Shyla Del-Aguila-Arcentales, M.Sc.

About the Authors, Reviewers, Editors, and Their Contributions

This book would not have been possible without the incredible and dedicated efforts and contributions of its many authors, who wrote chapters and often assisted in reviewing the chapters of others.

Camila Almanza-Cruz is the coauthor of Chap "Virtual Education: Carbon Footprint and Circularity".

Aldo Alvarez-Risco is coauthor of Chaps. "Measuring Circular Economy", "Better Students, Better Companies, Better Life: Circular Learning", "Leadership for Sustainability in Crisis Time", "Circular Economy for Food Loss Reduction and Water Foodprint", "3D Print, Circularity and Footprints", "Circular Economy for Packaging and Carbon Footprint", "Circular Economy for Waste Reduction and Carbon Footprint", "Food Loss Reduction and Carbon Footprint Practices Worldwide: A Benchmarking Approach of Circular Economy", "Fashion and Textile Circularity and Waste Footprint", "Material Selection for Circularity and Footprints", "Water Footprint in the Textile and Food Supply Chain Management: Trends to Become Circular and Sustainable", "Virtual Tourism, Carbon Footprint and Circularity", "Virtual Education: Carbon Footprint and Circularity". Also, he is one of the editors and reviewers of the book.

Maria de las Mercedes Anderson-Seminario is the coauthor of Chaps. "Better Students, Better Companies, Better Life: Circular Learning", "Circular Economy for Food Loss Reduction and Water foodprint", "3D Print, Circularity and Footprints", "Circular Economy for Packaging and Carbon Footprint", "Circular Economy for Waste Reduction and Carbon Footprint", "Food Loss Reduction and Carbon Footprint Practices Worldwide: A Benchmarking Approach of Circular Economy", "Fashion and Textile Circularity and Waste Footprint", "Material Selection for Circularity and Footprints", "Water Footprint in the Textile and Food Supply Chain Management: Trends to Become Circular and Sustainable", "Virtual Tourism, Carbon Footprint and Circularity", "Virtual Education: Carbon Footprint and Circularity".

Marián Arias-Meza is the coauthor of Chap "Fashion and Textile Circularity and Waste Footprint".

Marco Calle-Nole is the coauthor of Chap "Virtual Education: Carbon Footprint and Circularity".

Nilda Campos-Dávalos is the coauthor of Chaps "Water Footprint in the Textile and Food Supply Chain Management: Trends to Become Circular and Sustainable" and "Virtual Education: Carbon Footprint and Circularity".

Sarahit Castillo-Benancio is the coauthor of Chaps "Circular Economy for Packaging and Carbon Footprint" and "Material Selection for Circularity and Footprints".

Anguie Contreras-Taica is the coauthor of Chaps "Circular Economy for Food Loss Reduction and Water foodprint", and "Virtual Education: Carbon Footprint and Circularity".

Berdy Briggitte Cuya-Velásquez is the coauthor of Chaps "Circular Economy for Food Loss Reduction and Water foodprint" and "Fashion and Textile Circularity and Waste Footprint".

Myreya De-la-Cruz-Diaz is the coauthor of Chaps "3D Print, Circularity and Footprints" and "Virtual Tourism, Carbon Footprint and Circularity".

Shyla Del-Aguila-Arcentales is the coauthor of Chaps. "Measuring Circular Economy", "Better Students, Better Companies, Better Life: Circular Learning", "Leadership for Sustainability in Crisis Time", "Circular Economy for Food Loss Reduction and Water foodprint", "3D Print, Circularity and Footprints", "Circular Economy for Packaging and Carbon Footprint", "Circular Economy for Waste Reduction and Carbon Footprint", "Food Loss Reduction and Carbon Footprint Practices Worldwide: A Benchmarking Approach of Circular Economy", "Fashion and Textile Circularity and Waste Footprint", "Material Selection for Circularity and Footprints", "Water Footprint in the Textile and Food Supply Chain Management: Trends to Become Circular and Sustainable", "Virtual Tourism, Carbon Footprint and Circularity", "Virtual Education: Carbon Footprint and Circularity". Also, he is one of the editors and reviewers of the book.

Santiago Diaz-Risco is the coauthor of Chap "Leadership for Sustainability in Crisis Time".

Sharon Esquerre-Botton is the coauthor of Chaps "Circular Economy for Packaging and Carbon Footprint" and "Food Loss Reduction and Carbon Footprint Practices Worldwide: A Benchmarking Approach of Circular Economy".

Romina Gómez-Prado is the coauthor of Chaps "Circular Economy for Food Loss Reduction and Water foodprint" and "Circular Economy for Waste Reduction and Carbon Footprint".

Micaela Jaramillo-Arévalo is the coauthor of Chaps "3D Print, Circularity and Footprints" and "Virtual Tourism, Carbon Footprint and Circularity".

Luis Juarez-Rojas is the coauthor of Chaps "Circular Economy for Food Loss Reduction and Water foodprint" and "Water Footprint in the Textile and Food Supply Chain Management: Trends to Become Circular and Sustainable".

Luigi Leclercq-Machado is the coauthor of Chaps "Circular Economy for Packaging and Carbon Footprint" and "Food Loss Reduction and Carbon Footprint Practices Worldwide: A Benchmarking Approach of Circular Economy".

Maria F. Lenti-Dulong is the coauthor of Chap "Virtual Tourism, Carbon Footprint and Circularity".

Coralia Mesa-Gomez is the coauthor of Chap "Leadership for Sustainability in Crisis Time".

Flavio Morales-Ríos is the coauthor of Chaps "Circular Economy for Packaging and Carbon Footprint" and "Material Selection for Circularity and Footprints".

Jorge Sánchez-Palomino is the coauthor of Chap "Circular Economy for Waste Reduction and Carbon Footprint".

Subramanian Senthilkannan Muthu is the coauthor of Chap "Measuring Circular Economy".

Diego Villalobos-Alvarez is the coauthor of Chap "Leadership for Sustainability in Crisis Time".

Contents

Editors and Contributors

About the Editors

Dr. Aldo Alvarez-Risco is an associate professor at Universidad de Lima in Peru. He obtained his Ph.D. from the Universidad Autónoma de Nuevo León in Mexico. Also, he is a Doctor in Pharmacy and Biochemistry, Master in Pharmacology and Pharmacist at the Universidad Nacional Mayor de San Marcos in Peru, and Master in Pharmaceutical Care at Universidad de Granada in Spain. He has six years of regulatory affairs experience in pharmaceutical companies and eight years of experience in the Ministry of Health in Peru. He has published more than 50 research publications, written some book chapters, and authored/edited books in the areas of Pharmaceutical Care, Sustainability, Entrepreneurship, and Circular Economy. He is a lecturer in postgraduate programs since 2004 and a speaker in academic events in 21 countries.

Dr. Subramanian Senthilkannan Muthu currently works for SgT Group as Head of Sustainability and is based out of Hong Kong. He earned his Ph.D. from the Hong Kong Polytechnic University and is a renowned expert in the areas of Environmental Sustainability in Textiles & Clothing Supply Chain, Product Life Cycle Assessment (LCA), and Product Carbon Footprint Assessment (PCF) in various industrial sectors. He has five years of industrial experience in textile manufacturing, research and development and textile testing and over a decade's of experience in life cycle assessment (LCA), and carbon and ecological footprints assessment of various consumer products. He has published more than 100 research publications, written numerous book chapters, and authored/edited over 100 books in the areas of Carbon Footprint, Recycling, Environmental Assessment and Environmental Sustainability.

Prof. Shyla Del-Aguila-Arcentales is part of the staff in Escuela Nacional de Marina Mercante "Almirante Miguel Grau". She is also Master in Industrial Pharmacy at Universidad Nacional Mayor de San Marcos in Peru. Also, she is pharmacist at the

Universidad Nacional de la Amazonia Peruana in Peru. She has 12 years of experience in pharmaceutical companies' processes in manufacturing, regulatory affairs, and control quality. She has published more than 25 research publications, written some book chapters, and authored/edited books in the areas of Pharmaceutical Care, Management, Smart Cities, Sustainability, Entrepreneurship, and Circular Economy. Also, she is a lecturer in undergraduate and postgraduate programs since 2010.

Contributors

Camila Almanza-Cruz Universidad de Lima, Lima, Perú

Aldo Alvarez-Risco Universidad de Lima, Lima, Perú

Marian Arias-Meza Universidad de Lima, Lima, Perú

Marco Calle-Nole Universidad de Lima, Lima, Perú

Nilda Campos-Dávalos Universidad de Lima, Lima, Perú

Sarahit Castillo-Benancio Universidad de Lima, Lima, Perú

Anguie Contreras-Taica Universidad de Lima, Lima, Perú

Berdy Briggitte Cuya-Velásquez Universidad de Lima, Lima, Perú

Myreya De-la-Cruz-Diaz Universidad de Lima, Lima, Peru

Santiago Diaz-Risco Centro de Fertilidad Cajamarca, Cajamarca, Perú

Shyla Del-Aguila-Arcentales Escuela Nacional de Marina Mercante "Almirante Miguel Grau", Callao, Peru

Sharon Esquerre-Botton Universidad de Lima, Lima, Perú

Romina Gómez-Prado Universidad de Lima, Lima, Perú

Micaela Jaramillo-Arévalo Universidad de Lima, Lima, Perú

Luis Juarez-Rojas Universidad de Lima, Lima, Peru

Maria de las Mercedes Anderson-Seminario Universidad de Lima, Lima, Perú

Luigi Leclercq-Machado Universidad de Lima, Lima, Perú

Maria F. Lenti-Dulong Universidad de Lima, Lima, Perú

Coralia Mesa-Gomez Ministerio de Salud Pública de Cuba, La Habana, Cuba

Flavio Morales-Ríos Universidad de Lima, Lima, Perú

Subramanian Senthilkannan Muthu SgT & API, Hong Kong, Hong Kong

Arianne Ortiz-Guerra Universidad de Lima, Lima, Peru

Jorge Sánchez-Palomino Universidad de Lima, Lima, Perú

Diego Villalobos-Alvarez Universidad Tecnológica del Perú, Lima, Perú

Circular Economy: Strategies, Deglobalization and Leadership

Measuring Circular Economy

Shyla Del-Aguila-Arcentales, Aldo Alvarez-Risco, and Subramanian Senthilkannan Muthu

Abstract The circular economy seeks its development through various efforts. Specific action guidelines and indicators are required for different levels, types of organizations, regions, etc. Likewise, diverse experiences are required to build indicators for each reality. A review of some indicators is made, and specific evidence is presented for each material or country. Future research is needed to test various indicators for their importance and validity.

Keywords Circular economy · Footprint · Waste · Plastic · Indicators · Index

1 Introduction

Globally, there is a growing trend for organizations to develop projects that enable the shift from linear economy to circular economy-based activities [17, 21]. In organizations based on linear economics, the processes performed by workers are based on the use of materials in one direction only, where the raw materials that enter the process are used to obtain the final product, the resulting waste is thrown away without any further action. However, under the circular economy approach, two fundamental components underpin the circular management approach: recovery and valorization of waste. These approaches imply that particular materials can be reused in the supply chain. Several countries are already initiating the regulation and promotion of circular activities such as Vietnam [25], Canada [9], Russia [78], Latin America

S. Del-Aguila-Arcentales (✉)
Escuela Nacional de Marina Mercante "Almirante Miguel Grau", Callao, Perú
e-mail: sdelaguila@enamm.edu.pe

A. Alvarez-Risco
Universidad de Lima, Lima, Perú
e-mail: aralvare@ulima.edu.pe

S. S. Muthu
SgT & API, Hong Kong, Hong Kong

A. Alvarez-Risco et al. (eds.), *Circular Economy*, Environmental Footprints and Eco-design of Products and Processes,
https://doi.org/10.1007/978-981-19-0549-0_1

[85], New Zealand [58], Bolivia [15], Ghana [5], and others. The European Union [99] and China [14] are leading the way to implement the circular economy.

In the scientific literature, there are several reports, ranging from correlational or descriptive models to descriptive [44] or correlational models [2], but there is very little publication associated with measuring circularity, which is critical to monitoring implementation progress.

2 Measure of Circular Economy

When thinking about indicators, one needs to recognize the efforts that can be made to achieve monitoring and eco-efficient use at different levels. The first component that requires monitoring is electricity. Table 1 shows the different indicators that can be applied at different levels. By the eco-innovation action plan of the European Union [13], there are different indicators (Table 1).

Table 1 Indicators based on sustainable resource management, societal behaviors, and business operations

Type of indicators	Source
Sustainable resource management	
Material footprint	https://ec.europa.eu/eurostat/statistics-explained/index.php?title=Material_flow_accounts_statistics_-_material_footprints
Resource productivity	https://ec.europa.eu/eurostat/statistics-explained/index.php?title=Resource_productivity_statistics
Trends in the repair sector	https://appsso.eurostat.ec.europa.eu/nui/show.do?dataset=sbs_na_1a_se_r2&lang=en
Extended producer responsibility	https://circulareconomy.europa.eu/platform/sites/default/files/ecopreneur-circular-economy-update-report-2019.pdf
Recycling rates in Europe by waste stream	https://www.eea.europa.eu/data-and-maps/indicators/waste-recycling-1/assessment-1
Municipal solid waste	https://ec.europa.eu/eurostat/databrowser/view/env_wasmun/default/table?lang=en
Recycling of packaging waste	https://ec.europa.eu/eurostat/web/products-datasets/product?code=ten00063
Recycling of biowaste	https://ec.europa.eu/eurostat/cache/metadata/en/cei_wm030_esmsip2.htm
Recycling of construction	https://ec.europa.eu/eurostat/databrowser/view/cei_wm040/default/table?lang=en

(continued)

Table 1 (continued)

Type of indicators	Source
Municipal waste recycled	https://ec.europa.eu/eurostat/statistics-explained/index.php?title=Municipal_waste_statistics
Societal behaviors	
Citizens who have chosen alternatives to buying new products	https://data.europa.eu/data/datasets/s1102_388?locale=en
Coverage of circular economy	Scopus
Turnover in repair of computers and personal goods	https://ec.europa.eu/eurostat/statistics-explained/index.php?title=Computer_and_personal_and_household_goods_repair_statistics_-_NACE_Rev._2
Number of enterprises and employment in repair of computers	
Number of countries of enterprises and employment in repair of computers	
Business operations	
Difficulties implementing circular economy	https://data.europa.eu/data/datasets/s2110_441_eng?locale=en
Financing sources for circular economy	
Availability of information	
Share of enterprises that facilitated recycling of products	https://ec.europa.eu/eurostat/cache/metadata/en/inn_cis9_esms.htm
Enterprises that extended product life	
Number of eco labeled products and services	https://ec.europa.eu/environment/ecolabel/facts-and-figures.html

3 OECD Indicators

Another indicator relevant to consider is the circular economy indicator proposed by the OECD. The most critical indicators are presented below by category type, sector, indicator, unit of measurement, and source.

Specifically, Moraga et al. [56] described indicators such as self-sufficiency for raw materials, green public procurement, waste generation, food waste, recycling rates, recovery for specific waste streams, the contribution of recycled materials to raw materials demand, trade-in recyclable raw materials, private investments, jobs and gross value added, and patents related to recycling and secondary raw materials. Other authors have proposed different indicators such as Geng et al. [22] (Table 2), Smol et al. [91] (Table 3), and De Pascale et al. [70] (Table 4).

Also, it is recommended to review the proposal of indicators and footprints by Saidani et al. [83], Huysman et al. [31], de Oliveira et al. [65], Padilla-Rivera et al. [66], Rincón-Moreno et al. [79], Avdiushchenko and Zając [4], and Cayzer et al. [8]. More specifically, it is relevant focus on longevity and circularity [16], eight historic port cities [24], indicators in plastic, textile and electro-electronic cases [82], use of energy accounting method [87], material circularity and life cycle indicators [27, 30, 50, 55, 61, 75, 94, 103, 105], manufacturing network [37, 52], product families

Table 2 Indicators based in Geng et al. [22]

Calculation formula	Unit
Output of main mineral resource = GDP/total consumption of main mineral resource	10,000 U/ton
Output of energy = GDP/energy consumption	10,000 U/ton sce
Energy consumption per unit of GDP = energy consumption/GDP (unit: ton sce/10,000 U)	ton sce/10,000 U
Energy consumption per added industrial value = industrial energy consumption AVI	ton sce/10,000 U
Energy consumption of key industrial product = energy consumption of steel (copper, aluminum, cement, fertilizer, paper)/steel production	ton sce/ton
Water withdrawal per unit of GDP = total amount of water withdrawal/GDP	10,000 m^3/U
Water withdrawal per added industrial value = amount of industrial water withdrawal/AVI	10,000 m3/U
Water consumption of key industrial sector product = total amount of fresh water consumption/total amount of steel production	108 m^3/ton
Coefficient of irrigation water utilization = actual amount of irrigation water consumption/total amount of irrigation water consumption	
Recycling rate of industrial solid waste = (industrial solid waste integrated utilization Q/industrial solid waste generation) 100%	

Table 3 Indicators based in Smol et al. [91]

CE–Eco-innovation inputs	*Unit*
Regional authorities environmental and energy R&D for CE appropriations and outlays	% of GDP
Regional total value of green early stage investments	EURO per capita
CE–Eco-innovation activities	*Unit*
Firms having implemented CE–eco-innovation activities aiming at a reduction of material input per unit output	% of total firms in region
Firms having implemented CE–eco-innovation activities aiming at an increase of material recycling	% of total firms in region
CE–Eco-innovation outputs	*Unit*
Generated industrial waste	Amount of waste/person
Generated municipal waste	Amount of waste/person
Recycled industrial waste	Amount of waste/person
Recycled municipal waste	Amount of waste/person
Life cycle assessment of enterprises activity	Amount companies with LCA reports per regions
Number of companies with "zero waste" program	

(continued)

Table 3 (continued)

Resource efficiency outcomes	*Unit*
Material productivity	Regional GDP/domestic material consumption of region
Water productivity	Regional GDP/water footprint of region
Energy productivity	Regional GDP/gross inland energy consumption of region
GHG emissions intensity	CO_2e/regional GDP
Socio-economic outcomes	*Unit*
Employment in eco-industries and circular economy (% of total employment across all companies of region). Revenue in eco-industries and circular economy (% of total revenue across all companies of region)	% of total revenue across all companies of region
GHG emissions intensity	(CO_2e/regional GDP)

Table 4 Indicators based in De Pascale et al. [70]

Micro-level	Meso-level	Macro-level
Disassembly Effort Index	Energy-based Indices	Multi-scale integrated analysis of societal metabolism
End-of-Life Index	Resource Productivity	A comprehensive index of circular Economy
Reuse Potential Indicator	MEP Indicators System	Super-efficiency DEA model
Material Circularity Indicator	Best-Worst Method	CE monitoring framework
Recyclability Benefit Rate	Evaluation Index System	Regional indicators of eco-innovation
Longevity Indicator	Resource Productivity Indicator	Index system for evaluating the circular economy development
Material Reutilization Score	Eco-Efficiency Indicator	Circularity indicators based on the MFA approach
Recycling Indices	Wastewater Circonomics Index	The evaluation index system of circular economy development level

[49], cultural heritage buildings [18], BWM-DEMATEL approach [108], levels of innovation [42], waste [53, 54, 74, 77, 86, 92], relation with sustainability [37, 40, 47, 80], standard BS 8001:2017 [71], agriculture [72, 100], city level [1, 26, 39, 60, 67, 68, 88], alternative methods [10, 45, 11, 20, 33, 35, 36, 51, 62–64, 69, 76, 81, 84, 101, 102], in companies [34], plastic [32, 93], mobile phones [19], and supply chain [7]. Also, there is evidence of indicators by regions or countries such as Germany [29], China [14, 23, 48, 95, 106–107], Sweden [28], Croatia [43], and European Union [6, 46, 89, 90, 96, 97, 104] (Table 5).

Table 5 Indicators' contribution according to Kristensen and Mosgaard [41]

Indicator	Contribution
Disassembly Effort Index	Academic
Remanufacturing Product Profiles	Academic
Circular Economy Toolkit	Practical
End-of-life Index	Academic
Reuse Potential Indicator	Academic
Circular Economy Index Material Circularity Indicator	Practical
Circularity Calculator	Practical
Eco-cost/Value Ratio	Academic
Longevity Indicator	Academic
Material Reutilization Score	Practical

Some reports must be reviewed to obtain global information to develop successful strategies.

Source Morley et al. [57]

Source Tully [98]

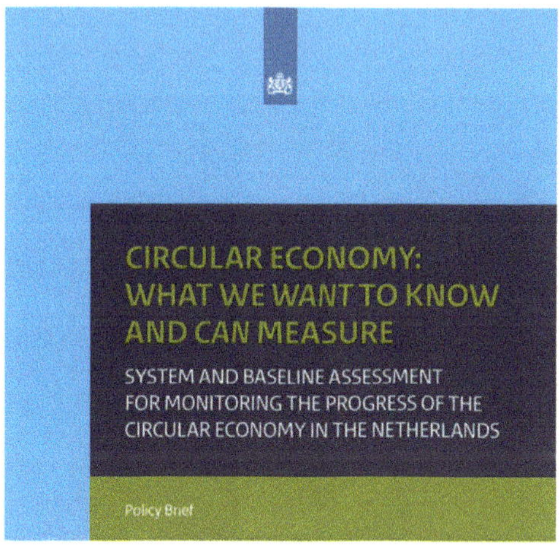

Source Potting et al. [73]

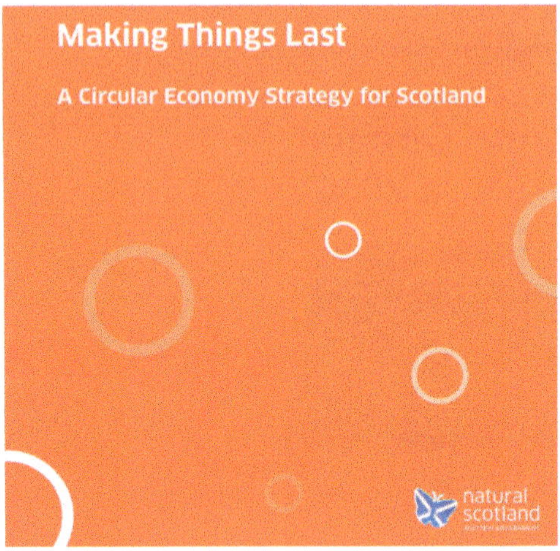

Source Natural Scotland [59]

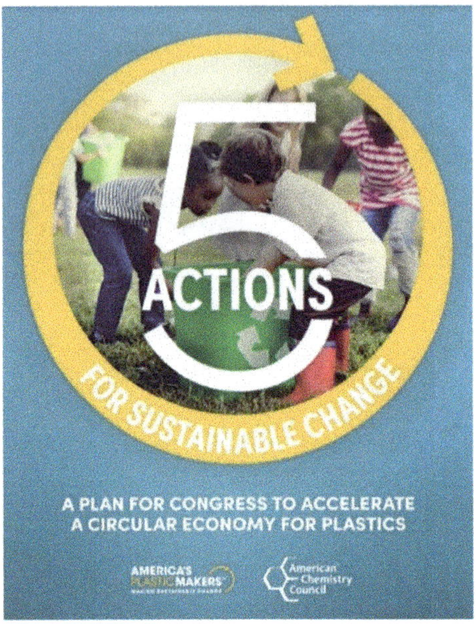

Source America's Plastic Makers [3]

Source European Circular Economy Stakeholder Platform [12]

Closing Remarks

In these times of the COVID-19 pandemic, resilience must be based on the eco-efficient use of materials and, therefore, requires building fundamental indicators to help monitor. The book presents the specific development of footprint certifications, focused on different materials. The evidence presented in this chapter should be used as inputs for future research to be carried out, testing the indicators from the governmental, business, and citizen point of view.

References

1. Alaerts L, Van Acker K, Rousseau S, De Jaeger S, Moraga G, Dewulf J, De Meester S, Van Passel S, Compernolle T, Bachus K, Vrancken K, Eyckmans J (2019) Towards a more direct policy feedback in circular economy monitoring via a societal needs perspective. Resour Conserv Recycl 149:363–371. https://doi.org/10.1016/j.resconrec.2019.06.004
2. Alvarez-Risco A, Estrada-Merino A, Rosen MA, Vargas-Herrera A, Del-Aguila-Arcentales S (2021) Factors for implementation of circular economy in firms in COVID-19 pandemic times: the case of Peru. Environ 8(9). https://doi.org/10.3390/environments8090095
3. America's Plastic Makers (2021) A circular economy for plastics. https://www.plasticmakers.org/files/d6b3a34b9a88b1a6ee4da0a73b24562d740f80e4.pdf
4. Avdiushchenko A, Zając P (2019) Circular economy indicators as a supporting tool for European regional development policies. Sustain 11(11). https://doi.org/10.3390/su11113025
5. Boon EK, Anuga SW (2020) Circular economy and its relevance for improving food and nutrition security in Sub-Saharan Africa: the case of Ghana. Mater Circ Econ 2(1):1–14
6. Busu M, Trica CL (2019) Sustainability of circular economy indicators and their impact on economic growth of the European Union. Sustain 11(19). https://doi.org/10.3390/su11195481

7. Calzolari T, Genovese A, Brint A (2022) Circular economy indicators for supply chains: a systematic literature review. Environ Sustain Indic 13:100160. https://doi.org/10.1016/j.indic.2021.100160
8. Cayzer S, Griffiths P, Beghetto V (2017) Design of indicators for measuring product performance in the circular economy. Int J Sustain Eng 10(4–5):289–298. https://doi.org/10.1080/19397038.2017.1333543
9. Cocker J, Graham K (2020) Circular economy in Canada. Circular economy: global perspective. Springer, Singapore, pp 87–122
10. Corona B, Shen L, Reike D, Rosales Carreón J, Worrell E (2019) Towards sustainable development through the circular economy—a review and critical assessment on current circularity metrics. Resour Conserv Recycl 151:104498. https://doi.org/10.1016/j.resconrec.2019.104498
11. Elia V, Gnoni MG, Tornese F (2017) Measuring circular economy strategies through index methods: a critical analysis. J Clean Prod 142:2741–2751. https://doi.org/10.1016/j.jclepro.2016.10.196
12. European Circular Economy Stakeholder Platform (2021) ECESP coordination group work plan 2021: delivering on the transition to a circular economy. https://circulareconomy.europa.eu/platform/en/news-and-events/all-news/ecesp-coordination-group-work-plan-2021-delivering-transition-circular-economy
13. European Commission (2021) Circular economy indicators. Retrieved 01 January 2022 from https://ec.europa.eu/environment/ecoap/indicators/circular-economy-indicators_en
14. Fan Y, Fang C (2020) Circular economy development in China-current situation, evaluation and policy implications. Environ Impact Assess Rev 84:106441. https://doi.org/10.1016/j.eiar.2020.106441
15. Ferronato N, Preziosi G, Portillo MAG, Lizarazu EGG, Torretta V (2020) Assessment of municipal solid waste selective collection scenarios with geographic information systems in Bolivia. Waste Manage 102:919–931
16. Figge F, Thorpe AS, Givry P, Canning L, Franklin-Johnson E (2018) Longevity and circularity as indicators of eco-efficient resource use in the circular economy. Ecol Econ 150:297–306. https://doi.org/10.1016/j.ecolecon.2018.04.030
17. Fogarassy C, Finger D (2020) Theoretical and practical approaches of circular economy for business models and technological solutions. Resour 9(6). https://doi.org/10.3390/resources9060076
18. Foster G, Kreinin H (2020) A review of environmental impact indicators of cultural heritage buildings: a circular economy perspective. Environ Res Lett 15(4):043003. https://doi.org/10.1088/1748-9326/ab751e
19. Franklin-Johnson E, Figge F, Canning L (2016) Resource duration as a managerial indicator for circular economy performance. J Clean Prod 133:589–598. https://doi.org/10.1016/j.jclepro.2016.05.023
20. Garcia-Bernabeu A, Hilario-Caballero A, Pla-Santamaria D, Salas-Molina F (2020) A process oriented MCDM approach to construct a circular economy composite index. Sustain 12(2). https://doi.org/10.3390/su12020618
21. Garcés-Ayerbe C, Rivera-Torres P, Suárez-Perales I, Leyva-de la Hiz DI (2019) Is it possible to change from a linear to a circular economy? An overview of opportunities and barriers for European small and medium-sized enterprise companies. Int J Environ Res Public Health 16(5). https://doi.org/10.3390/ijerph16050851
22. Geng Y, Fu J, Sarkis J, Xue B (2012) Towards a national circular economy indicator system in China: an evaluation and critical analysis. J Clean Prod 23(1):216–224. https://doi.org/10.1016/j.jclepro.2011.07.005
23. Geng Y, Sarkis J, Ulgiati S, Zhang P (2013) Measuring China's circular economy. Sci 339(6127):1526–1527. https://doi.org/10.1126/science.1227059
24. Gravagnuolo A, Angrisano M, Fusco Girard L (2019) Circular economy strategies in eight historic port cities: criteria and indicators towards a circular city assessment framework. Sustain 11(13). https://doi.org/10.3390/su11133512

25. Hai HT, Quang ND, Thang NT, Nam NH (2020) Circular economy in Vietnam. Circular economy: global perspective. Springer, Singapore, pp 423–452
26. Harris S, Martin M, Diener D (2021) Circularity for circularity's sake? Scoping review of assessment methods for environmental performance in the circular economy. Sustain Prod Consum 26:172–186. https://doi.org/10.1016/j.spc.2020.09.018
27. Haupt M, Hellweg S (2019) Measuring the environmental sustainability of a circular economy. Environ Sustain Indic 1–2:100005. https://doi.org/10.1016/j.indic.2019.100005
28. Haupt M, Vadenbo C, Hellweg S (2017) Do we have the right performance indicators for the circular economy? Insight into the Swiss waste management system. J Ind Ecol 21(3):615–627. https://doi.org/10.1111/jiec.12506
29. Helander H, Petit-Boix A, Leipold S, Bringezu S (2019) How to monitor environmental pressures of a circular economy: an assessment of indicators. J Ind Ecol 23(5):1278–1291. https://doi.org/10.1111/jiec.12924
30. Hu Y, He X, Poustie M (2018) Can legislation promote a circular economy? A material flow-based evaluation of the circular degree of the Chinese economy. Sustain 10(4). https://doi.org/10.3390/su10040990
31. Huysman S, De Schaepmeester J, Ragaert K, Dewulf J, De Meester S (2017) Performance indicators for a circular economy: a case study on post-industrial plastic waste. Resour Conserv Recycl 120:46–54. https://doi.org/10.1016/j.resconrec.2017.01.013
32. Huysveld S, Hubo S, Ragaert K, Dewulf J (2019) Advancing circular economy benefit indicators and application on open-loop recycling of mixed and contaminated plastic waste fractions. J Clean Prod 211:1–13. https://doi.org/10.1016/j.jclepro.2018.11.110
33. Hysa E, Kruja A, Rehman NU, Laurenti R (2020) Circular economy innovation and environmental sustainability impact on economic growth: an integrated model for sustainable development. Sustain 12(12). https://doi.org/10.3390/su12124831
34. Janik A, Ryszko A (2019) Circular economy in companies: an analysis of selected indicators from a managerial perspective. Multidiscip Asp Prod Eng 2(1):523–535. https://doi.org/10.2478/mape-2019-0053
35. Jerome A, Helander H, Ljunggren M, Janssen M (2022) Mapping and testing circular economy product-level indicators: a critical review. Resour Conserv Recycl 178:106080. https://doi.org/10.1016/j.resconrec.2021.106080
36. Kravchenko M, McAloone TC, Pigosso DCA (2019) Implications of developing a tool for sustainability screening of circular economy initiatives. Procedia CIRP 80:625–630. https://doi.org/10.1016/j.procir.2019.01.044
37. Kravchenko M, McAloone TC, Pigosso DCA (2020) To what extent do circular economy indicators capture sustainability? Procedia CIRP 90:31–36. https://doi.org/10.1016/j.procir.2020.02.118
38. Kravchenko M, Pigosso DCA, McAloone TC (2019) Towards the ex-ante sustainability screening of circular economy initiatives in manufacturing companies: consolidation of leading sustainability-related performance indicators. J Clean Prod 241:118318. https://doi.org/10.1016/j.jclepro.2019.118318
39. Kravchenko M, Pigosso DCA, McAloone TC (2020) A procedure to support systematic selection of leading indicators for sustainability performance measurement of circular economy initiatives. Sustain 12(3). https://doi.org/10.3390/su12030951
40. Kravchenko M, Pigosso DCA, McAloone TC (2021) A trade-off navigation framework as a decision support for conflicting sustainability indicators within circular economy implementation in the manufacturing industry. Sustain 13(1). https://doi.org/10.3390/su13010314
41. Kristensen HS, Mosgaard MA (2020) A review of micro level indicators for a circular economy—moving away from the three dimensions of sustainability? J Clean Prod 243:118531. https://doi.org/10.1016/j.jclepro.2019.118531
42. Kuzma EL, Sehnem S, Lopes de Sousa Jabbour AB, Campos LMS (2021) Circular economy indicators and levels of innovation: an innovative systematic literature review. Int J Prod Perform Manag (ahead-of-print). https://doi.org/10.1108/IJPPM-10-2020-0549

43. Luttenberger LR (2020) Waste management challenges in transition to circular economy—case of Croatia. J Clean Prod 256:120495. https://doi.org/10.1016/j.jclepro.2020.120495
44. Lüdeke-Freund F, Gold S, Bocken NMP (2019) A review and typology of circular economy business model patterns. J Ind Ecol 23(1):36–61. https://doi.org/10.1111/jiec.12763
45. Di Maio F, Rem PC, Baldé K, Polder M (2017) Measuring resource efficiency and circular economy: a market value approach. Resour Conserv Recycl 122:163–171. https://doi.org/10.1016/j.resconrec.2017.02.009
46. Marino A, Pariso P (2020) Comparing European countries' performances in the transition towards the circular economy. Sci Total Environ 729:138142. https://doi.org/10.1016/j.scitotenv.2020.138142
47. Martinho VJPD (2021) Insights into circular economy indicators: emphasizing dimensions of sustainability. Environ Sustain Indic 10:100119. https://doi.org/10.1016/j.indic.2021.100119
48. McDowall W, Geng Y, Huang B, Barteková E, Bleischwitz R, Türkeli S, Kemp R, Doménech T (2017) Circular economy policies in China and Europe. J Ind Ecol 21(3):651–661. https://doi.org/10.1111/jiec.12597
49. Mesa J, Esparragoza I, Maury H (2018) Developing a set of sustainability indicators for product families based on the circular economy model. J Clean Prod 196:1429–1442. https://doi.org/10.1016/j.jclepro.2018.06.131
50. Mesa J, González-Quiroga A, Maury H (2020) Developing an indicator for material selection based on durability and environmental footprint: a circular economy perspective. Resour Conserv Recycl 160:104887. https://doi.org/10.1016/j.resconrec.2020.104887
51. Mies A, Gold S (2021) Mapping the social dimension of the circular economy. J Clean Prod 321:128960. https://doi.org/10.1016/j.jclepro.2021.128960
52. Mishra S, Singh SP, Johansen J, Cheng Y, Farooq S (2019) Evaluating indicators for international manufacturing network under circular economy. Manag Decis 57(4):811–839. https://doi.org/10.1108/MD-05-2018-0565
53. Molina-Moreno V, Leyva-Díaz JC, Llorens-Montes FJ, Cortés-García FJ (2017) Design of indicators of circular economy as instruments for the evaluation of sustainability and efficiency in wastewater from pig farming industry. Water 9(9). https://doi.org/10.3390/w9090653
54. Molina-Sánchez E, Leyva-Díaz JC, Cortés-García FJ, Molina-Moreno V (2018) Proposal of sustainability indicators for the waste management from the paper industry within the circular economy model. Water 10(8). https://doi.org/10.3390/w10081014
55. Moraga G, Huysveld S, De Meester S, Dewulf J (2022) Resource efficiency indicators to assess circular economy strategies: a case study on four materials in laptops. Resour Conserv Recycl 178:106099. https://doi.org/10.1016/j.resconrec.2021.106099
56. . https://doi.org/10.1016/j.resconrec.2019.03.045
57. Morley A, Emelyne L, Zhao C (2018) Measuring the circular economy. http://www.futurepeterborough.com/wp-content/uploads/2018/08/Final-Report-Measuring-the-Circular-Economy-of-Peterborough-2-1.pdf
58. Munir MT, Mohaddespour A, Nasr AT, Carter S (2021) Municipal solid waste-to-energy processing for a circular economy in New Zealand. Renew Sustain Energy Rev 145:111080.
59. Natural Scotland (2016) Making things last: a circular economy strategy for Scotland. https://www.gov.scot/publications/making-things-last-circular-economy-strategy-scotland/
60. Ngan SL, How BS, Teng SY, Promentilla MAB, Yatim P, Er AC, Lam HL (2019) Prioritization of sustainability indicators for promoting the circular economy: the case of developing countries. Renew Sustain Energy Rev 111:314–331. https://doi.org/10.1016/j.rser.2019.05.001
61. Niero M, Kalbar PP (2019) Coupling material circularity indicators and life cycle based indicators: a proposal to advance the assessment of circular economy strategies at the product level. Resour Conserv Recycl 140:305–312. https://doi.org/10.1016/j.resconrec.2018.10.002
62. Nika CE, Vasilaki V, Expósito A, Katsou E (2020) Water cycle and circular economy: developing a circularity assessment framework for complex water systems. Water Res 187:116423. https://doi.org/10.1016/j.watres.2020.116423

63. Nika C-E, Expósito A, Kisser J, Bertino G, Oral HV, Dehghanian K, Vasilaki V, Iacovidou E, Fatone F, Atanasova N, Katsou E (2021) Validating circular performance indicators: the interface between circular economy and stakeholders. Water 13(16). https://doi.org/10.3390/w13162198
64. Nuñez-Cacho P, Górecki J, Molina-Moreno V, Corpas-Iglesias FA (2018) What gets measured, gets done: development of a circular economy measurement scale for building industry. Sustain 10(7). https://doi.org/10.3390/su10072340
65. de Oliveira CT, Dantas TET, Soares SR (2021) Nano and micro level circular economy indicators: assisting decision-makers in circularity assessments. Sustain Prod Consum 26:455–468. https://doi.org/10.1016/j.spc.2020.11.024
66. Padilla-Rivera A, do Carmo BBT, Arcese G, Merveille N (2021) Social circular economy indicators: selection through fuzzy delphi method. Sustain Prod Consum 26:101–110. https://doi.org/10.1016/j.spc.2020.09.015
67. Padilla-Rivera A, Russo-Garrido S, Merveille N (2020) Addressing the social aspects of a circular economy: a systematic literature review. Sustain 12(19). https://doi.org/10.3390/su12197912
68. Papageorgiou A, Henrysson M, Nuur C, Sinha R, Sundberg C, Vanhuyse F (2021) Mapping and assessing indicator-based frameworks for monitoring circular economy development at the city-level. Sustain Cities Soc 75:103378. https://doi.org/10.1016/j.scs.2021.103378
69. Parchomenko A, Nelen D, Gillabel J, Rechberger H (2019) Measuring the circular economy—a multiple correspondence analysis of 63 metrics. J Clean Prod 210:200–216. https://doi.org/10.1016/j.jclepro.2018.10.357
70. De Pascale A, Arbolino R, Szopik-Depczyńska K, Limosani M, Ioppolo G (2021) A systematic review for measuring circular economy: the 61 indicators. J Clean Prod 281:124942. https://doi.org/10.1016/j.jclepro.2020.124942
71. Pauliuk S (2018) Critical appraisal of the circular economy standard BS 8001:2017 and a dashboard of quantitative system indicators for its implementation in organizations. Resour Conserv Recycl 129:81–92. https://doi.org/10.1016/j.resconrec.2017.10.019
72. Poponi S, Arcese G, Pacchera F, Martucci O (2022) Evaluating the transition to the circular economy in the agri-food sector: selection of indicators. Resour Conserv Recycl 176:105916. https://doi.org/10.1016/j.resconrec.2021.105916
73. Potting J, Hanemaaijer A, Delahaye R, Ganzevles J, Hoekstra R, Lijzen J (2018) Circular economy: what we want to know and can measure. System and baseline assessment for monitoring the progress of the circular economy in The Netherlands. https://circulareconomy.europa.eu/platform/sites/default/files/pbl-2018-circular-economy-what-we-want-to-know-and-can-measure-3216.pdf
74. Preisner M, Smol M, Horttanainen M, Deviatkin I, Havukainen J, Klavins M, Ozola-Davidane R, Kruopienė J, Szatkowska B, Appels L, Houtmeyers S, Roosalu K (2022) Indicators for resource recovery monitoring within the circular economy model implementation in the wastewater sector. J Environ Manage 304:114261. https://doi.org/10.1016/j.jenvman.2021.114261
75. Primc K, Kalar B, Slabe-Erker R, Dominko M, Ogorevc M (2020) Circular economy configuration indicators in organizational life cycle theory. Ecol Ind 116:106532. https://doi.org/10.1016/j.ecolind.2020.106532
76. Rabta B (2020) An economic order quantity inventory model for a product with a circular economy indicator. Comput Ind Eng 140:106215. https://doi.org/10.1016/j.cie.2019.106215
77. Rashid MI, Shahzad K (2021) Food waste recycling for compost production and its economic and environmental assessment as circular economy indicators of solid waste management. J Clean Prod 317:128467. https://doi.org/10.1016/j.jclepro.2021.128467
78. Ratner, S, Lazanyuk, I, Revinova S, Gomonov K (2021) Barriers of consumer behavior for the development of the circular economy: empirical evidence from Russia. Appl Sci 11(1):46
79. Rincón-Moreno J, Ormazábal M, Álvarez MJ, Jaca C (2021) Advancing circular economy performance indicators and their application in Spanish companies. J Clean Prod 279:123605. https://doi.org/10.1016/j.jclepro.2020.123605

80. Rodriguez-Anton JM, Rubio-Andrada L, Celemín-Pedroche MS, Alonso-Almeida MDM (2019) Analysis of the relations between circular economy and sustainable development goals. Int J Sust Dev World 26(8):708–720. https://doi.org/10.1080/13504509.2019.1666754
81. Roos Lindgreen E, Mondello G, Salomone R, Lanuzza F, Saija G (2021) Exploring the effectiveness of grey literature indicators and life cycle assessment in assessing circular economy at the micro level: a comparative analysis. Int J Life Cycle Assess 26(11):2171–2191. https://doi.org/10.1007/s11367-021-01972-4
82. Rossi E, Bertassini AC, Ferreira CdS, Neves do Amaral WA, Ometto AR (2020) Circular economy indicators for organizations considering sustainability and business models: plastic, textile and electro-electronic cases. J Clean Prod 247:119137. https://doi.org/10.1016/j.jclepro.2019.119137
83. Saidani M, Yannou B, Leroy Y, Cluzel F, Kendall A (2019) A taxonomy of circular economy indicators. J Clean Prod 207:542–559. https://doi.org/10.1016/j.jclepro.2018.10.014
84. Saidani M, Yannou B, Leroy Y, Cluzel F (2017) How to assess product performance in the circular economy? Proposed requirements for the design of a circularity measurement framework. Recycl 2(1). https://doi.org/10.3390/recycling2010006
85. Salas DA, Criollo P, Ramirez AD (2021) The role of higher education institutions in the implementation of circular economy in Latin America. Sustain 13(17):9805
86. Salguero-Puerta L, Leyva-Díaz JC, Cortés-García FJ, Molina-Moreno V (2019) Sustainability indicators concerning waste management for implementation of the circular economy model on the University of Lome (Togo) campus. Int J Environ Res Public Health 16(12). https://doi.org/10.3390/ijerph16122234
87. Santagata R, Zucaro A, Viglia S, Ripa M, Tian X, Ulgiati S (2020) Assessing the sustainability of urban eco-systems through energy-based circular economy indicators. Ecol Ind 109:105859. https://doi.org/10.1016/j.ecolind.2019.105859
88. Schröder P, Lemille A, Desmond P (2020) Making the circular economy work for human development. Resour Conserv Recycl 156:104686. https://doi.org/10.1016/j.resconrec.2020.104686
89. Silvestri F, Spigarelli F, Tassinari M (2020) Regional development of circular economy in the European Union: a multidimensional analysis. J Clean Prod 255:120218. https://doi.org/10.1016/j.jclepro.2020.120218
90. Smol M (2021) Inventory and comparison of performance indicators in circular economy roadmaps of the European countries. Circ Econ Sustain. https://doi.org/10.1007/s43615-021-00127-9
91. Smol M, Kulczycka J, Avdiushchenko A (2017) Circular economy indicators in relation to eco-innovation in European regions. Clean Technol Environ Policy 19(3):669–678. https://doi.org/10.1007/s10098-016-1323-8
92. Smol M, Koneczna R (2021) Economic indicators in water and wastewater sector contributing to a circular economy (CE). Resour 10(12). https://doi.org/10.3390/resources10120129
93. Spierling S, Venkatachalam V, Behnsen H, Herrmann C, Endre, HJ (2019) Bioplastics and circular economy—performance indicators to identify optimal pathways. In: Schebek L, Herrmann C, Cerdas F (eds) Progress in life cycle assessment. Springer, pp 147–154. https://doi.org/10.1007/978-3-319-92237-9_16
94. Steinmann ZJN, Huijbregts MAJ, Reijnders L (2019) How to define the quality of materials in a circular economy? Resour Conserv Recycl 141:362–363. https://doi.org/10.1016/j.resconrec.2018.10.040
95. Su B, Heshmati A, Geng Y, Yu X (2013) A review of the circular economy in China: moving from rhetoric to implementation. J Clean Prod 42:215–227. https://doi.org/10.1016/j.jclepro.2012.11.020
96. Sverko Grdic Z, Krstinic Nizic M, Rudan E (2020) Circular economy concept in the context of economic development in EU countries. Sustain 12(7). https://doi.org/10.3390/su12073060
97. Tantau AD, Maassen MA, Fratila L (2018) Models for analyzing the dependencies between indicators for a circular economy in the European Union. Sustain 10(7). https://doi.org/10.3390/su10072141

98. Tully RC (2021) Green and Circular Extremadura 2030. https://www.interregeurope.eu/fileadmin/user_upload/plp_uploads/events/Webinars/2021_Low-carbon_strategies/4._LCStrategies_AGENEX.pdf
99. Ūsas J, Balezentis T, Streimikiene D (2021) Development and integrated assessment of the circular economy in the European Union: the outranking approach. J of Enterp Inf Manag
100. Velasco-Muñoz JF, Mendoza JMF, Aznar-Sánchez JA, Gallego-Schmid A (2021) Circular economy implementation in the agricultural sector: definition, strategies and indicators. Resour Conserv Recycl 170:105618. https://doi.org/10.1016/j.resconrec.2021.105618
101. Veleva V, Bodkin G, Todorova S (2017) The need for better measurement and employee engagement to advance a circular economy: lessons from Biogen's "zero waste" journey. J Clean Prod 154:517–529. https://doi.org/10.1016/j.jclepro.2017.03.177
102. Vinante C, Sacco P, Orzes G, Borgianni Y (2021) Circular economy metrics: literature review and company-level classification framework. J Clean Prod 288:125090. https://doi.org/10.1016/j.jclepro.2020.125090
103. Virtanen M, Manskinen K, Uusitalo V, Syvänne J, Cura K (2019) Regional material flow tools to promote circular economy. J Clean Prod 235:1020–1025. https://doi.org/10.1016/j.jclepro.2019.06.326
104. Völker T, Kovacic Z, Strand R (2020) Indicator development as a site of collective imagination? The case of European Commission policies on the circular economy. Cult Organ 26(2):103–120. https://doi.org/10.1080/14759551.2019.1699092
105. Walker S, Coleman N, Hodgson P, Collins N, Brimacombe L (2018) Evaluating the environmental dimension of material efficiency strategies relating to the circular economy. Sustain 10(3). https://doi.org/10.3390/su10030666
106. Wang N, Lee JCK, Zhang J, Chen H, Li H (2018) Evaluation of urban circular economy development: an empirical research of 40 cities in China. J Clean Prod 180:876–887. https://doi.org/10.1016/j.jclepro.2018.01.089
107. Wang H, Schandl H, Wang X, Ma F, Yue Q, Wang G, Wang Y, Wei Y, Zhang Z, Zheng R (2020) Measuring progress of China's circular economy. Resour Conserv Recycl 163:105070. https://doi.org/10.1016/j.resconrec.2020.105070
108. Yadav G, Mangla SK, Bhattacharya A, Luthra S (2020) Exploring indicators of circular economy adoption framework through a hybrid decision support approach. J Clean Prod 277:124186. https://doi.org/10.1016/j.jclepro.2020.124186

Better Students, Better Companies, Better Life: Circular Learning

Maria de las Mercedes Anderson-Seminario and Aldo Alvarez-Risco

Abstract New education programs related to the circular economy model are being developed and companies applying the circularity model and strategies according to the company and the context in which it is developed. This article aims to know what skills and knowledge new professionals require to minimize the risks in the transition from one model to another and the continuity of circularity. Throughout the literature review, it is shown that the knowledge and application of the circular economy model are focused on engineering careers but not on business careers, which through their acquired knowledge have significant contributions to competitiveness, consumer behavior, supply and demand, costs, and other variables that are important in a decision-making process of a circular economy. In the end, the competencies that are achieved through the different subjects and methodologies suggested are determined and the different actors that should be considered when developing a subject or a circular economy program.

Keywords Circular economy · Education · Circular economy · Circular competencies · Learning outcomes

1 Circularity and Its Application in Economies and Companies

Do you know the Tableau Economique? Many consider the circular economy as new processes, strategies, and ways of life. The circularity was born many years ago [188]. The multiplier effect considered surpluses in the form of income returns and expenditures as essential factors affecting the determination of the scale of the system as the axes of economic growth [50, 139, 151, 162, 188]. Circularity has changed under its different forms as an alternative to the linear model. Some examples include changes in its concept [114], its application [133, 147] or also the political initiatives

---

M. de las Mercedes Anderson-Seminario · A. Alvarez-Risco (✉)
Universidad de Lima, Lima, Perú
e-mail: aralvare@ulima.edu.pe

A. Alvarez-Risco et al. (eds.), *Circular Economy*, Environmental Footprints and Eco-design of Products and Processes,
https://doi.org/10.1007/978-981-19-0549-0_2

that have been taking place [73], and the different pilot programs to support the change of waste recirculation and for the benefit of climate change [39, 141, 142, 167, 173–174].

In summary, it is said that CE is an alternative model of production and consumption, a growth strategy that allows contributing to sustainable development [90, 92, 134, 151]. These results would be obtained through a cyclical flow for extraction, which allows for transformation, distribution and use and recovery through maintenance, repair of recyclable materials in products, and recovering raw materials from the waste stream [144, 161, 169, 176]. The generation of this economic benefit can coincide with creating social and environmental value, being adopted as a new sustainable economic principle [120], which is how we can say according to Ellen MacArthur Foundation [73] CE is mainly based on three fundamental principles: (1) conserve and optimize non-renewable resource stocks and the balance of renewable resource flows, (2) keep products and materials in use at the maximum in the biological and technical cycles, and (3) design out waste and negative environmental externalities such as pollution [94].

The world was transformed by COVID-19 pandemic (Table 1).

All this leads us to think about how companies and consumers take part in the development, and this is how in recent decades, the circular economy model has been seen as the option to overcome the current model of linear production and consumerism [92]. Thus, China and Japan were the first to adopt the model at the country level. China has been implementing circular economy policies since the 2000s, which has allowed the concept to evolve, becoming flow management and an innovation agenda [183]. It was mainly an industrial ecology program at its inception, being its objective the three "R's": Reduce, Reuse, and Recycle. It is worth noting that China published the Circular Economy Development Plan, integrated into the XIV Five-Year Plan (2021–2025). In 2017, the Circular Economic Policies were published, which contemplate eco-design (as a concept and as a policy) and extended producer responsibility, the latter being a crucial step in implementing the circular economy.

Meanwhile, due to its space and resource constraints, Japan asked itself: how can we better use material flow? As an economy that applies efficient industrialization models (e.g., lean manufacturing), it could visualize models that would improve efficiency by generating change in industrial processes and the application of the three Rs.

Considering the European Union as an economic bloc launched its first EC Action Plan in 2015 [80] and the new EC Action Plan in 2020 [81], it is one of the main blocks that make up the European Green Pact. In Europe, several countries have developed their initiatives, such as the Netherlands [96], France [82], the Energy Transition for Green Growth Act [95], and the Law against waste [97].

Italy is one of the countries that have a high value in the global circularity index, which was regulated by Law 221/2015 to promote green economy and sustainable development, and Law 216/2020 "Waste Decree," which incorporates the four European directives (851/2018 and 852/2018) contained in the EC Package. Other standards applied in Italy include the EMAS standard [76] and the European Union

Table 1 Impact of COVID-19 in different sectors

Education	Ali [9], Allen et al. [10], Alvarez-Risco, Estrada-Merino, et al. [24], Alvarez-Risco, Del-Aguila-Arcentales, Rosen, et al. [15], Alvarez-Risco, Del-Aguila-Arcentales, Yáñez, et al. [16], T. Chen et al. [60], Lackie et al. [119], Lashley et al. [121], Zollinger and DiCindio [187]
Entrepreneurship	Afshan et al. [4], Alvarez-Risco, Mlodzianowska, Zamora-Ramos, et al. [28], Alvarez-Risco, Mlodzianowska, García-Ibarra, et al. [27], Alvarez-Risco and Del-Aguila-Arcentales [16], Belitski et al. [42], Block et al. [45], Brown et al. [52], Bacq and Lumpkin [38], Chafloque-Cespedes et al. [58], Maritz et al. [132], Liguori and Winkler [126]
Health sector	Abiakam et al. [1], Alan et al. [8], Alsairafi et al. [11], Alvarez-Risco, Dawson, et al. [13], Alvarez-Risco, Del-Aguila-Arcentales and Yáñez [20], Barello et al. [40], Bozdağ and Ergün [49], X. Chen et al. [61], Deressa et al. [69], Koetter et al. [116], Raudenská et al. [150], Rojas Román et al. [154], Yáñez, Afshar Jahanshahi, et al. [179], Zhang et al. [185], Zhang et al. [186]
Hospitality	Duarte Alonso et al. [71], Gursoy and Chi [100], Ho et al. [105], Kaushal and Srivastava [111], Yan et al. [178]
Intellectual property	Altindis [12], Alvarez-Risco and Del-Aguila-Arcentales [14], del Castillo [67], Erfani et al. [78], Jecker and Atuire [109], Krishtel and Malpani [118], Okereke [143], Sekalala et al. [158], Zarocostas [184]
Population	Ahsan [6], Alvarez-Risco, Mejia, et al. [25], Cegarra-Navarro et al. [57], Hanzl [102], Lim and Prakash [127], Quispe-Cañari et al. [149], Sánchez-Clavijo et al. [156], Shaw et al. [159], Wen et al. [175], Wu et al. [177], Yáñez, Alvarez-Risco, et al. [179]
Prices	Apcho-Ccencho et al. [36], Chung et al. [62], Galanakis et al. [89], Hepburn et al. [103], Leiva-Martinez et al. [123], Obergassel et al. [140]
Tourism	Assaf and Scuderi [37], Baum and Hai [41], Brouder [51], Carvache-Franco, Alvarez-Risco, Carvache-Franco, Carvache-Franco, Estrada-Merino and Villalobos-Alvarez [54], Fotiadis et al. [88], Carvache-Franco, Alvarez-Risco, Carvache-Franco, Carvache-Franco, Estrada-Merino and Rosen [54], Sigala [160], Carvache-Franco, Carvache-Franco, et al. [56]
Trade	Abudureheman and Nilupaer [2], Acuña-Zegarra et al. [3], Aguirre et al. [5], Alvarez-Risco, Quipuzco-Chicata, et al. [29] Alvarez-Risco, Rosen, et al. [31], Borzée et al. [47], Cruz-Torres et al. [64], Kirk and Rifkin [115], Lopez-Odar et al. [129], Tran [164], Vidya and Prabheesh [171], Yasin Ar [181]
Violence against women	Amarillo [34], Chafloque-Cespedes et al. [59], Dominelli [70], Gulati and Kelly [99], Roesch et al. [152], Sánchez et al. [157], Viero et al. [172]

eco-labels [79]. Germany is a leader in waste management because it has developed policies concerning Sustainable Development Strategies [91], Resource Efficiency Program [85], and the National Program for Sustainable Consumption [86]. In Latin America, Mexico has applied the CE model through waste management [107], and Chile proposed a roadmap, with emphasis on innovation [137] as well as in Peru [72], a roadmap has released that drive and promote the transition from a linear economic model to a circular one in the manufacturing and industrial fish processing industries. Thus, different countries worldwide have been implementing policies, regulations, and the development of the circular economy in search of economic benefit and a better quality of life.

Many countries are applying political plans about the CE, which are gaining strength, and many institutions and people are involved in changing from a linear economy to a circular one. Are the state policies and the different promotion and innovative programs in the new circular economy model enough to achieve a better standard of living in societies?

One of the actors taking part in these state policies is companies, which must modify their resource management and organization practices (energy, water, raw materials) when considering a CE model where they must design, manufacture, distribute, and recover differently from what they have been developing [147]. Under the principles of CE, companies are expected to maintain above all the properties of the products and their components, which allow better management of resources, such as better maintenance [139]. Switching to a CE model implies adopting a new business model and applying new strategies, but these are always appropriate to the CE model applied. The strategies to be followed depend on each company, as shown in Table 2.

Companies must adapt to different organizational processes of the different links of a supply chain and the international physical distribution, seeking to generate value

Table 2 Strategies to be applied by companies in Circular Economy Models

"Resolve" * (2015)	"4R" ** (2017)	"10R" *** (2017)	Based in chain value **** (2018)
Regenerate	Reduce	Refuse	Procurement of materials
Share	Reuse	Rethink	Design
Optimize	Recycle	Reduce	Manufacturing
Make a loop	Retrieve	Reus	Distribution and sales
Virtualize		Repair	Consumption and use
Exchange		Refurbish	Collection and disposal
		Remanufacture	Recycling and recovery
		Repurpose	Remanufacturing
		Recycle	Circular entry
		Recover	

Source Adapted from: *: Ellen MacArthur Foundation [74], **: Kirchherr et al. [114], ***: Potting et al. [146], ****: Kalmykova et al. [110]

along with all the different links. These processes to be adopted seek the efficiency and effectiveness of each of these links for which it requires trained human capital and with the knowledge, skills, and competencies aligned with the EC principles. All areas of the company and the different positions in these different areas should require personnel at different levels who have the skills and competencies that can cope with change in the company and the market and its customer [182]. Thus, another question arises: Do university students receive the necessary training in SC to generate change in small and medium-sized enterprises?

2 Human Resources as a Factor of Change Toward a Circular Economy

As mentioned above, the contribution of companies and private sector organizations is important axis in implementing the new model. Change leads to a new way of seeing, thinking, developing, and doing business. The more significant the change in technological or product innovation, the more complex it is and the greater the knowledge requirements in business models, according to Bocken et al. [46], which enable the identification of strategies for business models that align with resource cycle closure approaches [46, 104]. The changes imply a staff with the required capabilities to take on a new challenge and the risk to change.

The managers of the different areas of the company that assume the change of model must adapt their skills and competencies to the new context. Likewise, organizational functions, such as production, purchasing, marketing and sales, supply chain, design, production, and logistics, are affected by the shift toward a CE model, for which new and updated skills and competencies aligned with the principles of CE are required [93].

The demand for skills and knowledge also varies between sectors and from the stages of circularity in which companies find themselves from the diversification and complexity of tasks, technological advances, reverse processes, and any process in general. Human resources at different levels must change mentality a shift toward greater responsibility about the product or service they offer by establishing an ecologically safe and socially fair operating space for society [63, 98, 101].

According to Burger et al. [53], the study conducted distinguishes 6 levels of competencies to be fulfilled, which are as follows: 1. basic skills (learning ability), 2. problem-solving skills,3. resource management skills (management efficiency skills); 4. social skills (ability to work in teams to achieve objectives); 5. systems skills (ability to understand, monitor, and improve socio-technical systems); and 6. technical skills (developing the ability to design, assemble, operate, and correct difficulties in machine applications or technological systems). It is important to stress the importance of skills to connect and co-create, such as assessing the systemic implications from design decisions until the product reaches the shelf, and then, this goes through a recycling or reuse process [48].

The shift toward a circular economy depends on the skills available, and these likely involve a combination of traditional skills, such as soft skills and service-related skills, and not neglect hard skills [53, 101]. However, it is essential to recognize that CE is a heterogeneous sector composed of different subsectors with different knowledge bases regarding the skills, education, and experience they require, given that activities, duties, and tasks vary across these sectors [170].

The last study conducted by Burger et al. [53] indicates that according to CE jobs, that is, jobs that contribute and are directly related to CE are not only increasing in companies and sectors that develop CE but also in sectors where CE is not applied but that in some way collaborate and support companies that apply CE (provide goods and services). They observed that companies that apply CE processes require more manual and technological skills than companies that support them, requiring more complex cognitive skills.

A company that applies circularity and good CE management generates benefits such as job opportunities and reduces inequality gaps through global redistribution of value. An increasing number of organizations show interest in implementing new goods and services that meet the requirements of a CE model [155], but it is worth highlighting that the application of circular business models is still scarce [128, 165].

3 Without Education, Change Toward a Circular Economy Cannot Be Generated

Education is how a change in values, behaviors, and lifestyles can be generated [43]. It requires leadership on teachers and more knowledge of new tools and approaches [125, 130]. According to Ellen MacArthur Foundation [74] and Kirchherr et al. [113], university professors have not been trained with the latest necessary knowledge as it is a radical innovation. However, universities have incorporated subjects related to sustainability in their curricula since 1990 [35, 173], but it seems that they have not presented an integration between the contents of their curricula and their relationship with sustainability and how to contribute to sustainable societies and the transition to change toward these [87, 106, 117, 131].

Environmentally oriented universities see the need to provide tools and strategies to support, manage, and solve pioneering situations to address urgent issues in waste, renewable energy, and sustainable agriculture [44], which hints that the focus goes toward the improvement of waste management and an improvement an increase in what refers to recycling [92, 161]. At the same time, others relate CE to design thinking and others from a product design perspective [124, 145].

According to Geissdoerfer et al. [90], Türkeli and Schophuizen [166], and Bocken et al. [46], this shift toward a CE model is based on complex tasks and is an integrated process with ecological, social, economic, and political trends. Thus, the circular economy model is not only referred to as waste management and recycling, since the model also allows minimizing risks such as price volatility, improving competitiveness, differentiation of products and services offered, and a reduction in operating costs, which allow a better profit. These points are little studied and are essential factors in the decision-making process of the production and internationalization of a product.

The change from one model to another varies depending on the application and the product. Different teaching methodologies are being applied nowadays, ranging from theory, applied cases, and gamification. However, Türkeli and Schophuizen [166] state that unidirectional education (teacher-students) is no longer competent for teaching CE and even less so if one wishes to raise awareness of the search for tangible solutions for sustainable production and consumption. The change from a linear model to CE is complex due to the lack of knowledge of the processes. For this reason, purely theoretical teaching does not allow the development of competencies and adequate learning for the application of a circular economy in a systemic way.

It is possible to generate a methodology that allows the development of competencies and knowledge to transition change and a circular economy. In this way, a better understanding of its importance can be achieved through CE education, leading students to become factors of change by becoming responsible designers, producers, entrepreneurs, politicians, and consumers.

Unlike other articles, the methodology and knowledge to be developed below are specifically developed for business schools where disciplines other than engineering are taught. Figure 1 shows how the circular economy in its analysis considers the country of destination of the product generated through an innovative eco-friendly model for which it has developed through circular processes to serve a market which

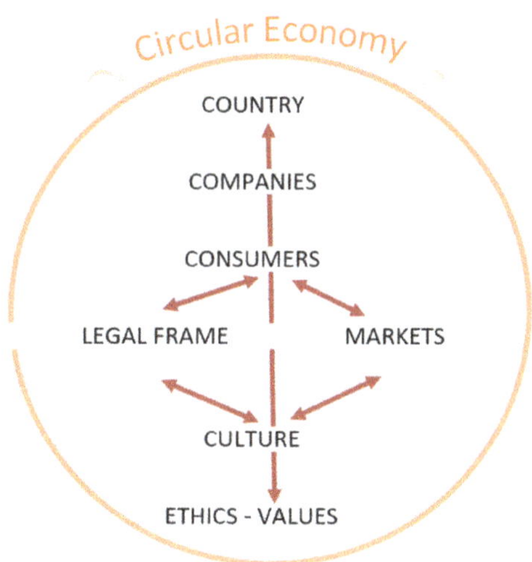

Fig. 1 Macro- and Microstructure analysis when producing with a circular economy model *Source* Adapted from Fang et al. [84], Kirchherr and Piscicelli [112]

is considered by the characteristics of its consumers to be an attractive, necessary, and differentiated product, thus satisfying the needs of the consumer. The rules and regulations, such as certifications requested for entry to the new destination, should be considered the national market existing to achieve the internationalization of the eco-friendly product. Finally, it is essential to consider the ethics and values of the market as those of the company we are dealing with in this operation [112].

Micro- and macro-analysis at the national and international level allows the evaluation of the different objectives either by economic sectors, business models, material flow, long-term results of CE, product life cycle, and CE metrics and indicators [138]. From the management perspective, the analysis of objectives is necessary to generate the required strategies covering national, regional, international, business, and consumer levels [92, 110, 114] and to analyze and verify how the objectives are converted into organizational policies and models [7].

The management of CE needs a redefinition of consumption and production systems that lead to a change in the behaviors of consumers. Therefore, we can say that a mindset change is needed on both the supply and demand sides [122]. Also, entrepreneurs must support their efforts [138].

Considering what has been developed, it can be defined as it is necessary to prepare future professionals, company managers, and create in them a mindset that allows them to develop and apply the appropriate strategies for the generation of change toward a CE that encompasses everything that involves production and consumption and not only the adoption of objectives that allow meeting environmental objectives [138]. Table 3 shows the suggested competencies to be developed by universities to generate a process of change toward a sustainable circular economy model. It should be clarified that it is possible to change toward circularity, but it does not necessarily refer to sustainability.

4 How Do We Achieve These Competencies?

Circular economy is an interdisciplinary discipline and as a subject or topic to be developed is relatively new in university institutions. For this reason, there are no adequate teaching materials, and there is no adequate coordination for its inclusion in the curricula [148]. According to Ellen MacArthur Foundation [75], all undergraduate and graduate students should follow at least one subject, which allows them to know and learn about the circular economy and how unsustainable the linear model is. Regardless of their job placement within the company, making better decisions for society and the environment is essential.

It is essential to consider that the fundamentals of circularity may be the same for all students worldwide. However, its application depends on each economy and

Table 3 Competences to be developed in EC students

Competences required in EC Janssens et al. [108]	**Sustainability competences** UNESCO [168]	**Circular competencies in design** Sumter et al. [163]
a. Competencias técnicas • Professional knowledge and accuracy directed toward sustainability • STEM-skills (Science + Technology + Mathematics) • STEAM-skills (out-of-the-box thinking and creativity	**a.** Systems thinking competency	Systems thinking and holistic thinking
	b. Anticipatory competency	Offering circle products achieve multiple life cycles
	c. Normative competency	• Setting circular criteria • Assessing circular solutions • Analyze and determine the environmental impact (over multiple life cycles)
b. Competencias valorativas • Critically contextualizing of knowledge • Knowledge to engage social, economic, and environmental problems • Transformation of knowledge into action • Environmental awareness and the application of ethical and sustainable principles	**d.** Strategic competency	• Analyze consumer value perception during the design process • Consider circular logistics, inverse logistics, and distribution process
	e. Collaboration competency	
	f. Critical thinking competency	
	g. Self-awareness competency	
c. Competencias transversales • Entrepreneurship competences • Collaboration, creative thinking, and flexibility	**h.** Integrated problem-solving competency	

Source Adapted from Janssens et al. [108], UNESCO [168], andSumter et al. [163]

its business sector where the knowledge of CE is transferred and where companies should support with the technical part and experience to participate in the development, organization of the subjects and their evaluation [65]. We can say that there is a clear opportunity for students to learn from experiences in a circular economy, generating little by little an impact on society.

Universities can contribute to the promotion of CE through activities at the university level as well as through their curriculum [148]. Nowadays, despite the progress about the CE message, it is still not enough,what does exist is a comprehensive development in subjects under the term “sustainability,” which has been adapted to different curricula and subjects. Table 3 shows the different possible subjects to be part of a curriculum for students studying careers related to business sciences.

The study centers and the different programs evaluate how to initiate CE learning, whether it is through a subject, a group of subjects, a specialization, or postgraduate studies. Table 4 also shows the learning methodologies used throughout the subjects using one more of these throughout the course. The last column shows us the different actors that would participate in virtual or face-to-face visits to inform the students about their experiences and knowledge acquired through the course, depending on the subject related to the theory or the activity in progress.

Considering the list of suggested subjects and depending on the level of learning granted by the university and the competencies developed, a CE manager or strategist could be generated, with knowledge, skills, and competencies different from those acquired in the classic careers of engineering, administration, and economics. The professional specializing in CE can develop policies and business strategies based on CE, mitigate the risks in the transition of change, and be a visionary, detecting business opportunities in CE. Likewise, they see the efficiency of resources and organizational models applied to CE [83].

5 Closing Remarks

The circular economy is gradually gaining importance in different countries where policies, legislation, and standards are being applied to improve society’s quality of life and the world. However, it is not enough; the process is prolonged and requires the knowledge and application of the model, for which it is considered that the educational centers are the ones that could develop the knowledge and be through the students to support the transition or continuity of a circular model. The innovative educational approach facilitates the change toward a more sustainable society in CE and allows the development of professionals with new knowledge, attitudes, and competencies to apply and manage the new model.

Table 4 Courses, methodologies, and actors related to circular economy

Courses related to circular economy	Methodology	Actors
CE for business	Practical workshops on circular economy	Policymakers
Sustainability management	Business Cases	Innovative manufactures
Management	Practical problems	Designers
Efficient resource management	Solve problems collaboratively	Environmental organizations
Design and development	Stimulating critical reflection	EC companies
Implications and opportunities for business and policymakers	Learning to value experience, positive and negative experience	International organizations
The consumer perception and behavior		
Micro- and macroeconomics foundations	Creating opportunities for participatory learning	
Innovation and tech innovation	Workshops on challenges and opportunities	
Marketing	Several lectures of CE	
Sustainable firm	Simulation software, gamming	
Elements for agri-food technology	Learning from errors	
Manufacturing, logistics, and inverse logistics	Written reports that include describing results and decisions taken	
Circular entrepreneurship and business models	Real-life cases using "ideal." Framework principles	
Health and environment	Good and bad practices	
Eco-Innovation as an opportunity	Presentation of group work results	
Circular models of production and consumption		
Risks with the CE transition		
Life cycle thinking		

Source Adapted from Giannoccaro et al. [94], Minguez et al. [136], Lanz et al. [120], Kirchherr and Piscicelli [112]

References

1. Abiakam N, Worsley P, Jayabal H, Mitchell K, Jones M, Fletcher J, Spratt F, Bader D (2021) Personal protective equipment related skin reactions in healthcare professionals during COVID-19. Int Wound J 18(3):312–322. https://doi.org/10.1111/iwj.13534
2. Abudureheman A, Nilupaer A (2021) Optimization model design of cross-border e-commerce transportation path under the background of prevention and control of COVID-19 pneumonia. Soft Comput 25(18):12007–12015. https://doi.org/10.1007/s00500-021-05685-6
3. Acuña-Zegarra MA, Santana-Cibrian M, Velasco-Hernandez JX (2020) Modeling behavioral change and COVID-19 containment in Mexico: a trade-off between lockdown and compliance. Math Biosci 325:108370. https://doi.org/10.1016/j.mbs.2020.108370
4. Afshan G, Shahid S, Tunio MN (2021) Learning experiences of women entrepreneurs amidst COVID-19. Int J Gend Entrep 13(2):162–186. https://doi.org/10.1108/IJGE-09-2020-0153
5. Aguirre AA, Catherina R, Frye H, Shelley L (2020) Illicit wildlife trade, wet markets, and COVID-19: preventing future pandemics. World Med & Health Policy 12(3):256–265. https://doi.org/10.1002/wmh3.348
6. Ahsan MM (2020) Strategic decisions on urban built environment to pandemics in Turkey: lessons from COVID-19. J Urban Manag 9(3):281–285. https://doi.org/10.1016/j.jum.2020.07.001
7. Akenji L, Bengtsson M, Bleischwitz R, Tukker A, Schandl H (2016) Ossified materialism: introduction to the special volume on absolute reductions in materials throughput and emissions. J Clean Prod 132:1–12. https://doi.org/10.1016/j.jclepro.2016.03.071
8. Alan H, Eskin Bacaksiz F, Tiryaki Sen H, Taskiran Eskici G, Gumus E, Harmanci Seren AK (2021) "I'm a hero, but...": an evaluation of depression, anxiety, and stress levels of frontline healthcare professionals during COVID-19 pandemic in Turkey. Perspect Psychiatr Care 57(3):1126–1136. https://doi.org/10.1111/ppc.12666
9. Ali W (2020) Online and remote learning in higher education institutes: a necessity in light of COVID-19 pandemic. High Educ Stud 10(3):16–25
10. Allen J, Rowan L, Singh P (2020) Teaching and teacher education in the time of COVID-19. Asia-Pac J Teach Educ 48(3):233–236. https://doi.org/10.1080/1359866X.2020.1752051
11. Alsairafi Z, Naser AY, Alsaleh FM, Awad A, Jalal Z (2021) Mental health status of healthcare professionals and students of health sciences faculties in Kuwait during the COVID-19 pandemic. Int J Environ Res Public Health 18(4). https://doi.org/10.3390/ijerph18042203
12. Altindis E (2021) Inequitable COVID-19 vaccine distribution and the intellectual property rights prolong the pandemic. Expert Rev Vaccines, null-null. https://doi.org/10.1080/14760584.2022.2014819
13. Alvarez-Risco A, Dawson J, Johnson W, Conteh-Barrat M, Aslani P, Del-Aguila-Arcentales S, Diaz-Risco S (2021) Ebola virus disease outbreak: a global approach for health systems [Ebola; health professionals; global approach; pharmacist: pharmaceutical care; COVID-19.]. 2021 53(4). http://www.revfarmacia.sld.cu/index.php/far/article/view/491
14. Alvarez-Risco A, Del-Aguila-Arcentales S (2015) Errores de prescripción como barrera para la Atención Farmacéutica en establecimientos de salud públicos: Experiencia Perú. Pharmaceutical Care España 17(6):725–731. https://www.pharmcareesp.com/index.php/PharmaCARE/article/view/246
15. Alvarez-Risco A, Del-Aguila-Arcentales S (2021a) A note on changing regulation in international business: the World Intellectual Property Organization (WIPO) and artificial intelligence. In: Verbeke A, van Tulder R, Rose EL, Wei Y (eds) The multiple dimensions of institutional complexity in international business research, vol 15. Emerald Publishing Limited, pp 363–371. https://doi.org/10.1108/S1745-886220210000015020
16. Alvarez-Risco A, Del-Aguila-Arcentales S (2021b) Public policies and private efforts to increase women entrepreneurship based on STEM background. In: Mari M, Poggesi S, Foss L (eds) Women's entrepreneurship in STEM disciplines: issues and perspectives. Springer International Publishing, pp 75–87. https://doi.org/10.1007/978-3-030-83792-1_5

17. Alvarez-Risco A, Del-Aguila-Arcentales S, Delgado-Zegarra J, Yáñez JA, Diaz-Risco S (2019) Doping in sports: findings of the analytical test and its interpretation by the public. Sport Sci Health 15(1):255–257. https://doi.org/10.1007/s11332-018-0484-8
18. Alvarez-Risco A, Del-Aguila-Arcentales S, Diaz-Risco S (2018a) Pharmacovigilance as a tool for sustainable development of healthcare in Peru. PharmacoVigilance Rev 10(2):4–6
19. Alvarez-Risco A, Del-Aguila-Arcentales S, Rosen MA, García-Ibarra V, Maycotte-Felkel S, Martínez-Toro GM (2021b) Expectations and interests of university students in COVID-19 times about sustainable development goals: evidence from Colombia, Ecuador, Mexico, and Peru. Sustain 13(6):3306. https://doi.org/10.3390/su13063306
20. Alvarez-Risco A, Del-Aguila-Arcentales S, Yáñez JA (2021c) Telemedicine in Peru as a result of the COVID-19 pandemic: perspective from a country with limited internet access. Am J Trop Med Hyg 105(1):6–11. https://doi.org/10.4269/ajtmh.21-0255
21. Alvarez-Risco A, Del-Aguila-Arcentales S, Yáñez JA, Rosen MA, Mejia CR (2021d) Influence of technostress on academic performance of university medicine students in Peru during the COVID-19 pandemic. Sustain 13(16):8949. https://doi.org/10.3390/su13168949
22. Alvarez-Risco A, Del-Aguila-Arcentales S, Stevenson JG (2015) Pharmacists and mass communication for implementing pharmaceutical care. Am J Pharm Benefits 7(3):e125–e126
23. Alvarez-Risco A, Delgado-Zegarra J, Yáñez JA, Diaz-Risco S, Del-Aguila-Arcentales S (2018b) Predation risk by gastronomic boom - Case Peru. J Landsc Ecol 11(1):100–103. https://doi.org/10.2478/jlecol-2018-0003
24. Alvarez-Risco A, Estrada-Merino A, Anderson-Seminario M. d. l. M, Mlodzianowska S, García-Ibarra V, Villagomez-Buele C, Carvache-Franco M (2021) Multitasking behavior in online classrooms and academic performance: case of university students in Ecuador during COVID-19 outbreak. Interact Technol Smart Educ 18(3):422–434. https://doi.org/10.1108/ITSE-08-2020-0160
25. Alvarez-Risco A, Mejia CR, Delgado-Zegarra J, Del-Aguila-Arcentales S, Arce-Esquivel AA, Valladares-Garrido MJ, Del Portal MR, Villegas LF, Curioso WH, Sekar MC, Yáñez JA (2020) The Peru approach against the COVID-19 infodemic: insights and strategies. Am J Trop Med Hyg 103(2):583–586. https://doi.org/10.4269/ajtmh.20-053
26. Alvarez-Risco A, Mil JWFv (2007) Pharmaceutical care in community pharmacies: Practice and research in Peru. Ann Pharmacother 41(12):2032–2037. https://doi.org/10.1345/aph.1K117
27. Alvarez-Risco A, Mlodzianowska S, García-Ibarra V, Rosen MA, Del-Aguila-Arcentales S (2021f) Factors affecting green entrepreneurship intentions in business university students in COVID-19 pandemic times: case of Ecuador. Sustain 13(11):6447. https://doi.org/10.3390/su13116447
28. Alvarez-Risco A, Mlodzianowska S, Zamora-Ramos U, Del-Aguila S (2021) Green entrepreneurship intention in university students: the case of Peru. Entrep Bus Econ Rev 9(4):85–100. https://doi.org/10.15678/EBER.2021.090406
29. Alvarez-Risco A, Quipuzco-Chicata L, Escudero-Cipriani C (2021) Determinantes de la intención de recompra en línea en tiempos de COVID-19: evidencia de una economía emergente. Lecturas de Economía (96). https://doi.org/10.17533/udea.le.n96a342638
30. Alvarez-Risco A, Quiroz-Delgado D, Del-Aguila-Arcentales S (2016a) Pharmaceutical care in hypertension patients in a peruvian hospital. Indian J Public Health Res Dev 7(3):183–188
31. Alvarez-Risco A, Rosen MA, Del-Aguila-Arcentales S (2020b) A new regulation for supporting a circular economy in the plastic industry: the case of Peru (short communication). J Landsc Ecol 13(1):1–3. https://doi.org/10.2478/jlecol-2020-0004
32. Alvarez-Risco A, Turpo-Cama A, Ortiz-Palomino L, Gongora-Amaut N, Del-Aguila-Arcentales S (2016) Barreras para la implementación de la Atención Farmacéutica en establecimientos farmacéuticos de Cusco, Perú. Pharmaceutical Care España 18(5):194–205. https://www.pharmcareesp.com/index.php/PharmaCARE/article/view/326
33. Álvarez-Risco A, Zegarra Arellano E, Matos Valerio E, Mejía Acosta N, Solis Tarazona Z (2013) Campaña de atención farmacéutica como estrategia de implementación de los servicios farmacéuticos: Experiencia Perú. Pharmaceutical Care España 15(1):35. https://www.pharmcareesp.com/index.php/PharmaCARE/article/view/105

34. Amarillo CR (2020) Feminism on lockdown. NACLA Rep Am 52(3):274–281. https://doi.org/10.1080/10714839.2020.1809084
35. Anderberg E, Nordén B, Hansson B (2009) Global learning for sustainable development in higher education: recent trends and a critique. Int J Sustain High Educ 10(4):368–378. https://doi.org/10.1108/14676370910990710
36. Apcho-Ccencho L-V, Cuya-Velásquez B-B, Alvarado Rodríguez D, Anderson-Seminario MdlM, Alvarez-Risco A, Estrada-Merino A, Mlodzianowska S (2021) The impact of international price on the technological industry in the United States and China during times of crisis: commercial war and COVID-19. In: Lawrence KD, Klimberg RK (eds) Advances in business and management forecasting, vol 14. Emerald Publishing Limited, pp 149–160. https://doi.org/10.1108/S1477-407020210000014010
37. Assaf A, Scuderi R (2020) COVID-19 and the recovery of the tourism industry. Tour Econ 26(5):731–733. https://doi.org/10.1177/1354816620933712
38. Bacq S, Lumpkin GT (2021) Social entrepreneurship and COVID-19. J Manage Stud 58(1):285–288. https://doi.org/10.1111/joms.12641
39. Banco Interamericano de Desarrollo (2021) IDB Group Climate Change Action Plan 2021–2025 [Plan de Acción del Grupo BID en Materia de Cambio Climático 2021–2025]. Retrieved 01/02/2022 from https://publications.iadb.org/publications/spanish/document/Plan-de-accion-del-Grupo-BID-en-materia-de-cambio-climatico-2021-2025.pdf
40. Barello S, Caruso R, Palamenghi L, Nania T, Dellafiore F, Bonetti L, Silenzi A, Marotta C, Graffigna G (2021) Factors associated with emotional exhaustion in healthcare professionals involved in the COVID-19 pandemic: an application of the job demands-resources model. Int Arch Occup Environ Health 94(8):1751–1761. https://doi.org/10.1007/s00420-021-01669-z
41. Baum T, Hai NTT (2020) Hospitality, tourism, human rights and the impact of COVID-19. Int J Contemp Hosp Manag 32(7):2397–2407. https://doi.org/10.1108/IJCHM-03-2020-0242
42. Belitski M, Guenther C, Kritikos AS, Thurik R (2021) Economic effects of the COVID-19 pandemic on entrepreneurship and small businesses. Small Bus Econ. https://doi.org/10.1007/s11187-021-00544-y
43. Berryman T, Sauvé L (2016) Ruling relationships in sustainable development and education for sustainable development. J Environ Educ 47(2):104–117. https://doi.org/10.1080/00958964.2015.1092934
44. Beynaghi A, Trencher G, Moztarzadeh F, Mozafari M, Maknoon R, Leal Filho W (2016) Future sustainability scenarios for universities: moving beyond the United Nations Decade of Education for Sustainable Development. J Clean Prod 112:3464–3478. https://doi.org/10.1016/j.jclepro.2015.10.117
45. Block JH, Fisch C, Hirschmann M (2021) The determinants of bootstrap financing in crises: evidence from entrepreneurial ventures in the COVID-19 pandemic. Small Bus Econ. https://doi.org/10.1007/s11187-020-00445-6
46. Bocken NMP, Olivetti EA, Cullen JM, Potting J, Lifset R (2017) Taking the circularity to the next level: a special issue on the circular economy. J Ind Ecol 21(3):476–482. https://doi.org/10.1111/jiec.12606
47. Borzée A, McNeely J, Magellan K, Miller JR, Porter L, Dutta T, Kadinjappalli KP, Sharma S, Shahabuddin G, Aprilinayati F, Ryan GE (2020) COVID-19 highlights the need for more effective wildlife trade legislation. Trends Ecol & Evol 35(12):1052–1055. https://doi.org/10.1016/j.tree.2020.10.001
48. Bovea MD, Pérez-Belis V (2018) Identifying design guidelines to meet the circular economy principles: a case study on electric and electronic equipment. J Environ Manage 228:483–494. https://doi.org/10.1016/j.jenvman.2018.08.014
49. Bozdağ F, Ergün N (2020) Psychological resilience of healthcare professionals during COVID-19 pandemic. Psychol Rep 124(6):2567–2586. https://doi.org/10.1177/0033294120965477
50. Brewer, A. (2005). Cantillon, Quesnay, and the Tableau Economique. Bristol Economics Discussion Papers, 5, 577.

51. Brouder P (2020) Reset redux: possible evolutionary pathways towards the transformation of tourism in a COVID-19 world. Tour Geogr 22(3):484–490. https://doi.org/10.1080/14616688.2020.1760928
52. Brown R, Rocha A, Cowling M (2020) Financing entrepreneurship in times of crisis: exploring the impact of COVID-19 on the market for entrepreneurial finance in the United Kingdom. Int Small Bus J 38(5):380–390. https://doi.org/10.1177/0266242620937464
53. Burger M, Stavropoulos S, Ramkumar S, Dufourmont J, van Oort F (2019) The heterogeneous skill-base of circular economy employment. Res Policy 48(1):248–261. https://doi.org/10.1016/j.respol.2018.08.015
54. Carvache-Franco, M., Alvarez-Risco, A., Carvache-Franco, O., Carvache-Franco, W., Estrada-Merino, A., & Villalobos-Alvarez, D. (2021). Perceived value and its influence on satisfaction and loyalty in a coastal city: a study from Lima, Peru. J Policy Res Tour, Leis Events, 1–16. https://doi.org/10.1080/19407963.2021.1883634
55. Carvache-Franco M, Alvarez-Risco A, Carvache-Franco W, Carvache-Franco O, Estrada-Merino A, Rosen MA (2021b) Coastal cities seen from loyalty and their tourist motivations: a study in Lima Peru. Sustain 13(21):11575. https://doi.org/10.3390/su132111575
56. Carvache-Franco M, Carvache-Franco O, Carvache-Franco W, Alvarez-Risco A, Estrada-Merino A (2021c) Motivations and segmentation of the demand for coastal cities: a study in Lima Peru. Int J Tour Res 23(4):517–531. https://doi.org/10.1002/jtr.2423
57. Cegarra-Navarro J-G, Vătămănescu E-M, Martínez-Martínez A (2021) A context-driven approach on coping with COVID-19: from hiding knowledge toward citizen engagement. Knowl Process Manag 28(2):134–140. https://doi.org/10.1002/kpm.1662
58. Chafloque-Cespedes R, Alvarez-Risco A, Robayo-Acuña PV, Gamarra-Chavez CA, Martinez-Toro GM, Vicente-Ramos W (2021) Effect of sociodemographic factors in entrepreneurial orientation and entrepreneurial intention in university students of Latin American business schools. In Jones P, Apostolopoulos N, Kakouris A, Moon C, Ratten V, Walmsley A (eds) Universities and entrepreneurship: meeting the educational and social challenges, vol 11. Emerald Publishing Limited, pp 151–165. https://doi.org/10.1108/S2040-724620210000011010
59. Chafloque-Céspedes R, Vara-Horna A, Asencios-Gonzales Z, López-Odar D, Álvarez-Risco A, Quipuzco-Chicata L, Schulze C, Sánchez-Villagomez M (2020) Presentismo académico y violencia contra las mujeres en escuelas de negocios e ingeniería en universidades peruanas. Lecturas de Economía 0(93):127–153. https://doi.org/10.17533/udea.le.n93a340726
60. Chen T, Peng L, Yin X, Rong J, Yang J, Cong G (2020) Analysis of user satisfaction with online education platforms in China during the COVID-19 pandemic. Healthcare 8(3). https://doi.org/10.3390/healthcare8030200
61. Chen X, Zhang SX, Jahanshahi AA, Alvarez-Risco A, Dai H, Li J, Ibarra VG (2020b) Belief in a COVID-19 conspiracy theory as a predictor of mental health and well-being of health care workers in Ecuador: cross-sectional survey study. JMIR Public Health Surveill 6(3):e20737. https://doi.org/10.2196/20737
62. Chung SA, Olivera S, Román BR, Alanoca E, Moscoso S, Terceros BL, Alvarez-Risco A, Yáñez JA (2021) Temáticas de la producción científica de la Revista Cubana de Farmacia indizada en Scopus (1967–2020) [Revista Cubana de Farmacia; farmacia; Cuba; evidencia]. 2021 54(1). http://www.revfarmacia.sld.cu/index.php/far/article/view/511
63. Circle Economy (2019) Why employers should care about the circular economy. Retrieved 01/01/2022 from https://medium.com/circleeconomy/why-employers-should-care-about-the circular-economy-2c0740372eec
64. Cruz-Torres W, Alvarez-Risco A, Del-Aguila-Arcentales S (2021) Impact of Enterprise Resource Planning (ERP) implementation on performance of an education enterprise: a Structural Equation Modeling (SEM). Stud Bus Econ 16(2):37–52. https://doi.org/10.2478/sbe-2021-0023
65. Cui X, Di Y, Nan N, Liu X, Zhou C (2008) Practice and effectiveness of the talent training model of RRSE under the combination of industry and education and college-enterprise thinking. Renew Resour Recycl Econ 11(7):7–10

66. Del-Aguila-Arcentales S, Alvarez-Risco A (2013) Human error or burnout as explanation for mistakes in pharmaceutical laboratories. Accred Qual Assur 18(5):447–448. https://doi.org/10.1007/s00769-013-1000-0
67. del Castillo FA (2021) Temporary waiver of intellectual property on Covid-19 vaccines: toward the creation of a better, post-pandemic society. J Public Health 43(3):e559–e560. https://doi.org/10.1093/pubmed/fdab184
68. Delgado-Zegarra J, Alvarez-Risco A, Yáñez JA (2018) Uso indiscriminado de pesticidas y ausencia de control sanitario para el mercado interno en Perú. Rev Panam Salud Publica 42:e3. https://doi.org/10.26633/RPSP.2018.3
69. Deressa W, Worku A, Abebe W, Gizaw M, Amogne W (2021) Risk perceptions and preventive practices of COVID-19 among healthcare professionals in public hospitals in Addis Ababa Ethiopia. PLoS ONE 16(6):e0242471. https://doi.org/10.1371/journal.pone.0242471
70. Dominelli L (2021) A green social work perspective on social work during the time of COVID-19. Int J Soc Welf 30(1):7–16. https://doi.org/10.1111/ijsw.12469
71. Duarte Alonso A, Kok SK, Bressan A, O'Shea M, Sakellarios N, Koresis A, Solis MA, Santoni LJ (2020) COVID-19, aftermath, impacts, and hospitality firms: an international perspective. Int J Hosp Manag 91:102654. https://doi.org/10.1016/j.ijhm.2020.102654
72. El Peruano (2021) Ordinance incorporating the principles of circular economy in local management [Ordenanza que incorpora a la gestión local los principios de economía circular]. Retrieved 01/01/2022 from https://busquedas.elperuano.pe/normaslegales/ordenanza-que-incorpora-a-la-gestion-local-los-principios-de-ordenanza-no-2367-2021-1975306-1
73. Ellen MacArthur Foundation (2015) Delivering the CE. A Toolkit for Policymakers. Retrieved 12/30/2021 from https://ellenmacarthurfoundation.org/a-toolkit-for-policymakers
74. Ellen MacArthur Foundation (2020) Towards the circular economy Vol. 1: an economic and business rationale for an accelerated transition. Retrieved 01/01/2022 from https://ellenmacarthurfoundation.org/towards-the-circular-economy-vol-1-an-economic-and-business-rationale-for-an
75. Ellen MacArthur Foundation (2021) What is a circular economy? Retrieved 12/22/2021 from https://ellenmacarthurfoundation.org/topics/circular-economy-introduction/overview
76. EMAS (2021) EMAS Regulation. Retrieved 01/01/2022 from https://ec.europa.eu/environment/emas/emas_publications/policy_en.htm
77. Enciso-Zarate A, Guzmán-Oviedo J, Sánchez-Cardona F, Martínez-Rohenes D, Rodríguez-Palomino JC, Alvarez-Risco A, Del-Aguila-Arcentales S, Diaz-Risco S (2016) Evaluación de la contaminación con agentes citotóxicos en hospitales en Colombia. Pharmaceutical Care España 18(6):241–250. https://www.pharmcareesp.com/index.php/PharmaCARE/article/view/354
78. Erfani P, Binagwaho A, Jalloh MJ, Yunus M, Farmer P, Kerry V (2021) Intellectual property waiver for covid-19 vaccines will advance global health equity. BMJ 374:n1837. https://doi.org/10.1136/bmj.n1837
79. European Commission - EU Ecolabel (2021) Ecolabel. Retrieved 01/01/2022 from https://ec.europa.eu/environment/ecolabel
80. European Commission (2015) Closing the Loop—An EU Action Plan for the Circular Economy; European Commission: Brussels, Belgium. Retrieved 01/01/2022 from https://www.eea.europa.eu/policy-documents/com-2015-0614-final
81. European Commission (2020) Circular Economy Action Plan: For a Cleaner and More Competitive Europe; European Commission: Brussels, Belgium. Retrieved 01/01/2022 from https://ec.europa.eu/environment/strategy/circular-economy-action-plan_es
82. European Union (2018) Circular Economy roadmap of France: 50 measures for a 100% circular economy. Retrieved 01/01/2022 from https://circulareconomy.europa.eu/platform/en/strategies/circular-economy-roadmap-france-50-measures-100-circular-economy
83. Evans S, Fernando L, Yang M (2017) Sustainable value creation—From concept towards implementation. In: Stark R, Seliger G, Bonvoisin J (eds), Sustainable manufacturing: challenges, solutions and implementation perspectives. Springer International Publishing, pp 203–220. https://doi.org/10.1007/978-3-319-48514-0_13

84. Fang Y, Côté RP, Qin R (2007) Industrial sustainability in China: Practice and prospects for eco-industrial development. J Environ Manage 83(3):315–328. https://doi.org/10.1016/j.jenvman.2006.03.007
85. Federal Ministry for The Environment, N. C., Nuclear Safety and Consumer Protection (2021a) German Resource Efficiency Programme (ProgRess) – an overview. Retrieved 01/01/2022 from https://www.bmu.de/en/topics/water-resources-waste/resource-efficiency/german-resource-efficiency-programme-progress-an-overview
86. Federal Ministry for The Environment, N. C., Nuclear Safety and Consumer Protection (2021b) National Programme on Sustainable Consumption. Retrieved 01/01/2022 from https://www.bmu.de/en/publication/national-programme-on-sustainable-consumption
87. Filho WL (2010) Climate change at universities: results of a world survey. In: Leal Filho W (ed), Universities and Climate Change: Introducing Climate Change to University Programmes. Springer Berlin Heidelberg, pp 1–19. https://doi.org/10.1007/978-3-642-10751-1_1
88. Fotiadis A, Polyzos S, Huan T-CTC (2021) The good, the bad and the ugly on COVID-19 tourism recovery. Ann Tour Res 87:103117. https://doi.org/10.1016/j.annals.2020.103117
89. Galanakis CM, Rizou M, Aldawoud TMS, Ucak I, Rowan NJ (2021) Innovations and technology disruptions in the food sector within the COVID-19 pandemic and post-lockdown era. Trends Food Sci Technol 110:193–200. https://doi.org/10.1016/j.tifs.2021.02.002
90. Geissdoerfer M, Savaget P, Bocken NMP, Hultink EJ (2017) The circular economy – A new sustainability paradigm? J Clean Prod 143:757–768. https://doi.org/10.1016/j.jclepro.2016.12.048
91. German Council for Sustainable Development (2021) German Sustainable Development Strategy. Retrieved 01/01/2022 from https://www.nachhaltigkeitsrat.de/en/german-sustainablility-strategy/?cn-reloaded=1
92. Ghisellini P, Cialani C, Ulgiati S (2016) A review on circular economy: the expected transition to a balanced interplay of environmental and economic systems. J Clean Prod 114:11–32. https://doi.org/10.1016/j.jclepro.2015.09.007
93. Giannoccaro I (2018) Centralized vs. decentralized supply chains: the importance of decision maker's cognitive ability and resistance to change. Ind Mark Manage 73:59–69. https://doi.org/10.1016/j.indmarman.2018.01.034
94. Giannoccaro I, Ceccarelli G, Fraccascia L (2021) Features of the higher education for the circular economy: the case of Italy. Sustain 13(20). https://doi.org/10.3390/su132011338
95. Gouvernement (2015) Energy transition. Retrieved 01/01/2022 from https://www.gouvernement.fr/en/energy-transition
96. Government of the Netherlands (2016) A circular economy in the Netherlands by 2050. Retrieved 01/01/2022 from https://www.government.nl/documents/policy-notes/2016/09/14/a-circular-economy-in-the-netherlands-by-2050
97. Grantham Research Institute on Climate Change and the Environment (2020) LAW n ° 2020-105 of 10 February 2020 relating to the fight against waste and the circular economy. Retrieved 01/01/2022 from https://climate-laws.org/geographies/france/laws/law-n-2020-105-of-10-february-2020-relating-to-the-fight-against-waste-and-the-circular-economy
98. Gregson N, Crang M, Botticello J, Calestani M, Krzywoszynska A (2014) Doing the 'dirty work' of the green economy: resource recovery and migrant labour in the EU. Eur Urban RegNal Stud 23(4):541–555. https://doi.org/10.1177/0969776414554489
99. Gulati G, Kelly BD (2020) Domestic violence against women and the COVID-19 pandemic: what is the role of psychiatry? Int J Law Psychiatry 71:101594. https://doi.org/10.1016/j.ijlp.2020.101594
100. Gursoy D, Chi CG (2020) Effects of COVID-19 pandemic on hospitality industry: review of the current situations and a research agenda. J Hosp Market Manag 29(5):527–529. https://doi.org/10.1080/19368623.2020.1788231
101. Guyot Phung C (2019) Implications of the circular economy and digital transition on skills and green jobs in the plastics industry. Field Actions Sci Rep. The journal of field actions(Special Issue 19):100–107

102. Hanzl M (2020) Urban forms and green infrastructure – the implications for public health during the COVID-19 pandemic. Cities & Health, 1–5. https://doi.org/10.1080/23748834.2020.1791441
103. Hepburn C, O'Callaghan B, Stern N, Stiglitz J, Zenghelis D (2020) Will COVID-19 fiscal recovery packages accelerate or retard progress on climate change? Oxf Rev Econ Policy, 36(Supplement_1):S359-S381. https://doi.org/10.1093/oxrep/graa015
104. Heshmati A (2017) A review of the circular economy and its implementation. Int J Green Econ 11(3–4):251–288. https://doi.org/10.1504/IJGE.2017.089856
105. Ho C-Y, Tsai B-H, Chen C-S, Lu M-T (2021) Exploring green marketing orientations toward sustainability the hospitality industry in the COVID-19 pandemic. Sustain 13(8). https://doi.org/10.3390/su13084348
106. Huckle J, Wals AEJ (2015) The UN Decade of Education for Sustainable Development: business as usual in the end. Environ Educ Res 21(3):491–505. https://doi.org/10.1080/13504622.2015.1011084
107. INECC (2020) Assessment of the current situation of the circular economy for the development of a roadmap for Brazil, Chile, Mexico and Uruguay [Evaluación de la situación actual de la economía circular para el desarrollo de una hoja de ruta para Brasil, Chile, México y Uruguay]. Retrieved 01/01/2022 from https://www.gob.mx/cms/uploads/attachment/file/641380/VF_version_ejecutiva_Economia_Circular_2.pdf
108. Janssens L, Kuppens T, Van Schoubroeck S (2021) Competences of the professional of the future in the circular economy: evidence from the case of Limburg Belgium. J Clean Prod 281:125365. https://doi.org/10.1016/j.jclepro.2020.125365
109. Jecker NS, Atuire CA (2021) What's yours is ours: waiving intellectual property protections for COVID-19 vaccines. J Med Ethics 47(9):595. https://doi.org/10.1136/medethics-2021-107555
110. Kalmykova Y, Sadagopan M, Rosado L (2018) Circular economy – From review of theories and practices to development of implementation tools. Resour, Conserv Recycl 135:190–201. https://doi.org/10.1016/j.resconrec.2017.10.034
111. Kaushal V, Srivastava S (2021) Hospitality and tourism industry amid COVID-19 pandemic: Perspectives on challenges and learnings from India. Int J Hosp Manag 92:102707. https://doi.org/10.1016/j.ijhm.2020.102707
112. Kirchherr J, Piscicelli L (2019) Towards an education for the circular economy (ECE): five teaching principles and a case study. Resour Conserv Recycl 150:104406. https://doi.org/10.1016/j.resconrec.2019.104406
113. Kirchherr J, Piscicelli L, Bour R, Kostense-Smit E, Muller J, Huibrechtse-Truijens A, Hekkert M (2018) Barriers to the circular economy: evidence from the European Union (EU). Ecol Econ 150:264–272. https://doi.org/10.1016/j.ecolecon.2018.04.028
114. Kirchherr J, Reike D, Hekkert M (2017) Conceptualizing the circular economy: an analysis of 114 definitions. Resour Conserv Recycl 127:221–232. https://doi.org/10.1016/j.resconrec.2017.09.005
115. Kirk CP, Rifkin LS (2020) I'll trade you diamonds for toilet paper: consumer reacting, coping and adapting behaviors in the COVID-19 pandemic. J Bus Res 117:124–131. https://doi.org/10.1016/j.jbusres.2020.05.028
116. Koetter P, Pelton M, Gonzalo J, Du P, Exten C, Bogale K, Buzzelli L, Connolly M, Edel K, Hoffman A, Legro NR (2020) Implementation and process of a COVID-19 contact tracing initiative: leveraging health professional students to extend the workforce during a pandemic. Am J Infect Control 48(12):1451–1456. https://doi.org/10.1016/j.ajic.2020.08.012
117. Kopnina H (2012) Education for sustainable development (ESD): the turn away from 'environment' in environmental education? Environ Educ Res 18(5):699–717. https://doi.org/10.1080/13504622.2012.658028
118. Krishtel P, Malpani R (2021) Suspend intellectual property rights for covid-19 vaccines. BMJ 373:n1344. https://doi.org/10.1136/bmj.n1344
119. Lackie K, Najjar G, El-Awaisi A, Frost J, Green C, Langlois S, Lising D, Pfeifle AL, Ward H, Xyrichis A, Khalili H (2020). Interprofessional education and collaborative practice research

during the COVID-19 pandemic: Considerations to advance the field. J Interprofessional Care 34(5):583–586. https://doi.org/10.1080/13561820.2020.1807481
120. Lanz M, Nylund H, Lehtonen T, Juuti T, Rättyä K (2019) Circular economy in integrated product and production development education. Procedia Manuf 33:470–476. https://doi.org/10.1016/j.promfg.2019.04.058
121. Lashley MA, Acevedo M, Cotner S, Lortie CJ (2020) How the ecology and evolution of the COVID-19 pandemic changed learning. Ecol Evol 10(22):12412–12417. https://doi.org/10.1002/ece3.6937
122. Laurenti R, Singh J, Frostell B, Sinha R, Binder CR (2018) The socio-economic embeddedness of the circular economy: an integrative framework. Sustain 10(7). https://doi.org/10.3390/su10072129
123. Leiva-Martinez M-A, Anderson-Seminario MdlM, Alvarez-Risco A, Estrada-Merino A, Mlodzianowska S (2021) Price variation in lower goods as of previous economic crisis and the contrast of the current price situation in the context of COVID-19 in Peru. In: Lawrence KD, Klimberg RK (eds) Advances in business and management forecasting, vol 14. Emerald Publishing Limited, pp 161–166. https://doi.org/10.1108/S1477-407020210000014011
124. Leube M, Walcher D (2017). Designing for the next (Circular) ECONOMY. An appeal to renew the Curricula of Design Schools. Des J 20(sup1):S492-S501. https://doi.org/10.1080/14606925.2017.1352999
125. Lidgren A, Rodhe H, Huisingh D (2006) A systemic approach to incorporate sustainability into university courses and curricula. J Clean Prod 14(9):797–809. https://doi.org/10.1016/j.jclepro.2005.12.011
126. Liguori E, Winkler C (2020) From offline to online: challenges and opportunities for entrepreneurship education following the COVID-19 pandemic. Entrepreneurship Education and Pedagogy 3(4):346–351. https://doi.org/10.1177/2515127420916738
127. Lim S, Prakash A (2021) Pandemics and citizen perceptions about their country: did COVID-19 increase national pride in South Korea? Nations Natl 27(3):623–637. https://doi.org/10.1111/nana.12749
128. Linder M, Williander M (2017) Circular business model innovation: inherent uncertainties. Bus Strateg Environ 26(2):182–196. https://doi.org/10.1002/bse.1906
129. Lopez-Odar D, Alvarez-Risco A, Vara-Horna A, Chafloque-Cespedes R, Sekar MC (2020) Validity and reliability of the questionnaire that evaluates factors associated with perceived environmental behavior and perceived ecological purchasing behavior in Peruvian consumers. Soc Responsib J 16(3):403–417. https://doi.org/10.1108/SRJ-08-2018-0201
130. Lozano R (2006) Incorporation and institutionalization of SD into universities: breaking through barriers to change. J Clean Prod 14(9):787–796. https://doi.org/10.1016/j.jclepro.2005.12.010
131. Lozano R, Lukman R, Lozano FJ, Huisingh D, Lambrechts W (2013) Declarations for sustainability in higher education: becoming better leaders, through addressing the university system. J Clean Prod 48:10–19. https://doi.org/10.1016/j.jclepro.2011.10.006
132. Maritz A, Perenyi A, de Waal G, Buck C (2020) Entrepreneurship as the unsung hero during the current COVID-19 economic crisis: Australian perspectives. Sustain 12(11). https://doi.org/10.3390/su12114612
133. McDonough W, Braungart M (2010) Cradle to cradle: remaking the way we make things. North point press.
134. McKinsey & Company (2015) Europe's Circular Economy Opportunity. Retrieved 01/02/2022 from https://www.mckinsey.com/business-functions/sustainability/our-insights/europes-circular-economy-opportunity#:~:text=It%20was%20first%20published%20in,up%20to%203%20percent%20annually
135. Mejía-Acosta N, Alvarez-Risco A, Solís-Tarazona Z, Matos-Valerio E, Zegarra-Arellano E, Del-Aguila-Arcentales S (2016) Reacciones Adversas a Medicamentos reportadas como resultado de la implementación de Atención Farmacéutica en la Farmacia Institucional DIGEMID - Ministerio de Salud de Perú. Pharmaceutical Care España 18(2):67–74. https://www.pharmcareesp.com/index.php/PharmaCARE/article/view/311

136. Minguez R, Lizundia E, Iturrondobeitia M, Akizu-Gardoki O, Saez-de-Camara E (2020). Education in Circular Economy: Focusing on Life Cycle Thinking at the University of the Basque Country. International Joint Conference on Mechanics, Design Engineering & Advanced Manufacturing
137. Ministerio del Medio Ambiente (2021) What is circular economy? ¿Qué es economia circular? Retrieved 01/01/2022 from https://economiacircular.mma.gob.cl/#:~:text=Economia%C3%ADa%20Circular%20%E2%80%93%20Ministerio%20del%20Medio,la%20contaminaci%C3%B3n%20con%20dise%C3%B1o%20inteligente
138. Morseletto P (2020) Targets for a circular economy. Resour Conserv Recycl 153:104553. https://doi.org/10.1016/j.resconrec.2019.104553
139. Murray A, Skene K, Haynes K (2017) The circular economy: an interdisciplinary exploration of the concept and application in a global context. J Bus Ethics 140(3):369–380. https://doi.org/10.1007/s10551-015-2693-2
140. Obergassel W, Hermwille L, Oberthür S (2021) Harnessing international climate governance to drive a sustainable recovery from the COVID-19 pandemic. Climate Policy 21(10):1298–1306. https://doi.org/10.1080/14693062.2020.1835603
141. OCED. (2016). Organization for Economic Cooperation and Development [Organización para la Cooperación y el Desarrollo Económico]. Retrieved 01/02/2022 from https://www.oecd.org/stories/clima-25-acciones
142. OECD. (2019). Boosting climate action. Policy reformulation from a welfare perspective [Impulso a la acción climática. Reformulación de las políticas desde la óptica del bienestar]. Retrieved 01/02/2022 from https://www.oecd.org/environment/cc/Impulso-a-la-acci%C3%B3n-clim%C3%A1tica-reformulaci%C3%B3n-de-las-pol%C3%ADticas-desde-la-%C3%B3ptica-del-bienestar-Extracto.pdf
143. Okereke M (2021) Towards vaccine equity: should big pharma waive intellectual property rights for COVID-19 vaccines? Public Health Pract 2:100165. https://doi.org/10.1016/j.puhip.2021.100165
144. Park J, Sarkis J, Wu Z (2010) Creating integrated business and environmental value within the context of China's circular economy and ecological modernization. J Clean Prod 18(15):1494–1501. https://doi.org/10.1016/j.jclepro.2010.06.001
145. Pitt J, Heinemeyer C (2015) Introducing ideas of a circular economy. In: Environment, ethics and cultures. Brill Sense, pp 245–260
146. Potting J, Hekkert M, Worrell E, Hanemaaijer A (2017) Circular economy: measuring innovation in the product chain. PBL Publishers.
147. Prieto-Sandoval V, Jaca García C, Ormazabal Goenaga M (2016) Circular Economy: An economic and industrial model to achieve the sustainability of society. Proceedings of the 22nd Annual International Sustainable Development Research Society Conference. Rethinking Sustainability Models and Practices: Challenges for the New and Old World Contexts
148. Qu D, Shevchenko T, Yan X (2020) University curriculum education activities towards circular economy implementation. Int J Sci Technol Res 9(5):200–206
149. Quispe-Cañari JF, Fidel-Rosales E, Manrique D, Mascaró-Zan J, Huamán-Castillón KM, Chamorro–Espinoza SE, Garayar–Peceros H, Ponce–López VL, Sifuentes-Rosales J, Alvarez-Risco A, Yáñez JA (2021) Self-medication practices during the COVID-19 pandemic among the adult population in Peru: a cross-sectional survey. Saudi Pharm J 29(1):1–11. https://doi.org/10.1016/j.jsps.2020.12.001
150. Raudenská J, Steinerová V, Javůrková A, Urits I, Kaye AD, Viswanath O, Varrassi G (2020) Occupational burnout syndrome and post-traumatic stress among healthcare professionals during the novel coronavirus disease 2019 (COVID-19) pandemic. Best Pract Res Clin Anaesthesiol 34(3):553–560. https://doi.org/10.1016/j.bpa.2020.07.008
151. Reike D, Vermeulen WJV, Witjes S (2018) The circular economy: new or refurbished as CE 3.0?—Exploring controversies in the conceptualization of the circular economy through a focus on history and resource value retention options. Resour Conserv Recycl 135:246–264. https://doi.org/10.1016/j.resconrec.2017.08.027

152. Roesch E, Amin A, Gupta J, García-Moreno C (2020) Violence against women during covid-19 pandemic restrictions. BMJ 369:m1712. https://doi.org/10.1136/bmj.m1712
153. Rojas-Osorio M, Alvarez-Risco A (2019) Intention to use smartphones among Peruvian university students. Int J Interact Mob Technol (iJIM) 13(03):13. https://doi.org/10.3991/ijim.v13i03.9356
154. Rojas Román B, Moscoso S, Chung SA, Limpias Terceros B, Álvarez-Risco A, Yáñez JA (2020) Tratamiento de la COVID-19 en Perú y Bolivia y los riesgos de la automedicación [COVID-19; tratamiento; estrategia; fármacos; automedicación; Perú; Bolivia.]. 2020 53(2). http://www.revfarmacia.sld.cu/index.php/far/article/view/435/351
155. Salvador R, Barros MV, Luz LMd, Piekarski CM, de Francisco AC (2020) Circular business models: current aspects that influence implementation and unaddressed subjects. J Clean Prod 250:119555. https://doi.org/10.1016/j.jclepro.2019.119555
156. Sánchez-Clavijo LM, Martínez-Callejas SJ, Acevedo-Charry O, Diaz-Pulido A, Gómez-Valencia B, Ocampo-Peñuela N, Ocampo D, Olaya-Rodríguez MH, Rey-Velasco JC, Soto-Vargas C, Ochoa-Quintero JM (2021) Differential reporting of biodiversity in two citizen science platforms during COVID-19 lockdown in Colombia. Biol Conserv 256:109077. https://doi.org/10.1016/j.biocon.2021.109077
157. Sánchez OR, Vale DB, Rodrigues L, Surita FG (2020) Violence against women during the COVID-19 pandemic: an integrative review. Int J Gynecol Obstet 151(2):180–187. https://doi.org/10.1002/ijgo.13365
158. Sekalala S, Forman L, Hodgson T, Mulumba M, Namyalo-Ganafa H, Meier BM (2021) Decolonising human rights: how intellectual property laws result in unequal access to the COVID-19 vaccine. BMJ Glob Health 6(7):e006169. https://doi.org/10.1136/bmjgh-2021-006169
159. Shaw R, Kim Y-K, Hua J (2020) Governance, technology and citizen behavior in pandemic: lessons from COVID-19 in East Asia. Prog Disaster Sci 6:100090. https://doi.org/10.1016/j.pdisas.2020.100090
160. Sigala M (2020) Tourism and COVID-19: impacts and implications for advancing and resetting industry and research. J Bus Res 117:312–321. https://doi.org/10.1016/j.jbusres.2020.06.015
161. Stahel WR (2016) The circular economy. Nature 531(7595):435–438. https://doi.org/10.1038/531435a
162. Steenge AE, van den Berg R (2008) Impact studies without Multipliers: Lessons from Quesnay's Tableau Economique. International Input-Output Meeting on Managing the Environment
163. Sumter D, de Koning J, Bakker C, Balkenende R (2020) Circular economy competencies for design. Sustaina 12(4). https://doi.org/10.3390/su12041561
164. Tran LTT (2021) Managing the effectiveness of e-commerce platforms in a pandemic. J Retail Consum Serv 58:102287. https://doi.org/10.1016/j.jretconser.2020.102287
165. Tura N, Hanski J, Ahola T, Ståhle M, Piiparinen S, Valkokari P (2019) Unlocking circular business: a framework of barriers and drivers. J Clean Prod 212:90–98. https://doi.org/10.1016/j.jclepro.2018.11.202
166. Türkeli S, Schophuizen M (2019) Decomposing the complexity of value: integration of digital transformation of education with circular economy transition. Soc Sci 8(8). https://doi.org/10.3390/socsci8080243
167. UNEP (2021) Environmental Moments: A UNEP@50 timeline. Retrieved 01/02/2022 from https://www.unep.org/who-we-are/about-us/environmental-moments-unep50-timeline
168. UNESCO (2017) Education for Sustainable Development Goals: Learning Objectives. Retrieved 01/01/2022 from https://www.unesco.de/sites/default/files/2018-08/unesco_education_for_sustainable_development_goals.pdf
169. Van Buren N, Demmers M, Van der Heijden R, Witlox F (2016). Towards a circular economy: the role of Dutch logistics industries and governments. Sustain 8(7). https://doi.org/10.3390/su8070647
170. van Oort F, Ramkumar S, Stavropoulos S, Burger M, Dufourmont J, Thissen M (2018) Putting the Circular Economy to Work: Measuring Gross Employment Effects in the Netherlands, 1996–2015

171. Vidya CT, Prabheesh KP (2020) Implications of COVID-19 pandemic on the global trade networks. Emerg Mark Financ Trade 56(10):2408–2421. https://doi.org/10.1080/1540496X.2020.1785426
172. Viero A, Barbara G, Montisci M, Kustermann K, Cattaneo C (2021) Violence against women in the Covid-19 pandemic: a review of the literature and a call for shared strategies to tackle health and social emergencies. Forensic Sci Int 319:110650. https://doi.org/10.1016/j.forsciint.2020.110650
173. WEF (2020) Schools of the Future Defining New Models of Education for the Fourth Industrial Revolution. Retrieved 01/02/2022 from http://www3.weforum.org/docs/WEF_Schools_of_the_Future_Report_2019.pdf
174. WEF (2021) Climate Governance Initiative. Retrieved 01/02/2022 from https://www.weforum.org/projects/climate-governance-initiative
175. Wen J, Kozak M, Yang S, Liu F (2021) COVID-19: potential effects on Chinese citizens' lifestyle and travel. Tour Rev 76(1):74–87. https://doi.org/10.1108/TR-03-2020-0110
176. Whalen KA, Milios L, Nussholz J (2018) Bridging the gap: Barriers and potential for scaling reuse practices in the Swedish ICT sector. Resour Conserv Recycl 135:123–131. https://doi.org/10.1016/j.resconrec.2017.07.029
177. Wu T-Y, Ford O, Rainville AJ, Bessire R (2020) COVID-19 care package distribution for senior citizens and families in Detroit and Hamtramck, Michigan. J Hunger & Environ Nutr 15(4):585–587. https://doi.org/10.1080/19320248.2020.1799586
178. Yan J, Kim S, Zhang SX, Foo M-D, Alvarez-Risco A, Del-Aguila-Arcentales S, Yáñez JA (2021) Hospitality workers' COVID-19 risk perception and depression: a contingent model based on transactional theory of stress model. Int J Hosp Manag 95:102935. https://doi.org/10.1016/j.ijhm.2021.102935
179. Yáñez JA, Afshar Jahanshahi A, Alvarez-Risco A, Li J, Zhang SX (2020a) Anxiety, distress, and turnover intention of healthcare workers in Peru by their distance to the epicenter during the COVID-19 crisis. Am J Trop Med Hyg 103(4):1614–1620. https://doi.org/10.4269/ajtmh.20-0800
180. Yáñez JA, Alvarez-Risco A, Delgado-Zegarra J (2020) Covid-19 in Peru: from supervised walks for children to the first case of Kawasaki-like syndrome. BMJ 369:m2418. https://doi.org/10.1136/bmj.m2418
181. Yasin Ar A (2020) Managing e-commerce during a pandemic: lessons from GrubHub during COVID-19. In: George B, Mahar Q (eds) International case studies in the management of disasters. Emerald Publishing Limited, pp 155–167. https://doi.org/10.1108/978-1-83982-186-820201010
182. Young W, Hwang K, McDonald S, Oates CJ (2010) Sustainable consumption: green consumer behaviour when purchasing products. Sustain Dev 18(1):20–31. https://doi.org/10.1002/sd.394
183. Yuan Z, Jiang W, Liu B, Bi J (2008) Where will China go? A viewpoint based on an analysis of the challenges of resource supply and pollution. Environ Prog 27(4):503–514. https://doi.org/10.1002/ep.10300
184. Zarocostas J (2021) What next for a COVID-19 intellectual property waiver? The Lancet 397(10288):1871–1872. https://doi.org/10.1016/S0140-6736(21)01151-X
185. Zhang SX, Chen J, Afshar Jahanshahi A, Alvarez-Risco A, Dai H, Li J, Patty-Tito RM (2021) Succumbing to the COVID-19 pandemic—healthcare workers not satisfied and intend to leave their jobs. Int J Ment Heal Addict. https://doi.org/10.1007/s11469-020-00418-6
186. Zhang SX, Sun S, Afshar Jahanshahi A, Alvarez-Risco A, Ibarra VG, Li J, Patty-Tito RM (2020) Developing and testing a measure of COVID-19 organizational support of healthcare workers – results from Peru, Ecuador, and Bolivia. Psychiatry Res 291:113174. https://doi.org/10.1016/j.psychres.2020.113174
187. Zollinger R, DiCindio C (2021) Community ecology: Museum education and the digital divide during and after COVID-19. J Mus Educ 46(4):481–492. https://doi.org/10.1080/10598650.2021.1983711
188. ヒシヤマ, イ (1960) The tableau economique of quesnay-its analysis, reconstruction, and application. Kyoto Univ Econ Rev 30(1):1–46

Leadership for Sustainability in Crisis Time

Aldo Alvarez-Risco, Shyla Del-Aguila-Arcentales, Diego Villalobos-Alvarez, and Santiago Diaz-Risco

Abstract Leadership has been a crucial topic of moving forward in times of COVID-19, and as can be seen in the chapter, it forces having to address different needs, from the death of workers, providing good teams, maintaining commitment, maintaining leadership image, avoiding harmful habits in closing, and mainly, ensuring smooth and effective communication in all teams locally and globally. Many people have passed away, many companies have remained, and that more than ever should drive the focus of academic and research efforts on leadership in times of crisis. About resilience, there is a need to lead processes that have sustainability at their core, specifically with projects that include sustainable indicators and certifications such as footprinting.

Keywords Circular economy · Leadership · Leader · Sustainable development goals · SDG

1 Introduction

Leadership is an urgent requirement in times of crisis, such as the current COVID-19 pandemic for both face-to-face and remote work [115]. World leaders in each country and global institutions should reflect and ask themselves whether this type of leadership is expected to move a country or the world forward with a pandemic ahead. As of December 31, 2021, 281,808,270 people have been infected worldwide,

A. Alvarez-Risco (✉)
Universidad de Lima, Lima, Perú
e-mail: aralvare@ulima.edu.pe

S. Del-Aguila-Arcentales
Escuela Nacional de Marina Mercante "Almirante Miguel Grau", Callao, Perú

D. Villalobos-Alvarez
Universidad Tecnológica del Perú, Lima, Perú

S. Diaz-Risco
Centro de Fertilidad Cajamarca, Cajamarca, Perú

A. Alvarez-Risco et al. (eds.), *Circular Economy*, Environmental Footprints and Eco-design of Products and Processes,
https://doi.org/10.1007/978-981-19-0549-0_3

and 5,411,759 have died [203]. When we speak of global leadership in times of crisis, we are talking about the leadership necessary for transnational companies to be able to continue with optimal supply chain management to supply their customers in the world with their products, likewise, leadership is expected from governments to take the most balanced and effective measures to control the pandemic in each country. On the other hand, leadership is expected in the medical environment to feel supported and protect patients. Finally, leadership is expected in education, which by moving to the online format became genuinely global, generating that access to education can be from and to any part of the world. Leadership is required in times of crisis to have a vision of sustainability to achieve sustainable results over time.

We present the leadership actions that have been recognized in the world and reported in the academic literature. We also present a pilot leadership outcome and its effects on job performance, and finally, the pending leadership actions required, emphasizing the communication of information, which is the basis for decision making.

2 Before and Current Situation About COVID-19

The world has changed, and people's habits regarding sustainability have changed forever. The companies and people are working hard to reestablish the world they usually live in. The world has changed and people's habits regarding sustainability have changed, depending on the characteristics of pre-pandemic time. Sustainable education is the cornerstone for leadership that can guide the companies and habitants for efficient use of sources as water and food. In Table 1, it is presented different sectors that have been damaged by COVID-19.

3 The Employee Needs Sustainable Leadership in Times of Crisis

Today's world is characterized by VUCAT (Volatility, Uncertainty, Complexity, Ambiguity, and Technological). These elements cause workers to need leadership that must be addressed continuously [76]. MNEs have to dedicate efforts to ensure that their employees in all their branches can be led in a homogeneous way to have a strengthened global team [66, 205]. Despite the difficulties that workers of all levels have had to adapt to remote work, based on communication technology, an adaptation has been made in general terms, but in many cases, human resources' typical performance has been altered. It must be considered that remote work is a permanent feature for more organizations and companies, which means that institutional leaders at a global and local level design new rules to safeguard the quality of life of workers and at the same time contribute to their higher performance.

Table 1 Impact of COVID-19 in different sectors

Education	Ali [9], Allen et al. [10], Alvarez-Risco, Estrada-Merino, et al. [20], Alvarez-Risco, Del-Aguila-Arcentales, Rosen, et al. [17], Alvarez-Risco, Del-Aguila-Arcentales, Yáñez, et al. [18], T. Chen et al. [57], Lackie et al. [122], Lashley et al. [123], Zollinger and DiCindio [213]
Entrepreneurship	Afshan et al. [4], Alvarez-Risco, Mlodzianowska, Zamora-Ramos, et al. [23], Alvarez-Risco, Mlodzianowska, García-Ibarra, et al. [22], Alvarez-Risco and Del-Aguila-Arcentales [16], Belitski et al. [34], Block et al. [38], Brown et al. [46], Bacq and Lumpkin [29], Chafloque-Cespedes et al. [54], Maritz et al. [135], Liguori and Winkler [129]
Health sector	Abiakam et al. [1], Alan et al. [8], Alsairafi et al. [12], Alvarez-Risco, Dawson, et al. [14], Alvarez-Risco, Del-Aguila-Arcentales and Yáñez [19], Barello et al. [31], Bozdağ and Ergün [43], X. Chen et al. [58], Deressa et al. [79], Koetter et al. [117], Raudenská et al. [159], Rojas Román et al. [164], Yáñez, Afshar Jahanshahi, et al. [207], Zhang et al. [211], Zhang et al. [212]
Hospitality	Duarte Alonso et al. [85], Gursoy and Chi [97], Ho et al. [103], Kaushal and Srivastava [112], Yan et al. [206]
Intellectual property	Altindis [13], Alvarez-Risco and Del-Aguila-Arcentales [15], del Castillo [77], Erfani et al. [89], Jecker and Atuire [110], Krishtel and Malpani [120], Okereke [144], Sekalala et al. [174], Zarocostas [210]
Population	Ahsan [6], Alvarez-Risco, Mejia, et al. [21], Cegarra-Navarro et al. [53], Hanzl [98], Lim and Prakash [130], Quispe-Cañari et al. [153], Sánchez-Clavijo et al. [169], Shaw et al. [176], Wen et al. [202], Wu et al. [204], Yáñez, Alvarez-Risco, et al. [208]
Prices	Apcho-Ccencho et al. [27], Chung et al. [61], Galanakis et al. [92], Hepburn et al. [101], Leiva-Martinez et al. [127], Obergassel et al. [143]
Tourism	Assaf and Scuderi [28], Baum and Hai [32], Brouder [45], Carvache-Franco et al. [50], Fotiadis et al. [91], Carvache-Franco et al. [51], Sigala [178], Carvache-Franco, Carvache-Franco, et al. [52]
Trade	Abudureheman and Nilupaer [2], Acuña-Zegarra et al. [3], Aguirre et al. [5], Alvarez-Risco, Quipuzco-Chicata, et al. [24] Alvarez-Risco, Rosen, et al. [25], Borzée et al. [40], Cruz-Torres et al. [67], Kirk and Rifkin [116], Lopez-Odar et al. [133], Tran [193], Vidya and Prabheesh [197], Yasin Ar [209]
Violence against women	Amarillo [26], Chafloque-Cespedes et al. [55], Dominelli [82], Gulati and Kelly [96], Roesch et al. [163], Sánchez et al. [170], Viero et al. [198]

For this, mainly in multinationals, the CEO and regional and local managers in the areas of Human Resources, Finance, and IT must propose strategies to take care of the companies' human capital in all their premises. Some of the main components that must be secured include:

Fig. 1 Activities, type of communication, and alternatives of software

1. Ensure that managers can successfully convey work rules, including annual goals, indicators, activities, and planned schedules. Additionally, it means communication for formal communication [126]. It is relevant that global leadership provides efficient communication in times of pandemic and that it serves after the pandemic to be the efficient way to carry it out [80]. Some concrete examples of standardization of the forms of communication are shown in Fig. 1.

   One example of communication leadership must be the following example:

   a. Do not repeat the myth
   Wrong message
   "It is said that when we are with trusted people in the office, we can stop using masks."
   Correct message
   "You always wear the mask during all the time that I am in contact with people in the office."

2. Ensure everyone has the equipment and the correct connectivity to be carried out at the scheduled times. Everyone is expected to have the same minimal technological tools: webcam, headphones, microphone, monitor, printer, and scanner. Leadership must make workers feel part of the same team with the same weapons. This lack of identification has been a problem in the health field where there have been health workers in the first line of care who have not had the necessary protective equipment for many months and, nevertheless, have continued to save the lives of patients with COVID-19 [71, 125], 131].
3. Evaluate the most suitable descriptions for remote jobs so that effective recruitment can occur for both part-time and full-time workers. Another leadership element that workers tend to manifest is the control they may feel from the company, represented by their bosses, of meeting specific hours in a

manufacturing model. Do people have to be sitting the same hours they were in the office because they are supposed to be productive? Do you think that they should not make the same effort because people are at home? What is the impact of control devices on the image of a leader built in front of the workers? How can a worker feel like a leader, a person who controls them millimetrically through the use of software?

Some current ways to overcontrol employees and do not allow them to create a leadership image include [36]:

- Taking screenshots of employees' screens, making video recordings
- Invisible installs and stealth monitoring features
- Keyloggers
- Instant Messaging app monitoring
- Remote desktop control
- Spying on employees' mobile devices
- Complete communication logs
- VOIP calls spying
- Internet monitoring
- Mobile Keylogger
- Geofencing alerts
- GPS tracking
- Remote control of the infected device along with viewing and blocking specific apps
- Access to the calendar, notes, and reminders
- Surroundings audio recording capability
- Taking over the phone's camera, making screenshots.

4. Ensure that people keep their work, assisting that they can continue to develop their activities despite all the mental effects of the pandemic and also, motivate them to remain focused on institutional objectives [64, 100]. Leadership in the MNE has been reflected in various strategies and commitments, but it is still pending to know what workers expect from their bosses [107]. The urgent need to generate research that allows us to know the impact of leadership and what workers expect from their local and corporate leaders has been recognized [49]. It can be started by conducting studies that link the three main variables: leadership, work environment, and satisfaction. One of the strategies that have been seen throughout the companies is the empowerment of staff skills through the increase and variety of training [86, 166]. The pandemic has also facilitated the global proliferation of Webinars. At a global level, the leaders of companies have understood in most cases that they must retain their talents and must ensure that they can be optimally developed in times of crisis. This distance training also has the advantage of being carried out in the home's comfort, contributing to the work-family balance to achieve employee engagement [56]. Due to all experienced in the global health environment, the urgent need for leadership can be recognized [99, 181]. One aspect that global leaders must prevent and address

is to know the pattern of distraction that employees have due to lockdown. Thus, technostress presence has been recognized [138, 142, 146, 152]. Technostress generates various forms of burnout, which can also lead to increased alcohol consumption at home [182], with all the implications that this entails and an increase in gambling. The institutional leadership must address all these circumstances, achieving that habits are changed, promoting activities in the company that can distract the busy work schedule [44].

Business leadership has responded to the challenge of the pandemic with responsibility and innovation. These attributes are essential for a social actor to operate in an environment of multiple crises: health, economic, environmental, trust, and social. The world continues to undergo rapid transformations. For years, it has faced sustainability-related problems such as biodiversity loss, water scarcity, air pollution, frequent migration crises, and social inequality. When setting these business goals, we often look at the limitations to be overcome, focusing primarily on the profit approach but ignoring the great urgency of addressing environmental, social, and economic issues.

Offering green products and services carries great responsibility as it is expected that the green component of the offer is such. Scientific literature shows that certain companies and individuals have a green expectation of a product when it is not: greenwashing. These previous studies show results from the last 20 years, such as the factors [78], policies [158], evaluation of greenwashing [84], strategies [173], ethical commitment [148], relation with corporate social responsibility [41, 134], impact [74, 149, 185], taxonomy [177], the role of stakeholders [191], the effect of purchase intention [7], and consumer reaction [156]. More recently, a systematic review presented the most recent evidence [73]. Some certifications can help recognize products that deliver on their environmental promise to guide companies and customers. There have been various efforts to encourage companies to obtain sustainable certifications to be standardized. Thus, there are the following certifications:

1. **ENERGY STAR** is the symbol of energy efficiency endorsed by the U.S. government. Thousands of organizations work with the U.S. Environmental Protection Agency (EPA) to deliver energy efficiency solutions that save costs and protect the climate. ENERGY STAR was created in 1992, and over the life of the program, every dollar the EPA has spent on ENERGY STAR has translated into $350 in energy cost savings. A review of the following articles specific to this certification is recommended [42, 47, 105, 147, 155, 160, 184, 201].

2. **Water Sense** is an EPA-sponsored consortia program that seeks to promote efficient water management and bring water-efficient products and services to the marketplace [59, 94, 95, 186, 196].

3. **Cradle to cradle** is a certification that verifies that a product uses safe and environmentally sound materials and is designed for material reuse, recycling, or composting [30, 37, 65, 68, 72, 75, 118–119, 121, 132, 167, 179–180, 192, 200].

4. **Forest Stewardship Council (FSC)** accredits forest managers, manufacturing companies, and controlled wood products that exhibit responsible consumption of forest products. A review of the following articles specific to this certification is recommended [35, 128, 136, 139, 141, 150, 151, 154, 165, 183, 187, 188.

5. **Leadership in Energy and Environmental Design** is a certification that focuses on green buildings. A review of the following articles specific to this certification is recommended [48, 60, 62, 69, 70, 81, 93, 102, 108, 111, 113, 114, 161, 168, 175].

4 Research in Sustainability and Consumers

The development of research that allows management decisions is an urgent issue, given the need for economic reactivation after the COVID-19 pandemic. The authors describe variables that could be part of future studies, which should be quantitative and correlational in order to predict the influence between variables and to be able to organize more efficient strategies. Thus, several studies could be carried out. The following items can be used in Likert scales.

4.1 Behavior of Circular Economy

In Fig. 2, it is presented the items that can be used to test models to explain the behavior of the circular economy.

4.2 Intention of e-commerce

Website Quality

The online purchasing platforms that I use give me detailed information.
The online purchasing platforms that I use give me complete information.
The online purchasing platforms that I use are interesting.
The online purchasing platforms I use have innovative designs.
Information can be easily found on online platforms.

Customer Satisfaction

I enjoy making purchases online.
Interaction is convenient when I use online shopping platforms.
Purchasing online is a good decision to.
Purchasing online is enjoyable.
I am satisfied with the whole experience of purchasing online.

Customer trust

I believe that the companies where I am purchasing online protect their customers.
I consider that the companies that I purchase from online are honest when doing business.
I feel safe when purchasing online.
I believe that the online shopping platforms I use can do business online.
I am sure that the online shopping platforms that I use are reliable.

Attitudes
Recycling plastic waste is good
Recycling plastic waste is useful
Recycling plastic waste is rewarding
Recycling plastic waste is sensible
Recycling plastic waste would give our organization great satisfaction
It is our organization's responsibility to recycle plastic waste
Subjective Norms
Most people who influence our decisions think that we should recycle plastic waste
Most people inside our organization think that we should participate in recycling plastic waste
Most people outside our organization think that we should participate in recycling plastic waste
Many organizations similar to our organization participate in recycling plastic waste
Our neighboring organizations participate in recycling plastic waste
Perceived Behavioral Control
Our organization knows what items of plastic waste can be recycled
Our organization knows where to take plastic waste for recycling
Our organization knows how to recycle plastic waste
Whether our organization recycles plastic waste is entirely up to us
Whether our organization recycles plastic waste effectively is entirely within our control
Intention of Circular Economy
Our organization intends to recycle plastic waste
Our organization plans to recycle plastic waste
Our organization is willing to put efforts to recycle plastic waste
Our organization is willing to participate or continue plastic recycling
Pressures
International trade regulations
Regional regulations on plastic recycling
Local regulations on plastic recycling
Threat of future environmental regulations
Green strategies of competitors
Environmental awareness of customers
Behavior of Circular Economy
Utilizing eco-friendly packaging
Segregating plastics from other waste
Handing over generated plastic waste to a waste management company
Selling generated plastic waste to other organizations
Reusing generated plastic waste within our organization

Fig. 2 Items to explain Behavior of Circular Economy

Online repurchase intention

When the quarantine ends, I intend to continue purchasing online.
When the quarantine ends, I intend to increase my online purchases.
I intend to buy instead of traditional (physical) purchasing when the quarantine ends.
When the quarantine ends, I intend to use my credit card to make online purchases.

When the quarantine ends, I will recommend my friends/family/acquaintances make online purchases.
When the quarantine ends, if my friends/family/acquaintances ask me for advice, I would recommend purchasing online.
When the quarantine ends, I will not purchase online.

4.3 Intention to Purchase Sustainable Clothing

In Fig. 3, it is presented the items that can be used to test models to explain the intention to purchase sustainable clothing.

Although sustainability professionals have been advocating for years for environmental impact to be integrated into corporate management, with the advent of the pandemic, significant changes have put these issues on the agenda of the CEOs of major companies. Currently, more significant regulatory pressure and activism can be evidenced by an increase in companies' fiduciary responsibility and their leaders.

Perceived environmental knowledge
I know how to behave sustainably
I know how I could lower the ecological harm with my behavior
I understand how I could reduce the negative environmental consequences of my behavior
I understand how to protect the environment in the long-term
Environmental concern
I am concerned about the environmental development
I am concerned about the long-term consequences of unsustainable behavior
I often think about the potential negative development of the environmental situation
I am concerned that humanity will cause lasting damage to the environment
Attitude
Generally, I have a favorable attitude towards the sustainable version of clothes
I am positive-minded towards buying secondhand clothes
I like the idea of buying sustainable clothes instead of conventional clothes to contribute to environmental protection
Subjective norms
My friends expect me to buy sustainable clothes
My family expect me to buy sustainable clothes
People who are important to me expect me to buy sustainable clothes
Purchase intention
I consider purchasing sustainable clothes
I intend to buy sustainable clothes instead of conventional clothes in the future
I might buy sustainable clothes in the future
I would consider buying sustainable clothes if I happen to see them in an online store
Purchase behavior
I choose to buy exclusively sustainable clothes
I buy sustainable clothes instead of conventional clothes if the quality is comparable
I purchase sustainable clothes even if they are more expensive than conventional clothes
When buying clothes, I pay attention that they are sustainable

Fig. 3 Items to explain Intention to purchase sustainable clothing

Fig. 4 Organic recycling process

Globally, leadership is a basis for successful planning. In this period of global resilience, thinking sustainability is the key to making efficient use of the resources available to countries. Various legislative initiatives are being promoted and implemented. Figure 4 shows the organic recycling cycle that must be planned, executed, and monitored. These practices are part of the most recent regulations increasingly demanded by the population.

The regulatory pressure undoubtedly leads to the transformation of sectors and business models that have to comply with their legal obligations and anticipate change to survive in the new paradigm. If sustainability is not made transversal to the business, not only will it be possible for the company to incur non-compliance with the law, but it lose business opportunities, investment, innovation, and competitiveness for the future.

There are several global initiatives to achieve eco-efficient plastic management: the Global Commitment Report [87], the Plastics Pact Network [88], Alliance To End Plastic Waste [11], Commonwealth Clean Ocean Alliance [63], and the UK Plastics Pact [145].

Specifically, the Paris Agreement has been under investigation since 2016, when it was signed [39, 83, 90, 104, 106, 109, 124, 137, 157, 162, 171, 195]. However, recent information shows that the goals set by the Paris Agreement are not being achieved [33, 140, 189, 190, 194, 199] (Fig. 5).

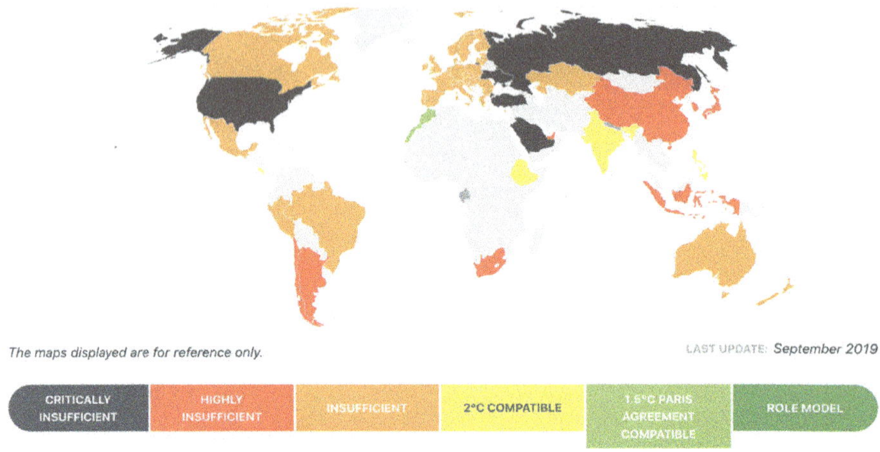

Fig. 5 Compliance with the Paris Agreement

5 Closing Remarks

The current pandemic that is plaguing the world will remain in the memory of each person and company. Each activity developed in this pandemic time is vital because it is the image that remains in the memory of all the people in an organization and is a predictor of the engagement that workers have concerning their leaders. Companies need to work on leadership at all levels so that all the directory measures are reflected in the most peripheral workers since they are direct with direct customers and finished products. Let us look at the future we want to create from global leadership. All the latter shows that leadership is required to achieve the goals set globally.

References

1. Abiakam N, Worsley P, Jayabal H, Mitchell K, Jones M, Fletcher J, Spratt F, Bader D (2021) Personal protective equipment related skin reactions in healthcare professionals during COVID-19. Int Wound J 18(3):312–322. https://doi.org/10.1111/iwj.13534
2. Abudureheman A, Nilupaer A (2021) Optimization model design of cross-border e-commerce transportation path under the background of prevention and control of COVID-19 pneumonia. Soft Comput 25(18):12007–12015. https://doi.org/10.1007/s00500-021-05685-6
3. Acuña-Zegarra MA, Santana-Cibrian M, Velasco-Hernandez JX (2020) Modeling behavioral change and COVID-19 containment in Mexico: a trade-off between lockdown and compliance. Math Biosci 325:108370. https://doi.org/10.1016/j.mbs.2020.108370
4. Afshan G, Shahid S, Tunio MN (2021) Learning experiences of women entrepreneurs amidst COVID-19. Int J Gend Entrep 13(2):162–186. https://doi.org/10.1108/IJGE-09-2020-0153
5. Aguirre AA, Catherina R, Frye H, Shelley L (2020) Illicit wildlife trade, wet markets, and COVID-19: preventing future pandemics. World Med & Health Policy 12(3):256–265. https://doi.org/10.1002/wmh3.348

6. Ahsan MM (2020) Strategic decisions on urban built environment to pandemics in Turkey: lessons from COVID-19. J Urban Manag 9(3):281–285. https://doi.org/10.1016/j.jum.2020.07.001
7. Akturan U (2018) How does greenwashing affect green branding equity and purchase intention? An empirical research. Mark Intell Plan 36(7):809–824. https://doi.org/10.1108/MIP-12-2017-0339
8. Alan H, Eskin Bacaksiz F, Tiryaki Sen H, Taskiran Eskici G, Gumus E, Harmanci Seren AK (2021) "I'm a hero, but...": an evaluation of depression, anxiety, and stress levels of frontline healthcare professionals during COVID-19 pandemic in Turkey. Perspect Psychiatr Care 57(3):1126–1136. https://doi.org/10.1111/ppc.12666
9. Ali W (2020) Online and remote learning in higher education institutes: a necessity in light of COVID-19 pandemic. High Educ Stud 10(3):16–25
10. Allen J, Rowan L, Singh P (2020) Teaching and teacher education in the time of COVID-19. Asia-Pac J Teach Educ 48(3):233–236. https://doi.org/10.1080/1359866X.2020.1752051
11. Alliance To End Plastic Waste (2021) The Alliance Progress Report 2021. Retrieved 01/01/2022 from https://endplasticwaste.org/en
12. Alsairafi Z, Naser AY, Alsaleh FM, Awad A, Jalal Z (2021) Mental health status of healthcare professionals and students of health sciences faculties in Kuwait during the COVID-19 pandemic. Int J Environ Res Public Health 18(4). https://doi.org/10.3390/ijerph18042203
13. Altindis E (2021) Inequitable COVID-19 vaccine distribution and the intellectual property rights prolong the pandemic. Expert Rev Vaccines, null-null. https://doi.org/10.1080/14760584.2022.2014819
14. Alvarez-Risco A, Dawson J, Johnson W, Conteh-Barrat M., Aslani P, Del-Aguila-Arcentales, S, Diaz-Risco S (2021) Ebola virus disease outbreak: a global approach for health systems [Ebola; health professionals; global approach; pharmacist: pharmaceutical care; COVID-19]. 2021 53(4). http://www.revfarmacia.sld.cu/index.php/far/article/view/491
15. Alvarez-Risco A, Del-Aguila-Arcentales S (2021a) A note on changing regulation in international business: the World Intellectual Property Organization (WIPO) and artificial intelligence. In: Verbeke A, van Tulder R, Rose EL, Wei Y (eds) The multiple dimensions of institutional complexity in International Business Research, vol 15. Emerald Publishing Limited, pp 363–371. https://doi.org/10.1108/S1745-886220210000015020
16. Alvarez-Risco A, Del-Aguila-Arcentales S (2021b) Public policies and private efforts to increase women entrepreneurship based on STEM background. In: Mari M, Poggesi S, Foss L (eds) Women's entrepreneurship in STEM disciplines: issues and perspectives. Springer International Publishing, pp 75–87. https://doi.org/10.1007/978-3-030-83792-1_5
17. Alvarez-Risco A, Del-Aguila-Arcentales S, Rosen MA, García-Ibarra V, Maycotte-Felkel S, Martínez-Toro GM (2021) Expectations and interests of university students in COVID-19 times about Sustainable Development Goals: evidence from Colombia, Ecuador, Mexico, and Peru. Sustain 13(6):3306. https://doi.org/10.3390/su13063306
18. Alvarez-Risco A, Del-Aguila-Arcentales S, Yáñez JA (2021) Telemedicine in Peru as a result of the COVID-19 pandemic: perspective from a country with limited internet access. Am J Trop Med Hyg 105(1):6–11. https://doi.org/10.4269/ajtmh.21-0255
19. Alvarez-Risco A, Del-Aguila-Arcentales S, Yáñez JA, Rosen MA, Mejia CR (2021) Influence of technostress on academic performance of university medicine students in Peru during the COVID-19 pandemic. Sustain 13(16):8949. https://doi.org/10.3390/su13168949
20. Alvarez-Risco A, Estrada-Merino A, Anderson-Seminario MdlM, Mlodzianowska S, García-Ibarra V, Villagomez-Buele C, Carvache-Franco M (2021) Multitasking behavior in online classrooms and academic performance: case of university students in Ecuador during COVID-19 outbreak. Interact Technol Smart Educ 18(3):422–434. https://doi.org/10.1108/ITSE-08-2020-0160
21. Alvarez-Risco A, Mejia CR, Delgado-Zegarra J, Del-Aguila-Arcentales S, Arce-Esquivel AA, Valladares-Garrido MJ, Del Portal MR, Villegas LF, Curioso WH, Sekar MC, Yáñez JA (2020) The Peru approach against the COVID-19 infodemic: insights and strategies. Am J Trop Med Hyg 103(2):583–586. https://doi.org/10.4269/ajtmh.20-0536

22. Alvarez-Risco A, Mlodzianowska S, García-Ibarra V, Rosen MA, Del-Aguila-Arcentales S (2021) Factors affecting green entrepreneurship intentions in business university students in COVID-19 pandemic times: Case of Ecuador. Sustain 13(11):6447. https://doi.org/10.3390/su13116447
23. Alvarez-Risco A, Mlodzianowska S, Zamora-Ramos U, Del-Aguila S (2021) Green entrepreneurship intention in university students: the case of Peru. Entrep Bus Econ Rev 9(4):85–100. https://doi.org/10.15678/EBER.2021.090406
24. Alvarez-Risco A, Quipuzco-Chicata L, Escudero-Cipriani C (2021) Determinantes de la intención de recompra en línea en tiempos de COVID-19: evidencia de una economía emergente. Lecturas de Economía (96). https://doi.org/10.17533/udea.le.n96a342638
25. Alvarez-Risco A, Rosen MA, Del-Aguila-Arcentales S (2020) A new regulation for supporting a circular economy in the plastic industry: the case of Peru (short communication). J Landsc Ecol 13(1):1–3. https://doi.org/10.2478/jlecol-2020-0004
26. Amarillo CR (2020) Feminism on lockdown. NACLA Rep Am 52(3):274–281. https://doi.org/10.1080/10714839.2020.1809084
27. Apcho-Ccencho L-V, Cuya-Velásquez B-B, Alvarado Rodríguez D, Anderson-Seminario MdlM, Alvarez-Risco A, Estrada-Merino A, Mlodzianowska S (2021) The impact of international price on the technological industry in the United States and China during times of crisis: commercial war and COVID-19. In: Lawrence KD, Klimberg RK (eds) Advances in business and management forecasting, vol 14. Emerald Publishing Limited, pp 149–160. https://doi.org/10.1108/S1477-407020210000014010
28. Assaf A, Scuderi R (2020) COVID-19 and the recovery of the tourism industry. Tour Econ 26(5):731–733. https://doi.org/10.1177/1354816620933712
29. Bacq S, Lumpkin GT (2021) Social entrepreneurship and COVID-19. J Manage Stud 58(1):285–288. https://doi.org/10.1111/joms.12641
30. Bakker CA, Wever R, Teoh C, De Clercq S (2010) Designing cradle-to-cradle products: a reality check. Int J Sustain Eng 3(1):2–8. https://doi.org/10.1080/19397030903395166
31. Barello S, Caruso R, Palamenghi L, Nania T, Dellafiore F, Bonetti L, Silenzi A, Marotta C, Graffigna G (2021) Factors associated with emotional exhaustion in healthcare professionals involved in the COVID-19 pandemic: an application of the job demands-resources model. Int Arch Occup Environ Health 94(8):1751–1761. https://doi.org/10.1007/s00420-021-01669-z
32. Baum T, Hai NTT (2020) Hospitality, tourism, human rights and the impact of COVID-19. Int J Contemp Hosp Manag 32(7):2397–2407. https://doi.org/10.1108/IJCHM-03-2020-0242
33. BBC (2021) Why is the Paris climate agreement important for COP26? Retrieved 01/01/2022 from https://www.bbc.com/news/science-environment-35073297
34. Belitski M, Guenther C, Kritikos AS, Thurik R (2021) Economic effects of the COVID-19 pandemic on entrepreneurship and small businesses. Small Bus Econ. https://doi.org/10.1007/s11187-021-00544-y
35. Bell S, Hindmoor A (2012) Governance without government? The case of the Forest Stewardship Council [https://doi.org/10.1111/j.1467-9299.2011.01954.x]. Public Adm 90(1):144–159. https://doi.org/10.1111/j.1467-9299.2011.01954.x
36. Berzinya A (2020) 16 worst and most extreme ways employers are spying on their people. Retrieved 01/01/2022 from https://turtler.io/news/16-worst-and-most-extreme-ways-employers-are-spying-on-their-people
37. Bjørn A, Hauschild MZ (2013) Absolute versus relative environmental sustainability [https://doi.org/10.1111/j.1530-9290.2012.00520.x]. J Ind Ecol 17(2):321–332. https://doi.org/10.1111/j.1530-9290.2012.00520.x
38. Block JH, Fisch C, Hirschmann M (2021) The determinants of bootstrap financing in crises: evidence from entrepreneurial ventures in the COVID-19 pandemic. Small Bus Econ. https://doi.org/10.1007/s11187-020-00445-6
39. Bodansky D (2016) The legal character of the Paris Agreement [https://doi.org/10.1111/reel.12154]. Rev Eur, Comp & Int Environ Law 25(2):142–150. https://doi.org/10.1111/reel.12154
40. Borzée A, McNeely J, Magellan K, Miller JRB, Porter L, Dutta T, … Zhang L (2020) COVID-19 highlights the need for more effective wildlife trade legislation. Trends Ecol & Evol 35(12):1052–1055. https://doi.org/10.1016/j.tree.2020.10.001

41. Bowen F, Aragon-Correa JA (2014) Greenwashing in corporate environmentalism research and practice: the importance of what we say and do. Organ Environ 27(2):107–112. https://doi.org/10.1177/1086026614537078
42. Boyd G, Dutrow E, Tunnessen W (2008) The evolution of the ENERGY STAR® energy performance indicator for benchmarking industrial plant manufacturing energy use. J Clean Prod 16(6):709–715. https://doi.org/10.1016/j.jclepro.2007.02.024
43. Bozdağ F, Ergün N (2020) Psychological resilience of healthcare professionals during COVID-19 pandemic. Psychol Rep 124(6):2567–2586. https://doi.org/10.1177/0033294120965477
44. Brennan F (2021) Technostress and leadership: a case study in higher education during the COVID-19 crisis. Tampere University of Applied Sciences]. https://www.theseus.fi/handle/10024/380031
45. Brouder P (2020) Reset redux: possible evolutionary pathways towards the transformation of tourism in a COVID-19 world. Tour Geogr 22(3):484–490. https://doi.org/10.1080/14616688.2020.1760928
46. Brown R, Rocha A, Cowling M (2020) Financing entrepreneurship in times of crisis: exploring the impact of COVID-19 on the market for entrepreneurial finance in the United Kingdom. Int Small Bus J 38(5):380–390. https://doi.org/10.1177/0266242620937464
47. Brown R, Webber C, Koomey JG (2002) Status and future directions of the Energy Star program. Energy 27(5):505–520. https://doi.org/10.1016/S0360-5442(02)00004-X
48. Cai H, Wang X, Kim J-H, Gowda A, Wang M, Mlade J, Farbman S, Leung, L (2022) Whole-building life-cycle analysis with a new GREET® tool: embodied greenhouse gas emissions and payback period of a LEED-Certified library. Build Environ 209:108664. https://doi.org/10.1016/j.buildenv.2021.108664
49. Caligiuri P, De Cieri H, Minbaeva D, Verbeke A, Zimmermann A (2020) International HRM insights for navigating the COVID-19 pandemic: implications for future research and practice. In: vol 51. Springer, pp 697–713
50. Carvache-Franco M, Alvarez-Risco A, Carvache-Franco O, Carvache-Franco W, Estrada-Merino A, Villalobos-Alvarez D (2021) Perceived value and its influence on satisfaction and loyalty in a coastal city: a study from Lima, Peru. J Policy Res Tour, Leis Events, 1–16. https://doi.org/10.1080/19407963.2021.1883634
51. Carvache-Franco M, Alvarez-Risco A, Carvache-Franco W, Carvache-Franco O, Estrada-Merino A, Rosen MA (2021) Coastal cities seen from loyalty and their tourist motivations: a study in Lima Peru. Sustain 13(21):11575. https://doi.org/10.3390/su132111575
52. Carvache-Franco M, Carvache-Franco O, Carvache-Franco W, Alvarez-Risco A, Estrada-Merino A (2021) Motivations and segmentation of the demand for coastal cities: a study in Lima Peru. Int J Tour Res 23(4):517–531. https://doi.org/10.1002/jtr.2423
53. Cegarra-Navarro J-G, Vătămănescu E-M, Martínez-Martínez A (2021) A context-driven approach on coping with COVID-19: from hiding knowledge toward citizen engagement. Knowl Process Manag 28(2):134–140. https://doi.org/10.1002/kpm.1662
54. Chafloque-Cespedes R, Alvarez-Risco A, Robayo-Acuña, P-V, Gamarra-Chavez C-A, Martinez-Toro G-M, Vicente-Ramos W (2021) Effect of sociodemographic factors in entrepreneurial orientation and entrepreneurial intention in university students of Latin American business schools. In: Jones P, Apostolopoulos N, Kakouris A, Moon C, Ratten V, Walmsley A (eds) Universities and entrepreneurship: meeting the educational and social challenges, vol 11. Emerald Publishing Limited, pp 151–165. https://doi.org/10.1108/S2040-724620210000011010
55. Chafloque-Céspedes R, Vara-Horna A, Asencios-Gonzales Z, López-Odar D, Álvarez-Risco A, Quipuzco-Chicata L, Schulze C, Sánchez-Villagomez M (2020) Presentismo académico y violencia contra las mujeres en escuelas de negocios e ingeniería en universidades peruanas. *Lecturas de Economía*, *0*(93), 127–153. https://doi.org/10.17533/udea.le.n93a340726
56. Chanana N, Sangeeta (2021) Employee engagement practices during COVID-19 lockdown [https://doi.org/10.1002/pa.2508]. J Public Aff 21(4):e2508. https://doi.org/10.1002/pa.2508

57. Chen T, Peng L, Yin X, Rong J, Yang J, Cong G (2020) Analysis of user satisfaction with online education platforms in China during the COVID-19 pandemic. Healthcare 8(3). https://doi.org/10.3390/healthcare8030200
58. Chen X, Zhang SX, Jahanshahi AA, Alvarez-Risco A, Dai H, Li J, Ibarra VG (2020) Belief in a COVID-19 conspiracy theory as a predictor of mental health and well-being of health care workers in Ecuador: cross-sectional survey study. JMIR Public Health Surveill 6(3):e20737. https://doi.org/10.2196/20737
59. Chen Y, Fuchs H, Schein J, Franco V, Stratton H, Burke T, Dunham C (2021) Water heating energy use reductions from EPA WaterSense lavatory plumbing fittings. Resour, Conserv Recycl 174:105781. https://doi.org/10.1016/j.resconrec.2021.105781
60. Chi B, Lu W, Ye M, Bao Z, Zhang X (2020) Construction waste minimization in green building: a comparative analysis of LEED-NC 2009 certified projects in the US and China. J Clean Prod 256:120749. https://doi.org/10.1016/j.jclepro.2020.120749
61. Chung SA, Olivera S, Rojas Román B, Alanoca E, Moscoso S, Limpias Terceros B, Alvarez-Risco A, Yáñez JA (2021) Temáticas de la producción científica de la Revista Cubana de Farmacia indizada en Scopus (1967–2020) [Revista Cubana de Farmacia; farmacia; Cuba; evidencia]. 2021 54(1). http://www.revfarmacia.sld.cu/index.php/far/article/view/511
62. Cidell J (2009) Building green: the emerging geography of LEED-certified buildings and professionals. Prof Geogr 61(2):200–215. https://doi.org/10.1080/00330120902735932
63. Commonwealth Clean Ocean Alliance (2021) Commonwealth clean ocean alliance. Retrieved 01/01/2022 from https://bluecharter.thecommonwealth.org/action-groups/marine-plastic-pollution
64. Company M (2020) COVID-19 and the employee experience: How leaders can seize the moment. Retrieved 01/01/2022 from https://www.mckinsey.com/business-functions/organization/our-insights/covid-19-and-the-employee-experience-how-leaders-can-seize-the-moment
65. Contreras-Lisperguer R, Muñoz-Cerón E, Aguilera J, Casa Jdl (2017) Cradle-to-cradle approach in the life cycle of silicon solar photovoltaic panels. J Clean Prod 168:51–59. https://doi.org/10.1016/j.jclepro.2017.08.206
66. Contreras F, Baykal E, Abid G (2020) E-Leadership and teleworking in times of COVID-19 and beyond: what we know and where do we go [Review]. Front Psychol 11(3484). https://doi.org/10.3389/fpsyg.2020.590271
67. Cruz-Torres W, Alvarez-Risco A, Del-Aguila-Arcentales S (2021) Impact of Enterprise Resource Planning (ERP) implementation on performance of an education enterprise: a Structural Equation Modeling (SEM). Studies in Business and Economics 16(2):37–52. https://doi.org/10.2478/sbe-2021-0023
68. Cruz Rios F, Grau D, Chong WK (2019) Reusing exterior wall framing systems: a cradle-to-cradle comparative life cycle assessment. Waste Manag 94:120–135. https://doi.org/10.1016/j.wasman.2019.05.040
69. Dall G, Bruni E, Panza A (2013) Improvement of the sustainability of existing school buildings according to the leadership in Energy and Environmental Design (LEED)® protocol: a case study in Italy. Energies 6(12). https://doi.org/10.3390/en6126487
70. Dall G, Galante A, Sanna N, Miller K (2013) On the integration of Leadership in Energy and Environmental Design (LEED)® ND protocol with the energy planning and management tools in Italy: strengths and weaknesses. Energies 6(11). https://doi.org/10.3390/en6115990
71. Daly J, Jackson D, Anders R, Davidson PM (2020) Who speaks for nursing? COVID-19 highlighting gaps in leadership [https://doi.org/10.1111/jocn.15305]. J Clin Nurs 29(15–16):2751–2752. https://doi.org/10.1111/jocn.15305
72. Daughton CG (2003) Cradle-to-cradle stewardship of drugs for minimizing their environmental disposition while promoting human health. I. Rationale for and avenues toward a green pharmacy. Environ Health Perspect 111(5):757–774. https://doi.org/10.1289/ehp.5947
73. de Freitas Netto SV, Sobral MFF, Ribeiro ARB, Soares GRdL (2020) Concepts and forms of greenwashing: a systematic review. Environ Sci Eur 32(1):19. https://doi.org/10.1186/s12302-020-0300-3

74. De Jong MDT, Harkink KM, Barth S (2017) Making green stuff? Effects of corporate greenwashing on consumers. J Bus Tech Commun 32(1):77–112. https://doi.org/10.1177/1050651917729863
75. de Pauw IC, Karana E, Kandachar P, Poppelaars F (2014) Comparing Biomimicry and Cradle to Cradle with Ecodesign: a case study of student design projects. J Clean Prod 78:174–183. https://doi.org/10.1016/j.jclepro.2014.04.077
76. Deepika, Chitranshi J (2021) Leader readiness of Gen Z in VUCA business environment. foresight 23(2):154–171. https://doi.org/10.1108/FS-05-2020-0048
77. del Castillo FA (2021) Temporary waiver of intellectual property on Covid-19 vaccines: toward the creation of a better, post-pandemic society. J Public Health 43(3):e559–e560. https://doi.org/10.1093/pubmed/fdab184
78. Delmas MA, Burbano VC (2011) The drivers of greenwashing. Calif Manage Rev 54(1):64–87. https://doi.org/10.1525/cmr.2011.54.1.64
79. Deressa W, Worku A, Abebe W, Gizaw M, Amogne W (2021) Risk perceptions and preventive practices of COVID-19 among healthcare professionals in public hospitals in Addis Ababa, Efthiopia. PLoS ONE 16(6):e0242471. https://doi.org/10.1371/journal.pone.0242471
80. Dirani KM, Abadi M, Alizadeh A, Barhate B, Garza RC, Gunasekara N, Ibrahim G, Majzun Z (2020) Leadership competencies and the essential role of human resource development in times of crisis: a response to Covid-19 pandemic. Hum Resour Dev Int 23(4):380–394. https://doi.org/10.1080/13678868.2020.1780078
81. Dobiáš J, Macek, D (2014) Leadership in Energy and Environmental Design (LEED) and its impact on building operational expenditures. Procedia Eng 85:132–139. https://doi.org/10.1016/j.proeng.2014.10.537
82. Dominelli L (2021) A green social work perspective on social work during the time of COVID-19. Int J Soc Welf 30(1):7–16. https://doi.org/10.1111/ijsw.12469
83. Doukas H, Nikas A, González-Eguino M, Arto I, Anger-Kraavi A (2018) From integrated to integrative: delivering on the Paris Agreement. Sustain 10(7). https://doi.org/10.3390/su10072299
84. Du X (2015) How the market values greenwashing? Evidence from China. J Bus Ethics 128(3):547–574
85. Duarte Alonso A, Kok SK, Bressan A, O'Shea M, Sakellarios N, Koresis A, Solis MA, Santoni LJ (2020) COVID-19, aftermath, impacts, and hospitality firms: an international perspective. Int J Hosp Manag 91:102654. https://doi.org/10.1016/j.ijhm.2020.102654
86. ElHawary H, Salimi A, Alam P, Gilardino MS (2020) Educational alternatives for the maintenance of educational competencies in surgical training programs affected by the COVID-19 pandemic. J Med Educ Curric Dev 7:2382120520951806. https://doi.org/10.1177/2382120520951806
87. Ellen MacArthur Foundation (2021a) The Global Commitment 2021 Progress Report. Retrieved 01/01/2021 from https://ellenmacarthurfoundation.org/global-commitment/overview
88. Ellen MacArthur Foundation (2021b) Plastics Pact Network. Retrieved 01/01/2021 from https://ellenmacarthurfoundation.org/the-plastics-pact-network
89. Erfani P, Binagwaho A, Jalloh MJ, Yunus M, Farmer P, Kerry V (2021) Intellectual property waiver for covid-19 vaccines will advance global health equity. BMJ 374:n1837. https://doi.org/10.1136/bmj.n1837
90. Falkner R (2016) The Paris Agreement and the new logic of international climate politics. Int Aff 92(5):1107–1125. https://doi.org/10.1111/1468-2346.12708
91. Fotiadis A, Polyzos S, Huan T-CTC (2021) The good, the bad and the ugly on COVID-19 tourism recovery. Ann Tour Res 87:103117. https://doi.org/10.1016/j.annals.2020.103117
92. Galanakis CM, Rizou M, Aldawoud TMS, Ucak I, Rowan NJ (2021) Innovations and technology disruptions in the food sector within the COVID-19 pandemic and post-lockdown era. Trends Food Sci Technol 110:193–200. https://doi.org/10.1016/j.tifs.2021.02.002
93. Gluszak M, Malkowska A, Marona B (2021) Green building adoption on office markets in Europe: An empirical investigation into LEED certification. Energies 14(7). https://doi.org/10.3390/en14071971

94. Gonzalez AM, Reynolds-Tylus T, Quick BL (2021) Clustering energy and water conservation behaviors as choices: examining the moderating roles of message elaboration and involvement. Appl Environ Educ Commun 20(2):139–154. https://doi.org/10.1080/1533015X.2020.1740116
95. Grumbles BH (2008) WaterSense® makes good sense [https://doi.org/10.1002/j.1551-8833.2008.tb09619.x]. Journal AWWA 100(5):34–36. https://doi.org/10.1002/j.1551-8833.2008.tb09619.x
96. Gulati G, Kelly BD (2020) Domestic violence against women and the COVID-19 pandemic: what is the role of psychiatry? Int J Law Psychiatry 71:101594. https://doi.org/10.1016/j.ijlp.2020.101594
97. Gursoy D, Chi CG (2020) Effects of COVID-19 pandemic on hospitality industry: review of the current situations and a research agenda. J Hosp Market Manag 29(5):527–529. https://doi.org/10.1080/19368623.2020.1788231
98. Hanzl M (2020) Urban forms and green infrastructure – the implications for public health during the COVID-19 pandemic. Cities & Health, 1–5. https://doi.org/10.1080/23748834.2020.1791441
99. Harris A, Jones M (2020) COVID 19 – School leadership in disruptive times. Sch Leadership & Manag 40(4):243–247. https://doi.org/10.1080/13632434.2020.1811479
100. HBS (2020) Keep your weary workers engaged and motivated. Retrieved 01/01/2022 from https://hbswk.hbs.edu/item/keep-covid-weary-employees-engaged-and-motivated
101. Hepburn C, O'Callaghan, B, Stern N, Stiglitz J, Zenghelis D (2020) Will COVID-19 fiscal recovery packages accelerate or retard progress on climate change? Oxf Rev Econ Policy 36(Supplement_1), S359–S381. https://doi.org/10.1093/oxrep/graa015
102. Hightower R (2013) Investigating the green Leadership in Energy and Environmental Design (LEED) servicescape scale in Brazil. Constr Innov 13(3):242–265. https://doi.org/10.1108/CI-Oct-2011-0045
103. Ho C-Y, Tsai B-H, Chen C-S, Lu M-T (2021) Exploring green marketing orientations toward sustainability the hospitality industry in the COVID-19 pandemic. Sustain 13(8). https://doi.org/10.3390/su13084348
104. Höhne N, Gidden MJ, den Elzen M, Hans F, Fyson C, Geiges A, Jeffery ML, Gonzales-Zuñiga S, Mooldijk S, Hare W, Rogelj J (2021) Wave of net zero emission targets opens window to meeting the Paris Agreement. Nat Clim Chang 11(10):820–822. https://doi.org/10.1038/s41558-021-01142-2
105. Hopkins EA, Carswell AT, Love KR (2020) Impacts of ENERGY STAR appliances on U.S. Multifamily rents [https://doi.org/10.1111/fcsr.12372]. Fam Consum Sci Res J 49(1):37–51. https://doi.org/10.1111/fcsr.12372
106. Horowitz CA (2016) Paris Agreement. Int Leg Mater 55(4):740–755. https://doi.org/10.1017/S0020782900004253
107. IFC (2020) Lessons for COVID-19-era CEOs. Retrieved 01/01/2022 from https://www.ifc.org/wps/wcm/connect/news_ext_content/ifc_external_corporate_site/news+and+events/news/insights/covid-19-era-ceo
108. Ismaeel WSE, Elsayed MA (2022) Sustainable sites in two generations of city development using GIS-MCDM and LEED LT and SS categories. J Clean Prod 330:129782. https://doi.org/10.1016/j.jclepro.2021.129782
109. Jackson RB, Friedlingstein P, Andrew RM, Canadell JG, Le Quéré C, Peters GP (2019) Persistent fossil fuel growth threatens the Paris Agreement and planetary health. Environ Res Lett 14(12):121001. https://doi.org/10.1088/1748-9326/ab57b3
110. Jecker NS, Atuire CA (2021) What's yours is ours: waiving intellectual property protections for COVID-19 vaccines. J Med Ethics 47(9):595. https://doi.org/10.1136/medethics-2021-107555
111. Johansson O (2012) The spatial diffusion of green building technologies: the case of Leadership in Energy and Environmental Design (LEED) in the United States. Int J Technol Manag Sustain Dev 10(3):251–266. https://doi.org/10.1386/tmsd.10.3.251_1

112. Kaushal V, Srivastava S (2021) Hospitality and tourism industry amid COVID-19 pandemic: perspectives on challenges and learnings from India. Int J Hosp Manag 92:102707. https://doi.org/10.1016/j.ijhm.2020.102707
113. Kern AP, Antoniolli CB, Wander PR, Mancio M, Stumpf González MA (2016) Energy and water consumption during the post-occupancy phase and the users' perception of a commercial building certified by Leadership in Energy and Environmental Design (LEED). J Clean Prod 133:826–834. https://doi.org/10.1016/j.jclepro.2016.05.081
114. Kientzel J, Kok G (2011) Environmental assessment methodologies for commercial buildings: an elicitation study of U.S. building professionals' beliefs on Leadership in Energy and Environmental Design (LEED). Sustain 3(12). https://doi.org/10.3390/su3122392
115. Kirchner K, Ipsen C, Hansen JP (2021) COVID-19 leadership challenges in knowledge work. Knowl Manag Res Pract 19(4):493–500. https://doi.org/10.1080/14778238.2021.1877579
116. Kirk CP, Rifkin LS (2020) I'll trade you diamonds for toilet paper: consumer reacting, coping and adapting behaviors in the COVID-19 pandemic. J Bus Res 117:124–131. https://doi.org/10.1016/j.jbusres.2020.05.028
117. Koetter P, Pelton M, Gonzalo J, Du P, Exten C, Bogale K, Buzzelli L, Connolly M, Edel K, Hoffman A, Legro NR, Sciamanna C (2020) Implementation and process of a COVID-19 contact tracing initiative: leveraging health professional students to extend the workforce during a pandemic. Am J Infect Control 48(12):1451–1456. https://doi.org/10.1016/j.ajic.2020.08.012
118. Kopnina H (2018) Circular economy and Cradle to Cradle in educational practice. J Integr Environ Sci 15(1):119–134. https://doi.org/10.1080/1943815X.2018.1471724
119. Kopnina H (2019) Green-washing or best case practices? Using circular economy and Cradle to Cradle case studies in business education. J Clean Prod 219:613–621. https://doi.org/10.1016/j.jclepro.2019.02.005
120. Krishtel P, Malpani R (2021) Suspend intellectual property rights for covid-19 vaccines. BMJ 373:n1344. https://doi.org/10.1136/bmj.n1344
121. Kumar S, Putnam V (2008) Cradle to cradle: reverse logistics strategies and opportunities across three industry sectors. Int J Prod Econ 115(2):305–315. https://doi.org/10.1016/j.ijpe.2007.11.015
122. Lackie K, Najjar G, El-Awaisi A, Frost J, Green C, Langlois S, Lising D, Pfeifle AL, Ward H, Xyrichis A, Khalili H (2020) Interprofessional education and collaborative practice research during the COVID-19 pandemic: considerations to advance the field. J Interprof Care 34(5):583–586. https://doi.org/10.1080/13561820.2020.1807481
123. Lashley MA, Acevedo M, Cotner S, Lortie CJ (2020) How the ecology and evolution of the COVID-19 pandemic changed learning. Ecol Evol 10(22):12412–12417. https://doi.org/10.1002/ece3.6937
124. Lawrence MG, Schäfer S, Muri H, Scott V, Oschlies A, Vaughan NE, Boucher O, Schmidt H, Haywood J, Scheffran J (2018) Evaluating climate geoengineering proposals in the context of the Paris Agreement temperature goals. Nat Commun 9(1):3734. https://doi.org/10.1038/s41467-018-05938-3
125. Lee DMM (2020) Covid-19: agnotology, inequality, and leadership. Hum Resour Dev Int 23(4):333–346. https://doi.org/10.1080/13678868.2020.1779544
126. Lee Y, Tao W, Li J-YQ, Sun R (2021) Enhancing employees' knowledge sharing through diversity-oriented leadership and strategic internal communication during the COVID-19 outbreak. J Knowl Manag 25(6):1526–1549. https://doi.org/10.1108/JKM-06-2020-0483
127. Leiva-Martinez M-A, Anderson-Seminario MdlM, Alvarez-Risco A, Estrada-Merino A, Mlodzianowska S (2021) Price variation in lower goods as of previous economic crisis and the contrast of the current price situation in the context of COVID-19 in Peru. In: Lawrence KD, Klimberg RK (eds.) Advances in business and management forecasting, vol 14. Emerald Publishing Limited, pp 161–166. https://doi.org/10.1108/S1477-407020210000014011
128. Lemes PG, Zanuncio JC, Jacovine LAG, Wilcken CF, Lawson SA (2021) Forest Stewardship Council and Responsible Wood certification in the integrated pest management in Australian forest plantations. For Policy Econ 131:102541. https://doi.org/10.1016/j.forpol.2021.102541

129. Liguori E, Winkler C (2020) From offline to online: challenges and opportunities for entrepreneurship education following the COVID-19 pandemic. Entrep Educ Pedagog 3(4):346–351. https://doi.org/10.1177/2515127420916738
130. Lim S, Prakash A (2021) Pandemics and citizen perceptions about their country: did COVID-19 increase national pride in South Korea? Nations Natl 27(3):623–637. https://doi.org/10.1111/nana.12749
131. Livingston E, Desai A, Berkwits M (2020) Sourcing personal protective equipment during the COVID-19 pandemic. JAMA 323(19):1912–1914. https://doi.org/10.1001/jama.2020.5317
132. Llorach-Massana P, Farreny R, Oliver-Solà J (2015) Are Cradle to Cradle certified products environmentally preferable? Analysis from an LCA approach. J Clean Prod 93:243–250. https://doi.org/10.1016/j.jclepro.2015.01.032
133. Lopez-Odar D, Alvarez-Risco A, Vara-Horna A, Chafloque-Cespedes R, Sekar MC (2020) Validity and reliability of the questionnaire that evaluates factors associated with perceived environmental behavior and perceived ecological purchasing behavior in Peruvian consumers. Soc Responsib J 16(3):403–417. https://doi.org/10.1108/SRJ-08-2018-0201
134. Mahoney LS, Thorne L, Cecil L, LaGore W (2013) A research note on standalone corporate social responsibility reports: signaling or greenwashing? Crit Perspect Account 24(4):350–359. https://doi.org/10.1016/j.cpa.2012.09.008
135. Maritz A, Perenyi A, de Waal G, Buck C (2020) Entrepreneurship as the unsung hero during the current COVID-19 economic crisis: Australian perspectives. Sustain 12(11). https://doi.org/10.3390/su12114612
136. Marx A, Cuypers D (2010) Forest certification as a global environmental governance tool: what is the macro-effectiveness of the Forest Stewardship Council? [https://doi.org/10.1111/j.1748-5991.2010.01088.x]. Regul & Gov 4(4):408–434. https://doi.org/10.1111/j.1748-5991.2010.01088.x
137. McCollum DL, Zhou W, Bertram C, de Boer H-S, Bosetti V, Busch S, Després J, Drouet L, Emmerling J, Fay M, Fricko O (2018) Energy investment needs for fulfilling the Paris Agreement and achieving the Sustainable Development Goals. Nat Energy 3(7):589–599. https://doi.org/10.1038/s41560-018-0179-z
138. Molino M, Ingusci E, Signore F, Manuti A, Giancaspro ML, Russo V, Russo V, Zito M, Cortese CG (2020) Wellbeing costs of technology use during Covid-19 remote working: an investigation using the Italian translation of the technostress creators scale. Sustain 12(15). https://doi.org/10.3390/su12155911
139. Moog S, Spicer A, Böhm S (2015) The politics of multi-stakeholder initiatives: the crisis of the Forest Stewardship Council. J Bus Ethics 128(3):469–493. https://doi.org/10.1007/s10551-013-2033-3
140. National Geographic (2021) The world is still falling short of meeting its climate goals. Retrieved 01/01/2022 from https://www.nationalgeographic.com/environment/article/the-world-is-still-falling-short-of-meeting-its-climate-goals
141. Niedziałkowski K, Shkaruba A (2018) Governance and legitimacy of the Forest Stewardship Council certification in the national contexts – A comparative study of Belarus and Poland. For Policy Econ 97:180–188. https://doi.org/10.1016/j.forpol.2018.10.005
142. Nimrod G (2020) Technostress in a hostile world: older internet users before and during the COVID-19 pandemic. Aging & Ment Health, 1–8. https://doi.org/10.1080/13607863.2020.1861213
143. Obergassel W, Hermwille L, Oberthür S (2021) Harnessing international climate governance to drive a sustainable recovery from the COVID-19 pandemic. Climate Policy 21(10):1298–1306. https://doi.org/10.1080/14693062.2020.1835603
144. Okereke M (2021) Towards vaccine equity: should big pharma waive intellectual property rights for COVID-19 vaccines? Public Health Pract 2:100165. https://doi.org/10.1016/j.puhip.2021.100165
145. Pact TUP (2021) What is The UK Plastics Pact? Retrieved 01/01/2022 from https://wrap.org.uk/taking-action/plastic-packaging/the-uk-plastics-pact

146. Panisoara IO, Lazar I, Panisoara G, Chirca R, Ursu AS (2020) Motivation and continuance intention towards online instruction among teachers during the COVID-19 pandemic: the mediating effect of burnout and technostress. Int J Environ Res Public Health 17(21). https://doi.org/10.3390/ijerph17218002
147. Papadopoulos S, Kontokosta CE (2019) Grading buildings on energy performance using city benchmarking data. Appl Energy 233–234:244–253. https://doi.org/10.1016/j.apenergy.2018.10.053
148. Parguel B, Benoît-Moreau F, Larceneux F (2011) How sustainability ratings might deter 'greenwashing': a closer look at ethical corporate communication. J Bus Ethics 102(1):15. https://doi.org/10.1007/s10551-011-0901-2
149. Parguel B, Benoit-Moreau F, Russell CA (2015) Can evoking nature in advertising mislead consumers? The power of 'executional greenwashing'. Int J Advert 34(1):107–134. https://doi.org/10.1080/02650487.2014.996116
150. Pattberg P (2005) What role for private rule-making in global environmental governance? Analysing the Forest Stewardship Council (FSC). Int Environ Agreem: Polit, Law Econ 5(2):175–189. https://doi.org/10.1007/s10784-005-0951-y
151. Pattberg PH (2005) The Forest Stewardship Council: risk and potential of private forest governance. J Environ & Dev 14(3):356–374. https://doi.org/10.1177/1070496505280062
152. Penado Abilleira M, Rodicio-García M-L, Ríos-de Deus MP, Mosquera-González MJ (2021) Technostress in Spanish University Teachers During the COVID-19 Pandemic [Original Research]. Front Psychol 12(496). https://doi.org/10.3389/fpsyg.2021.617650
153. Quispe-Cañari JF, Fidel-Rosales E, Manrique D, Mascaró-Zan J, Huamán-Castillón KM, Chamorro-Espinoza SE, Garayar-Peceros H, Ponce-López VL, Sifuentes-Rosales J, Alvarez-Risco A, Yáñez JA (2021) Self-medication practices during the COVID-19 pandemic among the adult population in Peru: a cross-sectional survey. Saudi Pharm J 29(1):1–11. https://doi.org/10.1016/j.jsps.2020.12.001
154. Rafael GC, Fonseca A, Jacovine LAG (2018) Non-conformities to the Forest Stewardship Council (FSC) standards: empirical evidence and implications for policy-making in Brazil. For Policy Econ 88:59–69. https://doi.org/10.1016/j.forpol.2017.12.013
155. Rahaman F, Maulick R, Yadav AK, Ray S, Sharma R (2012) Singularity-free dark energy star. Gen Relativ Gravit 44(1):107–124. https://doi.org/10.1007/s10714-011-1262-y
156. Rahman I, Park J, Chi CG-Q (2015) Consequences of "greenwashing." Int J Contemp Hosp Manag 27(6):1054–1081. https://doi.org/10.1108/IJCHM-04-2014-0202
157. Rajamani L (2016) Ambition and differentiation in the 2015 Paris Agreement: interpretative possibilities and underlying politics. Int Comp Law Q 65(2):493–514. https://doi.org/10.1017/S0020589316000130
158. Ramus CA, Montiel I (2005) When are corporate environmental policies a form of greenwashing? Bus Soc 44(4):377–414. https://doi.org/10.1177/0007650305278120
159. Raudenská J, Steinerová V, Javůrková A, Urits I, Kaye AD, Viswanath O, Varrassi G (2020) Occupational burnout syndrome and post-traumatic stress among healthcare professionals during the novel coronavirus disease 2019 (COVID-19) pandemic. Best Pract Res Clin Anaesthesiol 34(3):553–560. https://doi.org/10.1016/j.bpa.2020.07.008
160. Ren Z, Chen D (2018) Modelling study of the impact of thermal comfort criteria on housing energy use in Australia. Appl Energy 210:152–166. https://doi.org/10.1016/j.apenergy.2017.10.110
161. Retzlaff RC (2009) The Use of LEED in planning and development regulation: an exploratory analysis. J Plan Educ Res 29(1):67–77. https://doi.org/10.1177/0739456X09340578
162. Robiou du Pont Y, Meinshausen M (2018) Warming assessment of the bottom-up Paris Agreement emissions pledges. Nat Commun 9(1):4810. https://doi.org/10.1038/s41467-018-07223-9
163. Roesch E, Amin A, Gupta J, García-Moreno C (2020) Violence against women during covid-19 pandemic restrictions. BMJ 369:m1712. https://doi.org/10.1136/bmj.m1712

164. Rojas Román B, Moscoso S, Chung SA, Limpias Terceros B, Álvarez-Risco A, Yáñez JA (2020) Tratamiento de la COVID-19 en Perú y Bolivia y los riesgos de la automedicación [COVID-19; tratamiento; estrategia; fármacos; automedicación; Perú; Bolivia.]. 2020 53(2). http://www.revfarmacia.sld.cu/index.php/far/article/view/435/351
165. Romero C, Putz FE (2018) Theory-of-Change Development for the Evaluation of Forest Stewardship Council Certification of Sustained Timber Yields from Natural Forests in Indonesia. Forests 9(9). https://doi.org/10.3390/f9090547
166. Rosen Geoffrey H, Murray Katie S, Greene Kirsten L, Pruthi Raj S, Richstone L, Mirza M (2020) Effect of COVID-19 on urology residency training: a nationwide survey of program directors by the society of academic urologists. J Urol 204(5):1039–1045. https://doi.org/10.1097/JU.0000000000001155
167. Rossi M, Charon S, Wing G, Ewell J (2006) Design for the next generation: incorporating Cradle-to-Cradle design into Herman Miller products [https://doi.org/10.1162/jiec.2006.10.4.193]. J Ind Ecol 10(4):193–210. https://doi.org/10.1162/jiec.2006.10.4.193
168. Sabory NR, Senjyu T, Momand AH, Waqfi H, Saboor N, Mobarez R, Razeqi F (2021) LEED scores of residential buildings in poor cities: Kabul City Case. Sustain 13(12). https://doi.org/10.3390/su13126959
169. Sánchez-Clavijo LM, Martínez-Callejas SJ, Acevedo-Charry O, Diaz-Pulido A, Gómez-Valencia B, Ocampo-Peñuela N, Ocampo D, Olaya-Rodríguez MH, Rey-Velasco JC, Soto-Vargas C, Ochoa-Quintero JM (2021) Differential reporting of biodiversity in two citizen science platforms during COVID-19 lockdown in Colombia. Biol Cons 256:109077. https://doi.org/10.1016/j.biocon.2021.109077
170. Sánchez OR, Vale DB, Rodrigues L, Surita FG (2020) Violence against women during the COVID-19 pandemic: an integrative review. Int J Gynecol Obstet 151(2):180–187. https://doi.org/10.1002/ijgo.13365
171. Savaresi A (2016) The Paris Agreement: a new beginning? J Energy & Nat Resour Law 34(1):16–26. https://doi.org/10.1080/02646811.2016.1133983
172. Schepers DH (2010) Challenges to Legitimacy at the Forest Stewardship Council. J Bus Ethics 92(2):279–290. https://doi.org/10.1007/s10551-009-0154-5
173. Seele P, Gatti L (2017) Greenwashing revisited: in search of a typology and accusation-based definition incorporating legitimacy strategies [https://doi.org/10.1002/bse.1912]. Bus Strat Environ 26(2):239–252. https://doi.org/10.1002/bse.1912
174. Sekalala S, Forman L, Hodgson T, Mulumba M, Namyalo-Ganafa H, Meier BM (2021) Decolonising human rights: how intellectual property laws result in unequal access to the COVID-19 vaccine. BMJ Glob Health 6(7):e006169. https://doi.org/10.1136/bmjgh-2021-006169
175. Senaratne S, Hewamanage PR (2015) The role of team leadership in achieving LEED certification in a green building project. Built Environ Proj Asset Manag 5(2):170–183. https://doi.org/10.1108/BEPAM-09-2013-0036
176. Shaw R, Kim Y-K, Hua J (2020) Governance, technology and citizen behavior in pandemic: lessons from COVID-19 in East Asia. Prog Disaster Sci 6:100090. https://doi.org/10.1016/j.pdisas.2020.100090
177. Siano A, Vollero A, Conte F, Amabile S (2017) "More than words": expanding the taxonomy of greenwashing after the Volkswagen scandal. J Bus Res 71:27–37. https://doi.org/10.1016/j.jbusres.2016.11.002
178. Sigala M (2020) Tourism and COVID-19: impacts and implications for advancing and resetting industry and research. J Bus Res 117:312–321. https://doi.org/10.1016/j.jbusres.2020.06.015
179. Silvestre JD, de Brito J, Pinheiro MD (2013) From the new European Standards to an environmental, energy and economic assessment of building assemblies from cradle-to-cradle (3E-C2C). Energy Build 64:199–208. https://doi.org/10.1016/j.enbuild.2013.05.001
180. Silvestre JD, de Brito J, Pinheiro MD (2014) Environmental impacts and benefits of the end-of-life of building materials – calculation rules, results and contribution to a "cradle to cradle" life cycle. J Clean Prod 66:37–45. https://doi.org/10.1016/j.jclepro.2013.10.028

181. Singh A, Haynes M (2020) The challenges of COVID-19 in nursing education: the time for faculty leadership training is now. Nurse Educ Pract 47:102831. https://doi.org/10.1016/j.nepr.2020.102831
182. Spagnoli P, Molino M, Molinaro D, Giancaspro ML, Manuti A, Ghislieri C (2020) Workaholism and technostress during the COVID-19 emergency: the crucial role of the leaders on remote working [Brief Research Report]. Front Psychol 11(3714). https://doi.org/10.3389/fpsyg.2020.620310
183. Sugiura K, Oki Y (2018) Reasons for Choosing Forest Stewardship Council (FSC) and Sustainable Green Ecosystem Council (SGEC) schemes and the effects of certification acquisition by forestry enterprises in Japan. Forests 9(4). https://doi.org/10.3390/f9040173
184. Sun B (2018) Heterogeneous direct rebound effect: theory and evidence from the Energy Star program. Energy Econ 69:335–349. https://doi.org/10.1016/j.eneco.2017.11.025
185. Szabo S, Webster J (2021) Perceived greenwashing: the effects of green marketing on environmental and product perceptions. J Bus Ethics 171(4):719–739. https://doi.org/10.1007/s10551-020-04461-0
186. Tanner S (2009) How does WaterSense® select products for specification development? [https://doi.org/10.1002/j.1551-8833.2009.tb09827.x]. J AWWA 101(2):37–39. https://doi.org/10.1002/j.1551-8833.2009.tb09827.x
187. Taylor C, Lindenmayer DB (2021) Stakeholder engagement in a Forest Stewardship Council Controlled Wood assessment. Environ Sci & Policy 120:204–212. https://doi.org/10.1016/j.envsci.2021.03.014
188. Taylor PL (2005) In the market but not of it: fair trade coffee and Forest Stewardship Council Certification as market-based social change. World Dev 33(1):129–147. https://doi.org/10.1016/j.worlddev.2004.07.007
189. The Guardian (2021) Governments falling woefully short of Paris climate pledges, study finds. Retrieved 01/01/2022 from https://www.theguardian.com/science/2021/sep/15/governments-falling-short-paris-climate-pledges-study
190. The Wall Street Journal (2021) U.N. Finds Nations' Climate Plans Fall Short of Paris Accord. Retrieved 01/01/2022 from https://www.wsj.com/articles/u-n-finds-nations-climate-plans-fall-short-of-paris-accord-11635259447
191. Torelli R, Balluchi F, Lazzini A (2020) Greenwashing and environmental communication: Effects on stakeholders' perceptions [https://doi.org/10.1002/bse.2373]. Bus Strat Environ 29(2):407–421. https://doi.org/10.1002/bse.2373
192. Toxopeus ME, de Koeijer BLA, Meij AGGH (2015) Cradle to Cradle: effective vision vs. efficient practice? Procedia CIRP 29:384–389. https://doi.org/10.1016/j.procir.2015.02.068
193. Tran LTT (2021) Managing the effectiveness of e-commerce platforms in a pandemic. J Retail Consum Serv 58:102287. https://doi.org/10.1016/j.jretconser.2020.102287
194. UN (2021) COP26 ends with agreement but falls short on climate action. Retrieved 01/01/2022 from https://www.unep.org/news-and-stories/story/cop26-ends-agreement-falls-short-climate-action
195. Urpelainen J, Van de Graaf T (2018) United States non-cooperation and the Paris agreement. Clim Policy 18(7):839–851. https://doi.org/10.1080/14693062.2017.1406843
196. Vickers A, Bracciano D (2014) Low-volume plumbing fixtures achieve water savings [https://doi.org/10.5991/OPF.2014.40.0047]. Opflow 40(7):8–9. https://doi.org/10.5991/OPF.2014.40.0047
197. Vidya CT, Prabheesh KP (2020) Implications of COVID-19 pandemic on the global trade networks. Emerg Mark Financ Trade 56(10):2408–2421. https://doi.org/10.1080/1540496X.2020.1785426
198. Viero A, Barbara G, Montisci M, Kustermann K, Cattaneo C (2021) Violence against women in the Covid-19 pandemic: a review of the literature and a call for shared strategies to tackle health and social emergencies. Forensic Sci Int 319:110650. https://doi.org/10.1016/j.forsciint.2020.110650
199. VOX (2019) The Paris climate agreement is at risk of falling apart in the 2020s. Retrieved 01/01/2022 from https://www.vox.com/energy-and-environment/2019/11/5/20947289/paris-climate-agreement-2020s-breakdown-trump

200. Wang P, Li W, Kara S (2017) Cradle-to-cradle modeling of the future steel flow in China. Resour, Conserv Recycl 117:45–57. https://doi.org/10.1016/j.resconrec.2015.07.009
201. Ward DO, Clark CD, Jensen KL, Yen ST, Russell CS (2011) Factors influencing willingness-to-pay for the ENERGY STAR® label. Energy Policy 39(3):1450–1458. https://doi.org/10.1016/j.enpol.2010.12.017
202. Wen J, Kozak M, Yang S, Liu F (2021) COVID-19: potential effects on Chinese citizens' lifestyle and travel. Tour Rev 76(1):74–87. https://doi.org/10.1108/TR-03-2020-0110
203. WHO (2022) WHO Coronavirus (COVID-19) Dashboard. Retrieved 01/01/2022 from https://covid19.who.int/
204. Wu T-Y, Ford O, Rainville AJ, Bessire R (2020) COVID-19 care package distribution for senior citizens and families in Detroit and Hamtramck, Michigan. J Hunger & Environ Nutr 15(4):585–587. https://doi.org/10.1080/19320248.2020.1799586
205. Yadav S, Hazarika L (2021) Managing employee relations in multinational firms: challenges and opportunities. In Critical issues on changing dynamics in employee relations and workforce diversity. IGI Global, pp 18–44
206. Yan J, Kim S, Zhang SX, Foo M-D, Alvarez-Risco A, Del-Aguila-Arcentales S, Yáñez JA (2021) Hospitality workers' COVID-19 risk perception and depression: a contingent model based on transactional theory of stress model. Int J Hosp Manag 95:102935. https://doi.org/10.1016/j.ijhm.2021.102935
207. Yáñez JA, Afshar Jahanshahi A, Alvarez-Risco A, Li J, Zhang SX (2020) Anxiety, distress, and turnover intention of healthcare workers in Peru by their distance to the epicenter during the COVID-19 crisis. Am J Trop Med Hyg 103(4):1614–1620. https://doi.org/10.4269/ajtmh.20-0800
208. Yáñez JA, Alvarez-Risco A, Delgado-Zegarra J (2020) Covid-19 in Peru: from supervised walks for children to the first case of Kawasaki-like syndrome. BMJ 369:m2418. https://doi.org/10.1136/bmj.m2418
209. Yasin Ar A (2020) Managing e-commerce during a pandemic: lessons from GrubHub during COVID-19. In: George B, Mahar Q (eds) International case studies in the management of disasters. Emerald Publishing Limited, pp 155–167. https://doi.org/10.1108/978-1-83982-186-820201010
210. Zarocostas J (2021) What next for a COVID-19 intellectual property waiver? The Lancet 397(10288):1871–1872. https://doi.org/10.1016/S0140-6736(21)01151-X
211. Zhang SX, Chen J, Afshar Jahanshahi A, Alvarez-Risco A, Dai H, Li J, Patty-Tito RM (2021) Succumbing to the COVID-19 pandemic—Healthcare workers not satisfied and intend to leave their jobs. Int J Ment Heal Addict. https://doi.org/10.1007/s11469-020-00418-6
212. Zhang SX, Sun S, Afshar Jahanshahi A, Alvarez-Risco A, Ibarra VG, Li J, Patty-Tito RM (2020) Developing and testing a measure of COVID-19 organizational support of healthcare workers – results from Peru, Ecuador, and Bolivia. Psychiatry Res 291:113174. https://doi.org/10.1016/j.psychres.2020.113174
213. Zollinger R, DiCindio C (2021) Community ecology: Museum education and the digital divide during and after COVID-19. J Mus Educ 46(4):481–492. https://doi.org/10.1080/10598650.2021.1983711

Circular Economy for Food Loss Reduction and Water Footprint

Berdy Briggitte Cuya-Velásquez, Aldo Alvarez-Risco, Romina Gomez-Prado, Luis Juarez-Rojas, Anguie Contreras-Taica, Arianne Ortiz-Guerra, Maria de las Mercedes Anderson-Seminario, and Shyla Del-Aguila-Arcentales

Abstract Food loss is a global problem with significant social, economic, and ecological impacts. According to the Food and Agriculture Organization (FAO), a third of the food produced globally is lost or wasted along the food supply chain (FSC). Food requires water for its production. This water consumption can be measured under the WF indicator's Water Footprint (WF). This indicator makes it possible to measure fresh water's direct and indirect use throughout the food supply chain. Estimates of the water footprint of crop production range between 5,938 and 8,508 km^3/year. The greater the loss or waste of food, the greater the water consumption. Therefore, applying strategies related to the circular economy is an alternative solution. This chapter provides the relationship between food loss and the water footprint to provide alternative solutions under the circular economy model.

Keywords Circular economy · Food loss · Water footprint · Global trends · Impacts

1 Introduction

Globally, firms and citizens try to reestablish the world they usually live in, but it is not possible. Even recent news about new variants of COVID-19 pushes the governments and people to keep their lives according to the protocols. So what has been the damage due to COVID-19? The world has changed and people's habits regarding sustainability have changed, depending on the characteristics of pre-pandemic life [14, 17, 18, 22, 23, 26, 30, 32, 33, 64, 66, 73, 112, 129]. The world has changed and people's habits regarding sustainability have changed, depending on the characteristics of pre-pandemic life.

B. B. Cuya-Velásquez · A. Alvarez-Risco (✉) · R. Gomez-Prado · L. Juarez-Rojas · A. Contreras-Taica · A. Ortiz-Guerra · M. de las Mercedes Anderson-Seminario
Universidad de Lima, Lima, Perú
e-mail: aralvare@ulima.edu.pe

S. Del-Aguila-Arcentales
Escuela Nacional de Marina Mercante "Almirante Miguel Grau", Callao, Perú

A. Alvarez-Risco et al. (eds.), *Circular Economy*, Environmental Footprints and Eco-design of Products and Processes,
https://doi.org/10.1007/978-981-19-0549-0_4

Table 1 Impact of COVID-19 in different sectors

Education	[9, 10, 24 19, 20, 59, 103, 104, 166]
Entrepreneurship	[5, 16, 27, 28, 38, 41, 42, 47, 55, 107, 110]
Health sector	[1, 8, 11, 13 21, 39, 44, 60, 67, 99, 125, 130, 157, 163, 164]
Hospitality	[71, 87, 91, 95, 156]
Intellectual property	[12, 15, 65, 74, 94, 100, 122, 135, 162]
Population	[7, 25, 53, 88, 108, 123, 132, 136, 154, 155, 158]
Prices	[35, 62, 82, 90, 105, 121]
Tourism	[36, 40, 46, 50, 51, 52, 80, 137]
Trade	[2, 3, 6, 29, 31, 43, 63, 98, 109, 145, 150, 159]
Violence against women	[34, 56, 70, 86, 128, 133, 151]

Sustainable education is the cornerstone for firms and the population to ensure the efficient use of sources as water and food. In Table 1, it is presented different sectors that have been damaged by COVID-19.

According to the Food Loss and Waste Protocol, a global partnership, FLW refers to the decrease of food during the whole food supply chain. On the one hand, Food Loss (FL) decreases food during the following food supply chain stages: production, post-harvest, and processing. On the other hand, Food Waste (FW) is food in good condition to be consumed, discarded at the end of the food supply chain [152].

As part of the productive process of food harvesting, the water footprint is a factor to consider when talking about the FL and FW; it refers to the quantity of water used for people's consumption in any aspect. Water footprints of nations are the water used by inhabitants as a function of their consumption pattern. A solution to counteract these problems, the concept of circular economy (CE), has become a trend and is increasingly attracting the interest of academics and professionals because it is taken as a form of operationalization for companies to implement the concept of sustainable growth [83, 120]. Information has been compiled on the amount of FL and FW worldwide, information was added on the places where they are primarily generated and the factors surrounding each one, the amount of FL generated by one person per year and what is generated during the production process, distribution, sale, and consumption by people. Moreover, the environmental, social, and economic impacts of FL and FW were analyzed.

2 A Review of Circular Economy

The classic linear model that we have today of extraction-production-use-discharge of material and energy flow of the economic system turns out to be unsustainable [81]. As a complement and to facilitate the reader's understanding, the linear economy model is shown in Fig. 1. The materials cycle initiative has been from the beginning

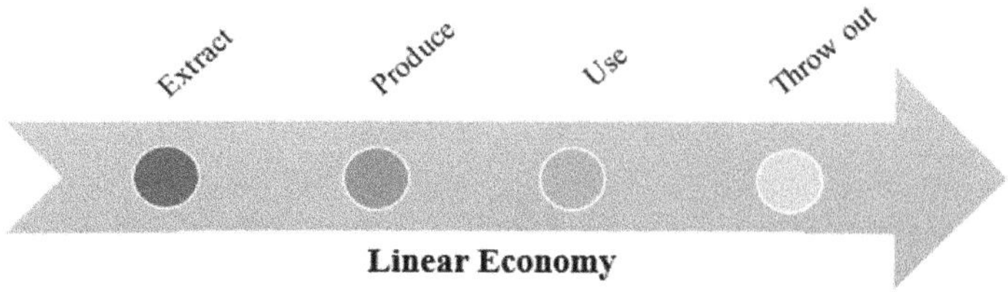

Fig. 1 Linear economy model (*Source* REPSOL [127])

of industrialization and has been put into practice linked to the idea that it reduces the negative impact on the environment and promotes novel trade opportunities throughout the origin of industrialization [68, 69].

However, this linear performance model has established itself in the business world by taking control of general development and causing dangerous damage to the environment. Many people make, use and dispose of the products; in fact, a third of the plastic waste in the world is not collected or managed. Nevertheless, the concern for the ethics and the stability of the resources and the reduction of greenhouse gases begins to be more noticeable and changes the way of seeing the materials. People believe that materials should be preserved rather than constantly consumed and thrown away [138].

According to Keijer et al. [96], awareness of the finite characteristic of many resources and limited environmental tolerance by the chemical industry has increased highly during the last decades. Therefore, they argue that the current linear economy approach of "take, make and dispose of" should be replaced by a new approach to a circular economy (CE). This business-oriented approach emphasizes reusing products, materials, and components, applying remanufacturing, repair, renovation, and improvement processes, and promoting the use of solar, wind, biomass, and other sources derived from waste obtained from the entire product manufacturing process [45, 114, 124]. Zacarías Farah [161] points out that changing the way it is produced and consumed is essential, moving toward the CE model in which production cycles are closed, and an uninterrupted flow of natural resources is maintained.

CE is born primarily in the literature through 3 main "actions," which are called the Principles of the 3Rs: Reduction, Reuse, and Recycling [58, 106, 126, 131, 142, 160, 165].

Charonis [57] establishes his definition of CE based on the vision of the Ellen MacArthur Foundation [72], explaining it as a system made to restore and regenerate. Likewise, it is defined as an economic system that replaces the idea of "end of life" with reduction, the option of reuse, recycling, and recovery of materials in the production, distribution, or consumption processes [97]. According to a current report from the United Nations Environment Program in 2018, titled Re-defining Value—The Manufacturing Revolution, the CE could minimize industrial waste between 80 and 99% in some sectors. In the same way, it would reduce its emissions between 79 and 99%, with optimal results that drive its application [161].

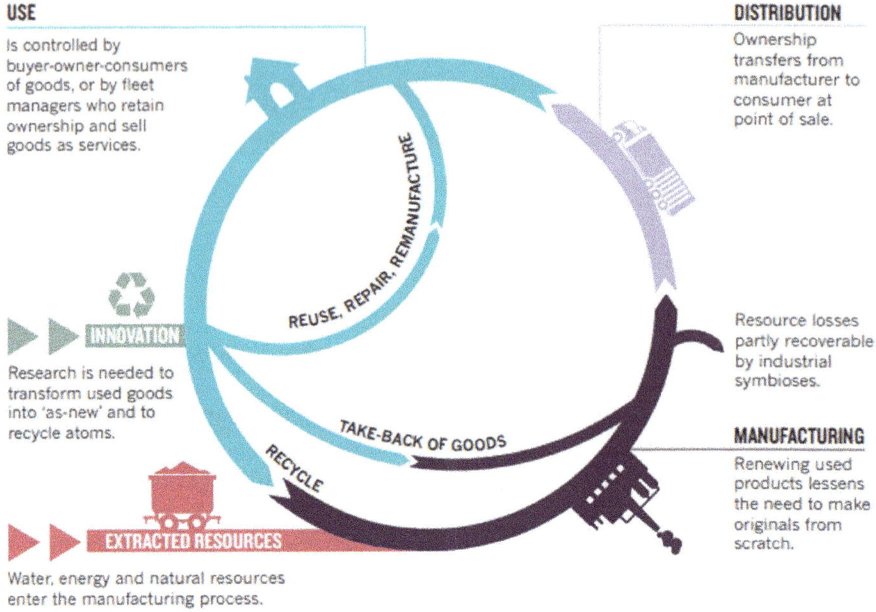

Fig. 2 Closing loops model (*Source* Stahel [138])

A balance would be achieved between the environment, the economy, and society. The CE serves as a tool to develop and create policies related to environmental care and bottom-up stewardship [83]. Thus, by applying this new economic model to achieve sustainable practices, companies contribute to the care and preservation of the environment and cause positive impacts on a social and economic level. Following this line, Fig. 2 shows a CE model called "Closing loops," which applies those above, focusing on transforming the goods in the last stage of their useful life into resources for others, uniting and closing ties in industrial ecosystems decreasing amounts of waste.

3 Food Loss: Data Analysis and Impact Study

3.1 Global Food Loss Trends and Data

According to the Food and Agriculture Organization of the United Nations [76], one-third of the food produced globally is lost or wasted throughout the FSC. It is essential to understand the difference is between food loss and food waste (PDA). Both concepts refer to decreased mass or nutritional value of foods during FSC. According to the United Nations Environment Program [146], the main difference is that food loss refers to food that spills, spoils, or loses its quality before its final product stage. The concept of food waste refers to those products that are discarded or not consumed in the form of a final product when it is fit for consumption.

According to the "Food Waste Index 2021," in 2019, around 931 million tons of food waste were generated, which indicates that almost 17% of world food production was wasted. This figure indicates the ineffective or non-existent residual treatment that food receives when lost or wasted. In Spain, for example, a small part of the FLW is used for aerobic composting systems or receives heat treatment. However, most FLW is sent to landfills [92, p. 3]. These strategies are essential to consider since it is estimated that between 8 and 10% of greenhouse gases are produced due to lost or wasted food. The UNEP study found that most of the food waste comes from households, followed by food services, and finally, retail Fig. 3 shows the distribution of food waste by sector in percentage and metric tons.

Data indicates that 61% of food wasted comes from households, which is explained by various situations, for example, those foods that were bought but not consumed or those that lost quality and nutritional properties because they expired. On the other hand, 26% of food is wasted in food services (i.e., places consumed outside the home) because the end-user does not consume all the food he asks for. 13% of wasted food is generated in supermarkets due to improper handling, distribution, or the disposal of products that have not been sold. Figure 4 shows the amount of food (in kilograms) discarded per person in each sector in 2019.

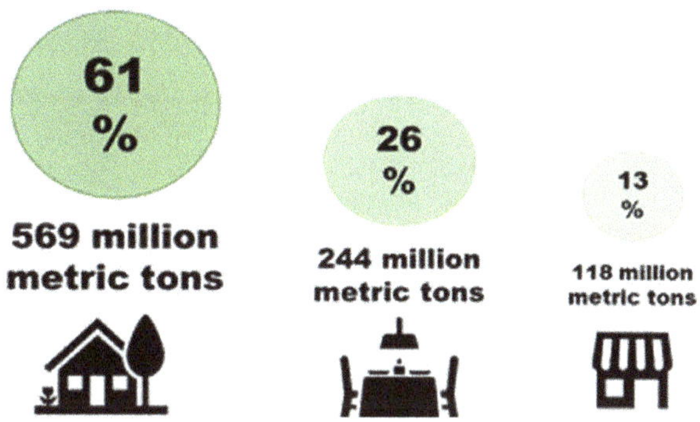

Fig. 3 Total food waste generated by the sector (*Source* UNEP [146])

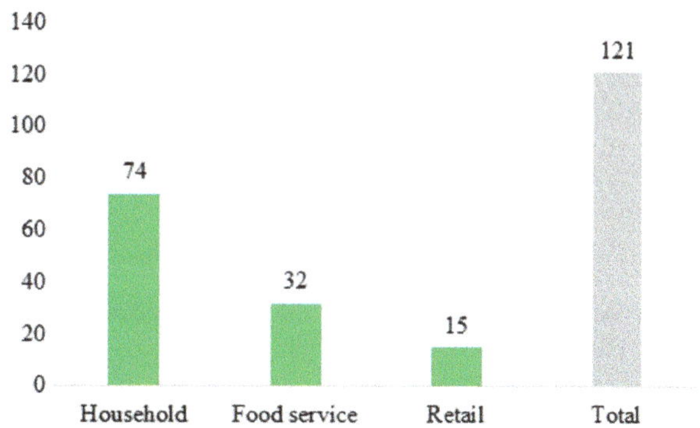

Fig. 4 Per capita food waste generated by households worldwide (*Source* Statista [139])

As indicated in the figure, approximately 121 kg of food was discarded per person in 2019. It is essential to be able to reduce this amount of food loss. We could relate this data to Sustainable Development Goal 12.3, which states the following: "By 2030, halve per capita global food waste at the retail and consumer levels and reduce food losses along production and supply chains, including post-harvest losses" [76].

It is necessary to achieve reducing the amount of wasted food can be achieved by using different strategies not only at the end consumer level but by being more efficient and controlling the food distribution processes throughout the FSC. Policies and strategies are highly dependent on FSC processes. Figure 5 summarizes the loss levels at the FSC level.

During the FSC, we could apply strategies to achieve the SDGs. Compliance with the Sustainable Development Goals is ideal for reducing food loss or waste during the different FSC processes. On the other hand, the per capita food waste and loss trend generated mainly by households show an exciting characteristic to analyze at the regional and country level. Figure 6 shows the food waste generated by households in 2020; the units are expressed in millions of metric tons.

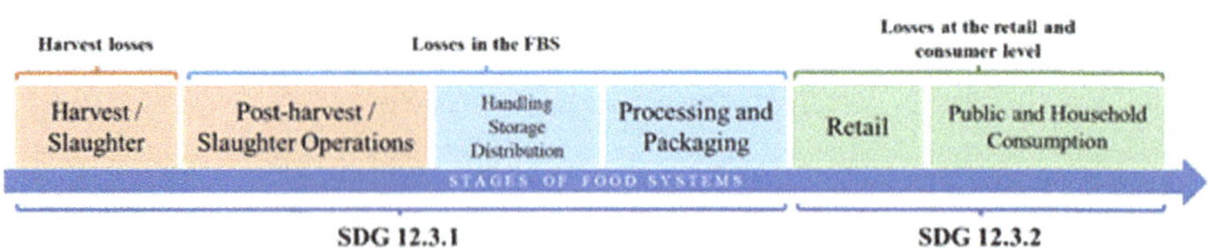

Fig. 5 Levels of food loss throughout the FSC (*Source* United Nations [147])

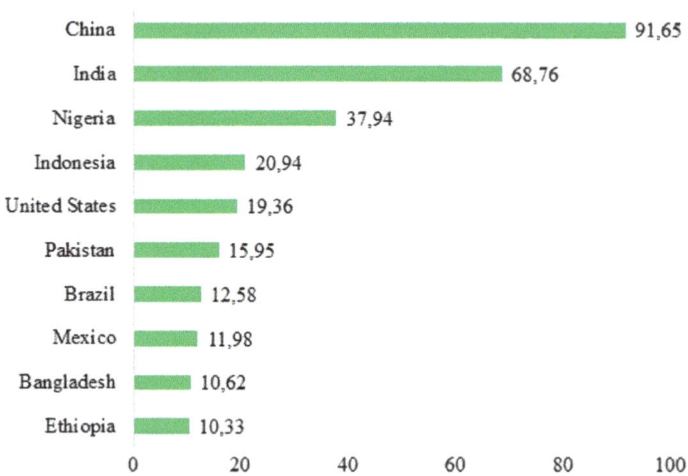

Fig. 6 Annual household waste produced by selected countries around the world in 2020 (million metric tons) (*Source* Statista [140])

As shown in Fig. 6, the levels of waste at the household level express similar amounts in high-, middle-upper-, and middle-low-income countries [146]. For example, food waste in Nigeria is 37.94 million metric tons. By way of analysis, Nigeria has a lower middle income than China or India,in addition, the prevalence level of undernourishment in that country in 2019 was 14.6% [144]. West Asia and sub-Saharan Africa countries have the highest per capita food waste at the domestic level [146]. Therefore, it suggests that the problem of food waste is not an exclusive problem in developed countries but also in developing countries that impact a social, economic, and environmental level.

3.2 *Impacts of Food Loss*

The social impact of food loss is related to hunger, and malnutrition is the primary social consequence. The world food production is increasing, malnutrition, far from decreasing, continues to grow. To understand this relationship, we define food loss as the decrease in volume of produced food fit for human consumption because it is discarded during the production, distribution, and marketing stages, while food waste refers to food fit for consumption discarded on purpose.

Likewise, hunger is defined as the deprivation of food, and the concept of malnutrition refers to disorders resulting from an imbalance of nutrients in the body. For the present book, three forms of malnutrition were considered: stunting in children under 5 years of age, anemia in women of reproductive age (15–49 years), and overweight (including obesity) in adult women (18 years and older), because they are the most common forms in the countries of the world. As previously mentioned, the Food Waste Index report [146] indicates that around 931 million tons of food waste were wasted in 2019. Further, Fig. 7 shows that 61% came from households, 26% from food services, and 13% from commercial establishments.

The Food Waste Index report [146] also reports that between 17 and 18% of global food production fit for consumption (plus inedible parts such as shells, seeds, and bones) ended up in a household, retail, restaurant, and other foodservice landfills. Figure 8 shows that 11% of the waste comes from households, 5% from food services, and 2% from commercial establishments.

The FAO Undernourishment Prevalence Indicator calculates a range of the population suffering from food energy sufficiency. It serves to control and monitor hunger at the global level (view Fig. 9) and even regionally since the data cannot be disaggregated sufficiently to track vulnerable populations within each region. For this report, the averages presented each year are used.

According to this indicator, it is estimated that between 766 million people suffer from food insufficiency (9.9% of the world population). Also, it is essential to highlight the significant difference between the average estimate of the range of 2019 and 2020, people suffering from food insufficiency increased from 650 million people in 2019 to 766 million in 2020; that is, in just one year, there was an increase of approximately 116 million people suffering from food insufficiency. Moreover, taking the

Fig. 7 Global Food Loss in 2019 (*Source* UNEP [146])

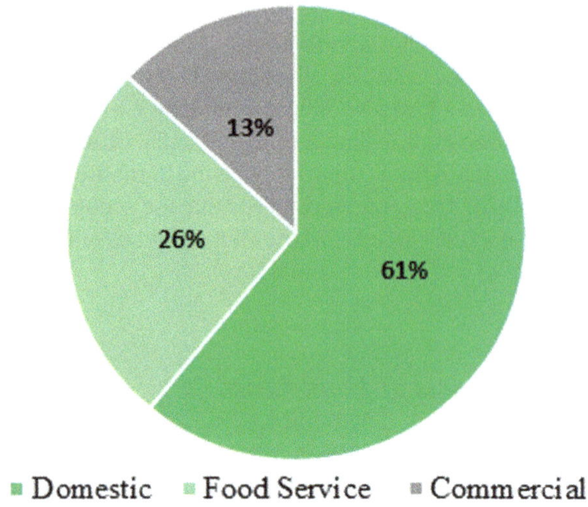

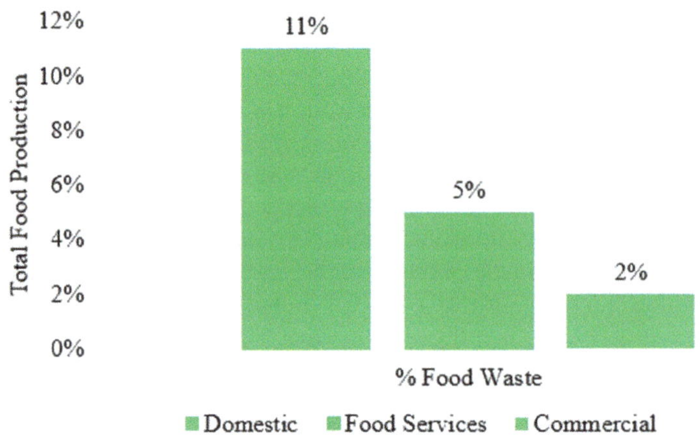

Fig. 8 Global Food Waste in 2019 (*Source* UNEP [146])

higher value of the 2020 range, 811 million people, the difference increases to 161 million people.

Considering that 933 million tons of food are lost, and 18% of total food production is wasted annually, it is inevitable to mention the relationship between food loss and waste and the hunger suffered by 766 million people (9.9% of the world population). In addition, approximately 2.3 billion people (30% of the world population) do not have access to a healthy diet, so it is essential to reduce these amounts to help people with fewer resources.

Finally, another point to consider as part of the social impact is food insecurity. Food security exists when people have physical and economic access to nutritious and safe food to have an active and healthy life [75]. Food insecurity involves a high

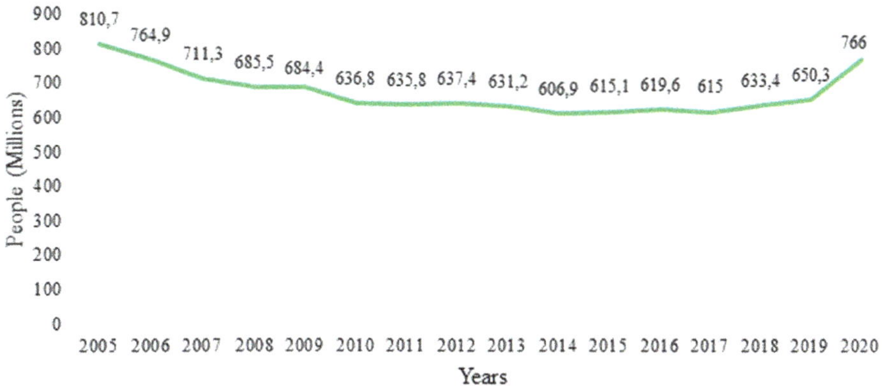

Fig. 9 Indicator of Prevalence of Undernourishment (Millions of people) (*Source* FAO [79])

degree of social impact since it arises from violence against women, climate change, and informality related to mass migration [149]. We could add that the loss of food also generates food insecurity. Around 690 million people suffer from hunger,despite that high-impact figure, food continues to be lost and wasted throughout the FSC. Food loss and waste must be reduced to improve food security levels.

The economic impact of food loss generates a series of economic costs, which impact each country's economy. According to FAO [78], the costs in the environmental sphere were the US$ 700,000 million, while in the social areas, they were US$ 900,000 million. The costs of food loss increase every year. In 2018, the amount of food lost was 2,100 tons, equivalent to US$ 1.6 billion [89]. However, companies and institutions' collaboration helped reduce costs significantly by $ 700,000 million. Table 2 shows the costs incurred for years according to social and environmental impacts.

Insignificant changes in society's thoughts and the behavior of institutions about food loss have influenced economic failures [102]. For Morone et al. [119], the

Table 2 The economic impact of food loss in US$ (billions per year)

Impacts	Costs
Emissions of greenhouse gases	US$ 394
Water scarcity in arid and dry regions	US$ 164
Soil erosion (due to the dismissal of nutrients, reduced yield, and biological deuterium)	US$ 35
Loss of resources due to soil erosion	US$ 32
Health damage from pesticide use	US$ 396
Repercussions on biodiversity due to pesticides, phosphates, and losses of pollinators	US$ 333
Risk of conflicts due to soil erosion and use of natural resources	US$ 153

Source FAO [78]

valorization of food loss requires the help of public and private institutions. Morone et al. [118] consider that a change in ecological behavior, economic awareness, and improvements in collaborative behaviors is necessary (p. 14). In this sense, Hegnsholt et al. [89] mention that five factors cause food loss, but proper management offers to save each aspect. Factors include lack of awareness that US$ 260 billion can be saved, supply chain infrastructure US$ 150 billion, chain efficiency US$ 120 billion, food chain collaboration in US$ 60 billion, and regulations US$ 110 billion.

The savings in each factor are based on the expenses made in recent years. The amount of food lost increases by around 1.9% per year until 2030, and the dollar is 1.8% per year [89]. In other words, the increase in the environmental loss of food and the increase in the dollar are related, generating a more significant impact on the economic, social, and social spheres. Figure 10 shows the costs incurred due to food loss in recent years and projections until 2030.

The environmental impact of food loss occurs in agricultural production because various natural resources are used, such as water, soil, air, among others. Chaboud and Daviron [54] mention that FLW produces an effective use of essential resources for food that are not used (p. 5). The waste of resources with food that was never consumed is evident throughout the supply chain. According to FAO [79], the environmental impacts of food loss are calculated from the life cycle analysis based on calculations of greenhouse gas (GHG) emissions and water and land use footprints. The environmental impacts of food loss cover four main aspects: biodiversity, carbon footprint, water use, and land use [79].

Biodiversity is an essential aspect that producers require to give good harvests, but the loss of food causes environmental damage that affects food crops for society [79]. For this reason, large companies consider the environmental factor as the greatest motivation to reduce food losses [102]. According to Chinie et al. [61], the mitigation of food loss can be done through reduction, reuse, recovery, and elimination (p. 2).

The land is one of the essential resources for the harvest and production of food. The degradation of this resource due to the loss of food causes damage to food production and the yield of fertile fields. According to Kummu et al. [101], the farmland used to grow these lost foods is 198 million hectares per year. The foods that most use the land for its cultivation are meats and dairy products, representing

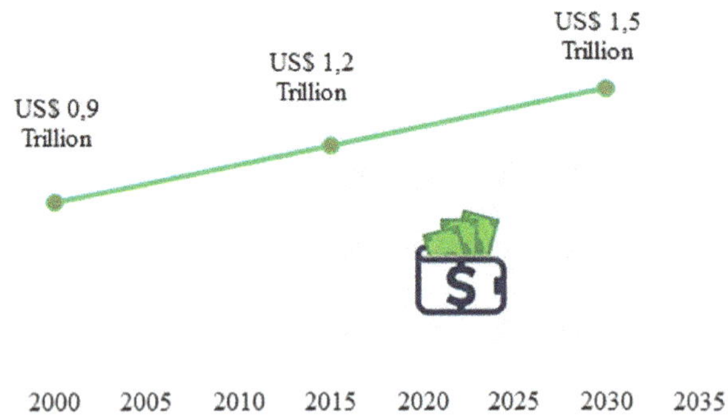

Fig. 10 Increase and behavior of waste loss costs in trillions of dollars from 2000 to 2030 (*Source* Hegnsholt et al. [89])

78%, while the loss of food represents 11% of the total surface [77]. The carbon footprint is another aspect affected since it relates to the number of greenhouse gases caused by food production during the life cycle. According to Statista [141], 10% of global greenhouse gas emissions are caused by food loss. For Mbow et a. [111], 8 to 10% is related to uneaten food (p. 200). The food loss index concerning the agricultural phase generates emissions of 16%, the post-harvest 16%, while in the processing, it produces 14% of emissions compared to the total generated by food production [77]. Finally, water is used in agricultural production worldwide, representing 92% of the available water in the world. According to FAO [79], the water used for the lost food is abundant since the representativeness of this resource exceeds the averages of water use compared to countries such as India and China who have a large population water consumption is high. In this sense, food loss substantially impacts water [79].

4 The Connection Between Food Loss and the Water Footprint

Water is the essential natural resource for food production after soil. However, this resource's indiscriminate use makes it increasingly scarce and overexploited worldwide [92]. To assess the impact of water consumption on food production, Hoekstra and Hung initially quantified virtual water (VA) flows from international trade in crops. These authors popularized the concept of the Water Footprint (WF) in 2002. As a concept, the water footprint is an indicator that measures the direct and indirect use of freshwater throughout the entire supply chain. The water footprint is divided into three components classified by color: blue (surface and groundwater), green (rainwater stored on the ground), and gray (contaminated water in the production process) [93]. It should be noted that the water footprint is not an indicator that measures the severity of the local impact on water consumption, but rather, it is a volumetric measure of water consumption.

Many economic sectors use water resources for their production processes. Agriculture has become the economic activity with the highest water consumption. By 2050, agricultural production is expected to increase by 50% compared to 2012 [113]. Increased agricultural activity could lead to higher water consumption. Agricultural products have a great demand for water resources [49].

Broadly, the food supply chain (FSC) can be divided into five main stages: (1) agricultural production, (2) post-harvest handling, (3) processing, (4) distribution/retail, and (5) consumption [4]. These stages involve harvesting the product under certain environmental conditions to consumption by the end-user of the supply chain. During FSC processes, much food is wasted, causing food loss problems worldwide. However, there are several ways to quantify food loss during FSC, the most widely used method is mass flow analysis that measures variations in the volume of certain products during FSC processes [48]. Applying methods that quantify food loss is ideal for establishing FSC processes policies. Some studies have focused on estimating the water footprint based on crop production. The differences in the estimates are mainly due to quantification techniques and modeling approaches, as well as externalities related to the environment in which the food is grown, including climate, number of crops, treatment, among others. In general, estimates of the water footprint of crop production range between 5,938 and 8,508 km^3/year [113]. Many investigations that estimate the water footprint in agricultural products, derivatives, and animal products have expanded the subject's knowledge. One of the most widely used studies for water footprint research was developed by Mekonnen and Gerbens-Leenes [113]. Hoekstra et al. [93] used a grid-based water balance model to calculate water use during a period for different crops. According to FAO [76], there are seven essential products between agriculture and animals, lost or wasted on a smaller or larger scale depending on the region under study. The groups are divided into Cereals, roots and tubers, oilseeds and legumes, fruits and vegetables, meat, fish, and dairy products. Table 3 is helpful to analyze the relationship between food loss and water footprint in more detail.

The data indicates that farm animal products require higher water consumption per kilogram production. The table shows that beef has a water footprint of 15,415 l/kg. This product is highly commercial and used by large restaurant chains worldwide; not only that but its consumption is linked to its importance as a source of protein, which has increased its demand [85]. However, as we have been pointing out, during the FSC, beef also suffers losses and waste that are not used efficiently. Table 4 shows lost or wasted beef during the five main FSC stages.

Table 3 The global average water footprint of agricultural and farm animal products

Product	Global Average Water Footprint (l/kg)			
	Green	Blue	Gray	Total
Agricultural products				
Beans, dry	3,945	125	983	**5,053**
Wheat	1,277	342	207	**1,827**
Rice	1,146	341	187	**1,673**
Barley	1,213	79	131	**1,423**
Apples	561	133	127	**822**
Bananas	660	97	33	**790**
Potato	191	33	63	**287**
Farm Animal Products				
Beef	14,414	550	451	**15,415**
Pork	4,907	459	622	**5,988**
Chicken meat	3,545	313	467	**4,325**
Egg	2,592	244	429	**3,265**
Milk	863	86	72	**1,020**

Source Hoekstra et al. [93] and Gerbens-Leenes [113]

Table 4 Activities that generate the loss or waste of beef throughout the FSC

Stages of the FSC	Activities that generate loss or waste
Production	• Animal deaths during breeding • Animals that are discarded for presenting diseases • Loss or misplacement of cattle
Handling and storage	• Deaths during transport to the slaughterhouse
Processing	• Industrial processing to produce derivatives (e.g., production of sausages) • Unused animal waste
Distribution	• Loss or waste in the market system. Unsold products that are thrown away
Consumption	• Loss or waste in household consumption • Product exposed to suboptimal conditions that cause rotting of the meat • Uneaten meat waste

Source FAO [77]

The activities described suggest that beef suffers loss and waste throughout the FSC. Strategies to reduce the amount of beef lost during the first two stages (i.e., production and handling and storage) may be challenging to implement because both stages require knowing the health status of the animals and whether they are suitable for consumption. However, numerous strategies could be applied at the final consumer level to reduce beef consumption and, therefore, the water levels required to produce this food. An alternative—among many others—could be to generate changes in eating habits. The impact would be significant in the long-term considering that the average meat consumer requires 16 Olympic pools of water to produce the meat they eat. At the country level, the annual water footprint for the crop sector (i.e., agricultural products except for farm animal products) reaches up to $1*10^{12}$ m^3/year. Figure 11 represents the annual green, blue, and gray water footprint for agricultural products in different countries.

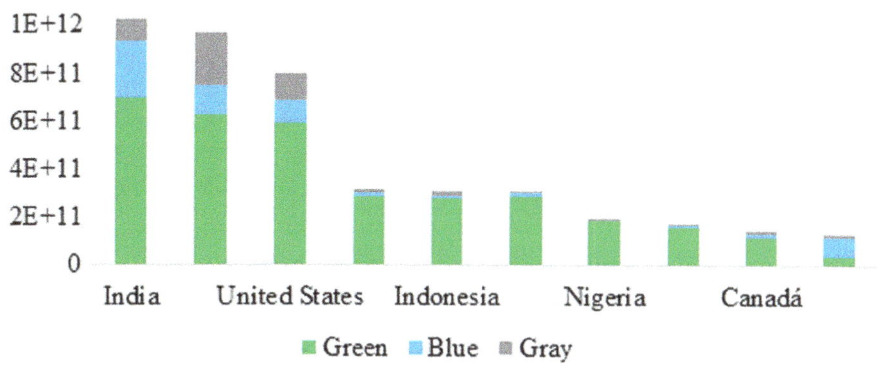

Fig. 11 Green, blue, and gray annual water footprint of agriculture (all crops) in the world by country (*Source* Water Footprint Network [153])

Based on the data recorded in the previous figure, the green water footprint is predominant in most selected countries, reaching levels of up to $7*10^{11}$ m^3/year. It is interesting to show that those countries that lose or waste food in more significant quantities have a higher annual water footprint record. In general, policies and strategies must be applied to reduce food loss and, therefore, the water footprint.

5 Food Loss Prevention and Management Options

Regarding the reduction or prevention of food loss, it is essential to create a zero hunger world and to achieve the Sustainable Development Goals (SDGs), especially SDG 2 (Zero Hunger) and SDG 12 (Ensure sustainable consumption and production patterns). Many people on the planet take food for granted, but the food is not guaranteed for the staggering 820 million-plus people who go hungry. FAO aims to increase respect for food, as well as for the farmers who produce it, the natural resources that are used to produce it, and the people who do not have access to it [79].

In Argentina, with Ministerial Resolution 392 of 2015, warned against the opportunity cost left by not making use of food and turning it into waste, establishing basic guidelines to combat problems, such as research, production of initiatives, technology immersion, communication, practical training, and partnerships [115]. Similarly, as in the case of the reduction in salt consumption, augur good results [37]. Argentina adopted the Sustainable Development Goals (SDGs) to the national context and, based on a diagnosis of the territory, was directed toward the quantification of losses and waste [134].

In Chile, it was approved Law 20.920, the Framework for the Management of Waste, Extended Producer Responsibility and Promotion of Recycling, granted powers to the Ministry of the Environment to prevent the generation of waste that, according to PASO balance sheets, are entirely avoidable [117].

Recycling Law (Universal Recycling Law) seeks the involvement of the most significant actors in the process. For the United States Department of Agriculture (USDA), there is great importance in the climatic variability and food refrigeration, as this helps stop microbial growth [148]. Among the proposals of the recent North American legislation, the one of Vermont with its Universal Law stands out. First, it expressly prohibits the incineration or landfilling of materials commercial organic, which includes food and other plant materials generated by all kinds of entities, in an amount more significant than one ton per week [143], which directly involves large producers of waste such as marketplaces, supermarkets, and the food industry.

6 Water Footprint Management Options

It is essential to indicate that water's responsible use and management represents a concern worldwide regarding the water footprint. According to the United Nations, the water footprint directly impacts the fulfillment of the Sustainable Development Goals of the 2030 Agenda of the international organization. It is worth noting that SDG 6 mentioned the following: "the sustainable management of water and sanitation and the objective is to increase the efficiency of water use in every sector and guarantee the sustainable harvest and allocation of water resources to face scarcity of water…." In this sense, water footprint policies and initiatives are closely related to the objectives agreed by governments to make fair and sustainable use of water in various sectors.

In several Latin American and Caribbean countries, governments have shown great interest in conducting water footprint assessments to monitor compliance with their mandates for the sustainable management of water resources. The Economic Commission for Latin America and the Caribbean (ECLAC or CEPAL) has identified outstanding initiatives in Peru, Colombia, Chile, Brazil, Argentina, Mexico, and Ecuador. Next, the proposals of some of the countries above are explained in detail.

In Peru, the National Water Authority (ANA) began operations in 2009 as the governing body of the National Water Resources Management System. Thanks to the joint work between the ANA and the Ministry of Water and Irrigation, several studies related to the concept of water footprint have been carried out in various sectors. It is noteworthy that, in 2018, the standard promotes the measurement and voluntary reduction of the water footprint and the shared value in the hydrographic basins with scope throughout the country. The latter aims to measure and reduce the water footprint and implement actions of shared value in the basins. The so-called Water Footprint Program comprises activities that seek to reduce consumption and impacts resulting from water resources and the development of social responsibility actions. The program led by the Directorate of the National Water Resources Information System of the National Water Authority is voluntary and, to identify the participating users, they are awarded a so-called Blue Certificate. In other words, the volunteers are recognized with a certificate as hydraulically responsible and supportive users, for which they indicate the operating unit, process, or product based on which their water footprint was measured and reduced.

In Colombia, the Ministry of Environment and Sustainable Development (MINAMBIENTE) and, more specifically, its Institute of Hydrology, Meteorology and Environmental Studies (IDEAM) has investigated and applied the concept of water footprint since 2009. In 2014, it carried out multisectoral research on the blue and green water footprints in the agricultural, industrial, energy, and oil sectors. In addition, it is worth noting that the domestic sector was included. As a result, in the agricultural sector, it was obtained that the most important crop in the country, in terms of water footprint (HH), is coffee. It is also important to note that the study above followed the methodology and definitions presented by the Water Footprint Manual [93], published by the Water Footprint Network.

In Chile, the General Water Directorate (DGA) evaluated HH and a pilot study in the Rapel river basin in a joint effort with Fundación Chile [116]. As a result, based on socioeconomic and environmental indicators, the areas in which water consumption is not sustainable were obtained. In this way, the DGA can prioritize actions to reduce HH in those regions. The evaluation of water consumption included all production sectors, such as domestic, forestry, agriculture, mining, energy, and industrial, to have a complete vision of the situation in the territory under analysis. Therefore, it represents a significant contribution to improving sustainable water management. In Table 5, a summary of the essential tools that can be used concerning the water footprint is presented.

Table 5 Most essential tools that can be used concerning the water footprint

Tool	Description
AWS	A water management platform that provides a detailed description of sustainable water management at the local level
CDP Water Disclosure	A platform to "reveal" corporate water. It allows corporations with the most significant potential to be impacted by water management to show their actions and impacts on this resource
Ceres Aqua Gauge	The dashboard allows investors to rate the water management activities of a company under sustainability and water security criteria
CEO Water Mandate	A framework that offers standard corporate water disclosure metrics and provides guidance to determine the relevance of the report content
Water Risk Filter	Water risk assessment tool, including mapping, mitigation responses, case studies, and country profiles. The latest version includes a water risk assessment for agriculture with specific indicators, questionnaires, and responses
WBCSD—Global water tool	Excel application available for companies. It allows ordering the information for different indices, such as CDP Water, Bloomberg, GRI Report, and Dow Jones, depending on the location, use of water, and the characteristics of the user's wastewater
WRI Aqueduct	A tool that allows knowing the water risk online with a robust geo-referenced data functionality. The key indicator is the "water stress at baseline" indicator developed by WRI
SWAN system tool	The system is used to provide a sustainability indicator for farms. The first time, it forms the SWAN sustainability score according to the farm. Process updates and measurements are compared to this baseline, showing farmers their progress over time

Source Gomez and Inthamoussu [84]

7 Closing Remarks

The implementation of a circular economic model helps companies achieve that change toward sustainability that is so sought after today, at the same time improving the corporate image and promoting their growth in the market both locally and internationally. In this way, they contribute and promote the proper use of resources that exist in nature in a limited way, such as water. All foods, in general, require the use of water for their production. Beef has been determined to be one of the foods with a higher water footprint than agricultural products (approximately 15,415 l / kg). The loss of food throughout the FSC is directly related to the levels of water footprint in each product when food is lost or wasted, water is lost, and the energy that the product can provide. The application of strategies and policies should not focus only on the final stage of the FSC but rather on the initial stages such as harvest, post-harvest, handling, and distribution. The application of the circular economy is ideal for solving the problem under study. The loss and waste of food generate a moral dispute in society since the lack or waste of a resource that is difficult to access for millions of people is caused to reflect on the importance of reducing food loss and optimizing its use, especially in households, since these are the places where the most significant food loss is generated. Thus, a new approach to global food production through food optimization was feeding more people what is already produced is more relevant than maximizing food production.

The money spent on food that is not consumed ranges between the US$ 35 billion and the US$ 1.6 billion, showing a tremendous economic loss. In the same way, the environmental environment suffers damage because the agricultural sector is one of the industries that most use natural resources to harvest, produce, and process the food consumed and not consumed. However, the union between society and institutions could help reduce these impacts related to food loss for greater sustainability. To do this, it must start from a change in awareness about the damage caused by food loss through its supply chain and changes in regulations and improved chain efficiency.

Therefore, there are various initiatives in many countries to combat food loss and adequately manage the water footprint that has been carried out. It is exciting that many systems can support large corporations and companies to control their sustainable water management about managing the water footprint. All of this marks an important precedent and a source of inspiration for further similar or innovative initiatives to address concepts relevant to all the world's inhabitants.

References

1. Abiakam N, Worsley P, Jayabal H, Mitchell K, Jones M, Fletcher J et al (2021) Personal protective equipment related skin reactions in healthcare professionals during COVID-19. Int Wound J 18(3):312–322. https://doi.org/10.1111/iwj.13534
2. Abudureheman A, Nilupaer A (2021) Optimization model design of cross-border e-commerce transportation path under the background of prevention and control of COVID-19 pneumonia.

Soft Comput 25(18):12007–12015. https://doi.org/10.1007/s00500-021-05685-6
3. Acuña-Zegarra MA, Santana-Cibrian M, Velasco-Hernandez JX (2020) Modeling behavioral change and COVID-19 containment in Mexico: a trade-off between lockdown and compliance. Math Biosci 325:108370. https://doi.org/10.1016/j.mbs.2020.108370
4. Aday S, Aday MS (2020) Impact of COVID-19 on the food supply chain. Food Qual Saf 4(4):167–180. https://doi.org/10.1093/fqsafe/fyaa024
5. Afshan G, Shahid S, Tunio MN (2021) Learning experiences of women entrepreneurs amidst COVID-19. Int J Gend Entrep 13(2):162–186. https://doi.org/10.1108/IJGE-09-2020-0153
6. Aguirre AA, Catherina R, Frye H, Shelley L (2020) Illicit wildlife trade, wet markets, and COVID-19: preventing future pandemics. World Med Health Policy 12(3):256–265. https://doi.org/10.1002/wmh3.348
7. Ahsan MM (2020) Strategic decisions on urban built environment to pandemics in Turkey: lessons from COVID-19. J Urban Manag 9(3):281–285. https://doi.org/10.1016/j.jum.2020.07.001
8. Alan H, Eskin Bacaksiz F, Tiryaki Sen H, Taskiran Eskici G, Gumus E, Harmanci Seren AK (2021) "I'm a hero, but...": an evaluation of depression, anxiety, and stress levels of frontline healthcare professionals during COVID-19 pandemic in Turkey. Perspect Psychiatr Care 57(3):1126–1136. https://doi.org/10.1111/ppc.12666
9. Ali W (2020) Online and remote learning in higher education institutes: a necessity in light of COVID-19 pandemic. High Educ Stud 10(3):16–25
10. Allen J, Rowan L, Singh P (2020) Teaching and teacher education in the time of COVID-19. Asia-Pacific Journal of Teacher Education 48(3):233–236. https://doi.org/10.1080/1359866X.2020.1752051
11. Alsairafi Z, Naser AY, Alsaleh FM, Awad A, Jalal Z (2021) Mental health status of healthcare professionals and students of health sciences faculties in Kuwait during the COVID-19 pandemic. Int J Environ Res Public Health 18(4). https://doi.org/10.3390/ijerph18042203
12. Altindis E (2021) Inequitable COVID-19 vaccine distribution and the intellectual property rights prolong the pandemic. Expert Rev Vaccines. https://doi.org/10.1080/14760584.2022.2014819
13. Alvarez-Risco A, Dawson J, Johnson W, Conteh-Barrat M, Aslani P, Del-Aguila-Arcentales S, Diaz-Risco S (2021) Ebola virus disease outbreak: a global approach for health systems [Ebola; health professionals; global approach; pharmacist: pharmaceutical care; COVID-19] 53(4). http://www.revfarmacia.sld.cu/index.php/far/article/view/491
14. Alvarez-Risco A, Del-Aguila-Arcentales S (2015) Errores de prescripción como barrera para la Atención Farmacéutica en establecimientos de salud públicos: Experiencia Perú. Pharmaceutical Care España 17(6):725–731. https://www.pharmcareesp.com/index.php/PharmaCARE/article/view/246
15. Alvarez-Risco A, Del-Aguila-Arcentales S (2021a) A note on changing regulation in international business: the World Intellectual Property Organization (WIPO) and artificial intelligence. In: Verbeke A, van Tulder R, Rose EL, Wei Y (eds) The multiple dimensions of institutional complexity in international business research, vol 15. Emerald Publishing Limited, pp 363–371. https://doi.org/10.1108/S1745-886220210000015020
16. Alvarez-Risco A, Del-Aguila-Arcentales S (2021b) Public policies and private efforts to increase women entrepreneurship based on STEM background. In: Mari M, Poggesi S, Foss L (eds) Women's entrepreneurship in STEM disciplines: issues and perspectives. Springer, pp 75–87. https://doi.org/10.1007/978-3-030-83792-1_5
17. Alvarez-Risco A, Del-Aguila-Arcentales S, Delgado-Zegarra J, Yáñez JA, Diaz-Risco S (2019) Doping in sports: findings of the analytical test and its interpretation by the public. Sport Sci Health 15(1):255–257. https://doi.org/10.1007/s11332-018-0484-8
18. Alvarez-Risco A, Del-Aguila-Arcentales S, Diaz-Risco S (2018) Pharmacovigilance as a tool for sustainable development of healthcare in Peru. PharmacoVigilance Review 10(2):4–6
19. Alvarez-Risco A, Del-Aguila-Arcentales S, Rosen MA, García-Ibarra V, Maycotte-Felkel S, Martínez-Toro GM (2021) Expectations and interests of university students in COVID-19 times about Sustainable Development Goals: evidence from Colombia, Ecuador, Mexico, and Peru. Sustainability 13(6):3306. https://doi.org/10.3390/su13063306

20. Alvarez-Risco A, Del-Aguila-Arcentales S, Yáñez JA (2021) Telemedicine in Peru as a result of the COVID-19 pandemic: perspective from a country with limited internet access. Am J Trop Med Hyg 105(1):6–11. https://doi.org/10.4269/ajtmh.21-0255
21. Alvarez-Risco A, Del-Aguila-Arcentales S, Yáñez JA, Rosen MA, Mejia CR (2021) Influence of technostress on academic performance of university medicine students in Peru during the COVID-19 pandemic. Sustainability 13(16):8949. https://doi.org/10.3390/su13168949
22. Alvarez-Risco A, Del-Aguila-Arcentales S, Stevenson JG (2015) Pharmacists and mass communication for implementing pharmaceutical care. Am J Pharm Benefits 7(3):e125–e126
23. Alvarez-Risco A, Delgado-Zegarra J, Yáñez JA, Diaz-Risco S, Del-Aguila-Arcentales S (2018) Predation risk by gastronomic boom—Case Peru. J Landsc Ecol 11(1):100–103. https://doi.org/10.2478/jlecol-2018-0003
24. Alvarez-Risco A, Estrada-Merino A, Anderson-Seminario MdlM, Mlodzianowska S, García-Ibarra V, Villagomez-Buele C, Carvache-Franco M (2021) Multitasking behavior in online classrooms and academic performance: case of university students in Ecuador during COVID-19 outbreak. Interact Technol Smart Educ 18(3):422–434. https://doi.org/10.1108/ITSE-08-2020-0160
25. Alvarez-Risco A, Mejia CR, Delgado-Zegarra J, Del-Aguila-Arcentales S, Arce-Esquivel AA, Valladares-Garrido MJ et al (2020) The Peru approach against the COVID-19 infodemic: insights and strategies. Am J Trop Med Hyg 103(2):583–586. https://doi.org/10.4269/ajtmh.20-0536
26. Alvarez-Risco A, Mil JWFv (2007) Pharmaceutical care in community pharmacies: Practice and research in Peru. Ann Pharmacother 41(12):2032–2037. https://doi.org/10.1345/aph.1K117
27. Alvarez-Risco A, Mlodzianowska S, García-Ibarra V, Rosen MA, Del-Aguila-Arcentales S (2021) Factors affecting green entrepreneurship intentions in business university students in COVID-19 pandemic times: case of Ecuador. Sustainability 13(11):6447. https://doi.org/10.3390/su13116447
28. Alvarez-Risco A, Mlodzianowska S, Zamora-Ramos U, Del-Aguila S (2021) Green entrepreneurship intention in university students: the case of Peru. Entrep Bus Econ Rev 9(4), 85–100. https://doi.org/10.15678/EBER.2021.090406
29. Alvarez-Risco A, Quipuzco-Chicata L, Escudero-Cipriani C (2021) Determinantes de la intención de recompra en línea en tiempos de COVID-19: evidencia de una economía emergente. Lecturas de Economía (96). https://doi.org/10.17533/udea.le.n96a342638
30. Alvarez-Risco A, Quiroz-Delgado D, Del-Aguila-Arcentales S (2016) Pharmaceutical care in hypertension patients in a Peruvian hospital. Indian J Public Health Res Dev 7(3):183–188
31. Alvarez-Risco A, Rosen MA, Del-Aguila-Arcentales S (2020) A new regulation for supporting a circular economy in the plastic industry: the case of Peru (short communication). J Landsc Ecol 13(1):1–3. https://doi.org/10.2478/jlecol-2020-0004
32. Alvarez-Risco A, Turpo-Cama A, Ortiz-Palomino L, Gongora-Amaut N, Del-Aguila-Arcentales S (2016) Barreras para la implementación de la Atención Farmacéutica en establecimientos farmacéuticos de Cusco, Perú. Pharmaceutical Care España 18(5):194–205. https://www.pharmcareesp.com/index.php/PharmaCARE/article/view/326
33. Álvarez-Risco A, Zegarra Arellano E, Matos Valerio E, Mejía Acosta N, Solis Tarazona Z (2013) Campaña de atención farmacéutica como estrategia de implementación de los servicios farmacéuticos: Experiencia Perú. Pharmaceutical Care España 15(1):35. https://www.pharmcareesp.com/index.php/PharmaCARE/article/view/105
34. Amarillo CR (2020) Feminism on lockdown. NACLA Rep Am 52(3):274–281. https://doi.org/10.1080/10714839.2020.1809084
35. Apcho-Ccencho L-V, Cuya-Velásquez B-B, Alvarado Rodríguez D, Anderson-Seminario MdlM, Alvarez-Risco A, Estrada-Merino A, Mlodzianowska S (2021) The impact of international price on the technological industry in the United States and China during times of crisis: commercial war and COVID-19. In: Lawrence KD, Klimberg RK (eds) Advances in business and management forecasting, vol. 14. Emerald Publishing Limited, pp 149–160. https://doi.org/10.1108/S1477-407020210000014010

36. Assaf A, Scuderi R (2020) COVID-19 and the recovery of the tourism industry. Tour Econ 26(5):731–733. https://doi.org/10.1177/1354816620933712
37. Aulicino, C. (2012). Una primera aproximación a las políticas de Educación Nutricional en las provincias argentinas. Documento de trabajo N°90. Programa de Protección Social, CIPPEC. Retrieved 12/22/2021 from http://municipios.unq.edu.ar/modules/mislibros/archivos/201205 nutricion.pdf
38. Bacq S, Lumpkin GT (2021) Social entrepreneurship and COVID-19. J Manage Stud 58(1):285–288. https://doi.org/10.1111/joms.12641
39. Barello S, Caruso R, Palamenghi L, Nania T, Dellafiore F, Bonetti L et al (2021) Factors associated with emotional exhaustion in healthcare professionals involved in the COVID-19 pandemic: an application of the job demands-resources model. Int Arch Occup Environ Health 94(8):1751–1761. https://doi.org/10.1007/s00420-021-01669-z
40. Baum T, Hai NTT (2020) Hospitality, tourism, human rights and the impact of COVID-19. Int J Contemp Hosp Manag 32(7):2397–2407. https://doi.org/10.1108/IJCHM-03-2020-0242
41. Belitski M, Guenther C, Kritikos AS, Thurik R (2021) Economic effects of the COVID-19 pandemic on entrepreneurship and small businesses. Small Bus Econ. https://doi.org/10.1007/s11187-021-00544-y
42. Block JH, Fisch C, Hirschmann M (2021) The determinants of bootstrap financing in crises: evidence from entrepreneurial ventures in the COVID-19 pandemic. Small Bus Econ. https://doi.org/10.1007/s11187-020-00445-6
43. Borzée A, McNeely J, Magellan K, Miller JRB, Porter L, Dutta T et al (2020) COVID-19 highlights the need for more effective wildlife trade legislation. Trends Ecol Evol 35(12):1052–1055. https://doi.org/10.1016/j.tree.2020.10.001
44. Bozdağ F, Ergün N (2020) Psychological resilience of healthcare professionals during COVID-19 pandemic. Psychol Rep 124(6):2567–2586. https://doi.org/10.1177/003329412 0965477
45. Braungart M, McDonough W, Kälin A, Bollinger A (2012) Cradle-to-cradle design: creating healthy emissions—a strategy for eco-effective product and system design. Birkhäuser.
46. Brouder P (2020) Reset redux: possible evolutionary pathways towards the transformation of tourism in a COVID-19 world. Tour Geogr 22(3):484–490. https://doi.org/10.1080/146 16688.2020.1760928
47. Brown R, Rocha A, Cowling M (2020) Financing entrepreneurship in times of crisis: exploring the impact of COVID-19 on the market for entrepreneurial finance in the United Kingdom. Int Small Bus J 38(5):380–390. https://doi.org/10.1177/0266242620937464
48. Caldeira C, De Laurentiis V, Corrado S, van Holsteijn F, Sala S (2019) Quantification of food waste per product group along the food supply chain in the European Union: a mass flow analysis. Resour Conserv Recycl 149:479–488. https://doi.org/10.1016/j.resconrec.2019.06.011
49. Caldeira C, Quinteiro P, Castanheira E, Boulay A-M, Dias AC, Arroja L, Freire F (2018) Water footprint profile of crop-based vegetable oils and waste cooking oil: comparing two water scarcity footprint methods. J Clean Prod 195:1190–1202. https://doi.org/10.1016/j.jcl epro.2018.05.221
50. Carvache-Franco M, Alvarez-Risco A, Carvache-Franco O, Carvache-Franco W, Estrada-Merino A, Villalobos-Alvarez D (2021) Perceived value and its influence on satisfaction and loyalty in a coastal city: a study from Lima, Peru. Journal of Policy Research in Tourism, Leisure and Events, 1–16. https://doi.org/10.1080/19407963.2021.1883634
51. Carvache-Franco M, Alvarez-Risco A, Carvache-Franco W, Carvache-Franco O, Estrada-Merino A, Rosen MA (2021) Coastal cities seen from loyalty and their tourist motivations: a study in Lima, Peru. Sustainability 13(21):11575. https://doi.org/10.3390/su132111575
52. Carvache-Franco M, Carvache-Franco O, Carvache-Franco W, Alvarez-Risco A, Estrada-Merino A (2021) Motivations and segmentation of the demand for coastal cities: a study in Lima, Peru. Int J Tour Res 23(4):517–531. https://doi.org/10.1002/jtr.2423
53. Cegarra-Navarro J-G, Vătămănescu E-M, Martínez-Martínez A (2021) A context-driven approach on coping with COVID-19: from hiding knowledge toward citizen engagement. Knowl Process Manag 28(2):134–140. https://doi.org/10.1002/kpm.1662

54. Chaboud G, Daviron B (2017) Food losses and waste: navigating the inconsistencies. Glob Food Sec 12:1–7. https://doi.org/10.1016/j.gfs.2016.11.004
55. Chafloque-Cespedes R, Alvarez-Risco A, Robayo-Acuña P-V, Gamarra-Chavez C-A, Martinez-Toro G-M, Vicente-Ramos W (2021) Effect of sociodemographic factors in entrepreneurial orientation and entrepreneurial intention in university students of Latin American business schools. In Jones P, Apostolopoulos N, Kakouris A, Moon C, Ratten V, Walmsley A (eds) Universities and entrepreneurship: meeting the educational and social challenges, vol 11. Emerald Publishing Limited, pp 151–165. https://doi.org/10.1108/S2040-724620210000011010
56. Chafloque-Cespedes R, Vara-Horna A, Asencios-Gonzales Z, Lopez-Odar DR, Alvarez-Risco A, Quipuzco-Chicata L, Schulze C, Sanchez-Villagomez M (2020) Presentismo académico y violencia contra las mujeres en escuelas de negocios e ingeniería en universidades peruanas. Lecturas de Economía 0(93):127–153. https://doi.org/10.17533/udea.le.n93a340726
57. Charonis G-K (2012) Degrowth, steady state economics and the circular economy: three distinct yet increasingly converging alternative discourses to economic growth for achieving environmental sustainability and social equity. World Economic Association Sustainability Conference
58. Chatham House (2012) A global redesign? Shaping the circular economy. Retrieved 12/22/2021 from https://www.chathamhouse.org/2012/03/global-redesign-shaping-circular-economy
59. Chen T, Peng L, Yin X, Rong J, Yang J, Cong G (2020) Analysis of user satisfaction with online education platforms in China during the COVID-19 pandemic. Healthcare 8(3). https://doi.org/10.3390/healthcare8030200
60. Chen X, Zhang SX, Jahanshahi AA, Alvarez-Risco A, Dai H, Li J, Ibarra VG (2020) Belief in a COVID-19 conspiracy theory as a predictor of mental health and well-being of health care workers in Ecuador: cross-sectional survey study. JMIR Public Health Surveill 6(3):e20737. https://doi.org/10.2196/20737
61. Chinie C, Biclesanu I, Bellini F (2021) The impact of awareness campaigns on combating the food wasting behavior of consumers. Sustainability 13(20). https://doi.org/10.3390/su132011423
62. Chung SA, Olivera S, Rojas Román B, Alanoca E, Moscoso S, Limpias Terceros B, Alvarez-Risco A, Yáñez JA (2021) Temáticas de la producción científica de la Revista Cubana de Farmacia indizada en Scopus (1967–2020) [Revista Cubana de Farmacia; farmacia; Cuba; evidencia] 54(1). http://www.revfarmacia.sld.cu/index.php/far/article/view/511
63. Cruz-Torres W, Alvarez-Risco A, Del-Aguila-Arcentales S (2021) Impact of Enterprise Resource Planning (ERP) implementation on performance of an education enterprise: a Structural Equation Modeling (SEM). Stud Bus Econ 16(2):37–52. https://doi.org/10.2478/sbe-2021-0023
64. Del-Aguila-Arcentales S, Alvarez-Risco A (2013) Human error or burnout as explanation for mistakes in pharmaceutical laboratories. Accred Qual Assur 18(5):447–448. https://doi.org/10.1007/s00769-013-1000-0
65. del Castillo FA (2021) Temporary waiver of intellectual property on Covid-19 vaccines: toward the creation of a better, post-pandemic society. J Public Health 43(3):e559–e560. https://doi.org/10.1093/pubmed/fdab184
66. Delgado-Zegarra J, Alvarez-Risco A, Yáñez JA (2018) Uso indiscriminado de pesticidas y ausencia de control sanitario para el mercado interno en Perú. Rev Panam Salud Publica 42:e3. https://doi.org/10.26633/RPSP.2018.3
67. Deressa W, Worku A, Abebe W, Gizaw M, Amogne W (2021) Risk perceptions and preventive practices of COVID-19 among healthcare professionals in public hospitals in Addis Ababa, Ethiopia. PLoS ONE 16(6):e0242471. https://doi.org/10.1371/journal.pone.0242471
68. Desrochers P (2002) Regional development and inter-industry recycling linkages: some historical perspectives. Entrep Reg Dev 14(1):49–65. https://doi.org/10.1080/08985620110096627

69. Desrochers P (2004) Industrial symbiosis: the case for market coordination. J Clean Prod 12(8):1099–1110. https://doi.org/10.1016/j.jclepro.2004.02.008
70. Dominelli L (2021) A green social work perspective on social work during the time of COVID-19. Int J Soc Welf 30(1):7–16. https://doi.org/10.1111/ijsw.12469
71. Duarte Alonso A, Kok SK, Bressan A, O'Shea M, Sakellarios N, Koresis A et al (2020) COVID-19, aftermath, impacts, and hospitality firms: an international perspective. Int J Hosp Manag 91:102654. https://doi.org/10.1016/j.ijhm.2020.102654
72. Ellen MacArthur Foundation (2021) What is a circular economy? Retrieved 12/22/2021 from https://ellenmacarthurfoundation.org/topics/circular-economy-introduction/overview
73. Enciso-Zarate A, Guzmán-Oviedo J, Sánchez-Cardona F, Martínez-Rohenes D, Rodríguez-Palomino JC, Alvarez-Risco A, et al (2016) Evaluación de la contaminación con agentes citotóxicos en hospitales en Colombia. Pharmaceutical Care España 18(6):241–250. https://www.pharmcareesp.com/index.php/PharmaCARE/article/view/354
74. Erfani P, Binagwaho A, Jalloh MJ, Yunus M, Farmer P, Kerry V (2021) Intellectual property waiver for covid-19 vaccines will advance global health equity. BMJ 374:n1837. https://doi.org/10.1136/bmj.n1837
75. FAO (2006) Food security. Retrieved 12/22/2021 from https://www.fao.org/fileadmin/templates/faoitaly/documents/pdf/pdf_Food_Security_Cocept_Note.pdf
76. FAO (2012) Food loss and waste in the world—scope, causes and prevention. Retrieved 12/22/2021 from https://www.fao.org/3/i2697s/i2697s.pdf
77. FAO (2013) Loss and waste of food in the context of sustainable food systems. Retrieved 12/22/2021 from https://www.fao.org/3/i3901s/i3901s.pdf
78. FAO (2014) Food wastage footprint, full—cost final report. Retrieved 12/22/2021 from https://www.wavespartnership.org/en/knowledge-center/food-wastage-footprint-full-cost-accounting-final-report
79. FAO (2021) Food loss and waste must be reduced to increase food security and environmental sustainability. Retrieved 12/22/2021 from https://www.fao.org/news/story/es/item/1310444/icode
80. Fotiadis A, Polyzos S, Huan T-CTC (2021) The good, the bad and the ugly on COVID-19 tourism recovery. Ann Tour Res 87:103117. https://doi.org/10.1016/j.annals.2020.103117
81. Frosch RA, Gallopoulos NE (1989) Strategies for manufacturing. Sci Am 261(3):144–153
82. Galanakis CM, Rizou M, Aldawoud TMS, Ucak I, Rowan NJ (2021) Innovations and technology disruptions in the food sector within the COVID-19 pandemic and post-lockdown era. Trends Food Sci Technol 110:193–200. https://doi.org/10.1016/j.tifs.2021.02.002
83. Ghisellini P, Cialani C, Ulgiati S (2016) A review on circular economy: the expected transition to a balanced interplay of environmental and economic systems. J Clean Prod 114:11–32. https://doi.org/10.1016/j.jclepro.2015.09.007
84. Gomez X, Inthamoussu A (2019) Water footprint in Uruguay. Potential impact on agro-industrial export sectors [Huella hídrica en Uruguay. Potencial impacto en los sectores agro-industriales exportadores]. Retrieved 12/22/2021 from https://www.uruguayxxi.gub.uy/uploads/informacion/77ed3ef6cc0b8f3e9ef5fdeaa0211828caa389cb.pdf
85. Gotoh T, Nishimura T, Kuchida K, Mannen H (2018) The Japanese Wagyu beef industry: current situation and future prospects—a review. Asian-Australas J Anim Sci 31(7):933–950. https://doi.org/10.5713/ajas.18.0333
86. Gulati G, Kelly BD (2020) Domestic violence against women and the COVID-19 pandemic: what is the role of psychiatry? Int J Law Psychiatry 71:101594. https://doi.org/10.1016/j.ijlp.2020.101594
87. Gursoy D, Chi CG (2020) Effects of COVID-19 pandemic on hospitality industry: review of the current situations and a research agenda. J Hosp Market Manag 29(5):527–529. https://doi.org/10.1080/19368623.2020.1788231
88. Hanzl M (2020) Urban forms and green infrastructure—the implications for public health during the COVID-19 pandemic. Cities & Health, 1–5. https://doi.org/10.1080/23748834.2020.1791441

89. Hegnsholt, E., Unnikrishnan, S., Pollmann-Larsen, M., Askelsdottir, B., & Gerard, M. (2018). *Tackling the 1.6-billion-ton food loss and waste crisis*. Retrieved 01/01/2022 from https://www.bcg.com/publications/2018/tackling-1.6-billion-ton-food-loss-and-waste-crisis
90. Hepburn C, O'Callaghan B, Stern N, Stiglitz J, Zenghelis D (2020) Will COVID-19 fiscal recovery packages accelerate or retard progress on climate change? Oxf Rev Econ Policy 36(Suppl 1):S359–S381. https://doi.org/10.1093/oxrep/graa015
91. Ho C-Y, Tsai B-H, Chen C-S, Lu M-T (2021) Exploring green marketing orientations toward sustainability the hospitality industry in the COVID-19 pandemic. Sustainability 13(8). https://doi.org/10.3390/su13084348
92. Hoehn D, Margallo M, Laso J, Ruiz-Salmón I, Fernández-Ríos A, Campos C, et al (2021) Water footprint assessment of food loss and waste management strategies in Spanish regions. Sustainability 13(14). https://doi.org/10.3390/su13147538
93. Hoekstra AY, Chapagain AK, Mekonnen MM, Aldaya MM (2011) The water footprint assessment manual: Setting the global standard. Retrieved 12/22/2021 from https://waterfootprint.org/media/downloads/TheWaterFootprintAssessmentManual_2.pdf
94. Jecker NS, Atuire CA (2021) What's yours is ours: waiving intellectual property protections for COVID-19 vaccines. J Med Ethics 47(9):595. https://doi.org/10.1136/medethics-2021-107555
95. Kaushal V, Srivastava S (2021) Hospitality and tourism industry amid COVID-19 pandemic: perspectives on challenges and learnings from India. Int J Hosp Manag 92:102707. https://doi.org/10.1016/j.ijhm.2020.102707
96. Keijer T, Bakker V, Slootweg JC (2019) Circular chemistry to enable a circular economy. Nat Chem 11(3):190–195. https://doi.org/10.1038/s41557-019-0226-9
97. Kirchherr J, Reike D, Hekkert M (2017) Conceptualizing the circular economy: an analysis of 114 definitions. Resour Conserv Recycl 127:221–232. https://doi.org/10.1016/j.resconrec.2017.09.005
98. Kirk CP, Rifkin LS (2020) I'll trade you diamonds for toilet paper: consumer reacting, coping and adapting behaviors in the COVID-19 pandemic. J Bus Res 117:124–131. https://doi.org/10.1016/j.jbusres.2020.05.028
99. Koetter P, Pelton M, Gonzalo J, Du P, Exten C, Bogale K et al (2020) Implementation and process of a COVID-19 contact tracing initiative: leveraging health professional students to extend the workforce during a pandemic. Am J Infect Control 48(12):1451–1456. https://doi.org/10.1016/j.ajic.2020.08.012
100. Krishtel P, Malpani R (2021) Suspend intellectual property rights for covid-19 vaccines. BMJ 373:n1344. https://doi.org/10.1136/bmj.n1344
101. Kummu M, de Moel H, Porkka M, Siebert S, Varis O, Ward PJ (2012) Lost food, wasted resources: global food supply chain losses and their impacts on freshwater, cropland, and fertiliser use. Sci Total Environ 438:477–489. https://doi.org/10.1016/j.scitotenv.2012.08.092
102. Kurthy G, Darvasne Ordog E, Dudas G, Radoczne Kocsis T, Szekelyhidi K, Takacs E (2021) Food loss in the Hungarian food industry—examining quantification problems and attitudes of managers and their impact on policy. Waste Manage 135:11–19. https://doi.org/10.1016/j.wasman.2021.08.029
103. Lackie K, Najjar G, El-Awaisi A, Frost J, Green C, Langlois S et al (2020) Interprofessional education and collaborative practice research during the COVID-19 pandemic: considerations to advance the field. J Interprof Care 34(5):583–586. https://doi.org/10.1080/13561820.2020.1807481
104. Lashley MA, Acevedo M, Cotner S, Lortie CJ (2020) How the ecology and evolution of the COVID-19 pandemic changed learning. Ecol Evol 10(22):12412–12417. https://doi.org/10.1002/ece3.6937
105. Leiva-Martinez M-A, Anderson-Seminario MdlM, Alvarez-Risco A, Estrada-Merino A, Mlodzianowska S (2021) Price variation in lower goods as of previous economic crisis and the contrast of the current price situation in the context of COVID-19 in Peru. In: Lawrence KD, Klimberg RK (eds) Advances in business and management forecasting, vol 14. Emerald Publishing Limited, pp 161–166. https://doi.org/10.1108/S1477-407020210000014011

106. Lett LA (2014) Las amenazas globales, el reciclaje de residuos y el concepto de economía circular. Rev Argent Microbiol 46(1):1–2
107. Liguori E, Winkler C (2020) From offline to online: challenges and opportunities for entrepreneurship education following the COVID-19 pandemic. Entrepreneurship Education and Pedagogy 3(4):346–351. https://doi.org/10.1177/2515127420916738
108. Lim S, Prakash A (2021) Pandemics and citizen perceptions about their country: did COVID-19 increase national pride in South Korea? Nations and Nationalism 27(3):623–637. https://doi.org/10.1111/nana.12749
109. Lopez-Odar D, Alvarez-Risco A, Vara-Horna A, Chafloque-Cespedes R, Sekar MC (2020) Validity and reliability of the questionnaire that evaluates factors associated with perceived environmental behavior and perceived ecological purchasing behavior in Peruvian consumers. Soc Responsib J 16(3):403–417. https://doi.org/10.1108/SRJ-08-2018-0201
110. Maritz A, Perenyi A, de Waal G, Buck C (2020) Entrepreneurship as the unsung hero during the current COVID-19 economic crisis: Australian perspectives. Sustainability 12(11). https://doi.org/10.3390/su12114612
111. Mbow C, Rosenzweig C, Barioni L, Benton T, Herrero M, Krishnapillai M, Liwenga E, Pradhan P, Rivera-Ferre MG, Sapkota T, Tubiello F, Xu Y (2019) Food security. Retrieved 12/22/2021 from https://www.ipcc.ch/site/assets/uploads/sites/4/2021/02/08_Chapter-5_3.pdf
112. Mejía-Acosta N, Alvarez-Risco A, Solís-Tarazona Z, Matos-Valerio E, Zegarra-Arellano E, Del-Aguila-Arcentales S (2016) Reacciones Adversas a Medicamentos reportadas como resultado de la implementación de Atención Farmacéutica en la Farmacia Institucional DIGEMID - Ministerio de Salud de Perú. Pharmaceutical Care España 18(2):67–74. https://www.pharmcareesp.com/index.php/PharmaCARE/article/view/311
113. Mekonnen MM, Gerbens-Leenes W (2020) The water footprint of global food production. Water 12(10). https://doi.org/10.3390/w12102696
114. Mihelcic JR, Crittenden JC, Small MJ, Shonnard DR, Hokanson DR, Zhang Q et al (2003) Sustainability Science and Engineering: the Emergence of a New Metadiscipline. Environ Sci Technol 37(23):5314–5324. https://doi.org/10.1021/es034605h
115. Ministerio de Agricultura GyP (2015) Programa Nacional de Reducción de perdida y desperdicio de alimentos - creación. Retrieved 12/22/2021 from https://www.argentina.gob.ar/normativa/nacional/resoluci%C3%B3n-392-2015-248599
116. Ministerio de Obras Públicas (2015) DGA and Fundación Chile conduct second training on water footprinting [DGA y Fundación Chile realizan segunda capacitación sobre huella hídrica]. Retrieved 12/22/2021 from https://dga.mop.gob.cl/noticias/Paginas/DetalledeNoticias.aspx?item=371
117. Ministerio del Ambiente de Chile (2016) Retrieved 12/22/2021 from https://www.bcn.cl/leychile/navegar?idNorma=1090894
118. Morone P, Falcone PM, Imbert E, Morone A (2018) Does food sharing lead to food waste reduction? An experimental analysis to assess challenges and opportunities of a new consumption model. J Clean Prod 185:749–760. https://doi.org/10.1016/j.jclepro.2018.01.208
119. Morone P, Falcone PM, Tartiu VE (2019) Food waste valorisation: assessing the effectiveness of collaborative research networks through the lenses of a COST action. J Clean Prod 238:117868. https://doi.org/10.1016/j.jclepro.2019.117868
120. Murray A, Skene K, Haynes K (2017) The circular economy: an interdisciplinary exploration of the concept and application in a global context. J Bus Ethics 140(3):369–380. https://doi.org/10.1007/s10551-015-2693-2
121. Obergassel W, Hermwille L, Oberthür S (2021) Harnessing international climate governance to drive a sustainable recovery from the COVID-19 pandemic. Climate Policy 21(10):1298–1306. https://doi.org/10.1080/14693062.2020.1835603
122. Okereke M (2021) Towards vaccine equity: should big pharma waive intellectual property rights for COVID-19 vaccines? Public Health in Practice 2:100165. https://doi.org/10.1016/j.puhip.2021.100165

123. Quispe-Cañari JF, Fidel-Rosales E, Manrique D, Mascaró-Zan J, Huamán-Castillón KM, Chamorro-Espinoza SE et al (2021) Self-medication practices during the COVID-19 pandemic among the adult population in Peru: a cross-sectional survey. Saudi Pharm J 29(1):1–11. https://doi.org/10.1016/j.jsps.2020.12.001
124. Rashid A, Asif FMA, Krajnik P, Nicolescu CM (2013) Resource Conservative Manufacturing: an essential change in business and technology paradigm for sustainable manufacturing. J Clean Prod 57:166–177. https://doi.org/10.1016/j.jclepro.2013.06.012
125. Raudenská J, Steinerová V, Javůrková A, Urits I, Kaye AD, Viswanath O, Varrassi G (2020) Occupational burnout syndrome and post-traumatic stress among healthcare professionals during the novel coronavirus disease 2019 (COVID-19) pandemic. Best Pract Res Clin Anaesthesiol 34(3):553–560. https://doi.org/10.1016/j.bpa.2020.07.008
126. Reh L (2013) Process engineering in circular economy. Particuology 11(2):119–133. https://doi.org/10.1016/j.partic.2012.11.001
127. REPSOL (2021) Cirular economy [Economia circular]. Retrieved 12/22/2021 from https://www.repsol.com/es/sostenibilidad/economia-circular/index.cshtml
128. Roesch E, Amin A, Gupta J, García-Moreno C (2020) Violence against women during covid-19 pandemic restrictions. BMJ 369:m1712. https://doi.org/10.1136/bmj.m1712
129. Rojas-Osorio M, Alvarez-Risco A (2019) Intention to use smartphones among Peruvian university students. International Journal of Interactive Mobile Technologies (iJIM) 13(03):13. https://doi.org/10.3991/ijim.v13i03.9356
130. Rojas Román, B., Moscoso, S., Chung, S. A., Limpias Terceros, B., Álvarez-Risco, A., & Yáñez, J. A. (2020). Tratamiento de la COVID-19 en Perú y Bolivia y los riesgos de la automedicación [COVID-19; tratamiento; estrategia; fármacos; automedicación; Perú; Bolivia.]. *2020, 53*(2). http://www.revfarmacia.sld.cu/index.php/far/article/view/435/351
131. Sakai S-I, Yoshida H, Hirai Y, Asari M, Takigami H, Takahashi S et al (2011) International comparative study of 3R and waste management policy developments. J Mater Cycles Waste Manage 13(2):86–102. https://doi.org/10.1007/s10163-011-0009-x
132. Sánchez-Clavijo LM, Martínez-Callejas SJ, Acevedo-Charry O, Diaz-Pulido A, Gómez-Valencia B, Ocampo-Peñuela N et al (2021) Differential reporting of biodiversity in two citizen science platforms during COVID-19 lockdown in Colombia. Biol Cons 256:109077. https://doi.org/10.1016/j.biocon.2021.109077
133. Sánchez OR, Vale DB, Rodrigues L, Surita FG (2020) Violence against women during the COVID-19 pandemic: an integrative review. Int J Gynecol Obstet 151(2):180–187. https://doi.org/10.1002/ijgo.13365
134. Schein L (2018) Argentina: food losses and waste study linked to the Sustainable Development Goal to ensure responsible responsible production and consumption (SDG 12) [Argentina: Estudio Pérdidas y desperdicios de alimentos vinculado al Objetivo Desarrollo Sostenible para garantizar producción y consumo responsable (ODS 12)]. Retrieved 12/22/2021 from http://www.alimentosargentinos.gob.ar/HomeAlimentos/ValoremoslosAlimentos/documentos/Resumen_Propuesta_Cuantificacion_PDA_ODS.pdf
135. Sekalala S, Forman L, Hodgson T, Mulumba M, Namyalo-Ganafa H, Meier BM (2021) Decolonising human rights: how intellectual property laws result in unequal access to the COVID-19 vaccine. BMJ Glob Health 6(7):e006169. https://doi.org/10.1136/bmjgh-2021-006169
136. Shaw R, Kim Y-K, Hua J (2020) Governance, technology and citizen behavior in pandemic: lessons from COVID-19 in East Asia. Progress in Disaster Science 6:100090. https://doi.org/10.1016/j.pdisas.2020.100090
137. Sigala M (2020) Tourism and COVID-19: impacts and implications for advancing and resetting industry and research. J Bus Res 117:312–321. https://doi.org/10.1016/j.jbusres.2020.06.015
138. Stahel WR (2016) The circular economy. Nature 531(7595):435–438. https://doi.org/10.1038/531435a
139. Statista (2021a) Annual food waste produced per capita worldwide in 2019, by sector (in kilograms). Retrieved 12/22/2021 from https://www.statista.com/statistics/1219850/global-food-waste-by-sector-per-capita

140. Statista (2021b) Annual household food waste produced in selected countries worldwide as of 2020 (in million metric tons per year). Retrieved 12/22/2021 from https://www.statista.com/statistics/933083/food-waste-of-selected-countries
141. Statista (2021c) Total food waste produced worldwide in 2019, by sector (in million metric tons). Retrieved 12/22/2021 from https://www.statista.com/statistics/1219836/global-food-waste-by-sector
142. Su B, Heshmati A, Geng Y, Yu X (2013) A review of the circular economy in China: moving from rhetoric to implementation. J Clean Prod 42:215–227. https://doi.org/10.1016/j.jclepro.2012.11.020
143. Teixidó LS (2019) Las landfinll bans estadounidenses como modelo en la lucha contra el desperdicio de alimentos tras la reforma de la Directiva Marco de residuos. *Revista General de Derecho Administrativo* 17(52).
144. The World Bank (2019) Prevalence of undernourishment (% of population)—Nigeria. Retrieved 12/22/2021 from https://data.worldbank.org/indicator/SN.ITK.DEFC.ZS?locations=NG
145. Tran LTT (2021) Managing the effectiveness of e-commerce platforms in a pandemic. J Retail Consum Serv 58:102287. https://doi.org/10.1016/j.jretconser.2020.102287
146. UNEP (2021) Food Waste Index Report 2021. Nairobi. Retrieved 12/22/2021 from https://wedocs.unep.org/bitstream/handle/20.500.11822/35280/FoodWaste.pdf
147. United Nations (2015) Goal 12: Ensure sustainable consumption and production patterns. Retrieved 12/23/2021 from https://www.un.org/sustainabledevelopment/sustainable-consumption-production
148. USDA (2017) USDA innovations to reduce food waste help the farmers' bounty go farther. Retrieved 12722/2021 from https://www.usda.gov/media/blog/2016/02/02/usda-innovations-reduce-food-waste-help-farmers-bounty-go-farther
149. Vara-Horna A, Alvarez-Risco A (2020) Food insecurity. In: Alvarez-Risco A, Rosen M, Del-Aguila-Arcentales S, Marinova D (eds) Building sustainable cities: social, economic and environmental factors. Springer, pp 105–122. https://doi.org/10.1007/978-3-030-45533-0_9
150. Vidya CT, Prabheesh KP (2020) Implications of COVID-19 pandemic on the global trade networks. Emerg Mark Financ Trade 56(10):2408–2421. https://doi.org/10.1080/1540496X.2020.1785426
151. Viero A, Barbara G, Montisci M, Kustermann K, Cattaneo C (2021) Violence against women in the Covid-19 pandemic: a review of the literature and a call for shared strategies to tackle health and social emergencies. Forensic Sci Int 319:110650. https://doi.org/10.1016/j.forsciint.2020.110650
152. Vilariño MV, Franco C, Quarrington C (2017) Food loss and waste reduction as an integral part of a circular economy [mini review]. Frontiers in Environmental Science 5(21). https://doi.org/10.3389/fenvs.2017.00021
153. Water Footprint Network (2020) Water footprint assessment tool. https://www.waterfootprintassessmenttool.org
154. Wen J, Kozak M, Yang S, Liu F (2021) COVID-19: potential effects on Chinese citizens' lifestyle and travel. Tourism Review 76(1):74–87. https://doi.org/10.1108/TR-03-2020-0110
155. Wu T-Y, Ford O, Rainville AJ, Bessire R (2020) COVID-19 care package distribution for senior citizens and families in Detroit and Hamtramck, Michigan. J Hunger Environ Nutr 15(4):585–587. https://doi.org/10.1080/19320248.2020.1799586
156. Yan J, Kim S, Zhang SX, Foo M-D, Alvarez-Risco A, Del-Aguila-Arcentales S, Yáñez JA (2021) Hospitality workers' COVID-19 risk perception and depression: a contingent model based on transactional theory of stress model. Int J Hosp Manag 95:102935. https://doi.org/10.1016/j.ijhm.2021.102935
157. Yáñez JA, Afshar Jahanshahi A, Alvarez-Risco A, Li J, Zhang SX (2020) Anxiety, distress, and turnover intention of healthcare workers in Peru by their distance to the epicenter during the COVID-19 crisis. Am J Trop Med Hyg 103(4):1614–1620. https://doi.org/10.4269/ajtmh.20-0800

158. Yáñez JA, Alvarez-Risco A, Delgado-Zegarra J (2020) Covid-19 in Peru: from supervised walks for children to the first case of Kawasaki-like syndrome. BMJ 369:m2418. https://doi.org/10.1136/bmj.m2418
159. Yasin Ar A (2020) Managing e-commerce during a pandemic: lessons from GrubHub during COVID-19. In: George B, Mahar Q (eds) International case studies in the management of disasters. Emerald Publishing Limited, pp 155–167. https://doi.org/10.1108/978-1-83982-186-820201010
160. Yong R (2007) The circular economy in China. J Mater Cycles Waste Manage 9(2):121–129. https://doi.org/10.1007/s10163-007-0183-z
161. Zacarías Farah A (2018) What is the circular economy and how does it care for the environment? [¿Qué es la economía circular y cómo cuida del medio ambiente?]. https://news.un.org/es/interview/2018/12/1447801
162. Zarocostas J (2021) What next for a COVID-19 intellectual property waiver? The Lancet 397(10288):1871–1872. https://doi.org/10.1016/S0140-6736(21)01151-X
163. Zhang SX, Chen J, Afshar Jahanshahi A, Alvarez-Risco A, Dai H, Li J, Patty-Tito RM (2021) Succumbing to the COVID-19 pandemic—healthcare workers not satisfied and intend to leave their jobs. Int J Ment Heal Addict. https://doi.org/10.1007/s11469-020-00418-6
164. Zhang SX, Sun S, Afshar Jahanshahi A, Alvarez-Risco A, Ibarra VG, Li J, Patty-Tito RM (2020) Developing and testing a measure of COVID-19 organizational support of healthcare workers – results from Peru, Ecuador, and Bolivia. Psychiatry Res 291:113174. https://doi.org/10.1016/j.psychres.2020.113174
165. Zhijun F, Nailing Y (2007) Putting a circular economy into practice in China. Sustain Sci 2(1):95–101. https://doi.org/10.1007/s11625-006-0018-1
166. Zollinger R, DiCindio C (2021) Community ecology: museum education and the digital divide during and after COVID-19. Journal of Museum Education 46(4):481–492. https://doi.org/10.1080/10598650.2021.1983711

3D Print, Circularity, and Footprints

Myreya De-la-Cruz-Diaz, Aldo Alvarez-Risco, Micaela Jaramillo-Arévalo, Maria de las Mercedes Anderson-Seminario, and Shyla Del-Aguila-Arcentales

Abstract 3D printing is part of the advanced technology that is beneficial in many areas such as construction, packaging, and medicine because it can create objects that can offer a better understanding of specific prototypes or models that you want to study. In addition, it also allows a smaller-scale production with lower costs for companies, so its use represents a comparative advantage and allows creating complex and more personalized designs to offer to its customers. In general, three-dimensional printing represents an efficient form of production that has a wide range of possibilities for its use. However, it is essential to know its impact at the environmental level to manage better and control its use. In the present work, we seek to know the relationship between circularity and three-dimensional impressions and the carbon footprint on the planet. Different sources and comparisons were made, and examples were proposed to explain and describe each previously proposed relationship to meet the research objective. The analysis results concluded that the impact was undoubtedly less than traditional production methods but that there is much room for improvement to reduce further the footprint left by 3D printing, such as a change in the use of materials. On the other hand, it is possible to achieve 3D prints that are circular and sustainable over time by making the necessary efforts. Additionally, the requirements for sustainable production through 3D printing must be not only feasible but also accessible to all to make a real change starting from the smallest companies and individuals that use this technology to the mass productions made by large companies.

Keywords 3D-printing · Circularity · Footprints · Sustainability · Three-dimensional · Additive manufacturing

---

M. De-la-Cruz-Diaz · A. Alvarez-Risco (✉) · M. Jaramillo-Arévalo · M. de las Mercedes Anderson-Seminario
Universidad de Lima, Lima, Perú
e-mail: aralvare@ulima.edu.pe

S. Del-Aguila-Arcentales
Escuela Nacional de Marina Mercante "Almirante Miguel Grau", Callao, Perú

A. Alvarez-Risco et al. (eds.), *Circular Economy*, Environmental Footprints and Eco-design of Products and Processes,
https://doi.org/10.1007/978-981-19-0549-0_5

1 Introduction

The world has changed, and people's habits regarding sustainability have changed, depending on the characteristics of pre-pandemic life. Due to limited sources by COVID-19, new ways to build new materials are essential in the reactivation of the global economies. The 3D print is a significant opportunity to create new materials for sustainable factories. The world changed and people's habits about sustainability have changed too, related to characteristics before COVID-19 pandemic mainly in health regulation [100], [24], [30], [27], [124], [29], healthcare service [137], [23], [26], [25], [81], [121], healthcare management [74], education [134] and sustainability ([10], [32], [31], [75]. Sustainable efforts are needed for companies and the inhabitants to promote the efficient management of 3D printing. Table 1, present the different sectors that the COVID-19 pandemic generated.

For decades, three-dimensional printing (3-D printing), also known as additive manufacturing (AM) or rapid prototyping, has existed due to the development of the 3D printer in 1984 [119]. In the beginning, the technology was prohibitively expensive and unavailable to the ordinary public [43]; now, due to technological advances over time, it is possible to create prototypes and objects quickly and easily for visualization and interaction. It is one of the most disruptive technologies of these times that could even cause a third world war [160].

Previously, a combination of texts and images or videos was used to understand a process or model; nowadays, because of the 3D printer, people can create 3D objects as well, which are used in different areas and can bring different benefits [43, 116]. The operation of 3D printers consists of creating three-dimensional pieces through the use of raw materials that are put layer by layer until the previously detailed object is created on the machine [141].

According to Silva et al. [141], Cartesian FFF is the most widely used 3D printing process, which employs a thermoplastic filament as a raw material to create 3D structures. Three axes (X, Y, and Z) regulate the object's position in the print bed in printers that use this technology. The Cartesian coordinate system always moves

Table 1 Effects of COVID-19 in different sectors

Education	[13, 14, 17, 19, 35, 69, 110, 112, 163]
Entrepreneurship	[6, 21, 34, 36, 46, 50, 52, 57, 65, 114, 120]
Health sector	[3, 12, 15, 18, 28, 48, 54, 68, 77, 106, 132, 135, 154, 158, 159]
Hospitality	[80, 91, 95, 102, 152]
Intellectual property	[16, 33, 62, 82, 98, 109, 126, 138, 156]
Population	[9, 20, 63, 93, 115, 130, 139, 143, 148, 151, 155]
Prices	[39, 70, 84, 94, 113, 125]
Tourism	[42, 49, 56, 58–60, 83, 140]
Trade	[4, 5, 7, 22, 37, 53, 72, 105, 118, 144, 145, 153]
Violence against women	[38, 66, 79, 90, 133, 142, 146]

the X, Y, and Z axes. The table can also do these as it moves through the extruder in any direction. Every FFF printer, on the other hand, has an extruder, which is the part that builds the 3D items. It's further divided into a cold tip that draws in the tough filament and a hot portion that fuses until its fluid enough to escape the nozzle and land on the print bed. Most modern 3D printers have hotbeds, which have the advantage of allowing the use of filaments that require a hot surface to adhere to the layers on top of each other. Figure 1 explains the most basic process to perform the three-dimensional printing process.

The behavior of industries has also changed in terms of their modes of production and management. In this way, the creation of 3D printing represented a technological advance that should be beneficial in many areas [141] (Fig. 2).

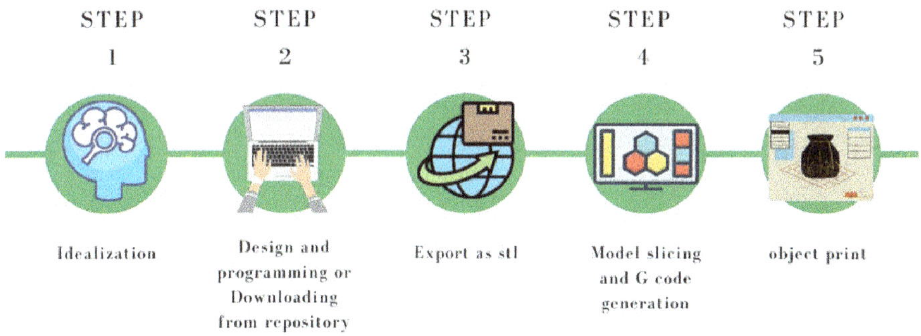

Fig. 1 Printing process (*Source* Adapted from Silva et al. [141])

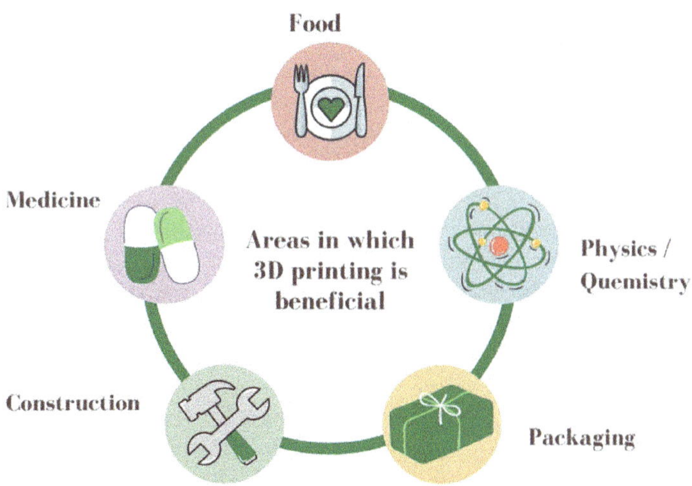

Fig. 2 Benefited areas (*Source* Adapted from Lipson et al. [1], Silva et al. [116], Otcu et al. [127], and Abdalla et al. [141])

For physics, it can be helpful to create prototypes and observe their behavior in real life, for example, kinematic mechanisms [116]. On the other hand, chemistry has also benefited from advances in 3D printing, especially in electrochemistry. In this area, research managers often look for standard yet customized ways to produce electrochemical devices that are sustainable, cost-saving, have a long-life cycle, and are produced efficiently. Three-dimensional printing has been quite beneficial to meet the requirements of preparing these chemical devices with all the characteristics mentioned above [141].

The application of the 3D printer in the industrial field allows making the manufacturing processes of certain products shorter, reducing costs and decreasing the demand for materials. In turn, this constitutes a comparative advantage for companies and allows greater customization of the products offered by them. The application in this field has many possibilities and reduces costs for manufacturing companies by offering greater control with the help of advanced technology in each production process [147].

In medicine, the existence of especially medical 3D printers is quite helpful for specific medical procedures, allowing at the same time a different and more efficient medical care for hospitals and doctors. These printers have specific characteristics to be used in this area, as they are certified as sterilized and biocompatible. In addition, its size was also designed to fit in spaces such as laboratories [55]. A specific case where 3D print has been used has been in eye care, where its earliest use was training and planning models, which later evolved to the creation of prostheses, surgical instruments, glasses, among others. In addition, it is essential to emphasize that there was an increase in demand for surgical instruments during the pandemic that led many hospitals to manufacture their instruments using 3D printers [111]. On the other hand, 3D prints also serve to recreate human body parts that can serve as training for medical personnel, providing them with a greater understanding of human anatomy [67]. 3D prints are accompanied by improvements in other technologies such as artificial intelligence that allows it to be combined with prints to create higher quality objects at lower costs and saving material, which reduces the waste of its use and creates more sustainable models [136].

In the case of the food industry, lately, the market has aided the inclusion of many AMTs in producing a wide range of meals, including chocolate, pasta, and pizza. The layer-by-layer material deposition mechanism of 3D Food Printing (3DFP) technology, based directly on a predesigned CAD file, is characterized by the treated material being food [129]. This revolutionary method of using 3D food printing can change the way food is produced, formulated, prepared, cooked, served, eaten, and even digested [127]. Because of the many disciplines of applications that 3D Food Printing brings together, this technology's foundation holds the promise of being able to "print" highly tailored foods, allowing for the creation of special meals [73, 127].

This chapter aims to identify the relationship between 3D printing and its effect on current environmental conditions according to the principle of circularity and these processes' footprint. Therefore, the positive or negative impact of additive manufacturing is presented.

2 Relationship Between Circularity and 3D Printing

The circular economy model replaces the concept of "end of life" in the manufacturing, distribution, and consumption processes in all industries by reducing the possibility of reuse, recycling, and recovery of resources [104]. The circular economy relies on the recycling of parts and resources. It reduces the usage of raw materials by including waste as a participant in the manufacturing process. Materials and components can be repaired, reused, reconditioned, and recycled in the technological cycle [86].

In the case of 3D printing, circularity is mainly achieved by reusing waste materials from different industries to convert them into the filaments necessary for final printing products, which can be seen in many examples, primarily plastics and, lately, in the food industry. Plastic is a trendy material used in different production methods to make various products, and the 3D printer is no exception. It also uses this material to create its three dimensions [123]. Plastic seems to go on and on increasing over the years, and it is estimated that around 850 million tons have been produced by 2050 [108]. Because of this exponential level of plastic production that generates negative impacts on the environment, new measures are being generated to counteract these effects [122].

Distributed Recycling for Additive Manufacturing is one of many ways to achieve a circular economy for plastic (DRAM). Consumers have an economic incentive to recycle under the DRAM technique because they can utilize their garbage as feedstock for a wide range of consumer products that can be produced at a fraction of the cost of comparable products [71, 149, 161]. PET has long been considered the holy grail of DRAM since it is one of the most easily identified polymer waste sources for consumers and is widely recycled through centralized procedures, though not at the rate required to support a circular economy [45]. PET is an excellent water and moisture barrier, and its use in the packaging sector for consumable packaging of water, soft beverages, and meals is predicted to expand at a rate of 4.5% per year [101]. PET water bottles are simple to recycle since they are already clean and available in vast quantities worldwide,this is why a few companies, such as Refil and B-PET, sell recycled PET filament [117].

Upcycling plastic waste into 3-D printer filament with a recyclebot, an open-source waste plastic extruder, is another technique of distributed plastic recycling [150]. Previous research on the life cycle analysis (LCA) or recyclebot method using post-consumer polymers rather than raw materials found that the energy performance of the filament from extracting, refining natural resources, and synthesizing was reduced by 90% when opposed to conventional manufacturing [47, 107, 108].

3D printing technology may be one such solution to address the challenge of plastic contamination as it is a low-cost and accessible technology, which uses plastics as its primary manufacturing material [122]. The EcoPrinting system is a compact nanogrid-style system with a solar-charging battery to switchgears and designed as an excellent example of how to turn plastic wastes into finished goods through this kind of 3D technology. Hand-operated granulation equipment, a low-power filament

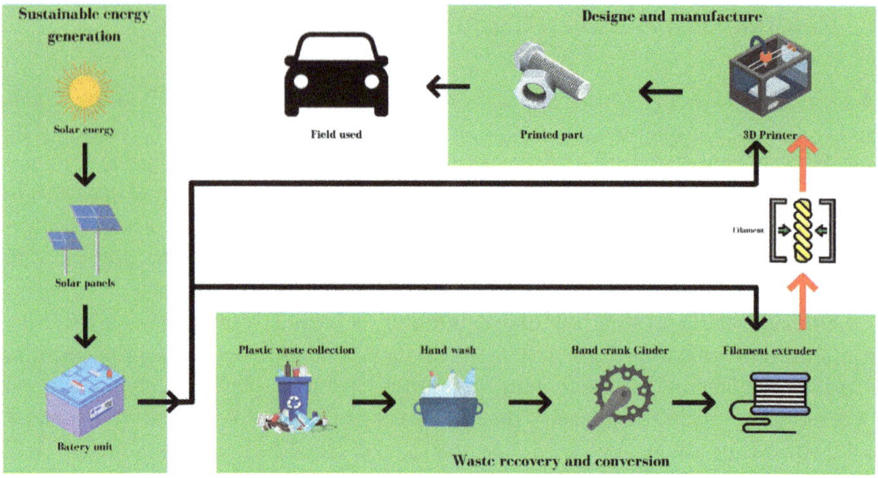

Fig. 3 The EcoPrinting system (*Source* Mohammed et al. [123])

extrusion system, a customized 3D printer capable of using recycled plastics, a laptop for design creation, and 3D printer interfacing are among the low-power gadgets created to achieve this. As a result, the recycled plastics could be formed into filaments with tolerances equal to commercial-grade filaments [123] (Fig. 3).

Similarly, another material that can be reused as a 3D printing filament is food. Several researchers and practitioners have advocated for using technology's characteristics in the food sector to address sustainability, increased regulations, complex supply chains, and so on [87], [128]. Compared to standard AM solutions, 3D food printing is still in its early stages of development and implementation. Yet, it broadens people's perspectives by giving them unrestricted design flexibility, allowing them to create more reliable products with more economic and social impact and fewer negative environmental consequences [127].

The developing method of three-dimensional printing (3D printing) should be evaluated as a feasible solution considering the mounting hazards linked with health, food shortages, and the other issues mentioned before. Food three-dimensional printing could be a viable means to assure food supply by minimizing the cultivation and stockbreeding process and a practical way to avoid being exposed to unhealthy food [162]. Powder bed AM systems are helpful in food printing for any scenarios where the input material's shape is input as a powder [129]. This 3D-printing technology was utilized to make nutrient-dense noodles from potato peel waste from processing plants. According to the research, the fine potato peel fractions were better printable than the coarse fractions. According to Pallottino et al. [128], it is possible that, per the current trends, strategies for effective food waste utilization become a key field of research and market soon.

Likewise, [61] show the potential of the salmon gelatine gels to flow as an independent filament and create 3D shapes was demonstrated at various doses. Because of its good printability, shape retention, and dimensional stability, the sample with a salmon gelatine concentration of 8% was the best suitable for the 3D printing procedure. As a result of this research, a foundation for studying food matrices based on marine gelatine has been established, which could be valuable in preparing diets for people with specific nutritional needs utilizing 3D printing.

A final example of circularity within 3D printing in the food industry can be seen through industry waste such as grape pomace and broken wheat as primary materials in a 3D printing material supply to create functional cookies [97]. The addition of grape pomace to the cookies increased their nutritional content and antioxidant capabilities leading to a final product that was high in both proteins and dietary fiber [97]. This method emphasizes the value addition potential of industrial waste streams, with a high consumer preference. As a result, additive manufacturing's unrivaled degrees of customization can be perfectly complemented by personalizing foods in terms of nutritional value while allowing for cleaner production processes and enhanced resource recovery from food processing wastes.

3 3D Print and Footprints

As mentioned above, the materials used in the three-dimensional printing process impact the environment. However, at the same time, this process can be compared with the impact of other more traditional methods and reach a conclusion about which is more efficient and leaves a lower carbon footprint. Next, the comparisons between methods and the different efforts that have been made to reduce the environmental impact of 3D printing are mentioned.

Following the plastic use in 3D printing mentioned before, according to Mohammed et al. [123], this technology has revealed that it operates at low average power (about 60Whr continuous use), perhaps allowing it to be powered by a small renewable energy system. Thus, using EcoPrinting to reprocess waste plastics into usable and functional things would generate a minimal carbon footprint and sustainable energy generation.

On the other hand, the construction sector is responsible for 46% of energy consumption annually [78]. It is also responsible for 38% of the global emission of greenhouse gasses [8]; as the sustainable objectives became relevant, the demand for a new, more eco-friendly system for this sector was growing [1]. In addition, this industry faces other problems such as the high number of accidents at work and low productivity. Three-dimensional printing is a potential solution to improve conventional construction methods that improve workers' quality of life and the work environment to face these problems. The use of 3D printing and its environmental and social impacts have been studied. A case study in China demonstrated the potential of this method through the construction of many houses of approximately 200 m2 with quality materials [103].

It was possible to conclude through the study that the results of the scenario where 3D printing was used were much more favorable than traditional methods, with a

Table 2 Environmental results

Impact category	3D printing	Conventional Construction
Global warming (Kg CO_2 eq)	608.55	1154.2
Land occupation (m^2 a crop eq)	0.4	6.8
Water consumption (m^3)	183.95	233.35
Human Health (Pt)	5.3	18.63
Ecosystems (Pt)	0.64	1.3
Resources (Pt)	0.05	0.2

Source Adapted from Abdalla et al. [1]

performance of up to 50% better since it was possible to reduce the emissions related to many processes [1]. On the other hand, other constructions under investigation have shown that implementing these 3D prints allowed the use of materials efficiently, thus reducing waste. And it could also reduce up to 5% of emissions by 2025 in larger projects.

In Table 2, you can see the results of the environmental impacts according to some specific indicators. It can be observed that comparing 3D printing and conventional construction, the impact of 3D printing is significantly less. The impacts were calculated in [1] investigation using SimaPro in four stages.

On the other hand, it is essential to mention that sustainability is crucial for the construction industry [51] and the packaging industry. The industry consumes 38% of the plastics derived from petroleum produced annually [131]. By 2050, it is estimated that plastic waste reach 12 billion tons that end up in landfills [85] which is how alternatives have emerged to use less polluting raw materials such as biomass composed of mushrooms that can significantly reduce the negative impact on both industries [44, 89].

The biomass that has been derived from agricultural residues fulfills the role of substrate and acts as a source of nutrition for the fungi that later grow as white filaments known as mycelia appointment [51]. The mycelia bind to the biomass since it has properties similar to wood and cork [41]. The benefits of using these materials are that they constitute lower costs [99], are biodegradable, and have a low environmental impact [2, 92], thus reducing the carbon footprint.

Different studies have been carried out on the characteristics of biomass-fungi, which has materials that may present different characteristics depending on their source of nutrition, species of fungi, and the method of processing [40, 92]. It is possible that even environmental factors such as light and carbon dioxide concentration also impact fungal density [41]. It has experimented on the factors and how it affects biomass-fungi since it is sought that this can be part of other productive processes of 3D printing [51].

One of the advantages of 3D printing is that it is cheaper for smaller-scale productions; it allows reducing costs that would otherwise be higher [51]. It is possible to produce complex parts in small quantities and create complex figures [11, 88]. To produce these objects, using methods based on molds are involved [96]; these molds are commonly made of plastic, that is why it is necessary to change this part of the production process to reduce the environmental impact of total production further and become more sustainable [44].

A wide range of materials meets the printer's requirements to be processed and used to find the most optimal and viable material and sustainable over time. These proposals have been given and investigated because 3D technologies, especially those with plastic as the primary material, produce a significant amount of waste that later ends up in landfills. Due to the pandemic, plastic use has increased due to the high demand for masks and protectors [157]. Although the environmental impact of prints is not as significant as energy consumption, added to the growth of plastic production, prints that continue to use this material after being discarded also need to use energy to be disposed of Zgodavová et al. [157]. The impact of prints depends on the volume produced, the materials used in the process, and the printer's technology [157]. According to the study conducted by Zgodavová et al. [157], the most suitable and sustainable material for production concerning face shields was a filament called PHA BioWOOD Rosa 3D due to its reduced environmental impact.

Just as technology is available to everyone, sustainable alternatives should also be. As mentioned before, the combination of 3D printing and other technologies such as artificial intelligence and computational intelligence [136]. It is necessary to have good management and use these advanced technologies correctly. As mentioned before, to achieve sustainable 3D printing, the material used must be considered, but other factors such as energy, emitted particles, and the waste they generate must also be considered. In addition, they are considering that these printers are increasingly accessible to the public, the waste generated by the easy manufacturing done by individuals increase [136]. To reduce this waste and the emissions generated by 3D printers that increase the carbon footprint they generate on the planet, the use of the correct technologies that can convert data into information and that later that information becomes learning is what is necessary to be able to propose changes in the current model and thus reduce the impact [76]. The correct use of the information supposes a comparative advantage for the companies since it allows them to raise much more viable hypotheses and make changes in their processes that reduce costs and increase efficiency. It also allows them to improve in environments where a problem had probably not been correctly identified [64].

Ci-based software does not replace printers but rather enhances and improves them to make them more sustainable [136]. On the other hand, being and breathing the chemicals produced by 3D printers is harmful to human health, specifically in their respiratory system, so it is essential to gather information through technologies to have better control and create safer protocols for personnel exposed to working with these printers. All this is to protect employees' health and have a better safe work environment for everyone.

4 Closing Remarks

3D printing, also known as additive manufacturing, quickly makes a model/prototype of a component or finished product. It is a very innovative and influential technology in today's world, which has gained relevance in many industries, from medicine and chemistry to the food industry, thanks to its relative ease of creating different and unique products adapted to each situation.

One relevant advantage is the possibility of contributing to the improvement of circularity in the industries to which it belongs. First, this technology can make it possible to use plastic waste by recycling it and converting it into base filaments to create final products. In the first place, this technology can make it possible to use plastic waste through its recycling and conversion into base filaments to create final products, with waste being one of the pillars of the circular economy. Moreover, recent experiments and research on the reuse of other waste such as food make the potential of 3d printing a problem-solving tool that most industries have to face, such as sustainability. In that case, chemical processes added to the 3D printing could reduce hazards related to food production due to the enhancement of food properties such as its durability and even protein and nutrients.

In addition to this possible reuse of resources, 3D printing is increasingly contributing to reducing the carbon footprints of industries. Taking the EcoPrinting System as an example, this technology can provide significant benefits to the neediest communities through sustainable energy sources and the production of necessary parts, thus avoiding the waste of resources which can be demonstrated in the same way in other sectors such as construction. The production of 3D printed parts helps one of the most polluting industries in terms of carbon emissions to reduce its footprint on the environment. Where the use of materials other than plastic is proposed for a reduction in environmental impact.

Although the production of goods in 3D can often be positive for both people and environmental issues, it should be noted that there are still difficulties in the use of 3D printers, such as the fact that many of the filaments currently used are made of materials such as metals and non-recycled plastic. Even if there are new experiments and tests to maximize the resources used in 3D printing, there are still many limitations.

Despite all the benefits that 3D printing generates for people and companies, there are still specific challenges to solve for this technology. Such as chemicals that are harmful to the health of those who work near these printers. On the other hand, since they are accessible, their use could become massive using non-biodegradable materials, negatively impacting the environment. It is also essential to consider that these 3D printers generally do not always occur directly but through the entire process of use which is how the objects created pollute even after being discarded since their recycling also consumes energy and other chemicals are needed.

It can then be concluded that efforts should continue to be made to continue improving 3D printing and reducing its impact on the planet. On the other hand, it is necessary to rely on the newly proposed models and advanced technologies that emerge over time and support them in making 3D printing even more efficient and sustainable in the long term.

References

1. Abdalla H, Fattah KP, Abdallah M, Tamimi AK (2021) Environmental footprint and economics of a full-scale 3D-printed house. Sustainability 13(21). https://doi.org/10.3390/su132111978
2. Abhijith R, Ashok A, Rejeesh CR (2018) Sustainable packaging applications from mycelium to substitute polystyrene: a review. Materials Today: Proceedings 5(1, Part 2):2139–2145. https://doi.org/10.1016/j.matpr.2017.09.211
3. Abiakam N, Worsley P, Jayabal H, Mitchell K, Jones M, Fletcher J, Spratt F, Bader D (2021) Personal protective equipment related skin reactions in healthcare professionals during COVID-19. Int Wound J 18(3):312–322. https://doi.org/10.1111/iwj.13534
4. Abudureheman A, Nilupaer A (2021) Optimization model design of cross-border e-commerce transportation path under the background of prevention and control of COVID-19 pneumonia. Soft Comput 25(18):12007–12015. https://doi.org/10.1007/s00500-021-05685-6
5. Acuña-Zegarra MA, Santana-Cibrian M, Velasco-Hernandez JX (2020) Modeling behavioral change and COVID-19 containment in Mexico: a trade-off between lockdown and compliance. Math Biosci 325:108370. https://doi.org/10.1016/j.mbs.2020.108370
6. Afshan G, Shahid S, Tunio MN (2021) Learning experiences of women entrepreneurs amidst COVID-19. Int J Gend Entrep 13(2):162–186. https://doi.org/10.1108/IJGE-09-2020-0153
7. Aguirre AA, Catherina R, Frye H, Shelley L (2020) Illicit wildlife trade, wet markets, and COVID-19: preventing future pandemics. World Med Health Policy 12(3):256–265. https://doi.org/10.1002/wmh3.348
8. Agustí-Juan I, Habert G (2017) Environmental design guidelines for digital fabrication. J Clean Prod 142:2780–2791. https://doi.org/10.1016/j.jclepro.2016.10.190
9. Ahsan MM (2020) Strategic decisions on urban built environment to pandemics in Turkey: lessons from COVID-19. J Urban Manag 9(3):281–285. https://doi.org/10.1016/j.jum.2020.07.001
10. Albareda-Tiana S, Vidal-Raméntol S, Pujol-Valls M, Fernández-Morilla M (2018) Holistic approaches to develop sustainability and research competencies in pre-service teacher training. Sustainability 10:3698. https://doi.org/10.3390/su10103698
11. Al-Ketan O, Rowshan R, Abu Al-Rub RK (2018) Topology-mechanical property relationship of 3D printed strut, skeletal, and sheet based periodic metallic cellular materials. Addit Manuf 19:167–183. https://doi.org/10.1016/j.addma.2017.12.006
12. Alan H, Eskin Bacaksiz F, Tiryaki Sen H, Taskiran Eskici G, Gumus E, Harmanci Seren AK (2021) "I'm a hero, but…": An evaluation of depression, anxiety, and stress levels of frontline healthcare professionals during COVID-19 pandemic in Turkey. Perspectives in Psychiatric Care 57(3):1126–1136 https://doi.org/10.1111/ppc.12666
13. Ali W (2020) Online and remote learning in higher education institutes: a necessity in light of COVID-19 pandemic. High Educ Stud 10(3):16–25
14. Allen J, Rowan L, Singh P (2020) Teaching and teacher education in the time of COVID-19. Asia-Pacific Journal of Teacher Education 48(3):233–236. https://doi.org/10.1080/1359866X.2020.1752051
15. Alsairafi Z, Naser AY, Alsaleh FM, Awad A, Jalal Z (2021) Mental health status of health-care professionals and students of health sciences faculties in Kuwait during the COVID-19 pandemic. Int J Environ Res Public Health 18(4). https://doi.org/10.3390/ijerph18042203
16. Altindis E (2021) Inequitable COVID-19 vaccine distribution and the intellectual property rights prolong the pandemic. Expert Rev. Vaccines. https://doi.org/10.1080/14760584.2022.2014819

17. Alvarez-Risco A, Del-Aguila-Arcentales S, Rosen MA, García-Ibarra V, Maycotte-Felkel S, Martínez-Toro GM (2021) Expectations and interests of university students in COVID-19 times about sustainable development goals: evidence from Colombia, Ecuador, Mexico, and Peru. Sustainability 13(6):3306. https://doi.org/10.3390/su13063306
18. Alvarez-Risco A, Del-Aguila-Arcentales S, Yáñez JA (2021) Telemedicine in Peru as a result of the COVID-19 pandemic: perspective from a country with limited internet access. Am J Trop Med Hyg 105(1):6–11. https://doi.org/10.4269/ajtmh.21-0255
19. Alvarez-Risco A, Del-Aguila-Arcentales S, Yáñez JA, Rosen MA, Mejia CR (2021) Influence of technostress on academic performance of university medicine students in Peru during the COVID-19 pandemic. Sustainability 13(16):8949. https://doi.org/10.3390/su13168949
20. Alvarez-Risco A, Mejia CR, Delgado-Zegarra J, Del-Aguila-Arcentales S, Arce-Esquivel AA, Valladares-Garrido MJ, del Portal MR, Villegas LF, Curioso WH, Sekar MC, Yáñez JA (2020) The Peru approach against the COVID-19 infodemic: insights and strategies. Am J Trop Med Hyg 103(2):583–586. https://doi.org/10.4269/ajtmh.20-0536
21. Alvarez-Risco A, Mlodzianowska S, García-Ibarra V, Rosen MA, Del-Aguila-Arcentales S (2021) Factors affecting green entrepreneurship intentions in business university students in COVID-19 pandemic times: case of Ecuador. Sustainability 13(11):6447. https://doi.org/10.3390/su13116447
22. Alvarez-Risco A, Rosen MA, Del-Aguila-Arcentales S (2020) A new regulation for supporting a circular economy in the plastic industry: the case of Peru (short communication). J Landsc Ecol 13(1):1–3. https://doi.org/10.2478/jlecol-2020-0004
23. Álvarez-Risco A, Zegarra Arellano E, Matos Valerio E, Mejía Acosta N, Solis Tarazona Z (2013) Campaña de atención farmacéutica como estrategia de implementación de los servicios farmacéuticos: experiencia Perú. Pharm Care España 15(1):35. https://www.pharmcareesp.com/index.php/PharmaCARE/article/view/105
24. Alvarez-Risco A, Mil JWFv (2007) Pharmaceutical care in community pharmacies: practice and research in Peru. Ann Pharmacother 41(12):2032–2037. https://doi.org/10.1345/aph.1K117
25. Alvarez-Risco A, Del-Aguila-Arcentales S, Delgado-Zegarra J, Yáñez JA, Diaz-Risco S (2019) Doping in sports: findings of the analytical test and its interpretation by the public. Sport Sci Health 15(1):255–257. https://doi.org/10.1007/s11332-018-0484-8
26. Alvarez-Risco A, Turpo-Cama A, Ortiz-Palomino L, Gongora-Amaut N, Del-Aguila-Arcentales S (2016) Barreras para la implementación de la Atención Farmacéutica en establecimientos farmacéuticos de Cusco, Perú. Pharm Care España 18(5):194–205. https://www.pharmcareesp.com/index.php/PharmaCARE/article/view/326
27. Alvarez-Risco A, Quiroz-Delgado D, Del-Aguila-Arcentales S (2016) Pharmaceutical care in hypertension patients in a peruvian hospital. Indian J Public Health Res Dev 7(3):183–188
28. Alvarez-Risco A, Dawson J, Johnson W, Conteh-Barrat M, Aslani P, Del-Aguila-Arcentales S, Diaz-Risco S (2021) Ebola virus disease outbreak: a global approach for health systems [Ebola; health professionals; global approach; pharmacist: pharmaceutical care; COVID-19] 53(4). http://www.revfarmacia.sld.cu/index.php/far/article/view/491
29. Alvarez-Risco A, Del-Aguila-Arcentales S (2015) Errores de prescripción como barrera para la Atención Farmacéutica en establecimientos de salud públicos: experiencia Perú. Pharm Care España 17(6):725–731
30. Alvarez-Risco A, Del-Aguila-Arcentales S, Stevenson JG (2015) Pharmacists and mass communication for implementing pharmaceutical care. Am J Pharm Benefits 7(3):e125–e126
31. Alvarez-Risco A, Del-Aguila-Arcentales S, Diaz-Risco S (2018) Pharmacovigilance as a tool for sustainable development of healthcare in Peru. PharmacoVigil Rev 10(2):4–6
32. Alvarez-Risco A, Delgado-Zegarra J, Yáñez JA, Diaz-Risco S, Del-Aguila-Arcentales S (2018) Predation risk by gastronomic boom-case Peru. J Lands Ecol 11(1):100–103. https://doi.org/10.2478/jlecol-2018-0003

33. Alvarez-Risco A, Del-Aguila-Arcentales S (2021a) A note on changing regulation in international business: the World Intellectual Property Organization (WIPO) and artificial intelligence. In: Verbeke A, van Tulder R, Rose EL, Wei Y (eds) The multiple dimensions of institutional complexity in international business research, vol 15. Emerald Publishing Limited, pp 363–371. https://doi.org/10.1108/S1745-886220210000015020
34. Alvarez-Risco A, Del-Aguila-Arcentales S (2021b) Public policies and private efforts to increase women entrepreneurship based on STEM background. In: Mari M, Poggesi S, Foss L (eds) Women's entrepreneurship in STEM disciplines: issues and perspectives. Springer, pp 75–87. https://doi.org/10.1007/978-3-030-83792-1_5
35. Alvarez-Risco A, Estrada-Merino A, Anderson-Seminario MdlM, Mlodzianowska S, García-Ibarra V, Villagomez-Buele C, Carvache-Franco M (2021) Multitasking behavior in online classrooms and academic performance: case of university students in Ecuador during COVID-19 outbreak. Interact Technol Smart Educ 18(3):422–434. https://doi.org/10.1108/ITSE-08-2020-0160
36. Alvarez-Risco A, Mlodzianowska S, Zamora-Ramos U, Del-Aguila S (2021) Green entrepreneurship intention in university students: the case of Peru. Entrep Bus Econ Rev 9(4), 85–100. https://doi.org/10.15678/EBER.2021.090406
37. Alvarez-Risco A, Quipuzco-Chicata L, Escudero-Cipriani C (2021) Determinantes de la intención de recompra en línea en tiempos de COVID-19: evidencia de una economía emergente. Lecturas de Economía (96). https://doi.org/10.17533/udea.le.n96a342638
38. Amarillo CR (2020) Feminism on Lockdown. NACLA Rep Am 52(3):274–281. https://doi.org/10.1080/10714839.2020.1809084
39. Apcho-Ccencho L-V, Cuya-Velásquez B-B, Alvarado Rodríguez D, Anderson-Seminario MdlM, Alvarez-Risco A, Estrada-Merino A, Mlodzianowska S (2021) The impact of international price on the technological industry in the United States and China during times of crisis: commercial war and COVID-19. In: Lawrence KD, Klimberg RK (eds) Advances in business and management forecasting, vol. 14. Emerald Publishing Limited, pp 149–160. https://doi.org/10.1108/S1477-407020210000014010
40. Appels FVW, Camere S, Montalti M, Karana E, Jansen KMB, Dijksterhuis J, Krijgsheld P, Wösten HAB (2019) Fabrication factors influencing mechanical, moisture- and water-related properties of mycelium-based composites. Mater Des 161:64–71. https://doi.org/10.1016/j.matdes.2018.11.027
41. Appels FVW, Dijksterhuis J, Lukasiewicz CE, Jansen KMB, Wösten HAB, Krijgsheld P (2018) Hydrophobin gene deletion and environmental growth conditions impact mechanical properties of mycelium by affecting the density of the material. Sci Rep 8(1):4703. https://doi.org/10.1038/s41598-018-23171-2
42. Assaf A, Scuderi R (2020) COVID-19 and the recovery of the tourism industry. Tour Econ 26(5):731–733. https://doi.org/10.1177/1354816620933712
43. Attaran M (2017) The rise of 3-D printing: the advantages of additive manufacturing over traditional manufacturing. Bus Horiz 60(5):677–688. https://doi.org/10.1016/j.bushor.2017.05.011
44. Attias N, Danai O, Abitbol T, Tarazi E, Ezov N, Pereman I, Grobman YJ (2020) Mycelium bio-composites in industrial design and architecture: comparative review and experimental analysis. J Clean Prod 246:119037. https://doi.org/10.1016/j.jclepro.2019.119037
45. Awaja F, Pavel D (2005) Recycling of PET. Eur Polymer J 41(7):1453–1477. https://doi.org/10.1016/j.eurpolymj.2005.02.005
46. Bacq S, Lumpkin GT (2021) Social entrepreneurship and COVID-19. J Manage Stud 58(1):285–288. https://doi.org/10.1111/joms.12641
47. Baechler C, DeVuono M, Pearce JM (2013) Distributed recycling of waste polymer into RepRap feedstock. Rapid Prototyp J 19(2):118–125. https://doi.org/10.1108/13552541311302978
48. Barello S, Caruso R, Palamenghi L, Nania T, Dellafiore F, Bonetti L, Silenzi A, Marotta C, Graffigna G (2021) Factors associated with emotional exhaustion in healthcare professionals involved in the COVID-19 pandemic: an application of the job demands-resources model. Int Arch Occup Environ Health 94(8):1751–1761. https://doi.org/10.1007/s00420-021-01669-z

49. Baum T, Hai NTT (2020) Hospitality, tourism, human rights and the impact of COVID-19. Int J Contemp Hosp Manag 32(7):2397–2407. https://doi.org/10.1108/IJCHM-03-2020-0242
50. Belitski M, Guenther C, Kritikos AS, Thurik R (2021) Economic effects of the COVID-19 pandemic on entrepreneurship and small businesses. Small Bus Econ. https://doi.org/10.1007/s11187-021-00544-y
51. Bhardwaj A, Rahman AM, Wei X, Pei Z, Truong D, Lucht M, Zou N (2021) 3D printing of biomass–fungi composite material: Effects of mixture composition on print quality. J Manuf Mater Process 5(4). https://doi.org/10.3390/jmmp5040112
52. Block JH, Fisch C, Hirschmann M (2021) The determinants of bootstrap financing in crises: evidence from entrepreneurial ventures in the COVID-19 pandemic. Small Bus Econ. https://doi.org/10.1007/s11187-020-00445-6
53. Borzée A, McNeely J, Magellan K, Miller JRB, Porter L, Dutta T et al (2020) COVID-19 highlights the need for more effective wildlife trade legislation. Trends Ecol Evol 35(12):1052–1055. https://doi.org/10.1016/j.tree.2020.10.001
54. Bozdağ F, Ergün N (2020) Psychological resilience of healthcare professionals during COVID-19 pandemic. Psychol Rep 124(6):2567–2586. https://doi.org/10.1177/0033294120965477
55. British Dental Association (2021, 2021/06/01) All-in-one medical 3D printer. Retrieved 12 from https://doi.org/10.1038/s41415-021-3196-4
56. Brouder P (2020) Reset redux: possible evolutionary pathways towards the transformation of tourism in a COVID-19 world. Tour Geogr 22(3):484–490. https://doi.org/10.1080/14616688.2020.1760928
57. Brown R, Rocha A, Cowling M (2020) Financing entrepreneurship in times of crisis: exploring the impact of COVID-19 on the market for entrepreneurial finance in the United Kingdom. Int Small Bus J 38(5):380–390. https://doi.org/10.1177/0266242620937464
58. Carvache-Franco M, Alvarez-Risco A, Carvache-Franco W, Carvache-Franco O, Estrada-Merino A, Rosen MA (2021) Coastal cities seen from loyalty and their tourist motivations: a study in Lima, Peru. Sustainability 13(21):11575. https://doi.org/10.3390/su132111575
59. Carvache-Franco M, Carvache-Franco O, Carvache-Franco W, Alvarez-Risco A, Estrada-Merino A (2021) Motivations and segmentation of the demand for coastal cities: a study in Lima, Peru. Int J of Tourism Res 23(4):517–531. https://doi.org/10.1002/jtr.2423
60. Carvache-Franco M, Alvarez-Risco A, Carvache-Franco O, Carvache-Franco W, Estrada-Merino A, Villalobos-Alvarez D (2021) Perceived value and its influence on satisfaction and loyalty in a coastal city: a study from Lima, Peru. Journal of Policy Research in Tourism, Leisure and Events, 1–16. https://doi.org/10.1080/19407963.2021.1883634
61. Carvajal-Mena N, Tabilo-Munizaga G, Pérez-Won M, Lemus-Mondaca R (2022) Valorization of salmon industry by-products: evaluation of salmon skin gelatin as a biomaterial suitable for 3D food printing. LWT 155:112931. https://doi.org/10.1016/j.lwt.2021.112931
62. del Castillo FA (2021) Temporary waiver of intellectual property on Covid-19 vaccines: toward the creation of a better, post-pandemic society. J Public Health 43(3):e559–e560. https://doi.org/10.1093/pubmed/fdab184
63. Cegarra-Navarro J-G, Vătămănescu E-M, Martínez-Martínez A (2021) A context-driven approach on coping with COVID-19: from hiding knowledge toward citizen engagement. Knowl Process Manag 28(2):134–140. https://doi.org/10.1002/kpm.1662
64. Chadha K, Tian Y, Bocher P, Spray JG, Aranas C (2020) Microstructure evolution, mechanical properties and deformation behavior of an additively manufactured maraging steel. Materials 13(10). https://doi.org/10.3390/ma13102380
65. Chafloque-Cespedes R, Alvarez-Risco A, Robayo-Acuña P-V, Gamarra-Chavez C-A, Martinez-Toro G-M, Vicente-Ramos W (2021) Effect of sociodemographic factors in entrepreneurial orientation and entrepreneurial intention in university students of Latin American business schools. In Jones P, Apostolopoulos N, Kakouris A, Moon C, Ratten V, Walmsley A (eds) Universities and entrepreneurship: meeting the educational and social challenges, vol 11. Emerald Publishing Limited, pp 151–165. https://doi.org/10.1108/S2040-724620210000011010

66. Chafloque-Cespedes R, Vara-Horna A, Asencios-Gonzales Z, Lopez-Odar DR, Alvarez-Risco A, Quipuzco-Chicata L, Schulze C, Sanchez-Villagomez M (2020) Presentismo académico y violencia contra las mujeres en escuelas de negocios e ingeniería en universidades peruanas. Lecturas de Economía 0(93):127–153. https://doi.org/10.17533/udea.le.n93a340726
67. Chaudhary B, Anand U, Kumari V, Agrawal P, Kumar P, Priyadarshi R (2021) Feasibility and adaptation of three-dimensional model for surgical planning and training: a pilot study [Original Article]. Natl J Clin Anat 10(4):220–225. https://doi.org/10.4103/2277-4025.329493
68. Chen X, Zhang SX, Jahanshahi AA, Alvarez-Risco A, Dai H, Li J, Ibarra VG (2020) Belief in a COVID-19 conspiracy theory as a predictor of mental health and well-being of health care workers in Ecuador: cross-sectional survey study. JMIR Public Health Surveill 6(3):e20737. https://doi.org/10.2196/20737
69. Chen T, Peng L, Yin X, Rong J, Yang J, Cong G (2020) Analysis of user satisfaction with online education platforms in China during the COVID-19 pandemic. Healthcare 8(3). https://doi.org/10.3390/healthcare8030200
70. Chung SA, Olivera S, Rojas Román B, Alanoca E, Moscoso S, Limpias Terceros B, Alvarez-Risco A, Yáñez JA (2021) Temáticas de la producción científica de la Revista Cubana de Farmacia indizada en Scopus (1967–2020) [Revista Cubana de Farmacia; farmacia; Cuba; evidencia] 54(1). http://www.revfarmacia.sld.cu/index.php/far/article/view/511
71. Cruz Sanchez FA, Boudaoud H, Camargo M, Pearce JM (2020) Plastic recycling in additive manufacturing: a systematic literature review and opportunities for the circular economy. J Clean Prod 264:121602. https://doi.org/10.1016/j.jclepro.2020.121602
72. Cruz-Torres W, Alvarez-Risco A, Del-Aguila-Arcentales S (2021) Impact of Enterprise Resource Planning (ERP) implementation on performance of an education enterprise: a Structural Equation Modeling (SEM). Stud Bus Econ 16(2):37–52. https://doi.org/10.2478/sbe-2021-0023
73. Dankar I, Haddarah A, Omar FEL, Sepulcre F, Pujolà M (2018) 3D printing technology: the new era for food customization and elaboration. Trends Food Sci Technol 75:231–242. https://doi.org/10.1016/j.tifs.2018.03.018
74. Del-Aguila-Arcentales S, Alvarez-Risco A (2013) Human error or burnout as explanation for mistakes in pharmaceutical laboratories. Accreditation Qual Assur 18(5):447–448. https://doi.org/10.1007/s00769-013-1000-0
75. Delgado-Zegarra J, Alvarez-Risco A, Yáñez JA (2018) Uso indiscriminado de pesticidas y ausencia de control sanitario para el mercado interno en Perú. Rev Panam Salud Publica 42:e3. https://doi.org/10.26633/RPSP.2018.3
76. Deng Y, Mao Z, Yang N, Niu X, Lu X (2020) Collaborative optimization of density and surface roughness of 316L stainless steel in selective laser melting. Materials 13(7). https://doi.org/10.3390/ma13071601
77. Deressa W, Worku A, Abebe W, Gizaw M, Amogne W (2021) Risk perceptions and preventive practices of COVID-19 among healthcare professionals in public hospitals in Addis Ababa, Ethiopia. PLoS ONE 16(6):e0242471. https://doi.org/10.1371/journal.pone.0242471
78. Dixit MK (2019) 3-D printing in building construction: a literature review of opportunities and challenges of reducing life cycle energy and carbon of buildings. IOP Conference Series: Earth and Environmental Science 290(1):012012. https://doi.org/10.1088/1755-1315/290/1/012012
79. Dominelli L (2021) A green social work perspective on social work during the time of COVID-19. Int J Soc Welf 30(1):7–16. https://doi.org/10.1111/ijsw.12469
80. Duarte Alonso A, Kok SK, Bressan A, O'Shea M, Sakellarios N, Koresis A et al (2020) COVID-19, aftermath, impacts, and hospitality firms: an international perspective. Int J Hosp Manag 91:102654. https://doi.org/10.1016/j.ijhm.2020.102654
81. Enciso-Zarate A, Guzmán-Oviedo J, Sánchez-Cardona F, Martínez-Rohenes D, Rodríguez-Palomino JC, Alvarez-Risco A, ... Diaz-Risco SS (2016) Evaluación de la contaminación con agentes citotóxicos en hospitales en Colombia. J Pharm Care 18(6):241–250. https://www.pharmcareesp.com/index.php/PharmaCARE/article/view/354

82. Erfani P, Binagwaho A, Jalloh MJ, Yunus M, Farmer P, Kerry V (2021) Intellectual property waiver for covid-19 vaccines will advance global health equity. BMJ 374:n1837. https://doi.org/10.1136/bmj.n1837
83. Fotiadis A, Polyzos S, Huan T-CTC (2021) The good, the bad and the ugly on COVID-19 tourism recovery. Ann Tour Res 87:103117. https://doi.org/10.1016/j.annals.2020.103117
84. Galanakis CM, Rizou M, Aldawoud TMS, Ucak I, Rowan NJ (2021) Innovations and technology disruptions in the food sector within the COVID-19 pandemic and post-lockdown era. Trends Food Sci Technol 110:193–200. https://doi.org/10.1016/j.tifs.2021.02.002
85. Geyer R, Jambeck Jenna R, Law Kara L (2021) Production, use, and fate of all plastics ever made. Sci Adv 3(7):e1700782. https://doi.org/10.1126/sciadv.1700782
86. Gil Muñoz V, Muneta LM, Carrasco-Gallego R, de Juanes Marquez J, Hidalgo-Carvajal D (2020) Evaluation of the circularity of recycled PLA filaments for 3D printers. Appl Sci 10(24). https://doi.org/10.3390/app10248967
87. Godoi FC, Prakash S, Bhandari BR (2016) 3d printing technologies applied for food design: status and prospects. J Food Eng 179:44–54. https://doi.org/10.1016/j.jfoodeng.2016.01.025
88. Goh GD, Neo SJC, Dikshit V, Yeong WY (2021) Quasi-static indentation and sound-absorbing properties of 3D printed sandwich core panels. J Sandwich Struct Mater. https://doi.org/10.1177/10996362211037015
89. Grimm D, Wösten HAB (2018) Mushroom cultivation in the circular economy. Appl Microbiol Biotechnol 102(18):7795–7803. https://doi.org/10.1007/s00253-018-9226-8
90. Gulati G, Kelly BD (2020) Domestic violence against women and the COVID-19 pandemic: what is the role of psychiatry? Int J Law Psychiatry 71:101594. https://doi.org/10.1016/j.ijlp.2020.101594
91. Gursoy D, Chi CG (2020) Effects of COVID-19 pandemic on hospitality industry: review of the current situations and a research agenda. J Hosp Market Manag 29(5):527–529. https://doi.org/10.1080/19368623.2020.1788231
92. Haneef M, Ceseracciu L, Canale C, Bayer IS, Heredia-Guerrero JA, Athanassiou A (2017) Advanced materials from fungal mycelium: fabrication and tuning of physical properties. Sci Rep 7(1):41292. https://doi.org/10.1038/srep41292
93. Hanzl M (2020) Urban forms and green infrastructure—the implications for public health during the COVID-19 pandemic. Cities & Health, 1–5. https://doi.org/10.1080/23748834.2020.1791441
94. Hepburn C, O'Callaghan B, Stern N, Stiglitz J, Zenghelis D (2020) Will COVID-19 fiscal recovery packages accelerate or retard progress on climate change? Oxf Rev Econ Policy 36(Suppl 1):S359–S381. https://doi.org/10.1093/oxrep/graa015
95. Ho C-Y, Tsai B-H, Chen C-S, Lu M-T (2021) Exploring green marketing orientations toward sustainability the hospitality industry in the COVID-19 pandemic. Sustainability 13(8). https://doi.org/10.3390/su13084348
96. Holt GA, McIntyre G, Flagg D, Bayer E, Wanjura JD, Pelletier MG (2012) Fungal mycelium and cotton plant materials in the manufacture of biodegradable molded packaging material: evaluation study of select blends of cotton byproducts. J Biobased Mater Bioenergy 6(4):431–439. https://doi.org/10.1166/jbmb.2012.1241
97. Jagadiswaran B, Alagarasan V, Palanivelu P, Theagarajan R, Moses JA, Anandharamakrishnan C (2021) Valorization of food industry waste and by-products using 3D printing: a study on the development of value-added functional cookies. Future Foods 4:100036. https://doi.org/10.1016/j.fufo.2021.100036
98. Jecker NS, Atuire CA (2021) What's yours is ours: waiving intellectual property protections for COVID-19 vaccines. J Med Ethics 47(9):595. https://doi.org/10.1136/medethics-2021-107555
99. Jones M, Mautner A, Luenco S, Bismarck A, John S (2020) Engineered mycelium composite construction materials from fungal biorefineries: a critical review. Mater Des 187:108397. https://doi.org/10.1016/j.matdes.2019.108397
100. Jones EJM, MacKinnon NJ, Tsuyuki RT (2005) Pharmaceutical care in community pharmacies: practice and research in Canada. Ann Pharmacother 39(9):1527–1533. https://doi.org/10.1345/aph.1E456

101. Karayannidis GP, Achilias DS (2007) Chemical recycling of poly(ethylene terephthalate). Macromol Mater Eng 292(2):128–146. https://doi.org/10.1002/mame.200600341
102. Kaushal V, Srivastava S (2021) Hospitality and tourism industry amid COVID-19 pandemic: perspectives on challenges and learnings from India. Int J Hosp Manag 92:102707. https://doi.org/10.1016/j.ijhm.2020.102707
103. Kietzmann J, Pitt L, Berthon P (2015) Disruptions, decisions, and destinations: enter the age of 3-D printing and additive manufacturing. Bus Horiz 58(2):209–215. https://doi.org/10.1016/j.bushor.2014.11.005
104. Kirchherr J, Reike D, Hekkert M (2017) Conceptualizing the circular economy: an analysis of 114 definitions. Resour Conserv Recycl 127:221–232. https://doi.org/10.1016/j.resconrec.2017.09.005
105. Kirk CP, Rifkin LS (2020) I'll trade you diamonds for toilet paper: consumer reacting, coping and adapting behaviors in the COVID-19 pandemic. J Bus Res 117:124–131. https://doi.org/10.1016/j.jbusres.2020.05.028
106. Koetter P, Pelton M, Gonzalo J, Du P, Exten C, Bogale K et al (2020) Implementation and process of a COVID-19 contact tracing initiative: leveraging health professional students to extend the workforce during a pandemic. Am J Infect Control 48(12):1451–1456. https://doi.org/10.1016/j.ajic.2020.08.012
107. Kreiger M, Anzalone GC, Mulder ML, Glover A, Pearce JM (2012) Distributed recycling of post-consumer plastic waste in rural areas. MRS Online Proc Libr 1492(1):101–106. https://doi.org/10.1557/opl.2013.258
108. Kreiger MA, Mulder ML, Glover AG, Pearce JM (2014) Life cycle analysis of distributed recycling of post-consumer high density polyethylene for 3-D printing filament. J Clean Prod 70:90–96. https://doi.org/10.1016/j.jclepro.2014.02.009
109. Krishtel P, Malpani R (2021) Suspend intellectual property rights for covid-19 vaccines. BMJ 373:n1344. https://doi.org/10.1136/bmj.n1344
110. Lackie K, Najjar G, El-Awaisi A, Frost J, Green C, Langlois S et al (2020) Interprofessional education and collaborative practice research during the COVID-19 pandemic: considerations to advance the field. J Interprof Care 34(5):583–586. https://doi.org/10.1080/13561820.2020.1807481
111. Larochelle RD, Mann SE, Ifantides C (2021) 3D printing in eye care. Ophthalmol Therapy 10(4):733–752. https://doi.org/10.1007/s40123-021-00379-6
112. Lashley MA, Acevedo M, Cotner S, Lortie CJ (2020) How the ecology and evolution of the COVID-19 pandemic changed learning. Ecol Evol 10(22):12412–12417. https://doi.org/10.1002/ece3.6937
113. Leiva-Martinez M-A, Anderson-Seminario MdlM, Alvarez-Risco A, Estrada-Merino A, Mlodzianowska S (2021) Price variation in lower goods as of previous economic crisis and the contrast of the current price situation in the context of COVID-19 in Peru. In: Lawrence KD, Klimberg RK (eds) Advances in business and management forecasting, vol 14. Emerald Publishing Limited, pp 161–166. https://doi.org/10.1108/S1477-407020210000014011
114. Liguori E, Winkler C (2020) From offline to online: challenges and opportunities for entrepreneurship education following the COVID-19 pandemic. Entrepreneurship Education and Pedagogy 3(4):346–351. https://doi.org/10.1177/2515127420916738
115. Lim S, Prakash A (2021) Pandemics and citizen perceptions about their country: did COVID-19 increase national pride in South Korea? Nations and Nationalism 27(3):623–637. https://doi.org/10.1111/nana.12749
116. Lipson H, Moon FC, Hai J, Paventi C (2004) 3-D printing the history of mechanisms. J Mech Des 127(5):1029–1033. https://doi.org/10.1115/1.1902999
117. Little HA, Tanikella NG, Reich MJ, Fiedler MJ, Snabes SL, Pearce JM (2020) Towards distributed recycling with additive manufacturing of PET flake feedstocks. Materials, 13(19). https://doi.org/10.3390/ma13194273
118. Lopez-Odar D, Alvarez-Risco A, Vara-Horna A, Chafloque-Cespedes R, Sekar MC (2020) Validity and reliability of the questionnaire that evaluates factors associated with perceived environmental behavior and perceived ecological purchasing behavior in Peruvian consumers. Soc Responsib J 16(3):403–417. https://doi.org/10.1108/SRJ-08-2018-0201

119. Manju R, Deepika R, Gokulakrishnan T, Srinithi K, Mohamed M (2019) A research on 3D printing concrete. Int J Recent Technol Eng 8:1691–1693
120. Maritz A, Perenyi A, de Waal G, Buck C (2020) Entrepreneurship as the unsung hero during the current COVID-19 economic crisis: Australian perspectives. Sustainability 12(11). https://doi.org/10.3390/su12114612
121. Mejía-Acosta N, Alvarez-Risco A, Solís-Tarazona Z, Matos-Valerio E, Zegarra-Arellano E, Del-Aguila-Arcentales S (2016) Reacciones adversas a medicamentos reportadas como resultado de la implementación de Atención Farmacéutica en la Farmacia Institucional DIGEMID-Ministerio de Salud de Perú. Pharm Care España 18(2):67–74. https://www.pharmcareesp.com/index.php/PharmaCARE/article/view/311
122. Mohammed M, Das A, Gomez-Kervin E, Wilson D, Gibson I (2017) EcoPrinting: investigating the use of 100% recycled acrylonitrile butadiene styrene (ABS) for additive manufacturing. https://hdl.handle.net/2134/12627581.v1
123. Mohammed MI, Wilson D, Gomez-Kervin E, Rosson L, Long J (2018, November 11–13) EcoPrinting: investigation of solar powered plastic recycling and additive manufacturing for enhanced waste management and sustainable manufacturing. 2018 IEEE conference on technologies for sustainability (SusTech).
124. Morillo-Verdugo R, Calleja-Hernández MÁ, de las Aguas Robustillo-Cortés M (2019) A new pharmaceutical care concept: more capable, motivated, and timely. Hosp Pharm 54(6):348–350. https://doi.org/10.1177/0018578719867657
125. Obergassel W, Hermwille L, Oberthür S (2021) Harnessing international climate governance to drive a sustainable recovery from the COVID-19 pandemic. Climate Policy 21(10):1298–1306. https://doi.org/10.1080/14693062.2020.1835603
126. Okereke M (2021) Towards vaccine equity: should big pharma waive intellectual property rights for COVID-19 vaccines? Public Health in Practice 2:100165. https://doi.org/10.1016/j.puhip.2021.100165
127. Otcu GB, Ramundo L, Terzi S (2019, June, 17–19) State of the art of sustainability in 3D food printing. 2019 IEEE international conference on engineering, technology and innovation (ICE/ITMC)
128. Pallottino F, Hakola L, Costa C, Antonucci F, Figorilli S, Seisto A, Menesatti P (2016) Printing on food or food printing: a review. Food Bioprc Tech 9(5): 725–733. https://doi.org/10.1007/s11947-016-1692-3
129. Pinna C, Ramundo L, Sisca FG, Angioletti CM, Taisch M, Terzi S (2016) Additive manufacturing applications within food industry: an actual overview and future opportunities. 21st Summer School Francesco Turco 2016
130. Quispe-Cañari JF, Fidel-Rosales E, Manrique D, Mascaró-Zan J, Huamán-Castillón KM, Chamorro-Espinoza SE et al (2021) Self-medication practices during the COVID-19 pandemic among the adult population in Peru: a cross-sectional survey. Saudi Pharm J 29(1):1–11. https://doi.org/10.1016/j.jsps.2020.12.001
131. Rabnawaz M, Wyman I, Auras, R, Cheng S (2017) A roadmap towards green packaging: the current status and future outlook for polyesters in the packaging industry. Green Chem 19(20):4737–4753. https://doi.org/10.1039/C7GC02521A
132. Raudenská J, Steinerová V, Javůrková A, Urits I, Kaye AD, Viswanath O, Varrassi G (2020) Occupational burnout syndrome and post-traumatic stress among healthcare professionals during the novel coronavirus disease 2019 (COVID-19) pandemic. Best Pract Res Clin Anaesthesiol 34(3):553–560. https://doi.org/10.1016/j.bpa.2020.07.008
133. Roesch E, Amin A, Gupta J, García-Moreno C (2020) Violence against women during covid-19 pandemic restrictions. BMJ 369:m1712. https://doi.org/10.1136/bmj.m1712
134. Rojas-Osorio M, Alvarez-Risco A (2019) Intention to use smartphones among peruvian university students [intention to use; Peruvian students; technology acceptance model; smartphone] 13(03):3. https://doi.org/10.3991/ijim.v13i03.9356
135. Rojas Román B, Moscoso S, Chung SA, Limpias Terceros B, Álvarez-Risco A, Yáñez JA. (2020) Tratamiento de la COVID-19 en Perú y Bolivia y los riesgos de la automedicación [COVID-19; tratamiento; estrategia; fármacos; automedicación; Perú; Bolivia] 2020 53(2). http://www.revfarmacia.sld.cu/index.php/far/article/view/435/351

136. Rojek I, Mikołajewski D, Kopowski J, Kotlarz P, Piechowiak M, Dostatni E. (2021). Reducing waste in 3D printing using a neural network based on an own elbow exoskeleton. Materials 14(17). https://doi.org/10.3390/ma14175074
137. Scott PA, Harvey C, Felzmann H, Suhonen R, Habermann M, Halvorsen K, ... Papastavrou E (2018) Resource allocation and rationing in nursing care: a discussion paper. Nurs Ethics 26(5):1528–1539. https://doi.org/10.1177/0969733018759831
138. Sekalala S, Forman L, Hodgson T, Mulumba M, Namyalo-Ganafa H, Meier BM (2021) Decolonising human rights: how intellectual property laws result in unequal access to the COVID-19 vaccine. BMJ Glob Health 6(7):e006169. https://doi.org/10.1136/bmjgh-2021-006169
139. Shaw R, Kim Y-K, Hua J (2020) Governance, technology and citizen behavior in pandemic: lessons from COVID-19 in East Asia. Progress in Disaster Science 6:100090. https://doi.org/10.1016/j.pdisas.2020.100090
140. Sigala M (2020) Tourism and COVID-19: impacts and implications for advancing and resetting industry and research. J Bus Res 117:312–321. https://doi.org/10.1016/j.jbusres.2020.06.015
141. Silva AL, Salvador GMdS, Castro SVF, Carvalho NMF, Munoz RAA (2021) A 3D printer guide for the development and application of electrochemical cells and devices [Mini Review]. Front. Chem 9(439). https://doi.org/10.3389/fchem.2021.684256
142. Sánchez OR, Vale DB, Rodrigues L, Surita FG (2020) Violence against women during the COVID-19 pandemic: an integrative review. Int J Gynecol Obstet 151(2):180–187. https://doi.org/10.1002/ijgo.13365
143. Sánchez-Clavijo LM, Martínez-Callejas SJ, Acevedo-Charry O, Diaz-Pulido A, Gómez-Valencia B, Ocampo-Peñuela N et al (2021) Differential reporting of biodiversity in two citizen science platforms during COVID-19 lockdown in Colombia. Biol Cons 256:109077. https://doi.org/10.1016/j.biocon.2021.109077
144. Tran LTT (2021) Managing the effectiveness of e-commerce platforms in a pandemic. J Retail Consum Serv 58:102287. https://doi.org/10.1016/j.jretconser.2020.102287
145. Vidya CT, Prabheesh KP (2020) Implications of COVID-19 pandemic on the global trade networks. Emerg Mark Financ Trade 56(10):2408–2421. https://doi.org/10.1080/1540496X.2020.1785426
146. Viero A, Barbara G, Montisci M, Kustermann K, Cattaneo C (2021) Violence against women in the Covid-19 pandemic: a review of the literature and a call for shared strategies to tackle health and social emergencies. Forensic Sci Int 319:110650. https://doi.org/10.1016/j.forsciint.2020.110650
147. Wang J (2021) The design research analysis of 3D printer based on FDM. J Phys: Conf Ser 1992(2):022143. https://doi.org/10.1088/1742-6596/1992/2/022143
148. Wen J, Kozak M, Yang S, Liu F (2021) COVID-19: potential effects on Chinese citizens' lifestyle and travel. Tourism Review 76(1):74–87. https://doi.org/10.1108/TR-03-2020-0110
149. Wittbrodt BT, Glover AG, Laureto J, Anzalone GC, Oppliger D, Irwin JL, Pearce JM (2013) Life-cycle economic analysis of distributed manufacturing with open-source 3-D printers. Mechatronics 23(6):713–726. https://doi.org/10.1016/j.mechatronics.2013.06.002
150. Woern AL, McCaslin JR, Pringle AM, Pearce JM (2018) RepRapable Recyclebot: open source 3-D printable extruder for converting plastic to 3-D printing filament. HardwareX 4. https://doi.org/10.1016/j.ohx.2018.e00026
151. Wu T-Y, Ford O, Rainville AJ, Bessire R (2020) COVID-19 care package distribution for senior citizens and families in Detroit and Hamtramck, Michigan. J Hunger Environ Nutr 15(4):585–587. https://doi.org/10.1080/19320248.2020.1799586
152. Yan J, Kim S, Zhang SX, Foo M-D, Alvarez-Risco A, Del-Aguila-Arcentales S, Yáñez JA (2021) Hospitality workers' COVID-19 risk perception and depression: a contingent model based on transactional theory of stress model. Int J Hosp Manag 95:102935. https://doi.org/10.1016/j.ijhm.2021.102935
153. Yasin Ar A (2020) Managing e-commerce during a pandemic: lessons from GrubHub during COVID-19. In: George B, Mahar Q (eds) International case studies in the management of

disasters. Emerald Publishing Limited, pp 155–167. https://doi.org/10.1108/978-1-83982-186-820201010
154. Yáñez JA, Afshar Jahanshahi A, Alvarez-Risco A, Li J, Zhang SX (2020) Anxiety, distress, and turnover intention of healthcare workers in Peru by their distance to the epicenter during the COVID-19 crisis. Am J Trop Med Hyg 103(4):1614–1620. https://doi.org/10.4269/ajtmh.20-0800
155. Yáñez JA, Alvarez-Risco A, Delgado-Zegarra J (2020) Covid-19 in Peru: from supervised walks for children to the first case of Kawasaki-like syndrome. BMJ 369:m2418. https://doi.org/10.1136/bmj.m2418
156. Zarocostas J (2021) What next for a COVID-19 intellectual property waiver? The Lancet 397(10288):1871–1872. https://doi.org/10.1016/S0140-6736(21)01151-X
157. Zgodavová K, Lengyelová K, Bober P, Eguren JA, Moreno A (2021) 3D printing optimization for environmental sustainability: experimenting with materials of protective face shield frames. Materials 14(21). https://doi.org/10.3390/ma14216595
158. Zhang SX, Chen J, Afshar Jahanshahi A, Alvarez-Risco A, Dai H, Li J, Patty-Tito RM (2021) Succumbing to the COVID-19 pandemic—healthcare workers not satisfied and intend to leave their jobs. Int J Ment Heal Addict. https://doi.org/10.1007/s11469-020-00418-6
159. Zhang SX, Sun S, Afshar Jahanshahi A, Alvarez-Risco A, Ibarra VG, Li J, Patty-Tito RM (2020) Developing and testing a measure of COVID-19 organizational support of healthcare workers—results from Peru, Ecuador, and Bolivia. Psychiatry Res 291:113174. https://doi.org/10.1016/j.psychres.2020.113174
160. Zhang F, Li C, Wang Z, Zhang J, Wang Y (2019) Multimaterial 3D printing for arbitrary distribution with nanoscale resolution. Nanomaterials 9(8). https://doi.org/10.3390/nano9081108
161. Zhong S, Pearce JM (2018) Tightening the loop on the circular economy: coupled distributed recycling and manufacturing with recyclebot and RepRap 3-D printing. Resour Conserv Recycl 128:48–58. https://doi.org/10.1016/j.resconrec.2017.09.023
162. Zhu X, Yuan X, Zhang Y, Liu H, Wang J, Sun B (2022) The global concern of food security during the COVID-19 pandemic: impacts and perspectives on food security. Food Chem 370:130830. https://doi.org/10.1016/j.foodchem.2021.130830
163. Zollinger R, DiCindio C (2021) Community ecology: museum education and the digital divide during and after COVID-19. Journal of Museum Education 46(4):481–492. https://doi.org/10.1080/10598650.2021.1983711

Circular Economy: Business Applications

Circular Economy for Packaging and Carbon Footprint

Sarahit Castillo-Benancio, Aldo Alvarez-Risco, Sharon Esquerre-Botton, Luigi Leclercq-Machado, Marco Calle-Nole, Flavio Morales-Ríos, María de las Mercedes Anderson-Seminario, and Shyla Del-Aguila-Arcentales

Abstract The academic and scientific literature agreed that the current economic, social, and environmental contexts require it to be treated immediately, especially in the packaging industry. Even though countries, organizations, and consumers took action to solve these problems, a change in the economic situation impacts the environmental and social ones and vice versa. To address all of them simultaneously, developing an economic model considering the three factors is necessary. This chapter discusses how a circular economy can lead to economic and sustainable growth, focusing on the packaging industry and carbon footprint. Analyzing what is currently done can also show a general picture of the current circular economy practices worldwide.

Keywords Circular economy · Circularity · Carbon footprint · Climate change · Environmental impact · Sustainable development goals · Packaging

1 Introduction

Climate concern is not new as several actions have been carried out, such as the Kyoto Protocol, which focused on covering emissions of the six leading greenhouse gases: carbon dioxide (CO_2), methane (CH_4), nitrous oxide (N_2O), hydrofluorocarbons (HFCs), perfluorocarbons (PFCs), and sulfur hexafluoride (SF_6) [137]. The world was already facing several environmental problems that required prompt attention to find solutions. Although several efforts were made, these have been altered by the arrival of COVID-19, which has generated several changes. Life as we knew it was gone forever. There may be many human and business activities that maintain some

S. Castillo-Benancio (✉) · A. Alvarez-Risco · S. Esquerre-Botton · L. Leclercq-Machado · M. Calle-Nole · F. Morales-Ríos · M. de las Mercedes Anderson-Seminario
Universidad de Lima, Lima, Perú
e-mail: aralvare@ulima.edu.pe

S. Del-Aguila-Arcentales
Escuela Nacional de Marina Mercante "Almirante Miguel Grau", Callao, Perú

A. Alvarez-Risco et al. (eds.), *Circular Economy*, Environmental Footprints and Eco-design of Products and Processes,
https://doi.org/10.1007/978-981-19-0549-0_6

pre-pandemic components, but ultimately, they now have a mix that comes from the virtual world due to the isolation of the population.

Environmental education is the basis for the population and companies to have an eco-efficient use of the different products, mainly to generate ecological preferences for ecological packaging. However, education was affected by COVID-19, creating several effects in students [9, 10, 14, 16, 51, 91, 93, 161]. These effects on people's education are seen in the coming years, and environmental education plays an active role in society.

During tourist activities, people use different types of containers and, usually, the containers used are disposable. However, several elements are used during tourism activities. With the arrival of COVID-19, tourism also underwent a change, generating various impacts [31, 36, 39, 41–43, 66, 126].

When violence against women is analyzed, a crucial issue that prevents people from having a stable life is touched. Social conflicts generate different changes in social dynamics, generating preferences and rejections. Social upheaval generates that more specific aspects such as environmental behaviors are not prioritized. Thus, it has been reported that due to COVID, several negative effects were generated [28, 47, 59, 72, 119, 133, 145].

The prices of products and services have changed significantly due to the pandemic. Thus, airline flights had different fluctuations, going from costing at least 3 times the usual price at the beginning of the pandemic to costing 50% of the price to generate the promotion of vaccination flights. Likewise, as currently observed, there are problems with containers, which has caused a tremendous logistical problem due to delays in the delivery of products, impacting prices. This impact on prices due to COVID-19 has also been reported [30, 95]. Also, it was generated the need for further research to solve the problems that COVID-19 has generated [53, 68, 77, 86, 112] and aspects such as intellectual property [22, 81, 89, 123, 155].

Due to the pandemic, the use of single-use plastics to avoid contagion increased exponentially due to the necessary protection for health professionals; at the same time, this has generated an environmental problem that already existed and has worsened. COVID-19 also generated other damage in the health sector [2, 8, 11, 20, 34, 38, 51, 58, 87, 118, 121, 154, 156–158].

Entrepreneurship has increased as a survival mechanism in the face of the crisis, generating various types of entrepreneurships, focused on solving many everyday problems of social isolation and, at the same time, based on environmental and digital issues. Ventures focused on food delivery have had to increase single-use plastics and quick-disposal packaging, with increased pollution. Entrepreneurship efforts have been diverse, including general interest in entrepreneurship from university students [23, 17, 46]. In addition, there have been other negative impacts due to COVID-19 such as in trade [4–6, 17, 19, 37, 55, 85, 100, 136, 144, 152, 159], hospitality [60, 74, 78, 84, 150], and citizens behavior [6, 24, 44, 76, 98, 117, 134, 124, 147, 148, 153, 160].

The world has changed and people's habits regarding sustainability have changed, depending on the characteristics of pre-pandemic life [13, 12, 15, 21, 25, 26, 18, 27, 56, 57, 62, 107, 120]. The world has changed and people's habits regarding sustainability have changed, depending on the characteristics of pre-pandemic life.

A circular economy intervenes as a sustainable development model that changes how we consume resources [125]. Make-use-dispose approach has been used for centuries to assume natural resources scarcity [128]. This economic model does not consider the environmental impact of the traditional production process, resulting in the continuous reduction of non-renewable natural resources [105]. Moreover, the current inexistence of effective waste management of the disposed of materials such as packaging destabilizes the planet and foments the environmental degradation of its ecosystem. The closing loop strategy behind this business model assumes the scarcity of natural resources and, consequently, looks for the production design of output such as packaging that can be reused [128], which changes the make-use-dispose model into a take-make-consume-reuse approach. Even though it sounds like a viable solution, another concern regarding performance metrics on how well countries, companies and consumers are doing arised. The Sustainable Development Goals (SDGs) became the benchmarks to ensure the environmental, economic, and societal improvement [125]. As a result, ensuring economic growth while minimizing environmental footprint requires a new business model to maximize value [69, 125]. This chapter analyzes how countries approach sustainable growth through circular economy practices based on SDGs. In this case, we focus on the packaging industry and carbon footprint, two of the essential headaches for micro- and macro-level institutions and individuals. Despite the current solid waste management strategies and policies, the closing loop production chain is still a difficult task to achieve [1].

2 A Review of the Circular Economy Framework

2.1 *Circular Economy Model Approach*

A circular economy (CE) is an economic system that is characterized by the sustainable growth of production, distribution, and consumption phases. This mechanism seeks to integrate definitions such as "sustainable" and "green" to reduce the waste generated in a linear system within the phases [92, 99, 103]. The method of application of the circular system performs not only among the technologies employed by companies, governments, and individuals, "but also changes in regulation, laws and infrastructures, industrial networks, consumption cultures, etc." [99]. Thus, technologies are constantly evolving to become increasingly resource-effective and environment-friendly,for this, the already existing linear economy is being developed [92]. However, with the reorganization of processes in companies, with an approach based on the 3 R's, "the key point here is that waste of output could be reused so energy and material could be arranged most efficiently" (reduce, recycle and reuse) [103].

The CE, in simple terms, "involves and integrates the production and consumption sides of our societies" [103]. In other words, it "is a simple, but convincing, strategy, which aims at reducing both inputs of virgin materials and output of wastes by closing economic and ecological loops of resource flows" [75]. The "model aimed at the efficient use of resources" and plans the "value retention, reduction of primary resources, and closed loops of products, product parts, and materials" in a long-term period [103]. Even if the CE offers diverse solutions that drive less invasive processes concerning the environment, there are several criticisms, which should not be understood as a method of highlighting the negative aspects, but instead that there are features that are supported and others that are not [54]. First, [143] refer that one of the main obstacles to achieving a higher level of resource recovery is the limitations present in material properties and modern manufacturing and reprocessing technologies, they proposed that "more effective reprocessing technologies is necessary to contribute to recovery value and closing material loops." Another critique made by [79] mentions that the CE defines strategies and solutions proposed by the CE "focus on large-scale manufacturing and industrial processes," besides, "while the actors, scales and sometimes sectors are identified, the practicalities and potential circular activities are not."

2.2 *Incorporating Sustainable Development Goals in the Circular Economy Model*

As part of the Sustainable Development Goals (SDGs), several indicators may be related to Circular Economy principles and strategies. Global standards are known as SDGs that promote, for instance, sustainable production and re-utilization of resources [139]. It also allows continuous monitoring and evaluation of countries' performance during a period [105] (Table 1).

The purpose of a circular economy is to encourage environmental and social prosperity and the long-term economic and sustainable growth of nations [132]. Moreover, the idea of recycling, remanufacturing, reusing, among other activities, is becoming necessary and allows the transformation of waste into a valuable asset [80]. Since the increasing environmental problems, the public and private sectors are continuously looking for new solutions [88]. Those are the main reasons why implementing the CE practices approach can lead to SDGs' accomplishment [122].

Table 1 SDGs examples related to circular economy

Sustainable Development Goal	Sub Goals Examples	Sustainable Development Goal	Sub Goals Examples
6 CLEAN WATER AND SANITATION	Improve Water Quality, Wastewater Treatment and Safe Reuse Increase Water-Use Efficiency and Ensure Freshwater Supplies Protect and Restore Water-Related Ecosystems	12 RESPONSIBLE CONSUMPTION AND PRODUCTION	Sustainable Management and use of Natural Resources Substantially Reduce Waste Generation Responsible Management of Chemicals and Waste Promote Universal Understanding of Sustainable Lifestyles Encourage Companies to Adopt Sustainable Practices
7 AFFORDABLE AND CLEAN ENERGY	Increase Global Percentage of Renewable Energy Double the Improvement in Energy Efficiency	13 CLIMATE ACTION	Build Knowledge and Capacity to meet Climate Change
8 DECENT WORK AND ECONOMIC GROWTH	Sustainable Economic Growth Improve Resource Efficiency in Consumption and Production	14 LIFE BELOW WATER	Reduce Marine Pollution Protect and Restore Ecosystems
9 INDUSTRY, INNOVATION AND INFRASTRUCTURE	Promote Inclusive and Sustainable Industrialization Upgrade All Industries and Infrastructures for Sustainability Enhance Research and Upgrade Industrial Technologies	15 LIFE ON LAND	End Deforestation and Restore Degraded Forests Protect Biodiversity and Natural Habitats Conserve and Restore Terrestrial and Freshwater Ecosystems Increase Financial Resources to Conserve and Sustainably Use Ecosystems and Biodiversity
11 SUSTAINABLE CITIES AND COMMUNITIES	Protect the World's Cultural and Natural Heritage Reduce the Environmental Impact of Cities	17 PARTNERSHIPS FOR THE GOALS	Enhance Global Partnership for Sustainable Development Promote Sustainable Technologies to Developing Countries

3 Circular Economy and Packaging

3.1 Circular Economy and Its Relationship with Packaging

The term packaging refers to a structure applied for protection, distribution, storage, presentation, use, and disuse of the enclosed goods. Packaging systems can be classified in different ways. Starting from characteristics such as the materials of its composition, manufacturing design, to its purpose, different products require different levels of handling coupled with different techniques and standards to ensure the correct shipment of the merchandise to the users [67]. Focusing on the materials, we classify them into the following categories:

- **Artificial packaging** is produced from synthetic materials obtained or processed by men whose resources may or may not be natural derivatives; examples include the production of glass, polymers (such as plastic), and/or fabrics [127].
- **Natural or ecological packaging** involves using the organic and biological matter to craft and elaborate like artisan products. This type of packaging is characterized as highly eco-friendly as they degrade quickly and do not pose a threat of becoming permanent pollutants in our environment [127].

The linear economic model, which continues to be enforced today, is based upon the principles of "take, make, waste," relying on the disposition of large, cheap, and easily accessible quantities of materials and energy [113]. However, this approach has proven not sustainable for countries that look forward to reducing their environmental footprint. Moreover, it has shown a continuous deterioration and harm toward the environment on a long-term scale. Those issues require implementing new business models such as a circular economy, which aims to optimize resource employment, reduce raw material consumption, and restate waste through a renewable flow [48]. Therefore, when applying this model to the packaging sector, we are not only focusing on the reduction of waste, but we are also switching from the traditional development processes of a product's life cycle by implementing new and more functional designs, means of distribution, and general practice, as well as the recycle and recovery of material [67].

There is an increasing need to implement a methodology on the design of industrial products where eco-design takes an impactful role, allowing to reach new opportunities with the firm's purpose of reducing production costs, the resource consumption of products, improving the quality, and increasing the shelf life of products [135]. The concept of eco-efficiency is proving to be an effective business strategy, guiding companies to assume a much more responsible role toward society and serving as a motivation for their activity to have greater competitiveness, adapting and remodeling existing production systems to the needs of the market and the environment, thus ensuring horizons of economic and social development [35]. Ecological packaging is the contribution of responsible companies to reducing waste that pollutes the environment, but they are also increasingly a requirement of consumers. A package serves to contain a product and gives it an identity. If it is made with ecological materials, it can be a compelling sales argument among new generations of environmentally conscious consumers [135].

3.2 Current Situation of Packaging in the World

After discussing the importance of the circular economy for packaging, we must understand the current situation of this industry worldwide. Let's look at two essential components of the packaging industry: plastic, paper, and cardboard.

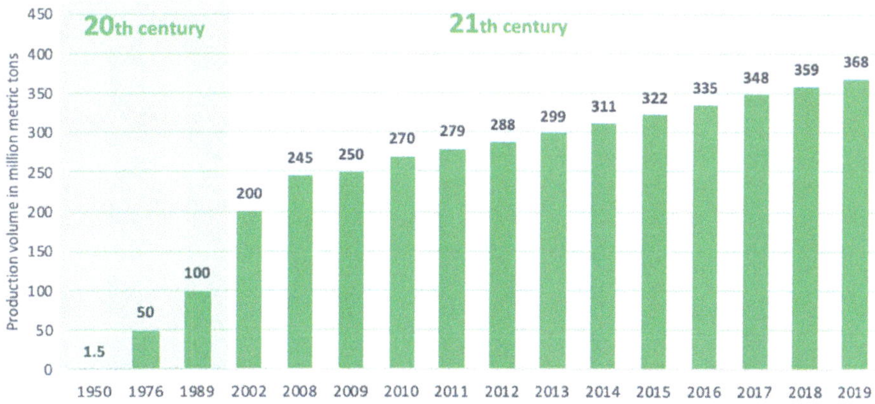

Fig. 1 Global plastic production since 1950 (*Source* [130])

3.2.1 Plastic

Plastic is one of the most common materials used for packaging nowadays. According to [130], its production has increased exponentially since 1950, rising to 368 million metric tons in 2019, and it is forecasted that by 2050 it register approximately 600 million. Geographically, the concentration of this production in 2019 was in China (30%), NAFTA (18%), the Rest of Asia (17%), and Europe (17%) [116]. If we go back to 2017, 146 million metric tons were spent on the packaging industry, making it "the biggest consumer" of the plastic industry [130] (Fig. 1).

From all these volumes, we should notice that in 2018, 54% of the 359 million metric tons of plastic produced have been discarded, 22% have been incinerated, and only 21.3% went to the recycling process [131]. As we can see, the globe has a severe problem regarding waste management. The packaging industry represented 46% of the total plastic waste generation in 2018, representing approximately 164 million metric tons of volume. For instance, the case of the European Union is fascinating. In 2018, they registered 17.2 million metric tons of plastic packaging waste, 17% more than in 2005 [131].

3.2.2 Paper and Cardboard

Focusing on the paper and cardboard industry represents a significant part of the packaging materials being used. In 2018, its production rose to 420 million metric tons [131]. 48% was produced in Asia, 26% in Europe, and 19% in North America (Fig. 2).

If we look at the paper and cardboard waste internationally, we should take the following examples:

- Singapore: Around 712 thousand tons were disposed of in 2020 [111].

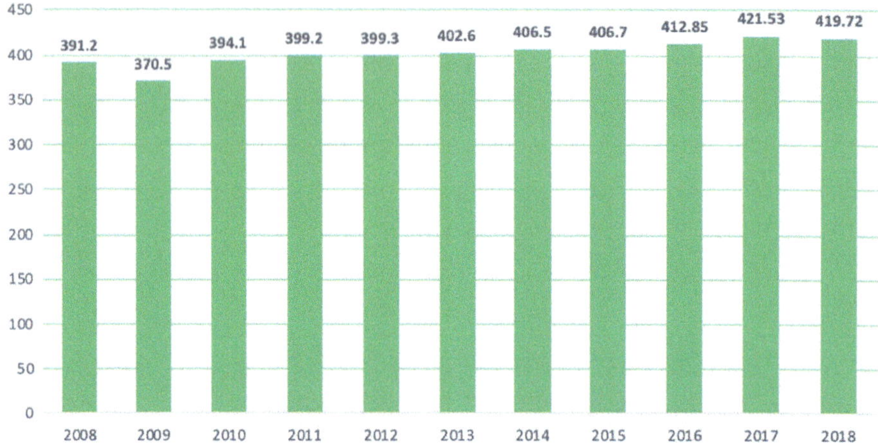

Fig. 2 Production volume of paper and cardboard worldwide 2008–2018 (in a million metric tons) (*Source* [131])

- United States: 17,220 thousand tons were landfilled in 2018 [141].
- United Kingdom: 1260 thousand tons were not recycled in 2018 (Department for Environment Food & Rural Affairs, 2018).

Waste management has several challenges, such as the increased disposal and waste generation of packaging materials such as plastic, glass, and paper [33]. Since only a single part of the million metric tons of waste is being recycled, it is significant that countries promote a sustainable development approach and reduce their environmental footprint. This economic approach, known as Circular Economy, aims to reduce the dependence on natural resources through waste output management [99].

3.3 Circular Economy for Packaging in the World

Governments can play a significant role in promoting and developing a circular economy. For instance, several countries changed their perspective from waste management to circular economy principles [64, 104]. That is mainly because the traditional linear business model is no longer an option and is inefficient [104]. We look at some examples of how countries are implementing this concept.

3.3.1 China

To begin with, we should first talk about China's perspective. This country represents 31% of the world's plastic production [130]. As a result, implementing circular economy principles as mandatory objectives is complex. One answer to this dilemma is starting to move from an environmental policy to an economic development one, including these last [104]. For instance, the government decided to practice the "National Sword" policy, which banned imports of several recyclable materials such as plastic [7]. This step was aimed to reduce the current flow of waste into the country.

Other's policies implemented by the government are the following:

- Law for Environmental Pollution of Solid Waste
- Amended Law on Pollution Prevention and Control of Solid Waste
- Laws and Regulations for Reuse and Recycling Specific Solid Waste
- Amended Law of the Prevention and Control of Environmental Pollution by Solid Waste
- Environmental Protection Tax Law [113].

In this last, circular economy is focused on 3 R activities: Reducing resources consumption, Reuse, and recycling waste materials [97]. All those initiatives are based on the Circular Economy Promotion Law that came into force in 2009 [106]. The aim of the implementation of the Circular Economy policy is mainly due to the scarcity of natural resources and emissions of gases such as the greenhouse one. All the issues facing this country are caused by its rapid industrial and economic development in the last decades [113]. Focusing now on performance results over time, we can see that China's paper and cardboard recycling rate increased from 42.9% in 2009 to 49% in 2019 [129]. The plastic recycling rate has decreased from 28% in 2017 to 17.6% in 2020, mainly due to the import ban of plastic materials [131].

3.3.2 Germany

Germany consumes 2019 around 19 million tons of packaging, of which 44% and 18% are for paper and plastic, respectively [131]. This number of materials can generate significant waste if not managed adequately. That is one of the reasons for the implementation of the Circular Economy Act in 1996. This Act allows continuous packaging collection from industrial and end consumers. It was aimed to foment and implement a sustainable production system where producers develop products that minimize waste, facilitate its recovery for further recycling, and allow re-utilization of this last [113]. Other policies implemented by the German government are the following:

- Waste Disposal Act in 1992
- Producer Responsibility for Packaging waste in 1991
- Packaging Ordinance in 1998
- Ordinance on managing the municipal waste of commercial origin and specific construction and demolition waste in 2002 [113].

According to [131], since 1999, the country has experienced excellent collection practices. In 2019, it registered that almost 12 million metric tons of packaging were collected. Also, its recycling rate has risen from 81% in 2000 up to 97% in 2018

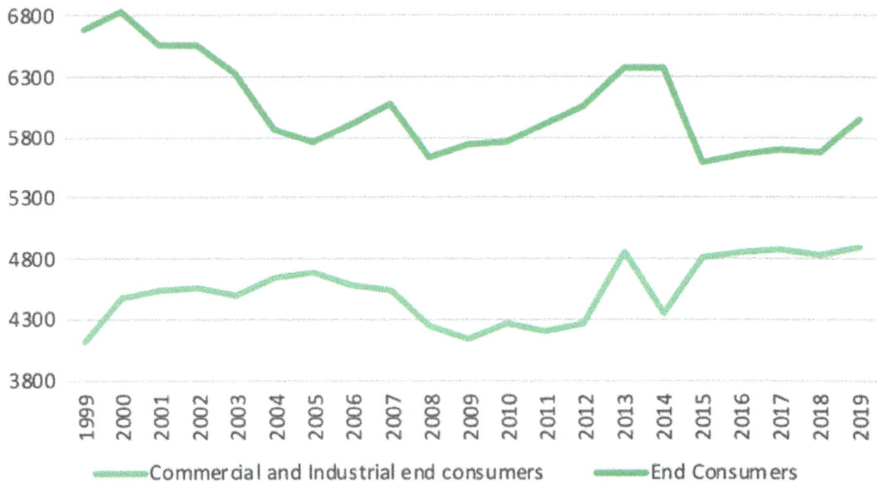

Fig. 3 Volume of packaging collected by type of consumers in Germany (in thousands of metric tons) (*Source* [131])

[131]. Specifically for each type of packaging, the recycling rate was the following: plastic packaging with 47.7%, glass packaging with 83%, and 99,7% for paper and cardboard packaging [131] (Fig. 3).

3.3.3 Other Countries

Around the world, we have evidenced similar actions implemented by other countries (see Table 2). As we can see, several strategies adopted worldwide were aimed to regulate the inefficient use of resources and waste [65]. Consequently, the traditional linear economy model is ineffective and focuses on the short term [105]. The idea of developing circular economy practices is to ensure the implementation of a *closed-loop system*, which means the continuous circulation of materials to avoid waste [132]. The minimization of the input of virgin raw materials and output of waste is the principle of this sustainable model, allowing the protection of the environment and simultaneously obtaining the desired performance metrics to achieve SDGs [75].

To sum up, a circular economy can indeed enhance sustainable development. Nonetheless, each government should implement politics to achieve economic and sustainable growth. Moreover, several authors explained the current difficulties in developing circular economy practices. Even though governments continuously issue new regulations, the current limitations are also related to organizations and consumers [54]. For instance, several business models appear with a circular approach but only focus on specific activities, which can significantly reduce the effectiveness of the total circular economy. It may be related to the unclear enactments from countries.

Table 2 Strategies adopted by other countries worldwide

Country	Propositions to achieve circular economy	Sources
Japan	Waste disposal Law in 1970 Resource Efficient Law in 1991 Environmental Law in 1993 Recycling Based Society Law in 2002	[82, 113]
Australia	National Waste Policy in 2009 Environment Protection Act in 2002 Australian Packaging Covenant National Environmental Protection Measure in 2011	[32, 73]
Peru	Law 3084—Regulation of single-use plastics and disposable containers	[19]
Brazil	Brazilian Policy of Solid Waste in 2010	[71]
South Korea	Framework Act on Resource Circulation 2016	[65]
United Kingdom	Producer Responsibility Obligations (Packaging Waste) in 2007 Packaging Regulations in 2015 Environmental Permitting Regulation in 2016	CMS [40, 94]
Colombia	Environmental Management Plans for Container and Packaging Waste (Amended in 2020)	[29]
European Union	European Green Deal	European Commission (n.d.)

4 Circular Economy for Carbon Footprint

4.1 Circular Economy and Its Relationship with Carbon Footprint

In recent years, conferences have been held on topics of global concern such as environmental pollution and ways to combat it. It should be noted that the Kyoto Protocol was approved in 1997, but it has been in force since 2005, and 192 parties have signed up to it; industrialized countries have committed themselves to limiting and reducing their GHG emissions [140]. It is also important to note that the first commitment period for all parties lasted from 2008 to 2012, a second period was signed for 2013–2020, the Doha amendment in 2012, although it has not yet entered into force, which is because there is global concern about the negative consequences that harm not only society but also business and the economy. Therefore, to calculate the amount of greenhouse gas emissions that accelerate climate change, the carbon footprint was created, which affects the economic development of nations [63]. In addition, the 6 GHGs established in the Kyoto Protocol are also considered for measurement. The main objective is to reduce emissions to slow down global warming and reduce pollution [61]. For example, China is a highly developed nation, but its cities are on the world's worst air quality lists due to their high pollution rates [101]. Since the nation entered into the World Trade Organization and its early economic recovery, $C0_2$ emissions have only increased [50]. Since 2013, the country has been conducting an anti-smog campaign, which has begun to bear fruit, such as in Beijing, which introduces blue skies [96].

According to [102], a new concept for all people was born out of the 2012 publication by the Ellen MacArthur Foundation in the United States on how to integrate environmental sustainability into the framework of economic development, which promoted the circular economy as supporting research for businesses and government entities. The old model was the linear economy model in which the extraction of natural resources, production of goods and services, consumption, and subsequent waste production is paramount, and it is in this last part that the result is pollution and an increase in GHGs [49]. On the other hand, the new CE model proposes a reconversion of waste by promoting innovation in the production process, thus converting it into a new reusable resource through renewable energy and product design [45]. Nowadays, companies that monitor their carbon emission levels have a more significant advantage over their competitors, as they can plan and take measures to reduce their measures, which results in high energy efficiency, which in the long run increases productivity and makes them more competitive not only in the national but also in the international market. In addition, companies with "green" certificates are more accepted by end customers [61].

Finally, something important to consider is that incorporating SDGs as strategic objectives by countries and companies can reduce contamination such as greenhouse gas emissions [110]. Since the circular economy approach is also based on a low-carbon strategy, this ensures economic and sustainable growth by meeting SDGs regarding carbon footprint [149].

4.2 Current Situation of Carbon Footprint in the World

Climate change is currently one of the main problems being faced globally, a problem originating from the product of human activities and the subsequent pressure exerted on surrounding ecosystems. Global warming is one of the significant issues, denoting a rapid development due to the continuous increase in CO_2 emissions globally. In 2019, CO_2 emissions reached 36,45 million tons [114]. Regionally, the United States (14.48%), China (27.9%), European Union (8%), and India (7.1%) are the leading territories emitting this component. Nonetheless, the current change regarding growth was around 0.6%, lower compared to 2018 (Fig. 4).

Focusing on plastic, a packaging raw material, carbon dioxide emissions related to GHG were around 0.86 billion tons in 2019 [129]. It was forecasted that by 2030 and 2050, it would reach 1.34 billion and 2.8 billion tons of carbon dioxide, respectively. Nonetheless, this increase is not seen everywhere. Taking a step back, specifically in 1990, the UK registered around 4,4 million tons of carbon dioxide in plastic product manufacturing. Nowadays, in 2019, it fell to 2,5 million tons of this last [129]. As

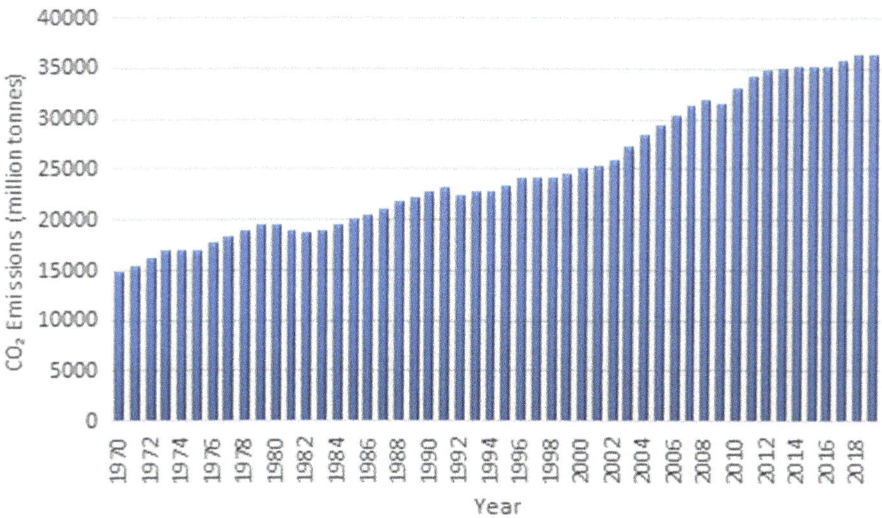

Fig. 4 CO_2 emissions in the world (million tons) (*Source* [114])

we can see, we can see an increase in GHG emissions, and in some countries, we can evidence a reduction in carbon footprint. It is then necessary to see what is happening worldwide and how countries address this problem.

4.3 Circular Economy for Carbon Footprint in the World

Governments can also play a significant role in the carbon footprint by implementing different laws and strategies. Policies made by this last ensure new sustainable practices that may reduce, for instance, greenhouse gas emissions [115]. Regulations must be clear, strong, and achievable so companies can enhance their processes, add value, take a long-term perspective, among other elements. The adoption of CE practices, as a result, requires the participation of the governments by enacting acts, strategies, regulations, laws, and policy such as the followings (Table 3).

Focusing now on consumers and companies, according to the United Nations—Department of Economic and Social Affairs (2019), the global population is estimated to continue growing to about 9.7 billion in the year 2025, which simultaneously incentivizes companies to produce more and expand their production capacity. This growth generates more significant pressure on the use of natural resources and a greater demand for consumer goods and services, causing a greater volume of waste to be cast into the environment. However, evaluating greenhouse gas emissions associated with products is rather complex. To define the footprint in its entire dimension, it is also necessary to consider consumers' responsibility in this process through their purchasing power, which could be considered one of the main causes of

Table 3 Strategies adopted by other countries worldwide regarding carbon footprint

Country	Propositions for carbon footprint	Sources
USA	Federal Water Pollution Control Act Amendments in 1972 Clean Air Act Amendments in 1990 The Federal Insecticide, Fungicide, and Rodenticide Act Amendments in 2003 Energy Independence and Security Act in 2007 Climate Adaptation Plan in 2021	[141]
China	Renewable Energy Law in 2006 Energy Conservation Law in 2007 National Plan for Combating Climate Change in 2007	[83, 151]
India	The Water (Prevention and Control of Pollution) Act in 1974 The Air (Prevention and Control of Pollution) Act in 1981 The Environment Protection Act in 1986 The National Green Tribunal Act in 2010 Hazardous Waste Management Rules in 2016	[142]
Russia	Environmental Protection Law in 2002 Environmental Doctrine in 2002 Climatic Doctrine in 2009 National Climate Change Adaptation Plan in 2019 Climate Project Concept 2020 Presidential Decree No. 666 to limit GHG emissions in 2020	[108, 109]
Japan	Energy Conservation Act in 1979 Joint Crediting Mechanism with India in 2013 Law Concerning Promotion of Measures to cope with Global Warming in 1998	[52, 70, 90]

the carbon footprint generated by certain goods or services. Therefore, developing countries find it more challenging to meet the demand for goods and services, to which export markets traditionally agreed, since these markets are raising environmental demands, circular, low carbon and could, if not anticipated, be transformed into new trade barriers [138, 146].

However, during the first half of 2020, the world suffered the severe effects of the health crisis caused by the COVID-19 virus, which has transcended a deep global economic crisis after the collapse of markets for weeks of inactivity in various sectors [138, 146]. So, in compensation for the difficulties that may be faced in the current markets, demand is expected to be significantly expanded in emerging markets which poses significant opportunities in the long term. The improvements in developing countries' productivity via the circular economy allow addressing these growing demands and increasing trade flows between developing countries, without the implications of a correlative increment in pollution, degradation of the environment, growth in greenhouse gas emissions, and potential depletion of resources in those countries.

In other words, the world is currently going through a period of profound transformation caused by the combination of a climate crisis, a latent social crisis exacerbated by the current health emergency, and an economic crisis from which we are estimated to recover for another decade. Therefore, it is urgent to lay the foundations and chart the route toward the transition to a circular economy that ensures sustainable development for the long term and economic recovery for years to come. Thus, the industrial sector should aim to accelerate its transition in implementing innovation, clean and cost-efficient technologies, and the design of ensuing professional opportunities. At the same time, markets have to reconfigure themselves in the creation of enabling mechanisms for this new economy, which, on the one hand, promote and decouple economic growth from environmental impact, and on the other hand, contribute to the closing of social inequality gaps that today play a determining role in a reality of insecurity and social precariousness for various countries in the world [138, 146].

4.3.1 Closing Remarks

There is no clear evidence about the correct development of a circular economy. Many countries adopted a specific approach according to their needs and strategic goals. Thus, implementing a Circular Economy can be done either at the micro- or macro-level. Nonetheless, it requires severe planning and continuous improvement of processes that may take years. Also, not every strategy can be implemented in each country, making it crucial to enact clear and achievable goals to ensure the scalability of the circular economy. Sustainable Development Goals regarding packaging are an efficient way to measure the environmental performance of countries and companies. These last accomplishments lead to economic and sustainable growth, resulting in a more excellent quality of life and well-being. A circular economy model shows us the importance of designing packaging to be reused, recycled, repaired, recollected, redesigned, among other strategies, which ensure the minimization of the linear model, which is harming the biosphere with its take-make-waste approach. Also, it increases the efficient use of natural resources, reduces waste generation, keeps long-term added value, and, as a result, a closed-loop system.

References

1. Aarnio T, Hämäläinen A (2008) Challenges in packaging waste management in the fast food industry. Resour Conserv Recycl 52(4):612–621. https://doi.org/10.1016/j.resconrec.2007.08.002
2. Abiakam N, Worsley P, Jayabal H, Mitchell K, Jones M, Fletcher J et al (2021) Personal protective equipment related skin reactions in healthcare professionals during COVID-19. Int. Wound J. 18 (3): 312–322. https://doi.org/10.1111/iwj.13534
3. Abudureheman A, Nilupaer A (2021) Optimization model design of cross-border e-commerce transportation path under the background of prevention and control of COVID-19 pneumonia. Soft Comput 25(18):12007–12015. https://doi.org/10.1007/s00500-021-05685-6
4. Acuña-Zegarra MA, Santana-Cibrian M, Velasco-Hernandez JX (2020) Modeling behavioral change and COVID-19 containment in Mexico: a trade-off between lockdown and compliance. Math Biosci 325. https://doi.org/10.1016/j.mbs.2020.108370

5. Aguirre AA, Catherina R, Frye H, Shelley L (2020) Illicit wildlife trade, wet markets, and COVID-19: preventing future pandemics. World Med & Health Policy 12(3):256–265. https://doi.org/10.1002/wmh3.348
6. Ahsan MM (2020) Strategic decisions on urban built environment to pandemics in Turkey: lessons from COVID-19. J Urban Manag 9(3):281–285. https://doi.org/10.1016/j.jum.2020.07.001
7. Alademi R (2020) The national sword policy of China and its effect on global policy. Retrieved 12/09/2021 from https://www.researchgate.net/publication/340439578_The_National_Sword_Policy_of_China_and_its_Effect_on_Global_Policy
8. Alan H, Eskin Bacaksiz F, Tiryaki Sen H, Taskiran Eskici G, Gumus E, Harmanci Seren AK (2021) "I'm a hero, but…": an evaluation of depression, anxiety, and stress levels of frontline healthcare professionals during COVID-19 pandemic in Turkey. Perspect Psychiatr Care 57(3):1126–1136. https://doi.org/10.1111/ppc.12666
9. Ali W (2020) Online and remote learning in higher education institutes: a necessity in light of COVID-19 pandemic. High Educ Stud 10(3):16–25
10. Allen J, Rowan L, Singh P (2020) Teaching and teacher education in the time of COVID-19. Asia-Pac J Teach Educ 48(3):233–236. https://doi.org/10.1080/1359866X.2020.1752051
11. Alsairafi Z, Naser AY, Alsaleh FM, Awad A, Jalal Z (2021) Mental health status of healthcare professionals and students of health sciences faculties in Kuwait during the COVID-19 pandemic. Int J Environ Res Public Health 18 (4). https://doi.org/10.3390/ijerph18042203
12. Alvarez-Risco A, Del-Aguila-Arcentales S, Delgado-Zegarra J, Yáñez JA, Diaz-Risco S (2019) Doping in sports: findings of the analytical test and its interpretation by the public. Sport Sci Health 15(1):255–257. https://doi.org/10.1007/s11332-018-0484-8
13. Alvarez-Risco A, Del-Aguila-Arcentales S, Stevenson JG (2015) Pharmacists and mass communication for implementing pharmaceutical care. Am J Pharm Benefits 7(3):e125–e126
14. Alvarez-Risco A, Del-Aguila-Arcentales S, Yáñez JA, Rosen MA, Mejia CR (2021) Influence of technostress on academic performance of university medicine students in Peru during the covid-19 pandemic. Sustainability 13(16):8949. https://doi.org/10.3390/su13168949
15. Alvarez-Risco A, Delgado-Zegarra J, Yáñez JA, Diaz-Risco S, Del-Aguila-Arcentales S (2018) Predation risk by gastronomic boom—case Peru. J Landsc Ecol 11(1):100–103. https://doi.org/10.2478/jlecol-2018-0003
16. Alvarez-Risco A, Estrada-Merino A, Rosen MA, Vargas-Herrera A, Del-Aguila-Arcentales S (2021) Factors for implementation of circular economy in firms in covid-19 pandemic times: the case of Peru. Environments 8(9):95. https://doi.org/10.3390/environments8090095
17. Alvarez-Risco A, Mlodzianowska S, García-Ibarra V, Rosen MA, Del-Aguila-Arcentales S (2021) Factors affecting green entrepreneurship intentions in business university students in covid-19 pandemic times: case of ecuador. Sustainability 13(11):6447. https://doi.org/10.3390/su13116447
18. Alvarez-Risco A, Quiroz-Delgado D, Del-Aguila-Arcentales S (2016) Pharmaceutical care in hypertension patients in a peruvian hospital. Indian J Public Health Res Dev 7(3):183–188
19. Alvarez-Risco A, Rosen MA, Del-Aguila-Arcentales S (2020) A new regulation for supporting a circular economy in the plastic industry: the case of Peru (short communication). J Landsc Ecol 13(1):1–3. https://doi.org/10.2478/jlecol-2020-0004
20. Alvarez-Risco A, Dawson J, Johnson W, Conteh-Barrat M, Aslani P, Del-Aguila-Arcentales S, Diaz-Risco S (2021) Ebola virus disease outbreak: a global approach for health systems [Ebola; health professionals; global approach; pharmacist: pharmaceutical care; COVID-19] 53 (4). http://www.revfarmacia.sld.cu/index.php/far/article/view/491
21. Alvarez-Risco A, Del-Aguila-Arcentales S (2015) Errores de prescripción como barrera para la Atención Farmacéutica en establecimientos de salud públicos: Experiencia Perú. Pharmaceutical Care España 17 (6): 725–731. https://www.pharmcareesp.com/index.php/PharmaCARE/article/view/246

22. Alvarez-Risco A, Del-Aguila-Arcentales S (2021a) A note on changing regulation in international business: the World Intellectual Property Organization (WIPO) and artificial intelligence. In: Verbeke A, van Tulder R, Rose EL, Wei Y (eds) The multiple dimensions of institutional complexity in international business research, vol 15. Emerald Publishing Limited, pp 363–371. https://doi.org/10.1108/S1745-886220210000015020
23. Alvarez-Risco A, Del-Aguila-Arcentales S (2021b) Public policies and private efforts to increase women entrepreneurship based on stem background. In Mari M, Poggesi S, Foss L (eds) *Women's entrepreneurship in STEM disciplines: issues and perspectives*. Springer International Publishing, pp 75–87. https://doi.org/10.1007/978-3-030-83792-1_5
24. Alvarez-Risco A, Mejia CR, Delgado-Zegarra J, Del-Aguila-Arcentales S, Arce-Esquivel AA, Valladares-Garrido MJ, Yáñez JA (2020) The Peru approach against the COVID-19 infodemic: insights and strategies. Am J Trop Med Hyg 103(2):583–586. https://doi.org/10.4269/ajtmh.20-0536
25. Alvarez-Risco A, Mil JWFv (2007) Pharmaceutical care in community pharmacies: practice and research in Peru. Ann Pharmacother 41(12): 2032–2037. https://doi.org/10.1345/aph.1K117
26. Alvarez-Risco A, Turpo-Cama A, Ortiz-Palomino L, Gongora-Amaut N, Del-Aguila-Arcentales S (2016) Barreras para la implementación de la Atención Farmacéutica en establecimientos farmacéuticos de Cusco, Perú. Pharmaceutical Care España 18 (5): 194–205. https://www.pharmcareesp.com/index.php/PharmaCARE/article/view/326
27. Álvarez-Risco A, Zegarra Arellano E, Matos Valerio E, Mejía Acosta N, Solis Tarazona Z (2013) Campaña de atención farmacéutica como estrategia de implementación de los servicios farmacéuticos: Experiencia Perú. Pharmaceutical Care España 15 (1): 35. https://www.pharmcareesp.com/index.php/PharmaCARE/article/view/105
28. Amarillo CR (2020) Feminism on lockdown. NACLA Rep Am 52(3):274–281. https://doi.org/10.1080/10714839.2020.1809084
29. ANLA (2021) Container and packaging waste environmental management plans [Planes de Gestión Ambiental de Residuos de Envases y Empaques]. Retrieved 12/09/2021 from https://www.anla.gov.co/permiso-y-autorizacion-plan-gestion-ambiental-residuos-envaces-y-empaques
30. Apcho-Ccencho L-V, Cuya-Velásquez B-B, Alvarado Rodríguez D, Anderson-Seminario MdlM, Alvarez-Risco A, Estrada-Merino A, Mlodzianowska S (2021) The impact of international price on the technological industry in the united states and china during times of crisis: commercial war and COVID-19. In: Lawrence KD, Klimberg RK (eds) Advances in business and management forecasting, vol. 14. Emerald Publishing Limited, pp. 149–160. https://doi.org/10.1108/S1477-407020210000014010
31. Assaf A, Scuderi R (2020) COVID-19 and the recovery of the tourism industry. Tour Econ 26(5):731–733. https://doi.org/10.1177/1354816620933712
32. Australian Capital Territory (2020) Waste management and resource recovery (Environment protection—used packaging materials) Code of practice 2020. Retrieved 12/09/2021 from https://www.legislation.act.gov.au/DownloadFile/di/2020-256/current/PDF/2020-256.PDF
33. Balwada J, Samaiya S, Mishra RP (2021) Packaging plastic waste management for a circular economy and identifying a better waste collection system using Analytical Hierarchy Process (AHP). Procedia CIRP 98:270–275. https://doi.org/10.1016/j.procir.2021.01.102
34. Barello S, Caruso R, Palamenghi L, Nania T, Dellafiore F, Bonetti L et al. (2021) Factors associated with emotional exhaustion in healthcare professionals involved in the COVID-19 pandemic: an application of the job demands-resources model. Int Arch Occup Environ Health 94 (8): 1751–1761. https://doi.org/10.1007/s00420-021-01669-z
35. Basuki B (2015) Eco-efficiency and sustainable development as efforts to produce environmentally friendly product: an exploratory case study. Issues Soc Environ Account 9(3):199–218
36. Baum T, Hai NTT (2020) Hospitality, tourism, human rights and the impact of COVID-19. Int J Contemp Hosp Manag 32(7):2397–2407. https://doi.org/10.1108/IJCHM-03-2020-0242

37. Borzée A, McNeely J, Magellan K, Miller J RB, Porter L, Dutta T et al. (2020) COVID-19 Highlights the need for more effective wildlife trade legislation. Trends Ecol & Evol 35 (12): 1052–1055. https://doi.org/10.1016/j.tree.2020.10.001
38. Bozdağ F, Ergün N (2020) Psychological resilience of healthcare professionals during covid-19 pandemic. Psychol Rep 124(6):2567–2586. https://doi.org/10.1177/0033294120965477
39. Brouder P (2020) Reset redux: possible evolutionary pathways towards the transformation of tourism in a COVID-19 world. Tour Geogr 22(3):484–490. https://doi.org/10.1080/14616688.2020.1760928
40. CMS LAW (2020) *Plastic and packaging waste laws in the United Kingdom.* Retrieved 12/08/2021 from https://cms.law/en/int/expert-guides/plastics-and-packaging-laws/united-kingdom
41. Carvache-Franco M, Alvarez-Risco A, Carvache-Franco W, Carvache-Franco O, Estrada-Merino A, Rosen MA (2021) Coastal cities seen from loyalty and their tourist motivations: a study in lima Peru. Sustainability 13(21):11575. https://doi.org/10.3390/su132111575
42. Carvache-Franco M, Carvache-Franco O, Carvache-Franco W, Alvarez-Risco A, Estrada-Merino A (2021) Motivations and segmentation of the demand for coastal cities: a study in lima Peru. Int J Tour Res 23(4):517–531. https://doi.org/10.1002/jtr.2423
43. Carvache-Franco M, Alvarez-Risco A, Carvache-Franco O, Carvache-Franco W, Estrada-Merino A, Villalobos-Alvarez D (2021) Perceived value and its influence on satisfaction and loyalty in a coastal city: a study from Lima, Peru. Journal of Policy Research in Tourism, Leisure and Events 1–16. https://doi.org/10.1080/19407963.2021.1883634
44. Cegarra-Navarro J-G, Vătămănescu E-M, Martínez-Martínez A (2021) A context-driven approach on coping with COVID-19: from hiding knowledge toward citizen engagement. Knowl Process Manag 28(2):134–140. https://doi.org/10.1002/kpm.1662
45. Cerdá E, Khalilova A (2016) Economía Circular, estrategia y competitividad empresarial. Retrieved 12/09/2021 from https://www.mincotur.gob.es/Publicaciones/Publicacionesperiodicas/EconomiaIndustrial/RevistaEconomiaIndustrial/401/CERD%C3%81%20y%20KHALILOVA.pdf
46. Chafloque-Cespedes R, Alvarez-Risco A, Robayo-Acuña P-V, Gamarra-Chavez C-A, Martinez-Toro G-M, Vicente-Ramos W (2021) Effect of sociodemographic factors in entrepreneurial orientation and entrepreneurial intention in university students of Latin American business schools. In: Jones P, Apostolopoulos N, Kakouris A, Moon C, Ratten V, Walmsley A (eds) Universities and entrepreneurship: meeting the educational and social challenges, vol. 11. Emerald Publishing Limited, pp 151–165. https://doi.org/10.1108/S2040-724620210000011010
47. Chafloque-Cespedes R, Vara-Horna A, Asencios-Gonzales Z, Lopez-Odar DR, Alvarez-Risco A, Quipuzco-Chicata, L et al. (2020) Presentismo académico y violencia contra las mujeres en escuelas de negocios e ingeniería en universidades peruanas. Lecturas de Economía 0 (93): 127–153. https://doi.org/10.17533/udea.le.n93a340726
48. Chaniotaki KG (2020) Food packaging and circular economy in the Netherlands: challenges and policy solutions.
49. Chaves Ávila R, Monzón Campos JL (2018) La economía social ante los paradigmas económicos emergentes: innovación social, economía colaborativa, economía circular, responsabilidad social empresarial, economía del bien común, empresa social y economía solidaria. CIRIEC-España Revista de economía pública, social y cooperativa 93: 5–50
50. Chen S, Lin F, Yao X, Zhang P (2020) WTO accession, trade expansion, and air pollution: evidence from China's county-level panel data. Rev Int Econ 28(4):1020–1045. https://doi.org/10.1111/roie.12480
51. Chen T, Peng L, Yin X, Rong J, Yang J, Cong G (2020) Analysis of user satisfaction with online education platforms in China during the COVID-19 pandemic. Healthcare 8 (3). https://doi.org/10.3390/healthcare8030200
52. Chotimah HC, Arisanto PT, Pratiwi TS (2020) Bilateral agreement of Indonesia-Japan for low carbon growth cooperation: an analysis of the effectiveness and the compliance level. Nation State: Journal of International Studies 3(2):98–113

53. Chung SA, Olivera S, Rojas Román B, Alanoca E, Moscoso S, Limpias Terceros B (2021) Temáticas de la producción científica de la Revista Cubana de Farmacia indizada en Scopus (1967–2020) [Revista Cubana de Farmacia; farmacia; Cuba; evidencia] 54 (1). http://www.revfarmacia.sld.cu/index.php/far/article/view/511
54. Corvellec H, Stowell AF, Johansson N (2021) Critiques of the circular economy. *Journal of Industrial Ecology*, *n/a*(n/a). https://doi.org/10.1111/jiec.13187
55. Cruz-Torres W, Alvarez-Risco A, Del-Aguila-Arcentales S (2021) Impact of Enterprise Resource Planning (ERP) implementation on performance of an education enterprise: a Structural Equation Modeling (SEM). Stud Bus Econ 16(2):37–52. https://doi.org/10.2478/sbe-2021-0023
56. Del-Aguila-Arcentales S, Alvarez-Risco A (2013) Human error or burnout as explanation for mistakes in pharmaceutical laboratories. Accred Qual Assur 18(5):447–448. https://doi.org/10.1007/s00769-013-1000-0
57. Delgado-Zegarra J, Alvarez-Risco A, Yáñez, JA (2018) Uso indiscriminado de pesticidas y ausencia de control sanitario para el mercado interno en Perú. Rev Panam Salud Publica 42: e3. https://doi.org/10.26633/RPSP.2018.3
58. Deressa W, Worku A, Abebe W, Gizaw M, Amogne W (2021) Risk perceptions and preventive practices of COVID-19 among healthcare professionals in public hospitals in Addis Ababa Ethiopia. PLoS ONE 16(6). https://doi.org/10.1371/journal.pone.0242471
59. Dominelli L (2021) A green social work perspective on social work during the time of COVID-19. Int J Soc Welf 30(1):7–16. https://doi.org/10.1111/ijsw.12469
60. Duarte Alonso A, Kok SK, Bressan A, O'Shea M, Sakellarios N, Koresis A et al. (2020) COVID-19, aftermath, impacts, and hospitality firms: An international perspective. Int J Hosp Manag 91: 102654. https://doi.org/10.1016/j.ijhm.2020.102654
61. Economic Commission for Latin America and the Caribbean (2013) Economic Commission for Latin America and the Caribbean (ECLAC): biennial report (2010–2011). Retrieved 12/09/2021 from https://digitallibrary.un.org/record/755139
62. Enciso-Zarate A, Guzmán-Oviedo J, Sánchez-Cardona F, Martínez-Rohenes D, Rodríguez-Palomino JC, Alvarez-Risco A et al. (2016) Evaluación de la contaminación con agentes citotóxicos en hospitales en Colombia. Pharmaceutical Care España 18 (6): 241–250. https://www.pharmcareesp.com/index.php/PharmaCARE/article/view/354
63. Estrada-Merino A (2020) La huella de carbono y las empresas: Reflexiones en la actual coyuntura. Retrieved 12/09/2021 from https://repositorio.ulima.edu.pe/bitstream/handle/20.500.12724/11613/Estrada_huella_de_carbono_y_empresas.pdf?sequence=1&isAllowed=y
64. Ezeudu OB, Ezeudu TS (2019) Implementation of circular economy principles in industrial solid waste management: case studies from a developing economy (Nigeria). Recycling 4 (4). https://doi.org/10.3390/recycling4040042
65. Fitch-Roy O, Benson D, Monciardini D (2021) All around the world: Assessing optimality in comparative circular economy policy packages. J Clean Prod 286. https://doi.org/10.1016/j.jclepro.2020.125493
66. Fotiadis A, Polyzos S, Huan T-CTC (2021) The good, the bad and the ugly on COVID-19 tourism recovery. Ann Tour Res 87. https://doi.org/10.1016/j.annals.2020.103117
67. French Packaging Council (2014) Packaging & circular economy: a case study of the circular economy model. Paris: CNE. Retrieved 12/09/2021 from https://circulareconomy.europa.eu/platform/sites/default/files/packaging-and-circular-economy-final-report-en-september-2014.pdf
68. Galanakis CM, Rizou M, Aldawoud TMS, Ucak I, Rowan NJ (2021) Innovations and technology disruptions in the food sector within the COVID-19 pandemic and post-lockdown era. Trends Food Sci Technol 110:193–200. https://doi.org/10.1016/j.tifs.2021.02.002
69. Ghosh SK (2020) Circular economy: global perspective. Springer
70. Gokhale H (2021) Japan's carbon tax policy: limitations and policy suggestions. Curr Res Environ Sustain 3. https://doi.org/10.1016/j.crsust.2021.100082
71. Guarnieri P, Cerqueira-Streit JA, Batista LC (2020) Reverse logistics and the sectoral agreement of packaging industry in Brazil towards a transition to circular economy. Resour Conserv Recycl 153. https://doi.org/10.1016/j.resconrec.2019.104541

72. Gulati G, Kelly BD (2020) Domestic violence against women and the COVID-19 pandemic: what is the role of psychiatry? Int J Law Psychiatry 71. https://doi.org/10.1016/j.ijlp.2020.101594
73. Gumley W (2014) An analysis of regulatory strategies for recycling and re-use of metals in Australia. Resources 3 (2). https://doi.org/10.3390/resources3020395
74. Gursoy D, Chi CG (2020) Effects of COVID-19 pandemic on hospitality industry: review of the current situations and a research agenda. J Hosp Market Manag 29(5):527–529. https://doi.org/10.1080/19368623.2020.1788231
75. Haas W, Krausmann F, Wiedenhofer D, Heinz M (2015) How circular is the global economy?: an assessment of material flows, waste production, and recycling in the European Union and the World in 2005. J Ind Ecol 19(5):765–777. https://doi.org/10.1111/jiec.12244
76. Hanzl M (2020) Urban forms and green infrastructure—the implications for public health during the COVID-19 pandemic. Cities & Health 1–5. https://doi.org/10.1080/23748834.2020.1791441
77. Hepburn C, O'Callaghan B, Stern N, Stiglitz J, Zenghelis D (2020) Will COVID-19 fiscal recovery packages accelerate or retard progress on climate change? Oxf Rev Econ Policy 36 (Supplement_1): S359–S381. https://doi.org/10.1093/oxrep/graa015
78. Ho C-Y, Tsai B-H, Chen C-S, Lu M-T (2021) Exploring green marketing orientations toward sustainability the hospitality industry in the COVID-19 pandemic. Sustainability 13 (8). https://doi.org/10.3390/su13084348
79. Holmes H, Wieser H, Kasmire J (2021) Critical approaches to circular economy research: time, space and evolution. In Bali Swain R, Sweet S (eds), Sustainable consumption and production, volume ii: circular economy and beyond. Springer International Publishing, pp 55–74. https://doi.org/10.1007/978-3-030-55285-5_4
80. Jawahir IS, Bradley R (2016) Technological elements of circular economy and the principles of 6r-based closed-loop material flow in sustainable manufacturing. Procedia CIRP 40:103–108. https://doi.org/10.1016/j.procir.2016.01.067
81. Jecker NS, Atuire CA (2021) What's yours is ours: waiving intellectual property protections for COVID-19 vaccines. J Med Ethics 47(9):595. https://doi.org/10.1136/medethics-2021-107555
82. Ji X, Zhang Y, Hao L (2012) Analyses of Japanese circular economy mode and its inspiration significance for China. Advances in Asian Social Science (AASS), 725.
83. Jiang B, Sun Z, Liu M (2010) China's energy development strategy under the low-carbon economy. Energy 35(11):4257–4264. https://doi.org/10.1016/j.energy.2009.12.040
84. Kaushal V, Srivastava S (2021) Hospitality and tourism industry amid COVID-19 pandemic: perspectives on challenges and learnings from India. Int J Hosp Manag 92. https://doi.org/10.1016/j.ijhm.2020.102707
85. Kirk CP, Rifkin LS (2020) I'll trade you diamonds for toilet paper: consumer reacting, coping and adapting behaviors in the COVID-19 pandemic. J Bus Res 117:124–131. https://doi.org/10.1016/j.jbusres.2020.05.028
86. Kitz R, Walker T, Charlebois S, Music J (2021) Food packaging during the COVID-19 pandemic: Consumer perceptions. Int J Consum Stud n/a (n/a). https://doi.org/10.1111/ijcs.12691
87. Koetter P, Pelton M, Gonzalo J, Du P, Exten C, Bogale K et al. (2020). Implementation and process of a Covid-19 contact tracing initiative: leveraging health professional students to extend the workforce during a pandemic. Am J Infect Control 48 (12): 1451–1456. https://doi.org/10.1016/j.ajic.2020.08.012
88. Kozik N (2020) Sustainable packaging as a tool for global sustainable development. SHS Web Conf. 74:04012. https://doi.org/10.1051/shsconf/20207404012
89. Krishtel P, Malpani R (2021) Suspend intellectual property rights for covid-19 vaccines. BMJ 373. https://doi.org/10.1136/bmj.n1344
90. Kuriyama A, Tamura K, Kuramochi T (2019) Can Japan enhance its 2030 greenhouse gas emission reduction targets? assessment of economic and energy-related assumptions in Japan's NDC. Energy Policy 130:328–340. https://doi.org/10.1016/j.enpol.2019.03.055

91. Lackie K, Najjar G, El-Awaisi A, Frost J, Green C, Langlois S et al. (2020) Interprofessional education and collaborative practice research during the COVID-19 pandemic: considerations to advance the field. J Interprofessional Care 34 (5): 583–586. https://doi.org/10.1080/13561820.2020.1807481
92. Larsson M (2018) Circular business models: developing a sustainable future. Springer
93. Lashley MA, Acevedo M, Cotner S, Lortie CJ (2020) How the ecology and evolution of the COVID-19 pandemic changed learning. Ecol Evol 10(22):12412–12417. https://doi.org/10.1002/ece3.6937
94. Legislation.gov.uk (2016) The environmental permitting (England and wales) regulations 2016. Retrieved 12/09/2021 from https://www.legislation.gov.uk/uksi/2016/1154/contents/made
95. Leiva-Martinez M-A, Anderson-Seminario MdlM, Alvarez-Risco A, Estrada-Merino A, Mlodzianowska S (2021) Price variation in lower goods as of previous economic crisis and the contrast of the current price situation in the context of COVID-19 in Peru. In: Lawrence KD, Klimberg RK (eds) Advances in Business and Management Forecasting, vol. 14. Emerald Publishing Limited, pp 161–166. https://doi.org/10.1108/S1477-407020210000014011
96. Li X, Tilt B (2019) Public engagements with smog in urban China: Knowledge, trust, and action. Environ Sci Policy 92:220–227. https://doi.org/10.1016/j.envsci.2018.12.008
97. Li W, Lin W (2016) Circular economy policies in China. Economic Research Institute for ASEAN and East Asia. https://www.eria.org/ERIA_RPR_FY2014_44.pdf
98. Lim S, Prakash A (2021) Pandemics and citizen perceptions about their country: did COVID-19 increase national pride in South Korea? Nations Natl 27(3):623–637. https://doi.org/10.1111/nana.12749
99. Liu JY-S (2012) Circular economy and environmental efficiency-the case of traditional hakka living system. Procedia Soc Behav Sci 57:255–260. https://doi.org/10.1016/j.sbspro.2012.09.1183
100. Lopez-Odar D, Alvarez-Risco A, Vara-Horna A, Chafloque-Cespedes R, Sekar MC (2020) Validity and reliability of the questionnaire that evaluates factors associated with perceived environmental behavior and perceived ecological purchasing behavior in Peruvian consumers. Soc Responsib J 16(3):403–417. https://doi.org/10.1108/SRJ-08-2018-0201
101. Lu Z-N, Chen H, Hao Y, Wang J, Song X, Mok TM (2017) The dynamic relationship between environmental pollution, economic development and public health: evidence from China. J Clean Prod 166:134–147. https://doi.org/10.1016/j.jclepro.2017.08.010
102. Malagón-Vélez LE (2021) Social and solidarity economy conceptual contributions to the circular economy. Cuadernos de Administración (Universidad del Valle), 37 (70).
103. Manniche J, Topsø Larsen K, Brandt Broegaard R, Holland E (2017) Destination: a circular tourism economy: a handbook for transitioning toward a circular economy within the tourism and hospitality sectors in the South Baltic Region. In: Centre for Regional and Tourism Research.
104. Mathews JA, Tan H (2011) Progress toward a circular economy in China. J Ind Ecol 15(3):435–457. https://doi.org/10.1111/j.1530-9290.2011.00332.x
105. Mazur-Wierzbicka E (2021) Towards circular economy—a comparative analysis of the countries of the European Union. Resources 10 (5). https://doi.org/10.3390/resources10050049
106. McDowall W, Geng Y, Huang B, Barteková E, Bleischwitz R, Türkeli S, ... Doménech T (2017) Circular economy policies in China and Europe. J Ind Ecol 21(3):651–661.
107. Mejía-Acosta N, Alvarez-Risco A, Solís-Tarazona Z, Matos-Valerio E, Zegarra-Arellano E, Del-Aguila-Arcentales S (2016) Reacciones Adversas a Medicamentos reportadas como resultado de la implementación de Atención Farmacéutica en la Farmacia Institucional DIGEMID—Ministerio de Salud de Perú. Pharmaceutical Care España 18 (2): 67–74. https://www.pharmcareesp.com/index.php/PharmaCARE/article/view/311
108. Milutin S (2021) The environment and climate change law review: Russia. Retrieved 12/09/2021 from https://thelawreviews.co.uk/title/the-environment-and-climate-change-law-review/russia

109. Ministry of Economic Development Russian Federation (2020) State report on the state of energy saving and increasing energy efficiency in Russian federations. Retrieved 12/09/2021 from https://www.economy.gov.ru/material/news/ilya_torosov_aktivnaya_rabota_po_povysheniyu_energoeffektivnosti_polozhitelno_povliyaet_na_snizhenie_antropogennogo_vozdeystviya_na_klimat.html
110. Mostert C, Sameer H, Glanz D, Bringezu S (2021) Climate and resource footprint assessment and visualization of recycled concrete for circular economy. Resour Conserv Recycl 174. https://doi.org/10.1016/j.resconrec.2021.105767
111. National Environment Agency (2020) Waste statistics and overall recycling. Retrieved 12/08/2021 from https://www.nea.gov.sg/our-services/waste-management/waste-statistics-and-overall-recycling
112. Obergassel W, Hermwille L, Oberthür S (2021) Harnessing international climate governance to drive a sustainable recovery from the COVID-19 pandemic. Climate Policy 21(10):1298–1306. https://doi.org/10.1080/14693062.2020.1835603
113. Ogunmakinde OE (2019) A review of circular economy development models in China, Germany and Japan. Recycling 4 (3). https://doi.org/10.3390/recycling4030027
114. Our World in Data (2019) CO2 emissions. Retrieved 12/09/2021 from https://ourworldindata.org/co2-emissions#co2-emissions-by-region
115. Patwa N, Sivarajah U, Seetharaman A, Sarkar S, Maiti K, Hingorani K (2021) Towards a circular economy: an emerging economies context. J Bus Res 122:725–735. https://doi.org/10.1016/j.jbusres.2020.05.015
116. PlasticsEurope (2019) Plastics—the facts 2019. An analysis of European plastics production, demand and waste data. Retrieved 12/09/2021 from https://www.plasticseurope.org/application/files/9715/7129/9584/FINAL_web_version_Plastics_the_facts2019_14102019.pdf
117. Quispe-Cañari JF, Fidel-Rosales E, Manrique D, Mascaró-Zan J, Huamán-Castillón KM, Chamorro–Espinoza SE et al. (2021) Self-medication practices during the COVID-19 pandemic among the adult population in Peru: a cross-sectional survey. Saudi Pharm J 29 (1): 1–11. https://doi.org/10.1016/j.jsps.2020.12.001
118. Raudenská J, Steinerová V, Javůrková A, Urits I, Kaye AD, Viswanath O, Varrassi G (2020) Occupational burnout syndrome and post-traumatic stress among healthcare professionals during the novel coronavirus disease 2019 (COVID-19) pandemic. Best Pract Res Clin Anaesthesiol 34(3):553–560. https://doi.org/10.1016/j.bpa.2020.07.008
119. Roesch E, Amin A, Gupta J, García-Moreno C (2020) Violence against women during covid-19 pandemic restrictions. BMJ 369. https://doi.org/10.1136/bmj.m1712
120. Rojas-Osorio M, Alvarez-Risco A (2019) Intention to use smartphones among Peruvian university students [intention to use; Peruvian students; technology acceptance model; smartphone]. 13 (03): 13. https://doi.org/10.3991/ijim.v13i03.9356
121. Rojas Román B, Moscoso S, Chung SA, Limpias Terceros B, Álvarez-Risco A, Yáñez JA (2020) Tratamiento de la COVID-19 en Perú y Bolivia y los riesgos de la automedicación [COVID-19; tratamiento; estrategia; fármacos; automedicación; Perú; Bolivia] 53 (2). http://www.revfarmacia.sld.cu/index.php/far/article/view/435/351
122. Schroeder P, Anggraeni K, Weber U (2019) The relevance of circular economy practices to the sustainable development goals. J Ind Ecol 23(1):77–95. https://doi.org/10.1111/jiec.12732
123. Sekalala S, Forman L, Hodgson T, Mulumba M, Namyalo-Ganafa H, Meier BM (2021) Decolonising human rights: how intellectual property laws result in unequal access to the COVID-19 vaccine. BMJ Glob Health 6(7). https://doi.org/10.1136/bmjgh-2021-006169
124. Shaw R, Kim Y-K, Hua J (2020) Governance, technology and citizen behavior in pandemic: lessons from COVID-19 in East Asia. Prog Disaster Sci 6. https://doi.org/10.1016/j.pdisas.2020.100090
125. Shpak N, Kuzmin O, Melnyk O, Ruda M, Sroka W (2020) Implementation of a circular economy in Ukraine: the context of European integration. Resources 9 (8). https://doi.org/10.3390/resources9080096
126. Sigala M (2020) Tourism and COVID-19: Impacts and implications for advancing and resetting industry and research. J Bus Res 117:312–321. https://doi.org/10.1016/j.jbusres.2020.06.015

127. Siracusa V, Rosa MD (2018) 8—sustainable packaging. In Galanakis, CM (ed) *Sustainable food systems from agriculture to industry*. Academic Press, pp 275–307. https://doi.org/10.1016/B978-0-12-811935-8.00008-1
128. Stahel WR (2016) The circular economy. Nature 531(7595):435–438. https://doi.org/10.1038/531435a
129. Statista (2019) Forecast of greenhouse gas emissions from the plastic life cycle worldwide from 2019 to 2050. Retrieved 12/09/2021 from https://www.statista.com/statistics/1011906/forecast-ghg-plastic-lifecycle-globally
130. Statista (2020) The life cycle of plastics. Retrieved 12/09/2021 from https://www.statista.com/study/83215/the-life-cycle-of-plastics
131. Statista (2021) Paper industry worldwide. Retrieved 12/09/2021 from https://www.statista.com/study/18116/paper-industry--statista-dossier
132. Sverko Grdic Z, Krstinic Nizic M, Rudan E (2020) Circular economy concept in the context of economic development in EU countries. Sustainability 12 (7). https://doi.org/10.3390/su12073060
133. Sánchez OR, Vale DB, Rodrigues L, Surita FG (2020) Violence against women during the COVID-19 pandemic: an integrative review. Int J Gynecol Obstet 151(2):180–187. https://doi.org/10.1002/ijgo.13365
134. Sánchez-Clavijo LM, Martínez-Callejas SJ, Acevedo-Charry O, Diaz-Pulido A, Gómez-Valencia B, Ocampo-Peñuela N et al. (2021) Differential reporting of biodiversity in two citizen science platforms during COVID-19 lockdown in Colombia. Biological Conservation 256: 109077. https://doi.org/10.1016/j.biocon.2021.109077
135. The Consumer Goods Forum (2020) Building a circular economy for packaging: a view from the consumer goods industry on optimal extended producer responsibility. Retrieved 12/08/2021 from https://www.theconsumergoodsforum.com/wp-content/uploads/EPR-Building-a-Circular-Economy-for-Plastic-Packaging-cgf-plastic-waste.pdf
136. Tran LTT (2021) Managing the effectiveness of e-commerce platforms in a pandemic. J Retail Consum Serv 58. https://doi.org/10.1016/j.jretconser.2020.102287
137. UNFCC (2020) Kyoto protocol—targets for the first commitment period. Retrieved 12/09/2021 from https://unfccc.int/process-and-meetings/the-kyoto-protocol/what-is-the-kyoto-protocol/kyoto-protocol-targets-for-the-first-commitment-period
138. Udara Willhelm Abeydeera LH, Wadu Mesthrige J, Samarasinghalage TI (2019) Global research on carbon emissions: a scientometric review. Sustainability 11 (14). https://doi.org/10.3390/su11143972
139. United Nations (2021) Take action for the sustainable development goals. Retrieved 12/09/2021 from https://www.un.org/sustainabledevelopment/sustainable-development-goals
140. United Nations Climate Change (2021) ¿Qué es el protocolo de Kyoto? Retrieved 12/09/2021 from https://unfccc.int/es/kyoto_protocol
141. United States Environmental Protection Agency (2018) Paper and paperboard: material-specific data. https://www.epa.gov/facts-and-figures-about-materials-waste-and-recycling/paper-and-paperboard-material-specific-data
142. Vaish V, Mehta H (2017) India: environment laws In India. octubre 30, 2021, de Vaish Associates Advocates. Retrieved 12/08/2021 from https://www.mondaq.com/india/waste-management/624836/environment-laws-in-india
143. Velis CA, Vrancken KC (2015) Which material ownership and responsibility in a circular economy? Waste Manage Res 33(9):773–774. https://doi.org/10.1177/0734242X15599305
144. Vidya CT, Prabheesh KP (2020) Implications of COVID-19 pandemic on the global trade networks. Emerg Mark Financ Trade 56(10):2408–2421. https://doi.org/10.1080/1540496X.2020.1785426
145. Viero A, Barbara G, Montisci M, Kustermann K, Cattaneo C (2021) Violence against women in the Covid-19 pandemic: a review of the literature and a call for shared strategies to tackle health and social emergencies. Forensic Sci Int 319. https://doi.org/10.1016/j.forsciint.2020.110650

146. Wandana S, Wadanambi R, Preethika P, Dassanayake N, Chathumini L, Arachchige U (2021) Carbon footprint analysis: promoting sustainable development. J Res Technol Eng 2 (1): 73–80. https://www.jrte.org/2021/01/11/carbon-footprint-analysis-promoting-sustainable-development
147. Wen J, Kozak M, Yang S, Liu F (2021) COVID-19: potential effects on Chinese citizens' lifestyle and travel. Tour Rev 76(1):74–87. https://doi.org/10.1108/TR-03-2020-0110
148. Wu T-Y, Ford O, Rainville AJ, Bessire R (2020) COVID-19 care package distribution for senior citizens and families in Detroit and Hamtramck, Michigan. J Hunger & Environ Nutr 15(4):585–587. https://doi.org/10.1080/19320248.2020.1799586
149. Xue Y-N, Luan W-X, Wang H, Yang Y-J (2019) Environmental and economic benefits of carbon emission reduction in animal husbandry via the circular economy: case study of pig farming in Liaoning China. J Clean Prod 238. https://doi.org/10.1016/j.jclepro.2019.117968
150. Yan J, Kim S, Zhang SX, Foo M-D, Alvarez-Risco A, Del-Aguila-Arcentales S, Yáñez JA (2021) Hospitality workers' COVID-19 risk perception and depression: a contingent model based on transactional theory of stress model. Int J Hosp Manag 95. https://doi.org/10.1016/j.ijhm.2021.102935
151. Yang Y, Meng G (2020) The evolution and research framework of carbon footprint: based on the perspective of knowledge mapping. Ecol Ind 112. https://doi.org/10.1016/j.ecolind.2020.106125
152. Yasin Ar A (2020) Managing e-commerce during a pandemic: lessons from grubhub during coVID-19. In George B, Mahar Q (eds) International case studies in the management of disasters. Emerald Publishing Limited, pp 155–167. https://doi.org/10.1108/978-1-83982-186-820201010
153. Yáñez JA, Afshar Jahanshahi A, Alvarez-Risco A, Li J, Zhang SX (2020) Anxiety, distress, and turnover intention of healthcare workers in Peru by their distance to the epicenter during the COVID-19 crisis. Am J Trop Med Hyg 103(4):1614–1620. https://doi.org/10.4269/ajtmh.20-0800
154. Yáñez JA, Alvarez-Risco A, Delgado-Zegarra J (2020) Covid-19 in Peru: from supervised walks for children to the first case of Kawasaki-like syndrome. BMJ 369. https://doi.org/10.1136/bmj.m2418
155. Zarocostas J (2021) What next for a COVID-19 intellectual property waiver? The Lancet 397(10288):1871–1872. https://doi.org/10.1016/S0140-6736(21)01151-X
156. Zhang SX, Chen J, Afshar Jahanshahi A, Alvarez-Risco A, Dai H, Li J, Patty-Tito RM (2021) Succumbing to the COVID-19 pandemic—healthcare workers not satisfied and intend to leave their Jobs. Int J Ment Heal Addict. https://doi.org/10.1007/s11469-020-00418-6
157. Zhang SX, Chen J, Jahanshahi AA, Alvarez-Risco A, Dai H, Li J, Patty-Tito RM (2021) Correction to: succumbing to the COVID-19 pandemic—healthcare workers not satisfied and intend to leave their jobs. Int J Ment Heal Addict. https://doi.org/10.1007/s11469-021-00502-5
158. Zhang SX, Sun S, Afshar Jahanshahi A, Alvarez-Risco A, Ibarra VG, Li J, Patty-Tito RM (2020) Developing and testing a measure of COVID-19 organizational support of healthcare workers—results from Peru, Ecuador, and Bolivia. Psychiatry Res 291. https://doi.org/10.1016/j.psychres.2020.113174
159. Zhang Y, Li J, Lei Y, Yan T (2021) Explore the approaches to corporate social responsibility implemented by e-commerce platforms in china during the early stage of covid-19: a mixed-methods content analysis. Disaster Med Public Health Prep 1–23. https://doi.org/10.1017/dmp.2021.346
160. Zhu J, Xu C (2021) Sina microblog sentiment in Beijing city parks as measure of demand for urban green space during the COVID-19. Urban Forestry & Urban Greening 58. https://doi.org/10.1016/j.ufug.2020.126913
161. Zollinger R, DiCindio C (2021) Community Ecology: Museum Education and the Digital Divide During and After COVID-19. Journal of Museum Education 46(4):481–492. https://doi.org/10.1080/10598650.2021.1983711

Circular Economy for Waste Reduction and Carbon Footprint

Romina Gómez-Prado, Aldo Alvarez-Risco, Jorge Sánchez-Palomino, Maria de las Mercedes Anderson-Seminario, and Shyla Del-Aguila-Arcentales

Abstract The anthropogenic activities developed in the manufacturing and distribution of products that are carried out at greater intensity to meet the growing demand and consumption worldwide produce a more significant amount of greenhouse gases (GHG), which impacts global warming. In the same way, the problem of the scarcity of natural resources, increasingly growing, creates the need to seek better alternatives with a sustainable approach to protect the state of the planet in which we live. For this, the support of governments, companies, and even consumers is required. This chapter takes the perspective of the circular economy (CE) as an opportunity to implement sustainable methods based on the management and reduction of waste, giving them a new function. Adding the use of the carbon footprint (CF) as a tool that allows knowing the impact of anthropogenic activities on the environment is added. In addition, confirmed cases aim to reduce GHG emissions by implementing a more efficient system that originates based on the CE principles.

Keywords Circular economy · Carbon footprint · Environmental · Waste · Global warming · Greenhouse gases

1 Introduction

Today's high growth of world consumerism has brought significant consequences to the environment, making its effects irreversible for the future. The application of the linear economy has not been favorable in waste management; on the contrary, the waste of final products threatens the planet both in the short and long term [74].

On the other hand, this problem has taken more significant consideration in recent years. One of the primary triggers has been climate change and biodiversity loss, so

R. Gómez-Prado · A. Alvarez-Risco (✉) · J. Sánchez-Palomino ·
M. de las Mercedes Anderson-Seminario
Universidad de Lima, Lima, Perú
e-mail: aralvare@ulima.edu.pe

S. Del-Aguila-Arcentales
Escuela Nacional de Marina Mercante "Almirante Miguel Grau", Callao, Perú

A. Alvarez-Risco et al. (eds.), *Circular Economy*, Environmental Footprints and Eco-design of Products and Processes,
https://doi.org/10.1007/978-981-19-0549-0_7

there is now a greater awareness of the environmental lobby and its purpose [104]. For their part, some irresponsible consumers create a gap between regular consumption and sustainable consumption, preventing such waste from being minimized and achieving sustainable consumption. In addition, this model proposes to generate greater consumer interest toward sustainable questions [46].

The CF and its implication to the environment are crucial in analyzing the environmental impact. The CF is probably the best-known indicator for measuring environmental impact. It is responsible for measuring the amount of CH_4 and CO_2 emissions on a chosen population or activity; that is, it measures the impact of certain activities that threaten the Earth's climate [93]. Among the main emitters of these gases, we find large companies, both manufacturing and non-manufacturing, and developing countries as primary threats to the environment [53, 117]. In addition, there is work on institutional organizations (OI) to ensure climate risks, increasingly complex challenges due to political factors, adaptability, interests, and others. Many regional governments tend to have a strong fragmentation between policy communities, which leads to the non-application of the actions proposed by OI. A similar case is experienced internationally [72].

The world has changed, and people's habits regarding sustainability have changed, depending on the characteristics of pre-pandemic life. Due to limited sources by COVID-19, new ways to build new materials are essential in the reactivation of the global economies.

In Table 1 shows the different negative impacts that the COVID-19 pandemic generated.

The CE has been widely accepted as a mechanism that seeks to balance economic, social, and environmental outcomes [65]. In response, the concept of CE brings together a set of activities that improve resource management [52]. This chapter also mentions the circular economy as a basis for creating ideal models to mitigate environmental impact, which is born apart from the obsolescence of the linear economy, characterized by the waste of reusable inputs [32].

Table 1 Changes created by COVID-19 in different sectors

Education	[8, 9, 19, 13, 12, 47, 81, 82, 149]
Entrepreneurship	[4, 21, 14, 18, 26, 30, 31, 36, 44, 86, 91]
Health sector	[1, 7, 10, 16, 12, 28, 34, 48, 55, 79, 106, 111, 144, 147, 148]
Hospitality	[58, 67, 71, 77, 142]
Intellectual property	[11, 17, 42, 59, 75, 80, 97, 116, 146]
Population	[6, 20, 43, 68, 87, 105, 130, 120, 139, 141, 145]
Prices	[24, 50, 63, 70, 84, 96]
Tourism	[25, 29, 35, 39, 62, 40, 122, 41]
Trade	[2, 3, 5, 22, 15, 33, 51, 78, 89, 133, 137, 143]
Violence against women	[23, 45, 56, 66, 110, 129, 138]

2 Problematic Situation

2.1 Poor Waste Management

Many countries are adopting policies based on waste management [136]. This growing importance is due to the promotion of environmental management tools, which involve controlling inputs and managing construction waste [131]. In addition to this, another of the great motivations is the rise in raw material costs, stricter policies on waste, and high disposal costs [136].

One industry that has made the most changes for proper waste management has been the food industry. According to the Food and Agriculture Organization of the United Nations [135], approximately one-third of the food produced globally, a value of around 1,300 million tons per year, is thrown away by the trash. In addition, there is a per capita amount of food waste, which reaches 95–115 kilos in Europe and North America, while in Africa and South and Southeast Asia, it is around 6–11 kilos. This data highlights the food gap between industrialized countries compared to developing countries.

A distinction needs to be made in food waste as avoidable and unavoidable. It is considered avoidable to those edible foods before being discarded, while the unavoidable foods never came to be considered food. Food waste implies a financial loss and a waste of the resources used to produce the food: energy, land, fertilizer, transport, etc. This loss of resources negatively impacts the environment, so waste management policies seek to ensure both the use of resources and reduce environmental damage [132].

It is a challenge to precisely calculate the amount of food wasted globally. Even so, various IOs seek to quantify the magnitude of the damage in environmental indicators [76]. Regarding the balance of damages from food waste in terms of GHG emissions, the food industry emits approximately 4,400 million tons of CO_2, representing 8% of total GHG emissions [60]. On the other hand, the construction sector also adapts processes to improve waste management. The implementation of construction waste management tools is currently part of the agenda of this industry. In recent years, there has been a greater awareness of the impacts on the environment [131].

The growing awareness of concerns about managing construction and demolition waste has led to the development of waste management as an essential function of construction project management, the overwhelming promotion of waste management tools, and sustainable development activities. As a result, there is a growing awareness of waste management issues and potential problems related to negative impacts on the environment [131]. Those responsible for regulating environmental policies in the industry have been external institutions, which provided solutions even to reuse waste. Thanks to technological advances, products made from precast steel can be seen, such as polymer recycling and safe combustion. Other responsible characters are global customers, who ultimately demand quality service, responsible for using resources without wasting them [123]. Still, many companies do not implement sustainable practices.

A survey of construction companies in Hong Kong held that minimizing the cost of the construction project is the most crucial objective, while concern for the environment is the least important. On the other hand, the lack of experience of many construction companies is the main reason why recycling construction materials is not very encouraging. In response, the Hong Kong government-imposed regulations to minimize waste and implement the WMP method for waste management. However, the companies claimed that their productivity was significantly affected [131]. For this reason, the companies themselves must invest and develop their recycling or waste reuse programs adapted to their conditions in order not to be harmed [113].

2.2 *Carbon Footprint*

Human activity has been responsible for the emissions of harmful gases such as carbon dioxide (CO_2), methane (CH_4), nitrous oxide (N_2O), and other gases that threaten the planet [119, 140]. Also, the concept of CF has become part of the global agenda by various IOs, governments, and institutions as GHG emissions have constantly been increasing in recent years, thus generating great global concern [57]. All the tasks carried out by a company, company, or specific group to be evaluated must be considered to measure CF. From the extraction of the raw material, transportation, production, and final disposal, each of these stages must be analyzed on how much they compromise the environment [69].

There is great difficulty in measuring CF; it is necessary to calculate both direct and indirect emissions. It is called direct to those consuming domestic energy, while indirect seeks to produce or eliminate products and services at home [49]. This lack of knowledge does not allow people to apply the necessary strategies to reduce carbon levels [92]. For this reason, the scientific community seeks to inform and raise awareness among the population by showing little encouraging indicators about the accelerated rate of degradation of the planet and guiding companies toward adequate environmental policies [61, 101]. According to the Fifth Assessment Report of the Intergovernmental Panel on Climate Change, the most notorious changes caused by environmental degradation are extreme weather, scarcity of resources, air pollution, increased sea levels, and multiple epidemics [126]. That evidence that not enough is being done to compensate for the damage caused.

In response, international interest has been aroused in quantifying the magnitude of the environmental impact [69, 125, 128]. Corporate carbon calculators are valuable tools that measure business CF, an indicator that allows comparisons of other business footprints. Another tool is the Greenhouse Gas Protocol (GGP), considered the most widely used global standard tool for measuring and reporting GHG levels of a particular activity [69].

Carrying out a correct measurement of CF brings with it significant benefits. At the business level, a correct application of measurement tools generates cost savings, taking advantage of these saved resources to synergies from other areas of the company, eco-friendly innovations, corporate image, among others [121]. On the

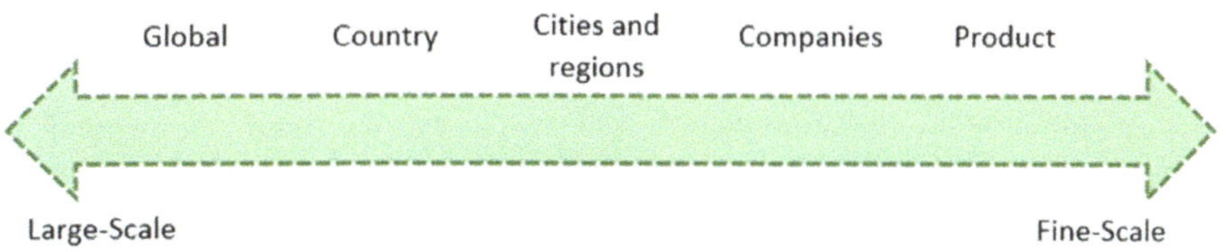

Fig. 1 Scheme of applications of the carbon footprint on a large and fine-scale (*Note* Adapted from [101])

social side, prioritizing a green agenda is vital for promoting environmental protection and global sustainable development, in addition to generating environmental awareness [69, 121].

Increased GHG emissions have been the biggest driver of climate change [73], leading to global action. Companies such as Hewlett Packard, Patagonia, and Walmart, concerned about having their current processes damage the environment, have sought to reverse this situation by managing the ecological supply chain (GrSCM). This environmental management tool ranges from designing products ecological to managing their reverse logistics [128]. In this way, the GrSCM allows reducing waste, saving energy, conserving natural resources, and, above all, reducing carbon emissions, generating both operational and financial savings for the company [37].

Like the companies mentioned, many other IOs and developed countries contribute significantly to GHG emissions [107, 109, 114]. Their respective supply chains can even be considered a threat to the environment if they do not adapt their stages to an environmental design. It is necessary to mention that, despite the current environmentalist trend, it is still a challenge for organizations, who must rigorously analyze new implementations of their processes without having some general framework that guarantees the success of their implementation [100, 114]. Operations redesign tends to focus on the efficiency of logistics networks and processes for more excellent added value [100]. That is why, regardless of being an IO, entity, or company of greater or lesser development, reducing the CF is a responsibility that all those who take advantage of the planet's resources must assume. Companies have numerous actions to take to benefit from the adaptation of their policies to more environmental ones, as well as cities and regions can implement local policies toward general national objectives [101]. In Fig. 1, we can see the different agents that, in addition to contributing to GHG emissions, are part of the large groups that must take care of environmental challenges to safeguard scarce and limited natural resources.

3 Theoretical Framework

Currently, the socioeconomic system focuses on the linear economy, which produces and consumes products that are discarded at the end of their useful life [94]. This system causes unnecessary losses of natural resources, which, year after year, lies in their existing depletion on Earth [94].

It was found in the literature that CE is considered as the trend that responds to the mismanagement of resources of the traditional linear economy system. However, changing the traditional system to a circular one entails several challenges that must be overcome [94, 112]. Therefore, it is necessary to know and investigate the subject and its representation.

The term CE is increasingly approached by scientists, whose central element is the "restorative use" of resources; in other words, it focuses on transforming the waste that remains after having consumed a product [64, 94]. For this reason, the CE concept is mainly considered a solution for insufficient resources, waste production, and increased economic advantages [85].

Given that environmental care is of great importance for the industrial business sector and its good performance, it is essential to determine the stakeholder's part of the waste management [85, 95]. Governments and companies search for a new, different, and sustainable business model (BM). However, to adopt a model attached to the CE concept, it is necessary to have a collective effort, which is made up of governments, companies, and even consumers [94, 99].

Before delving into the meaning of a CE model, it is necessary to understand the bases of a BM, which is made up of a sequence of patterns that allow it to evolve and innovate from time to time from previous methods that have already been successful [103, 108]. Similarly, the BMs that implement innovative tools to solve problems are more valuable [103].

The continuous search for commercial patterns that have circular improvement effects in the BM explains the creation of solid circular economy business models (CEBM) to rethink companies' general sense of value creation. Moreover, the different factors, whether internal or external, that affect the competitive environment or the set of laws that govern the market must also be taken into consideration since they could influence changes in the BM [32, 85, 103].

In recent years, sustainability has been an increasingly popular factor in the business world, meaning a competitive advantage for companies due to the excellent response from consumers by providing them with greater value. Therefore, the principles of sustainability were added to the BM, forming part of the model's design. Incorporating these sustainable models achieves a triple result due to its social, environmental, and economical approach [102].

Going from a linear economy BM based on the "manufacturing-consumption-waste" system, which is shown in greater detail in Fig. 2, to a CEBM implies that it

Fig. 2 Linear economy model (*Note* Adapted from [38])

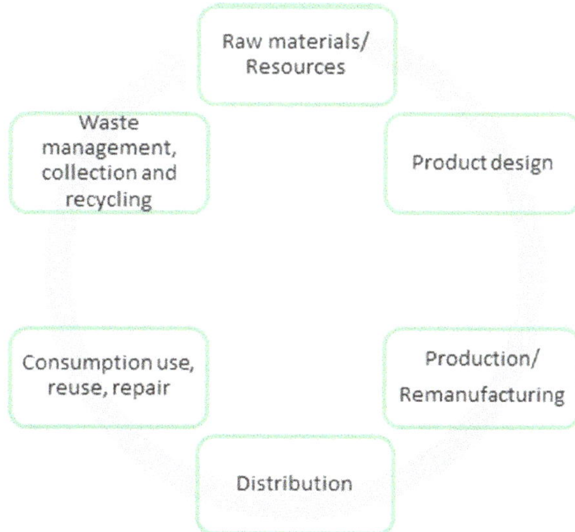

Fig. 3 Closed production circular economy model (*Note* Adapted from [38])

includes sustainable principles and maintains the purpose of increasing the productivity of natural resources. The CEBM is specified through strategies for reducing or closing resource flows [103, 115, 127]. In this way, it proves to be a model that can replace the traditional model and reduce the negative impact on the environment [118].

The CEBM is an option that proposes a path toward a more sustainable economic development, which proposes different solutions to face environmental challenges both locally and globally, taking care of the availability of resources and the preservation of the planet. The CEBM shown in Fig. 3 represents the closed production model, in which the raw material constantly travels in a loop [38, 115].

3.1 The Closed-Loop Model Focused on the 6Rs

The first of the 6 is "Reduce" which is the act that focuses on our commitment to reducing the number of products we consume, to avoid excesses to reduce the number of items to be discarded. "Reuse" refers to seeking to get the most out of the goods, even if they are not in their best condition or if they have already passed their useful life. "Recycle," probably the most prominent term in the 6Rs, is responsible for segmenting waste according to its nature to give it a new function. Another task is to "Redistribute" assets that are not being fully used and, failing that, maybe more valuable in other hands. Finally, "Repair" the goods that have supposedly been out of the use or with some damage. To not encourage unnecessary consumerism, it seeks to give them a "second life" [74].

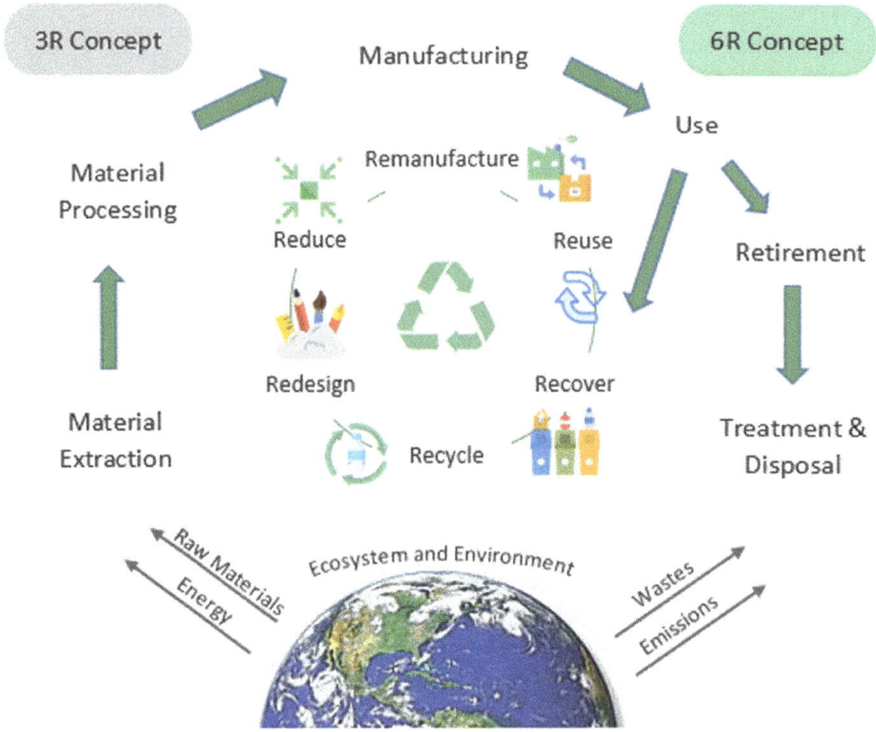

Fig. 4 The closed-loop model focused on the 6Rs (*Note* Adapted from [74])

The following model shown in Fig. 4 graphically represents a closed-loop system that follows the 6R concept, which proposes a sustainable system that confers a flow of very close to imperishable materials. Therefore, a model is established that optimizes processes and promotes GHG emission reduction [74].

3.2 Reverse Cycle Circular Economy Model

The closed-loop supply chain (CLSC) models are schemes adapted to the demands of today's market: a cost-effective model that in turn meets the environmental requirements proposed by the IOs. Ellen MacArthur Foundation proposes a CE model where each of the stages of the CLSC seeks to make the most of resources [94]. It consists of 5 stages, where each product can be repaired, reused, redistributed, reconditioned, or remanufactured [90].

It also proposes to reinvent supply chains, ensuring more optimization of systems than components, eradicating waste in an organized manner to maximize profitability while, at the same time, minimizing environmental impact and maximizing social

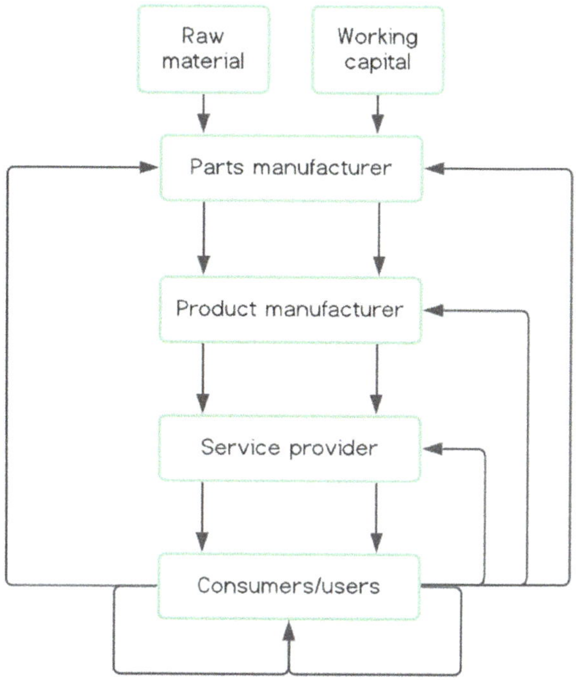

Fig. 5 Reverse cycles for the circular economy (*Note* Adapted from [90])

well-being [90]. In Fig. 5, you can see the model explained above, which includes the stages of a usual supply chain, adding reverse logistics actions in each of these.

3.3 Cases

If we talk about the industries in which there is the most significant amount of GHG emissions, the food production sector can be taken as an example due to its significant participation in anthropogenic GHG emissions, generating between 10 and 14% of these types of emissions in the world [124, 134].

During 2000–2014, the United States presented a decrease in CO_2 emissions regarding agribusiness. While, the case of the countries of China and India was the opposite, showing a pronounced increase in these emissions [27] (Fig. 6 below). It should be noted that China was the country with the highest record of carbon emissions, going from 280 to 475 Mt at the end of the period studied.

On the other hand, the closest way to measure CF levels is by segmenting them into three types: Direct, Indirect, and Global Indirect. The first classification is defined as those actions that lead to the use of products or services that directly generate consumption of natural resources, while indirect emissions refer to consumption without expending natural resources. Regarding the significant countries that emit

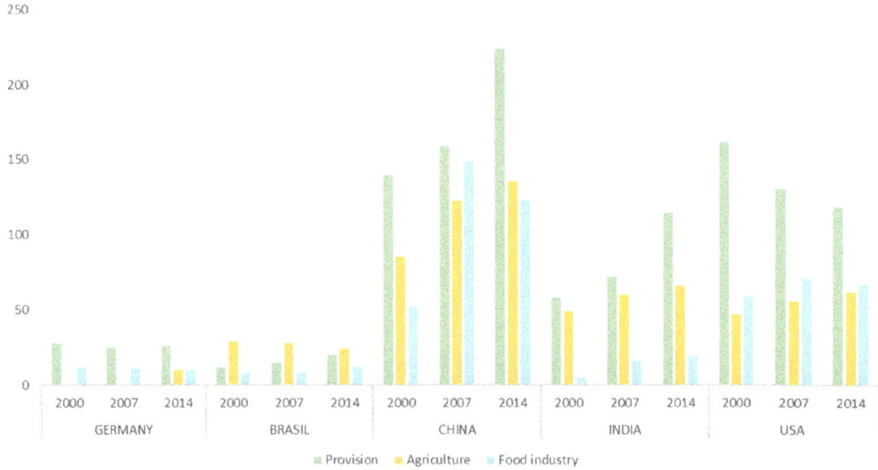

Fig. 6 Carbon emissions from agribusiness in 5 central countries (*Note* Adapted from [27])

GHG, we find China and India the countries with the highest proportion of indirect gas emissions, with 90% and 82%, respectively. Also, global indirect emissions are led by Canada and Japan, with 44% and 36%, respectively. Direct emissions are not usually that high, with Japan as the largest emitter, with 11% [98] (Fig. 6).

Larger companies are those that cause a more significant impact on the environment from their activities, compared to small companies. For this reason, large companies are the most potential candidates to look for reports, articles, and chapters and to implement new practices that have a more sustainable approach [52]. Therefore, below are some cases of countries adopting a CEBM and using CF as a support tool to manage GHG emissions.

3.4 China

This case analyzes China and different cities to determine whether the Chinese government's decision to implement a circular economy model was favorable or not. Its application was essential to overcome the problem of environmental degradation and shortage of sources. Another great success was the development of a comprehensive legal framework for the benefit of recycling. It is necessary to mention the complications and the risk assumed by implementing something different, especially in a country as influential as China. Aspects such as rapid industrialization and urbanization were significant obstacles, but it was not until the Asian country acted, such as relocations and regulations. In addition to this, the cities chosen had the necessary conditioning and resources to make this implementation more user-friendly [127].

The main reason behind the potential success of the upgrade would be the government's decision to focus its efforts on those sectors, which produce the most significant short-term impact, such as the relocation of heavy industry and the enforcement of instrument regulations that are likely to be the most influential sectors. Second, those four pilot cities are more modernized, more prosperous, and have more advanced energy efficiency technologies and equipment than other regions in China, making it relatively easier for these four cities to increase their energy efficiency [127].

3.5 *United States*

This new trend for using recyclable material applying CE strategies has been an excellent opportunity for the United States to enter the reused plastic or recycled PET (rPET) industry to represent a significant saving in raw materials while giving a message of care for the environment. This CE campaign seeks to unify the government, companies, and society to reduce the use of natural resources and, in contrast, prioritize recycled products [88].

A significant issue in implementing rPET bottles is its contribution to the GrSCM, giving more excellent circulation to final products instead of spending natural resources to manufacture more bottles. This action includes significantly saving time, money, resources and mitigating the environmental impact. However, these same bottles could be used for many other products such as fibers, sheets, and footwear [88].

3.6 *Germany*

The author argues that the current dominant linear economy causes a total waste of natural resources. He proposes a change from the traditional model toward a circular carbon one was, in addition to reducing GHG emissions, business efficiency increases. The competitiveness of the chemical sector in Germany is highly competitive, having total revenues of up to € 185 billion as of 2016, plus 447,000 jobs and investments of more than € 7 billion. For more than 30 years in the chemical industry, reduced GHG emissions by up to 50% [83].

Developing renewable energy from domestic lignite is essential; the European country is favorable, with the necessary raw material and fine chemicals originating from national carbon resources. For its part, Germany has more than 72.7 kg billion tons of brown coal, of which 50% is considered exploitable. Given this, the author adds that the energy sectors have a circular carbon economy [83].

3.7 Closing Remarks

Given the increase of the world's inhabitants, the global demand for resources is also considered increasing amounts of GHGs to the environment and waste [85]. The problems caused by the current linear economy model become an increasingly vital cause to seek more friendly options with the environment and society. Consequently, closed-loop models are adopted for manufacturing and consumption, that is, models focusing on the CE [54]. In this way, the waste that remains after consuming a product is given a new use instead of simply discarding it and polluting the environment.

The CE model must stop being seen as an alternative and necessary in everyday life. The misuse of natural resources caused by poor supply chain management is a short-term and long-term threat [100]. Economic prosperity and the ecological balance must coexist harmoniously to guarantee sustainable development [65, 74].

Faced with this, the implementation of tools for measuring environmental indicators, such as CF, is decisive when measuring the CE model's results. CF as a tool to manage and control GHG emissions due to anthropogenic actions can impact a global level in the same framework. However, it should be considered that it is more difficult to achieve its full implementation in practice [101].

Finally, it is expected that models continue to be applied that support the predisposition of caring for the planet, taking advantage of goods in a sustainable way. CE business models are an essential tool in a context where resources are increasingly scarce to safeguard the environment. The next generations also deserve to enjoy these goods, so implementing these models is no longer an alternative but a necessity in the business world [74].

References

1. Abiakam, N., Worsley, P., Jayabal, H., Mitchell, K., Jones, M., Fletcher J (2021) Personal protective equipment related skin reactions in healthcare professionals during COVID-19. Int Wound J 18 (3): 312–322. https://doi.org/10.1111/iwj.13534
2. Abudureheman A, Nilupaer A (2021) Optimization model design of cross-border e-commerce transportation path under the background of prevention and control of COVID-19 pneumonia. Soft Comput 25(18):12007–12015. https://doi.org/10.1007/s00500-021-05685-6
3. Acuña-Zegarra MA, Santana- M, Velasco-Hernandez JX (2020) Modeling behavioral change and COVID-19 containment in Mexico: a trade-off between lockdown and compliance. Math Biosci 325. https://doi.org/10.1016/j.mbs.2020.108370
4. Afshan G, Shahid S, Tunio MN (2021) Learning experiences of women entrepreneurs amidst COVID-19. Int J Gend Entrep 13(2):162–186. https://doi.org/10.1108/IJGE-09-2020-0153
5. Aguirre AA, Catherina R, Frye H, Shelley L (2020) Illicit wildlife trade, wet markets, and COVID-19: preventing future pandemics. World Medical & Health Policy 12(3):256–265. https://doi.org/10.1002/wmh3.348
6. Ahsan MM (2020) Strategic decisions on urban built environment to pandemics in Turkey: lessons from COVID-19. Journal of Urban Management 9(3):281–285. https://doi.org/10.1016/j.jum.2020.07.001
7. Alan H, Eskin F, Tiryaki H, Taskiran G, Gumus E, Harmanci AK (2021) "I'm a hero, but...": an evaluation of depression, anxiety, and stress levels of frontline healthcare professionals

during COVID-19 pandemic in Turkey. Perspect Psychiatr Care 57(3):1126–1136. https://doi.org/10.1111/ppc.12666
8. Ali W (2020) Online and remote learning in higher education institutes: a necessity in light of COVID-19 pandemic. High Educ Stud 10(3):16–25
9. Allen J, Rowan L, Singh P (2020) Teaching and teacher education in the time of COVID-19. Asia-Pac J Teach Educ 48(3):233–236. https://doi.org/10.1080/1359866X.2020.1752051
10. Alsairafi Z, Naser AY, Alsaleh FM, Awad A, Jalal Z (2021) Mental health status of healthcare professionals and students of health sciences faculties in Kuwait during the COVID-19 pandemic. Int J Environ Res Public Health 18 (4). https://doi.org/10.3390/ijerph18042203
11. Altindis E (2021) Inequitable COVID-19 vaccine distribution and the intellectual property rights prolong the pandemic. Expert Rev Vaccines, null-null. https://doi.org/10.1080/14760584.2022.2014819
12. Alvarez- A, Del- S, Yáñez JA (2021) Telemedicine in Peru as a result of the COVID-19 pandemic: perspective from a country with limited internet access. Am J Trop Med Hyg 105(1):6–11. https://doi.org/10.4269/ajtmh.21-0255
13. Alvarez- A, Del- S, Yáñez JA, Rosen MA, Mejia CR (2021) Influence of technostress on academic performance of university medicine students in Peru during the COVID-19 pandemic. Sustainability 13(16):8949. https://doi.org/10.3390/su13168949
14. Alvarez- A, Mlodzianowska S, García V, Rosen MA, Del- S (2021) Factors affecting green entrepreneurship intentions in business university students in COVID-19 pandemic times: Case of Ecuador. Sustainability 13(11):6447. https://doi.org/10.3390/su13116447
15. Alvarez- A, Rosen MA, Del- S (2020) A new regulation for supporting a circular economy in the plastic industry: the case of Peru (short communication). J Landsc Ecol 13(1):1–3. https://doi.org/10.2478/jlecol-2020-0004
16. Alvarez-Risco A, Dawson J, Johnson W, Conteh-Barrat M, Aslani P, Del-Aguila-Arcentales S, Diaz-Risco S (2021) Ebola virus disease outbreak: A global approach for health systems [Ebola; health professionals; global approach; pharmacist: pharmaceutical care; COVID-19.] 53 (4). http://www.revfarmacia.sld.cu/index.php/far/article/view/491
17. Alvarez-Risco A, Del-Aguila-Arcentales S (2021a) A note on changing regulation in international business: the World Intellectual Property Organization (WIPO) and artificial intelligence. In: Verbeke A, van Tulder R, Rose EL, Wei Y (eds), The multiple dimensions of institutional complexity in international business research, vol. 15. Emerald Publishing Limited, pp 363–371. https://doi.org/10.1108/S1745-886220210000015020
18. Alvarez-Risco A, Del-Aguila-Arcentales S (2021b) Public policies and private efforts to increase women entrepreneurship based on STEM background. In Mari M, Poggesi S, Foss L (eds), Women's entrepreneurship in STEM disciplines: issues and perspectives. Springer International Publishing, pp 75–87. https://doi.org/10.1007/978-3-030-83792-1_5
19. Alvarez-Risco A, Estrada-Merino A, Anderson-Seminario MdlM, Mlodzianowska S, García-Ibarra V, Villagomez-Buele C, Carvache-Franco M (2021) Multitasking behavior in online classrooms and academic performance: case of university students in Ecuador during COVID-19 outbreak. Interact Technol Smart Educ 18 (3): 422–434. https://doi.org/10.1108/ITSE-08-2020-0160
20. Alvarez-Risco A, Mejia CR, Delgado-Zegarra J, Del-Aguila-Arcentales S, Arce-Esquivel AA, Valladares-Garrido MJ et al. (2020) The Peru approach against the COVID-19 infodemic: insights and strategies. Am J Trop Med Hyg 103 (2): 583–586. https://doi.org/10.4269/ajtmh.20-0536
21. Alvarez-Risco A, Mlodzianowska S, Zamora-Ramos U, Del-Aguila S (2021) Green entrepreneurship intention in university students: the case of Peru. Entrep Bus Econ Rev 9 (4): 85–100. https://doi.org/10.15678/EBER.2021.090406
22. Alvarez-Risco A, Quipuzco-Chicata L, Escudero-Cipriani C (2021) Determinantes de la intención de recompra en línea en tiempos de COVID-19: evidencia de una economía emergente. Lecturas de Economía (96). https://doi.org/10.17533/udea.le.n96a342638
23. Amarillo CR (2020) Feminism on lockdown. NACLA Rep Am 52(3):274–281. https://doi.org/10.1080/10714839.2020.1809084

24. Apcho-Ccencho L-V, Cuya-Velásquez B-B, Alvarado Rodríguez D, Anderson-Seminario MdlM, Alvarez-Risco A, Estrada-Merino A, Mlodzianowska S (2021) The impact of international price on the technological industry in the United States and China during times of crisis: commercial war and COVID-19. In Lawrence KD, Klimberg RK (eds), Advances in business and management forecasting, vol. 14. Emerald Publishing Limited, pp 149–160. https://doi.org/10.1108/S1477-407020210000014010
25. Assaf A, Scuderi R (2020) COVID-19 and the recovery of the tourism industry. Tour Econ 26(5):731–733. https://doi.org/10.1177/1354816620933712
26. Bacq S, Lumpkin GT (2021) Social entrepreneurship and COVID-19. J Manage Stud 58(1):285–288. https://doi.org/10.1111/joms.12641
27. Bajan B, Mrówczyńska-Kamińska A (2020) Carbon footprint and environmental performance of agribusiness production in selected countries around the world. J Clean Prod 276. https://doi.org/10.1016/j.jclepro.2020.123389
28. Barello S, Caruso R, Palamenghi L, Nania T, Dellafiore F, Bonetti L et al. (2021) Factors associated with emotional exhaustion in healthcare professionals involved in the COVID-19 pandemic: an application of the job demands-resources model. Int Arch Occup Environ Health 94 (8): 1751–1761. https://doi.org/10.1007/s00420-021-01669-z
29. Baum T, Hai NTT (2020) Hospitality, tourism, human rights and the impact of COVID-19. Int J Contemp Hosp Manag 32(7):2397–2407. https://doi.org/10.1108/IJCHM-03-2020-0242
30. Belitski M, Guenther C, Kritikos AS, Thurik R (2021) Economic effects of the COVID-19 pandemic on entrepreneurship and small businesses. Small Bus Econ. https://doi.org/10.1007/s11187-021-00544-y
31. Block JH, Fisch C, Hirschmann M (2021) The determinants of bootstrap financing in crises: evidence from entrepreneurial ventures in the COVID-19 pandemic. Small Bus Econ. https://doi.org/10.1007/s11187-020-00445-6
32. Blomsma F, Pieroni M, Kravchenko M, Pigosso DCA, Hildenbrand J, Kristinsdottir AR et al. (2019) Developing a circular strategies framework for manufacturing companies to support circular economy-oriented innovation. J Clean Prod 241: 118271. https://doi.org/10.1016/j.jclepro.2019.118271
33. Borzée A, McNeely J, Magellan K, Miller JRB, Porter L, Dutta T et al. (2020) COVID-19 highlights the need for more effective wildlife trade legislation. Trends in Ecology & Evolution 35 (12): 1052–1055. https://doi.org/10.1016/j.tree.2020.10.001
34. Bozdağ F, Ergün N (2020) Psychological resilience of healthcare professionals during COVID-19 pandemic. Psychol Rep 124(6):2567–2586. https://doi.org/10.1177/0033294120965477
35. Brouder P (2020) Reset redux: possible evolutionary pathways towards the transformation of tourism in a COVID-19 world. Tour Geogr 22(3):484–490. https://doi.org/10.1080/14616688.2020.1760928
36. Brown R, Rocha A, Cowling M (2020) Financing entrepreneurship in times of crisis: Exploring the impact of COVID-19 on the market for entrepreneurial finance in the United Kingdom. Int Small Bus J 38(5):380–390. https://doi.org/10.1177/0266242620937464
37. Busch T, Hoffmann VH (2007) Emerging carbon constraints for corporate risk management. Ecol Econ 62(3):518–528. https://doi.org/10.1016/j.ecolecon.2006.05.022
38. Calvo-Porral C, Lévy-Mangin J-P (2020) The circular economy business model: examining consumers' acceptance of recycled goods. Adm Sci 10 (2). https://doi.org/10.3390/admsci10020028
39. Carvache M, Alvarez- A, Carvache O, Carvache W, Estrada-Merino A, Villalobos-Alvarez D (2021) Perceived value and its influence on satisfaction and loyalty in a coastal city: a study from Lima, Peru. J Policy Res Tour, Leis Events 1–16. https://doi.org/10.1080/19407963.2021.1883634
40. Carvache M, Alvarez- A, Carvache W, Carvache O, Estrada-Merino A, Rosen MA (2021) Coastal cities seen from loyalty and their tourist motivations: a study in Lima Peru. Sustainability 13(21):11575. https://doi.org/10.3390/su132111575

41. Carvache M, Carvache O, Carvache W, Alvarez- A, Estrada-Merino A (2021) Motivations and segmentation of the demand for coastal cities: A study in Lima Peru. Int J Tour Res 23(4):517–531. https://doi.org/10.1002/jtr.2423
42. del Castillo FA (2021) Temporary waiver of intellectual property on Covid-19 vaccines: toward the creation of a better, post-pandemic society. J Public Health 43(3):e559–e560. https://doi.org/10.1093/pubmed/fdab184
43. Cegarra J-G, Vătămănescu E-M, Martínez-Martínez A (2021) A context-driven approach on coping with COVID-19: From hiding knowledge toward citizen engagement. Knowl Process Manag 28(2):134–140. https://doi.org/10.1002/kpm.1662
44. Chafloque-Cespedes R, Alvarez-Risco A, Robayo-Acuña P-V, Gamarra-Chavez C-A, Martinez-Toro G-M, Vicente-Ramos W (2021) Effect of sociodemographic factors in entrepreneurial orientation and entrepreneurial intention in university students of Latin American business schools. In Jones P, Apostolopoulos N, Kakouris A, Moon C, Ratten V, Walmsley A (eds), Universities and entrepreneurship: meeting the educational and social challenges, vol. 11. Emerald Publishing Limited, pp 151–165. https://doi.org/10.1108/S2040-724620210000011010
45. Chafloque-Cespedes R, Vara-Horna A, Asencios-Gonzales Z, Lopez-Odar DR, Alvarez-Risco A, Quipuzco-Chicata L et al. (2020) Presentismo académico y violencia contra las mujeres en escuelas de negocios e ingeniería en universidades peruanas. Lecturas de Economía 0 (93): 127–153. https://doi.org/10.17533/udea.le.n93a340726
46. Chen C-C, Sujanto RY, Tseng M-L, Fujii M, Lim MK (2021) Sustainable consumption transition model: social concerns and waste minimization under willingness-to-pay in Indonesian food industry. Resour Conserv Recycl 170. https://doi.org/10.1016/j.resconrec.2021.105590
47. Chen T, Peng L, Yin X, Rong J, Yang J, Cong G (2020) Analysis of user satisfaction with online education platforms in China during the COVID-19 pandemic. Healthcare 8 (3). https://doi.org/10.3390/healthcare8030200
48. Chen X, Zhang SX, Jahanshahi AA, Alvarez- A, Dai H, Li J, Ibarra VG (2020) Belief in a COVID-19 conspiracy theory as a predictor of mental health and well-being of health care workers in Ecuador: cross-sectional survey study. JMIR Public Health Surveill 6(3). https://doi.org/10.2196/20737
49. Cheng M, Chen G, Wiedmann T, Hadjikakou M, Xu L, Wang Y (2020) The sharing economy and sustainability—assessing Airbnb's direct, indirect and induced carbon footprint in Sydney. J Sustain Tour 28(8):1083–1099. https://doi.org/10.1080/09669582.2020.1720698
50. Chung SA, Olivera S, Rojas Román B, Alanoca E, Moscoso S, Limpias Terceros B et al. (2021) Temáticas de la producción científica de la Revista Cubana de Farmacia indizada en Scopus (1967–2020) [Revista Cubana de Farmacia; farmacia; Cuba; evidencia] 54 (1). http://www.revfarmacia.sld.cu/index.php/far/article/view/511
51. Cruz-Torres W, Alvarez- A, Del- S (2021) Impact of Enterprise Resource Planning (ERP) implementation on performance of an education enterprise: a Structural Equation Modeling (SEM). Studies in Business and Economics 16(2):37–52. https://doi.org/10.2478/sbe-2021-0023
52. Dagiliene L, Frendzel M, Sutiene K, Wnuk-Pel T (2020) Wise managers think about circular economy, wiser report and analyze it. Research of environmental reporting practices in EU manufacturing companies. J Clean Prod 274: 121968. https://doi.org/10.1016/j.jclepro.2020.121968
53. Dagilienė L, Varaniūtė V, Bruneckienė J (2021) Local governments' perspective on implementing the circular economy: a framework for future solutions. J Clean Prod 310. https://doi.org/10.1016/j.jclepro.2021.127340
54. Daou A, Mallat C, Chammas G, Cerantola N, Kayed S, Saliba NA (2020) The Ecocanvas as a business model canvas for a circular economy. J Clean Prod 258. https://doi.org/10.1016/j.jclepro.2020.120938
55. Deressa W, Worku A, Abebe W, Gizaw M, Amogne W (2021) Risk perceptions and preventive practices of COVID-19 among healthcare professionals in public hospitals in Addis Ababa Ethiopia. PLoS ONE 16(6). https://doi.org/10.1371/journal.pone.0242471

56. Dominelli L (2021) A green social work perspective on social work during the time of COVID-19. Int J Soc Welf 30(1):7–16. https://doi.org/10.1111/ijsw.12469
57. Dong H, Ohnishi S, Fujita T, Geng Y, Fujii M, Dong L (2014) Achieving carbon emission reduction through industrial & urban symbiosis: a case of Kawasaki. Energy 64:277–286. https://doi.org/10.1016/j.energy.2013.11.005
58. Duarte Alonso A, Kok SK, Bressan A, O'Shea M, Sakellarios N, Koresis A et al. (2020) COVID-19, aftermath, impacts, and hospitality firms: An international perspective. Int J Hosp Manag, 91: 102654. https://doi.org/10.1016/j.ijhm.2020.102654
59. Erfani P, Binagwaho A, Jalloh MJ, Yunus M, Farmer P, Kerry V (2021) Intellectual property waiver for covid-19 vaccines will advance global health equity. BMJ 374. https://doi.org/10.1136/bmj.n1837
60. FAO (2015) Food wastage footprint & climate change. Retrieved 12/22/2021 from https://www.fao.org/3/bb144e/bb144e.pdf
61. Fang K, Heijungs R, de Snoo GR (2014) Theoretical exploration for the combination of the ecological, energy, carbon, and water footprints: overview of a footprint family. Ecol Ind 36:508–518. https://doi.org/10.1016/j.ecolind.2013.08.017
62. Fotiadis A, Polyzos S, Huan T-CTC (2021) The good, the bad and the ugly on COVID-19 tourism recovery. Ann Tour Res 87. https://doi.org/10.1016/j.annals.2020.103117
63. Galanakis CM, Rizou M, Aldawoud TMS, Ucak I, Rowan NJ (2021) Innovations and technology disruptions in the food sector within the COVID-19 pandemic and post-lockdown era. Trends Food Sci Technol 110:193–200. https://doi.org/10.1016/j.tifs.2021.02.002
64. Geisendorf S, Pietrulla F (2018) The circular economy and circular economic concepts—a literature analysis and redefinition. Thunderbird Int Bus Rev 60(5):771–782. https://doi.org/10.1002/tie.21924
65. Ghisellini P, Cialani C, Ulgiati S (2016) A review on circular economy: the expected transition to a balanced interplay of environmental and economic systems. J Clean Prod 114:11–32. https://doi.org/10.1016/j.jclepro.2015.09.007
66. Gulati G, Kelly BD (2020) Domestic violence against women and the COVID-19 pandemic: what is the role of psychiatry? Int J Law Psychiatry 71. https://doi.org/10.1016/j.ijlp.2020.101594
67. Gursoy D, Chi CG (2020) Effects of COVID-19 pandemic on hospitality industry: review of the current situations and a research agenda. J Hosp Market Manag 29(5):527–529. https://doi.org/10.1080/19368623.2020.1788231
68. Hanzl M (2020) Urban forms and green infrastructure—the implications for public health during the COVID-19 pandemic. Cities & Health 1–5. https://doi.org/10.1080/23748834.2020.1791441
69. Harangozo G, Szigeti C (2017) Corporate carbon footprint analysis in practice—with a special focus on validity and reliability issues. J Clean Prod 167:1177–1183. https://doi.org/10.1016/j.jclepro.2017.07.237
70. Hepburn, C., O'Callaghan, B., Stern, N., Stiglitz, J., & Zenghelis, D. (2020). Will COVID-19 fiscal recovery packages accelerate or retard progress on climate change? *Oxford Review of Economic Policy*, *36*(Supplement_1), S359-S381. https://doi.org/10.1093/oxrep/graa015
71. Ho C-Y, Tsai B-H, Chen C-S, Lu M-T (2021) Exploring green marketing orientations toward sustainability the hospitality industry in the COVID-19 pandemic. Sustainability 13 (8). https://doi.org/10.3390/su13084348
72. Ivanova M (2012) Institutional design and UNEP reform: historical insights on form, function and financing. Int Aff 88(3):565–584. https://doi.org/10.1111/j.1468-2346.2012.01089.x
73. Jantke K, Hartmann MJ, Rasche L, Blanz B, Schneider UA (2020) Agricultural greenhouse gas emissions: Knowledge and positions of German farmers. Land 9 (5). https://doi.org/10.3390/land9050130
74. Jawahir IS, Bradley R (2016) Technological elements of circular economy and the principles of 6R-based closed-loop material flow in sustainable manufacturing. Procedia CIRP 40:103–108. https://doi.org/10.1016/j.procir.2016.01.067

75. Jecker NS, Atuire CA (2021) What's yours is ours: waiving intellectual property protections for COVID-19 vaccines. J Med Ethics 47(9):595. https://doi.org/10.1136/medethics-2021-107555
76. Jeswani HK, Figueroa-Torres G, Azapagic A (2021) The extent of food waste generation in the UK and its environmental impacts. Sustain Prod Consum 26:532–547. https://doi.org/10.1016/j.spc.2020.12.021
77. Kaushal V, Srivastava S (2021) Hospitality and tourism industry amid COVID-19 pandemic: perspectives on challenges and learnings from India. Int J Hosp Manag 92. https://doi.org/10.1016/j.ijhm.2020.102707
78. Kirk CP, Rifkin LS (2020) I'll trade you diamonds for toilet paper: consumer reacting, coping and adapting behaviors in the COVID-19 pandemic. J Bus Res 117:124–131. https://doi.org/10.1016/j.jbusres.2020.05.028
79. Koetter P, Pelton M, Gonzalo J, Du P, Exten C, Bogale K et al (2020) Implementation and process of a COVID-19 contact tracing initiative: leveraging health professional students to extend the workforce during a pandemic. Am J Infect Control 48(12):1451–1456. https://doi.org/10.1016/j.ajic.2020.08.012
80. Krishtel P, Malpani R (2021) Suspend intellectual property rights for covid-19 vaccines. BMJ 373. https://doi.org/10.1136/bmj.n1344
81. Lackie K, Najjar G, El- A, Frost J, Green C, Langlois S et al (2020) Interprofessional education and collaborative practice research during the COVID-19 pandemic: considerations to advance the field. J Interprof Care 34(5):583–586. https://doi.org/10.1080/13561820.2020.1807481
82. Lashley MA, Acevedo M, Cotner S, Lortie CJ (2020) How the ecology and evolution of the COVID-19 pandemic changed learning. Ecol Evol 10(22):12412–12417. https://doi.org/10.1002/ece3.6937
83. Lee RP, Keller F, Meyer B (2017) A concept to support the transformation from a linear to circular carbon economy: net zero emissions, resource efficiency and conservation through a coupling of the energy, chemical and waste management sectors. Clean Energy 1(1):102–113. https://doi.org/10.1093/ce/zkx004
84. Leiva-Martinez M-A, Anderson-Seminario MdlM, Alvarez-Risco A, Estrada-Merino A, Mlodzianowska S (2021) Price variation in lower goods as of previous economic crisis and the contrast of the current price situation in the context of COVID-19 in Peru. In Lawrence KD, Klimberg RK (eds), Advances in business and management forecasting, vol. 14. Emerald Publishing Limited, pp 161–166. https://doi.org/10.1108/S1477-407020210000014011
85. Lieder M, Rashid A (2016) Towards circular economy implementation: a comprehensive review in context of manufacturing industry. J Clean Prod 115:36–51. https://doi.org/10.1016/j.jclepro.2015.12.042
86. Liguori E, Winkler C (2020) From offline to online: challenges and opportunities for entrepreneurship education following the COVID-19 pandemic. Entrep Educ Pedagog 3(4):346–351. https://doi.org/10.1177/2515127420916738
87. Lim S, Prakash A (2021) Pandemics and citizen perceptions about their country: Did COVID-19 increase national pride in South Korea? Nations Natl 27(3):623–637. https://doi.org/10.1111/nana.12749
88. Lonca G, Lesage P, Majeau-Bettez G, Bernard S, Margni M (2020) Assessing scaling effects of circular economy strategies: a case study on plastic bottle closed-loop recycling in the USA PET market. Resour Conserv Recycl 162. https://doi.org/10.1016/j.resconrec.2020.105013
89. Lopez- D, Alvarez- A, Vara-Horna A, Chafloque-Cespedes R, Sekar MC (2020) Validity and reliability of the questionnaire that evaluates factors associated with perceived environmental behavior and perceived ecological purchasing behavior in Peruvian consumers. Soc Responsib J 16(3):403–417. https://doi.org/10.1108/SRJ-08-2018-0201
90. Lüdeke F, Gold S, Bocken NMP (2019) A review and typology of circular economy business model patterns. J Ind Ecol 23(1):36–61. https://doi.org/10.1111/jiec.12763
91. Maritz A, Perenyi A, de Waal G, Buck C (2020) Entrepreneurship as the unsung hero during the current COVID-19 economic crisis: Australian perspectives. Sustainability 12(11). https://doi.org/10.3390/su12114612

92. Matthews HS, Hendrickson CT, Weber CL (2008) The importance of carbon footprint estimation boundaries. Environ Sci Technol 42(16):5839–5842. https://doi.org/10.1021/es703112w
93. Matuštík J, Kočí V (2021) What is a footprint? a conceptual analysis of environmental footprint indicators. J Clean Prod 285. https://doi.org/10.1016/j.jclepro.2020.124833
94. Michelini G, Moraes RN, Cunha RN, Costa JMH, Ometto AR (2017) From linear to circular economy: PSS conducting the transition. Procedia CIRP 64:2–6. https://doi.org/10.1016/j.procir.2017.03.012
95. Mohamed Sultan AA, Mativenga PT (2019) Sustainable Location Identification Decision Protocol (SuLIDeP) for determining the location of recycling centres in a circular economy. J Clean Prod 223:508–521. https://doi.org/10.1016/j.jclepro.2019.03.104
96. Obergassel W, Hermwille L, Oberthür S (2021) Harnessing international climate governance to drive a sustainable recovery from the COVID-19 pandemic. Clim Policy 21(10):1298–1306. https://doi.org/10.1080/14693062.2020.1835603
97. Okereke M (2021) Towards vaccine equity: should big pharma waive intellectual property rights for COVID-19 vaccines? Public Health Pract 2. https://doi.org/10.1016/j.puhip.2021.100165
98. Onat NC, Kucukvar M (2020) Carbon footprint of construction industry: a global review and supply chain analysis. Renew Sustain Energy Rev 124. https://doi.org/10.1016/j.rser.2020.109783
99. Parajuly K, Fitzpatrick C, Muldoon O, Kuehr R (2020) Behavioral change for the circular economy: a review with focus on electronic waste management in the EU. Resour, Conserv & Recycl X (6): 100035. https://doi.org/10.1016/j.rcrx.2020.100035
100. Parashar S, Sood G, Agrawal N (2020) Modelling the enablers of food supply chain for reduction in carbon footprint. J Clean Prod 275. https://doi.org/10.1016/j.jclepro.2020.122932
101. Peters GP (2010) Carbon footprints and embodied carbon at multiple scales. Curr Opin Environ Sustain 2(4):245–250. https://doi.org/10.1016/j.cosust.2010.05.004
102. Pieroni MPP, McAloone TC, Pigosso DCA (2019) Business model innovation for circular economy and sustainability: a review of approaches. J Clean Prod 215:198–216. https://doi.org/10.1016/j.jclepro.2019.01.036
103. Pieroni MPP, McAloone TC, Pigosso DCA (2021) Circular economy business model innovation: Sectorial patterns within manufacturing companies. J Clean Prod 286. https://doi.org/10.1016/j.jclepro.2020.124921
104. Pullin AS, Knight TM (2009) Doing more good than harm—building an evidence-base for conservation and environmental management. Biol Cons 142(5):931–934. https://doi.org/10.1016/j.biocon.2009.01.010
105. Quispe-Cañari JF, Fidel-Rosales E, Manrique D, Mascaró-Zan J, Huamán-Castillón KM, Chamorro–Espinoza SE et al. (2021) Self-medication practices during the COVID-19 pandemic among the adult population in Peru: A cross-sectional survey. Saudi Pharm J 29 (1): 1–11. https://doi.org/10.1016/j.jsps.2020.12.001
106. Raudenská J, Steinerová V, Javůrková A, Urits I, Kaye AD, Viswanath O, Varrassi G (2020) Occupational burnout syndrome and post-traumatic stress among healthcare professionals during the novel coronavirus disease 2019 (COVID-19) pandemic. Best Pract Res Clin Anaesthesiol 34(3):553–560. https://doi.org/10.1016/j.bpa.2020.07.008
107. Rees WE (1992) Ecological footprints and appropriated carrying capacity: what urban economics leaves out. Environ Urban 4(2):121–130. https://doi.org/10.1177/095624789200400212
108. Remane G, Hanlet A, Tesch JF, Kolbe LM (2017) The business model pattern database—a tool for systematic model innovation. Int J Innov Manag 21(01):1750004. https://doi.org/10.1142/s1363919617500049
109. Robinson OJ, Tewkesbury A, Kemp S, Williams ID (2018) Towards a universal carbon footprint standard: a case study of carbon management at universities. J Clean Prod 172:4435–4455. https://doi.org/10.1016/j.jclepro.2017.02.147

110. Roesch E, Amin A, Gupta J, García C (2020) Violence against women during covid-19 pandemic restrictions. BMJ 369. https://doi.org/10.1136/bmj.m1712
111. Rojas Román B, Moscoso S, Chung SA, Limpias Terceros B, Álvarez-Risco A, Yáñez JA (2020) Tratamiento de la COVID-19 en Perú y Bolivia y los riesgos de la automedicación [COVID-19; tratamiento; estrategia; fármacos; automedicación; Perú; Bolivia] 53 (2). http://www.revfarmacia.sld.cu/index.php/far/article/view/435/351
112. Ryen EG, Gaustad G, Babbitt CW, Babbitt G (2018) Ecological foraging models as inspiration for optimized recycling systems in the circular economy. Resour Conserv Recycl 135:48–57. https://doi.org/10.1016/j.resconrec.2017.08.006
113. Sarkis J, Dijkshoorn J (2007) Relationships between solid waste management performance and environmental practice adoption in Welsh small and medium-sized enterprises (SMEs). Int J Prod Res 45(21):4989–5015. https://doi.org/10.1080/00207540600690529
114. Schanes K, Giljum S, Hertwich E (2016) Low carbon lifestyles: A framework to structure consumption strategies and options to reduce carbon footprints. J Clean Prod 139:1033–1043. https://doi.org/10.1016/j.jclepro.2016.08.154
115. Schröder P, Lemille A, Desmond P (2020) Making the circular economy work for human development. Resour Conserv Recycl 156. https://doi.org/10.1016/j.resconrec.2020.104686
116. Sekalala S, Forman L, Hodgson T, Mulumba M, Namyalo-Ganafa H, Meier BM (2021) Decolonising human rights: how intellectual property laws result in unequal access to the COVID-19 vaccine. BMJ Glob Health 6(7). https://doi.org/10.1136/bmjgh-2021-006169
117. Sekito T, Prayogo TB, Dote Y, Yoshitake T, Bagus I (2013) Influence of a community-based waste management system on people's behavior and waste reduction. Resour Conserv Recycl 72:84–90. https://doi.org/10.1016/j.resconrec.2013.01.001
118. Selvefors A, Rexfelt O, Renström S, Strömberg H (2019) Use to use—a user perspective on product circularity. J Clean Prod 223:1014–1028. https://doi.org/10.1016/j.jclepro.2019.03.117
119. Shakoor A, Shahbaz M, Farooq TH, Sahar NE, Shahzad SM, Altaf MM, Ashraf M (2021) A global meta-analysis of greenhouse gases emission and crop yield under no-tillage as compared to conventional tillage. Sci Total Environ 750. https://doi.org/10.1016/j.scitotenv.2020.142299
120. Shaw R, Kim Y-K, Hua J (2020) Governance, technology and citizen behavior in pandemic: lessons from COVID-19 in East Asia. Prog Disaster Sci 6. https://doi.org/10.1016/j.pdisas.2020.100090
121. Shi S, Yin J (2021) Global research on carbon footprint: a scientometric review. Environ Impact Assess Rev 89. https://doi.org/10.1016/j.eiar.2021.106571
122. Sigala M (2020) Tourism and COVID-19: impacts and implications for advancing and resetting industry and research. J Bus Res 117:312–321. https://doi.org/10.1016/j.jbusres.2020.06.015
123. Simpson D (2012) Institutional pressure and waste reduction: the role of investments in waste reduction resources. Int J Prod Econ 139(1):330–339. https://doi.org/10.1016/j.ijpe.2012.05.020
124. Smith P, Martino D, Cai Z, Gwary D, Janzen H, Kumar P et al. (2007) Policy and technological constraints to implementation of greenhouse gas mitigation options in agriculture. Agric, Ecosyst & Environ 118 (1): 6–28. https://doi.org/10.1016/j.agee.2006.06.006
125. Sovacool BK, Brown MA (2010) Twelve metropolitan carbon footprints: a preliminary comparative global assessment. Energy Policy 38(9):4856–4869. https://doi.org/10.1016/j.enpol.2009.10.001
126. Stocker T (2014) Climate change 2013: the physical science basis: working Group I contribution to the fifth assessment report of the intergovernmental panel on climate change. Cambridge university press.
127. Su B, Heshmati A, Geng Y, Yu X (2013) A review of the circular economy in China: moving from rhetoric to implementation. J Clean Prod 42:215–227. https://doi.org/10.1016/j.jclepro.2012.11.020

128. Sundarakani B, de Souza R, Goh M, Wagner SM, Manikandan S (2010) Modeling carbon footprints across the supply chain. Int J Prod Econ 128(1):43–50. https://doi.org/10.1016/j.ijpe.2010.01.018
129. Sánchez OR, Vale DB, Rodrigues L, Surita FG (2020) Violence against women during the COVID-19 pandemic: an integrative review. Int J Gynecol Obstet 151(2):180–187. https://doi.org/10.1002/ijgo.13365
130. Sánchez-Clavijo LM, Martínez-Callejas SJ, Acevedo-Charry O, Diaz-Pulido A, Gómez-Valencia B, Ocampo-Peñuela N et al. (2021) Differential reporting of biodiversity in two citizen science platforms during COVID-19 lockdown in Colombia. Biol Conserv 256: 109077. https://doi.org/10.1016/j.biocon.2021.109077
131. Tam VWY (2008) On the effectiveness in implementing a waste-management-plan method in construction. Waste Manage 28(6):1072–1080. https://doi.org/10.1016/j.wasman.2007.04.007
132. Tonini D, Albizzati PF, Astrup TF (2018) Environmental impacts of food waste: Learnings and challenges from a case study on UK. Waste Manage 76:744–766. https://doi.org/10.1016/j.wasman.2018.03.032
133. Tran LTT (2021) Managing the effectiveness of e-commerce platforms in a pandemic. J Retail Consum Serv 58. https://doi.org/10.1016/j.jretconser.2020.102287
134. Tubiello FN, Salvatore M, Rossi S, Ferrara A, Fitton N, Smith P (2013) The FAOSTAT database of greenhouse gas emissions from agriculture. Environ Res Lett 8(1). https://doi.org/10.1088/1748-9326/8/1/015009
135. United Nations (2013) UN report: one-third of world's food wasted annually, at great economic, environmental cost. Retrieved 12/28/2021 from https://news.un.org/en/story/2013/09/448652
136. Viau S, Majeau-Bettez G, Spreutels L, Legros R, Margni M, Samson R (2020) Substitution modelling in life cycle assessment of municipal solid waste management. Waste Manage 102:795–803. https://doi.org/10.1016/j.wasman.2019.11.042
137. Vidya CT, Prabheesh KP (2020) Implications of COVID-19 pandemic on the global trade networks. Emerg Mark Financ Trade 56(10):2408–2421. https://doi.org/10.1080/1540496X.2020.1785426
138. Viero A, Barbara G, Montisci M, Kustermann K, Cattaneo C (2021) Violence against women in the Covid-19 pandemic: a review of the literature and a call for shared strategies to tackle health and social emergencies. Forensic Sci Int 319. https://doi.org/10.1016/j.forsciint.2020.110650
139. Wen J, Kozak M, Yang S, Liu F (2021) COVID-19: potential effects on Chinese citizens' lifestyle and travel. Tour Rev 76(1):74–87. https://doi.org/10.1108/TR-03-2020-0110
140. Wiedmann T, Minx J (2008) A definition of 'carbon footprint.' Ecol Econ Res Trends 1:1–11
141. Wu T-Y, Ford O, Rainville AJ, Bessire R (2020) COVID-19 care package distribution for senior citizens and families in Detroit and Hamtramck, Michigan. J Hunger & Environ Nutr 15(4):585–587. https://doi.org/10.1080/19320248.2020.1799586
142. Yan J, Kim S, Zhang SX, Foo M-D, Alvarez- A, Del- S, Yáñez JA (2021) Hospitality workers' COVID-19 risk perception and depression: a contingent model based on transactional theory of stress model. Int J Hosp Manag 95. https://doi.org/10.1016/j.ijhm.2021.102935
143. Yasin Ar A (2020) Managing e-commerce during a pandemic: lessons from GrubHub during COVID-19. In George B, Mahar Q (eds), International case studies in the management of disasters. Emerald Publishing Limited, pp 155–167. https://doi.org/10.1108/978-1-83982-186-820201010
144. Yáñez JA, Afshar A, Alvarez- A, Li J, Zhang SX (2020) Anxiety, distress, and turnover intention of healthcare workers in Peru by their distance to the epicenter during the COVID-19 crisis. Am J Trop Med Hyg 103(4):1614–1620. https://doi.org/10.4269/ajtmh.20-0800
145. Yáñez JA, Alvarez- A, Delgado- J (2020) Covid-19 in Peru: from supervised walks for children to the first case of Kawasaki-like syndrome. BMJ 369. https://doi.org/10.1136/bmj.m2418
146. Zarocostas J (2021) What next for a COVID-19 intellectual property waiver? The Lancet 397(10288):1871–1872. https://doi.org/10.1016/S0140-6736(21)01151-X

147. Zhang SX, Chen J, Afshar A, Alvarez- A, Dai H, Li J, Patty-Tito RM (2021) Succumbing to the COVID-19 pandemic—healthcare workers not satisfied and intend to leave their jobs. Int J Ment Heal Addict. https://doi.org/10.1007/s11469-020-00418-6
148. Zhang SX, Sun S, Afshar A, Alvarez- A, Ibarra VG, Li J, Patty-Tito RM (2020) Developing and testing a measure of COVID-19 organizational support of healthcare workers—results from Peru, Ecuador, and Bolivia. Psychiatry Res 291. https://doi.org/10.1016/j.psychres.2020.113174
149. Zollinger R, DiCindio C (2021) Community ecology: museum education and the digital divide during and after COVID-19. J MusM Educ 46(4):481–492. https://doi.org/10.1080/10598650.2021.1983711

Food Loss Reduction and Carbon Footprint Practices Worldwide: A Benchmarking Approach of Circular Economy

Sharon Esquerre-Botton, Aldo Alvarez-Risco, Luigi Leclercq-Machado, Maria de las Mercedes Anderson-Seminario, and Shyla Del-Aguila-Arcentales

Abstract Unsustainable food supply chains are negatively affecting the environment. As a result, this topic has been debated worldwide, and several actors came to the same conclusions: intervention and actions are a must. Nonetheless, this approach was not standardized, and countries and companies are implementing circular economies differently. This chapter aims to outline the current circular economy approach taken internationally and discuss the barriers of this last. A qualitative analysis was conducted by collecting several case studies and understanding the identified practices and challenges. In the findings, the authors discovered that countries' policies remain weak globally, though the demand is increasing. Drawbacks and limitations such as economic resources, technological innovation, and incentives are evidenced.

Keywords Circular economy · Food waste · Carbon footprint · Policies · Sustainable development

1 Introduction

In recent times, the term Circular Economy (CE) has been spreading, not only from generation to generation but also throughout different nations; this proposal works as a solution to different challenges such as waste generation, shortage of resources, and sustaining economic benefits [106]. The specifications of the CE concept consist of are implemented in different aspects to be regulated to configure new models in production and consumption systems [70]. Governmental and non-governmental entities have identified the need to improve the food industry, CE is modified to be

S. Esquerre-Botton · A. Alvarez-Risco (✉) · L. Leclercq-Machado ·
M. de las Mercedes Anderson-Seminario
Universidad de Lima, Lima, Perú
e-mail: aralvare@ulima.edu.pe

S. Del-Aguila-Arcentales
Escuela Nacional de Marina Mercante "Almirante Miguel Grau", Callao, Perú

A. Alvarez-Risco et al. (eds.), *Circular Economy*, Environmental Footprints and Eco-design of Products and Processes,
https://doi.org/10.1007/978-981-19-0549-0_8

used in agricultural processes, food chain, and alimentary systems, to reduce food loss and waste (FLW) and other adverse effects of traditional production [1, 59, 70]. For the development of the CE, [80] establish the existence of three circles: the corporate level (this means that the responsibility lies with the manufacturers and their production processes); the interfirm level (created to increase the commercialization of by-products). Finally, social level (which refers to eco-friendly activities that encourage a change in society's attitude toward intelligent consumption).

Today's high growth of world consumerism has brought significant consequences to the environment, making its effects irreversible for the future. The application of the linear economy has not been favorable in waste management; on the contrary, the waste of final products threatens the planet both in the short and long term [93].

On the other hand, this problem has taken more significant consideration in recent years. One of the primary triggers has been climate change and biodiversity loss, so there is now a greater awareness of the environmental lobby and its purpose [128]. For their part, some irresponsible consumers create a gap between regular consumption and sustainable consumption, preventing such waste from being minimized and achieving sustainable consumption. In addition, this model proposes to generate greater consumer interest toward sustainable questions [54].

The CF and its implication to the environment are crucial in analyzing the environmental impact. The CF is probably the best-known indicator for measuring environmental impact. It is responsible for measuring the amount of CH_4 and CO_2 emissions on a chosen population or activity; that is, it measures the impact of certain activities that threaten the Earth's climate [113]. Among the main emitters of these gases, we find large companies, both manufacturing and non-manufacturing, and developing countries as primary threats to the environment [61, 138]. In addition, there is work on institutional organizations (OI) to ensure climate risks, increasingly complex challenges due to political factors, adaptability, interests, and others. Many regional governments tend to have a strong fragmentation between policy communities, which leads to the non-application of the actions proposed by OI. A similar case is experienced internationally [92].

Human life has changed, and people's habits regarding sustainability have changed, depending on the characteristics of pre-pandemic life. Due to limited sources by COVID-19, new ways to build new materials are essential in the reactivation of the global economies. In Table 1, detail the sectors impacted by COVID-19 pandemic.

The purpose of this chapter is to develop an approach related to the importance of the CE in the face of alimentary systems within four sections; of which the influence of the CE in the reduction of food waste is exposed, in addition to the policies implemented by different countries, the strategies generated by companies and the actions carried out by individual actors, finally the barriers that interpose the application and appreciation of the Circular Economy.

Table 1 Changes created by COVID-19 in different sectors

Education	[10, 11, 14, 16, 22, 56, 102, 103, 166]
Entrepreneurship	[6, 17, 24] [21, 30, 34, 36, 42, 52, 107, 112]
Health sector	[2, 9, 12, 19, 15, 32, 40, 55, 63, 100, 130, 134, 161, 164, 165]
Hospitality	[66, 86, 90, 97, 159]
Intellectual property	[13, 20, 49, 67, 94, 101, 123, 137, 163]
Population	[8, 23, 51, 87, 108, 129, 148, 141, 156, 157, 162]
Prices	[27, 57, 75, 89, 105, 122]
Tourism	[29, 33, 41, 44–48, 74, 142]
Trade	[3, 4, 7, 25] [18, 39, 60, 98, 109, 150, 153, 160]
Violence against women	[26, 53, 65, 85, 133, 147, 154]

2 Circular Economy Approach for Food Loss Reduction and Carbon Footprint

As global food waste increases, changes along the food supply chain are necessary [5]. Circular economy (CE) has been defined as a strategy that provides a business perspective to achieve sustainability and increase brand positioning and profits [79, 124]. CE initiatives help companies reduce food waste and pollution by ensuring the maximum value utilization of resources and outputs [31, 81]. Industries and policymakers design regulations and standards so that businesses adopt CE principles to reduce food waste and carbon footprint [124]. According to [155], policy measures can be classified as the following (Fig. 1):

Despite the development of policies aimed at reducing food waste and carbon footprint, the difficulty is evidenced in the actual transformation of these lasts into laws, regulations, standards, awareness campaigns, fiscal incentives, among others [131]. Businesses must adapt and deal with several obstacles such as financial aspects, regulations, corporate culture, and others [5]. One of the reasons for such change is that the Circular Economy requires consumers and companies to change behaviors,

Fig. 1 Classification of policy measures (*Source* Adapted from [155])

consumption patterns, manufacturing practices, etc. [64, 127]. This situation can result in a more challenging transition toward a circular economy approach, which may discourage organizations from acting [78]. Nonetheless, the correct implementation of circular economy practices in the food supply chain can give products a "second life" and be used for new purposes [127]. A circular economy may imply the development of strategies for waste reduction through innovation and technological development [38].

A circular economy allows the incorporation of the "waste = food" concept, meaning industries and organizations can use food dispositions as reusable materials [38, 82]. As it can be seen, though CE offers several opportunities to reduce waste by adopting several plans of action, it is not easy to achieve [152]. Governments and organizations can help companies achieve CE goals in various ways. Based on the Fig. 2, three types of variables influence food waste and carbon footprint reduction.

For actors, non-lucrative organizations can collect food surplus from companies such as supermarkets and donate this last as part of the interventions on food waste policy [104]. Supermarkets, as a result, are crucial for the decrease in food waste. These lasts were already influenced by different kinds of legislation, awareness campaigns, and incentives to foment food surplus donations [72]. Other exciting participants are schools. These can adopt several initiatives to reduce waste, such as catering staff training, recompositing of menus, and reduction of portion sizes [37, 151]. Finally, on the policy-initiatives side, governments can foment the private sector contribution with fiscal incentives, awareness campaigns, waste quotas regulations, among others, which result in the reduction of food waste and carbon footprint.

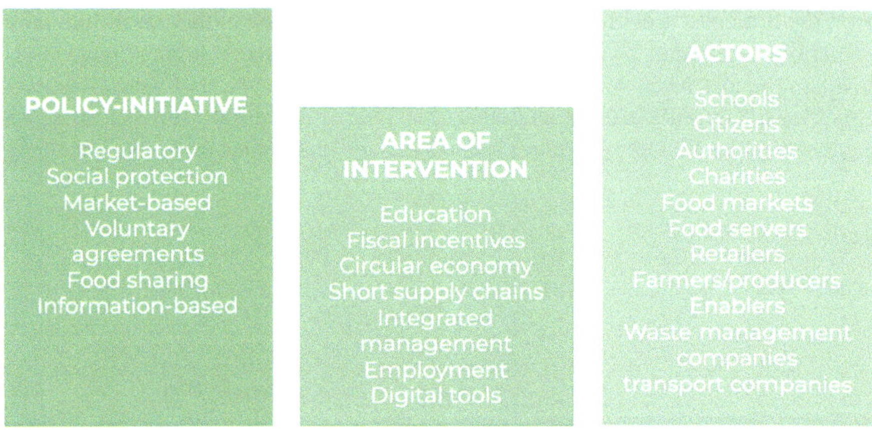

Fig. 2 Urban food waste policy (*Source* Adapted from [71])

3 Food Loss Reduction Practices Worldwide

3.1 Countries' Policy

During countries' development, governments adopted several strategies to reduce food waste. For instance, France adopted the Law 2016/138 to promote a donation from supermarkets' unsold food to organizations [116]. Others, such as Romania, are promoting voluntary commitment from companies in the agri-food sector to participate in awareness campaigns conducted by national authorities [126] (Table 2).

A government approach that promotes private sector contribution reduces food waste and carbon footprint. Indeed, supportive infrastructure and collaboration between actors are crucial for creating awareness campaigns and promoting food loss reduction initiatives [143]. For instance, in the United Kingdom, the "Love Food Hate Waste" campaign was raised in 2007 to increase consciousness from consumers and provide tips and measures to reduce their food waste [135]. Therefore, laws that encourage closed-loop business models result in a higher probability of sustainable operations implementation behaviors from companies [143]. Campaigns, as mentioned above, are not the only way to reduce food waste. Countries adopt different approaches.

An example is France, which implemented penalties if companies are not complying with the Law 2016/138 [62]. This last mandated supermarket to sign

Table 2 Food loss reduction policies practices

Country	Food loss reduction policies	Sources
Italia	National Waste Prevention Program, 2006 Law against food waste, Law No 166/2016	Giordano et al. [83], Arcuri [28]
France	Law for fighting food waste, Law No 2016-138	Condamine [58], Bunting Eubanks [43]
United States	Bill Emerson Food Donation Law, 1996 Federal Food Donation Act, 2008 Food Recovery Act, 2016	Bunting Eubanks [43]
Romania	Romanian Law 217-2016 against food waste	[126]
Spain	"More Food, Less Waste" Strategy, 2013	Magrini et al. [111]
Sri Lanka	National Agriculture Research Policy and Strategy	[131]
Czech Republic	Amendment 180/2016 on Food Act 110/1997	[135]

a non-profit organization to collect unsold food. In the case of Italy, the 166/2016 legislation offers to donators tax exemption, new funds, among other benefits [77]. For the United States, the American law before the Bill Emerson Donation Law followed the following approach. If a person gets sick from the food consumed, the direct distributor is automatically liable and responsible [77]. As fear of getting sued, companies, people, and organizations were not donating, though their desires were the opposite. It is worth emphasizing that national and supranational policies focus on achieving Sustainable Development Goals (SDGs), such as the number 12, focused on reducing and minimizing food waste by 2030 [88]. As a result, the growing international and national policies' development regarding circular economy practices is becoming more critical over the years [78, 144].

3.2 Companies and Actors Strategies

We can observe that several organizations of different sectors are adopting CE initiatives to improve economic-social-environmental development [35, 140]. Taking the example of charitable food aid in Finland, those organizations aimed at providing a solution for food poverty became one of the strategies to embrace sustainability as they reduce food waste disposal from retail sectors [149]. One of the problems evidenced in this industry is the shelf life, which increases food waste [50]. In Indonesia, an organic fertilizer company sends their output waste to other sectors such as the poultry and fish supply chain and reduces, consequently, final disposal [121]. Previous studies such as the one made by [70] identified 40 case studies of companies worldwide which implemented circular economy principles. Though they belong to different industries, they achieved several sustainable development goals by adopting circular economy principles and initiatives. For instance, the agri-food sector is considered one of the significant sectors of greenhouse emissions, and specific companies are redesigning their supply chain to reduce this indicator [136]. An example is 9 agri-food companies based in Italy, which developed corporate social responsibility initiatives as circular economy practices such as transforming waste into resources [73].

4 Carbon Footprint Reduction Practices Worldwide

4.1 Countries' Policy

As for food waste, countries are also implementing carbon footprint reduction practices to meet several sustainable development goals (Table 3).

Table 3 Carbon footprint reduction practices policies

Country	Carbon footprint reduction policies	Sources
Norway	Extended Producers Responsibilities policy (EPR) (2004)	Karstensen et al. [96]
Japan	Promotion of Effective Utilization of Resources (2010?)	[145]
China	12th five-year plan (2011–2015)	Murray et al. [117]
Singapore	Zero Waste Masterplan (2019–2030)	[132]
Portugal	Strategy and Action Plan for Combating Food Waste (2018)	[76]
Malaysia	Movement Control Order (MCO) (2020)	Naderipour et al. [119]

The environment is a topic of global concern that has been transmitted from generation to generation, and that gave rise to the concern of individuals and companies for the environment, generating concrete actions such as management of carbon emissions, waste and, carbon footprint others [76, 110, 158]. Table 3 shows only a few examples of policies implemented by certain countries, but it should be noted that these same countries have established a more significant number of policies and plans for the development of the Circular Economy and other countries that have not been mentioned. Thus, following a production and consumption model that is conscious of the importance of the 3Rs to minimize environmental impact [118].

Several actors are making efforts to meet carbon neutrality [44]. In the case of the European Union states members, they are required to implement waste management programs based on the Waste Framework Directive (2008/98/EC) [45]. As in Europe, practices such as recycling have been promoted [110], governments should enact restrictive laws and policies to increase carbon footprint reduction [44]. Such measures are often developed to manage waste and carbon emissions [145], increase recycling and improve composting [110], reducing packaging [132], in addition to other policies.

The solutions proposed by governments and various entities are based on CE [76]. Lack of fiscal incentives for food donation, lower VAT for recycled materials, among other initiatives, are broadly implemented by the European Union [45]. Working in FLW leads to trying to improve each of the steps in the food supply chain, from farmers, manufacturers, convenience stores, and large markets, such as restaurants, workplaces, colleges, universities; as well as working in the mind of the end consumer [158]. For these efforts to have the desired results, the actions to be taken must be worked on jointly, since the production of food integrates the use of resources such as non-renewable mineral resources, water from groundwater reserves, and fossil fuel [118, 158].

In this regard, policymakers are promoting low-carbon economic incentives to encourage companies to develop eco-friendly technological innovations [120]. It is proposed the creation of eco-friendly packaging design, the promotion of clean and conscious consumption of materials, and the adaptation of rules that allow buyers to adapt their consumption to an eco-friendlier one [114].

4.2 Companies and Actors Strategies

In this case, we can also observe different actors implementing CE to reduce the carbon footprint in different industries, as Greenhouse emissions keep growing and consumers are looking forward to modifying their choices [125]. For example, Spain has the largest agro-ecosystem that can be observed in Europe, which is facing climate change and the decrease of resources at the same time, this problem is being faced with the modification and improvement of farming activities, thus developing more sustainable systems from the beginning of operations [91]. The building sector makes efforts to reduce carbon emissions and energy consumption.

In Metropolitan Melbourne, Australia, the law created the "Building Code of Australia," which ensures that the construction of new houses is carried out in compliance with energy regulations [139]. Similarly, in La Abundancia-Florencia highway, Costa Rica, the need for a conscious construction in the mentality of public and private groups [68].

5 Barriers to Achieve Effective Circular Economy

The circular economy policy seems attractive for countries and companies. Nonetheless, several factors can make this objective difficult. Authors such as Stangherlin and de Barcellos (2018) mentioned that external influences might impact directly or indirectly individuals. In this sense, we can find the following aspects: historical, supply chain factors, and regulatory issues. For example, food waste can be generated when it is not adequately preserved [95], which requires suppliers' and consumers' contribution to initiatives to reduce food disposal [78]. Focusing on an organizational perspective, the lack of circular economy indicators reflecting food loss and carbon footprint reduction, lack of scalability of initiatives, and network support is also present [78].

It is possible to classify barriers to circularity in the food supply chain into 7 types: cultural, business and business finance, regulatory and governmental, technological, managerial, supply chain management, and knowledge/skills [5]. The academic field validates that these lasts affect businesses. For instance, awareness, capabilities, resources, and infrastructure for implementing green supply chain practices are not homogenous in all organizations [84, 140]. Furthermore, aspects such as government contribution in establishing regulations remain weak in a significant part of the globe [69]. It has been argued that their governments are not pushing the transition toward circular economy practices in the food sector because of the scarcity of economic resources and adequate infrastructure [99].

6 Closing Remarks

Over the last decades, food waste and carbon footprint awareness increased exponentially [121]. As a result, a circular economy, as a solution to these issues, can be implemented for diverse purposes in the supply chain to increase waste management. To promote this trend, governments have recommended implementing food waste policies [71]. Nonetheless, several challenges were identified. For instance, a lack of indicators, information, supply chain deficit analysis, among others, are present [50]. Moreover, food waste varies along the entire supply chain and can be affected by climate conditions, shelf life, manufacturing practices, fiscal incentives, etc. The private and public sectors must adopt initiatives to reduce the environmental impact [50, 146]. A possible solution can be the circular economy based on the interconnection between industries, focusing on delivering disposed output to other companies [70]. This proposal is challenging since it breaks the traditional linear model, requiring technological investment, redesigning the supply chain, among other strategic decisions [70]. Despite these drawbacks, food waste mitigation can be achievable and monitored efficiently [131] Nowadays, the current food production system is unsustainable [95]. Consequently, countries are developing programs and policies to increase and fasten the transition toward a circular economy but remain weak worldwide [71]. Also, actors such as NGOs, charitable organizations, among others, are strategic partners for the effective implementation of circular economy activities to reduce food loss [71, 115].

References

1. Abiad MG, Meho LI (2018) Food loss and food waste research in the Arab world: a systematic review. Food Secur 10(2):311–322. https://doi.org/10.1007/s12571-018-0782-7
2. Abiakam N, Worsley P, Jayabal H, Mitchell K, Jones M, Fletcher J, Spratt F, Bader D (2021) Personal protective equipment related skin reactions in healthcare professionals during COVID-19. Int Wound J 18(3):312–322. https://doi.org/10.1111/iwj.13534
3. Abudureheman A, Nilupaer A (2021) Optimization model design of cross-border e-commerce transportation path under the background of prevention and control of COVID-19 pneumonia. Soft Comput 25(18):12007–12015. https://doi.org/10.1007/s00500-021-05685-6
4. Acuña-Zegarra MA, Santana-Cibrian M, Velasco-Hernandez JX (2020) Modeling behavioral change and COVID-19 containment in Mexico: a trade-off between lockdown and compliance. Math Biosci 325:108370. https://doi.org/10.1016/j.mbs.2020.108370
5. Ada N, Kazancoglu Y, Sezer MD, Ede-Senturk C, Ozer I, Ram M (2021) Analyzing barriers of circular food supply chains and proposing industry 4.0 solutions. Sustainability 13(12). https://doi.org/10.3390/su13126812
6. Afshan G, Shahid S, Tunio MN (2021) Learning experiences of women entrepreneurs amidst COVID-19. Int J Gend Entrep 13(2):162–186. https://doi.org/10.1108/IJGE-09-2020-0153
7. Aguirre AA, Catherina R, Frye H, Shelley L (2020) Illicit wildlife trade, wet markets, and COVID-19: preventing future pandemics. World Med Health Policy 12(3):256–265. https://doi.org/10.1002/wmh3.348

8. Ahsan MM (2020) Strategic decisions on urban built environment to pandemics in Turkey: lessons from COVID-19. J Urban Manag 9(3):281–285. https://doi.org/10.1016/j.jum.2020.07.001
9. Alan H, Eskin Bacaksiz F, Tiryaki Sen H, Taskiran Eskici G, Gumus E, Harmanci Seren AK (2021) "I'm a hero, but…": an evaluation of depression, anxiety, and stress levels of frontline healthcare professionals during COVID-19 pandemic in Turkey. Perspect Psychiatr Care 57(3):1126–1136. https://doi.org/10.1111/ppc.12666
10. Ali W (2020) Online and remote learning in higher education institutes: a necessity in light of COVID-19 pandemic. High Educ Stud 10(3):16–25
11. Allen J, Rowan L, Singh P (2020) Teaching and teacher education in the time of COVID-19. Asia-Pac J Teach Educ 48(3):233–236. https://doi.org/10.1080/1359866X.2020.1752051
12. Alsairafi Z, Naser AY, Alsaleh FM, Awad A, Jalal Z (2021) Mental health status of healthcare professionals and students of health sciences faculties in Kuwait during the COVID-19 pandemic. Int J Environ Res Public Health 18(4). https://doi.org/10.3390/ijerph18042203
13. Altindis E (2021) Inequitable COVID-19 vaccine distribution and the intellectual property rights prolong the pandemic. Expert Rev Vaccines, null–null. https://doi.org/10.1080/14760584.2022.2014819
14. Alvarez-Risco A, Del-Aguila-Arcentales S, Rosen MA, García-Ibarra V, Maycotte-Felkel S, Martínez-Toro GM (2021) Expectations and interests of university students in COVID-19 times about sustainable development goals: evidence from Colombia, Ecuador, Mexico, and Peru. Sustainability 13(6):3306. https://doi.org/10.3390/su13063306
15. Alvarez-Risco A, Del-Aguila-Arcentales S, Yáñez JA (2021) Telemedicine in Peru as a result of the COVID-19 pandemic: perspective from a country with limited internet access. Am J Trop Med Hyg 105(1):6–11. https://doi.org/10.4269/ajtmh.21-0255
16. Alvarez-Risco A, Del-Aguila-Arcentales S, Yáñez JA, Rosen MA, Mejia CR (2021) Influence of technostress on academic performance of university medicine students in Peru during the COVID-19 pandemic. Sustainability 13(16):8949. https://doi.org/10.3390/su13168949
17. Alvarez-Risco A, Mlodzianowska S, García-Ibarra V, Rosen MA, Del-Aguila-Arcentales S (2021) Factors affecting green entrepreneurship intentions in business university students in COVID-19 pandemic times: case of Ecuador. Sustainability 13(11):6447. https://doi.org/10.3390/su13116447
18. Alvarez-Risco A, Rosen MA, Del-Aguila-Arcentales S (2020) A new regulation for supporting a circular economy in the plastic industry: the case of Peru (short communication). J Landsc Ecol 13(1):1–3. https://doi.org/10.2478/jlecol-2020-0004
19. Alvarez-Risco A, Dawson J, Johnson W, Conteh-Barrat M, Aslani P, Del-Aguila-Arcentales S, Diaz-Risco S (2021) Ebola virus disease outbreak: a global approach for health systems [Ebola; health professionals; global approach; pharmacist: pharmaceutical care; COVID-19.]. Rev Cuba Farm 53(4). http://www.revfarmacia.sld.cu/index.php/far/article/view/491
20. Alvarez-Risco A, Del-Aguila-Arcentales S (2021a) A note on changing regulation in international business: The World Intellectual Property Organization (WIPO) and artificial intelligence. In: Verbeke A, van Tulder R, Rose EL, Wei Y (eds) The multiple dimensions of institutional complexity in international business research, vol 15. Emerald Publishing Limited, p. 363–371. https://doi.org/10.1108/S1745-886220210000015020
21. Alvarez-Risco A, Del-Aguila-Arcentales S (2021b) Public policies and private efforts to increase women entrepreneurship based on STEM background. In: Mari M, Poggesi S, Foss L (eds) Women's entrepreneurship in STEM disciplines: issues and perspectives. Springer International Publishing, p 75–87. https://doi.org/10.1007/978-3-030-83792-1_5
22. Alvarez-Risco A, Estrada-Merino A, de las Mercedes Anderson-Seminario M, Mlodzianowska S, García-Ibarra V, Villagomez-Buele C, Carvache-Franco M (2021) Multitasking behavior in online classrooms and academic performance: case of university students in Ecuador during COVID-19 outbreak. Interact Technol Smart Educ 18(3):422–434. https://doi.org/10.1108/ITSE-08-2020-0160
23. Alvarez-Risco A, Mejia CR, Delgado-Zegarra J, Del-Aguila-Arcentales S, Arce-Esquivel AA, Valladares-Garrido MJ, del Portal MR, Villegas LF, Curioso WH, Chandra Sekar M, Yáñez

JA (2020) The Peru approach against the COVID-19 infodemic: insights and strategies. Am J Trop Med Hyg 103(2):583–586. https://doi.org/10.4269/ajtmh.20-0536
24. Alvarez-Risco A, Mlodzianowska S, Zamora-Ramos U, Del-Aguila S (2021) Green entrepreneurship intention in university students: the case of Peru. Entrep Bus Econ Rev 9(4):85–100. https://doi.org/10.15678/EBER.2021.090406
25. Alvarez-Risco A, Quipuzco-Chicata L, Escudero-Cipriani C (2021) Determinantes de la intención de recompra en línea en tiempos de COVID-19: evidencia de una economía emergente. Lect Econ (96). https://doi.org/10.17533/udea.le.n96a342638
26. Amarillo CR (2020) Feminism on lockdown. NACLA Rep Am 52(3):274–281. https://doi.org/10.1080/10714839.2020.1809084
27. Apcho-Ccencho L-V, Cuya-Velásquez B-B, Alvarado Rodríguez D, de las Mercedes Anderson-Seminario M, Alvarez-Risco A, Estrada-Merino A, Mlodzianowska S (2021) The impact of international price on the technological industry in the United States and China during times of crisis: commercial war and COVID-19. In: Lawrence KD, Klimberg RK (eds) Advances in business and management forecasting, vol 14. Emerald Publishing Limited, p 149–160. https://doi.org/10.1108/S1477-407020210000014010
28. Arcuri S (2019) Food poverty, food waste and the consensus frame on charitable food redistribution in Italy. Agric Hum Values 36(2):263–275.
29. Assaf A, Scuderi R (2020) COVID-19 and the recovery of the tourism industry. Tour Econ 26(5):731–733. https://doi.org/10.1177/1354816620933712
30. Bacq S, Lumpkin GT (2021) Social entrepreneurship and COVID-19. J Manage Stud 58(1):285–288. https://doi.org/10.1111/joms.12641
31. Bag S, Gupta S, Foropon C (2019) Examining the role of dynamic remanufacturing capability on supply chain resilience in circular economy. Manag Decis 57(4):863–885. https://doi.org/10.1108/MD-07-2018-0724
32. Barello S, Caruso R, Palamenghi L, Nania T, Dellafiore F, Bonetti L, Silenzi A, Marotta C, Graffigna G (2021) Factors associated with emotional exhaustion in healthcare professionals involved in the COVID-19 pandemic: an application of the job demands-resources model. Int Arch Occup Environ Health 94(8):1751–1761. https://doi.org/10.1007/s00420-021-01669-z
33. Baum T, Hai NTT (2020) Hospitality, tourism, human rights and the impact of COVID-19. Int J Contemp Hosp Manag 32(7):2397–2407. https://doi.org/10.1108/IJCHM-03-2020-0242
34. Belitski M, Guenther C, Kritikos AS, Thurik R (2021) Economic effects of the COVID-19 pandemic on entrepreneurship and small businesses. Small Bus Econ. https://doi.org/10.1007/s11187-021-00544-y
35. Bhattacharya A (2021) Achieving sustainability in supply chain operations in the interplay between circular economy and Industry 4.0. Prod Plan Control: 1–3. https://doi.org/10.1080/09537287.2021.1981032
36. Block JH, Fisch C, Hirschmann M (2021) The determinants of bootstrap financing in crises: evidence from entrepreneurial ventures in the COVID-19 pandemic. Small Bus Econ. https://doi.org/10.1007/s11187-020-00445-6
37. Bonomi S, Moggi S, Ricciardi F (2016, 2016//) Innovation for sustainable development by educating the local community: the case of an Italian project of food waste prevention. In: Exploring services science. Cham.
38. Borrello M, Caracciolo F, Lombardi A, Pascucci S, Cembalo L (2017) Consumers' perspective on circular economy strategy for reducing food waste. Sustainability 9(1). https://doi.org/10.3390/su9010141
39. Borzée A, McNeely J, Magellan K, Miller JRB, Porter L, Dutta T, Kadinjappalli KP, Sharma S, Shahabuddin G, Aprilinayati F, Ryan GE, Hughes A, Abd Mutalib AH, Wahab AZA, Bista D, Chavanich SA, Chong JL, Gale GA, Zhang L (2020). COVID-19 highlights the need for more effective wildlife trade legislation. Trends Ecol Evol 35(12):1052–1055. https://doi.org/10.1016/j.tree.2020.10.001
40. Bozdağ F, Ergün N (2020) Psychological resilience of healthcare professionals during COVID-19 pandemic. Psychol Rep 124(6):2567–2586. https://doi.org/10.1177/0033294120965477

41. Brouder P (2020) Reset redux: possible evolutionary pathways towards the transformation of tourism in a COVID-19 world. Tour Geogr 22(3):484–490. https://doi.org/10.1080/14616688.2020.1760928
42. Brown R, Rocha A, Cowling M (2020) Financing entrepreneurship in times of crisis: Exploring the impact of COVID-19 on the market for entrepreneurial finance in the United Kingdom. Int Small Bus J 38(5):380–390. https://doi.org/10.1177/0266242620937464
43. Bunting Eubanks L (2018) From A Culture Of Food Waste To A Culture Of Food Security: A Comparison Of Food Waste Law And Policy In France And In The United States Wm, Mary Envtl L, Pol'y Rev. 43: 667.
44. Caglar AE, Balsalobre-Lorente D, Akin CS (2021) Analysing the ecological footprint in EU-5 countries under a scenario of carbon neutrality: evidence from newly developed sharp and smooth structural breaks in unit root testing. J Environ Manage 295:113155. https://doi.org/10.1016/j.jenvman.2021.113155
45. Calisto Friant M, Vermeulen WJV, Salomone R (2021) Analysing European Union circular economy policies: words versus actions. Sustain Prod Consum 27:337–353. https://doi.org/10.1016/j.spc.2020.11.001
46. Carvache-Franco M, Alvarez-Risco A, Carvache-Franco W, Carvache-Franco O, Estrada-Merino A, Rosen MA (2021) Coastal cities seen from loyalty and their tourist motivations: a study in Lima, Peru. Sustainability 13(21):11575. https://doi.org/10.3390/su132111575
47. Carvache-Franco M, Alvarez-Risco A, Carvache-Franco O, Carvache-Franco W, Estrada-Merino A, Villalobos-Alvarez D (2021) Perceived value and its influence on satisfaction and loyalty in a coastal city: a study from Lima, Peru. J Policy Res Tour Leis Events 1–16. https://doi.org/10.1080/19407963.2021.1883634
48. Carvache-Franco M, Carvache-Franco O, Carvache-Franco W, Alvarez-Risco A, Estrada-Merino A (2021) Motivations and segmentation of the demand for coastal cities: a study in Lima, Peru. Int J Tour Res 23(4):517–531. https://doi.org/10.1002/jtr.2423
49. del Castillo FA (2021) Temporary waiver of intellectual property on COVID-19 vaccines: toward the creation of a better, post-pandemic society. J Public Health 43(3):e559–e560. https://doi.org/10.1093/pubmed/fdab184
50. Cattaneo A, Federighi G, Vaz S (2021) The environmental impact of reducing food loss and waste: a critical assessment. Food Policy 98:101890. https://doi.org/10.1016/j.foodpol.2020.101890
51. Cegarra-Navarro J-G, Vătămănescu E-M, Martínez-Martínez A (2021) A context-driven approach on coping with COVID-19: from hiding knowledge toward citizen engagement. Knowl Process Manag 28(2):134–140. https://doi.org/10.1002/kpm.1662
52. Chafloque-Cespedes R, Alvarez-Risco A, Robayo-Acuña P-V, Gamarra-Chavez C-A, Martinez-Toro G-M, Vicente-Ramos W (2021) Effect of sociodemographic factors in entrepreneurial orientation and entrepreneurial intention in university students of Latin American business schools. In: Jones P, Apostolopoulos N, Kakouris A, Moon C, Ratten V, Walmsley A (eds) Universities and entrepreneurship: meeting the educational and social challenges, vol 11. Emerald Publishing Limited, p 151–165. https://doi.org/10.1108/S2040-724620210000011010
53. Chafloque-Cespedes R, Vara-Horna A, Asencios-Gonzales Z, Lopez-Odar DR, Alvarez-Risco A, Quipuzco-Chicata L, Schulze C, Sanchez-Villagomez M (2020) Presentismo académico y violencia contra las mujeres en escuelas de negocios e ingeniería en universidades peruanas. Lect Econ 0(93):127–153. https://doi.org/10.17533/udea.le.n93a340726
54. Chen C-C, Sujanto RY, Tseng M-L, Fujii M, Lim MK (2021) Sustainable consumption transition model: social concerns and waste minimization under willingness-to-pay in Indonesian food industry. Resour Conserv Recycl 170:105590. https://doi.org/10.1016/j.resconrec.2021.105590
55. Chen X, Zhang SX, Jahanshahi AA, Alvarez-Risco A, Dai H, Li J, Ibarra VG (2020) Belief in a COVID-19 conspiracy theory as a predictor of mental health and well-being of health care workers in Ecuador: cross-sectional survey study. JMIR Public Health Surveill 6(3):e20737. https://doi.org/10.2196/20737

56. Chen T, Peng L, Yin X, Rong J, Yang J, Cong G (2020) Analysis of user satisfaction with online education platforms in China during the COVID-19 pandemic. Healthcare, 8(3). https://doi.org/10.3390/healthcare8030200
57. Chung SA, Olivera S, Román BR, Alanoca E, Moscoso S, Terceros BL, Alvarez-Risco A, Yáñez JA (2021) Temáticas de la producción científica de la Revista Cubana de Farmacia indizada en Scopus (1967–2020) Rev Cuba Farm 54(1). http://www.revfarmacia.sld.cu/index.php/far/article/view/511
58. Condamine P (2020) France's law for fighting food waste. Zero Waste Europe. https://www.zerowasteeurope.eu/wp-content/uploads/2020/11/zwe_11_2020_factsheet_france_en.pdf Accessed 22 September 2021
59. Corrado S, Caldeira C, Eriksson M, Hanssen OJ, Hauser H-E, van Holsteijn F, Liu G, Östergren K, Parry A, Secondi L, Stenmarck Å, Sala S (2019) Food waste accounting methodologies: challenges, opportunities, and further advancements. Glob Food Secur 20:93–100. https://doi.org/10.1016/j.gfs.2019.01.002
60. Cruz-Torres W, Alvarez-Risco A, Del-Aguila-Arcentales S (2021) Impact of enterprise resource planning (ERP) implementation on performance of an education enterprise: a structural equation modeling (SEM). Stud Bus Econ 16(2):37–52. https://doi.org/10.2478/sbe-2021-0023
61. Dagilienė L, Varaniūtė V, Bruneckienė J (2021) Local governments' perspective on implementing the circular economy: a framework for future solutions. J Clean Prod 310:127340. https://doi.org/10.1016/j.jclepro.2021.127340
62. DeLorenzo A, Parizeau K, von Massow M (2019) Regulating Ontario's circular economy through food waste legislation. Soc Bus Rev 14(2):200–216. https://doi.org/10.1108/SBR-12-2017-0115
63. Deressa W, Worku A, Abebe W, Gizaw M, Amogne W (2021) Risk perceptions and preventive practices of COVID-19 among healthcare professionals in public hospitals in Addis Ababa, Ethiopia. PLoS ONE 16(6):e0242471. https://doi.org/10.1371/journal.pone.0242471
64. do Canto NR, Grunert KG, De Barcellos MD (2021) Circular food behaviors: a literature review. Sustainability 13(4). https://doi.org/10.3390/su13041872
65. Dominelli L (2021) A green social work perspective on social work during the time of COVID-19. Int J Soc Welf 30(1):7–16. https://doi.org/10.1111/ijsw.12469
66. Duarte Alonso A, Kok SK, Bressan A, O'Shea M, Sakellarios N, Koresis A, Solis MAB, Santoni LJ (2020) COVID-19, aftermath, impacts, and hospitality firms: an international perspective. Int J Hosp Manag 91:102654. https://doi.org/10.1016/j.ijhm.2020.102654
67. Erfani P, Binagwaho A, Jalloh MJ, Yunus M, Farmer P, Kerry V (2021) Intellectual property waiver for COVID-19 vaccines will advance global health equity. BMJ 374:n1837. https://doi.org/10.1136/bmj.n1837
68. Espinoza M, Campos N, Yang R, Ozer H, Aguiar-Moya JP, Baldi A, Loría-Salazar LG, Al-Qadi IL (2019) Carbon footprint estimation in road construction: La Abundancia–Florencia case study. Sustainability 11(8). https://doi.org/10.3390/su11082276
69. Farooque M, Zhang A, Liu Y (2019) Barriers to circular food supply chains in China. Supply Chain Manag Int J 24(5):677–696. https://doi.org/10.1108/SCM-10-2018-0345
70. Fassio F, Tecco N (2019) Circular economy for food: a systemic interpretation of 40 case histories in the food system in their relationships with SDGs. Systems 7(3). https://doi.org/10.3390/systems7030043
71. Fattibene D, Recanati F, Dembska K, Antonelli M (2020) Urban food waste: a framework to analyse policies and initiatives. Resources 9(9). https://doi.org/10.3390/resources9090099
72. Ferrando T, Mansuy J (2018) The European action against food loss and waste: co-regulation and collisions on the way to the sustainable development goals. Yearb Eur Law 37:424–454. https://doi.org/10.1093/yel/yey015
73. Fortunati S, Morea D, Mosconi EM (2020) Circular economy and corporate social responsibility in the agricultural system: cases study of the Italian agri-food industry. Agric Econ 66(11):489–498

74. Fotiadis A, Polyzos S, Huan T-CTC (2021) The good, the bad and the ugly on COVID-19 tourism recovery. Ann Tour Res 87:103117. https://doi.org/10.1016/j.annals.2020.103117
75. Galanakis CM, Rizou M, Aldawoud TMS, Ucak I, Rowan NJ (2021) Innovations and technology disruptions in the food sector within the COVID-19 pandemic and post-lockdown era. Trends Food Sci Technol 110:193–200. https://doi.org/10.1016/j.tifs.2021.02.002
76. Galli A, Moreno Pires S, Iha K, Alves AA, Lin D, Mancini MS, Teles F (2020) Sustainable food transition in Portugal: assessing the footprint of dietary choices and gaps in national and local food policies. Sci Total Environ 749:141307. https://doi.org/10.1016/j.scitotenv.2020.141307
77. Garske B, Heyl K, Ekardt F, Weber LM, Gradzka W (2020) Challenges of food waste governance: an assessment of European legislation on food waste and recommendations for improvement by economic instruments. Land 9(7). https://doi.org/10.3390/land9070231
78. Gedam VV, Raut RD, de Sousa L, Jabbour AB, Tanksale AN, Narkhede BE (2021) Circular economy practices in a developing economy: barriers to be defeated. J Clean Prod 311:127670. https://doi.org/10.1016/j.jclepro.2021.127670
79. Geissdoerfer M, Savaget P, Bocken NMP, Hultink EJ (2017) The circular economy—a new sustainability paradigm? J Clean Prod 143:757–768. https://doi.org/10.1016/j.jclepro.2016.12.048
80. Geng Y, Doberstein B (2008) Developing the circular economy in China: challenges and opportunities for achieving "leapfrog development." Int J Sust Dev World 15(3):231–239. https://doi.org/10.3843/SusDev.15.3:6
81. Geng Y, Sarkis J, Ulgiati S, Zhang P (2013) Measuring China's circular economy. Science 339(6127):1526–1527. https://doi.org/10.1126/science.1227059
82. Ghisellini P, Cialani C, Ulgiati S (2016) A review on circular economy: the expected transition to a balanced interplay of environmental and economic systems. J Clean Prod 114:11–32. https://doi.org/10.1016/j.jclepro.2015.09.007
83. Giordano C, Falasconi, L, Cicatiello, C, Pancino B. (2020) The role of food waste hierarchy in addressing policy and research: A comparative analysis. J Clean Prod 252:119617.
84. Govindan K, Hasanagic M (2018) A systematic review on drivers, barriers, and practices towards circular economy: a supply chain perspective. Int J Prod Res 56(1–2):278–311. https://doi.org/10.1080/00207543.2017.1402141
85. Gulati G, Kelly BD (2020) Domestic violence against women and the COVID-19 pandemic: What is the role of psychiatry? Int J Law Psychiatry 71:101594. https://doi.org/10.1016/j.ijlp.2020.101594
86. Gursoy D, Chi CG (2020) Effects of COVID-19 pandemic on hospitality industry: review of the current situations and a research agenda. J Hosp Market Manag 29(5):527–529. https://doi.org/10.1080/19368623.2020.1788231
87. Hanzl M (2020) Urban forms and green infrastructure—the implications for public health during the COVID-19 pandemic. Cities Health 1–5. https://doi.org/10.1080/23748834.2020.1791441
88. Helander H, Bruckner M, Leipold S, Petit-Boix A, Bringezu S (2021) Eating healthy or wasting less? Reducing resource footprints of food consumption. Environ Res Lett 16(5):054033. https://doi.org/10.1088/1748-9326/abe673
89. Hepburn C, O'Callaghan B, Stern N, Stiglitz J, Zenghelis D (2020) Will COVID-19 fiscal recovery packages accelerate or retard progress on climate change? Oxf Rev Econ Policy 36(Supplement_1), S359–S381. https://doi.org/10.1093/oxrep/graa015
90. Ho C-Y, Tsai B-H, Chen C-S, Lu M-T (2021) Exploring green marketing orientations toward sustainability the hospitality industry in the COVID-19 pandemic. Sustainability 13(8). https://doi.org/10.3390/su13084348
91. Horrillo A, Gaspar P, Escribano M (2020) Organic farming as a strategy to reduce carbon footprint in Dehesa agroecosystems: a case study comparing different livestock products. Animals 10(1). https://doi.org/10.3390/ani10010162
92. Ivanova M (2012) Institutional design and UNEP reform: historical insights on form, function and financing. Int Aff 88(3):565–584. https://doi.org/10.1111/j.1468-2346.2012.01089.x

93. Jawahir IS, Bradley R (2016) Technological elements of circular economy and the principles of 6R-based closed-loop material flow in sustainable manufacturing. Procedia CIRP 40:103–108. https://doi.org/10.1016/j.procir.2016.01.067
94. Jecker NS, Atuire CA (2021) What's yours is ours: waiving intellectual property protections for COVID-19 vaccines. J Med Ethics 47(9):595. https://doi.org/10.1136/medethics-2021-107555
95. Jurgilevich A, Birge T, Kentala-Lehtonen J, Korhonen-Kurki K, Pietikäinen J, Saikku L, Schösler H (2016) Transition towards circular economy in the food system. Sustainability 8(1). https://doi.org/10.3390/su8010069
96. Karstensen KH, Engelsen CJ, Saha PK (2020) Circular economy initiatives in Norway. In Circular economy: Global perspective pp 299–316. Springer, Singapore
97. Kaushal V, Srivastava S (2021) Hospitality and tourism industry amid COVID-19 pandemic: perspectives on challenges and learnings from India. Int J Hosp Manag 92:102707. https://doi.org/10.1016/j.ijhm.2020.102707
98. Kirk CP, Rifkin LS (2020) I'll trade you diamonds for toilet paper: consumer reacting, coping and adapting behaviors in the COVID-19 pandemic. J Bus Res 117:124–131. https://doi.org/10.1016/j.jbusres.2020.05.028
99. Kirwan J, Maye D, Brunori G (2017) Acknowledging complexity in food supply chains when assessing their performance and sustainability. J Rural Stud 52:21–32. https://doi.org/10.1016/j.jrurstud.2017.03.008
100. Koetter P, Pelton M, Gonzalo J, Du P, Exten C, Bogale K, Buzzelli L, Connolly M, Edel K, Hoffman A, Legro NR, Medina D, Sood N, Blaker J, Kearcher K, Sciamanna C (2020) Implementation and process of a COVID-19 contact tracing initiative: leveraging health professional students to extend the workforce during a pandemic. Am J Infect Control, 48(12): 1451–1456. https://doi.org/10.1016/j.ajic.2020.08.012
101. Krishtel P, Malpani R (2021) Suspend intellectual property rights for COVID-19 vaccines. BMJ 373:n1344. https://doi.org/10.1136/bmj.n1344
102. Lackie K, Najjar G, El-Awaisi A, Frost J, Green C, Langlois S, Lising D, Pfeifle AL, Ward H, Xyrichiso A, Khalili H (2020) Interprofessional education and collaborative practice research during the COVID-19 pandemic: considerations to advance the field. J Interprofessional Care 34(5):583–586. https://doi.org/10.1080/13561820.2020.1807481
103. Lashley MA, Acevedo M, Cotner S, Lortie CJ (2020) How the ecology and evolution of the COVID-19 pandemic changed learning. Ecol Evol 10(22):12412–12417. https://doi.org/10.1002/ece3.6937
104. Leipold S, Weldner K, Hohl M (2021) Do we need a 'circular society'? Competing narratives of the circular economy in the French food sector. Ecol Econ 187:107086. https://doi.org/10.1016/j.ecolecon.2021.107086
105. Leiva-Martinez M-A, de las Mercedes Anderson-Seminario M, Alvarez-Risco A, Estrada-Merino A, Mlodzianowska S (2021) Price variation in lower goods as of previous economic crisis and the contrast of the current price situation in the context of COVID-19 in Peru. In: Lawrence KD, Klimberg RK (eds) Advances in business and management forecasting, vol 14. Emerald Publishing Limited, p 161–166. https://doi.org/10.1108/S1477-407020210000014011
106. Lieder M, Rashid A (2016) Towards circular economy implementation: a comprehensive review in context of manufacturing industry. J Clean Prod 115:36–51. https://doi.org/10.1016/j.jclepro.2015.12.042
107. Liguori E, Winkler C (2020) From offline to online: challenges and opportunities for entrepreneurship education following the COVID-19 pandemic. Entrep Educ Pedagog 3(4):346–351. https://doi.org/10.1177/2515127420916738
108. Lim S, Prakash A (2021) Pandemics and citizen perceptions about their country: Did COVID-19 increase national pride in South Korea? Nations Natl 27(3):623–637. https://doi.org/10.1111/nana.12749
109. Lopez-Odar D, Alvarez-Risco A, Vara-Horna A, Chafloque-Cespedes R, Sekar MC (2020) Validity and reliability of the questionnaire that evaluates factors associated with perceived

environmental behavior and perceived ecological purchasing behavior in Peruvian consumers. Soc Responsib J 16(3):403–417. https://doi.org/10.1108/SRJ-08-2018-0201
110. Magazzino C, Mele M, Schneider N, Sarkodie SA (2021) Waste generation, wealth and GHG emissions from the waste sector: Is Denmark on the path towards circular economy? Sci Total Environ 755:142510. https://doi.org/10.1016/j.scitotenv.2020.142510
111. Magrini C, D'Addato F, Bonoli A (2020) Municipal solid waste prevention: A review of market-based instruments in six European Union countries. Waste Manage Res 38:3–22.
112. Maritz A, Perenyi A, de Waal G, Buck C (2020) Entrepreneurship as the unsung hero during the current COVID-19 economic crisis: Australian perspectives. Sustainability 12(11). https://doi.org/10.3390/su12114612
113. Matuštík J, Kočí V (2021) What is a footprint? A conceptual analysis of environmental footprint indicators. J Clean Prod 285:124833. https://doi.org/10.1016/j.jclepro.2020.124833
114. McDowall W, Geng Y, Huang B, Barteková E, Bleischwitz R, Türkeli S, Kemp R, Doménech T (2017) Circular economy policies in China and Europe. J Ind Ecol 21(3):651–661. https://doi.org/10.1111/jiec.12597
115. Michelini G, Moraes RN, Cunha RN, Costa JMH, Ometto AR (2017) From linear to circular economy: PSS conducting the transition. Procedia CIRP 64:2–6. https://doi.org/10.1016/j.procir.2017.03.012
116. Mourad M (2016) Recycling, recovering and preventing "food waste": competing solutions for food systems sustainability in the United States and France. J Clean Prod 126:461–477. https://doi.org/10.1016/j.jclepro.2016.03.084
117. Murray A, Skene K, Haynes K (2017) The circular economy: An interdisciplinary exploration of the concept and application in a global context. J Bus Ethics 140(3): 369–380.
118. Muscio A, Sisto R (2020) Are agri-food systems really switching to a circular economy model? Implications for European research and innovation policy. Sustainability 12(14). https://doi.org/10.3390/su12145554
119. Naderipour A, Abdul-Malek Z, Ahmad NA., Kamyab H, Ashokkumar V, Ngamcharussrivichai C, Chelliapan S (2020) Effect of COVID-19 virus on reducing GHG emission and increasing energy generated by renewable energy sources: A brief study in Malaysian context. Environ Technol & innovation 20: 101151.
120. Nair M, Arvin MB, Pradhan RP, Bahmani S (2021) Is higher economic growth possible through better institutional quality and a lower carbon footprint? Evidence from developing countries. Renew Energy 167:132–145. https://doi.org/10.1016/j.renene.2020.11.056
121. Nattassha R, Handayati Y, Simatupang TM, Siallagan M (2020) Understanding circular economy implementation in the agri-food supply chain: the case of an Indonesian organic fertiliser producer. Agric Food Secur 9(1):10. https://doi.org/10.1186/s40066-020-00264-8
122. Obergassel W, Hermwille L, Oberthür S (2021) Harnessing international climate governance to drive a sustainable recovery from the COVID-19 pandemic. Clim Policy 21(10):1298–1306. https://doi.org/10.1080/14693062.2020.1835603
123. Okereke M (2021) Towards vaccine equity: Should big pharma waive intellectual property rights for COVID-19 vaccines? Public Health Pract 2:100165. https://doi.org/10.1016/j.puhip.2021.100165
124. de Oliveira MM, Lago A, Dal'Magro GP (2021) Food loss and waste in the context of the circular economy: a systematic review. J Clean Prod 294:126284. https://doi.org/10.1016/j.jclepro.2021.126284
125. Payet J (2021) Assessment of carbon footprint for the textile sector in France. Sustainability 13(5). https://doi.org/10.3390/su13052422
126. Pocol CB, Pinoteau M, Amuza A, Burlea-Schiopoiu A, Glogovețan A-I (2020) Food waste behavior among Romanian consumers: a cluster analysis. Sustainability 12(22). https://doi.org/10.3390/su12229708
127. Principato L, Secondi L, Cicatiello C, Mattia G (2020) Caring more about food: the unexpected positive effect of the COVID-19 lockdown on household food management and waste. Socioecon Plann Sci 100953. https://doi.org/10.1016/j.seps.2020.100953

128. Pullin AS, Knight TM (2009) Doing more good than harm—building an evidence-base for conservation and environmental management. Biol Conserv 142(5):931–934. https://doi.org/10.1016/j.biocon.2009.01.010
129. Quispe-Cañari JF, Fidel-Rosales E, Manrique D, Mascaró-Zan J, Huamán-Castillón KM, Chamorro-Espinoza SE, Garayar-Peceros H, Ponce-López VL, Sifuentes-Rosales J, Alvarez-Risco A, Yáñez JA, Mejia CR (2021) Self-medication practices during the COVID-19 pandemic among the adult population in Peru: a cross-sectional survey. Saudi Pharm J 29(1):1–11. https://doi.org/10.1016/j.jsps.2020.12.001
130. Raudenská J, Steinerová V, Javůrková A, Urits I, Kaye AD, Viswanath O, Varrassi G (2020) Occupational burnout syndrome and post-traumatic stress among healthcare professionals during the novel coronavirus disease 2019 (COVID-19) pandemic. Best Pract Res Clin Anaesthesiol 34(3):553–560. https://doi.org/10.1016/j.bpa.2020.07.008
131. Reitemeier M, Aheeyar M, Drechsel P (2021) Perceptions of food waste reduction in Sri Lanka's commercial capital, Colombo. Sustainability 13(2). https://doi.org/10.3390/su13020838
132. Rezvani Ghomi E, Khosravi F, Tahavori MA, Ramakrishna S (2021) Circular economy: a comparison between the case of Singapore and France. Mater Circ Econ 3(1):2. https://doi.org/10.1007/s42824-020-00016-w
133. Roesch E, Amin A, Gupta J, García-Moreno C (2020) Violence against women during COVID-19 pandemic restrictions. BMJ 369:m1712. https://doi.org/10.1136/bmj.m1712
134. Rojas Román B, Moscoso S, Chung SA, Limpias Terceros B, Álvarez-Risco A, Yáñez JA (2020) Tratamiento de la COVID-19 en Perú y Bolivia y los riesgos de la automedicación [COVID-19; tratamiento; estrategia; fármacos; automedicación; Perú; Bolivia.]. 53(2). http://www.revfarmacia.sld.cu/index.php/far/article/view/435/351
135. Schinkel J (2019) Review of policy instruments and recommendations for effective food waste prevention. Proceedings of the Institution of Civil Engineers-Waste and Resource Management.
136. Secondi L, Principato L, Ruini L, Guidi M (2019) Reusing food waste in food manufacturing companies: the case of the tomato-sauce supply chain. Sustainability 11(7). https://doi.org/10.3390/su11072154
137. Sekalala S, Forman L, Hodgson T, Mulumba M, Namyalo-Ganafa H, Meier BM (2021) Decolonising human rights: how intellectual property laws result in unequal access to the COVID-19 vaccine. BMJ Glob Health 6(7):e006169. https://doi.org/10.1136/bmjgh-2021-006169
138. Sekito T, Prayogo TB, Dote Y, Yoshitake T, Bagus I (2013) Influence of a community-based waste management system on people's behavior and waste reduction. Resour Conserv Recycl 72:84–90. https://doi.org/10.1016/j.resconrec.2013.01.001
139. Seo S, Foliente G (2021) Carbon footprint reduction through residential building stock retrofit: a metro Melbourne suburb case study. Energies 14(20). https://doi.org/10.3390/en14206550
140. Sharma YK, Mangla SK, Patil PP, Liu S (2019) When challenges impede the process. Manag Decis 57(4):995–1017. https://doi.org/10.1108/MD-09-2018-1056
141. Shaw R, Kim Y-K, Hua J (2020) Governance, technology and citizen behavior in pandemic: lessons from COVID-19 in East Asia. Prog Disaster Sci 6:100090. https://doi.org/10.1016/j.pdisas.2020.100090
142. Sigala M (2020) Tourism and COVID-19: impacts and implications for advancing and resetting industry and research. J Bus Res 117:312–321. https://doi.org/10.1016/j.jbusres.2020.06.015
143. do Carmo Stangherlin I, De Barcellos MD (2018) Drivers and barriers to food waste reduction. Br Food J, 120(10):2364–2387. https://doi.org/10.1108/BFJ-12-2017-0726
144. Stewart R, Niero M (2018) Circular economy in corporate sustainability strategies: a review of corporate sustainability reports in the fast-moving consumer goods sector. Bus Strateg Environ 27(7):1005–1022. https://doi.org/10.1002/bse.2048
145. Sun L, Li Z, Fujii M, Hijioka Y, Fujita T (2018) Carbon footprint assessment for the waste management sector: a comparative analysis of China and Japan. Front Energy 12(3):400–410. https://doi.org/10.1007/s11708-018-0565-z

146. Szulecka J, Strøm-Andersen N (2021) Norway's food waste reduction governance: From industry self-regulation to governmental regulation? Scand Polit Stud, n/a(n/a). https://doi.org/10.1111/1467-9477.12219
147. Sánchez OR, Vale DB, Rodrigues L, Surita FG (2020) Violence against women during the COVID-19 pandemic: an integrative review. Int J Gynecol Obstet 151(2):180–187. https://doi.org/10.1002/ijgo.13365
148. Sánchez-Clavijo LM, Martínez-Callejas SJ, Acevedo-Charry O, Diaz-Pulido A, Gómez-Valencia B, Ocampo-Peñuela N, Ocampo D, Olaya-Rodríguez MH, Rey-Velasco JC, Soto-Vargas C, Ochoa-Quintero JM (2021) Differential reporting of biodiversity in two citizen science platforms during COVID-19 lockdown in Colombia. Biol Conserv 256:109077. https://doi.org/10.1016/j.biocon.2021.109077
149. Tikka V (2019) Charitable food aid in Finland: from a social issue to an environmental solution. Agric Hum Values 36(2):341–352. https://doi.org/10.1007/s10460-019-09916-3
150. Tran LTT (2021) Managing the effectiveness of e-commerce platforms in a pandemic. J Retail Consum Serv 58:102287. https://doi.org/10.1016/j.jretconser.2020.102287
151. Tóth AJ, Koller Z, Illés CB, Bittsánszky A (2017) Development of conscious food handling in Hungarian school cafeterias. Food Control 73:644–649. https://doi.org/10.1016/j.foodcont.2016.09.011
152. Venkata Mohan S, Modestra JA, Amulya K, Butti SK, Velvizhi G (2016) A circular bioeconomy with biobased products from CO_2 sequestration. Trends Biotechnol 34(6):506–519. https://doi.org/10.1016/j.tibtech.2016.02.012
153. Vidya CT, Prabheesh KP (2020) Implications of COVID-19 pandemic on the global trade networks. Emerg Mark Financ Trade 56(10):2408–2421. https://doi.org/10.1080/1540496X.2020.1785426
154. Viero A, Barbara G, Montisci M, Kustermann K, Cattaneo C (2021) Violence against women in the COVID-19 pandemic: a review of the literature and a call for shared strategies to tackle health and social emergencies. Forensic Sci Int 319:110650. https://doi.org/10.1016/j.forsciint.2020.110650
155. Vittuari M, Azzurro P, Gaiani S, Gheoldus M, Burgos S, Aramyan L, Valeeva N, Rogers D, Östergren K, Timmermans T, Bos-Brouwers H (2016) *Recommendations and guidelines for a common European food waste policy framework*. Fusions. https://www.eu-fusions.org/phocadownload/Publications/D3.5%20recommendations%20and%20guidelines%20food%20waste%20policy%20FINAL.pdf
156. Wen J, Kozak M, Yang S, Liu F (2021) COVID-19: potential effects on Chinese citizens' lifestyle and travel. Tour Rev 76(1):74–87. https://doi.org/10.1108/TR-03-2020-0110
157. Wu T-Y, Ford O, Rainville AJ, Bessire R (2020) COVID-19 care package distribution for senior citizens and families in Detroit and Hamtramck, Michigan. J Hunger Environ Nutr 15(4):585–587. https://doi.org/10.1080/19320248.2020.1799586
158. Wunderlich SM, Martinez NM (2018) Conserving natural resources through food loss reduction: production and consumption stages of the food supply chain. Int Soil Water Conserv Res 6(4):331–339. https://doi.org/10.1016/j.iswcr.2018.06.002
159. Yan J, Kim S, Zhang SX, Foo M-D, Alvarez-Risco A, Del-Aguila-Arcentales S, Yáñez JA (2021) Hospitality workers' COVID-19 risk perception and depression: a contingent model based on transactional theory of stress model. Int J Hosp Manag 95:102935. https://doi.org/10.1016/j.ijhm.2021.102935
160. Yasin Ar A (2020). Managing e-commerce during a pandemic: lessons from GrubHub during COVID-19. In: George B, Mahar Q (eds) International case studies in the management of disasters. Emerald Publishing Limited, p 155–167. https://doi.org/10.1108/978-1-83982-186-820201010
161. Yáñez JA, Afshar Jahanshahi A, Alvarez-Risco A, Li J, Zhang SX (2020a) Anxiety, distress, and turnover intention of healthcare workers in Peru by their distance to the epicenter during the COVID-19 crisis. Am J Trop Med Hyg 103(4):1614–1620. https://doi.org/10.4269/ajtmh.20-0800

162. Yáñez JA, Alvarez-Risco A, Delgado-Zegarra J (2020) COVID-19 in Peru: from supervised walks for children to the first case of Kawasaki-like syndrome. BMJ 369:m2418. https://doi.org/10.1136/bmj.m2418
163. Zarocostas JJ (2021) What next for a COVID-19 intellectual property waiver? The Lancet 397(10288):1871–1872. https://doi.org/10.1016/S0140-6736(21)01151-X
164. Zhang SX, Chen J, Afshar Jahanshahi A, Alvarez-Risco A, Dai H, Li J, Patty-Tito RM (2021) Succumbing to the COVID-19 pandemic—healthcare workers not satisfied and intend to leave their jobs. Int J Ment Heal Addict. https://doi.org/10.1007/s11469-020-00418-6
165. Zhang SX, Sun S, Afshar Jahanshahi A, Alvarez-Risco A, Ibarra VG, Li J, Patty-Tito RM (2020) Developing and testing a measure of COVID-19 organizational support of healthcare workers—results from Peru, Ecuador, and Bolivia. Psychiatry Res 291:113174. https://doi.org/10.1016/j.psychres.2020.113174
166. Zollinger R, DiCindio C (2021) Community ecology: Museum education and the digital divide during and after COVID-19. J MusM Educ 46(4):481–492. https://doi.org/10.1080/10598650.2021.1983711

Fashion and Textile Circularity and Waste Footprint

Marián Arias-Meza, Aldo Alvarez-Risco, Berdy Briggitte Cuya-Velásquez, Maria de las Mercedes Anderson-Seminario, and Shyla Del-Aguila-Arcentales

Abstract Year by year, the waste footprint has been increasing significantly; thousands of tons of waste generated by various industry sectors is already bringing irreparable changes in the world. One of the principal industries that produce these effects is the fashion industry, which used to follow a traditional linear fashion promoted by marketing where consumption by demand becomes a "necessity." However, this perspective has been changing, where the final consumer nowadays is more conscious about the process and the negative aspects that fast fashion means for the environment and the social aspect. This current chapter analyzes the aspects that the fashion industry generates, such as economic, environmental, and social impact, and the solution that provides the fashion and textile circularity for the sector with practical cases being implemented.

Keywords Fashion industry · Circularity · Waste footprint · Recycling · Circular economy · Sustainability · Clothing

1 Introduction

It is crucial to know in detail the most relevant evidence of the impact of the COVID-19 pandemic and can develop more specific and successful plans for the recovery of the economy and commercial activities around the world (Table 1).

Worldwide, 80 billion new garments are sold, equivalent to $1.2 trillion annually for the fashion item. Garments are usually manufactured in higher proportions in China and Bangladesh, but Americans are the largest consumers of textile products [52]. The volume of garments purchased in the EU has increased by 40% compared to previous years and could triple by 2050 [80].

M. Arias-Meza · A. Alvarez-Risco (✉) · B. B. Cuya-Velásquez ·
M. de las Mercedes Anderson-Seminario
Universidad de Lima, Lima, Perú
e-mail: aralvare@ulima.edu.pe

S. Del-Aguila-Arcentales
Escuela Nacional de Marina Mercante "Almirante Miguel Grau", Callao, Perú

A. Alvarez-Risco et al. (eds.), *Circular Economy*, Environmental Footprints and Eco-design of Products and Processes,
https://doi.org/10.1007/978-981-19-0549-0_9

Table 1 Impact of COVID-19 in different sectors

Education	[10, 11, 17, 20, 31, 77, 126, 127, 205]
Entrepreneurship	[6, 30, 22, 33, 44, 51, 53, 60, 72, 133, 141]
Health sector	[2, 9, 13, 27, 19, 49, 58, 76, 85, 122, 162, 167, 199, 202, 203]
Hospitality	[87, 99, 106, 119, 195]
Intellectual property	[14, 29, 68, 91, 112, 125, 155, 171, 201]
Population	[8, 32, 70, 101, 135, 160, 181, 174, 193, 194, 200]
Prices	[38, 78, 95, 104, 129, 154]
Tourism	[41, 50, 59, 65–67, 94, 177]
Trade	[3, 5, 7, 34] [24, 56, 79, 121, 136, 186, 190, 197]
Violence against women	[37, 73, 86, 98, 165, 180, 191]

The growing demand for textile products increases pollution, as it is positioned as one of the most polluting industries [57, 175] with a generation of 5% of waste in the world [93]. It employs toxic chemicals [176], produces polluting carbon from waste pyrolysis [198], uses energy [149], and makes packaging waste [96]. It added that fashion businesses quickly bring out new collections to market [143], which is related to the term fast fashion, which caused the textile sector to be distinguished as an industry that discards goods quickly, reducing the useful life of textile products [61, 124]. About 90 million clothes go to landfills annually [157]. 20% are reused and recycled [156].

For this reason, there is an increased awareness of the environmental damages of this sector, causing the creation of new business models with a sustainable vision [156]. Companies in this industry have started to consider their social and environmental footprint [118]. For example, waste such as yarn can be employed for new fabric manufacturing [103]. Management performed within a textile and fashion circularity is related to the concept of economic circularity emerging in the mid-1960s [54], but for [178], it is a controversial form of business. However, since 2015, the European Commission reported that the circular economy is an important model to consider in business [120].

The application of the circular economy has a focus on reducing waste [40, 88]. For textile products that are not being reused but wasted [92]. Much research on this concept [89, 147]. Due to its sustainable contributions to business and the planet [107, 170]. However, [172] and [161] pointed out a slight increase in studies on applying this concept due to the different barriers that industries encounter when they intend to implement it within their business. Among the barriers are cultural, market, technological, and regulatory barriers [120].

In the same sense, the poor traceability related to the number of textiles available in commerce causes it to be complicated to quantify the number of textile materials that are wasted [150], and textile companies do not incorporate circularity strategies due to limited resources [164]. However, the importance of reusing and recycling textile inputs or the products themselves would allow introducing new products to the market, which could occur through incorporating the circular economy with strategies such as the return and reuse of products [117].

2 The Economic Impacts of the Textile Sector

Living in a globalized world has generated the development and economic growth of various industries present in the market [184], and the fashion industry is no exception. A study found that the footwear and apparel industry is equivalent to about 2% of the global economy, which is 3000 billion dollars [159, 176], which means that the industry being studied in this paper has a significant influence and what happens to it have a positive or negative impact. In commerce, the textile sector needs employees to remain in operation, so this industry employs many workers and generates income at a global level [69].

With time, the increase in sales volume of textile clothing has been observed because the amount of clothing a person consumes has increased; according to [159], in 1995, there was 7.6 kg of fiber per person; while for the year 2018 it was 13.8 kg of fiber per person, this being an increase of 47%. The significant increase is due to fast fashion companies executing a strategy of price reduction through offers, in addition to incentivizing seasonal or campaign fashion products, which has been driven mainly by the era of globalization [43]. An increase in volume brings the need to hire more staff, so the textile industry generates many jobs. For example, in Table 2, only H&M in 2010 had 59,440 employees globally. The growth for the company increased the number of workers from 104,634 in 2015 to 110,325 in 2020.

In the fashion industry, haute couture, fast fashion brands, and the ones that sell sports clothing are considered the biggest in the world. Also, [159] say that countries such as Australia, the United States, Canada, and several located in Western Europe tend to consume a more significant amount of fashion textiles and that there is also a relationship between purchasing power the amount purchased. Furthermore, [39] emphasize that women usually consume apparel just for fashion needs and no longer basic needs. Within the price of products sold in various market industries, taxes are included, which are taxes collected by the state to finance public works, health, infrastructure, education, and others [12].

[63] mentions that Nike is the biggest has a value of around 35 billion dollars, followed by Gucci, the Italian luxury fashion brand with 17.6 billion dollars,and then have placed two brands in the third position with 16.5 billion dollars, one is a sports

Table 2 Number of employees around the world of H&M and Macy's

Year	Number of employees
2020	110,325
2015	104,634
2010	59,440

Source [100]

Table 3 The most valuable clothing brands in the world

Brand	Country	Brand value	Market capitalization
Nike	United States	34.8	180.2
Gucci	Italy	17.6	79.0
Adidas	Germany	16.5	71.1
Louis Vuitton	France	16.5	327.5
Cartier	France	15.0	51.3
Zara	Spain	14.6	97.1
H&M	Sweden	13.9	32.6
Chanel	France	13.7	N/A
Uniqlo	Japan	12.9	103.8
Hermes	France	11.9	118.0

Source [63]

brand again which is Adidas and the other one is the French luxury brand, Louis Vuitton. In Table 3 appears the ten most valuable clothing brands around the world. Also, the brand which has the highest market capitalization is the French company Louis Vuitton, with $327 billion, followed by the American sports brand Nike with $180 billion; the French Hermes with $118 billion, and a Uniqlo, a not to known brand with $103 billion.

3 The Environmental Impacts of the Textile Sector and the Circular Economy

The textile industry has a high rate of pollution on the planet [114], which is because companies use materials such as nylon, elastane, and cotton where they develop a method of production and disposal without the vision of reuse during the manufacturing chain [74, 144, 189].

For example, cotton is linked to high consumption of water, soil, and chemicals [80, 179, 187], use of polluting chemical dyes [138]. In the face of this, [75] investigated a tool that allows assessing the level of water preservation as well as monitoring the amount of contamination in the manufacturing chain using a preservation technology that helps to manufacture with a higher level of cleanliness in the textile industry [130].

It generates 3% of GHGs, positioning it within the top 5 pollution-producing items in the world [153]. Each year, the textile sector produces about 110 million tons of clothing and fibers that generate textile waste globally [131]. These are incinerated, causing environmental damage such as emissions [48, 109, 111, 196]. It should be specified that the damage of fibers or fabrics depends on the type of garment [80]. At the same time, the thickness of textile yarn and fabric causes pollution [189]. Among the types of waste from the textile and fashion sector are cardboard/paper, plastics, chemical, fiber waste, spinning waste, weaving waste, knitting waste, and clothing waste, which are pollutants worldwide.

Therefore, today there is a growing awareness of the environmental footprint generated by the textile industry, such as waste and carbon emissions [152, 163]. In this regard, [71] mentioned that the incorporation of cleaner production practices promotes environmental benefits for the textile industry through the minimal use of toxic materials and reducing accidents related to the environment. Waste management involves all parts of the chain, from the manufacturer to the customers of textile products [131].

Within the circular economy, the circular supply chain (CSC) can be found, which provides a flow that has no end since companies perform reuse, remodeling, and recycling of materials [140]. It is used to study pollution problems by decreasing waste during the manufacturing process of textile items [132]. It aims to improve resource management [132]. Therefore, these negative impacts can be reduced by applying the circular economy.

4 The Social Impacts of the Textile Sector and the Circular Economy

In the textile and fashion sector, there is no unanimity about the social impacts to be considered [116]. However, it is considered an item with significant negative impacts due to many worker rotations [128]. This problem is related to a large number of informality that triggers the possibility that employees in this sector do not receive additional benefits from the company or the government [62]. In addition, companies recruit migrants to pay them less. And they hire cheap labor demonstrating the exploitation of workers [96].

SMEs account for at least 45% and 40% of industrial manufacturing and exports. They provide jobs to more than 42 million people where employees are engaged in dyeing, cutting, and sewing [158]. In addition, they provide 8,000 goods of excellent quality produced for international markets, but with a production where informality and minimal labor payment abound [138]. For example, India is a textile manufacturing country after China and Bangladesh.

Another problem is the benefits of employees in the entire garment chain and the population's welfare [52]. Workers can acquire respiratory diseases in the industry due to the constant use of chemicals [47]; they can also suffer different occupational incidences. For example, in 2017, Brazilian companies in the textile sector presented a high percentage of occupational accidents, being 16.4/1000 employees [146]. Table 4 shows an example of labor incidents occurring in textile companies located in Brazil between 2015 and 1017, where males suffered the most damage.

However, today, social pressures seek to improve Corporate Social Responsibility within fashion companies [96], which allows companies to be more competitive in

Table 4 Labor incidence in Brazilian companies in the textile and fashion sector

Variable	Amount	2015	2016	2017	Annual variation rate (%)
Textile production	490,540	1.1%	1.1%	1.1%	−9.1
Clothing production and accessories	1,064,874	0.5%	0.6%	0.5%	−8.6

Source [146]

international trade, as well as to improve the quality of work of their employees, which affects their efficient performance [192]. To do this, companies can train workers to learn how to collect waste and send it to recovery sites [131]. In this way, employees learn about sustainable practices and take care of their health. Against this, circularity can also be incorporated, focusing on reducing environmental and social problems [55].

5 Circular Fashion and Waste Footprint in the Textile and Fashion Industry

The increase in the world population, the increase in the socioeconomic level of the middle class, and the low prices for purchasing clothing items have brought with it a high demand for these products [123, 139], which has resulted in the fashion industry being one of the most polluting, so its carbon footprint is large [42, 148]. On the part of the offer, many clothing brands, with low prices, contributed with the manufacture of garments of poor quality but fashionable each season, in this sense, consumers would see the need to acquire new fashion apparel in a linear way which is called fast fashion to [123]. This sector generates a negative impact on the environment because different types of waste are produced, such as water, solid waste, and the disposal of textile garments once their life ends [105, 176], and the production of CO_2 caused by the supply chain [123].

Nowadays, especially the companies related to textile and fashion products are related to unfair conditions involving health problems and child exploitation [157]. Social impacts are also related to environmental damage because fashion companies are looking for solutions to generate less impact on the carbon footprint [42], which is because of the high volume of fashion waste that the brands in the industrial production of fashion [151].

It is necessary to consider that consumers are becoming more conscious about the process involving the production of one fashion item and its consequences on the environment [45, 185, 188]. Previous research made in the UK found that income, culture, and prices work as determinants when consumers make decisions [204]. Because of that, fashion companies are trying to improve their supply chain by rethinking and designing models which focus circularly and avoid the traditional linear fashion [97]. In that sense, the brands are going through challenges to reduce their environmental impact. [204] also, add that the support of governments is needed by improving policy interventions such as subsidies and taxes to promote sustainability in the fashion sector of each country.

For example, the European Union decades ago has started regulations and policies to reduce the carbon emissions that used to be produced from industrial processes, transportation of goods, the energy of the factories, and others [12]. Also, some of the policies involucrated incorporating the concept of recycling in commercialized products, including fashion textiles, to generate less impact on the waste footprint [142].

The linear economy model generates a high carbon footprint because it generates the depletion of non-renewable resources [64]; that is why it is proposed to use the circular economy in various industries [55, 82], especially in the fashion industry [110]. In addition, [64] add that for success in the circular economy, it is essential to select partners with whom trust can be developed with a common goal, a circular economy, and customers. However, the culture of fast fashion consumption by clients in the sector is considered one of the main factors preventing the fashion industry's fluid, circular economy model [81]. To add, fashion companies execute marketing strategies to increase demand, hiding the environmental damage caused by consuming fast fashion items [46].

In their research, the authors [81] mentioned that in fashion circularity. There are many barriers, initiatives depending on the material, and drivers in every statement. In Fig. 1, the process is shown in a simplified way. The design stage is related to market circular economy initiatives about reduction, reuse, recycling, institutional regulation from governments, and the stakeholders are the conscientious consumers. Then in the manufacturing stage, technological initiatives are present, market organizations and many technical challenges are manifested. Then, the consumer usage focuses on the consumer culture and depends on them not to throw and not to buy constantly. And finally, the collection or recycling where technical challenges are involved as barriers and organizations. In the distribution, the authors found that circular initiatives are needed to reduce the carbon footprint because of the logistic process.

Some limitations of the fashion circularity in the literature according to [168], on their research, the main findings were that there are necessary more studies about

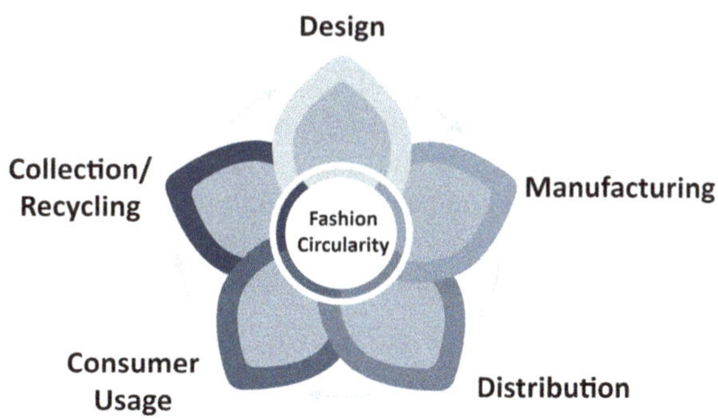

Fig. 1 Fashion circularity process (*Source* [81])

processes related to the collection, recycling, and classification. They add that fewer studies about wool and viscose, while the most studied are cotton and polyester. So is necessary more knowledge about these materials to process those to create new alternatives for the circularity process.

6 Case Studies of the Application of Circular Economy in the Textile Sector

The following paragraphs show various cases of companies and countries dedicated to the fashion industry implementing circular economy strategies and practices. The cases focus on countries such as the Netherlands, Brazil, and Australia, the incorporation of upcycling, and the company H&M, one of the companies that markets a large volume of textile products.

6.1 Garments Took Back by H&M

Companies in the fashion industry have implemented actions that involve returning clothing items in the store, by the end customer, through the collection in the same store, or working together with a social aid organization, among others [117]. With this innovative business model, brands make known their commitment to the associations with which they work together, in addition to their environmental responsibility [117].

Consumers' perception of this action has a positive influence because they capture the environmental and social awareness generated by the company in the garment collection campaign [113]. An example of a brand that is in the process of changing from traditional fast fashion to now opting for sustainability is H&M, with around 5020 stores worldwide, according to H&M 100, which seeks to exploit the garment waste [55]. It should be mentioned that it is essential to highlight the role played by the consumer since it depends a lot on them whether the fashion and textile circularity model works [148]. According to [115], the fourth industrial revolution in the fashion industry is a reshape of living and working. Both, in their research, highlight the importance of new technologies in the fashion companies and recycle the fashion textiles and give them another life with circularity and reduce the waste footprint that they would come with if the cloth ended in a dump. To carry out this process, the company mean high investment, but the implementation of this is considered as a tool of marketing that benefits the company [46] because of the consumer's conscience [45, 185, 188] as it was mentioned.

Table 5 Percentage approximately of fabric leftover while manufacturing

Fabric leftover	Large manufacturer (%)	Small manufacturer (%)
Cutting leftovers	12	21
End-pits and roll ends	2.5	4.2
Rejected fabric and garments	4.2	5.2
Excess made by the fabric	2.5	3.7
Over production	3.5	5.1
Total leftovers	24.7	39.2

Source [42]

6.2 Integrating Upcycling

The concept of upcycling refers to using textile leftovers that are considered waste for the creation of new textile garments [42], which works as a new method that many apparel brands can implement. Table 5 shows the high quantities of fabric leftover generated by large and small manufacturers where the main difference is cutting waste. In that sense, it demonstrates that when comparing both, the large manufacturing companies produce less leftover fabric because small companies can get an excess fabric. After all, their sales are in a smaller quantity.

6.3 Australia: Recycling Cotton from Denim

In the continent of Oceania, being specific in Australia, research made by [139] demonstrated a process of recycling where the denim waste goes through a process of dissolution to a binary solvent, and then while it is wet, it is spun. In that study, they show that using this process, the spun fiber, the color that belongs from the original textile, can keep on the new threads or, in the other case, can be regenerated if color is absent. Figure 2 shows the process that the authors propose.

6.4 Netherlands: Textile Materials in Circulation

In the Netherlands, research was developed to address the circular economy in the textile industry as an efficient method for the utilization of textile materials [147]. The approach was based on the analysis of institutional transformation in the circular economy transition [93] since no theory provides knowledge for organizations to develop new institutions to favor circular textile materials.

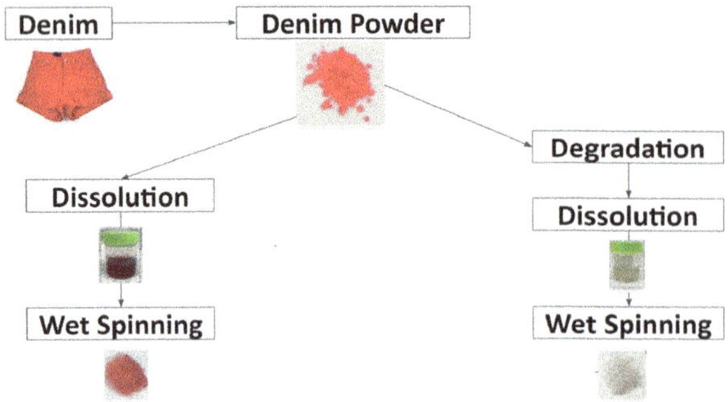

Fig. 2 Process for Denim recycle (*Source* [139])

Table 6 Strategic situations for a circular economy according to Dutch companies

Companies	Circular economy strategies	Strategic positions or situations
Mud Jeans	Rent a jeans	Develop a chain of coordination between all parties involved
Dutch aWEARness	Corporate rental of Dresses	
Lena Fashion Library	Fashion library	

Source [93]

The Netherlands and Nordics have a 61% rate of discarded clothes, go to landfills, or become incinerated [93, 111]. In some cases, they go to recycling, but recycling implies a downward life cycle of the materials after being reused. Therefore, Dutch apparel companies are founders of different companies that make transition processes toward circular development of materials, and this is executed with the generation of new organizational ways. For example, reconditioning and remanufacturing, sustainable strategies found in the circular economy [140] allow lengthening the utilization of materials by decreasing the application of harmful chemicals [176].

About institutions, new standards were created in this country regarding the use of textile materials to supervise the fair use of recycled fabrics and to develop cascade actions in the apparel chain. To this end, they created two agreements called PAS and SQ that are helping companies and institutions to produce positive impacts on the planet and society. The PAS agreement includes Mud Jeans, Lena Fashion Library, and Dutch aWEARness, which apply the circularity regulations proposed in this agreement. Table 6 shows the strategies implemented by these Dutch companies.

6.5 *Brazil: Textile Recycling*

Brazil ranks fifth among the countries involved in textile and fashion manufacturing and fourth in clothing manufacturing globally [146]. Within the state that produces the most polo in this country is São Paulo, followed by Santa Catarina. This second, in 2017, generated an income of more than 19.7% in the textile and fashion industry,

Table 7 Brazilian companies that recycle textile waste

Name	Raw material	Final product
Benefibras	Post-industrial waste of natural fibers	Fiber
Ecosimple		
Lonatex		
Superfios		
Ecofios	Post-industrial waste of natural fibers	Wires and strings
Benefios		
Eurofios		
Maxitex	Post-industrial waste of natural and chemical fiber	Yarns, fabrics, and clothing

Source [36]

Table 8 Cost of textile waste about recycling Brazilian companies

Components	Value per kilogram
Combined waste	R$ 0.05 < R$ 0.10
Cotton fabric with colors	R$ 0.70 < R$ 1.00
White cotton fabric	R$ 1.20 < R$ 1.70
Acrylic fiber or polyamide	R$ 0.70 < R$ 1.00

Source [36]

where companies provided jobs to 162,845 inhabitants since, in this state, there are 1,832 companies dedicated to this item [146].

In Brazil, companies use Jean's waste as inputs for the automotive industry, materials that are manufactured from 100% cotton and polyester are used to produce wires and ropes [36]. For this reason, it is essential to provide a practical example from this country related to sustainable textile waste practices based on waste recovery through recycling [131, 169]. Table 7 shows the name of Brazilian companies dedicated to recycling textile waste to improve their sustainability as a company to take care of the planet and society in general. While Table 8 shows the price of textile waste in companies located in Brazil.

7 Closing Remarks

The fashion industry influences the world economy due to the era of globalization. It is mainly due to the high demand of consumers to buy fashion garments driven by marketing [46], which has brought with it economic income for large companies that sell textile fashion, as well as for the GDP of these countries, it has also generated jobs for 45 million people at the international level [69, 173]. For example, in Bangladesh, 4.2 million job opportunities have been generated Hossain et al. [108], and 42 million for India [158], while the H&M company employed around 110,325 people in 2020 [100].

However, although this industry has been bringing a positive impact on the economy of the different countries, it has also brought with it a negative impact that harms the environment [137], despite having environmental regulations to control pollution in the sector [102]. Among the damages it causes are the high amount of "gray water" that causes the water footprint to be harmed and the climate due to accelerated variations in aquatic ecosystems [75, 182]. The waste is incinerated or thrown away [111]. The high levels of chemicals [187], among other problems detailed in items of environmental impacts of this industry.

Regarding social impacts, there is evidence of a high rate of labor incidents due to the activities carried out by employees to manufacture garments or textile accessories [146]. Added to this are stressful workloads [1] and the minimum payment for migrant labor [134]. Even though the institutions have implemented good employment conditions, they have not managed to ensure the well-being of the workers in this industry. For this reason, [4] states that the owners of international companies and local and international public authorities have to pay more attention to these problems related to confections. In the same sense, [183] pointed out that the industry needs an empowered workforce obtained through sustainable regulatory policies by international institutions and collaborations.

Because of that, to keep using the traditional linear fashion, [97] propose the implementation of the fashion circularity. The primary purpose of this is to avoid waste produced by the cloth garment, where basically the idea is that the process continues from the design, the manufacturing, the distribution, the consumer usage, then the collection of the apparel to recycle this, and then goes the design again and so forth [81].

The companies such as H&M have been implementing the clothing garments take back to reduce the waste solid created [55]. Also, innovative proposals such as integrating upcycling in Australian research [42], recycling cotton from denim [139], the recycling of textile materials in Dutch companies through the collaboration of institutions can later create new sustainability regulations [93], (Fischer et al. 2017). Finally, there is the recycling of textiles in Brazilian companies, where they develop transformations to waste by converting them into new products such as cables and ropes [36].

References

1. Abd-Elkader MR, Kabbash IA, El-Sallamy RM, El-Sawy H, Gad E-SA-H (2020) Tramadol abuse among workers in an industrial city in mid-Nile Delta region, Egypt. Environ Sci Pollut Res 27(30):37549–37556. https://doi.org/10.1007/s11356-020-08040-8
2. Abiakam N, Worsley P, Jayabal H, Mitchell K, Jones M, Fletcher J, Spratt F, Bader D (2021) Personal protective equipment related skin reactions in healthcare professionals during COVID-19. Int Wound J 18(3):312–322. https://doi.org/10.1111/iwj.13534

3. Abudureheman A, Nilupaer A (2021) Optimization model design of cross-border e-commerce transportation path under the background of prevention and control of COVID-19 pneumonia. Soft Comput 25(18):12007–12015. https://doi.org/10.1007/s00500-021-05685-6
4. Acharya AK (2021) Caste-based migration and exposure to abuse and exploitation: Dadan labour migration in India. Contemp Soc Sci 16(3):371–383. https://doi.org/10.1080/21582041.2020.1855467
5. Acuña-Zegarra MA, Santana-Cibrian M, Velasco-Hernandez JX (2020) Modeling behavioral change and COVID-19 containment in Mexico: a trade-off between lockdown and compliance. Math Biosci 325:108370. https://doi.org/10.1016/j.mbs.2020.108370
6. Afshan G, Shahid S, Tunio MN (2021) Learning experiences of women entrepreneurs amidst COVID-19. Int J Gend Entrep 13(2):162–186. https://doi.org/10.1108/IJGE-09-2020-0153
7. Aguirre AA, Catherina R, Frye H, Shelley L (2020) Illicit wildlife trade, wet markets, and COVID-19: preventing future pandemics. World Med Health Policy 12(3):256–265. https://doi.org/10.1002/wmh3.348
8. Ahsan MM (2020) Strategic decisions on urban built environment to pandemics in Turkey: lessons from COVID-19. J Urban Manag 9(3):281–285. https://doi.org/10.1016/j.jum.2020.07.001
9. Alan H, Eskin Bacaksiz F, Tiryaki Sen H, Taskiran Eskici G, Gumus E, Harmanci Seren AK (2021) "I'm a hero, but...": an evaluation of depression, anxiety, and stress levels of frontline healthcare professionals during COVID-19 pandemic in Turkey. Perspect Psychiatr Care 57(3):1126–1136. https://doi.org/10.1111/ppc.12666
10. Ali W (2020) Online and remote learning in higher education institutes: a necessity in light of COVID-19 pandemic. High Educ Stud 10(3):16–25
11. Allen J, Rowan L, Singh P (2020) Teaching and teacher education in the time of COVID-19. Asia-Pac J Teach Educ 48(3):233–236. https://doi.org/10.1080/1359866X.2020.1752051
12. Allevi E, Gnudi A, Konnov IV, Oggioni G (2018) Evaluating the effects of environmental regulations on a closed-loop supply chain network: a variational inequality approach. Ann Oper Res 261(1):1–43. https://doi.org/10.1007/s10479-017-2613-1
13. Alsairafi Z, Naser AY, Alsaleh FM, Awad A, Jalal Z (2021) Mental health status of healthcare professionals and students of health sciences faculties in Kuwait during the COVID-19 pandemic. Int J Environ Res Public Health 18(4). https://doi.org/10.3390/ijerph18042203
14. Altindis E (2021) Inequitable COVID-19 vaccine distribution and the intellectual property rights prolong the pandemic. Expert Rev Vaccines Null–null. https://doi.org/10.1080/14760584.2022.2014819
15. Alvarez-Risco A, Del-Aguila-Arcentales S, Delgado-Zegarra J, Yáñez JA, Diaz-Risco S (2019) Doping in sports: findings of the analytical test and its interpretation by the public. Sport Sci Health 15(1):255–257. https://doi.org/10.1007/s11332-018-0484-8
16. Alvarez-Risco A, Del-Aguila-Arcentales S, Diaz-Risco S (2018) Pharmacovigilance as a tool for sustainable development of healthcare in Peru. PharmacoVigilance Rev 10(2):4–6
17. Alvarez-Risco A, Del-Aguila-Arcentales S, Rosen MA, García-Ibarra V, Maycotte-Felkel S, Martínez-Toro GM (2021) Expectations and interests of university students in COVID-19 times about sustainable development goals: evidence from Colombia, Ecuador, Mexico, and Peru. Sustainability 13(6):3306. https://doi.org/10.3390/su13063306
18. Alvarez-Risco A, Del-Aguila-Arcentales S, Stevenson JG (2015) Pharmacists and mass communication for implementing pharmaceutical care. Am J Pharm Benefits 7(3):e125–e126
19. Alvarez-Risco A, Del-Aguila-Arcentales S, Yáñez JA (2021) Telemedicine in Peru as a result of the COVID-19 pandemic: perspective from a country with limited internet access. Am J Trop Med Hyg 105(1):6–11. https://doi.org/10.4269/ajtmh.21-0255
20. Alvarez-Risco A, Del-Aguila-Arcentales S, Yáñez JA, Rosen MA, Mejia CR (2021) Influence of technostress on academic performance of university medicine students in Peru during the COVID-19 pandemic. Sustainability 13(16):8949. https://doi.org/10.3390/su13168949
21. Alvarez-Risco A, Delgado-Zegarra J, Yáñez JA, Diaz-Risco S, Del-Aguila-Arcentales S (2018) Predation risk by gastronomic boom—case Peru. J Landsc Ecol 11(1):100–103. https://doi.org/10.2478/jlecol-2018-0003

22. Alvarez-Risco A, Mlodzianowska S, García-Ibarra V, Rosen MA, Del-Aguila-Arcentales S (2021) Factors affecting green entrepreneurship intentions in business university students in COVID-19 pandemic times: case of Ecuador. Sustainability 13(11):6447. https://doi.org/10.3390/su13116447
23. Alvarez-Risco A, Quiroz-Delgado D, Del-Aguila-Arcentales S (2016) Pharmaceutical care in hypertension patients in a peruvian hospital. Indian J Public Health Res Dev 7(3):183–188
24. Alvarez-Risco A, Rosen MA, Del-Aguila-Arcentales S (2020) A new regulation for supporting a circular economy in the plastic industry: the case of Peru (short communication). J Landsc Ecol 13(1):1–3. https://doi.org/10.2478/jlecol-2020-0004
25. Alvarez-Risco A, van Mil JWF (2007) Pharmaceutical care in community pharmacies: practice and research in Peru. Ann Pharmacother 41(12):2032–2037. https://doi.org/10.1345/aph.1K117
26. Álvarez-Risco A, Zegarra Arellano E, Matos Valerio E, Mejía Acosta N, Solis Tarazona Z (2013) Campaña de atención farmacéutica como estrategia de implementación de los servicios farmacéuticos: Experiencia Perú. Pharm Care España 15(1):35. https://www.pharmcareesp.com/index.php/PharmaCARE/article/view/105
27. Alvarez-Risco A, Dawson J, Johnson W, Conteh-Barrat M, Aslani P, Del-Aguila-Arcentales S, Diaz-Risco S (2021) Ebola virus disease outbreak: a global approach for health systems [Ebola; health professionals; global approach; pharmacist: pharmaceutical care; COVID-19.]. 53(4). http://www.revfarmacia.sld.cu/index.php/far/article/view/491
28. Alvarez-Risco A, Del-Aguila-Arcentales S (2015) Errores de prescripción como barrera para la Atención Farmacéutica en establecimientos de salud públicos: Experiencia Perú. Pharm Care España 17(6):725–731. https://www.pharmcareesp.com/index.php/PharmaCARE/article/view/246
29. Alvarez-Risco A, Del-Aguila-Arcentales S (2021a) A note on changing regulation in international business: The World Intellectual Property Organization (WIPO) and artificial intelligence. In: Verbeke A, van Tulder R, Rose EL, Wei Y (eds) The multiple dimensions of institutional complexity in international business research, vol 15. Emerald Publishing Limited, p. 363–371. https://doi.org/10.1108/S1745-886220210000015020
30. Alvarez-Risco A, Del-Aguila-Arcentales S (2021b) Public policies and private efforts to increase women entrepreneurship based on STEM background. In: Mari M, Poggesi S, Foss L (eds) Women's entrepreneurship in STEM disciplines: issues and perspectives. Springer International Publishing, pp 75–87. https://doi.org/10.1007/978-3-030-83792-1_5
31. Alvarez-Risco A, Estrada-Merino A, de las Mercedes Anderson-Seminario M, Mlodzianowska S, García-Ibarra V, Villagomez-Buele C, Carvache-Franco M (2021) Multitasking behavior in online classrooms and academic performance: case of university students in Ecuador during COVID-19 outbreak. Interact Technol Smart Educ 18(3):422–434. https://doi.org/10.1108/ITSE-08-2020-0160
32. Alvarez-Risco A, Mejia CR, Delgado-Zegarra J, Del-Aguila-Arcentales S, Arce-Esquivel AA, Valladares-Garrido MJ, del Portal MR, Villegas LF, Curioso WH, Chandra Sekar M, Yáñez JA (2020) The Peru approach against the COVID-19 infodemic: insights and strategies. Am J Trop Med Hyg 103(2):583–586. https://doi.org/10.4269/ajtmh.20-0536
33. Alvarez-Risco A, Mlodzianowska S, Zamora-Ramos U, Del-Aguila S (2021) Green entrepreneurship intention in university students: the case of Peru. Entrep Bus Econ Rev 9(4):85–100. https://doi.org/10.15678/EBER.2021.090406
34. Alvarez-Risco A, Quipuzco-Chicata L, Escudero-Cipriani C (2021) Determinantes de la intención de recompra en línea en tiempos de COVID-19: evidencia de una economía emergente. Lect Econ (96). https://doi.org/10.17533/udea.le.n96a342638
35. Alvarez-Risco A, Turpo-Cama A, Ortiz-Palomino L, Gongora-Amaut N, Del-Aguila-Arcentales S (2016) Barreras para la implementación de la Atención Farmacéutica en establecimientos farmacéuticos de Cusco, Perú. Pharm Care España 18(5):194–205. https://www.pharmcareesp.com/index.php/PharmaCARE/article/view/326
36. Amaral MCd, Zonatti WF, Silva KLd, Karam D, Amato J, Baruque-Ramos J (2018) Industrial textile recycling and reuse in Brazil: case study and considerations concerning the circular economy. Gestão & Produção 25:431–443

37. Amarillo CR (2020) Feminism on lockdown. NACLA Rep Am 52(3):274–281. https://doi.org/10.1080/10714839.2020.1809084
38. Apcho-Ccencho L-V, Cuya-Velásquez B-B, Alvarado Rodríguez D, de las Mercedes Anderson-Seminario M, Alvarez-Risco A, Estrada-Merino A, Mlodzianowska S (2021) The impact of international price on the technological industry in the United States and China during times of crisis: commercial war and COVID-19. In: Lawrence KD, Klimberg RK (eds) Advances in business and management forecasting, vol 14. Emerald Publishing Limited, pp 149–160. https://doi.org/10.1108/S1477-407020210000014010
39. Armstrong CM, Niinimäki K, Kujala S, Karell E, Lang C (2015) Sustainable product-service systems for clothing: exploring consumer perceptions of consumption alternatives in Finland. J Clean Prod 97:30–39. https://doi.org/10.1016/j.jclepro.2014.01.046
40. Ashby A (2018) Developing closed loop supply chains for environmental sustainability. J Manuf Technol Manag 29(4):699–722. https://doi.org/10.1108/JMTM-12-2016-0175
41. Assaf A, Scuderi R (2020) COVID-19 and the recovery of the tourism industry. Tour Econ 26(5):731–733. https://doi.org/10.1177/1354816620933712
42. Aus R, Moora H, Vihma M, Unt R, Kiisa M, Kapur S (2021) Designing for circular fashion: integrating upcycling into conventional garment manufacturing processes. Fash TextEs 8(1):34. https://doi.org/10.1186/s40691-021-00262-9
43. Backs S, Jahnke H, Lüpke L, Stücken M, Stummer C (2021) Traditional versus fast fashion supply chains in the apparel industry: an agent-based simulation approach. Ann Oper Res 305(1):487–512. https://doi.org/10.1007/s10479-020-03703-8
44. Bacq S, Lumpkin GT (2021) Social entrepreneurship and COVID-19. J Manage Stud 58(1):285–288. https://doi.org/10.1111/joms.12641
45. Bajaj S (2017) In support of the circular economy: an exercise in stakeholder mapping. Matériaux Tech 105(5–6):512
46. Baldassarre F, Campo R (2016) Sustainability as a marketing tool: To be or to appear to be? Bus Horiz 59(4):421–429. https://doi.org/10.1016/j.bushor.2016.03.005
47. Banerjee S, Goswami K (2019) The occupational illness of slum dwellers across industries: a case study in West Bengal. Int J Ind Ergon 73:102834. https://doi.org/10.1016/j.ergon.2019.102834
48. Barbero-Barrera MdM, Pombo O, Navacerrada, MdlÁ (2016) Textile fibre waste bindered with natural hydraulic lime. Compos Part B: Eng 94:26–33. https://doi.org/10.1016/j.compositesb.2016.03.013
49. Barello S, Caruso R, Palamenghi L, Nania T, Dellafiore F, Bonetti L, Silenzi A, Marotta C, Graffigna G (2021) Factors associated with emotional exhaustion in healthcare professionals involved in the COVID-19 pandemic: an application of the job demands-resources model. Int Arch Occup Environ Health 94(8):1751–1761. https://doi.org/10.1007/s00420-021-01669-z
50. Baum T, Hai NTT (2020) Hospitality, tourism, human rights and the impact of COVID-19. Int J Contemp Hosp Manag 32(7):2397–2407. https://doi.org/10.1108/IJCHM-03-2020-0242
51. Belitski M, Guenther C, Kritikos AS, Thurik R (2021) Economic effects of the COVID-19 pandemic on entrepreneurship and small businesses. Small Bus Econ. https://doi.org/10.1007/s11187-021-00544-y
52. Bick R, Halsey E, Ekenga CC (2018) The global environmental injustice of fast fashion. Environ Health 17(1):92. https://doi.org/10.1186/s12940-018-0433-7
53. Block JH, Fisch C, Hirschmann M (2021) The determinants of bootstrap financing in crises: evidence from entrepreneurial ventures in the COVID-19 pandemic. Small Bus Econ. https://doi.org/10.1007/s11187-020-00445-6
54. Blomsma F, Brennan G (2017) The emergence of circular economy: a new framing around prolonging resource productivity. J Ind Ecol 21(3):603–614. https://doi.org/10.1111/jiec.12603
55. Bocken NMP, Olivetti EA, Cullen JM, Potting J, Lifset R (2017) Taking the circularity to the next level: a special issue on the circular economy. J Ind Ecol 21(3):476–482. https://doi.org/10.1111/jiec.12606

56. Borzée A, McNeely J, Magellan K, Miller JRB, Porter L, Dutta T, Kadinjappalli KP, Sharma S, Shahabuddin G, Aprilinayati F, Ryan GE , Hughes A, Mutalib AHA, Wahab AZA, Bista D, Chavanich SA, Chong JL, Gale GA, Zhang, L (2020) COVID-19 highlights the need for more effective wildlife trade legislation. Trends Ecol Evol 35(12):1052–1055. https://doi.org/10.1016/j.tree.2020.10.001
57. Boström M, Micheletti M (2016) Introducing the sustainability challenge of textiles and clothing. J Consum Policy 39(4):367–375. https://doi.org/10.1007/s10603-016-9336-6
58. Bozdağ F, Ergün N (2020) Psychological resilience of healthcare professionals during COVID-19 pandemic. Psychol Rep 124(6):2567–2586. https://doi.org/10.1177/0033294120965477
59. Brouder P (2020) Reset redux: possible evolutionary pathways towards the transformation of tourism in a COVID-19 world. Tour Geogr 22(3):484–490. https://doi.org/10.1080/14616688.2020.1760928
60. Brown R, Rocha A, Cowling M (2020) Financing entrepreneurship in times of crisis: exploring the impact of COVID-19 on the market for entrepreneurial finance in the United Kingdom. Int Small Bus J 38(5):380–390. https://doi.org/10.1177/0266242620937464
61. Brydges T (2021) Closing the loop on take, make, waste: investigating circular economy practices in the Swedish fashion industry. J Clean Prod 293:126245. https://doi.org/10.1016/j.jclepro.2021.126245
62. Brydges T, Hanlon M (2020) Garment worker rights and the fashion industry's response to COVID-19. Dialogues Hum Geogr 10(2):195–198. https://doi.org/10.1177/2043820620933851
63. Buchholz K (2021) The world's most valuable apparel brands. Retrieved 22 Dec 2021 from https://www.statista.com/chart/24216/most-valuable-apparel-brands
64. Calicchio Berardi P, Peregrino de Brito R (2021) Supply chain collaboration for a circular economy—from transition to continuous improvement. J Clean Prod 328:129511. https://doi.org/10.1016/j.jclepro.2021.129511
65. Carvache-Franco M, Alvarez-Risco A, Carvache-Franco W, Carvache-Franco O, Estrada-Merino A, Rosen MA (2021) Coastal cities seen from loyalty and their tourist motivations: a study in Lima, Peru. Sustainability 13(21):11575. https://doi.org/10.3390/su132111575
66. Carvache-Franco M, Alvarez-Risco A, Carvache-Franco O, Carvache-Franco W, Estrada-Merino A, Villalobos-Alvarez D (2021) Perceived value and its influence on satisfaction and loyalty in a coastal city: a study from Lima, Peru. J Policy Res Tour, Leis Events. https://doi.org/10.1080/19407963.2021.1883634
67. Carvache-Franco M, Carvache-Franco O, Carvache-Franco W, Alvarez-Risco A, Estrada-Merino A (2021) Motivations and segmentation of the demand for coastal cities: a study in Lima, Peru. Int J Tour Res 23(4):517–531. https://doi.org/10.1002/jtr.2423
68. del Castillo FA (2021) Temporary waiver of intellectual property on COVID-19 vaccines: toward the creation of a better, post-pandemic society. J Public Health 43(3):e559–e560. https://doi.org/10.1093/pubmed/fdab184
69. Castro-López A, Iglesias V, Puente J (2021) Slow fashion trends: Are consumers willing to change their shopping behavior to become more sustainable? Sustainability 13(24). https://doi.org/10.3390/su132413858.
70. Cegarra-Navarro J-G, Vătămănescu E-M, Martínez-Martínez A (2021) A context-driven approach on coping with COVID-19: from hiding knowledge toward citizen engagement. Knowl Process Manag 28(2):134–140. https://doi.org/10.1002/kpm.1662
71. Cesar da Silva P, de Oliveira C, Neto G, Ferreira Correia JM, Pujol Tucci HN (2021) Evaluation of economic, environmental and operational performance of the adoption of cleaner production: survey in large textile industries. J Clean Prod 278:123855. https://doi.org/10.1016/j.jclepro.2020.123855
72. Chafloque-Cespedes R, Alvarez-Risco A, Robayo-Acuña P-V, Gamarra-Chavez C-A, Martinez-Toro G-M, Vicente-Ramos W (2021) Effect of sociodemographic factors in entrepreneurial orientation and entrepreneurial intention in university students of Latin American business schools. In: Jones P, Apostolopoulos N, Kakouris A, Moon C, Ratten V, Walmsley A (eds) Universities and entrepreneurship: meeting the educational and social challenges,

vol 11. Emerald Publishing Limited, pp 151–165. https://doi.org/10.1108/S2040-724620210000011010
73. Chafloque-Cespedes R, Vara-Horna A, Asencios-Gonzales Z, Lopez-Odar DR, Alvarez-Risco A, Quipuzco-Chicata L, Schulze C, Sanchez-Villagomez M (2020) Presentismo académico y violencia contra las mujeres en escuelas de negocios e ingeniería en universidades peruanas. Lect Econ 0(93):127–153. https://doi.org/10.17533/udea.le.n93a340726
74. Chaturvedi P, Kulshreshtha K, Tripathi V (2020) Investigating the determinants of behavioral intentions of generation Z for recycled clothing: an evidence from a developing economy. Young Consum 21(4):403–417. https://doi.org/10.1108/YC-03-2020-1110
75. Chen L, Wang L, Wu X, Ding X (2017) A process-level water conservation and pollution control performance evaluation tool of cleaner production technology in textile industry. J Clean Prod 143:1137–1143. https://doi.org/10.1016/j.jclepro.2016.12.006
76. Chen X, Zhang SX, Jahanshahi AA, Alvarez-Risco A, Dai H, Li J, Ibarra VG (2020) Belief in a COVID-19 conspiracy theory as a predictor of mental health and well-being of health care workers in Ecuador: cross-sectional survey study. JMIR Public Health Surveill 6(3):e20737. https://doi.org/10.2196/20737
77. Chen T, Peng L, Yin X, Rong J, Yang J, Cong G (2020) Analysis of user satisfaction with online education platforms in China during the COVID-19 pandemic. Healthcare 8(3). https://doi.org/10.3390/healthcare803020
78. Chung SA, Olivera S, Román BR, Alanoca E, Moscoso S, Terceros BL, Alvarez-Risco A, Yáñez JA (2021) Temáticas de la producción científica de la Revista Cubana de Farmacia indizada en Scopus (1967–2020) Rev Cuba Farm 54(1). http://www.revfarmacia.sld.cu/index.php/far/article/view/511
79. Cruz-Torres W, Alvarez-Risco A, Del-Aguila-Arcentales S (2021) Impact of enterprise resource planning (ERP) implementation on performance of an education enterprise: a structural equation modeling (SEM). Stud Bus Econ 16(2):37–52. https://doi.org/10.2478/sbe-2021-0023
80. Dahlbo H, Aalto K, Eskelinen H, Salmenperä H (2017) Increasing textile circulation—consequences and requirements. Sustain Prod Consum 9:44–57. https://doi.org/10.1016/j.spc.2016.06.005
81. de Aguiar Hugo A, de Nadae J, & da Silva Lima R (2021) Can fashion be circular? A literature review on circular economy barriers, drivers, and practices in the fashion industry's productive chain. Sustainability 13(21). https://doi.org/10.3390/su132112246
82. De los Rios IC, Charnley FJS (2017) Skills and capabilities for a sustainable and circular economy: the changing role of design. J Clean Prod 160:109–122. https://doi.org/10.1016/j.jclepro.2016.10.130
83. Del-Aguila-Arcentales S, Alvarez-Risco A (2013) Human error or burnout as explanation for mistakes in pharmaceutical laboratories. Accred Qual Assur 18(5):447–448. https://doi.org/10.1007/s00769-013-1000-0
84. Delgado-Zegarra J, Alvarez-Risco A, Yáñez JA (2018) Uso indiscriminado de pesticidas y ausencia de control sanitario para el mercado interno en Perú. Rev Panam Salud Publica 42:e3. https://doi.org/10.26633/RPSP.2018.3
85. Deressa W, Worku A, Abebe W, Gizaw M, Amogne W (2021) Risk perceptions and preventive practices of COVID-19 among healthcare professionals in public hospitals in Addis Ababa, Ethiopia. PLoS ONE 16(6):e0242471. https://doi.org/10.1371/journal.pone.0242471
86. Dominelli L (2021) A green social work perspective on social work during the time of COVID-19. Int J Soc Welf 30(1):7–16. https://doi.org/10.1111/ijsw.12469
87. Duarte Alonso A, Kok SK, Bressan A, O'Shea M, Sakellarios N, Koresis A, Solis MAB, Santoni LJ (2020) COVID-19, aftermath, impacts, and hospitality firms: an international perspective. Int J Hosp Manag 91:102654. https://doi.org/10.1016/j.ijhm.2020.102654
88. Dubey R, Gunasekaran A, Childe SJ, Papadopoulos T, Helo P (2019) Supplier relationship management for circular economy. Manag Decis 57(4):767–790. https://doi.org/10.1108/MD-04-2018-0396

89. Elia V, Gnoni MG, Tornese F (2017) Measuring circular economy strategies through index methods: a critical analysis. J Clean Prod 142:2741–2751. https://doi.org/10.1016/j.jclepro.2016.10.196
90. Enciso-Zarate A, Guzmán-Oviedo J, Sánchez-Cardona F, Martínez-Rohenes D, Rodríguez-Palomino JC, Alvarez-Risco A, Del-Aguila-Arcentales S, Diaz-Risco S (2016) Evaluación de la contaminación con agentes citotóxicos en hospitales en Colombia. Pharm Care España 18(6):241–250. https://www.pharmcareesp.com/index.php/PharmaCARE/article/view/354
91. Erfani P, Binagwaho A, Jalloh MJ, Yunus M, Farmer P, Kerry V (2021) Intellectual property waiver for COVID-19 vaccines will advance global health equity. BMJ 374:n1837. https://doi.org/10.1136/bmj.n1837
92. Esposito M, Tse T, Soufani K (2018) Introducing a circular economy: nw thinking with new managerial and policy implications. Calif Manage Rev 60(3):5–19. https://doi.org/10.1177/0008125618764691
93. Fischer A, Pascucci S (2017) Institutional incentives in circular economy transition: the case of material use in the Dutch textile industry. J Clean Prod 155:17–32. https://doi.org/10.1016/j.jclepro.2016.12.038
94. Fotiadis A, Polyzos S, Huan T-CTC (2021) The good, the bad and the ugly on COVID-19 tourism recovery. Ann Tour Res 87:103117. https://doi.org/10.1016/j.annals.2020.103117
95. Galanakis CM, Rizou M, Aldawoud TMS, Ucak I, Rowan NJ (2021) Innovations and technology disruptions in the food sector within the COVID-19 pandemic and post-lockdown era. Trends Food Sci Technol 110:193–200. https://doi.org/10.1016/j.tifs.2021.02.002
96. Garcia-Torres S, Rey-Garcia M, Albareda-Vivo L (2017) Effective disclosure in the fast-fashion industry: from sustainability reporting to action. Sustainability 9(12). https://doi.org/10.3390/su9122256
97. Gazzola P, Pavione E, Pezzetti R, Grechi D (2020) Trends in the fashion industry. The perception of sustainability and circular economy: a gender/generation quantitative approach. Sustainability 12(7). https://doi.org/10.3390/su12072809
98. Gulati G, Kelly BD (2020) Domestic violence against women and the COVID-19 pandemic: What is the role of psychiatry? Int J Law Psychiatry 71:101594. https://doi.org/10.1016/j.ijlp.2020.101594
99. Gursoy D, Chi CG (2020) Effects of COVID-19 pandemic on hospitality industry: review of the current situations and a research agenda. J Hosp Market Manag 29(5):527–529. https://doi.org/10.1080/19368623.2020.1788231
100. H&M (2021) Hennes and Mauritz full year report 2020. Retrieved Dec 22 2021 from https://hmgroup.com/wp-content/uploads/2021/02/H-M-Hennes-Mauritz-AB-Full-year-report-2020.pdf
101. Hanzl M (2020) Urban forms and green infrastructure—the implications for public health during the COVID-19 pandemic. Cities Health. https://doi.org/10.1080/23748834.2020.1791441
102. Haque N (2017) Exploratory analysis of fines for water pollution in Bangladesh. Water Resour Ind 18:1–8. https://doi.org/10.1016/j.wri.2017.05.001
103. Haslinger S, Hummel M, Anghelescu-Hakala A, Määttänen M, Sixta H (2019) Upcycling of cotton polyester blended textile waste to new man-made cellulose fibers. Waste Manage 97:88–96. https://doi.org/10.1016/j.wasman.2019.07.040
104. Hepburn C, O'Callaghan B, Stern N, Stiglitz J, Zenghelis D (2020) Will COVID-19 fiscal recovery packages accelerate or retard progress on climate change? Oxf Rev Econ Policy 36(Supplement_1), S359–S381. https://doi.org/10.1093/oxrep/graa015
105. Hiller Connell KY, Kozar JM (2017) Introduction to special issue on sustainability and the triple bottom line within the global clothing and textiles industry. Fash TextEs 4(1):16. https://doi.org/10.1186/s40691-017-0100-6
106. Ho C-Y, Tsai B-H, Chen C-S, Lu M-T (2021) Exploring green marketing orientations toward sustainability the hospitality industry in the COVID-19 pandemic. Sustainability 13(8). https://doi.org/10.3390/su13084348

107. Homrich AS, Galvão G, Abadia LG, Carvalho MM (2018) The circular economy umbrella: tends and gaps on integrating pathways. J Clean Prod 175:525–543. https://doi.org/10.1016/j.jclepro.2017.11.064
108. Hossain L, Khan, MS (2020) Water Footprint management for sustainable growth in the bangladesh apparel sector. Water 12(10): 2760.
109. Hu Y, Du C, Pensupa N, Lin CSK (2018) Optimisation of fungal cellulase production from textile waste using experimental design. Process Saf Environ Prot 118:133–142. https://doi.org/10.1016/j.psep.2018.06.009
110. Hultberg E, Pal R (2021) Lessons on business model scalability for circular economy in the fashion retail value chain: towards a conceptual model. Sustain Prod Consum 28:686–698. https://doi.org/10.1016/j.spc.2021.06.033
111. Istrate I-R, Galvez-Martos J-L, Dufour J (2021) The impact of incineration phase-out on municipal solid waste landfilling and life cycle environmental performance: case study of Madrid, Spain. Sci Total Environ 755:142537. https://doi.org/10.1016/j.scitotenv.2020.142537
112. Jecker NS, Atuire CA (2021) What's yours is ours: waiving intellectual property protections for COVID-19 vaccines. J Med Ethics 47(9):595. https://doi.org/10.1136/medethics-2021-107555
113. Jena SK, Sarmah SP (2015) Measurement of consumers' return intention index towards returning the used products. J Clean Prod 108:818–829. https://doi.org/10.1016/j.jclepro.2015.05.115
114. Jia F, Yin S, Chen L, Chen X (2020) The circular economy in the textile and apparel industry: a systematic literature review. J Clean Prod 259:120728. https://doi.org/10.1016/j.jclepro.2020.120728
115. Jin BE, Shin DC (2021) The power of 4th industrial revolution in the fashion industry: What, why, and how has the industry changed? Fash TextEs 8(1):31. https://doi.org/10.1186/s40691-021-00259-4
116. Joyce A, Paquin RL (2016) The triple layered business model canvas: a tool to design more sustainable business models. J Clean Prod 135:1474–1486. https://doi.org/10.1016/j.jclepro.2016.06.067
117. Kant Hvass K, Pedersen ERG (2019) Toward circular economy of fashion. J Fash Mark Manag: Int J 23(3):345–365. https://doi.org/10.1108/JFMM-04-2018-0059
118. Karaosman H, Morales-Alonso G, Brun A (2017) From a systematic literature review to a classification framework: sustainability Integration in fashion operations. Sustainability 9(1). https://doi.org/10.3390/su9010030
119. Kaushal V, Srivastava S (2021) Hospitality and tourism industry amid COVID-19 pandemic: perspectives on challenges and learnings from India. Int J Hosp Manag 92:102707. https://doi.org/10.1016/j.ijhm.2020.102707
120. Kirchherr J, Piscicelli L, Bour R, Kostense-Smit E, Muller J, Huibrechtse-Truijens A, Hekkert M (2018) Barriers to the circular economy: evidence from the European Union (EU). Ecol Econ 150:264–272. https://doi.org/10.1016/j.ecolecon.2018.04.028
121. Kirk CP, Rifkin LS (2020) I'll trade you diamonds for toilet paper: consumer reacting, coping and adapting behaviors in the COVID-19 pandemic. J Bus Res 117:124–131. https://doi.org/10.1016/j.jbusres.2020.05.028
122. Koetter P, Pelton M, Gonzalo J, Du P, Exten C, Bogale K, Buzzelli L, Connolly M, Edel K, Hoffman A, Legro NR, Medina D, Sood N, Blaker J, Kearcher K, Sciamanna C (2020) Implementation and process of a COVID-19 contact tracing initiative: leveraging health professional students to extend the workforce during a pandemic. Am J Infect Control 48(12):1451–1456. https://doi.org/10.1016/j.ajic.2020.08.012
123. Koszewska M (2018) Circular economy—challenges for the textile and clothing industry. Autex Res J 18(4):337–347. https://doi.org/10.1515/aut-2018-0023
124. Kozlowski A, Searcy C, Bardecki M (2015) Corporate sustainability reporting in the apparel industry. Int J Product Perform Manag 64(3):377–397. https://doi.org/10.1108/IJPPM-10-2014-0152

125. Krishtel P, Malpani R (2021) Suspend intellectual property rights for COVID-19 vaccines. BMJ 373:n1344. https://doi.org/10.1136/bmj.n1344
126. Lackie K, Najjar G, El-Awaisi A, Frost J, Green C, Langlois S, Lising D, Pfeifle AL, Ward H, Xyrichiso A, Khalili H (2020) Interprofessional education and collaborative practice research during the COVID-19 pandemic: considerations to advance the field. J Interprofessional Care 34(5):583–586. https://doi.org/10.1080/13561820.2020.1807481
127. Lashley MA, Acevedo M, Cotner S, Lortie CJ (2020) How the ecology and evolution of the COVID-19 pandemic changed learning. Ecol Evol 10(22):12412–12417. https://doi.org/10.1002/ece3.6937
128. Lee SH, Ha-Brookshire J (2017) Ethical climate and job attitude in fashion retail employees' turnover intention, and perceived organizational sustainability performance: a cross-sectional study. Sustainability 9(3). https://doi.org/10.3390/su9030465
129. Leiva-Martinez M-A, de las Mercedes Anderson-Seminario M, Alvarez-Risco A, Estrada-Merino A, Mlodzianowska S (2021) Price variation in lower goods as of previous economic crisis and the contrast of the current price situation in the context of COVID-19 in Peru. In: Lawrence KD, Klimberg RK (eds) Advances in business and management forecasting, vol 14. Emerald Publishing Limited, pp 161–166. https://doi.org/10.1108/S1477-407020210000014011
130. Li X, Ren J, Wu Z, Wu X, Ding X (2021) Development of a novel process-level water footprint assessment for textile production based on modularity. J Clean Prod 291:125884. https://doi.org/10.1016/j.jclepro.2021.125884
131. Li X, Wang L, Ding X (2021) Textile supply chain waste management in China. J Clean Prod 289:125147. https://doi.org/10.1016/j.jclepro.2020.125147
132. Lieder M, Rashid A (2016) Towards circular economy implementation: a comprehensive review in context of manufacturing industry. J Clean Prod 115:36–51. https://doi.org/10.1016/j.jclepro.2015.12.042
133. Liguori E, Winkler C (2020) From offline to online: challenges and opportunities for entrepreneurship education following the COVID-19 pandemic. Entrep Educ Pedagog 3(4):346–351. https://doi.org/10.1177/2515127420916738
134. Lilja F (2020) Global capital and local labour. Strategies and labour relations in the Hex River Textiles factory from the 1940s to the early 1990s. Labor Hist 61(1):48–59. https://doi.org/10.1080/0023656X.2019.1681644
135. Lim S, Prakash A (2021) Pandemics and citizen perceptions about their country: Did COVID-19 increase national pride in South Korea? Nations Natl 27(3):623–637. https://doi.org/10.1111/nana.12749
136. Lopez-Odar D, Alvarez-Risco A, Vara-Horna A, Chafloque-Cespedes R, Sekar MC (2020) Validity and reliability of the questionnaire that evaluates factors associated with perceived environmental behavior and perceived ecological purchasing behavior in Peruvian consumers. Soc Responsib J 16(3):403–417. https://doi.org/10.1108/SRJ-08-2018-0201
137. Luo Y, Song K, Ding X, Wu X (2021) Environmental sustainability of textiles and apparel: a review of evaluation methods. Environ Impact Assess Rev 86:106497. https://doi.org/10.1016/j.eiar.2020.106497
138. Luque A, Herrero-García N (2019) How corporate social (ir)responsibility in the textile sector is defined, and its impact on ethical sustainability: an analysis of 133 concepts. Corp Soc Responsib Environ Manag 26(6):1285–1306. https://doi.org/10.1002/csr.1747
139. Ma Y, Zeng B, Wang X, Byrne N (2019) Circular textiles: closed loop fiber to fiber wet spun process for recycling cotton from Denim. ACS Sustain Chem Eng 7(14):11937–11943. https://doi.org/10.1021/acssuschemeng.8b06166
140. Mangla SK, Luthra S, Mishra N, Singh A, Rana NP, Dora M, Dwivedi Y (2018) Barriers to effective circular supply chain management in a developing country context. Prod Plan Control 29(6):551–569. https://doi.org/10.1080/09537287.2018.1449265
141. Maritz A, Perenyi A, de Waal G, Buck C (2020) Entrepreneurship as the unsung hero during the current COVID-19 economic crisis: Australian perspectives. Sustainability 12(11). https://doi.org/10.3390/su12114612

142. McDowall W, Geng Y, Huang B, Barteková E, Bleischwitz R, Türkeli S, Kemp R, Doménech T (2017) Circular economy policies in China and Europe. J Ind Ecol 21(3):651–661. https://doi.org/10.1111/jiec.12597
143. McNeill L, Moore R (2015) Sustainable fashion consumption and the fast fashion conundrum: fashionable consumers and attitudes to sustainability in clothing choice. Int J Consum Stud 39(3):212–222. https://doi.org/10.1111/ijcs.12169
144. McNeill L, Venter B (2019) Identity, self-concept and young women's engagement with collaborative, sustainable fashion consumption models. Int J Consum Stud 43(4):368–378. https://doi.org/10.1111/ijcs.12516
145. Mejía-Acosta N, Alvarez-Risco A, Solís-Tarazona Z, Matos-Valerio E, Zegarra-Arellano E, Del-Aguila-Arcentales S (2016) Reacciones Adversas a Medicamentos reportadas como resultado de la implementación de Atención Farmacéutica en la Farmacia Institucional DIGEMID - Ministerio de Salud de Perú. *Pharm Care España* 18(2):67–74. https://www.pharmcareesp.com/index.php/PharmaCARE/article/view/311
146. Menegon LdS, Menegon FA, Maeno M, Kupek E (2021) Incidência e tendência temporal de acidentes de trabalho na indústria têxtil e de confecção: análise de Santa Catarina, Brasil, entre 2008 e 2017. Rev Bras Epidemiol 24
147. Murray A, Skene K, Haynes K (2017) The circular economy: an interdisciplinary exploration of the concept and application in a global context. J Bus Ethics 140(3):369–380. https://doi.org/10.1007/s10551-015-2693-2
148. Musova Z, Musa H, Drugdova J, Lazaroiu G, Alayasa J (2021) Consumer attitudes towards new circular models in the fashion industry. J Compet 13(3):111–128
149. Muthukumarana TT, Karunathilake HP, Punchihewa HKG, Manthilake MMID, Hewage KN (2018) Life cycle environmental impacts of the apparel industry in Sri Lanka: analysis of the energy sources. J Clean Prod 172:1346–1357. https://doi.org/10.1016/j.jclepro.2017.10.261
150. Naidoo M, Gasparatos A (2018) Corporate environmental sustainability in the retail sector: drivers, strategies and performance measurement. J Clean Prod 203:125–142. https://doi.org/10.1016/j.jclepro.2018.08.253
151. Nayak R, Houshyar S, Patnaik A, Nguyen LTV, Shanks RA, Padhye R, Fegusson M (2020) Sustainable reuse of fashion waste as flame-retardant mattress filing with ecofriendly chemicals. J Clean Prod 251:119620. https://doi.org/10.1016/j.jclepro.2019.119620
152. Niinimäki K, Peters G, Dahlbo H, Perry P, Rissanen T, Gwilt A (2020) The environmental price of fast fashion. Nat Rev Earth & Environ 1(4):189–200. https://doi.org/10.1038/s43017-020-0039-9
153. Nørup N, Pihl K, Damgaard A, Scheutz C (2018) Development and testing of a sorting and quality assessment method for textile waste. Waste Manage 79:8–21. https://doi.org/10.1016/j.wasman.2018.07.008
154. Obergassel W, Hermwille L, Oberthür S (2021) Harnessing international climate governance to drive a sustainable recovery from the COVID-19 pandemic. Clim Policy 21(10):1298–1306. https://doi.org/10.1080/14693062.2020.1835603
155. Okereke M (2021) Towards vaccine equity: Should big pharma waive intellectual property rights for COVID-19 vaccines? Public Health Pract 2:100165. https://doi.org/10.1016/j.puhip.2021.100165
156. Pal R, Gander J (2018) Modelling environmental value: an examination of sustainable business models within the fashion industry. J Clean Prod 184:251–263. https://doi.org/10.1016/j.jclepro.2018.02.001
157. Pedersen ERG, Gwozdz W, Hvass KK (2018) Exploring the relationship between business model innovation, corporate sustainability, and organisational values within the fashion industry. J Bus Ethics 149(2):267–284. https://doi.org/10.1007/s10551-016-3044-7
158. Pereira Neto ML, Ribeiro FAdS, Torres LPM (2020) Apresentação do dossiê "Estudos recentes sobre os mundos do trabalho têxtil no Brasil". Rev Tempo E Argum 12(30):e0100. https://doi.org/10.5965/2175180312302020e0100
159. Peters G, Li M, Lenzen M (2021) The need to decelerate fast fashion in a hot climate—a global sustainability perspective on the garment industry. J Clean Prod 295:126390. https://doi.org/10.1016/j.jclepro.2021.126390

160. Quispe-Cañari JF, Fidel-Rosales E, Manrique D, Mascaró-Zan J, Huamán-Castillón KM, Chamorro-Espinoza SE, Garayar-Peceros H, Ponce-López VL, Sifuentes-Rosales J, Alvarez-Risco A, Yáñez JA, Mejia CR (2021) Self-medication practices during the COVID-19 pandemic among the adult population in Peru: a cross-sectional survey. Saudi Pharm J 29(1):1–11. https://doi.org/10.1016/j.jsps.2020.12.001
161. Ranta V, Aarikka-Stenroos L, Ritala P, Mäkinen SJ (2018) Exploring institutional drivers and barriers of the circular economy: a cross-regional comparison of China, the US, and Europe. Resour Conserv Recycl 135:70–82. https://doi.org/10.1016/j.resconrec.2017.08.017
162. Raudenská J, Steinerová V, Javůrková A, Urits I, Kaye AD, Viswanath O, Varrassi G (2020) Occupational burnout syndrome and post-traumatic stress among healthcare professionals during the novel coronavirus disease 2019 (COVID-19) pandemic. Best Pract Res Clin Anaesthesiol 34(3):553–560. https://doi.org/10.1016/j.bpa.2020.07.008
163. Rausch TM, Kopplin CS (2021) Bridge the gap: consumers' purchase intention and behavior regarding sustainable clothing. J Clean Prod 278:123882. https://doi.org/10.1016/j.jclepro.2020.123882
164. Riemens J, Lemieux A-A, Lamouri S, Garnier L (2021) A Delphi-Régnier study addressing the challenges of textile recycling in Europe for the fashion and apparel industry. Sustainability 13(21). https://doi.org/10.3390/su132111700
165. Roesch E, Amin A, Gupta J, García-Moreno C (2020) Violence against women during COVID-19 pandemic restrictions. BMJ 369:m1712. https://doi.org/10.1136/bmj.m1712
166. Rojas-Osorio M, Alvarez-Risco A (2019) Intention to use smartphones among Peruvian university students. Int J Interact Mob Technol 13(03):13. https://doi.org/10.3991/ijim.v13i03.9356
167. Rojas Román B, Moscoso S, Chung SA, Limpias Terceros B, Álvarez-Risco A, Yáñez JA (2020) Tratamiento de la COVID-19 en Perú y Bolivia y los riesgos de la automedicación [COVID-19; tratamiento; estrategia; fármacos; automedicación; Perú; Bolivia.]. 53(2). http://www.revfarmacia.sld.cu/index.php/far/article/view/435/351
168. Sandin G, Peters GM (2018) Environmental impact of textile reuse and recycling—a review. J Clean Prod 184:353–365. https://doi.org/10.1016/j.jclepro.2018.02.266
169. Sandvik IM, Stubbs W (2019) Circular fashion supply chain through textile-to-textile recycling. J Fash Mark Manag Int J 23(3):366–381. https://doi.org/10.1108/JFMM-04-2018-0058
170. Schroeder P, Anggraeni K, Weber U (2019) The relevance of circular economy practices to the sustainable development goals. J Ind Ecol 23(1):77–95. https://doi.org/10.1111/jiec.12732
171. Sekalala S, Forman L, Hodgson T, Mulumba M, Namyalo-Ganafa H, Meier BM (2021) Decolonising human rights: how intellectual property laws result in unequal access to the COVID-19 vaccine. BMJ Glob Health 6(7):e006169. https://doi.org/10.1136/bmjgh-2021-006169
172. Shahbazi S, Wiktorsson M, Kurdve M, Jönsson C, Bjelkemyr M (2016) Material efficiency in manufacturing: Swedish evidence on potential, barriers and strategies. J Clean Prod 127:438–450. https://doi.org/10.1016/j.jclepro.2016.03.143
173. Shahi SK, Shiva A, Dia M (2021) Integrated sustainable supply chain management and firm performance in the Indian textile industry. Qual Res Organ Manag Int J 16(3/4):614–635. https://doi.org/10.1108/QROM-03-2020-1904
174. Shaw R, Kim Y-K, Hua J (2020) Governance, technology and citizen behavior in pandemic: lessons from COVID-19 in East Asia. Prog Disaster Sci 6:100090. https://doi.org/10.1016/j.pdisas.2020.100090
175. Shen B, Li Q, Dong C, Perry P (2017) Sustainability issues in textile and apparel supply chains. Sustainability 9(9). https://doi.org/10.3390/su9091592
176. Shirvanimoghaddam K, Motamed B, Ramakrishna S, Naebe M (2020) Death by waste: fashion and textile circular economy case. Sci Total Environ 718:137317. https://doi.org/10.1016/j.scitotenv.2020.137317
177. Sigala M (2020) Tourism and COVID-19: impacts and implications for advancing and resetting industry and research. J Bus Res 117:312–321. https://doi.org/10.1016/j.jbusres.2020.06.015

178. Skene KR (2018) Circles, spirals, pyramids and cubes: why the circular economy cannot work. Sustain Sci 13(2):479–492. https://doi.org/10.1007/s11625-017-0443-3
179. Subramanian K, Chopra SS, Cakin E, Li X, Lin CSK (2020) Environmental life cycle assessment of textile bio-recycling—valorizing cotton-polyester textile waste to pet fiber and glucose syrup. Resour Conserv Recycl 161:104989. https://doi.org/10.1016/j.resconrec.2020.104989
180. Sánchez OR, Vale DB, Rodrigues L, Surita FG (2020) Violence against women during the COVID-19 pandemic: an integrative review. Int J Gynecol Obstet 151(2):180–187. https://doi.org/10.1002/ijgo.13365
181. Sánchez-Clavijo LM, Martínez-Callejas SJ, Acevedo-Charry O, Diaz-Pulido A, Gómez-Valencia B, Ocampo-Peñuela N, Ocampo D, Olaya-Rodríguez MH, Rey-Velasco JC, Soto-Vargas C, Ochoa-Quintero JM (2021) Differential reporting of biodiversity in two citizen science platforms during COVID-19 lockdown in Colombia. Biol Conserv 256:109077. https://doi.org/10.1016/j.biocon.2021.109077
182. Sözen S, Dulkadiroglu H, Begum Yucel A, Insel G, Orhon D (2019) Pollutant footprint analysis for wastewater management in textile dye houses processing different fabrics. J Chem Technol Biotechnol 94(4):1330–1340. https://doi.org/10.1002/jctb.5891
183. Taifa IWR, Lushaju GG (2020) Establishing basic requirements for textile and garment mass production units in the Tanzanian context. Res J Text Appar 24(4):321–340. https://doi.org/10.1108/RJTA-11-2019-0054
184. Tang S, Wang Z, Yang G, Tang W (2020) What are the implications of globalization on sustainability?—a comprehensive study. Sustainability 12(8). https://doi.org/10.3390/su12083411
185. Todeschini BV, Cortimiglia MN, Callegaro-de-Menezes D, Ghezzi A (2017) Innovative and sustainable business models in the fashion industry: entrepreneurial drivers, opportunities, and challenges. Bus Horiz 60(6):759–770. https://doi.org/10.1016/j.bushor.2017.07.003
186. Tran LTT (2021) Managing the effectiveness of e-commerce platforms in a pandemic. J Retail Consum Serv 58:102287. https://doi.org/10.1016/j.jretconser.2020.102287
187. Ütebay B, Çelik P, Çay A (2019) Effects of cotton textile waste properties on recycled fibre quality. J Clean Prod 222:29–35. https://doi.org/10.1016/j.jclepro.2019.03.033
188. Vadakkepatt GG, Winterich KP, Mittal V, Zinn W, Beitelspacher L, Aloysius J, Ginger J, Reilman J (2021) Sustainable retailing. J Retail 97(1):62–80. https://doi.org/10.1016/j.jretai.2020.10.008
189. van der Velden NM, Patel MK, Vogtländer JG (2014) LCA benchmarking study on textiles made of cotton, polyester, nylon, acryl, or elastane. Int J Life Cycle Assess 19(2):331–356. https://doi.org/10.1007/s11367-013-0626-9
190. Vidya CT, Prabheesh KP (2020) Implications of COVID-19 pandemic on the global trade networks. Emerg Mark Financ Trade 56(10):2408–2421. https://doi.org/10.1080/1540496X.2020.1785426
191. Viero A, Barbara G, Montisci M, Kustermann K, Cattaneo C (2021) Violence against women in the COVID-19 pandemic: a review of the literature and a call for shared strategies to tackle health and social emergencies. Forensic Sci Int 319:110650. https://doi.org/10.1016/j.forsciint.2020.110650
192. Wang S, Liao Y-K, Wu, W-Y, Le, KB (2021).The role of corporate social responsibility perceptions in brand equity, brand credibility, brand reputation, and purchase intentions. Sustainability 13(21). https://doi.org/10.3390/su132111975
193. Wen J, Kozak M, Yang S, Liu F (2021) COVID-19: potential effects on Chinese citizens' lifestyle and travel. Tour Rev 76(1):74–87. https://doi.org/10.1108/TR-03-2020-0110
194. Wu T-Y, Ford O, Rainville AJ, Bessire R (2020) COVID-19 care package distribution for senior citizens and families in Detroit and Hamtramck, Michigan. J Hunger Environ Nutr 15(4):585–587. https://doi.org/10.1080/19320248.2020.1799586
195. Yan J, Kim S, Zhang SX, Foo M-D, Alvarez-Risco A, Del-Aguila-Arcentales S, Yáñez JA (2021) Hospitality workers' COVID-19 risk perception and depression: a contingent model based on transactional theory of stress model. Int J Hosp Manag 95:102935. https://doi.org/10.1016/j.ijhm.2021.102935

196. Yasin S, Sun D (2019) Propelling textile waste to ascend the ladder of sustainability: EOL study on probing environmental parity in technical textiles. J Clean Prod 233:1451–1464. https://doi.org/10.1016/j.jclepro.2019.06.009
197. Yasin Ar A (2020) Managing e-commerce during a pandemic: lessons from GrubHub during COVID-19. In: George B, Mahar Q (eds) International case studies in the management of disasters. Emerald Publishing Limited, pp 155–167. https://doi.org/10.1108/978-1-83982-186-820201010
198. Yousef S, Kalpokaitė-Dičkuvienė R, Baltušnikas A, Pitak I, Lukošiūtė SI (2021) A new strategy for functionalization of char derived from pyrolysis of textile waste and its application as hybrid fillers (CNTs/char and graphene/char) in cement industry. J Clean Prod 314:128058. https://doi.org/10.1016/j.jclepro.2021.128058
199. Yáñez JA, Afshar Jahanshahi A, Alvarez-Risco A, Li J, Zhang SX (2020) Anxiety, distress, and turnover intention of healthcare workers in Peru by their distance to the epicenter during the COVID-19 crisis. Am J Trop Med Hyg 103(4):1614–1620. https://doi.org/10.4269/ajtmh.20-0800
200. Yáñez JA, Alvarez-Risco A, Delgado-Zegarra J (2020) COVID-19 in Peru: from supervised walks for children to the first case of Kawasaki-like syndrome. BMJ 369:m2418. https://doi.org/10.1136/bmj.m2418
201. Zarocostas J (2021) What next for a COVID-19 intellectual property waiver? The Lancet 397(10288):1871–1872. https://doi.org/10.1016/S0140-6736(21)01151-X
202. Zhang SX, Chen J, Afshar Jahanshahi A, Alvarez-Risco A, Dai H, Li J, Patty-Tito RM (2021) Succumbing to the COVID-19 pandemic—healthcare workers not satisfied and intend to leave their jobs. Int J Ment Heal Addict. https://doi.org/10.1007/s11469-020-00418-6
203. Zhang SX, Sun S, Afshar Jahanshahi A, Alvarez-Risco A, Ibarra VG, Li J, Patty-Tito RM (2020) Developing and testing a measure of COVID-19 organizational support of healthcare workers—results from Peru, Ecuador, and Bolivia. Psychiatry Res 291:113174. https://doi.org/10.1016/j.psychres.2020.113174
204. Zhang B, Zhang Y, Zhou P (2021) Consumer attitude towards sustainability of fast fashion products in the UK. Sustainability 13(4). https://doi.org/10.3390/su13041646
205. Zollinger R, DiCindio C (2021) Community ecology: Museum education and the digital divide during and after COVID-19. J MusM Educ 46(4):481–492. https://doi.org/10.1080/10598650.2021.1983711

Material Selection for Circularity and Footprints

Flavio Morales-Ríos, Aldo Alvarez-Risco, Sarahit Castillo-Benancio, Maria de las Mercedes Anderson-Seminario, and Shyla Del-Aguila-Arcentales

Abstract Companies that offer a good have a significant carbon footprint due to the production of their products. In this way, a selection of more environmentally friendly materials is sought to reduce pollution, so that unused or no longer helpful raw materials can be reused in the production of other derived products. This research details a circular economy framework for carbon footprint reduction, focusing on material selection. Most of the articles reviewed date from 2017 to 2021, demonstrating that the topic is new to the research area. Based on the literature review, research on the circular economy in feedstock sorting has focused on the recovery and recycling of waste to facilitate circularity in future. The framework presented also allows analysis from an eco-efficiency point of view because it considers economic and environmental aspects that improve products and processes using technologies. This way provides professionals with a new approach to efficiently cost-effectively managing their waste. In addition, circularity can be especially useful for the long-term strategy work of various companies regardless of the sector they are in, but which are in the goods production sector.

Keywords Material selection · Circular economy · Environmental footprint · Green design · Sustainable development · Product durability

1 Introduction

Because of the exponential growth of technological advancement and human population, energy supply and environmental threats have become significant concerns worldwide [132]. Companies must examine their operations to see if there are any

F. Morales-Ríos · A. Alvarez-Risco (✉) · S. Castillo-Benancio ·
M. de las Mercedes Anderson-Seminario
Universidad de Lima, Lima, Perú
e-mail: aralvare@ulima.edu.pe

S. Del-Aguila-Arcentales
Escuela Nacional de Marina Mercante "Almirante Miguel Grau", Callao, Perú

A. Alvarez-Risco et al. (eds.), *Circular Economy*, Environmental Footprints and Eco-design of Products and Processes,
https://doi.org/10.1007/978-981-19-0549-0_10

opportunities to reduce wastage, energy demand, gas emissions, and resource exhaustion [45, 46, 113, 132]. It is no longer an issue of whether to include environmental management in business activities; instead, it is a question of how [46, 88]. From the standpoint of industrial sustainability, in which goods and wastes are created concurrently, it is attempted to identify how the industrial system may be constructed in such a manner that environmental consequences can be measured quickly and accurately, making it environmentally friendly [34, 46, 126]. Product designers must consider environmental consequences from the early phases of the product design lifecycle and the selection of essential materials [36, 55, 88, 109]. The selection of materials can guarantee that items adhere to environmental guidelines, make product maintenance and disassembly easier, and increase product reuse, recycling, and regeneration [93, 109].

Green product design not only plays a vital role in the manufacturing business, but it has also become a primary focal point in future, thanks to increased environmental consciousness and the implementation of environmental protection regulations [88, 123]. Consumers' needs are no longer satisfied just by practical and industrial design [88]; they now look for items that adhere to environmental principles and standards [45, 109].

In recent years, the circular economy as a concept has been part of the popular and scientific debate [90, 93, 138]. Because resources are mostly finite, the intensive extraction produced over time to obtain materials needed in a linear economic model for subsequent use and disposal has resulted in environmental pollution [75, 104, 138]. New laws and regulations have progressed with global development toward a sustainable one, in which environmental preservation is fundamental [100], although it is trendy in the public debate and still requires further refinement of its theory [93]. The costs arising from the disposal of waste have been increasing, thus with new technological advances, there has been an understanding of the potential value of waste, and an alternative for entrepreneurs has been to rethink waste materials into a renewable resource [93, 100, 135]. Because of this, it is imperative to know the necessary development for decision-making by companies that provide residual waste treatment [128].

A circular economy is viable for reducing environmental impact [93]. Its goal is to reduce waste through design while preserving financial and ecological value [69, 88, 132]. Product integrity is an important idea in the circular economy model [88], and keeping products functional takes precedence beyond material recovery [34, 109, 142]. Product value can be kept through durability, lifespan extension, and product recovery strategies, while the material value may be sustained through recycling [46, 93, 120, 134]. In addition, the quality of materials is critical in determining the economy's circularity [88, 109]. The functionality of compounds contained in materials is significant to downcycling [110], and this consideration is consistent with the concept that "as long as feasible" functionality and practicality conservation is vital for a circular economy [59, 69].

Worldwide, the population and companies work hard to reestablish the life that they enjoy daily before the COVID-19 pandemic. Initially, some activities have changed in differents areas [13, 16, 17, 21, 22, 25, 31, 29, 32, 62, 64, 70, 102, 118].

It is crucial to know in detail the most relevant evidence of the impact of the COVID-19 pandemic and can develop more specific and successful plans for the recovery of the economy and commercial activities around the world (Table 1).

2 Background and Environmental Context

Environmental issues have become more global in scope in recent decades, both in terms of their existence and repercussions, as well as the socioeconomic factors that cause them [79]. Upon sporadically acknowledging the rise in global awareness of climate change, this entry looks at the nature of environmental issues and their long-term implications, as well as evidence that humans are increasingly pushing against worldwide environmental boundaries [129] (Fig. 1).

Through the efforts of environmentalists, activist groups, and politicians, concern over air and water pollution quickly expanded to various other problems: climate change, deforestation, natural resource depletion, and waste production, among others [129]. The environment supplies us with the natural resources needed in industrial economies and production [68, 100]. The environment serves as a supply point for human cultures by offering what is referred to as the sustenance basis. It provides renewable and non-renewable resources [68]. Excessive use of natural resources, such as water, air, and biomass, may lead to shortages, while excessive use of non-renewable resources, such as coal, fossil fuels, and natural gas, might lead to scarcities [68, 79]. However, humans produce waste due to the utilization of natural resources [100]. Nowadays, it is quite conceivable to state that we live in a wastage-oriented culture, as a result of population expansion and increased production of waste, which causes landfills to become more frequent and deteriorate the biosphere [79, 100, 140] (Fig. 2).

Day after day, massive volumes of waste are created in cities and agricultural regions [100]. Millions of tons of waste material are delivered to the world's significant landfills [130]. The Sudokwon landfill facility in South Korea receives around 6.9 million tons of waste produced yearly, which is the most of any landfill site in the world [130]. The Apex Regional landfill in Las Vegas is the world's most considerable surface extension, receiving approximately 3.8 million tons of debris per year [130]. While escalating urbanization and industrialization have accelerated the rate of resource scarcity and environmental damage throughout the world, popular support for sustainability has risen [79, 84]. Now, green material selection, also known as sustainable material selection, is presented as an alternative model aiming to ensure product functionality while reducing the overall environmental and human health burden [109].

Table 1 Impact of COVID-19 in different sectors

Education	Ali [8], Allen et al. [9], Alvarez-Risco, Estrada-Merino et al. [23], Alvarez-Risco, Del-Aguila-Arcentales, Rosen et al. [18], Alvarez-Risco, Del-Aguila-Arcentales, Yáñez et al. [20], T. Chen et al. [56], Lackie et al. [94], Lashley et al. [95], Zollinger and DiCindio [151]
Entrepreneurship	Afshan et al. [4], Alvarez-Risco and Del-Aguila-Arcentales [15], Alvarez-Risco, Mlodzianowska, García-Ibarra et al. [26], Alvarez-Risco, Mlodzianowska, Zamora-Ramos et al. [27], Belitski et al. [38], Block et al. [41], Brown et al. [42], Bacq and Lumpkin [48], Chafloque-Cespedes et al. [53], Maritz et al. [97], Liguori and Winkler [101]
Health sector	Abiakam et al. [1], Alan et al. [7], Alsairafi et al. [10], Alvarez-Risco, Dawson et al. [12], Alvarez-Risco, Del-Aguila-Arcentales and Yáñez [19], Alvarez-Risco, Del-Aguila-Arcentales, Yáñez et al. [20], Barello et al. [39], Bozdağ and Ergün [44], X. Chen et al. [57], Deressa et al. [65], Koetter et al. [89], Raudenská et al. [116], Rojas Román et al. [119], Yáñez, Afshar Jahanshahi et al. [144], Zhang et al. [149], Zhang et al. [150]
Hospitality	Duarte Alonso et al. [67], Gursoy and Chi [78], Ho et al. [83], Kaushal and Srivastava [86], Yan et al. [143]
Intellectual property	Altindis [11], Alvarez-Risco and Del-Aguila-Arcentales [14], del Castillo [63], Erfani et al. [71], Jecker and Atuire [85], Krishtel and Malpani [92], Okereke [108], Sekalala et al. [124], Zarocostas [147]
Population	Ahsan [6], Alvarez-Risco, Mejia, et al. [24], Cegarra-Navarro et al. [52], Hanzl [80], Lim and Prakash [98], Quispe-Cañari et al. [111], Sánchez-Clavijo et al. [121], Shaw et al. [125], Wen et al. [139], Wu et al. [141], Yáñez, Alvarez-Risco et al. [145]
Prices	Apcho-Ccencho et al. [35], Chung et al. [58], Galanakis et al. [74], Hepburn et al. [81], Leiva-Martinez et al. [96], Obergassel et al. [107]
Tourism	Assaf and Scuderi [37], Baum and Hai [40], Brouder [47], Carvache-Franco, Alvarez-Risco, Carvache-Franco, Carvache-Franco, Estrada-Merino and Villalobos-Alvarez [49], [73], Carvache-Franco, Alvarez-Risco, Carvache-Franco, Carvache-Franco, Estrada-Merino and Rosen [50], [127], Carvache-Franco, Carvache-Franco et al. [51]
Trade	Abudureheman and Nilupaer [2], Acuña-Zegarra et al. [3], Aguirre et al. [5], Alvarez-Risco, Quipuzco-Chicata et al. [28], Alvarez-Risco, Rosen et al. [30], Borzée et al. [43], Cruz-Torres et al. [61], Kirk and Rifkin [87], Lopez-Odar et al. [99], Tran [133], Vidya and Prabheesh [136], Yasin Ar [146]
Violence against women	Amarillo [33], Chafloque-Cespedes et al. [54], Dominelli [66], Gulati and Kelly [77], Roesch et al. [117], Sánchez et al. [122], Viero et al. [137]

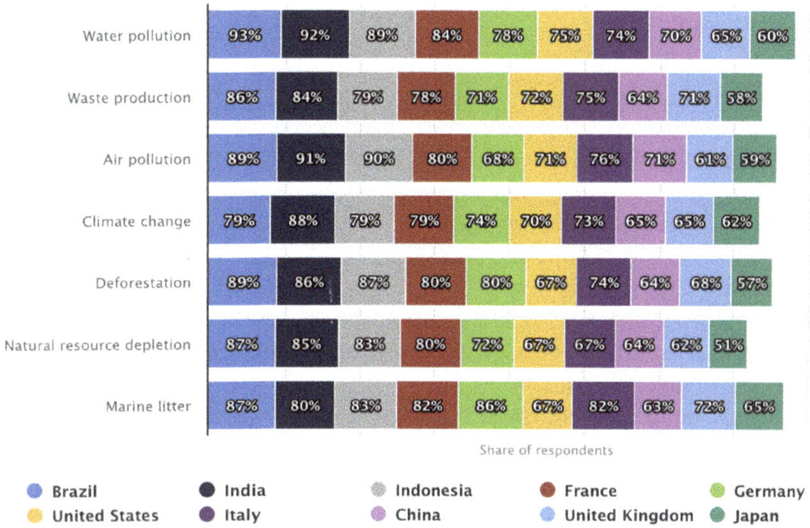

Fig. 1 Concerns about the environmental impacts of packaging worldwide 2020, by country (*Source* Statista [129])

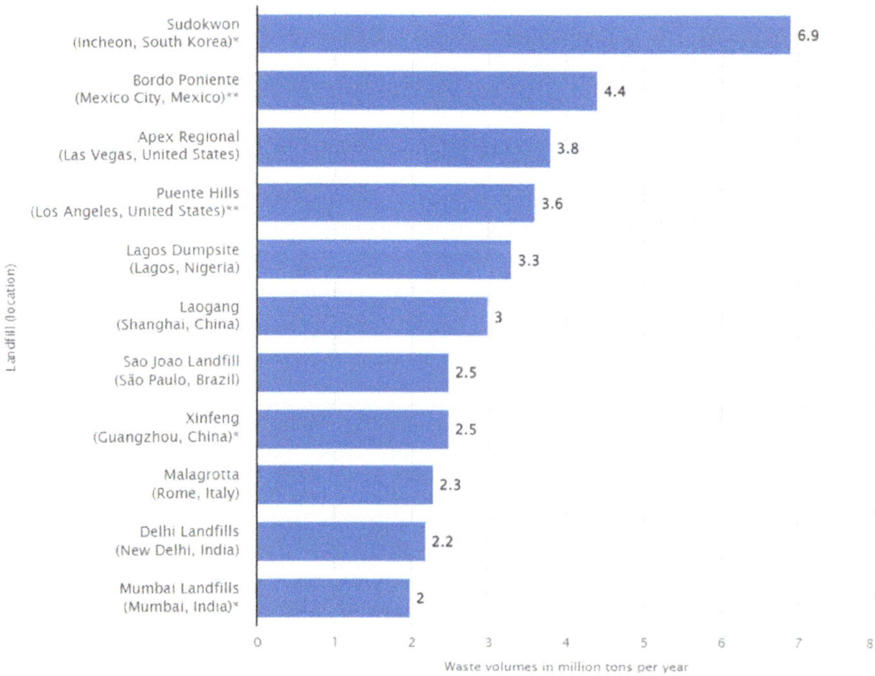

Fig. 2 Annual volumes of waste disposed at the largest landfills worldwide as of 2021 (in a million tons) (*Source* Statista [129])

3 Material Selection

The selection of a material for a new product or to improve an existing one is an essential part of the industrial design process [36]. The techniques for selecting materials are based on factors developed by physicists, mechanical engineers, thermal engineers, electrical engineers, and production engineers that define a material's technical utility [36, 132]. Previously, material selection approaches were mostly cost-driven [148]. Material selection is, however, becoming more critical in the context of the change from linear to the circular economy during the last decade, which has resulted in new aims to decrease ecological consequences, enhance process efficiency, and generate both social and economic advantages while retaining a product's usefulness, practicality, and value [109, 148].

The residues of unused materials can be reintroduced into the production chain and recycled as much as possible [72, 106, 115], which would be very beneficial for the environment [60]. Being paramount to reduce the extraction of raw material coming from natural resources so that the carbon emission is lower, which these measures are in favor of the environment, this is due to the circular production model [91], which tries to decrease land and marine pollution [140]. All plastic-related products are significant because they have a very short lifespan, the volumes in which they are produced are high, and their composition in many cases is harmful to the environment [84], which is why the management of these is of optional interest to governments that are investing in the EC [79].

3.1 Circular Model

The current linear economic model, based on the disposition of cheap and easily accessible large quantities of materials and energy, as well as cheap means to dispose of what no longer interests, has been at the heart of industrial development and has produced unprecedented levels of growth, is reaching its physical limits [72, 90]. The circular economy is a production and consumption framework that encourages people to share, reuse, repair, and recycle existing resources and products for as long as feasible, prolonging a product's life cycle [76, 91], which means reducing waste to the absolute minimum [90, 114]. When a product reaches the end of its useful life, its materials are reused as much as possible [76, 90]. These may be put to good use over again, resulting in increased value [72, 76, 91]. By design, a circular economy is restorative and regenerative, aiming to retain goods, components, and materials at their current use levels [72, 128]. The paucity of natural resources and high levels of pollution in the world have pushed productive systems toward the CE model, which emphasizes better material applications, longer product lifespans, and more innovative product creation and usage [84, 114]. Environmental implications are determined once the materials for each component are established, considering materials selection occurs during the product design process [109, 148]. Profits drive most manufacturers' production activities since any specific firm's existence, and financial state usually defines its operations [84, 148]. Enterprise decision-making must be directed by both market demand and the value toward a community based on environmental protection principles, resulting in more sustainable manufacturing processes, preparation procedures, and material selection schemes, based on the assumption of addressing economic issues [84, 109, 128]. As a result, the definition and selection of manufacturing materials have become a significant problem in CE approaches [72, 128].

Sustainable production should prioritize the use of renewable resources and the efficient use of non-renewable resources, and the avoidance of waste that the environment cannot absorb [88, 148]. Previous studies point out that materials must be chosen to ensure durability when it comes to material selection for increased circularity [103, 109, 148]. However, in terms of economic and marketing considerations, designers may consider durability a non-desirable feature [103, 115]. Durability and maintainability in product design have significant potential to improve and maximize product longevity [103, 132]. According to previous research, durability has been extensively investigated in the field of construction and buildings, where it has been supplemented with modular strategies to optimize structure assembly and disassembly as per product lifecycle prerequisites or the use of environmentally friendly materials [103, 112]. According to Rahla et al. [112], the purpose is to define the best product dependability and durability, allowing for extended periods of operation without failure. Increased longevity enhances resource efficiency by reducing material consumption and delaying demand for new products, allowing for shorter material cycles, according to Steinmann et al. [131]. Durability helps resource preservation and wastage mitigation, according to an examination of standardized tests and processes for goods and materials [103, 112, 131].

3.2 Environmental Footprint

The planet's ecological footprint is a measurement of human demand on the earth's natural ecosystems [123]. The amount of biologically productive land and marine area required to maintain current levels of resource consumption and generation of waste by the global population [105]. Choosing the suitable materials affects the product's functional performance and its pollutant emissions and energy consumption throughout its product life cycle [105, 109, 148]. The entire amount of carbon dioxide emissions, directly and indirectly, generated by activity or accumulated during the lifespan of a product is defined as the total amount of carbon dioxide emissions [105]. The total direct and indirect CO_2 emissions and the equivalent CO_2 emissions of other greenhouse gases are computed during the lifespan of a product, process, or activity [105, 123]. Moreover, the quantity of carbon accessible in various fuels and the efficiency of energy conversion determine the amount of carbon released [68, 79, 105].

With public awareness for sustainable development growing, material selection must adhere to green design principles, increase product quality, and expedite manufacturing processes, reducing negative environmental and human health consequences [109, 148]. Eco-design tools may enhance a product's environmental performance early in the design process, not only because they can cut manufacturing costs and materials, but also because they can minimize embodied energy and CO_2 emissions, as well as remove waste [34, 59, 109]. The manufacturing sector is dealing with building a product that has a lower environmental effect to create a more sustainable society, and green design plays a crucial part in this [105, 120, 123]. Green product design, also known as eco-design, can be defined as the use of elements that might lessen a product's environmental effect throughout the design phase [109, 123]. Eco-design provides businesses with a variety of benefits and prospects, not only in terms of the environment but also economic growth and social responsibility [46, 59, 123].

By optimizing the inputs and outputs of the manufacturing process, minimizing resource consumption, lowering environmental impact, and enhancing system efficiency, a product's environmental performance increases, lowering its overall ecological footprint [109, 123, 134]. Furthermore, the adoption of Environmental Management Systems (EMS) may be aided by green design approaches [109, 120]. Environmental data, as external factors, may be utilized for user communication and marketing while also enhancing the presence of the environment as a decision-making criterion affecting purchase intention [59, 120, 126]. Furthermore, better environmental profiles comply with existing standards and anticipate increasingly stringent norms in future [59, 82, 120, 126].

4 Closing Remarks

To validate the selection of materials in a CE context and reduce the carbon footprint, this is an issue of relevance for both the public and private sector, which can make decisions focused on a more circular measure using technologies. In a framework where production and consumption are designed to encourage people to share, reuse, repair, and recycle, extending the life cycle of products. In this way, waste is reduced to a minimum, with a positive impact on the environment, promoting a shift from linear to circular, increasing competitiveness in new markets with its end products being eco-efficient.

Knowing how to select a material that can be reused in the short and long term is paramount, as it becomes attractive in both the commercial and waste markets because it can be a potential raw material for a downstream product being demanded in different markets; therefore, good quality plays an important role. It is worth mentioning that an eco-design in a product in its early stages can reduce production costs, CO_2 emissions and eliminate waste, being more sustainable in its production. In addition, it offers companies very positive economic growth and projects a favorable image from a social responsibility point of view. In summary, we conclude that companies must adapt to the change to more environmentally friendly options, with the CE model being an alternative, and that regulatory frameworks are necessary for accordance with the evolving needs of the markets so that the private or public sector should also adopt environmental management systems (EMS). We hope that this document is part of future research and scientific studies related to the EC and the environmental footprint.

References

1. Abiakam N, Worsley P, Jayabal H, Mitchell K, Jones M, Fletcher J, Bader D (2021) Personal protective equipment related skin reactions in healthcare professionals during COVID-19. Int Wound J 18(3):312–322. https://doi.org/10.1111/iwj.13534
2. Abudureheman A, Nilupaer A (2021) Optimization model design of cross-border e-commerce transportation path under the background of prevention and control of COVID-19 pneumonia. Soft Comput 25(18):12007–12015. https://doi.org/10.1007/s00500-021-05685-6
3. Acuña-Zegarra MA, Santana-Cibrian M, Velasco-Hernandez JX (2020) Modeling behavioral change and COVID-19 containment in Mexico: a trade-off between lockdown and compliance. Math Biosci 325. https://doi.org/10.1016/j.mbs.2020.108370
4. Afshan G, Shahid S, Tunio MN (2021) Learning experiences of women entrepreneurs amidst COVID-19. Int J Gend Entrep 13(2):162–186. https://doi.org/10.1108/IJGE-09-2020-0153
5. Aguirre AA, Catherina R, Frye H, Shelley L (2020) Illicit wildlife trade, wet markets, and COVID-19: preventing future pandemics. World Medical & Health Policy 12(3):256–265. https://doi.org/10.1002/wmh3.348
6. Ahsan MM (2020) Strategic decisions on urban built environment to pandemics in Turkey: lessons from COVID-19. Journal of Urban Management 9(3):281–285. https://doi.org/10.1016/j.jum.2020.07.001
7. Alan H, Eskin Bacaksiz F, Tiryaki Sen H, Taskiran Eskici G, Gumus E, Harmanci Seren AK (2021) "I'm a hero, but...": an evaluation of depression, anxiety, and stress levels of frontline healthcare professionals during COVID-19 pandemic in Turkey. Perspect Psychiatr Care 57(3):1126–1136. https://doi.org/10.1111/ppc.12666
8. Ali W (2020) Online and remote learning in higher education institutes: a necessity in light of COVID-19 pandemic. High Educ Stud 10(3):16–25
9. Allen J, Rowan L, Singh P (2020) Teaching and teacher education in the time of COVID-19. Asia-Pac J Teach Educ 48(3):233–236. https://doi.org/10.1080/1359866X.2020.1752051
10. Alsairafi Z, Naser AY, Alsaleh FM, Awad A, Jalal Z (2021) Mental health status of healthcare professionals and students of health sciences faculties in Kuwait during the COVID-19 pandemic. Int J Environ Res Public Health 18(4). https://doi.org/10.3390/ijerph18042203
11. Altindis E (2021) Inequitable COVID-19 vaccine distribution and the intellectual property rights prolong the pandemic. Expert Rev Vaccines. https://doi.org/10.1080/14760584.2022.2014819
12. Alvarez-Risco A, Dawson J, Johnson W, Conteh-Barrat, M Aslani P, Del-Aguila-Arcentales S, Diaz-Risco S (2021) Ebola virus disease outbreak: a global approach for health systems [Ebola; health professionals; global approach; pharmacist: pharmaceutical care; COVID-19.] 53(4). http://www.revfarmacia.sld.cu/index.php/far/article/view/491
13. Alvarez-Risco A, Del-Aguila-Arcentales S (2015) Errores de prescripción como barrera para la Atención Farmacéutica en establecimientos de salud públicos: Experiencia Perú. Pharm Care España 17(6), 725–731. https://www.pharmcareesp.com/index.php/PharmaCARE/article/view/246
14. Alvarez-Risco A, Del-Aguila-Arcentales S (2021a) A note on changing regulation in international business: the World Intellectual Property Organization (WIPO) and artificial intelligence. In: Verbeke A, van Tulder R, Rose EL, Wei Y (eds) The Multiple Dimensions of

Institutional Complexity in International Business Research (Vol 15, pp 363–371). Emerald Publishing Limited. https://doi.org/10.1108/S1745-886220210000015020
15. Alvarez-Risco A, Del-Aguila-Arcentales S (2021b) Public policies and private efforts to increase women entrepreneurship based on STEM background. In: Mari M, Poggesi S, Foss L (eds) Women's Entrepreneurship in STEM Disciplines: Issues and Perspectives (pp. 75–87). Springer International Publishing. https://doi.org/10.1007/978-3-030-83792-1_5
16. Alvarez-Risco A, Del-Aguila-Arcentales S, Delgado-Zegarra J, Yáñez JA, Diaz-Risco S (2019) Doping in sports: findings of the analytical test and its interpretation by the public. Sport Sci Health 15(1):255–257. https://doi.org/10.1007/s11332-018-0484-8
17. Alvarez-Risco A, Del-Aguila-Arcentales S, Diaz-Risco S (2018) Pharmacovigilance as a tool for sustainable development of healthcare in Peru. PharmacoVigilance Rev 10(2):4–6
18. Alvarez-Risco A, Del-Aguila-Arcentales S, Rosen MA, García-Ibarra V, Maycotte-Felkel S, Martínez-Toro GM (2021) Expectations and interests of university students in COVID-19 times about Sustainable Development Goals: evidence from Colombia, Ecuador, Mexico, and Peru. Sustain 13(6):3306. https://doi.org/10.3390/su13063306
19. Alvarez-Risco A, Del-Aguila-Arcentales S, Yáñez JA (2021) Telemedicine in Peru as a result of the COVID-19 pandemic: perspective from a country with limited internet access. Am J Trop Med Hyg 105(1):6–11. https://doi.org/10.4269/ajtmh.21-0255
20. Alvarez-Risco A, Del-Aguila-Arcentales S, Yáñez JA, Rosen MA, Mejia CR (2021) Influence of technostress on academic performance of university medicine students in Peru during the COVID-19 pandemic. Sustain 13(16):8949. https://doi.org/10.3390/su13168949
21. Alvarez-Risco A, Del-Aguila-Arcentales S, Stevenson JG (2015) Pharmacists and mass communication for implementing pharmaceutical care. Am J Pharm Benefits 7(3):e125–e126
22. Alvarez-Risco A, Delgado-Zegarra J, Yáñez JA, Diaz-Risco S, Del-Aguila-Arcentales S (2018) Predation risk by gastronomic boom - case Peru. J Landsc Ecol 11(1):100–103. https://doi.org/10.2478/jlecol-2018-0003
23. Alvarez-Risco A, Estrada-Merino A, Anderson-Seminario MdlM, Mlodzianowska S, García-Ibarra V, Villagomez-Buele C, Carvache-Franco M (2021) Multitasking behavior in online classrooms and academic performance: case of university students in Ecuador during COVID-19 outbreak. Interact Technol Smart Educ 18(3):422–434. https://doi.org/10.1108/ITSE-08-2020-0160
24. Alvarez-Risco A, Mejia CR, Delgado-Zegarra J, Del-Aguila-Arcentales S, Arce-Esquivel AA, Valladares-Garrido MJ, Yáñez JA (2020) The Peru approach against the COVID-19 infodemic: insights and strategies. Am J Trop Med Hyg 103(2):583–586. https://doi.org/10.4269/ajtmh.20-0536
25. Alvarez-Risco A, Mil JWFv (2007) Pharmaceutical care in community pharmacies: practice and research in Peru. Ann Pharmacother 41(12):2032–2037. https://doi.org/10.1345/aph.1K117
26. Alvarez-Risco A, Mlodzianowska S, García-Ibarra V, Rosen MA, Del-Aguila-Arcentales S (2021) Factors affecting green entrepreneurship intentions in business university students in COVID-19 pandemic times: case of Ecuador. Sustain 13(11):6447. https://doi.org/10.3390/su13116447
27. Alvarez-Risco A, Mlodzianowska S, Zamora-Ramos U, Del-Aguila S (2021) Green entrepreneurship intention in university students: the case of Peru. Entrep Bus Econ Rev 9(4):85–100. https://doi.org/10.15678/EBER.2021.090406
28. Alvarez-Risco A, Quipuzco-Chicata L, Escudero-Cipriani C (2021) Determinantes de la intención de recompra en línea en tiempos de COVID-19: evidencia de una economía emergente. Lecturas de Economía (96). https://doi.org/10.17533/udea.le.n96a342638
29. Alvarez-Risco A, Quiroz-Delgado D, Del-Aguila-Arcentales S (2016) Pharmaceutical care in hypertension patients in a peruvian hospital. Indian J Public Health Res Dev 7(3):183–188
30. Alvarez-Risco A, Rosen MA, Del-Aguila-Arcentales S (2020) A new regulation for supporting a circular economy in the plastic industry: the case of Peru (short communication). J Landsc Ecol 13(1):1–3. https://doi.org/10.2478/jlecol-2020-0004

31. Alvarez-Risco A, Turpo-Cama A, Ortiz-Palomino L, Gongora-Amaut N, Del-Aguila-Arcentales S (2016) Barreras para la implementación de la Atención Farmacéutica en establecimientos farmacéuticos de Cusco, Perú. Pharm Care España 18(5):194–205. https://www.pharmcareesp.com/index.php/PharmaCARE/article/view/326
32. Álvarez-Risco A, Zegarra Arellano E, Matos Valerio E, Mejía Acosta N, Solis Tarazona Z (2013) Campaña de atención farmacéutica como estrategia de implementación de los servicios farmacéuticos: Experiencia Perú. Pharm Care España 15(1):35. https://www.pharmcareesp.com/index.php/PharmaCARE/article/view/105
33. Amarillo CR (2020) Feminism on lockdown. NACLA Rep Am 52(3):274–281. https://doi.org/10.1080/10714839.2020.1809084
34. Aoe T (2007) Eco-efficiency and ecodesign in electrical and electronic products. J Clean Prod 15(15):1406–1414. https://doi.org/10.1016/j.jclepro.2006.06.004
35. Apcho-Ccencho L-V, Cuya-Velásquez B-B, Alvarado Rodríguez D, Anderson-Seminario MdlM, Alvarez-Risco A, Estrada-Merino A, Mlodzianowska S (2021) The impact of international price on the technological industry in the United States and China during times of crisis: commercial war and COVID-19. In: Lawrence KD, Klimberg RK (eds) Advances in Business and Management Forecasting (Vol 14, pp 149–160). Emerald Publishing Limited. https://doi.org/10.1108/S1477-407020210000014010
36. Ashby MF, Johnson K (2013) Materials and design: the art and science of material selection in product design. Butterworth-Heinemann
37. Assaf A, Scuderi R (2020) COVID-19 and the recovery of the tourism industry. Tour Econ 26(5):731–733. https://doi.org/10.1177/1354816620933712
38. Bacq S, Lumpkin GT (2021) Social entrepreneurship and COVID-19. J Manage Stud 58(1):285–288. https://doi.org/10.1111/joms.12641
39. Barello S, Caruso R, Palamenghi L, Nania T, Dellafiore F, Bonetti L, Graffigna G (2021) Factors associated with emotional exhaustion in healthcare professionals involved in the COVID-19 pandemic: an application of the job demands-resources model. Int Arch Occup Environ Health 94(8):1751–1761. https://doi.org/10.1007/s00420-021-01669-z
40. Baum T, Hai NTT (2020) Hospitality, tourism, human rights and the impact of COVID-19. Int J Contemp Hosp Manag 32(7):2397–2407. https://doi.org/10.1108/IJCHM-03-2020-0242
41. Belitski M, Guenther C, Kritikos AS, Thurik R (2021) Economic effects of the COVID-19 pandemic on entrepreneurship and small businesses. Small Bus Econ. https://doi.org/10.1007/s11187-021-00544-y
42. Block JH, Fisch C, Hirschmann M (2021) The determinants of bootstrap financing in crises: evidence from entrepreneurial ventures in the COVID-19 pandemic. Small Bus Econ. https://doi.org/10.1007/s11187-020-00445-6
43. Borzée A, McNeely J, Magellan K, Miller JRB, Porter L, Dutta T, Zhang L (2020) COVID-19 highlights the need for more effective wildlife trade legislation. Trends Ecol Evol 35(12):1052–1055. https://doi.org/10.1016/j.tree.2020.10.001
44. Bozdağ F, Ergün N (2020) Psychological resilience of healthcare professionals during COVID-19 pandemic. Psychol Rep 124(6):2567–2586. https://doi.org/10.1177/0033294120965477
45. Bracke S, Yamada S, Kinoshita Y, Inoue M, Yamada T (2017) Decision making within the conceptual design phase of eco-friendly products. Procedia Manuf 8:463–470. https://doi.org/10.1016/j.promfg.2017.02.059
46. Brones F, de Carvalho MM, de Senzi Zancul E (2014) Ecodesign in project management: a missing link for the integration of sustainability in product development? J Clean Prod 80:106–118. https://doi.org/10.1016/j.jclepro.2014.05.088
47. Brouder P (2020) Reset redux: possible evolutionary pathways towards the transformation of tourism in a COVID-19 world. Tour Geogr 22(3):484–490. https://doi.org/10.1080/14616688.2020.1760928
48. Brown R, Rocha A, Cowling M (2020) Financing entrepreneurship in times of crisis: Exploring the impact of COVID-19 on the market for entrepreneurial finance in the United Kingdom. Int Small Bus J 38(5):380–390. https://doi.org/10.1177/0266242620937464

49. Carvache-Franco M, Alvarez-Risco A, Carvache-Franco O, Carvache-Franco W, Estrada-Merino A, Villalobos-Alvarez D (2021) Perceived value and its influence on satisfaction and loyalty in a coastal city: a study from Lima, Peru. J Policy Res Tour, Leis Events 1–16. https://doi.org/10.1080/19407963.2021.1883634
50. Carvache-Franco M, Alvarez-Risco A, Carvache-Franco W, Carvache-Franco O, Estrada-Merino A, Rosen MA (2021) Coastal cities seen from loyalty and their tourist motivations: a study in Lima, Peru. Sustain 13(21):11575. https://doi.org/10.3390/su132111575
51. Carvache M, Carvache O, Carvache W, Alvarez- A, Estrada-Merino A (2021) Motivations and segmentation of the demand for coastal cities: a study in Lima, Peru. Int J Tour Res 23(4):517–531. https://doi.org/10.1002/jtr.2423
52. Cegarra-Navarro J-G, Vătămănescu E-M, Martínez-Martínez A (2021) A context-driven approach on coping with COVID-19: from hiding knowledge toward citizen engagement. Knowl Process Manag 28(2):134–140. https://doi.org/10.1002/kpm.1662
53. Chafloque-Cespedes R, Alvarez-Risco A, Robayo-Acuña P-V, Gamarra-Chavez C-A, Martinez-Toro G-M, Vicente-Ramos W (2021) Effect of sociodemographic factors in entrepreneurial orientation and entrepreneurial intention in university students of Latin American business schools. In: Jones P, Apostolopoulos N, Kakouris A, Moon C, Ratten V, Walmsley A (eds) Universities and Entrepreneurship: Meeting the Educational and Social Challenges (Vol 11, pp 151–165). Emerald Publishing Limited. https://doi.org/10.1108/S2040-724620210000011010
54. Chafloque-Cespedes R, Vara-Horna A, Asencios-Gonzales Z, Lopez-Odar DR, Alvarez-Risco A, Quipuzco-Chicata L, . . . Sanchez-Villagomez M (2020) Presentismo académico y violencia contra las mujeres en escuelas de negocios e ingeniería en universidades peruanas. Lecturas de Economía 0(93):127–153. https://doi.org/10.17533/udea.le.n93a340726
55. Chauhan A, Vaish R (2012) Magnetic material selection using multiple attribute decision making approach. Mater Des 1980–2015(36):1–5. https://doi.org/10.1016/j.matdes.2011.11.021
56. Chen T, Peng L, Yin X, Rong J, Yang J, Cong G (2020) Analysis of user satisfaction with online education platforms in China during the COVID-19 pandemic. Healthcare 8(3). https://doi.org/10.3390/healthcare8030200
57. Chen X, Zhang SX, Jahanshahi AA, Alvarez-Risco A, Dai H, Li J, Ibarra VG (2020) Belief in a COVID-19 conspiracy theory as a predictor of mental health and well-being of health care workers in Ecuador: cross-sectional survey study. JMIR Public Health Surveill 6(3). https://doi.org/10.2196/20737
58. Chung SA, Olivera S, Rojas Román B, Alanoca E, Moscoso S, Limpias Terceros B, . . . Yáñez JA (2021) Temáticas de la producción científica de la Revista Cubana de Farmacia indizada en Scopus (1967–2020) [Revista Cubana de Farmacia; farmacia; Cuba; evidencia] 54(1). http://www.revfarmacia.sld.cu/index.php/far/article/view/511
59. Cicconi P (2020) Eco-design and Eco-materials: an interactive and collaborative approach. Sustain Mater Technol 23. https://doi.org/10.1016/j.susmat.2019.e00135
60. Conversio (2019) *Plastics - the Facts 2019*. Retrieved 01/01/2022 from https://www.conversio-gmbh.com/en/news/global-plastics-flow-studie-2019
61. Cruz-Torres W, Alvarez-Risco A, Del-Aguila-Arcentales S (2021) Impact of Enterprise Resource Planning (ERP) implementation on performance of an education enterprise: a Structural Equation Modeling (SEM). Stud Bus Econ 16(2):37–52. https://doi.org/10.2478/sbe-2021-0023
62. Del-Aguila-Arcentales S, Alvarez-Risco A (2013) Human error or burnout as explanation for mistakes in pharmaceutical laboratories. Accred Qual Assur 18(5):447–448. https://doi.org/10.1007/s00769-013-1000-0
63. del Castillo FA (2021) Temporary waiver of intellectual property on Covid-19 vaccines: toward the creation of a better, post-pandemic society. J Public Health 43(3):e559–e560. https://doi.org/10.1093/pubmed/fdab184
64. Delgado-Zegarra J, Alvarez-Risco A, Yáñez JA (2018) Uso indiscriminado de pesticidas y ausencia de control sanitario para el mercado interno en Perú. Rev Panam Salud Publica 42, e3. https://doi.org/10.26633/RPSP.2018.3

65. Deressa W, Worku A, Abebe W, Gizaw M, Amogne W (2021) Risk perceptions and preventive practices of COVID-19 among healthcare professionals in public hospitals in Addis Ababa, Ethiopia. PLoS ONE 16(6). https://doi.org/10.1371/journal.pone.0242471
66. Dominelli L (2021) A green social work perspective on social work during the time of COVID-19. Int J Soc Welf 30(1):7–16. https://doi.org/10.1111/ijsw.12469
67. Duarte Alonso A, Kok SK, Bressan A, O'Shea M, Sakellarios N, Koresis A, Santoni LJ (2020) COVID-19, aftermath, impacts, and hospitality firms: An international perspective. Int J Hosp Manag 91. https://doi.org/10.1016/j.ijhm.2020.102654
68. Dunlap RE, Jorgenson AK (2012) Environmental problems. The Wiley-Blackwell Encyclopedia of Globalization.
69. Dwek M (2017) Integration of material circularity in product design. Université Grenoble Alpes].
70. Enciso-Zarate A, Guzmán-Oviedo J, Sánchez-Cardona F, Martínez-Rohenes D, Rodríguez-Palomino JC, Alvarez-Risco A, . . . Diaz-Risco SS (2016) Evaluación de la contaminación con agentes citotóxicos en hospitales en Colombia. Pharm Care España 18(6), 241–250. https://www.pharmcareesp.com/index.php/PharmaCARE/article/view/354
71. Erfani P, Binagwaho A, Jalloh MJ, Yunus M, Farmer P, Kerry V (2021) Intellectual property waiver for covid-19 vaccines will advance global health equity. BMJ 374. https://doi.org/10.1136/bmj.n1837
72. Fellner J, Lederer J, Scharff C, Laner D (2017) Present potentials and limitations of a circular economy with respect to primary raw material demand. J Ind Ecol 21(3):494–496. https://doi.org/10.1111/jiec.12582
73. Fotiadis A, Polyzos S, Huan T-CTC (2021) The good, the bad and the ugly on COVID-19 tourism recovery. Ann Tour Res 87. https://doi.org/10.1016/j.annals.2020.103117
74. Galanakis CM, Rizou M, Aldawoud TMS, Ucak I, Rowan NJ (2021) Innovations and technology disruptions in the food sector within the COVID-19 pandemic and post-lockdown era. Trends Food Sci Technol 110:193–200. https://doi.org/10.1016/j.tifs.2021.02.002
75. Geissdoerfer M, Morioka SN, de Carvalho MM, Evans S (2018) Business models and supply chains for the circular economy. J Clean Prod 190:712–721. https://doi.org/10.1016/j.jclepro.2018.04.159
76. Geissdoerfer M, Savaget P, Bocken NMP, Hultink EJ (2017) The Circular Economy – a new sustainability paradigm? J Clean Prod 143:757–768. https://doi.org/10.1016/j.jclepro.2016.12.048
77. Gulati G, Kelly BD (2020) Domestic violence against women and the COVID-19 pandemic: what is the role of psychiatry? Int J Law Psychiatry 71. https://doi.org/10.1016/j.ijlp.2020.101594
78. Gursoy D, Chi CG (2020) Effects of COVID-19 pandemic on hospitality industry: review of the current situations and a research agenda. J Hosp Market Manag 29(5):527–529. https://doi.org/10.1080/19368623.2020.1788231
79. Gutberlet J, Bramryd T, Johansson M (2020) Expansion of the waste-based commodity frontier: insights from Sweden and Brazil. Sustain 12(7). https://doi.org/10.3390/su12072628
80. Hanzl M (2020) Urban forms and green infrastructure – the implications for public health during the COVID-19 pandemic. Cities & Health 1–5. https://doi.org/10.1080/23748834.2020.1791441
81. Hepburn C, O'Callaghan B, Stern N, Stiglitz J, Zenghelis D (2020) Will COVID-19 fiscal recovery packages accelerate or retard progress on climate change? Oxf Rev Econ Policy 36(Supplement_1), S359-S381. https://doi.org/10.1093/oxrep/graa015
82. Hinchliffe D, Akkerman F (2017) Assessing the review process of EU Ecodesign regulations. J Clean Prod 168:1603–1613. https://doi.org/10.1016/j.jclepro.2017.03.091
83. Ho C-Y, Tsai B-H, Chen C-S, Lu M-T (2021) Exploring green marketing orientations toward sustainability the hospitality industry in the COVID-19 pandemic. Sustain 13(8). https://doi.org/10.3390/su13084348
84. Huysman S, De Schaepmeester J, Ragaert K, Dewulf J, De Meester S (2017) Performance indicators for a circular economy: a case study on post-industrial plastic waste. Resour Conserv Recycl 120:46–54. https://doi.org/10.1016/j.resconrec.2017.01.013

85. Jecker NS, Atuire CA (2021) What's yours is ours: waiving intellectual property protections for COVID-19 vaccines. J Med Ethics 47(9):595. https://doi.org/10.1136/medethics-2021-107555
86. Kaushal V, Srivastava S (2021) Hospitality and tourism industry amid COVID-19 pandemic: perspectives on challenges and learnings from India. Int J Hosp Manag 92. https://doi.org/10.1016/j.ijhm.2020.102707
87. Kirk CP, Rifkin LS (2020) I'll trade you diamonds for toilet paper: consumer reacting, coping and adapting behaviors in the COVID-19 pandemic. J Bus Res 117:124–131. https://doi.org/10.1016/j.jbusres.2020.05.028
88. Ko Y-T (2020) Modeling an innovative green design method for sustainable products. Sustain 12(8). https://doi.org/10.3390/su12083351
89. Koetter P, Pelton M, Gonzalo J, Du P, Exten C, Bogale K, Sciamanna C (2020) Implementation and process of a COVID-19 contact tracing initiative: leveraging health professional students to extend the workforce during a pandemic. Am J Infect Control 48(12):1451–1456. https://doi.org/10.1016/j.ajic.2020.08.012
90. Korhonen J, Honkasalo A, Seppälä J (2018) Circular economy: The concept and its limitations. Ecol Econ 143:37–46. https://doi.org/10.1016/j.ecolecon.2017.06.041
91. Korhonen J, Nuur C, Feldmann A, Birkie SE (2018) Circular economy as an essentially contested concept. J Clean Prod 175:544–552. https://doi.org/10.1016/j.jclepro.2017.12.111
92. Krishtel P, Malpani R (2021) Suspend intellectual property rights for covid-19 vaccines. BMJ 373. https://doi.org/10.1136/bmj.n1344
93. Kyrö RK (2020) Share, Preserve, Adapt, Rethink – a focused framework for circular economy. IOP Conf Ser: Earth Environ Sci 588(4). https://doi.org/10.1088/1755-1315/588/4/042034
94. Lackie K, Najjar G, El-Awaisi A, Frost J, Green C, Langlois S, Khalili H (2020) Interprofessional education and collaborative practice research during the COVID-19 pandemic: considerations to advance the field. J Interprof Care 34(5):583–586. https://doi.org/10.1080/13561820.2020.1807481
95. Lashley MA, Acevedo M, Cotner S, Lortie CJ (2020) How the ecology and evolution of the COVID-19 pandemic changed learning. Ecol Evol 10(22):12412–12417. https://doi.org/10.1002/ece3.6937
96. Leiva-Martinez M-A, Anderson-Seminario MdlM, Alvarez-Risco A, Estrada-Merino A, Mlodzianowska S (2021) Price variation in lower goods as of previous economic crisis and the contrast of the current price situation in the context of COVID-19 in Peru. In: Lawrence KD, Klimberg RK (eds) Advances in Business and Management Forecasting (Vol 14, pp 161–166). Emerald Publishing Limited. https://doi.org/10.1108/S1477-407020210000014011
97. Liguori E, Winkler C (2020) From offline to online: challenges and opportunities for entrepreneurship education following the COVID-19 pandemic. Entrep Educ Pedagog 3(4):346–351. https://doi.org/10.1177/2515127420916738
98. Lim S, Prakash A (2021) Pandemics and citizen perceptions about their country: did COVID-19 increase national pride in South Korea? Nations Natl 27(3):623–637. https://doi.org/10.1111/nana.12749
99. Lopez-Odar D, Alvarez-Risco A, Vara-Horna A, Chafloque-Cespedes R, Sekar MC (2020) Validity and reliability of the questionnaire that evaluates factors associated with perceived environmental behavior and perceived ecological purchasing behavior in Peruvian consumers. Soc Responsib J 16(3):403–417. https://doi.org/10.1108/SRJ-08-2018-0201
100. Lu C, Pan X, Chen X, Mao J, Pang J, Xue B (2021) Modeling of waste flow in industrial symbiosis system at city-region level: a case study of Jinchang, China. Sustain, 13(2). https://doi.org/10.3390/su13020466
101. Maritz A, Perenyi A, de Waal G, Buck C (2020) Entrepreneurship as the unsung hero during the current COVID-19 economic crisis: Australian perspectives. Sustain, 12(11). https://doi.org/10.3390/su12114612
102. Mejía-Acosta N, Alvarez-Risco A, Solís-Tarazona Z, Matos-Valerio E, Zegarra-Arellano E, Del-Aguila-Arcentales S (2016) Reacciones Adversas a Medicamentos reportadas como resultado de la implementación de Atención Farmacéutica en la Farmacia Institucional

DIGEMID - Ministerio de Salud de Perú. Pharm Care España, 18(2):67–74. https://www.pharmcareesp.com/index.php/PharmaCARE/article/view/311
103. Mesa J, González-Quiroga A, Maury H (2020) Developing an indicator for material selection based on durability and environmental footprint: a circular economy perspective. Resour Conserv Recycl 160. https://doi.org/10.1016/j.resconrec.2020.104887
104. Millar N, McLaughlin E, Börger T (2019) The circular economy: swings and roundabouts? Ecol Econ 158:11–19. https://doi.org/10.1016/j.ecolecon.2018.12.012
105. Morini AA, Ribeiro MJ, Hotza D (2019) Early-stage materials selection based on embodied energy and carbon footprint. Mater Des 178. https://doi.org/10.1016/j.matdes.2019.107861
106. Nikanorova M, Stankevičienė J (2020) Development of environmental pillar in the context of circular economy assessment: Baltic Sea Region case. Entrep Sustain Issues 8(1):1209
107. Obergassel W, Hermwille L, Oberthür S (2021) Harnessing international climate governance to drive a sustainable recovery from the COVID-19 pandemic. Clim Policy 21(10):1298–1306. https://doi.org/10.1080/14693062.2020.1835603
108. Okereke M (2021) Towards vaccine equity: should big pharma waive intellectual property rights for COVID-19 vaccines? Public Health Pract 2. https://doi.org/10.1016/j.puhip.2021.100165
109. Peng C, Feng D, Guo S (2021) Material selection in green design: a method combining DEA and TOPSIS. Sustain 13(10). https://doi.org/10.3390/su13105497
110. Prasad K, Chakraborty S (2013) A quality function deployment-based model for materials selection. Mater Des 49:525–535. https://doi.org/10.1016/j.matdes.2013.01.035
111. Quispe-Cañari JF, Fidel-Rosales E, Manrique D, Mascaró-Zan J, Huamán-Castillón KM, Chamorro-Espinoza SE, Mejia CR (2021) Self-medication practices during the COVID-19 pandemic among the adult population in Peru: a cross-sectional survey. Saudi Pharm J 29(1):1–11. https://doi.org/10.1016/j.jsps.2020.12.001
112. Rahla KM, Mateus R, Bragança L (2021) Selection criteria for building materials and components in line with the circular economy principles in the built environment—a review of current trends. Infrastructures 6(4). https://doi.org/10.3390/infrastructures6040049
113. Ram VG, Kishore KC, Kalidindi SN (2020) Environmental benefits of construction and demolition debris recycling: evidence from an Indian case study using life cycle assessment. J Clean Prod 255. https://doi.org/10.1016/j.jclepro.2020.120258
114. Ranjbari M, Saidani M, Shams Esfandabadi Z, Peng W, Lam SS, Aghbashlo M, Tabatabaei M (2021) Two decades of research on waste management in the circular economy: insights from bibliometric, text mining, and content analyses. J Clean Prod 314. https://doi.org/10.1016/j.jclepro.2021.128009
115. Ranta V, Aarikka-Stenroos L, Ritala P, Mäkinen SJ (2018) Exploring institutional drivers and barriers of the circular economy: a cross-regional comparison of China, the US, and Europe. Resour Conserv Recycl 135:70–82. https://doi.org/10.1016/j.resconrec.2017.08.017
116. Raudenská J, Steinerová V, Javůrková A, Urits I, Kaye AD, Viswanath O, Varrassi G (2020) Occupational burnout syndrome and post-traumatic stress among healthcare professionals during the novel coronavirus disease 2019 (COVID-19) pandemic. Best Pract Res Clin Anaesthesiol 34(3):553–560. https://doi.org/10.1016/j.bpa.2020.07.008
117. Roesch E, Amin A, Gupta J, García-Moreno C (2020) Violence against women during covid-19 pandemic restrictions. BMJ 369. https://doi.org/10.1136/bmj.m1712
118. Rojas-Osorio M, Alvarez-Risco A (2019) Intention to use smartphones among Peruvian university students. Int J Interact Mob Technol (iJIM) 13(03):13. https://doi.org/10.3991/ijim.v13i03.9356
119. Rojas Román B, Moscoso S, Chung SA, Limpias Terceros B, Álvarez-Risco A, Yáñez JA (2020) Tratamiento de la COVID-19 en Perú y Bolivia y los riesgos de la automedicación [COVID-19; tratamiento; estrategia; fármacos; automedicación; Perú; Bolivia.] 53(2). http://www.revfarmacia.sld.cu/index.php/far/article/view/435/351
120. Rossi M, Germani M, Zamagni A (2016) Review of ecodesign methods and tools. Barriers and strategies for an effective implementation in industrial companies. J Clean Prod 129:361–373. https://doi.org/10.1016/j.jclepro.2016.04.051

121. Sánchez-Clavijo LM, Martínez-Callejas SJ, Acevedo-Charry O, Diaz-Pulido A, Gómez-Valencia B, Ocampo-Peñuela N, Ochoa-Quintero JM (2021) Differential reporting of biodiversity in two citizen science platforms during COVID-19 lockdown in Colombia. Biol Cons 256. https://doi.org/10.1016/j.biocon.2021.109077
122. Sánchez OR, Vale DB, Rodrigues L, Surita FG (2020) Violence against women during the COVID-19 pandemic: an integrative review. Int J Gynecol Obstet 151(2):180–187. https://doi.org/10.1002/ijgo.13365
123. Sanyé-Mengual E, Lozano RG, Farreny R, Oliver-Solà J, Gasol CM, Rieradevall J (2014) Introduction to the eco-design methodology and the role of product carbon footprint. In: Muthu SS (ed) Assessment of carbon footprint in different industrial sectors (Vol 1, pp 1–24). Springer Singapore. https://doi.org/10.1007/978-981-4560-41-2_1
124. Sekalala S, Forman L, Hodgson T, Mulumba M, Namyalo-Ganafa H, Meier BM (2021) Decolonising human rights: how intellectual property laws result in unequal access to the COVID-19 vaccine. BMJ Glob Health 6(7). https://doi.org/10.1136/bmjgh-2021-006169
125. Shaw R, Kim Y-K, Hua J (2020) Governance, technology and citizen behavior in pandemic: lessons from COVID-19 in East Asia. Prog Disaster Sci 6. https://doi.org/10.1016/j.pdisas.2020.100090
126. Short T, Lee-Mortimer A, Luttropp C, Johansson G (2012) Manufacturing, sustainability, ecodesign and risk: lessons learned from a study of Swedish and English companies. J Clean Prod 37:342–352. https://doi.org/10.1016/j.jclepro.2012.07.037
127. Sigala M (2020) Tourism and COVID-19: impacts and implications for advancing and resetting industry and research. J Bus Res 117:312–321. https://doi.org/10.1016/j.jbusres.2020.06.015
128. Snellinx S, Van Meensel J, Farahbakhsh S, Bourgeois L, Mertens A, Lauwers L, Buysse J (2021) Waste treatment company decision-making in a complex system of markets influenced by the circular economy. J Clean Prod 328. https://doi.org/10.1016/j.jclepro.2021.129672
129. Statista (2020) How concerned are you about product packaging and its impact on the following environmental issues? Retrieved 01/01/2022 from https://www.statista.com/statistics/1010173/packaging-environmental-impact-concerns-worldwide-by-country
130. Statista (2021) Annual volumes of waste disposed at the largest landfills worldwide as of 2021. Retrieved 01/01/2022 from https://www.statista.com/statistics/1252704/largest-dump-sites-by-annual-volume-worldwide
131. Steinmann ZJN, Huijbregts MAJ, Reijnders L (2019) How to define the quality of materials in a circular economy? Resour Conserv Recycl 141:362–363. https://doi.org/10.1016/j.resconrec.2018.10.040
132. Stoffels P, Kaspar J, Baehre D, Vielhaber M (2017) Holistic material selection approach for more sustainable products. Procedia Manufacturing 8:401–408. https://doi.org/10.1016/j.promfg.2017.02.051
133. Tran LTT (2021) Managing the effectiveness of e-commerce platforms in a pandemic. J Retail Consum Serv 58. https://doi.org/10.1016/j.jretconser.2020.102287
134. van der Velden NM, Kuusk K, Köhler AR (2015) Life cycle assessment and eco-design of smart textiles: the importance of material selection demonstrated through e-textile product redesign. Mater Des 84:313–324. https://doi.org/10.1016/j.matdes.2015.06.129
135. van Ewijk S, Stegemann JA, Ekins P (2021) Limited climate benefits of global recycling of pulp and paper. Nature Sustain 4(2):180–187. https://doi.org/10.1038/s41893-020-00624-z
136. Vidya CT, Prabheesh KP (2020) Implications of COVID-19 pandemic on the global trade networks. Emerg Mark Financ Trade 56(10):2408–2421. https://doi.org/10.1080/1540496X.2020.1785426
137. Viero A, Barbara G, Montisci M, Kustermann K, Cattaneo C (2021) Violence against women in the Covid-19 pandemic: a review of the literature and a call for shared strategies to tackle health and social emergencies. Forensic Sci Int 319. https://doi.org/10.1016/j.forsciint.2020.110650
138. Webster K (2017) The circular economy: a wealth of flows. Ellen MacArthur Foundation Publishing.

139. Wen J, Kozak M, Yang S, Liu F (2021) COVID-19: potential effects on Chinese citizens' lifestyle and travel. Tourism Rev 76(1):74–87. https://doi.org/10.1108/TR-03-2020-0110
140. Windsor FM, Durance I, Horton AA, Thompson RC, Tyler CR, Ormerod SJ (2019) A catchment-scale perspective of plastic pollution. Glob Change Biol 25(4):1207–1221. https://doi.org/10.1111/gcb.14572
141. Wu T-Y, Ford O, Rainville AJ, Bessire R (2020) COVID-19 care package distribution for senior citizens and families in Detroit and Hamtramck, Michigan. J Hunger & Environ Nutr 15(4):585–587. https://doi.org/10.1080/19320248.2020.1799586
142. Xie Y, Zhao Y, Chen Y, Allen C (2022) Green construction supply chain management: Integrating governmental intervention and public–private partnerships through ecological modernisation. J Clean Prod 331. https://doi.org/10.1016/j.jclepro.2021.129986
143. Yan J, Kim S, Zhang SX, Foo M-D, Alvarez-Risco A, Del-Aguila-Arcentales S, Yáñez JA (2021) Hospitality workers' COVID-19 risk perception and depression: a contingent model based on transactional theory of stress model. Int J Hosp Manag 95. https://doi.org/10.1016/j.ijhm.2021.102935
144. Yáñez JA, Afshar Jahanshahi A, Alvarez-Risco A, Li J, Zhang SX (2020) Anxiety, distress, and turnover intention of healthcare workers in Peru by their distance to the epicenter during the COVID-19 crisis. Am J Trop Med Hyg 103(4):1614–1620. https://doi.org/10.4269/ajtmh.20-0800
145. Yáñez JA, Alvarez-Risco A, Delgado-Zegarra J (2020) Covid-19 in Peru: from supervised walks for children to the first case of Kawasaki-like syndrome. BMJ 369. https://doi.org/10.1136/bmj.m2418
146. Yasin Ar A (2020) Managing e-commerce during a pandemic: lessons from GrubHub during COVID-19. In: George B, Mahar Q (eds) International Case Studies in the Management of Disasters (pp 155–167). Emerald Publishing Limited. https://doi.org/10.1108/978-1-83982-186-820201010
147. Zarocostas J (2021) What next for a COVID-19 intellectual property waiver? The Lancet 397(10288):1871–1872. https://doi.org/10.1016/S0140-6736(21)01151-X
148. Zhang H, Peng Y, Tian G, Wang D, Xie P (2017) Green material selection for sustainability: a hybrid MCDM approach. PLoS ONE 12(5). https://doi.org/10.1371/journal.pone.0177578
149. Zhang SX, Chen J, Afshar Jahanshahi A, Alvarez-Risco A, Dai H, Li J, Patty-Tito RM (2021) Succumbing to the COVID-19 pandemic—healthcare workers not satisfied and intend to leave their jobs. Int J Ment Heal Addict. https://doi.org/10.1007/s11469-020-00418-6
150. Zhang SX, Sun S, Afshar Jahanshahi A, Alvarez-Risco A, Ibarra VG, Li J, Patty-Tito RM (2020) Developing and testing a measure of COVID-19 organizational support of healthcare workers – results from Peru, Ecuador, and Bolivia. Psychiatry Res 291. https://doi.org/10.1016/j.psychres.2020.113174
151. Zollinger R, DiCindio C (2021) Community ecology: Museum education and the digital divide during and after COVID-19. J MusM Educ 46(4):481–492. https://doi.org/10.1080/10598650.2021.1983711

Application of Virtual Strategies

Water Footprint in the Textile and Food Supply Chain Management: Trends to Become Circular and Sustainable

Luis Juarez-Rojas, Aldo Alvarez-Risco, Nilda Campos-Dávalos, Maria de las Mercedes Anderson-Seminario, and Shyla Del-Aguila-Arcentales

Abstract The textile and food industries are two sectors whose water consumption generates a significant environmental footprint. The production of industries must be focused on sustainability if the organizations for these sectors are improved. Therefore, it is necessary to focus the analysis on the sustainable alternatives applied in the supply chain by applying circular strategies. This chapter presents water use and its water footprint in the industries above. In addition, some strategies applied to improve water footprint levels—which involve water consumption and pollution—through circular strategies are presented. Finally, some trends and external and internal factors that can determine the successful implementation of a sustainable supply chain are presented.

Keywords Textile industry · Food industry · Water footprint · Sustainability · Supply chain · Circularity · Sustainable strategies · Resource management

1 Introduction

Countries such as India, China, and Bangladesh generate between 2.8% and 3.3% of the world's wastewater from the textile sector. The wastewater generated by these countries is 640, 1840, and 1030 million m3/year, respectively [82]. Overall, water use in the textile industry accounts for a significant environmental footprint [113]. Along the supply chain, many stages require large amounts of resources to be carried out. Mainly, garment production requires higher water consumption, especially the dyeing stage. From this stage, large amounts of water resources are used and polluted [148]. There are various ways to apply circularity in the supply chain. Some authors suggest the application of technologies, while others propose changing

L. Juarez-Rojas · A. Alvarez-Risco (✉) · N. Campos-Dávalos ·
M. de las Mercedes Anderson-Seminario
Universidad de Lima, Lima, Perú
e-mail: aralvare@ulima.edu.pe

S. Del-Aguila-Arcentales
Escuela Nacional de Marina Mercante "Almirante Miguel Grau", Callao, Perú

A. Alvarez-Risco et al. (eds.), *Circular Economy*, Environmental Footprints and Eco-design of Products and Processes,
https://doi.org/10.1007/978-981-19-0549-0_11

and improving the stages of the supply chain. Whatever the alternative, it is necessary to be efficient in managing and controlling resources, especially water.

Overall, assuring a sustainable food supply for the world's fast-growing population is a significant challenge in the agricultural and food industry. The supply chain in this industry is one of the critical areas that require action, alongside food consumption, nutrition, and food security issues. The methods of global food production must change to minimize the impact on the environment and support the world's capacity to produce food in future. As well as other economic sectors, food production contributes to climate change, water scarcity, soil degradation, and the destruction of biodiversity. In this paper, the environmental effect we focus on is on the water footprint and scarcity. This research analyzes what type of immediate changes can be made to food production to make it more sustainable, with organizations such as the European Commission.

It is crucial to know in detail the most relevant evidence of the impact of the COVID-19 pandemic and can develop more specific and successful plans for the recovery of the economy and commercial activities around the world (Table 1).

For this chapter, information on the water footprint in the textile and food industry is discussed. Strategies for implementing a sustainable supply chain that focuses on the management of production resources, especially water, are proposed. Finally, some external and internal factors that improve the supply chain are presented. Conclusions are presented based on what has been developed.

2 Literature Background

2.1 Water Footprint in the Textile Industry

Reducing the water footprint in the textile industry is a concern for achieving sustainable development globally [137]. China, for example, has become the largest exporter of textiles in the world [121], which is reasonable considering that the Asian country has become the most significant global producer [54]. In this country, manufacturing industries are not eco-efficient and even transgress planetary boundaries, including freshwater use [146]. The study of the water footprint in the textile sector has focused on finding sustainable alternatives for the industry worldwide, considering that water consumption in this industry is high [137]. The water footprint provides valuable information regarding the consumption (in quantity and quality) of water in the production processes of different sectors, whether at industrial, domestic, or agricultural levels [83].

The water footprint has been high in the textile sector and can be supported by research from significant garment-producing countries. Wang et al. [133] calculated the direct blue and gray water footprint Chinese companies use for garment production. Three subsectors of the textile industry were studied: textile, footwear, cap manufacturing sector, and chemical fibers. It was concluded that the sector's water

Table 1 Impact of COVID-19 in different sectors

Education	Ali [10], Allen et al. [11], Alvarez-Risco, Estrada-Merino et al. [17], Alvarez-Risco, Del-Aguila-Arcentales, Rosen et al. [14], Alvarez-Risco, Del-Aguila-Arcentales, Yáñez et al. [16], T. Chen et al. [55], Lackie et al. [90], Lashley et al. [91], Zollinger and DiCindio [149]
Entrepreneurship	Afshan et al. [4], Alvarez-Risco, Mlodzianowska, Zamora-Ramos et al. [24], Alvarez-Risco, Mlodzianowska, García-Ibarra et al. [19], Alvarez-Risco and Del-Aguila-Arcentales [23], Belitski et al. [31], Block et al. [34], Brown et al. [36], Bacq and Lumpkin [43], Chafloque-Cespedes et al. [52], Maritz et al. [94], Liguori and Winkler [100]
Health sector	Abiakam et al. [1], Alan et al. [7], Alsairafi et al. [12], Alvarez-Risco, Dawson et al. [21], Alvarez-Risco, Del-Aguila-Arcentales and Yáñez [15], Barello et al. [32], Bozdağ and Ergün [40], X. Chen et al. [56], Deressa et al. [61], Koetter et al. [87], Raudenská et al. [112], Rojas Román et al. [115], Yáñez, Afshar Jahanshahi et al. [141], Zhang et al. [145], Zhang et al. [147]
Hospitality	Duarte Alonso et al. [64], Gursoy and Chi [73], Ho et al. [77], Kaushal and Srivastava [84], Yan et al. [139]
Intellectual property	Altindis [13], Alvarez-Risco and Del-Aguila-Arcentales [22], del Castillo [50], Erfani et al. [67], Jecker and Atuire [81], Krishtel and Malpani [88], Okereke [107], Sekalala et al. [118], Zarocostas [143]
Population	Ahsan [6], Alvarez-Risco, Mejia et al. [18], Cegarra-Navarro et al. [51], Hanzl [74], Lim and Prakash [95], Quispe-Cañari et al. [111], Sánchez-Clavijo et al. [123], Shaw et al. [119], Wen et al. [135], Wu et al. [136], Yáñez, Alvarez-Risco et al. [142]
Prices	Apcho-Ccencho et al. [28], Chung et al. [57], Galanakis et al. [69], Hepburn et al. [75], Leiva-Martinez et al. [92], Obergassel et al. [106]
Tourism	Assaf and Scuderi [30], Baum and Hai [33], Brouder [42], Carvache-Franco, Alvarez-Risco, Carvache-Franco, Carvache-Franco, Estrada-Merino and Villalobos-Alvarez [49], Fotiadis et al. [68], Carvache-Franco, Alvarez-Risco, Carvache-Franco, Carvache-Franco, Estrada-Merino and Rosen [47], Sigala [120], Carvache-Franco, Carvache-Franco et al. [48]
Trade	Abudureheman and Nilupaer [2], Acuña-Zegarra et al. [3], Aguirre et al. [5], Alvarez-Risco, Quipuzco-Chicata et al. [25], Alvarez-Risco, Rosen et al. [20], Borzée et al. [38], Cruz-Torres et al. [60], Kirk and Rifkin [86], Lopez-Odar et al. [97], Tran [127], Vidya and Prabheesh [129], Yasin Ar [140]
Violence against women	Amarillo [26], Chafloque-Cespedes et al. [53], Dominelli [62], Gulati and Kelly [72], Roesch et al. [114], Sánchez et al. [122], Viero et al. [130]

footprint was 1.8 billion m^3, which is a worrying figure as textile activities generate water pollution that could cause depletion of water sources and health problems for citizens. Despite this, the water footprint has decreased by applying water-saving technologies and encouraging it with government policies. Similarly, Hossain and Khan [78] found that the textile supply and production chain consumes and pollutes water in Bangladesh. In general, most research suggests a negative impact of the textile industry on the water footprint and water pollution [89].

Research has focused on determining the overall water footprint of a country's textile sector. The complexity of calculating the water footprint in the textile sector is the number of garments, materials, and processes involved in textile manufacturing [99]. Quian-Ulloa and Stange [110] investigated and calculated the water footprint of viscose staple fiber blouses and men's suits. It was determined that the stage whose water consumption is higher in the same viscose staple fiber production. This stage, in turn, has a more significant environmental burden on the water resources used. Thompson et al. [125] determined the water footprint throughout silk production in Thailand. It was evidenced that in 2019 the water footprint was the highest, with a record of 1,710 m^3/ton, which directly affects the textile industry since the water footprint of the production of a silk shirt is 376L. The authors propose alternatives based on the sustainability of the supply and production chain through the joint work of authorities and entrepreneurs. On a general level, not only this country but also the Asian region is going through a problem of water scarcity and wastewater pollution [93].

Water pollution contributes to 80% of the water footprint of the productive activities of the textile industry [134]. The water footprint measures not only water consumption but also water pollution. Different stages can be identified within the textile industry that increases the water footprint (Fig. 1).

Water consumption in the textile industry covers not only production processes but also previous stages such as obtaining raw materials and indirect consumption in the final stage. The amount of water used in the sector depends on the type of fiber, machinery used, the type of product sought to be made, and the stages applied [116]. Reducing water consumption by constantly evaluating the water footprint is an alternative to be more efficient with the water resource [93]. Achieving this goal

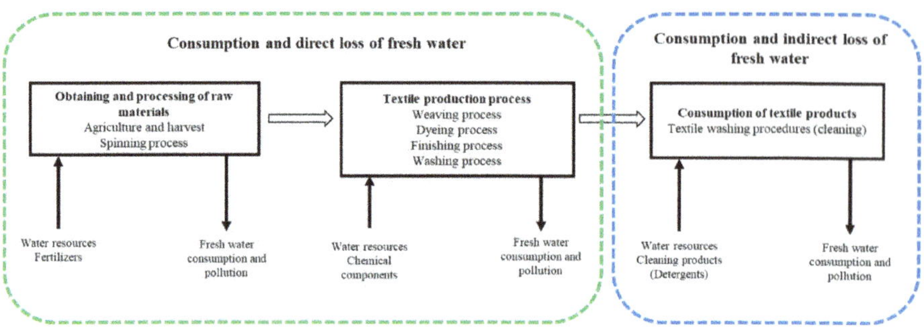

Fig. 1 Direct and indirect loss of freshwater (*Source* Adapted from Wang et al. [134])

is possible by becoming more sustainable along the supply and production chain of the textile industry.

3 Sustainability in the Textile Industry Supply Chain

To understand what sustainability alternatives have been applied to the textile supply chain, we must understand what activities comprise it. Table 2 summarizes the main stages of the sector's supply chain.

Many manufacturing industries seek to acquire technology to reduce production costs (i.e., labor hiring) and leverage the resources used to manufacture their products [63]. The supply chain has been characterized as long and complex [66]. For that reason, new tools have been integrated to achieve a more efficient and optimized supply chain.

Achieving a sustainable supply chain through the application of the circular economy requires textile industries to reduce their pollution levels, the number of resources used in the production of garments, and the reduction of greenhouse gas emissions [41]. Applying responsible measures such as using renewable energy to reduce the amount of CO_2 emitted or using eco-friendly fibers and dyes promotes the transition from a conventional supply chain to a sustainable one [110]. Saxena and Khare [117] applied the concept of green manufacturing within the textile industry, the main objective is to reduce the number of hazardous substances within the design and production of garments, which affects resources such as water, air, and land. It is wrong to think that all products generate positive impacts despite the above [76] demonstrated that within the bio-textile industry, bio-based leathers are used to produce shoes, clothing, and upholstery, which require a large amount of water for their production. It must be sustainable in terms of the materials used and the production processes.

Table 2 Textile industry supply chain

Primary Activities	Secondary Activities
Inbound Logistics: Activities to receive and store raw materials. It involves relationships with supply providers **Operations:** Activities of transformation and production of the final product **Outbound Logistics:** Activities to store and distribute the final products **Marketing and sales:** Activities to facilitate and promote the final product **Service:** Involves working effectively with the customer once the purchase has been made	**Procurement:** Acquisition of resources for the company **Human Resource Management:** Involves the recruitment and hiring of personnel for the company **Technological Development:** Involves the acquisition of technological equipment Other activities in the areas of accounting, finance, legal

Source Adapted from ElMessiry and ElMessiry [66]

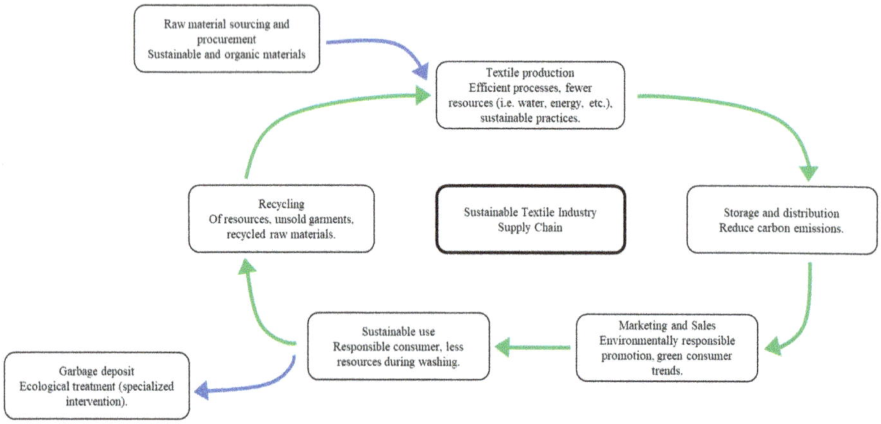

Fig. 2 The sustainable supply chain for the textile industry (*Source* Adapted from Wang et al. [134])

The leading sustainable practices in the textile industry focus on product design with longer lifetimes, energy efficiency and control strategies, implementation of circular strategies, and resource savings [80]. Water management, as a production resource, focuses on seeking alternatives to use various sources such as rainwater, seawater, wastewater, among others [82]. A priori, this would be a complex strategy considering that using this type of source requires having technology capable of processing such water and making it sound. It has been determined that water pollution occurs in different stages of garment production, mainly in washing and dyeing activities [109]. Some countries, mainly Europe, can recycle and reuse water from textile activities. For this, technological, economic, and environmental options are abundant and applicable at the industrial level [59]. However, this could be applied only by companies that can implement this strategy. Therefore, it is necessary to focus our strategies on small and medium-sized companies that do not have the resources to be sustainable. Figure 2 shows a circular strategy that can be sustainable and applicable at any scale.

4 Water Footprint in the Food Industry

With the food industry being one of the most demanded ones within the global economy, there are several environmental problems it relies on, which is an obstacle in the progress of the sustainable development goals. Even though there are guidelines on food, such as the food-based dietary guideline (FBDGs), the question is if it is being integrated into the practices [46]. Water footprint (WF) investigations are primarily directed to lower freshwater consumption. However, if biogas plants, photovoltaic panels, and other types of renewable energy are used in the agricultural sector, studies showed that not only carbon footprint can be reduced, but water footprint as well [37].

The water footprint is divided into three categories: green, blue, and gray waters, according to the Water Footprint Network [98]. For the type of water footprint, this industry is significant in, a study on the global water footprint in the food production of different crops shows that the more significant part is associated with green water and a small part on blue water, meaning that indeed green water is a significant factor in the crop production worldwide [101].

For instance, Blas et al. [35] studied two countries, Spain and the United States (US), and compared their water consumption due to the US having more water footprint than the Mediterranean diet. Overall, this paper concludes that the origin of products and the consumption pattern is prevalent and essential factors in reducing water footprint [35]. In the case of South America, a study based in Argentina showed that in the agricultural process, water footprint is lower, given that irrigation is not used commonly in the sector [131].

Over the years, there have been methodologies to save water and its consumption in the food supply chain, standing out both the water footprint assessment (WFA) approach and the life cycle analysis (LCA) one. These two methodologies have been compared and applied to numerous food products to develop the benefits and disadvantages each one may have [39]. It is essential to mention that the water used in the crop growing season is based on production. Using the accounting of water footprint would be an efficient measure to manage the relationship in terms of the use of water and crop yield [102].

The main reason for the increase in water footprint and environmental issues among the food industry is the evolution it had; nowadays, habits related to the diet have gone in the direction of processed food, meat, refined sugar, and less consumption of fruits and vegetables [71]. Studies have proved how diet-based choices significantly influence the food system's impact on the environment [132, 105, 126].

On the other hand, there are different levels of impact among each type of food, depending on the production system. For example, ruminant meats have tripled to tenth times the impact compared to plant-based food [58]. However, implementing projects to decrease the consumption of meat would be difficult in meat-loving countries like Argentina [29]. It is essential to evaluate the country's conditions, capabilities, and technologies because these are the significant factors regarding water volumes and water use [128].

5 Sustainability in the Food Industry Supply Chain

For sustainability, there have been actions taken to prioritize the environment's well-being and promote sustainable farming and processing among the food industry. First, water footprint is not the only indicator that can measure how sustainable food production is. Other considerations to be taken include the local, pedo-climatic, and technological factors [98].

For example, in the countries that belong to the EU, the European Commission posted an evaluation on the situation the EU is facing based on REFIT data. This program (REFIT) is an initiative in charge of improving the performance of EU businesses without creating burdens, especially in the framework of environmental issues [27]. It is aligned with the European Green deal, which is an institution focused on reducing the damage the food production impacts on the environment; it aims to transform the UE into a carbon–neutral sector by 2050 [9]. On the other hand, there is another method being used called Waste to Energy (WTE) for waste management (including water) that acts with the mindset of energy sustainability based on sustainable energy sources and sustainable energy systems [70].

Among the strategies promoted by the European Green deal, there is also the Farm-to-Fork one, which aims to create sustainable food systems in line with the United Nations Sustainable Development Goals (SDG). This strategy focuses on the food supply chains and looks at ways of reducing food waste and using it as a resource aligned with the food waste hierarchy [9]. In China, a recent foam system intends to shift from using inorganic materials to materials with a biological origin in stabilization, which is because inorganic materials in the food industry have shallow sustainable tendencies [144].

6 Trends to Become Circular in Sustainable

By 2030, 118 billion cubic meters are expected to be used for apparel production globally [113]. Therefore, companies must implement a sustainability approach through the SDGs that seek, among other things, to reduce the number of resources along the supply and production chain [44]. The massive water consumption of the textile industry is a problem with social, economic, and environmental impact in different countries. For this reason, it is necessary to apply the strategies above, such as water reuse [108], wastewater treatment, staged production system [124], use of ecological and sustainable fibers that consume fewer resources [8, 65, 96]. The solutions contribute to the improvement of the supply chain and improve the company's image considering that consumer behavior and their tendency toward environmentally friendly products is increasing [113].

For the food industry, there have been techniques to mitigate water footprint, such as the Micro-filtration (MF) one, which causes a significant reduction in chemical and biological oxygens (COD and BOD); additionally, it reduces the count of total bacteria in wastewater. In the case of the food processing industry, techniques such as the Microbial Fuel Cell (MFC) is an effective response to treat wastewater coming from various sectors because it reclaims water for applications in agriculture [79].

To achieve sustainability in terms of water consumption in the supply chain of the industries under study, companies should consider external and internal factors that can be applied. Table 3 indicates the factors that could contribute to achieving the circular economy and, therefore, becoming more sustainable.

Table 3 External and internal factors that enable sustainability

Factor	Level	Description
Globalization of competition (Differentiation)	External	Textile industry: Countries with lower production costs exert pressure on pricing and costs. Being responsible with the number of resources used to manufacture textiles. Therefore, it is ideal to differentiate our products Food Industry: Consumers' tendencies put pressure on businesses; nowadays, they look for traceability and organic food with an environmental focus and choose environmentally responsible companies
Change or improve the supply chain	Internal	Textile Industry: Companies must change their traditional supply chain to a sustainable one. Firms must apply strategies such as recycling and reusing resources Food Industry: The supply chain in this industry depends on each country's level of technology; there are also technological trends within this process, including blockchain, QR codes, and tracking systems to please the consumers' needs
Modify production processes (Automation)	Internal	Automating production processes is difficult depending on the size of the company. Small companies (Startups or SMEs) should make significant investments to acquire technology. However, companies should improve their processes to be more productive and efficient In the food industry, there needs to be an investment in innovation and technology. The change in the processes depends on each market's needs; for example, many companies have implemented omnichannel consumption, which gives the consumer more access to the information the product has driven by food safety and transparency
The role of government in policymaking	External	Governments should prioritize promoting support and incentive policies for sustainable textile and food companies. Companies, for their part, should embrace such policies, improve their processes, and make them more efficient

Source Adapted from Bressanelli et al. [41]

7 Closing Remarks

The application of the circular economy is presented as a sustainable alternative to the linear model that is mainly applied in the business environment [45]. Changes in product design and supply chain are required to achieve this [41].

On the other hand, in the case of the food industry, it is essential to acknowledge the methodologies used by already established organizations, given that they already have good results in reducing water footprint. Apart from that, there are also excellent and effective new technologies emerging in European and Asian countries. Overall, food industries must extend these technologies without losing the nutritional content that food has, which should be focused on the availability of their compounds; apart from that, promote and provide sustainability not only environmental but also economical. To achieve this, they need to embrace and take into consideration all the processing steps. Additionally, it would be efficient to have a global agricultural sustainability framework done by a multi-stakeholder congress that would secure the standardization of the certifications that farmers and users all over the food supply chain need to accomplish for their product to enter each market [103]. This result should still accomplish the final goal to fulfill consumers' needs in the case of sensory characteristics [103].

Sustainability focuses on efficiency [104]. Therefore, it is necessary to focus the organization of the supply chain in terms of the resources used in the sector, mainly water and energy. It is ideal to look for alternative sources and, above all, to promote innovation in the companies of the industry. For the most part, innovation focused on the acquisition of technology that allows us to facilitate and improve supply chain processes [138]. In general, improvements in the supply chain led to environmental differentiation that improves the performance of companies [85].

References

1. Abiakam N, Worsley P, Jayabal H, Mitchell K, Jones M, Fletcher J, Bader D (2021) Personal protective equipment related skin reactions in healthcare professionals during COVID-19. Int Wound J 18(3):312–322. https://doi.org/10.1111/iwj.13534
2. Abudureheman A, Nilupaer A (2021) Optimization model design of cross-border e-commerce transportation path under the background of prevention and control of COVID-19 pneumonia. Soft Comput 25(18):12007–12015. https://doi.org/10.1007/s00500-021-05685-6
3. Acuña-Zegarra MA, Santana-Cibrian M, Velasco-Hernandez JX (2020) Modeling behavioral change and COVID-19 containment in Mexico: trade-off between lockdown and compliance. Math Biosci 325. https://doi.org/10.1016/j.mbs.2020.108370
4. Afshan G, Shahid S, Tunio MN (2021) Learning experiences of women entrepreneurs amidst COVID-19. Int J Gend Entrep 13(2):162–186. https://doi.org/10.1108/IJGE-09-2020-0153
5. Aguirre AA, Catherina R, Frye H, Shelley L (2020) Illicit wildlife trade, wet markets, and COVID-19: preventing future pandemics. World Med & Health Policy 12(3):256–265. https://doi.org/10.1002/wmh3.348

6. Ahsan MM (2020) Strategic decisions on urban built environment to pandemics in Turkey: lessons from COVID-19. J Urban Manag 9(3):281–285. https://doi.org/10.1016/j.jum.2020.07.001
7. Alan H, Eskin Bacaksiz F, Tiryaki Sen H, Taskiran Eskici G, Gumus E, Harmanci Seren AK (2021) "I'm a hero, but…": An evaluation of depression, anxiety, and stress levels of frontline healthcare professionals during COVID-19 pandemic in Turkey. Perspect Psychiatr Care 57(3):1126–1136. https://doi.org/10.1111/ppc.12666
8. Alberghini M, Hong S, Lozano LM, Korolovych V, Huang Y, Signorato F, Boriskina SV (2021) Sustainable polyethylene fabrics with engineered moisture transport for passive cooling. Nat Sustain 4(8):715–724. https://doi.org/10.1038/s41893-021-00688-5
9. Albizzati PF, Tonini D, Astrup TF (2021) A quantitative sustainability assessment of food waste management in the European Union. Environ Sci Technol 55(23):16099–16109. https://doi.org/10.1021/acs.est.1c03940
10. Ali W (2020) Online and remote learning in higher education institutes: a necessity in light of COVID-19 pandemic. High Educ Stud 10(3):16–25
11. Allen J, Rowan L, Singh P (2020) Teaching and teacher education in the time of COVID-19. Asia-Pac J Teach Educ 48(3):233–236. https://doi.org/10.1080/1359866X.2020.1752051
12. Alsairafi Z, Naser AY, Alsaleh FM, Awad A, Jalal Z (2021) Mental health status of healthcare professionals and students of health sciences faculties in Kuwait during the COVID-19 pandemic. Int J Environ Res Public Health 18(4). https://doi.org/10.3390/ijerph18042203
13. Altindis E (2021) Inequitable COVID-19 vaccine distribution and the intellectual property rights prolong the pandemic. Expert Rev Vaccines. https://doi.org/10.1080/14760584.2022.2014819
14. Alvarez-Risco A, Del-Aguila-Arcentales S, Rosen MA, García-Ibarra V, Maycotte-Felkel S, Martínez-Toro GM (2021) Expectations and interests of university students in COVID-19 times about Sustainable Development Goals: evidence from Colombia, Ecuador, Mexico, and Peru. Sustain 13(6):3306. https://doi.org/10.3390/su13063306
15. Alvarez-Risco A, Del-Aguila-Arcentales S, Yáñez JA (2021) Telemedicine in Peru as a result of the COVID-19 pandemic: perspective from a country with limited internet access. Am J Trop Med Hyg 105(1):6–11. https://doi.org/10.4269/ajtmh.21-0255
16. Alvarez-Risco A, Del-Aguila-Arcentales S, Yáñez JA, Rosen MA, Mejia CR (2021) Influence of technostress on academic performance of university medicine students in Peru during the COVID-19 pandemic. Sustain 13(16):8949. https://doi.org/10.3390/su13168949
17. Alvarez-Risco A, Estrada-Merino A, Anderson-Seminario M, d. l. M., Mlodzianowska, S., García-Ibarra, V., Villagomez-Buele, C., & Carvache-Franco, M. (2021) Multitasking behavior in online classrooms and academic performance: case of university students in Ecuador during COVID-19 outbreak. Interact Technol Smart Educ 18(3):422–434. https://doi.org/10.1108/ITSE-08-2020-0160
18. Alvarez-Risco A, Mejia CR, Delgado-Zegarra J, Del-Aguila-Arcentales S, Arce-Esquivel AA, Valladares-Garrido MJ, Yáñez JA (2020) The Peru approach against the COVID-19 infodemic: insights and strategies. Am J Trop Med Hyg 103(2):583–586. https://doi.org/10.4269/ajtmh.20-0536
19. Alvarez-Risco A, Mlodzianowska S, García-Ibarra V, Rosen MA, Del-Aguila-Arcentales S (2021) Factors affecting green entrepreneurship intentions in business university students in COVID-19 pandemic times: case of Ecuador. Sustain 13(11):6447. https://doi.org/10.3390/su13116447
20. Alvarez-Risco A, Rosen MA, Del-Aguila-Arcentales S (2020) A new regulation for supporting a circular economy in the plastic industry: the case of Peru (short communication). J Landsc Ecol 13(1):1–3. https://doi.org/10.2478/jlecol-2020-0004
21. Alvarez-Risco A, Dawson J, Johnson W, Conteh-Barrat M, Aslani P, Del-Aguila-Arcentales S, Diaz-Risco S (2021) Ebola virus disease outbreak: a global approach for health systems [Ebola; health professionals; global approach; pharmacist: pharmaceutical care; COVID-19] 53(4). http://www.revfarmacia.sld.cu/index.php/far/article/view/491

22. Alvarez-Risco A, Del-Aguila-Arcentales S (2021a) A note on changing regulation in international business: the World Intellectual Property Organization (WIPO) and artificial intelligence. In: Verbeke A, van Tulder R, Rose EL, Wei Y (eds) The multiple dimensions of Institutional Complexity in International Business Research (Vol 15, pp 363–371). Emerald Publishing Limited. https://doi.org/10.1108/S1745-886220210000015020
23. Alvarez-Risco A, Del-Aguila-Arcentales S (2021b) Public policies and private efforts to increase women entrepreneurship based on STEM background. In: Mari M, Poggesi S, Foss L (eds) Women's entrepreneurship in STEM disciplines: issues and perspectives (pp 75–87). Springer International Publishing. https://doi.org/10.1007/978-3-030-83792-1_5
24. Alvarez-Risco A, Mlodzianowska S, Zamora-Ramos U, Del-Aguila S (2021) Green entrepreneurship intention in university students: the case of Peru. Entrep Bus Econ Rev 9(4):85–100. https://doi.org/10.15678/EBER.2021.090406
25. Alvarez-Risco A, Quipuzco-Chicata L, Escudero-Cipriani C (2021) Determinantes de la intención de recompra en línea en tiempos de COVID-19: evidencia de una economía emergente. Lecturas de Economía (96). https://doi.org/10.17533/udea.le.n96a342638
26. Amarillo CR (2020) Feminism on lockdown. NACLA Rep Am 52(3):274–281. https://doi.org/10.1080/10714839.2020.1809084
27. De Angelis F, Carreño I (2021) Achieving sustainability of the EU food chain with feed additives as a key tool: the European Commission intends to modernise the legal framework to foster innovation. Eur J Risk Regul 12(4):866–870. https://doi.org/10.1017/err.2021.10
28. Apcho-Ccencho L-V, Cuya-Velásquez B-B, Alvarado Rodríguez D, Anderson-Seminario MdlM, Alvarez-Risco A, Estrada-Merino A, Mlodzianowska S (2021) The impact of international price on the technological industry in the United States and China during times of crisis: commercial war and COVID-19. In: Lawrence KD, Klimberg RK (eds) Advances in business and management forecasting (Vol 14, pp 149–160). Emerald Publishing Limited. https://doi.org/10.1108/S1477-407020210000014010
29. Arrieta EM, Geri M, Coquet JB, Scavuzzo CM, Zapata ME, González AD (2021) Quality and environmental footprints of diets by socio-economic status in Argentina. Sci Total Environ 801. https://doi.org/10.1016/j.scitotenv.2021.149686
30. Assaf A, Scuderi R (2020) COVID-19 and the recovery of the tourism industry. Tour Econ 26(5):731–733. https://doi.org/10.1177/1354816620933712
31. Bacq S, Lumpkin GT (2021) Social entrepreneurship and COVID-19. J Manage Stud 58(1):285–288. https://doi.org/10.1111/joms.12641
32. Barello S, Caruso R, Palamenghi L, Nania T, Dellafiore F, Bonetti L, Graffigna G (2021) Factors associated with emotional exhaustion in healthcare professionals involved in the COVID-19 pandemic: an application of the job demands-resources model. Int Arch Occup Environ Health 94(8):1751–1761. https://doi.org/10.1007/s00420-021-01669-z
33. Baum T, Hai NTT (2020) Hospitality, tourism, human rights and the impact of COVID-19. Int J Contemp Hosp Manag 32(7):2397–2407. https://doi.org/10.1108/IJCHM-03-2020-0242
34. Belitski M, Guenther C, Kritikos AS, Thurik R (2021) Economic effects of the COVID-19 pandemic on entrepreneurship and small businesses. Small Bus Econ. https://doi.org/10.1007/s11187-021-00544-y
35. Blas A, Garrido A, Willaarts BA (2016) Evaluating the water footprint of the Mediterranean and American diets. Water 8(10). https://doi.org/10.3390/w8100448
36. Block JH, Fisch C, Hirschmann M (2021) The determinants of bootstrap financing in crises: evidence from entrepreneurial ventures in the COVID-19 pandemic. Small Bus Econ. https://doi.org/10.1007/s11187-020-00445-6
37. Borghesi G, Stefanini R, Vignali G (2022) Life cycle assessment of packaged organic dairy product: a comparison of different methods for the environmental assessment of alternative scenarios. J Food Eng 318. https://doi.org/10.1016/j.jfoodeng.2021.110902
38. Borzée A, McNeely J, Magellan K, Miller JRB, Porter L, Dutta T, Zhang L (2020) COVID-19 highlights the need for more effective wildlife trade legislation. Trends Ecol Evol 35(12):1052–1055. https://doi.org/10.1016/j.tree.2020.10.001

39. Boulay A-M, Hoekstra AY, Vionnet S (2013) Complementarities of water-focused life cycle assessment and water footprint assessment. Environ Sci Technol 47(21):11926–11927. https://doi.org/10.1021/es403928f
40. Bozdağ F, Ergün N (2020) Psychological resilience of healthcare professionals during COVID-19 pandemic. Psychol Rep 124(6):2567–2586. https://doi.org/10.1177/0033294120965477
41. Bressanelli G, Visintin F, Saccani N (2022) Circular Economy and the evolution of industrial districts: a supply chain perspective. Int J Prod Econ 243. https://doi.org/10.1016/j.ijpe.2021.108348
42. Brouder P (2020) Reset redux: possible evolutionary pathways towards the transformation of tourism in a COVID-19 world. Tour Geogr 22(3):484–490. https://doi.org/10.1080/14616688.2020.1760928
43. Brown R, Rocha A, Cowling M (2020) Financing entrepreneurship in times of crisis: exploring the impact of COVID-19 on the market for entrepreneurial finance in the United Kingdom. Int Small Bus J 38(5):380–390. https://doi.org/10.1177/0266242620937464
44. Cai Y-J, Choi T-M (2020) A United Nations' Sustainable Development Goals perspective for sustainable textile and apparel supply chain management. Transp Res Part E: Logist Transp Rev 141. https://doi.org/10.1016/j.tre.2020.102010
45. Calicchio Berardi P, Peregrino de Brito R (2021) Supply chain collaboration for a circular economy - from transition to continuous improvement. J Clean Prod 328. https://doi.org/10.1016/j.jclepro.2021.129511
46. Cambeses-Franco C, González-García S, Feijoo G, Moreira MT (2022) Driving commitment to sustainable food policies within the framework of American and European dietary guidelines. Sci Total Environ 807. https://doi.org/10.1016/j.scitotenv.2021.150894
47. Carvache-Franco M, Alvarez-Risco A, Carvache-Franco W, Carvache-Franco O, Estrada-Merino A, Rosen MA (2021) Coastal cities seen from loyalty and their tourist motivations: a study in Lima, Peru. Sustain 13(21):11575. https://doi.org/10.3390/su132111575
48. Carvache-Franco M, Carvache-Franco O, Carvache-Franco W, Alvarez-Risco A, Estrada-Merino A (2021) Motivations and segmentation of the demand for coastal cities: a study in Lima, Peru. Int J Tour Res 23(4):517–531. https://doi.org/10.1002/jtr.2423
49. Carvache-Franco M, Alvarez-Risco A, Carvache-Franco O, Carvache-Franco W, Estrada-Merino A, Villalobos-Alvarez D (2021) Perceived value and its influence on satisfaction and loyalty in a coastal city: a study from Lima, Peru. J Policy Res Tour, Leis Events 1–16. https://doi.org/10.1080/19407963.2021.1883634
50. del Castillo FA (2021) Temporary waiver of intellectual property on Covid-19 vaccines: toward the creation of a better, post-pandemic society. J Public Health 43(3):e559–e560. https://doi.org/10.1093/pubmed/fdab184
51. Cegarra-Navarro J-G, Vătămănescu E-M, Martínez-Martínez A (2021) A context-driven approach on coping with COVID-19: from hiding knowledge toward citizen engagement. Knowl Process Manag 28(2):134–140. https://doi.org/10.1002/kpm.1662
52. Chafloque-Cespedes R, Alvarez-Risco A, Robayo-Acuña P-V, Gamarra-Chavez C-A, Martinez-Toro G-M, Vicente-Ramos W (2021) Effect of sociodemographic factors in entrepreneurial orientation and entrepreneurial intention in university students of Latin American business schools. In: Jones P, Apostolopoulos N, Kakouris A, Moon C, Ratten V, Walmsley A (eds) Universities and Entrepreneurship: meeting the educational and social challenges (Vol 11, pp 151–165). Emerald Publishing Limited. https://doi.org/10.1108/S2040-724620210000011010
53. Chafloque-Cespedes R, Vara-Horna A, Asencios-Gonzales Z, Lopez-Odar DR, Alvarez-Risco A, Quipuzco-Chicata L, ... Sanchez-Villagomez M (2020) Presentismo académico y violencia contra las mujeres en escuelas de negocios e ingeniería en universidades peruanas. Lecturas de Economía 0(93):127–153. https://doi.org/10.17533/udea.le.n93a340726
54. Chen F, Shen Y, Liu S, Yang Y, Wang L (2021) Water footprint of textile industry: a case study of China. Environ Eng & Manag J (EEMJ) 20(2).

55. Chen T, Peng L, Yin X, Rong J, Yang J, Cong G (2020) Analysis of user satisfaction with online education platforms in China during the COVID-19 pandemic. Healthcare 8(3). https://doi.org/10.3390/healthcare8030200
56. Chen X, Zhang SX, Jahanshahi AA, Alvarez-Risco A, Dai H, Li J, Ibarra VG (2020) Belief in a COVID-19 conspiracy theory as a predictor of mental health and well-being of health care workers in Ecuador: cross-sectional survey study. JMIR Public Health Surveill 6(3). https://doi.org/10.2196/20737
57. Chung SA, Olivera S, Rojas Román B, Alanoca E, Moscoso S, Limpias Terceros B, . . . Yáñez JA (2021) Temáticas de la producción científica de la Revista Cubana de Farmacia indizada en Scopus (1967–2020) [Revista Cubana de Farmacia; farmacia; Cuba; evidencia] 54(1). http://www.revfarmacia.sld.cu/index.php/far/article/view/511
58. Clark M, Tilman D (2017) Comparative analysis of environmental impacts of agricultural production systems, agricultural input efficiency, and food choice. Environ Res Lett 12(6). https://doi.org/10.1088/1748-9326/aa6cd5
59. Colella M, Ripa M, Cocozza A, Panfilo C, Ulgiati S (2021) Challenges and opportunities for more efficient water use and circular wastewater management. The case of Campania Region, Italy. J Environ Manag 297:113171. https://doi.org/10.1016/j.jenvman.2021.113171
60. Cruz-Torres W, Alvarez-Risco A, Del-Aguila-Arcentales S (2021) Impact of Enterprise Resource Planning (ERP) implementation on performance of an education enterprise: a Structural Equation Modeling (SEM). Stud Bus Econ 16(2):37–52. https://doi.org/10.2478/sbe-2021-0023
61. Deressa W, Worku A, Abebe W, Gizaw M, Amogne W (2021) Risk perceptions and preventive practices of COVID-19 among healthcare professionals in public hospitals in Addis Ababa, Ethiopia. PLoS ONE 16(6). https://doi.org/10.1371/journal.pone.0242471
62. Dominelli L (2021) A green social work perspective on social work during the time of COVID-19. Int J Soc Welf 30(1):7–16. https://doi.org/10.1111/ijsw.12469
63. Dorduncu M, Olmus I, Rabczuk T (2022) A peridynamic approach for modeling of two dimensional functionally graded plates. Compos Struct 279. https://doi.org/10.1016/j.compstruct.2021.114743
64. Duarte Alonso A, Kok SK, Bressan A, O'Shea M, Sakellarios N, Koresis A, Santoni LJ (2020) COVID-19, aftermath, impacts, and hospitality firms: an international perspective. Int J Hosp Manag 91. https://doi.org/10.1016/j.ijhm.2020.102654
65. Duque Schumacher AG, Pequito S, Pazour J (2020) Industrial hemp fiber: a sustainable and economical alternative to cotton. J Clean Prod 268. https://doi.org/10.1016/j.jclepro.2020.122180
66. ElMessiry M, ElMessiry A (2018) Blockchain framework for textile supply chain management. Blockchain – ICBC 2018, Cham.
67. Erfani P, Binagwaho A, Jalloh MJ, Yunus M, Farmer P, Kerry V (2021) Intellectual property waiver for covid-19 vaccines will advance global health equity. BMJ 374. https://doi.org/10.1136/bmj.n1837
68. Fotiadis A, Polyzos S, Huan T-CTC (2021) The good, the bad and the ugly on COVID-19 tourism recovery. Ann Tour Res 87. https://doi.org/10.1016/j.annals.2020.103117
69. Galanakis CM, Rizou M, Aldawoud TMS, Ucak I, Rowan NJ (2021) Innovations and technology disruptions in the food sector within the COVID-19 pandemic and post-lockdown era. Trends Food Sci Technol 110:193–200. https://doi.org/10.1016/j.tifs.2021.02.002
70. Gil A (2022) Challenges on waste-to-energy for the valorization of industrial wastes: electricity, heat and cold, bioliquids and biofuels. Environ Nanotechnol, Monit & Manag 17. https://doi.org/10.1016/j.enmm.2021.100615
71. González-García S, Green RF, Scheelbeek PF, Harris F, Dangour AD (2020) Dietary recommendations in Spain –affordability and environmental sustainability? J Clean Prod 254. https://doi.org/10.1016/j.jclepro.2020.120125
72. Gulati G, Kelly BD (2020) Domestic violence against women and the COVID-19 pandemic: what is the role of psychiatry? Int J Law Psychiatry 71. https://doi.org/10.1016/j.ijlp.2020.101594

73. Gursoy D, Chi CG (2020) Effects of COVID-19 pandemic on hospitality industry: review of the current situations and a research agenda. J Hosp Market Manag 29(5):527–529. https://doi.org/10.1080/19368623.2020.1788231
74. Hanzl M (2020) Urban forms and green infrastructure – the implications for public health during the COVID-19 pandemic. Cities & Health 1–5. https://doi.org/10.1080/23748834.2020.1791441
75. Hepburn C, O'Callaghan B, Stern N, Stiglitz J, Zenghelis D (2020) Will COVID-19 fiscal recovery packages accelerate or retard progress on climate change? Oxf Rev Econ Policy 36(Supplement_1):S359-S381. https://doi.org/10.1093/oxrep/graa015
76. Hildebrandt J, Thrän D, Bezama A (2021) The circularity of potential bio-textile production routes: comparing life cycle impacts of bio-based materials used within the manufacturing of selected leather substitutes. J Clean Prod 287. https://doi.org/10.1016/j.jclepro.2020.125470
77. Ho C-Y, Tsai B-H, Chen C-S, Lu M-T (2021) Exploring green marketing orientations toward sustainability the hospitality industry in the COVID-19 pandemic. Sustain 13(8). https://doi.org/10.3390/su13084348
78. Hossain L, Khan MS (2020) Water footprint management for sustainable growth in the Bangladesh apparel sector. Water 12(10). https://doi.org/10.3390/w12102760
79. Iqbal M, Nauman S, Ghafari M, Parnianifard A, Gomes A, Gomes C (2021) Treatment of wastewater for agricultural applications in regions of water scarcity. Significance 16:17.
80. Islam MM, Perry P, Gill S (2021) Mapping environmentally sustainable practices in textiles, apparel and fashion industries: a systematic literature review. J Fash Mark Manag: Int J 25(2):331–353. https://doi.org/10.1108/JFMM-07-2020-0130
81. Jecker NS, Atuire CA (2021) What's yours is ours: waiving intellectual property protections for COVID-19 vaccines. J Med Ethics 47(9):595. https://doi.org/10.1136/medethics-2021-107555
82. Jegatheesan V, Shu L, Jegatheesan L (2021) Producing fit-for-purpose water and recovering resources from various sources: an overview. Environ Qual Manage 31(2):9–28. https://doi.org/10.1002/tqem.21780
83. Kalvani SR, Sharaai AH, Manaf L, Hamidian A (2020) Review on water footprint method In different sectors. Int J Adv Sci Technol 1778–1785.
84. Kaushal V, Srivastava S (2021) Hospitality and tourism industry amid COVID-19 pandemic: perspectives on challenges and learnings from India. Int J Hosp Manag 92. https://doi.org/10.1016/j.ijhm.2020.102707
85. Kirchoff JF, Falasca M (2022) Environmental differentiation from a supply chain practice view perspective. Int J Prod Econ 244. https://doi.org/10.1016/j.ijpe.2021.108365
86. Kirk CP, Rifkin LS (2020) I'll trade you diamonds for toilet paper: consumer reacting, coping and adapting behaviors in the COVID-19 pandemic. J Bus Res 117:124–131. https://doi.org/10.1016/j.jbusres.2020.05.028
87. Koetter P, Pelton M, Gonzalo J, Du P, Exten C, Bogale K, Sciamanna C (2020) Implementation and process of a COVID-19 contact tracing initiative: leveraging health professional students to extend the workforce during a pandemic. Am J Infect Control 48(12):1451–1456. https://doi.org/10.1016/j.ajic.2020.08.012
88. Krishtel P, Malpani R (2021) Suspend intellectual property rights for covid-19 vaccines. BMJ 373. https://doi.org/10.1136/bmj.n1344
89. Kumar PS, Prasanth SM, Harish S, Rishikesh M (2021) Industrial water footprint: Case study on textile industries. In: Muthu SS (ed) Water footprint: assessment and case studies (pp 35–60). Springer Singapore. https://doi.org/10.1007/978-981-33-4377-1_2
90. Lackie K, Najjar G, El-Awaisi A, Frost J, Green C, Langlois S, Khalili H (2020) Interprofessional education and collaborative practice research during the COVID-19 pandemic: considerations to advance the field. J Interprof Care 34(5):583–586. https://doi.org/10.1080/13561820.2020.1807481
91. Lashley MA, Acevedo M, Cotner S, Lortie CJ (2020) How the ecology and evolution of the COVID-19 pandemic changed learning. Ecol Evol 10(22):12412–12417. https://doi.org/10.1002/ece3.6937

92. Leiva-Martinez M-A, Anderson-Seminario MdlM, Alvarez-Risco A, Estrada-Merino A, Mlodzianowska S (2021) Price variation in lower goods as of previous economic crisis and the contrast of the current price situation in the context of COVID-19 in Peru. In: Lawrence KD, Klimberg RK (eds.) Advances in business and management forecasting (Vol 14, pp 161–166). Emerald Publishing Limited. https://doi.org/10.1108/S1477-407020210000014011
93. Li X, Ren J, Wu Z, Wu X, Ding X (2021) Development of a novel process-level water footprint assessment for textile production based on modularity. J Clean Prod 291. https://doi.org/10.1016/j.jclepro.2021.125884
94. Liguori E, Winkler C (2020) From offline to online: challenges and opportunities for entrepreneurship education following the COVID-19 pandemic. Entrep Educ Pedagog 3(4):346–351. https://doi.org/10.1177/2515127420916738
95. Lim S, Prakash A (2021) Pandemics and citizen perceptions about their country: did COVID-19 increase national pride in South Korea? Nations Natl 27(3):623–637. https://doi.org/10.1111/nana.12749
96. Liu H, Alam MK, He M, Liu Y, Wang L, Qin X, Yu J (2021) Sustainable cellulose aerogel from waste cotton fabric for high-performance solar steam generation. ACS Appl Mater Interfaces 13(42):49860–49867. https://doi.org/10.1021/acsami.1c13362
97. Lopez-Odar D, Alvarez-Risco A, Vara-Horna A, Chafloque-Cespedes R, Sekar MC (2020) Validity and reliability of the questionnaire that evaluates factors associated with perceived environmental behavior and perceived ecological purchasing behavior in Peruvian consumers. Soc Responsib J 16(3):403–417. https://doi.org/10.1108/SRJ-08-2018-0201
98. Lovarelli D, Bacenetti J, Fiala M (2016) Water footprint of crop productions: a review. Sci Total Environ 548–549:236–251. https://doi.org/10.1016/j.scitotenv.2016.01.022
99. Luo Y, Wu X, Ding X (2022) Carbon and water footprints assessment of cotton jeans using the method based on modularity: a full life cycle perspective. J Clean Prod 332. https://doi.org/10.1016/j.jclepro.2021.130042
100. Maritz A, Perenyi A, de Waal G, Buck C (2020) Entrepreneurship as the unsung hero during the current COVID-19 economic crisis: Australian perspectives. Sustain 12(11). https://doi.org/10.3390/su12114612
101. Mekonnen MM, Hoekstra AY (2011) The green, blue and grey water footprint of crops and derived crop products. Hydrol Earth Syst Sci 15(5):1577–1600. https://doi.org/10.5194/hess-15-1577-2011
102. Mohamed A, Mehaweed HS (2021) Water footprint as a tool of water resources management–review. Egypt J Chem 64(12):2–3
103. Nabi BG, Mukhtar K, Arshad RN, Radicetti E, Tedeschi P, Shahbaz MU, . . . Aadil RM (2021) High-pressure processing for sustainable food supply. Sustain 13(24). https://doi.org/10.3390/su132413908
104. Negri M, Cagno E, Colicchia C, Sarkis J (2021) Integrating sustainability and resilience in the supply chain: a systematic literature review and a research agenda. Bus Strateg Environ 30(7):2858–2886. https://doi.org/10.1002/bse.2776
105. Nijdam D, Rood T, Westhoek H (2012) The price of protein: review of land use and carbon footprints from life cycle assessments of animal food products and their substitutes. Food Policy 37(6):760–770. https://doi.org/10.1016/j.foodpol.2012.08.002
106. Obergassel W, Hermwille L, Oberthür S (2021) Harnessing international climate governance to drive a sustainable recovery from the COVID-19 pandemic. Climate Policy 21(10):1298–1306. https://doi.org/10.1080/14693062.2020.1835603
107. Okereke M (2021) Towards vaccine equity: should big pharma waive intellectual property rights for COVID-19 vaccines? Public Health Pract 2. https://doi.org/10.1016/j.puhip.2021.100165
108. Oliveira Neto GCd, Cesar da Silva P, Tucci HNP, Amorim M (2021) Reuse of water and materials as a cleaner production practice in the textile industry contributing to blue economy. J Clean Prod 305:127075.https://doi.org/10.1016/j.jclepro.2021.127075
109. Palacios-Mateo C, van der Meer Y, Seide G (2021) Analysis of the polyester clothing value chain to identify key intervention points for sustainability. Environ Sci Eur 33(1):2. https://doi.org/10.1186/s12302-020-00447-x

110. Quian-Ulloa R, Stange C (2021) Carotenoid biosynthesis and plastid development in plants: the role of light. Int J Mol Sci 22(3). https://doi.org/10.3390/ijms22031184
111. Quispe-Cañari JF, Fidel-Rosales E, Manrique D, Mascaró-Zan J, Huamán-Castillón KM, Chamorro-Espinoza SE, Mejia CR (2021) Self-medication practices during the COVID-19 pandemic among the adult population in Peru: a cross-sectional survey. Saudi Pharm J 29(1):1–11. https://doi.org/10.1016/j.jsps.2020.12.001
112. Raudenská J, Steinerová V, Javůrková A, Urits I, Kaye AD, Viswanath O, Varrassi G (2020) Occupational burnout syndrome and post-traumatic stress among healthcare professionals during the novel coronavirus disease 2019 (COVID-19) pandemic. Best Pract Res Clin Anaesthesiol 34(3):553–560. https://doi.org/10.1016/j.bpa.2020.07.008
113. Rausch TM, Kopplin CS (2021) Bridge the gap: consumers' purchase intention and behavior regarding sustainable clothing. J Clean Prod 278. https://doi.org/10.1016/j.jclepro.2020.123882
114. Roesch E, Amin A, Gupta J, García-Moreno C (2020) Violence against women during covid-19 pandemic restrictions. BMJ 369. https://doi.org/10.1136/bmj.m1712
115. Rojas Román B, Moscoso S, Chung SA, Limpias Terceros B, Álvarez-Risco A, Yáñez JA (2020) Tratamiento de la COVID-19 en Perú y Bolivia y los riesgos de la automedicación [COVID-19; tratamiento; estrategia; fármacos; automedicación; Perú; Bolivia] 53(2). http://www.revfarmacia.sld.cu/index.php/far/article/view/435/351
116. Samanta KK, Pandit P, Samanta P, Basak S (2019) 3 - Water consumption in textile processing and sustainable approaches for its conservation. In: Muthu SS (ed) Water in textiles and fashion (pp 41–59). Woodhead Publishing. https://doi.org/10.1016/B978-0-08-102633-5.00003-8
117. Saxena A, Khare AK (2015) Development of green manufacturing system in Indian apparel industry. Systems Thinking Approach for Social Problems, New Delhi.
118. Sekalala S, Forman L, Hodgson T, Mulumba M, Namyalo-Ganafa H, Meier BM (2021) Decolonising human rights: how intellectual property laws result in unequal access to the COVID-19 vaccine. BMJ Glob Health 6(7). https://doi.org/10.1136/bmjgh-2021-006169
119. Shaw R, Kim Y-K, Hua J (2020) Governance, technology and citizen behavior in pandemic: lessons from COVID-19 in East Asia. Prog Disaster Sci 6. https://doi.org/10.1016/j.pdisas.2020.100090
120. Sigala M (2020) Tourism and COVID-19: impacts and implications for advancing and resetting industry and research. J Bus Res 117:312–321. https://doi.org/10.1016/j.jbusres.2020.06.015
121. Statista (2021) Share in world exports of the leading clothing exporters in 2020, by country. Retrieved 01/01/2022 from https://www.statista.com/statistics/1094515/share-of-the-leading-global-textile-clothing-by-country/#:~:text=In%202020,%20China%20was%20the,textile%20exporters%20in%20the%20world
122. Sánchez OR, Vale DB, Rodrigues L, Surita FG (2020) Violence against women during the COVID-19 pandemic: an integrative review. Int J Gynecol Obstet 151(2):180–187. https://doi.org/10.1002/ijgo.13365
123. Sánchez-Clavijo LM, Martínez-Callejas SJ, Acevedo-Charry O, Diaz-Pulido A, Gómez-Valencia B, Ocampo-Peñuela N, Ochoa-Quintero JM (2021) Differential reporting of biodiversity in two citizen science platforms during COVID-19 lockdown in Colombia. Biol Cons 256. https://doi.org/10.1016/j.biocon.2021.109077
124. Tayyab M, Jemai J, Lim H, Sarkar B (2020) A sustainable development framework for a cleaner multi-item multi-stage textile production system with a process improvement initiative. J Clean Prod 246. https://doi.org/10.1016/j.jclepro.2019.119055
125. Thompson AL, Nicholas KM, Watson E, Terán E, Bentley ME (2020) Water, food, and the dual burden of disease in Galápagosm, Ecuador. Am J Hum Biol 32(1). https://doi.org/10.1002/ajhb.23344
126. Tilman D, Clark M (2014) Global diets link environmental sustainability and human health. Nature 515(7528):518–522. https://doi.org/10.1038/nature13959
127. Tran LTT (2021) Managing the effectiveness of e-commerce platforms in a pandemic. J Retail Consum Serv 58. https://doi.org/10.1016/j.jretconser.2020.102287

128. Vanham D, Bidoglio G (2013) A review on the indicator water footprint for the EU28. Ecol Ind 26:61–75. https://doi.org/10.1016/j.ecolind.2012.10.021
129. Vidya CT, Prabheesh KP (2020) Implications of COVID-19 pandemic on the global trade networks. Emerg Mark Financ Trade 56(10):2408–2421. https://doi.org/10.1080/1540496X.2020.1785426
130. Viero A, Barbara G, Montisci M, Kustermann K, Cattaneo C (2021) Violence against women in the Covid-19 pandemic: a review of the literature and a call for shared strategies to tackle health and social emergencies. Forensic Sci Int 319. https://doi.org/10.1016/j.forsciint.2020.110650
131. Viglizzo EF, Frank FC, CarreÑO LV, JobbÁGy EG, Pereyra H, Clatt J, Ricard MF (2011) Ecological and environmental footprint of 50 years of agricultural expansion in Argentina. Glob Change Biol 17(2):959–973. https://doi.org/10.1111/j.1365-2486.2010.02293.x
132. de Vries M, de Boer IJM (2010) Comparing environmental impacts for livestock products: a review of life cycle assessments. Livest Sci 128(1):1–11. https://doi.org/10.1016/j.livsci.2009.11.007
133. Wang L, Ding X, Wu X (2013) Blue and grey water footprint of textile industry in China. Water Sci Technol 68(11):2485–2491. https://doi.org/10.2166/wst.2013.532
134. Wang D, Hubacek K, Shan Y, Gerbens-Leenes W, Liu J (2021) A review of water stress and water footprint accounting. Water 13(2). https://doi.org/10.3390/w13020201
135. Wen J, Kozak M, Yang S, Liu F (2021) COVID-19: potential effects on Chinese citizens' lifestyle and travel. Tour Rev 76(1):74–87. https://doi.org/10.1108/TR-03-2020-0110
136. Wu T-Y, Ford O, Rainville AJ, Bessire R (2020) COVID-19 care package distribution for senior citizens and families in Detroit and Hamtramck, Michigan. J Hunger & Environ Nutr 15(4):585–587. https://doi.org/10.1080/19320248.2020.1799586
137. Wu G-H, Wu X-Y, Wang L, Liu S-Q, Ding X, Cuc S (2015) Water footprint and carbon footprint reduction in textile's waste recycling. DE REDACTIE 85.
138. Xing G, Duan Z, Yan W, Baykal-Gürsoy M (2021) Evaluation of "innovation chain + supply chain" fusion driven by blockchain technology under typical scenario. Int J Prod Econ 242. https://doi.org/10.1016/j.ijpe.2021.108284
139. Yan J, Kim S, Zhang SX, Foo M-D, Alvarez-Risco A, Del-Aguila-Arcentales S, Yáñez JA (2021) Hospitality workers' COVID-19 risk perception and depression: a contingent model based on transactional theory of stress model. Int J Hosp Manag 95. https://doi.org/10.1016/j.ijhm.2021.102935
140. Yasin Ar A (2020) Managing e-commerce during a pandemic: lessons from GrubHub during COVID-19. In: George B, Mahar Q (eds) International case studies in the management of disasters (pp 155–167). Emerald Publishing Limited. https://doi.org/10.1108/978-1-83982-186-820201010
141. Yáñez JA, Afshar Jahanshahi A, Alvarez-Risco A, Li J, Zhang SX (2020) Anxiety, distress, and turnover intention of healthcare workers in Peru by their distance to the epicenter during the COVID-19 crisis. Am J Trop Med Hyg 103(4):1614–1620. https://doi.org/10.4269/ajtmh.20-0800
142. Yáñez JA, Alvarez-Risco A, Delgado-Zegarra J (2020) Covid-19 in Peru: from supervised walks for children to the first case of Kawasaki-like syndrome. BMJ 369. https://doi.org/10.1136/bmj.m2418
143. Zarocostas J (2021) What next for a COVID-19 intellectual property waiver? The Lancet 397(10288):1871–1872. https://doi.org/10.1016/S0140-6736(21)01151-X
144. Zhan F, Youssef M, Shah BR, Li J, Li B (2022) Overview of foam system: Natural material-based foam, stabilization, characterization, and applications. Food Hydrocoll 125. https://doi.org/10.1016/j.foodhyd.2021.107435
145. Zhang SX, Chen J, Afshar Jahanshahi A, Alvarez-Risco A, Dai H, Li J, Patty-Tito RM (2021) Succumbing to the COVID-19 pandemic—healthcare workers not satisfied and intend to leave their jobs. Int J Ment Heal Addict. https://doi.org/10.1007/s11469-020-00418-6
146. Zhang Q, Wiedmann T, Fang K, Song J, He J, Chen X (2022) Bridging planetary boundaries and spatial heterogeneity in a hybrid approach: a focus on Chinese provinces and industries. Sci Total Environ 804. https://doi.org/10.1016/j.scitotenv.2021.150179

147. Zhang SX, Sun S, Afshar Jahanshahi A, Alvarez-Risco A, Ibarra VG, Li J, Patty-Tito RM (2020) Developing and testing a measure of COVID-19 organizational support of healthcare workers – results from Peru, Ecuador, and Bolivia. Psychiatry Res 291. https://doi.org/10.1016/j.psychres.2020.113174
148. Zhou Y, Ma J, Zhang Y, Qin B, Jeppesen E, Shi K, Gao G (2017) Improving water quality in China: environmental investment pays dividends. Water Res 118:152–159. https://doi.org/10.1016/j.watres.2017.04.035
149. Zollinger R, DiCindio C (2021) Community ecology: museum education and the digital divide during and after COVID-19. J MusM Educ 46(4):481–492. https://doi.org/10.1080/10598650.2021.1983711

Virtual Tourism, Carbon Footprint, and Circularity

Myreya De-la-Cruz-Diaz, Aldo Alvarez-Risco, Micaela Jaramillo-Arévalo, Maria F. Lenti-Dulong, Marco Calle-Nole, Maria de las Mercedes Anderson-Seminario, and Shyla Del-Aguila-Arcentales

Abstract This chapter aims to analyze the tourism sector globally and the creation of virtual tourism. It is presented along with the previous context of the sector and the damage made by the COVID-19 pandemic. This lens shows how virtual tourism emerged as an alternative to traditional tourism while people worldwide were forced to stay home during the uncertainty of the pandemic. Virtual tourism would be related to the arising environmental tendencies expected to be embraced by the industry companies to make tourism a more sustainable economic activity and, therefore, reduce its carbon footprint through circularity. Finally, it also analyses the possibility of its per durability once the consequences and lags of the pandemic are solved.

Keywords Virtual tourism · Circular economy · Circularity · Sustainability · Carbon footprint

1 Introduction

Due to COVID-19, tourism was one of the sectors most affected because of social distancing. Fortunately, it also became an opportunity for companies to innovate and rebuild themselves.

M. De-la-Cruz-Diaz · A. Alvarez-Risco (✉) · M. Jaramillo-Arévalo · M. F. Lenti-Dulong · M. Calle-Nole · M. de las Mercedes Anderson-Seminario
Universidad de Lima, Lima, Perú
e-mail: aralvare@ulima.edu.pe

S. Del-Aguila-Arcentales
Escuela Nacional de Marina Mercante "Almirante Miguel Grau", Callao, Perú
e-mail: sdelaguila@enamm.edu.pe

A. Alvarez-Risco et al. (eds.), *Circular Economy*, Environmental Footprints and Eco-design of Products and Processes,
https://doi.org/10.1007/978-981-19-0549-0_12

This chapter presents previous scientific articles about the impact of tourism on climate change, greenhouse gas emissions, and the circularity of said industry. Next, the most relevant antecedents to carry out this investigation are presented. The world has changed and people's habits regarding sustainability have changed, depending on the characteristics of pre-pandemic life [14, 17, 18, 20, 23, 26, 30, 32, 33, 58, 60, 64, 92, 103]. In Table 1, it is presented different sectors that have been damaged by COVID-19.

Rastati [100] analyses the impact of virtual tourism in times of COVID-19 in Indonesia and its impact in different areas. For this, a qualitative study was carried out to know the experiences of the people and workers of the sector participating in virtual tourism during the COVID-19 pandemic. The methodology used consisted of conducting interviews with these groups of people and asking them about their opinions about this new activity and what they considered to be the advantages and disadvantages of virtual tourism. Research results demonstrated that virtual tourism could be a starting point as an alternative to traditional tourism during and after the COVID-19 pandemic ends. In addition, the people interviewed stated that although some experiences of traditional tourism (such as trying local food) were lost, the virtual one turned out to be more accessible, motivated the interviewees to know the attractions after the end of the world situation, and generated less environmental impact than ordinary tourism. However, it was highlighted that for these activities to be carried out correctly, the government must provide advanced technology, such as virtual reality and panoramic images.

Kitamura et al. [80] investigated the carbon footprint in the tourism sector in Japan with a view to more sustainable tourism because this is a critical issue to act in all sectors, which also has relation with the goals of Japan to contribute with the Sustainable Development Goals. A life cycle assessment (LCA) based on processes and an input and output analysis was agreed upon; it was calculated based on tourism consumption using a Japanese input and output table, and the industry, in general, showed that total emissions were around 136 million tCO_2 per year. Regarding the

Table 1 Impact of COVID-19 in different sectors

Education	[9, 10, 19, 22, 24, 54, 83, 84, 130]
Entrepreneurship	[4, 16, 27, 28, 39, 42, 43, 47, 52, 86, 90]
Health sector	[1, 8, 11, 13, 21, 40, 45, 55, 61, 81, 101, 104, 125, 128, 129]
Hospitality	[63, 70, 73, 77, 124]
Intellectual property	[12, 15, 59, 65, 76, 82, 98, 107, 127]
Population	[6, 25, 51, 71, 87, 99, 105, 108, 120, 123, 125]
Prices	[36, 56, 68, 72, 85, 97]
Tourism	[38, 41, 46, 48–50, 67, 109]
Trade	[2, 3, 5, 29, 31, 44, 57, 79, 88, 113, 118, 126]
Violence against women	[34, 53, 62, 69, 102, 106, 119]

relationship by percentages according to sectors, it was determined that mainly transportation represents 56.3%, souvenirs 23.2%, gasoline 16.9% and accommodation 9.8%, food and beverages 7.5% and activities 3%.

Lu et al. [89] investigated the potential of virtual tourism in the tourism industry's recovery during the COVID-19 pandemic in China. For the development of the study, they used the Theory of Planned Behavior and qualitative and quantitative studies. The methodology used was questionnaires and interviews with samples of residents of China. Different statistical measures were used for the quantitative research, such as the Wilcoxon test, non-parametric methods, and binary logistic regression, and in the qualitative research, thirty interviews were conducted with Chinese residents. The findings of this study revealed that it is necessary to promote virtual tourism to museums and other indoor attractions to include this activity as part of their services and products offered to the public. In addition, to achieve a more significant amount of income from the tourism industry (especially during the pandemic), tourist places should develop new ways of using the internet to promote their new virtual tourism services. It should be noted that several interviewees mentioned that virtual tourism could help promote sustainability, accessibility, and diversity in the sector.

To determine the global impact of tourism activities, an environmental assessment that covers the entire life cycle of what is consumed by tourists has been identified as an appropriate methodology. WTTC [121] conducted a study to examine the potential for saving greenhouse gas emissions through 3 different strategies: prevention of food waste, reduction of single-use plastics, and recollection and recycling of waste. The general methodology used for the article was the analysis of databases regarding the waste of tourists and their accommodations, differentiated by their top-down or bottom-up approaches. The results were expressed in two measures: per kilogram of waste that could be avoided and per thousand tourists. The strategies presented demonstrated the potential savings in greenhouse gas emissions between 4 and 189 kg of CO_2 equivalents per 1000 tourists. The measures to reduce food waste and the collection and recycling of waste showed low emissions; however, the emissions associated with the single-use plastic reduction measure were relatively high.

O'Grady et al. [96] analyze how the construction industry generates the most significant percentage of waste. The methodology is based on an applied index to our purpose-built prefabricated building. The 3DR method illustrates the steps and decision-making that the research results show that building refurbishment is possible through the 83% of walls and ceilings that can be disassembled during the operation phase. The UNEP circularity platform provides an understanding of the circularity concept, its scope, and its contribution to promoting sustainable consumption and production patterns. As outlined by Inger Andersen, UNEP Executive Director, circularity and sustainable consumption and production are essential to delivering on every multilateral agreement, from the Sustainable Development Goals to the Paris Agreement to the post-2020 global biodiversity framework. They are essential to a sustainable recovery from the COVID-19 pandemic. Circularity builds upon value retention loops, as shown on the UNEP circularity.

2 Tourism Background

The tourism industry has been one of the most important contributors to the global economy for a long time. UNWTO [115] pointed out that as of 2018, this sector had nine consecutive years of sustained growth, improving people's lives all over the world. The importance of tourism is notable in certain countries, especially for those from America and the Asia–Pacific region. According to the WTTC [122], the total GDP contribution in 2019 was 10.4% generating about USD9,170 BN and 334 MN jobs. Moreover, during 2014–2019 about 1 in 4 new jobs were created by this industry alone.

The outburst of the coronavirus pandemic has dramatically changed the world. What once was a growing industry with high expectations was shocked by the sudden stop of people flowing worldwide. The measures taken by every government to reduce the spread of the virus were severe and progressively longer, such as the lockdown, close of borders, and curfew. Therefore, all nations experienced a period of uncertainty about when it would be safe to get back outside their houses and go back to how things were before the outbreak. Lots of lives were lost, and multitudes were fired from their jobs. The economy was severely wounded, some industries more than others, especially tourism, where interaction was crucial. This sector stopped all its activities in April 2020, leading just in said year to the loss of about 62 million jobs, representing a reduction of 18.5% [122].

The governments took several measures to reduce the impact, such as financial aid provided to tourist companies but still, the travel and tourism sector suffered a loss of 4.5 trillion dollars, a significant drop of about 49.1% compared to 2019. The countries whose economy depends more on tourism were the most affected. According to Statista [111], Fig. 1 represents the contribution of travel and tourism to the GDP of the world's major economies in 2019.

In addition, there has been a significant variation in the GDP contributed per country at a global level comparing 2020 and 2019. According to the WTTC [122], the average contribution per area during 2019 was 9.64%, however, during 2020, the average was 5.14%, nearly half of 2019's contribution. As we can convey from Fig. 1, it represents a decrease of 48.3% on average in the contribution of each geographical area. In 2019, the area with the smallest share in the whole GDP reached 6.6% (South Asia); during 2020, the smallest area (Africa) contributed only 3.7% of the GDP. It can also be noticed that the Caribbean region was the most affected, with a decrease of 58% from one year to another (Fig. 2).

A vivid example of the adverse effects of COVID-19 on tourism can be seen in the Dominican Republic, a country situated in the Caribbean, the most affected geographical area according to the WTTC. The country's strength in the services sector before the pandemic was an essential factor in the recovery of its economy. Nevertheless, it still affected the country by shrinking its GDP by 6.7% in 2020 [74]; with the vital measures applied by the government, many jobs were lost, especially those related to tourism. The authorities decided to implement social assistance programs for those who lost their jobs and others directly affected to face the situation.

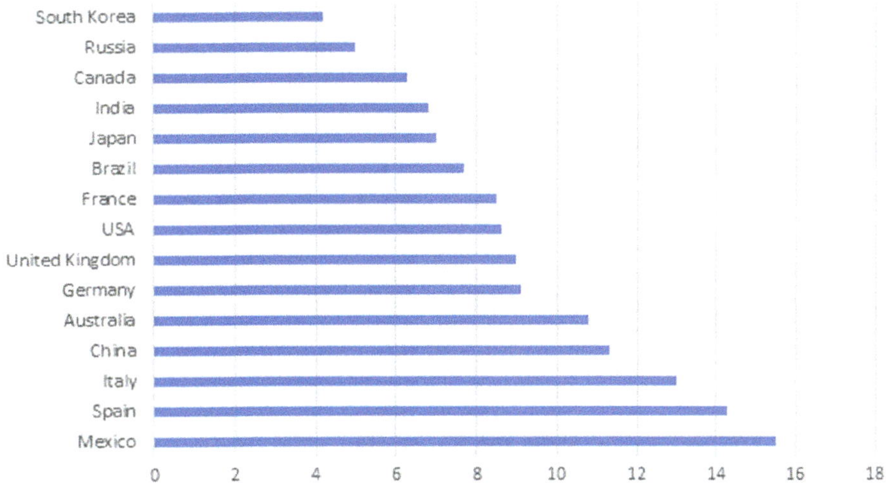

Fig. 1 Countries that are more dependent on the tourism industry (*Source* Adapted from Statista [111])

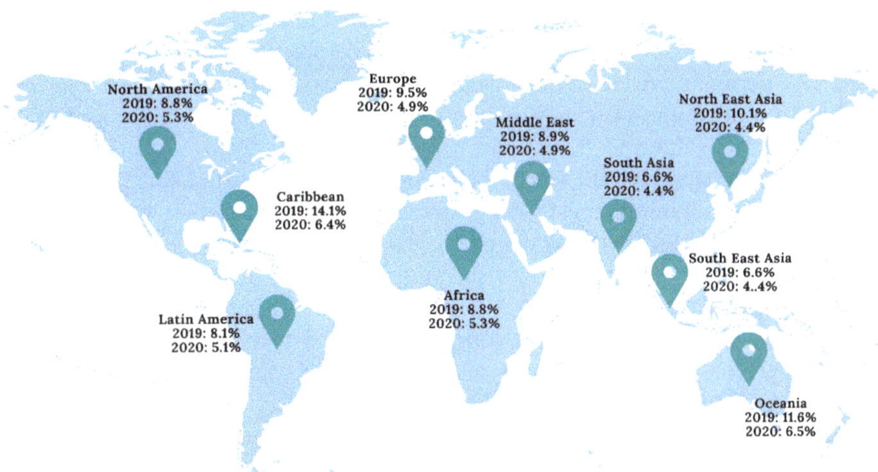

Fig. 2 Contribution in global GDP by geographical region (*Source* Adapted from WTTC [122])

Another exciting example is India. Before the pandemic, this country was an emblematic crowded touristic nation frequented because of religious and non-religious reasons that happened to be critically injured by the restrictions apply mainly for foreign tourists. Even though this country is in the south of Asia, the geographical area least affected by the pandemic, it was still severely shocked by the closure of borders and lockdown. Just in 2020, the government reduced its expenditure on travel and tourism by 30.7%, a significant sector in India's economy [122].

Luckily, in mid-October of 2021, the government decided it was time to reopen borders, and it is expected to recover quickly.

Despite the great efforts to go back to "normal" or the pre-pandemic habits, the pandemic is still far from over. The governments of all nations, especially those more dependent on tourism, face a significant challenge for recovering the industry and retaking the previously established objectives while creating new ways for travelers to experience the beautiful locations without endangering their health. With the help of foreign direct investment, government leaders are already implementing measures to recover the sector, such as inoculation programs and the progressive reopening of borders. These measures are showing their effectiveness and giving the expected results so far.

3 Virtual Tourism: A New Opportunity

Due to social distancing and people being restricted to stay at home, the already rising environmentalist tendencies increase, even more, influencing how people act and think while using a specific product or service. The tourism sector was no exception and started to notice the multiple ways it contributed to the world's contamination, starting to worry about their carbon footprint while preparing themselves to adapt to the new normality. A fair number of companies started to realize that it was time for them to make relevant changes that may be over the budget but would allow them to be sustainable and green in the long term. And although it is essential that companies are concerned and accept the necessity to take a sustainable path, people from the local communities must accept this too and push both tourists and companies to do their part, mainly because the changes would affect everyone. According to the UNWTO [116], sustainable tourism must include an optimum use of natural resources, respect host communities' social and cultural authenticity, and ensure that the activities are viable in the long term. Virtual tourism gives new options to consumers out of their new necessities and protects the world everyone lives in. According to Stainton [110], virtual tourism is the activity of "touring" various locations without traveling to them and using innovative technologies to feel as if you are there in real life. Many new technologies have been created to develop this venture, taking advantage of this tourist's new interest. It is related to the new tourism trends that have emerged from the COVID-19 pandemic, which created some restrictions in the industry, leading companies to innovate and try new ways to keep tourism going. Some of the technologies used by these businesses are virtual reality, augmented reality, 360°, and holograms (Fig. 3).

By using these new resources, tourism companies can provide new experiences and make virtual tourists feel they are in a foreign location, appreciating its beauty and features while being far away. To some people, virtual tourism may not be the best option since it may not give them the experience they are looking for when trying to meet different cultures. According to Mura et al. [94], people perceive virtual tourism lacks the authenticity factor because when participating in a virtual tour, you cannot

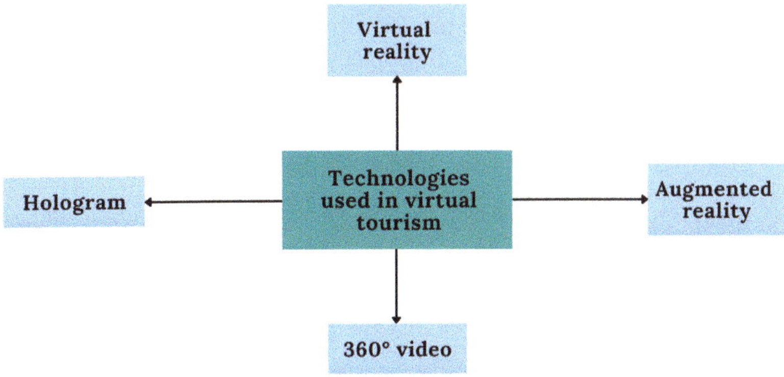

Fig. 3 Technologies used in virtual tourism (*Source* Adapted from Stainton [110])

become familiar with the atmosphere, the weather, the local language, and overall, all of the sensorial experience is lost.

Virtual tourism is mainly directed towards people that want to know more about the country but not necessarily travel there, probably because of the lack of means or time. This type of activity could be unconventional compared to natural tourism to some people and may bring skepticism; even so, this kind of service must be directed at people who want to appreciate, learn, and understand the cultural part. It can also be a terrific start to have an insight into the destination before a trip is booked, and based on the findings, plan the itinerary according to the locations and activities that appear to be the most compelling.

Virtual tourism can come in handy if you have a low budget and want to see various places. After having experienced some trips virtually, one could save money, smartly plan an itinerary, and finally book the actual trip of your dreams, knowing what to expect when you get there. In contrast, the segment of people that would not consume this type of tourism would be those interested in more adventurous activities and in an interpersonal connection with the locals of that country or space; for those people, a virtual tour would be boring and far from enough. Although these people only use virtual tours to complement their authentic tours, they may only cause more contamination, an action that would end up defeating the primary purpose.

Regardless of how much or less captivating the tourist might find virtual tourism, the reality is that it has changed the way tourists and tourism workers see this industry. For example, many touristic places are guided by the people who live there, so while implementing virtual tours would increase the number of people in each group, it would also decrease the tour guides needed in every expedition as a recording can be used in its place. It also would affect countless jobs generated by this sector. All the benefits obtained through virtual tourism can generate another way to contaminate and increase the carbon footprint. It's important to mention that being viable in the long term means that if virtual tourism is promoted, companies must ensure that this can be used after the pandemic, and it can sustain in the future because people do not accept this new way of traveling.

France is a strong example of the new opportunities provided by virtual tourism. The information reviewed by France Diplomacy [93], this country remains the first world destination with a record influx of between 88 and 89 million foreign visitors, and tourism is a crucial sector for the French economy with a presence of about 8% of GDP. It is at this point and in the wake of the COVID-19 pandemic that the European Union member country has sought new opportunities to be able to create tourism; clear examples of this are the virtual sessions of the museum of Le Havre, the Museum of Fine Arts in Rouen, the Castle Museum of Dieppe and many more. Italy is another example of countries whose industry has successfully adapted to virtual tourism. ItalyGuides [75] express that there are at least 20 different destinations that anyone can visit through the internet from the comfort of their sofas. These destinations include the eternal city, Roma, Florence, Verona, and Venice, all of them are possible to reach by audio guides and unique apps for every technological device. Blogs such as Headout include views from attractions such as the Colosseum and the Vatican with 360° tours.

Peru, despite being a developing country, according to Martínez [91], the Ministry of Foreign Trade and Tourism during the pandemic has promoted various activities so that companies and tour guides can use different technologies (virtual visitors through applications such as Zoom, Google Meet or 360º views, Google Maps, special effects, and even music) to develop tourism further but virtually. To give a more panoramic view, we can see that these new technologies are being developed in the Uros Islands, Manu National Park, Huascaran National Park, etc.

4 Virtual Tourism and Carbon Footprint

Therefore, with virtual tourism, another factor can be calculated. The Carbon Footprint is "the measure of the impact of all greenhouse gases produced by our activities (individual, collective, eventual, and from products) on the environment," according to the United Nations [114]. It is measured in tons of kilos of carbon dioxide equivalent to greenhouse gases. In the Earth's atmosphere, the main greenhouse gases are carbon dioxide (CO_2), water vapor (H_2O), nitrous oxide (N_2O), methane (CH_4), and ozone (O_3). Currently, companies carry out this calculation to identify and reduce contamination levels of the different production processes, engage employees in environmental issues and promote a product. It is a differentiating element that can satisfy the demands of a global society with greater environmental awareness. An analysis of the necessary activities to obtain the product must be carried out to know the carbon footprint, from the moment the materials are acquired for its preparation until its management to be reduced [114].

Tourism generates approximately 8% of the total carbon emissions in the world. They are mainly created due to boat rides and plane travels to the world's leading tourism hubs. The main activities that contribute to the total carbon footprint of tourism are Transport (49%), Goods (12%), Food and beverage (10%), Agriculture

(8%), Services (8%), Lodging (6%), Construction and mining (6%) and others (1%) [112].

Since the COVID-19 pandemic started, as mentioned before, all the tourism industries had to stop their activities for an unestablished period and wait until the cases started to reduce and find new ways to keep generating revenue. During this time that the routine activities stopped, the carbon footprint reduced its growth since travels by airplane or sea were stopped for a long time until the governments of the countries created new measures to do these activities in a safe way to prevent more contagions of COVID-19.

One sector that generates the most carbon emissions in tourism is transportation. Aviation is responsible for approximately 12% of the carbon emissions in all the transport sectors in the world. According to the Air Transport Action Group [7], in 2019, flights worldwide produced 915 million tons of carbon dioxide into the atmosphere. The tourism sector is tightly linked with the aviation sector because most people travel by plane to their destinations since it is the fastest way to do so. The COVID-19 pandemic has generated a considerable reduction in plane travels due to the restrictions imposed by governments of different countries around the world, which has had an enormous impact on the generation of carbon emissions. The airplanes generate many outflows of this chemical into the atmosphere, causing damage to the environment. However, with the reduction of traveling with this method, these emissions have been reduced considerably.

With the implementation of virtual tourism, many airline companies could see a decline in passengers because some people are still a bit scared of catching COVID-19 and do not want to be in a twelve-hour long plane and risk catching the virus rather than experience their destination from the safety of their homes. Also, for those who only want to know the place and see it, not necessarily in person, it would be much cheaper to pay for a virtual tour rather than a plane ticket. Another thing to keep in mind is that virtual tourism may affect local communities that work in tourism sites such as hotels, restaurants, and transport since people are now going straight to their virtual destination and do not need any of these services.

According to Aminy [35], countries such as Indonesia that have experienced substantial tourism lots (88.9% drop rate in tourism in 2020) have innovated and started offering virtual tours. Websites such as Traveloka have released more than 10,000 virtual tours that people can choose and take from their homes. However, being a country that is very dependent on tourism, these virtual tours could negatively impact the local communities since they are not receiving tourists that they can offer their services to, generating losses in these sectors. It is essential for governments, not only in Indonesia but also around the world, to find ways to try to incorporate local businesses that benefit from tourism into this new reality so that countries that depend heavily on tourist activities can still generate some income in this new global context.

However, not all these effects of virtual tourism are still in the future. With the COVID-19 pandemic coming to an end, it is hard to know if people still log into virtual tours when they can buy a plane ticket and head there in person. Another essential thing [95] is that virtual tours take away some aspects of regular travel, such as the foreign environment and culture that you experience for the first time when going to a new place. Some people may feel that if they take a virtual tour, it takes away all the illusion of seeing a new country or city in person since they have already experienced it virtually, and thus, the "wow" factor gets lost.

After the pandemic ends, many people usually start traveling again. The environmental benefits generated by the lack of international travel and the new implementation of virtual tours could be reversed and end up like the pre-pandemic times. People need to start to take conscience about the environmental impact of traveling and start taking actions to reduce their carbon footprint impact.

In recent years, many environmental-related topics such as climate change have been very prominent among people, and now there is more awareness of environmental care and what to do to prevent long-term effects. Because of this, virtual tourism may be one of the choices that people take when deciding to meet a new culture or country since they are aware of the benefits that this type of activity has on the environment and help reduce their carbon footprint in the long term, so is undeniable that some tourism companies may still have virtual tours as part of their services after the pandemic comes to an end.

Virtual tourism can be used as a new form for people to experience tourism through the internet. It is well known that while being abroad, there is hardly enough time to visit all the available cultural places. So instead of trying to visit all the places in one day, tourists can choose to visit it from their homes before or after their trips, this would save a lot of money, energy and most importantly it reduces the carbon footprint generated with transportation, food, and others. It could also be an excellent opportunity for travelers with a meager budget that want to know all the culture from the place but can't afford to visit; virtual tourism can be a new way to enjoy touristic experiences.

5 Virtual Tourism and Circularity

The tourism industry has been most affected by the COVID-19 pandemic. Mainly because mobility restrictions made it difficult to carry out the different activities that make up the tourism value chain and severely damaged tourism results, but it is in this context. With the growing awareness of creating a new economic system that helps the environment and does not exceed the global ecological limit, new ways of doing tourism have opened. In this change, virtual tourism emerges among the ways of adapting to the new reality. It should be noted that according to Aprende de Turismo [37], the economic activity model of virtual tourism is still linear, it is a model that depletes resources and generates too much waste, which does not make it environmentally sustainable. For this very reason, and to reduce the adverse effects of this model on tourism, it seems necessary to seek a new, more integrated production model, through the conversion of "systems," in which the cost of the production-environmental damage-product price equation is optimized, thus generating greater profits. This model would be based on circularity.

According to Vargas [117], circularity or circular economy applied to virtual tourism is a model created to replace the linear economy, based on the following sequence: obtain resources from nature, manufacture them, use or consume them and dispose of them. The circular model tries to keep resources in the economy if possible, preventing them from becoming waste, in this way, it contributes to the regeneration of the environment. Its method can be proactive (to avoid waste) or, if it fails, passive (to manage it by recycling). Due to the negative impact of waste—accumulated in landfills—on the environment, increasingly scarce resources, and rising costs.

The dilemma is how to apply circularity in virtual tourism; in other words, create circular tourism. As stated by Estévez [66], circular tourism is the tourism model that allows not only the protection of cultural and natural heritage through the reduction of resource extraction and the reduction of negative externalities but also the regeneration of the natural capital of the territory, which allows having leading tourist destinations in quality and innovation.

Virtual tourism is essential to help with this work since this type of tourism can be used as a lever to introduce the circular economy to the tourism market more efficiently, which is because virtuality is a way to avoid the wear and tear of the cultural heritage of different countries; it also avoids the transit and pollution of the mobilization of so many people that occurs in tourism. In addition, virtual tours can become a realistic experience where visitors can get to experience more about their destination and help people who cannot travel (time or financial constraints).

Closing Remarks

But despite all the benefits that can be achieved with circularity in virtual tourism, a few aspects may not be favorable for the circularity chain. Virtual tourism involves the constant use of new technology trends, whether computers, cell phones, tablets, virtual reality goggles, or others, which produce a large amount of electronic waste. According to a United Nations (UN) report, Kenny [78] quoted, the world recorded 53.6 million tons of e-waste in 2019. E-waste (discarded products that use a battery or a plug) increased by 21% in the last five years and is predicted to reach 74 million tons by 2030, almost double the e-waste of 16 years. For these reasons, applying circularity is imperative also in virtual tourism because it allows the industry to reduce the level of consumption of resources and waste products in the process of production, circulation, and consumption. Additionally, it is reusing products directly, using them as a new product after waste once modified, restored, or renewed, or using waste as a component of other products. And finally, recycling is where waste is used as a direct raw material or recycled waste.

References

1. Abiakam N, Worsley P, Jayabal H, Mitchell K, Jones M, Fletcher J, Spratt F, Bader D (2021) Personal protective equipment related skin reactions in healthcare professionals during COVID-19. Int Wound J 18(3):312–322. https://doi.org/10.1111/iwj.13534
2. Abudureheman A, Nilupaer A (2021) Optimization model design of cross-border e-commerce transportation path under the background of prevention and control of COVID-19 pneumonia. Soft Comput 25(18):12007–12015. https://doi.org/10.1007/s00500-021-05685-6
3. Acuña-Zegarra MA, Santana-Cibrian M, Velasco-Hernandez JX (2020) Modeling behavioral change and COVID-19 containment in Mexico: a trade-off between lockdown and compliance. Math Biosci 325:108370. https://doi.org/10.1016/j.mbs.2020.108370
4. Afshan G, Shahid S, Tunio MN (2021) Learning experiences of women entrepreneurs amidst COVID-19. Int J Gend Entrep 13(2):162–186. https://doi.org/10.1108/IJGE-09-2020-0153
5. Aguirre AA, Catherina R, Frye H, Shelley L (2020) Illicit wildlife trade, wet markets, and COVID-19: preventing future pandemics. World Med Health Policy 12(3):256–265. https://doi.org/10.1002/wmh3.348
6. Ahsan MM (2020). Strategic decisions on urban built environment to pandemics in Turkey: lessons from COVID-19. J Urban Manag 9(3):281–285. https://doi.org/10.1016/j.jum.2020.07.001
7. Air Transport Action Group (2021) Facts & figures. Retrieved 22 December 2021 from https://www.atag.org/facts-figures.html
8. Alan H, Eskin Bacaksiz F, Tiryaki Sen H, Taskiran Eskici G, Gumus E, Harmanci Seren AK (2021) "I'm a hero, but...": an evaluation of depression, anxiety, and stress levels of frontline healthcare professionals during COVID-19 pandemic in Turkey. Perspect Psychiatr Care 57(3):1126–1136. https://doi.org/10.1111/ppc.12666
9. Ali W (2020) Online and remote learning in higher education institutes: a necessity in light of COVID-19 pandemic. High Educ Stud 10(3):16–25
10. Allen J, Rowan L, Singh P (2020) Teaching and teacher education in the time of COVID-19. Asia-Pac J Teach Educ 48(3):233–236. https://doi.org/10.1080/1359866X.2020.1752051
11. Alsairafi Z, Naser AY, Alsaleh FM, Awad A, Jalal Z (2021) Mental health status of healthcare professionals and students of health sciences faculties in Kuwait during the COVID-19 pandemic. Int J Environ Res Public Health 18(4). https://doi.org/10.3390/ijerph18042203
12. Altindis E (2021) Inequitable COVID-19 vaccine distribution and the intellectual property rights prolong the pandemic. Expert Rev Vaccines. https://doi.org/10.1080/14760584.2022.2014819
13. Alvarez-Risco A, Dawson J, Johnson W, Conteh-Barrat M, Aslani P, Del-Aguila-Arcentales S, Diaz-Risco S (2021) Ebola virus disease outbreak: a global approach for health systems [Ebola; health professionals; global approach; pharmacist: pharmaceutical care; COVID-19]. 2021 53(4). http://www.revfarmacia.sld.cu/index.php/far/article/view/491
14. Alvarez-Risco A, Del-Aguila-Arcentales S (2015) Errores de prescripción como barrera para la Atención Farmacéutica en establecimientos de salud públicos: Experiencia Perú. Pharm Care España 17(6):725–731. https://www.pharmcareesp.com/index.php/PharmaCARE/article/view/246
15. Alvarez-Risco A, Del-Aguila-Arcentales S (2021a) A note on changing regulation in international business: the World Intellectual Property Organization (WIPO) and artificial intelligence. In: Verbeke A, van Tulder R, Rose EL, Wei Y (eds) The multiple dimensions of institutional complexity in international business research, vol. 15. Emerald Publishing Limited, pp 363–371. https://doi.org/10.1108/S1745-886220210000015020
16. Alvarez-Risco A, Del-Aguila-Arcentales S (2021b) Public policies and private efforts to increase women entrepreneurship based on STEM background. In: Mari M, Poggesi S, Foss L (eds) Women's entrepreneurship in STEM disciplines: issues and perspectives. Springer, pp 75–87. https://doi.org/10.1007/978-3-030-83792-1_5

17. Alvarez-Risco A, Del-Aguila-Arcentales S, Delgado-Zegarra J, Yáñez JA, Diaz-Risco S (2019) Doping in sports: findings of the analytical test and its interpretation by the public. Sport Sci Health 15(1):255–257. https://doi.org/10.1007/s11332-018-0484-8
18. Alvarez-Risco A, Del-Aguila-Arcentales S, Diaz-Risco S (2018) Pharmacovigilance as a tool for sustainable development of healthcare in Peru. PharmacoVigilance Rev 10(2):4–6
19. Alvarez-Risco A, Del-Aguila-Arcentales S, Rosen MA, García-Ibarra V, Maycotte-Felkel S, Martínez-Toro GM (2021) Expectations and interests of university students in COVID-19 times about Sustainable Development Goals: evidence from Colombia, Ecuador, Mexico, and Peru. Sustain 13(6):3306. https://doi.org/10.3390/su13063306
20. Alvarez-Risco A, Del-Aguila-Arcentales S, Stevenson JG (2015) Pharmacists and mass communication for implementing pharmaceutical care. Am J Pharm Benefits 7(3):e125–e126
21. Alvarez-Risco A, Del-Aguila-Arcentales S, Yáñez JA (2021) Telemedicine in Peru as a result of the COVID-19 pandemic: perspective from a country with limited internet access. Am J Trop Med Hyg 105(1):6–11. https://doi.org/10.4269/ajtmh.21-0255
22. Alvarez-Risco A, Del-Aguila-Arcentales S, Yáñez JA, Rosen MA, Mejia CR (2021) Influence of technostress on academic performance of university medicine students in Peru during the COVID-19 pandemic. Sustain 13(16):8949. https://doi.org/10.3390/su13168949
23. Alvarez-Risco A, Delgado-Zegarra J, Yáñez JA, Diaz-Risco S, Del-Aguila-Arcentales S (2018) Predation risk by gastronomic boom—case Peru. J Landsc Ecol 11(1):100–103. https://doi.org/10.2478/jlecol-2018-0003
24. Alvarez-Risco A, Estrada-Merino A, Anderson-Seminario MdlM, Mlodzianowska S, García-Ibarra V, Villagomez-Buele C, Carvache-Franco M (2021) Multitasking behavior in online classrooms and academic performance: case of university students in Ecuador during COVID-19 outbreak. Interact Technol Smart Educ 18(3):422–434. https://doi.org/10.1108/ITSE-08-2020-0160
25. Alvarez-Risco A, Mejia CR, Delgado-Zegarra J, Del-Aguila-Arcentales S, Arce-Esquivel AA, Valladares-Garrido MJ, Rosas Del Portal M, Villegas LF, Curioso WH, Sekar MC, Yáñez JA (2020) The Peru approach against the COVID-19 infodemic: insights and strategies. Am J Trop Med Hyg 103(2):583–586. https://doi.org/10.4269/ajtmh.20-0536
26. Alvarez-Risco A, Mil JWFv (2007) Pharmaceutical care in community pharmacies: practice and research in Peru. Ann Pharmacother 41(12):2032–2037. https://doi.org/10.1345/aph.1K117
27. Alvarez-Risco A, Mlodzianowska S, García-Ibarra V, Rosen MA, Del-Aguila-Arcentales S (2021) Factors affecting green entrepreneurship intentions in business university students in COVID-19 pandemic times: case of Ecuador. Sustain 13(11):6447. https://www.mdpi.com/2071-1050/13/11/6447
28. Alvarez-Risco A, Mlodzianowska S, Zamora-Ramos U, Del-Aguila S (2021) Green entrepreneurship intention in university students: the case of Peru. Entrep Bus Econ Rev 9(4):85–100. https://doi.org/10.15678/EBER.2021.090406
29. Alvarez-Risco A, Quipuzco-Chicata L, Escudero-Cipriani C (2021) Determinantes de la intención de recompra en línea en tiempos de COVID-19: evidencia de una economía emergente. Lect Econ (96). https://doi.org/10.17533/udea.le.n96a342638
30. Alvarez-Risco A, Quiroz-Delgado D, Del-Aguila-Arcentales S (2016) Pharmaceutical care in hypertension patients in a Peruvian hospital. Indian J Public Health Res Dev 7(3):183–188
31. Alvarez-Risco A, Rosen MA, Del-Aguila-Arcentales S (2020) A new regulation for supporting a circular economy in the plastic industry: the case of Peru (short communication). J Landsc Ecol 13(1):1–3. https://doi.org/10.2478/jlecol-2020-0004
32. Alvarez-Risco A, Turpo-Cama A, Ortiz-Palomino L, Gongora-Amaut N, Del-Aguila-Arcentales S (2016) Barreras para la implementación de la Atención Farmacéutica en establecimientos farmacéuticos de Cusco, Perú. Pharm Care España 18(5):194–205. https://www.pharmcareesp.com/index.php/PharmaCARE/article/view/326
33. Álvarez-Risco A, Zegarra Arellano E, Matos Valerio E, Mejía Acosta N, Solis Tarazona Z (2013) Campaña de atención farmacéutica como estrategia de implementación de los servicios farmacéuticos: Experiencia Perú. Pharm Care España 15(1):35. https://www.pharmcareesp.com/index.php/PharmaCARE/article/view/105

34. Amarillo CR (2020) Feminism on lockdown. NACLA Rep Am 52(3):274–281. https://doi.org/10.1080/10714839.2020.1809084
35. Aminy MM (2020) Can virtual tourism save local businesses in tourist destinations? Retrieved 22 December 2021 from https://theconversation.com/can-virtual-tourism-save-local-businesses-in-tourist-destinations-149537
36. Apcho-Ccencho L-V, Cuya-Velásquez B-B, Alvarado Rodríguez D, Anderson-Seminario MdlM, Alvarez-Risco A, Estrada-Merino A, Mlodzianowska S (2021) The impact of international price on the technological industry in the United States and China during times of crisis: commercial war and COVID-19. In: Lawrence KD, Klimberg, RK (eds) Advances in business and management forecasting, vol. 14. Emerald Publishing Limited, pp 149–160. https://doi.org/10.1108/S1477-407020210000014010
37. Aprende de Turismo (2021) Aprende de Turismo. Retrieved 22 December 2021 from https://www.aprendedeturismo.org
38. Assaf A, Scuderi R (2020) COVID-19 and the recovery of the tourism industry. Tour Econ 26(5):731–733. https://doi.org/10.1177/1354816620933712
39. Bacq S, Lumpkin GT (2021) Social entrepreneurship and COVID-19. J Manag Stud 58(1):285–288. https://doi.org/10.1111/joms.12641
40. Barello S, Caruso R, Palamenghi L, Nania T, Dellafiore F, Bonetti L, Silenzi A, Marotta C, Graffigna G (2021) Factors associated with emotional exhaustion in healthcare professionals involved in the COVID-19 pandemic: an application of the job demands-resources model. Int Arch Occup Environ Health 94(8):1751–1761. https://doi.org/10.1007/s00420-021-01669-z
41. Baum T, Hai NTT (2020) Hospitality, tourism, human rights and the impact of COVID-19. Int J Contemp Hosp Manag 32(7):2397–2407. https://doi.org/10.1108/IJCHM-03-2020-0242
42. Belitski M, Guenther C, Kritikos AS, Thurik R (2021) Economic effects of the COVID-19 pandemic on entrepreneurship and small businesses. Small Bus Econ. https://doi.org/10.1007/s11187-021-00544-y
43. Block JH, Fisch C, Hirschmann M (2021) The determinants of bootstrap financing in crises: evidence from entrepreneurial ventures in the COVID-19 pandemic. Small Bus Econ. https://doi.org/10.1007/s11187-020-00445-6
44. Borzée A, McNeely J, Magellan K, Miller JRB, Porter L, Dutta T, et al (2020) COVID-19 highlights the need for more effective wildlife trade legislation. Trends Ecol & Evol 35(12):1052–1055. https://doi.org/10.1016/j.tree.2020.10.001
45. Bozdağ F, Ergün N (2020) Psychological resilience of healthcare professionals during COVID-19 pandemic. Psychol Rep 124(6):2567–2586. https://doi.org/10.1177/0033294120965477
46. Brouder P (2020) Reset redux: possible evolutionary pathways towards the transformation of tourism in a COVID-19 world. Tour Geogr 22(3):484–490. https://doi.org/10.1080/14616688.2020.1760928
47. Brown R, Rocha A, Cowling M (2020) Financing entrepreneurship in times of crisis: exploring the impact of COVID-19 on the market for entrepreneurial finance in the United Kingdom. Int Small Bus J 38(5):380–390. https://doi.org/10.1177/0266242620937464
48. Carvache-Franco M, Alvarez-Risco A, Carvache-Franco O, Carvache-Franco W, Estrada-Merino A, Villalobos-Alvarez D (2021) Perceived value and its influence on satisfaction and loyalty in a coastal city: a study from Lima, Peru. J Policy Res Tour Leis Events:1–16. https://doi.org/10.1080/19407963.2021.1883634
49. Carvache-Franco M, Alvarez-Risco A, Carvache-Franco W, Carvache-Franco O, Estrada-Merino A, Rosen MA (2021) Coastal cities seen from loyalty and their tourist motivations: a study in Lima, Peru. Sustain 13(21):11575. https://doi.org/10.3390/su132111575
50. Carvache-Franco M, Carvache-Franco O, Carvache-Franco W, Alvarez-Risco A, Estrada-Merino A (2021) Motivations and segmentation of the demand for coastal cities: a study in Lima, Peru. Int J Tour Res 23(4):517–531. https://doi.org/10.1002/jtr.2423
51. Cegarra-Navarro J-G, Vătămănescu E-M, Martínez-Martínez A (2021) A context-driven approach on coping with COVID-19: from hiding knowledge toward citizen engagement. Knowl Process Manag 28(2):134–140. https://doi.org/10.1002/kpm.1662

52. Chafloque-Cespedes R, Alvarez-Risco A, Robayo-Acuña P-V, Gamarra-Chavez C-A, Martinez-Toro G-M, Vicente-Ramos W (2021) Effect of sociodemographic factors in entrepreneurial orientation and entrepreneurial intention in university students of Latin American business schools. In Jones P, Apostolopoulos N, Kakouris A, Moon C, Ratten V, Walmsley A (eds) Universities and entrepreneurship: meeting the educational and social challenges, vol. 11. Emerald Publishing Limited, pp 151–165. https://doi.org/10.1108/S2040-724620210000011010
53. Chafloque-Cespedes R, Vara-Horna A, Asencios-Gonzales Z, Lopez-Odar DR, Alvarez-Risco A, Quipuzco-Chicata L, et al (2020) Presentismo académico y violencia contra las mujeres en escuelas de negocios e ingeniería en universidades peruanas. Lect Econ (93):127–153. https://doi.org/10.17533/udea.le.n93a340726
54. Chen T, Peng L, Yin X, Rong J, Yang J, Cong G (2020) Analysis of user satisfaction with online education platforms in China during the COVID-19 pandemic. Healthc 8(3). https://doi.org/10.3390/healthcare8030200
55. Chen X, Zhang SX, Jahanshahi AA, Alvarez-Risco A, Dai H, Li J, Ibarra VG (2020) Belief in a COVID-19 conspiracy theory as a predictor of mental health and well-being of health care workers in Ecuador: cross-sectional survey study. JMIR Public Health Surveill 6(3):e20737. https://doi.org/10.2196/20737
56. Chung SA, Olivera S, Rojas Román B, Alanoca E, Moscoso S, Limpias Terceros B, et al (2021) Temáticas de la producción científica de la Revista Cubana de Farmacia indizada en Scopus (1967–2020) [Revista Cubana de Farmacia; farmacia; Cuba; evidencia]. 2021 54(1). http://www.revfarmacia.sld.cu/index.php/far/article/view/511
57. Cruz-Torres W, Alvarez-Risco A, Del-Aguila-Arcentales S (2021) Impact of Enterprise Resource Planning (ERP) implementation on performance of an education enterprise: a Structural Equation Modeling (SEM). Stud Bus Econ 16(2):37–52. https://doi.org/10.2478/sbe-2021-0023
58. Del-Aguila-Arcentales S, Alvarez-Risco A (2013) Human error or burnout as explanation for mistakes in pharmaceutical laboratories. Accred Qual Assur 18(5):447–448. https://doi.org/10.1007/s00769-013-1000-0
59. del Castillo FA (2021) Temporary waiver of intellectual property on Covid-19 vaccines: toward the creation of a better, post-pandemic society. J Public Health 43(3):e559–e560. https://doi.org/10.1093/pubmed/fdab184
60. Delgado-Zegarra J, Alvarez-Risco A, Yáñez JA (2018) Uso indiscriminado de pesticidas y ausencia de control sanitario para el mercado interno en Perú. Rev Panam Salud Publica 42:e3. https://doi.org/10.26633/RPSP.2018.3
61. Deressa W, Worku A, Abebe W, Gizaw M, Amogne W (2021) Risk perceptions and preventive practices of COVID-19 among healthcare professionals in public hospitals in Addis Ababa, Ethiopia. PLoS ONE 16(6):e0242471. https://doi.org/10.1371/journal.pone.0242471
62. Dominelli L (2021) A green social work perspective on social work during the time of COVID-19. Int J Soc Welf 30(1):7–16. https://doi.org/10.1111/ijsw.12469
63. Duarte Alonso A, Kok SK, Bressan A, O'Shea M, Sakellarios N, Koresis A et al (2020) COVID-19, aftermath, impacts, and hospitality firms: an international perspective. Int J Hosp Manag 91:102654. https://doi.org/10.1016/j.ijhm.2020.102654
64. Enciso-Zarate A, Guzmán-Oviedo J, Sánchez-Cardona F, Martínez-Rohenes D, Rodríguez-Palomino JC, Alvarez-Risco A, et al (2016) Evaluación de la contaminación con agentes citotóxicos en hospitales en Colombia. Pharm Care España 18(6):241–250. https://www.pharmcareesp.com/index.php/PharmaCARE/article/view/354
65. Erfani P, Binagwaho A, Jalloh MJ, Yunus M, Farmer P, Kerry V (2021) Intellectual property waiver for covid-19 vaccines will advance global health equity. BMJ 374:n1837. https://doi.org/10.1136/bmj.n1837
66. Estévez R (2020) What does circular tourism consist of? [¿En qué consiste el turismo circular?]. Retrieved 22 December 2021 from https://www.ecointeligencia.com/2020/08/turismo-circular

67. Fotiadis A, Polyzos S, Huan T-CTC (2021) The good, the bad and the ugly on COVID-19 tourism recovery. Ann Tour Res 87:103117. https://doi.org/10.1016/j.annals.2020.103117
68. Galanakis CM, Rizou M, Aldawoud TMS, Ucak I, Rowan NJ (2021) Innovations and technology disruptions in the food sector within the COVID-19 pandemic and post-lockdown era. Trends Food Sci Technol 110:193–200. https://doi.org/10.1016/j.tifs.2021.02.002
69. Gulati G, Kelly BD (2020) Domestic violence against women and the COVID-19 pandemic: what is the role of psychiatry? Int J Law Psychiatry 71:101594. https://doi.org/10.1016/j.ijlp.2020.101594
70. Gursoy D, Chi CG (2020) Effects of COVID-19 pandemic on hospitality industry: review of the current situations and a research agenda. J Hosp Market Manag 29(5):527–529. https://doi.org/10.1080/19368623.2020.1788231
71. Hanzl M (2020) Urban forms and green infrastructure—the implications for public health during the COVID-19 pandemic. Cities and Health:1–5. https://doi.org/10.1080/23748834.2020.1791441
72. Hepburn C, O'Callaghan B, Stern N, Stiglitz J, Zenghelis D (2020) Will COVID-19 fiscal recovery packages accelerate or retard progress on climate change? Oxf Rev Econ Policy 36(Suppl. 1):S359–S381. https://doi.org/10.1093/oxrep/graa015
73. Ho C-Y, Tsai B-H, Chen C-S, Lu M-T (2021) Exploring green marketing orientations toward sustainability the hospitality industry in the COVID-19 pandemic. Sustain 13(8). https://doi.org/10.3390/su13084348
74. International Monetary Fund (2021) Four charts on the Dominican Republic's crisis response and recovery. Retrieved 22 December 2021 from https://www.imf.org/en/News/Articles/2021/07/08/na071221-four-charts-on-the-dominican-republics-crisis-response-and-recovery
75. ItalyGuides (2021) Italy at its best. Retrieved 22 December 2021 from https://www.italyguides.it/en
76. Jecker NS, Atuire CA (2021) What's yours is ours: waiving intellectual property protections for COVID-19 vaccines. J Med Ethics 47(9):595. https://doi.org/10.1136/medethics-2021-107555
77. Kaushal V, Srivastava S (2021) Hospitality and tourism industry amid COVID-19 pandemic: perspectives on challenges and learnings from India. Int J Hosp Manag 92:102707. https://doi.org/10.1016/j.ijhm.2020.102707
78. Kenny P (2020) UN report reveals that global e-waste increased in recent years [Informe de la ONU revela que desechos electrónicos mundiales aumentaron en los últimos años]. Retrieved 22 December 2021 from https://www.aa.com.tr/es/mundo/informe-de-la-onu-revela-que-desechos-electr%C3%B3nicos-mundiales-aumentaron-en-los-%C3%BAltimos-a%C3%B1os/1899209
79. Kirk CP, Rifkin LS (2020) I'll trade you diamonds for toilet paper: consumer reacting, coping and adapting behaviors in the COVID-19 pandemic. J Bus Res 117:124–131. https://doi.org/10.1016/j.jbusres.2020.05.028
80. Kitamura Y, Ichisugi Y, Karkour S, Itsubo N (2020) Carbon footprint evaluation based on tourist consumption toward sustainable tourism in Japan. Sustain 12(6). https://doi.org/10.3390/su12062219
81. Koetter P, Pelton M, Gonzalo J, Du P, Exten C, Bogale K et al (2020) Implementation and process of a COVID-19 contact tracing initiative: leveraging health professional students to extend the workforce during a pandemic. Am J Infect Control 48(12):1451–1456. https://doi.org/10.1016/j.ajic.2020.08.012
82. Krishtel P, Malpani R (2021) Suspend intellectual property rights for covid-19 vaccines. BMJ 373:n1344. https://doi.org/10.1136/bmj.n1344
83. Lackie K, Najjar G, El-Awaisi A, Frost J, Green C, Langlois S et al (2020) Interprofessional education and collaborative practice research during the COVID-19 pandemic: considerations to advance the field. J Interprof Care 34(5):583–586. https://doi.org/10.1080/13561820.2020.1807481
84. Lashley MA, Acevedo M, Cotner S, Lortie CJ (2020) How the ecology and evolution of the COVID-19 pandemic changed learning. Ecol Evol 10(22):12412–12417. https://doi.org/10.1002/ece3.6937

85. Leiva-Martinez M-A, Anderson-Seminario MdlM, Alvarez-Risco A, Estrada-Merino A, Mlodzianowska S (2021) Price variation in lower goods as of previous economic crisis and the contrast of the current price situation in the context of COVID-19 in Peru. In: Lawrence KD, Klimberg RK (eds) Advances in business and management forecasting, vol. 14. Emerald Publishing Limited, pp 161–166. https://doi.org/10.1108/S1477-407020210000014011
86. Liguori E, Winkler C (2020) From offline to online: challenges and opportunities for entrepreneurship education following the COVID-19 pandemic. Entrep Educ Pedagog 3(4):346–351. https://doi.org/10.1177/2515127420916738
87. Lim S, Prakash A (2021) Pandemics and citizen perceptions about their country: did COVID-19 increase national pride in South Korea? Nations Natl 27(3):623–637. https://doi.org/10.1111/nana.12749
88. Lopez-Odar D, Alvarez-Risco A, Vara-Horna A, Chafloque-Cespedes R, Sekar MC (2020) Validity and reliability of the questionnaire that evaluates factors associated with perceived environmental behavior and perceived ecological purchasing behavior in Peruvian consumers. Soc Responsib J 16(3):403–417. https://doi.org/10.1108/SRJ-08-2018-0201
89. Lu J, Xiao X, Xu Z, Wang C, Zhang M, Zhou Y (2021) The potential of virtual tourism in the recovery of tourism industry during the COVID-19 pandemic. Curr Issues Tour:1–17. https://doi.org/10.1080/13683500.2021.1959526
90. Maritz A, Perenyi A, de Waal G, Buck C (2020) Entrepreneurship as the unsung hero during the current COVID-19 economic crisis: Australian perspectives. Sustain 12(11). https://doi.org/10.3390/su12114612
91. Martínez M (2021) Virtual tourism: a new way of "traveling" without leaving home [Turismo virtual: Una nueva forma de "viajar" sin salir de casa]. Retrieved 22 December 2021 from https://www.revistacasinoperu.com/turismo-virtual-una-nueva-forma-de-viajar-sin-salir-de-casa
92. Mejía-Acosta N, Alvarez-Risco A, Solís-Tarazona Z, Matos-Valerio E, Zegarra-Arellano E, Del-Aguila-Arcentales S (2016) Reacciones Adversas a Medicamentos reportadas como resultado de la implementación de Atención Farmacéutica en la Farmacia Institucional DIGEMID - Ministerio de Salud de Perú. Pharm Care España 18(2):67–74. https://www.pharmcareesp.com/index.php/PharmaCARE/article/view/311
93. Ministère de l'Europe et des Affaires étrangères (2018) Tourism, a key advantage for France [El turismo, una ventaja clave para Francia]. Retrieved 22 December 2021 from https://www.diplomatie.gouv.fr/es/politica-exterior/turismo/el-turismo-una-ventaja-clave-para
94. Mura P, Tavakoli R, Pahlevan Sharif S (2017) 'Authentic but not too much': exploring perceptions of authenticity of virtual tourism. Inf Technol Tour 17(2):145–159. https://doi.org/10.1007/s40558-016-0059-y
95. Negrao F (2020) Virtual and augmented reality: advantages and disadvantages [Realidad virtual y aumentada: Ventajas y desventajas]. Retrieved 22 December 2021 from https://www.encora.com/es/blog/realidad-virtual-y-aumentada-ventajas-y-desventajas
96. O'Grady T, Minunno R, Chong H-Y, Morrison GM (2021) Design for disassembly, deconstruction and resilience: a circular economy index for the built environment. Resour Conserv Recycl 175:105847. https://doi.org/10.1016/j.resconrec.2021.105847
97. Obergassel W, Hermwille L, Oberthür S (2021) Harnessing international climate governance to drive a sustainable recovery from the COVID-19 pandemic. Clim Policy 21(10):1298–1306. https://doi.org/10.1080/14693062.2020.1835603
98. Okereke M (2021) Towards vaccine equity: should big pharma waive intellectual property rights for COVID-19 vaccines? Public Health Pract 2:100165. https://doi.org/10.1016/j.puhip.2021.100165
99. Quispe-Cañari JF, Fidel-Rosales E, Manrique D, Mascaró-Zan J, Huamán-Castillón KM, Chamorro-Espinoza SE et al (2021) Self-medication practices during the COVID-19 pandemic among the adult population in Peru: a cross-sectional survey. Saudi Pharm J 29(1):1–11. https://doi.org/10.1016/j.jsps.2020.12.001
100. Rastati R (2020) Virtual tour: tourism in the time of corona. In: 6th International Conference on Social and Political Sciences (ICOSAPS 2020)

101. Raudenská J, Steinerová V, Javůrková A, Urits I, Kaye AD, Viswanath O, Varrassi G (2020) Occupational burnout syndrome and post-traumatic stress among healthcare professionals during the novel coronavirus disease 2019 (COVID-19) pandemic. Best Pract Res Clin Anaesthesiol 34(3):553–560. https://doi.org/10.1016/j.bpa.2020.07.008
102. Roesch E, Amin A, Gupta J, García-Moreno C (2020) Violence against women during covid-19 pandemic restrictions. BMJ 369:m1712. https://doi.org/10.1136/bmj.m1712
103. Rojas-Osorio M, Alvarez-Risco A (2019) Intention to use smartphones among Peruvian university students [intention to use; Peruvian students; technology acceptance model; smartphone]. 2019 13(03):13. https://doi.org/10.3991/ijim.v13i03.9356
104. Rojas Román B, Moscoso S, Chung SA, Limpias Terceros B, Álvarez-Risco A, Yáñez JA (2020) Tratamiento de la COVID-19 en Perú y Bolivia y los riesgos de la automedicación [COVID-19; tratamiento; estrategia; fármacos; automedicación; Perú; Bolivia]. 2020 53(2). http://www.revfarmacia.sld.cu/index.php/far/article/view/435/351
105. Sánchez-Clavijo LM, Martínez-Callejas SJ, Acevedo-Charry O, Diaz-Pulido A, Gómez-Valencia B, Ocampo-Peñuela N et al (2021) Differential reporting of biodiversity in two citizen science platforms during COVID-19 lockdown in Colombia. Biol Cons 256:109077. https://doi.org/10.1016/j.biocon.2021.109077
106. Sánchez OR, Vale DB, Rodrigues L, Surita FG (2020) Violence against women during the COVID-19 pandemic: an integrative review. Int J Gynecol Obstet 151(2):180–187. https://doi.org/10.1002/ijgo.13365
107. Sekalala S, Forman L, Hodgson T, Mulumba M, Namyalo-Ganafa H, Meier BM (2021) Decolonising human rights: how intellectual property laws result in unequal access to the COVID-19 vaccine. BMJ Glob Health 6(7):e006169. https://doi.org/10.1136/bmjgh-2021-006169
108. Shaw R, Kim Y-K, Hua J (2020) Governance, technology and citizen behavior in pandemic: lessons from COVID-19 in East Asia. Progress in Disaster Science 6:100090. https://doi.org/10.1016/j.pdisas.2020.100090
109. Sigala M (2020) Tourism and COVID-19: impacts and implications for advancing and resetting industry and research. J Bus Res 117:312–321. https://doi.org/10.1016/j.jbusres.2020.06.015
110. Stainton H (2020) Virtual tourism explained: what, why and where. Retrieved 22 December 2021 from https://tourismteacher.com/virtual-tourism
111. Statista (2020) Who's most vulnerable to COVID-19's impact on tourism? Retrieved 22 December 2021 from https://www.statista.com/chart/21391/total-contribution-of-travel-and-tourism-to-gdp-of-selected-countries
112. Sustainable Travel International (2021) Carbon footprint of tourism. Sustainable travel international. Retrieved 22 December 2021 from https://sustainabletravel.org/issues/carbon-footprint-tourism
113. Tran LTT (2021) Managing the effectiveness of e-commerce platforms in a pandemic. J Retail Consum Serv 58:102287. https://doi.org/10.1016/j.jretconser.2020.102287
114. United Nations (2021) Carbon footprint [Huella de carbono]. Retrieved 22 December 2021 from https://biblioguias.cepal.org/huellacarbono
115. UNWTO (2019) International tourism highlights. UNWTO. https://doi.org/10.18111/9789284421152
116. UNWTO (2021) Sustainable development [Desarrollo sostenible]. Retrieved 22 December 2021 from https://www.unwto.org/es/desarrollo-sostenible
117. Vargas A (2021) The future of tourism will be circular, or it will not [El futuro del turismo será circular o no será]. Retrieved 22 December 2021 from https://www.elmostrador.cl/generacion-m/2021/02/27/el-futuro-del-turismo-sera-circular-o-no-sera
118. Vidya CT, Prabheesh KP (2020) Implications of COVID-19 pandemic on the global trade networks. Emerg Mark Financ Trade 56(10):2408–2421. https://doi.org/10.1080/1540496X.2020.1785426
119. Viero A, Barbara G, Montisci M, Kustermann K, Cattaneo C (2021) Violence against women in the Covid-19 pandemic: a review of the literature and a call for shared strategies to tackle

health and social emergencies. Forensic Sci Int 319:110650. https://doi.org/10.1016/j.forsciint.2020.110650
120. Wen J, Kozak M, Yang S, Liu F (2021) COVID-19: potential effects on Chinese citizens' lifestyle and travel. Tourism Review 76(1):74–87. https://doi.org/10.1108/TR-03-2020-0110
121. WTTC (2019) Travel and tourism economic impact 2021. Retrieved 22 December 2021 from https://wttc.org/Portals/0/Documents/EIR/EIR2021%20Global%20Infographic.pdf?ver=2021-04-06-170951-897
122. WTTC (2021) Economic impact reports. Retrieved 22 December 2021 from https://wttc.org/Research/Economic-Impact
123. Wu T-Y, Ford O, Rainville AJ, Bessire R (2020) COVID-19 care package distribution for senior citizens and families in Detroit and Hamtramck, Michigan. J Hunger Environ Nutr 15(4):585–587. https://doi.org/10.1080/19320248.2020.1799586
124. Yan J, Kim S, Zhang SX, Foo M-D, Alvarez-Risco A, Del-Aguila-Arcentales S, Yáñez JA (2021) Hospitality workers' COVID-19 risk perception and depression: a contingent model based on transactional theory of stress model. Int J Hosp Manag 95:102935. https://doi.org/10.1016/j.ijhm.2021.102935
125. Yáñez JA, Afshar Jahanshahi A, Alvarez-Risco A, Li J, Zhang SX (2020) Anxiety, distress, and turnover intention of healthcare workers in Peru by their distance to the epicenter during the COVID-19 crisis. Am J Trop Med Hyg 103(4):1614–1620. https://doi.org/10.4269/ajtmh.20-0800
126. Yasin Ar A (2020) Managing e-commerce during a pandemic: lessons from GrubHub during COVID-19. In: George B, Mahar Q (eds) International case studies in the management of disasters. Emerald Publishing Limited, pp 155–167. https://doi.org/10.1108/978-1-83982-186-820201010
127. Zarocostas J (2021) What next for a COVID-19 intellectual property waiver? The Lancet 397(10288):1871–1872. https://doi.org/10.1016/S0140-6736(21)01151-X
128. Zhang SX, Chen J, Afshar Jahanshahi A, Alvarez-Risco A, Dai H, Li J, Patty-Tito RM (2021) Succumbing to the COVID-19 pandemic—healthcare workers not satisfied and intend to leave their jobs. Int J Ment Heal Addict. https://doi.org/10.1007/s11469-020-00418-6
129. Zhang SX, Sun S, Afshar Jahanshahi A, Alvarez-Risco A, Ibarra VG, Li J, Patty-Tito RM (2020) Developing and testing a measure of COVID-19 organizational support of healthcare workers—results from Peru, Ecuador, and Bolivia. Psychiatry Res 291:113174. https://doi.org/10.1016/j.psychres.2020.113174
130. Zollinger R, DiCindio C (2021) Community ecology: museum education and the digital divide during and after COVID-19. J Mus Educ 46(4):481–492. https://doi.org/10.1080/10598650.2021.1983711

Virtual Education: Carbon Footprint and Circularity

Anguie Contreras-Taica, Aldo Alvarez-Risco, Marian Arias-Meza, Nilda Campos-Dávalos, Marco Calle-Nole, Camila Almanza-Cruz, María de las Mercedes Anderson-Seminario, and Shyla Del-Aguila-Arcentales

Abstract Virtual education has appeared to upgrade traditional education in different dimensions, such as easing the learning process, developing new teaching methodologies, and eliminating distance barriers. Since the pandemic outbreak by COVID-19, this situation has intensified, forcing students and educators to adapt to this new scenario, which involves staying at home and using electronic devices for long hours, which seems to contribute positively to the environment, as there is no need to attend education centers where physical installations and different services are used. However, little has been said about the environmental impact of virtual learning. Nevertheless, to have a transparent approach and determine virtual education sustainability, it is essential to analyze its implications.

Keywords Education · Virtual education · Carbon footprint · Climate change · Circularity · Environmental impact · Circular economy

1 Introduction

The world changed dramatically by COVID-19. Some specific areas involved include education [5, 12, 17], tourism [35–37], violence against woman [39, 90], prices [27, 68], research [41], intellectual property [24], health care [22, 11, 40, 92, 116, 118, 119, 120], entrepreneurship [25, 13, 18, 38], commerce [10, 20, 15, 43, 70], hospitality [115], and citizens behavior [11, 84, 117].

The sustainability of many issues has changed since the usual behavior and characteristics [3, 4, 7, 9, 12, 14, 16, 21, 23, 45, 46, 49, 73, 91]. The changes are mainly

---

A. Contreras-Taica · A. Alvarez-Risco (✉) · M. Arias-Meza · N. Campos-Dávalos · M. Calle-Nole · C. Almanza-Cruz · M. de las Mercedes Anderson-Seminario
Universidad de Lima, Lima, Perú
e-mail: aralvare@ulima.edu.pe

S. Del-Aguila-Arcentales
Escuela Nacional de Marina Mercante "Almirante Miguel Grau", Callao, Perú
e-mail: sdelaguila@enamm.edu.pe

A. Alvarez-Risco et al. (eds.), *Circular Economy*, Environmental Footprints and Eco-design of Products and Processes,
https://doi.org/10.1007/978-981-19-0549-0_13

in communication, with increasing virtual behavior such as e-commerce, virtual tourism, and distance education, in different academic levels.

Due to the enormous concern over environmental care and reducing greenhouse gas emissions, it becomes relevant to analyze if virtual education is sustainable. Distance education can be considered a turning point in the education system. Therefore, it has always been compared with face-to-face instruction [1, 33, 60, 62, 83].

A comparison between face-to-face education and virtual education, from an environmental perspective, is presented by analyzing how carbon footprint and circularity work in both scenarios, which provide a clear overview of the direct and indirect effects of these two situations on sustainability. The innovations that the virtual world has brought to education and the environmental practices implemented in this sector are mentioned. Additionally, it is shown the initiatives and plans made by international organizations focused on reducing our carbon footprint and increasing circular sustainability. Finally, some conclusions and recommendations for educational institutions to implement excellent sustainable education programs, mainly in virtual education, presented after the current pandemic situation by COVID-19 are given.

All the information in the current chapter aims to create consciousness about the impact of virtual education by providing an environmental approach.

2 Face-to-Face Education

Every sector in the world generates greenhouse gases, whether it be agriculture [64, 72, 76, 79, 87, 89, 97, 100, 111], livestock [2, 48, 57, 112, 121], fishing [54, 58, 81], construction [28, 69, 94, 95], mining [29, 55, 61, 88], telecommunications [74, 85], and others. However, one sector cannot generate these gases. In the following lines, the concepts of carbon footprint and circularity and their applications are presented in the context of face-to-face education.

2.1 Carbon Footprint in Face-to-Face Education

Decades ago, sustainability issues did not consider in educational centers. However, after the Earth Summit in Rio de Janeiro in 1992, it generated awareness in European universities. Consequently, the Copernicus University Charter for Sustainable Development was drawn up in 1993 [56]. The carbon footprint is used and works as an environmental metric for calculating greenhouse gas emissions. According to Torres et al. [101], three scopes are considered:

Scope 1. Direct emissions produced by the educational center.
Scope 2. Indirect energy emissions.

Table 1 Description of the scope of the carbon footprint in face-to-face education

Scope 1	Scope 2	Scope 3
• Combustion of vehicles owned by the educational center • Combustion of canteen kitchens of the educational center	• Electricity consumption	• Private or public transportation of students and staff • Food consumed indoors • Waste generation • Services purchased from various companies

Source Adapted from Torres Ramos et al. [101]

Scope 3. Indirect emissions that may be generated.

In addition, the latter is usually the one that generates a higher percentage of participation, unlike scope 1 and 2. Table 1 details those above.

We can evidence that both variables decreased their participation in the carbon footprint impact. To better understand, as shown in Fig. 1, from 2018 to 2019, in scope 1, the variation was lower by 1%; scope 2 also had a reduction of 4%. However, in scope 3, where the student generates this carbon footprint, there is an increase of 5%, mainly due to the transportation factor (use of private vehicles instead of public transportation). The latter represents an entirely understandable preference in a pandemic context in which people seek to reduce the chances of contagion and spread of the virus.

In face-to-face education, according to the report by Filimonau et al. [52], scope 3 is the one that generates the most significant carbon footprint emissions, unlike scope 1 and scope 2. However, in 2020, the suspension of physical classes due to isolation to avoid contagion impacted the scope 3 because the student did not use public or private transportation, did not generate external solid waste, did not generate food waste at the university, among others, however, this gave rise to the following another point to consider: studying from home. Although the students do not very much consider it, it also negatively impacts the environment due to the frequent use

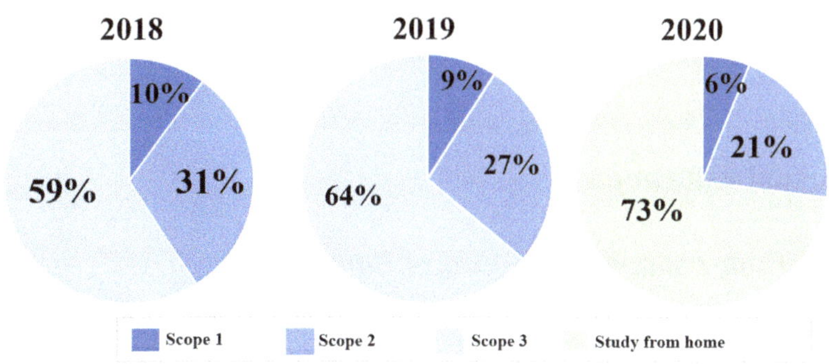

Fig. 1 Emissions per scope of assessment in the United Kingdom from 2018 to 2020 (*Source* Adapted from Filimonau et al. [52])

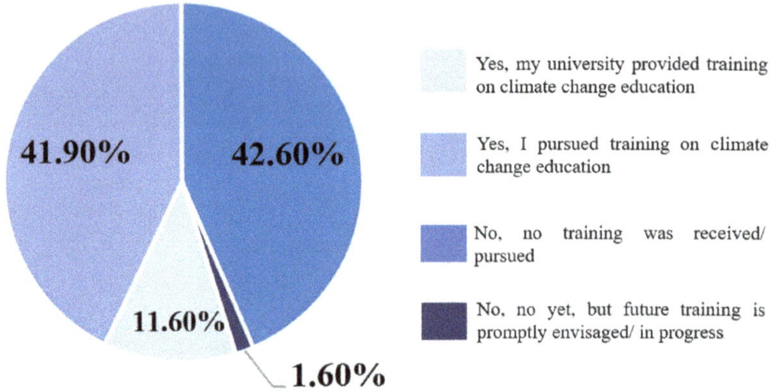

Fig. 2 Survey question: have you received or pursued training on matters related to climate change? (*Source* Leal Filho et al. [67])

of laptops from Monday to Friday, which increases energy consumption. These kinds of initiatives are discussed in more detail below.

2.2 Circularity in Face-to-Face Education

Many universities worldwide recognize their role concerning circularity to reduce climate change [65, 66, 75]. According to Leal Filho et al. [67], to have a minimal impact on the carbon footprint, they point out that universities are in search of becoming carbon neutral and are developing curricula and pedagogical approaches to educating about some topics like climate change mitigation and adaptation. Work is indeed being done, talks and activities have been started in many universities about climate change awareness, however, there is still a lack of knowledge about the student's carbon footprint. Figure 2 shows the survey results to know if students have received or pursued training on matters related to climate change. Around 44% of them have not yet received any information on the subject.

3 Virtual Education

3.1 Carbon Footprint and Virtual Education

Virtual education has brought numerous advantages for the environment, including a far lower carbon footprint emission [53, 82, 93, 110]. Students sometimes have to travel between one city to take classes because of distance learning. Less transportation is used, but the energy consumption is also increasing due to this now-standard

Table 2 Description of the scope of the carbon footprint in virtual education

Scope 1	Scope 2	Scope 3
• Direct emission from sources that are from the educational institution such as heating, cooling system, and electricity consumption	• Indirect emission from the institution provoked by the electricity consumption	• Indirect effects from the institution that are not controlled by it and are from other sources such as commute of staff, or the consequences of teleworking done by the professionals and students

Source Adapted from Versteijlen et al. [110]

method of education. Versteijlen et al. [110] indicate that "one of the great advantages of online education is a substantial decrease of carbon emissions" (p. 7). For their study, they interviewed professionals to ask them what they think about this fact, in which, surprisingly, most of them did not have an idea of the reducing carbon footprint that online education can bring, however, Filimonau et al. [52] explain that there is a possibility that virtual education might be less climate friendly than it seems. Considering that studying or working from home can generate significant emissions of the carbon footprint that the university generates, a significant share of the carbon footprint reduced from lengthy trips for the students' transportation can be effectively negated (p. 9) (Table 2).

3.2 *Virtual Education and Circularity*

As well as in face-to-face education, in online learning, there are also courses to give students environmental awareness and knowledge of the circular economy [63, 77, 113]. The inclusion of sustainability principles is needed in the plans of university studies as well as undergraduate teaching courses [26, 32, 86, 98]. Sites widely known, such as Coursera [42], edX [47] and Udemy [107], are teaching courses solely about this topic, and because they are fully online courses, more people can have access to its advantages and may apply its core in their professional life. Many universities globally have courses related to sustainability and are trying to boost awareness.

For example, student training projects seem to be an exciting and potential way to organize academic outreach. Those projects are challenges related to academic research protocols and provide new opportunities for universities, students, teachers, and community members. The latter dramatically benefits university students since volunteering as group consultants develop their hard and soft skills. In addition, the most important thing is that they contribute to the development of sustainable goals in their communities within the framework of the UN 2030 Agenda [80]. In this sense, the objective from now and on is for student training projects to be a promising method to integrate capacity building and empowerment in community sustainable development in higher education [44].

3.3 Virtual Education Innovations

Educational innovations in the context, based on international principles for sustainable education (see Table 3), implement novelties that allow students to develop their theoretical and practical knowledge about what they can do to contribute to sustainability.

Various ways of taking advantage of technological resources were found during the research, from sustainable computing to renewable energy sources applied to education. There is a tangible diversity of innovations in this field, which led us to

Table 3 International principles of education for sustainable development

Principles	Description
Future-oriented thinking	Thinking ahead involves people imagining preferential situations for the future. This future visualization process leads people to become aware of and take responsibility for a sustainable future. It allows people to understand the significant aspects of sustainable development and makes it possible to explore those assumptions
Creative and critical thinking	It allows people to explore new ways of thinking and acting, make decisions based on information, and create alternatives to present and choose wisely. It involves reflecting on how people interact, understanding cultural differences, and creating alternative ways of living together
Participation and participatory learning	Achieving the commitment of people is essential for building a shared future. Engaging diverse people and communities is essential as it broadens the knowledge system and perspectives. The participatory process is also important in creating a sense of ownership and empowerment
Partnerships	Those are motivating forces toward change. These organizations empower groups and individuals and allow them to act in taking action and making decisions and the processes that are part of sustainable development
Thinking about systems and networks	Thinking systematically is essential for sustainable development. Sustainable development requires approaches that go beyond problem-solving or cause of an effect

Source Adapted from Tilbury [99]

develop a greater awareness of the inordinate efforts of society and governments to provide excellent quality. The development of new technologies such as immersive augmented reality or the educational application of artificial intelligence shows a considerable improvement and a correct application at an educational level. The main innovation in the educational field is mainly found in digital immersion that connects and complements reality as we know it, the possibility of having a virtual campus, connecting with people from different parts of the world, and openness to a diversity of applied topics to the UNDP's Knowledge Management Strategy [102]. Visual communication in the digital age is seen as a convex source in the face of the diversified needs of people. Considering the energy consumption that virtual education implies, diversity in the electric field can be analyzed and on the other hand, the implementation of more and more sustainable sources.

4 Webinars and Talks to Reduce Your Carbon Footprint

As an example, it is possible to mention that the University of Lima, through its sustainability center, offers various conferences, awareness campaigns, recycling, and among many other activities that encourage university students to become interested in topics such as carbon footprint and circularity, which are the central focus of the current chapter, in a virtual education context. The last event developed, called "Development Opportunities with the UN Volunteers program," is especially noteworthy, in which the university offered the possibility of connecting with the world-renowned institution [108]. In the same way, the University of Cambridge brings solutions about educational procedures around new politics of business sustainability management and distance learning to manage better practices in the development of new technologies implemented in virtual educational systems [109].

Currently, in the Climate Neutral Now Events 2021 organized by United Nations Framework Convention on Climate Change [105], there are a series of virtual events with different actors sharing their experiences and inspiring others on their way for more sustainable and climate-neutral or net-zero emissions targets. The Path to Climate Neutrality: The basics for schools explained more about the importance of virtual education in the new generations and how it impacts the ecosystem of many organizations in how they interact with the environment. The discussion showed the increasing commitment and support for governments to the private sector, with policies to achieve more sustainable practices to reduce the footprint [105].

In the interaction with other stakeholders, the OECD Green Talks Live webinar series [78] starts the discussion with the implementation of new environmental efforts in the last decade. The relationship of environmental policies with productivity and sustainability systems management empowers the circular economy. Additionally, the UNEP experimented with new technology for reducing carbon footprint in VR in a talk about new tendencies in educational systems growing [103]. In the context of COP26, there is increasing knowledge about decarbonization and the role of

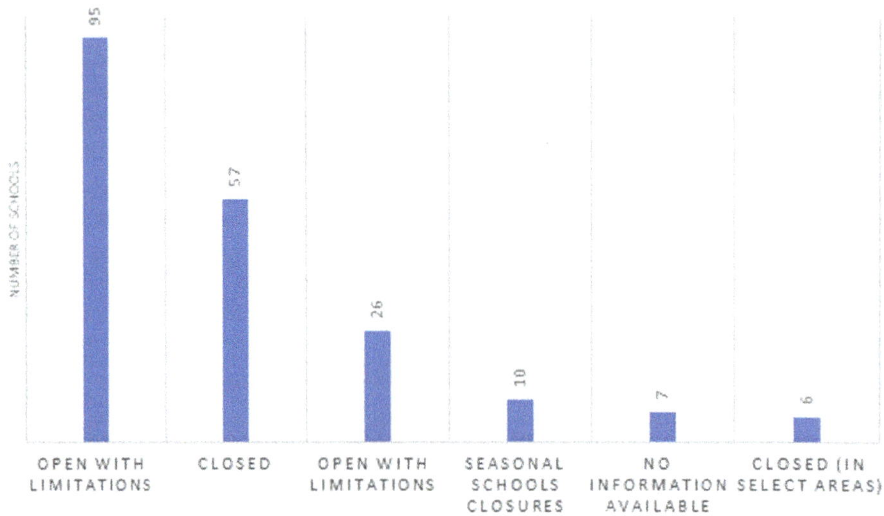

Fig. 3 Number of countries with close schools worldwide 2021, by status (*Source* Adapted from World Bank [114])

governments and capital markets to encourage the transition to a net-zero industry [106].

As shown in Fig. 3, there currently exists a massive variation for schools with high potential to return to face-to-face learning. Most of them, until this moment, wait to be restructured in a different industry or keep with less equipment for blended learning. In contrast to the established idea of events, virtual learning transforms into a different one with digital tools. In that context, webinars and online talks become the better improvement for virtual education.

5 International Organization Initiatives

In terms of sustainability, due to the UN's drive toward organizations promoting sustainable development, institutions such as the World Bank and the International Monetary Fund that are part of the United Nations system evaluate the initiatives or progress made to achieve the SDGs. In his case, these issues are more closely related to his financial and economic development functions.

In terms of education post-pandemic, Huang and Holotescu [59] explain that UNESCO, as a response to the pandemic, launched various initiatives such as releasing publications like "Guidance on Open Educational Practices during School Closures: Utilizing OER under COVID-19 Pandemic in line with UNESCO OER Recommendation" which has the objective to show the consequences if Open Educational Practices (OEP) and Open Educational Resources (OER) were to be applied on learning outcomes. Likewise, the Organization for Cooperation and Development promotes improved economic and social well-being policies. This guide explains innovative approaches when using both, and these experiences are lined up with UNESCO OER recommendations for five areas:

- "Building capacity of stakeholders to create access, use, adapt and redistribute OER
- Developing supportive policy
- Encouraging inclusive and equitable quality OER
- Nurturing the creation of sustainability models for OER
- Facilitating international cooperation." (p. 4)

Moreover, UNICEF, UNESCO, and the World Bank have been working together to monitor countries' responses to the COVID-19 pandemic in the education sector by analyzing the impact of school closures and learning losses worldwide. Following this line, UNESCO launched the Global Education Coalition to support educational recovery and assist countries in developing the best distance learning practices [104].

6 Efficient Environmental Practices in Education

6.1 Appropriate Use of the Email

Information and communication technologies have brought significant changes to society in various sectors, and education has not been the exception. These have become tools that help develop the student learning process in a didactic way, where technology is involved with education [30, 50, 51].

One of the most widely used services is email. In this line, Alvarez-Risco [19] proposes first, select and discard the inbox, i.e., delete emails that have no use, second is to sort; for this, you can use folders in the case of Outlook or labels in Gmail to be stored within these, so they will no longer be in the inbox. The third is to clean, that is to say, to be aware of the viruses that can arrive using the mails; the fourth is to standardize, develop habits so that the inbox shows what is answered, and have discipline be constant in all these steps.

6.2 Efficient Use of Laptops/Computers

It is a fact that traditional means are not indispensable to conduct classes; even the student's physical presence in the classroom is unnecessary. Technological devices have allowed classes to continue to be transmitted virtually, either live or recorded. One of the most indispensable devices is the laptop or computer. It is recommended to turn off the monitor, in case of a computer, when you are not using it, avoid opening many programs at the same time; reduce the brightness of the screen; unplug external devices such as USB and memory cards when you finish using them; and do not place the devices near sources of heat or cold.

6.3 Appropriate Use of Cloud Storage

Cloud storage appeared to reduce paper waste, saving files and having access to them from any device via the Internet, which has become an efficient tool in virtual learning environments as it provides students access to their educational materials anywhere and at any time. Cloud computing makes it possible to share, comment, visualize, and edit files in real time with other users, reducing the need for social gatherings. As well as in emails, it is recommended to use cloud storage appropriately to get the most benefit from it. Apart from organizing files in folders to better manage them, it is essential to periodically review the files that are no longer needed and delete them. The latter reduces the carbon emissions generated by cloud data centers [71].

7 Post-pandemic World Planners

Also, new practices seemed strange at first; however, it is part of everyday life. Likewise, according to Filimonau et al. [52], the return to classes brings with it the mandatory use of masks, an increase in the carbon footprint due to the limited capacity of public transportation, an increase in the use of water for disinfection, greater use of disposables, and greater use of classrooms or laboratories due to the fact of having more use of the facilities because the capacity in a class is reduced.

7.1 Use of Masks

Awareness of the use of single-use disposables was beginning to be raised in the world. However, with COVID-19, many of the practices to avoid the use of disposable have decreased. A clear example is masked as a form of protection against the virus. Selvaranjan et al. [96] showed that one person uses 5 masks per week. The authors mentioned that a biodegradable mask would be an option to reduce plastic waste. With the replacement of polypropylene with organic materials with similar chemical, mechanical, and physical properties, it can be inferred that the cost is higher due to the innovations that these masks would have.

7.2 Social Distancing on Public Transport

Because of the new restrictions about the social distancing on public transport, the carbon footprint would increase because more buses are needed and more use of petroleum or gas. In some cases, the student prefers to use cars to spread the virus, which means a higher carbon footprint per student. Consequently, the role of sustainable virtual education is fundamental to generating awareness about the efficient management of the carbon footprint without significantly affecting the environment [96].

7.3 Increased Water Use for Disinfection

Another variable that would affect the carbon footprint is the increase of water because it is frequently used for disinfection, according to Selvaranjan et al. [96]. So, what actions should universities take to have a minimum impact on climate change? On this point, it is pretty interesting to follow correct water management, and it is worth noting that SDG 6 mentioned the following: "the sustainable management of water and sanitation and the objective is to increase the efficiency of water use in every sector and guarantee the sustainable harvest and allocation of water resources to face scarcity of water…" In this sense, virtual education can lead to virtual training programs on efficient water use for all members of the various university communities. In this way, it is possible to face the fact that today it is necessary to consume more significant amounts of water resources.

7.4 Packaging and Other Solid Waste

Solid waste may be another problem for the environment while going back to face-to-face classes, as more single-use disposables are used to minimize the spread of the virus. Especially in campus dining services, some protocols are added to prevent COVID-19 contagions. For instance, it might be useful to give students packaged food and disposable utensils as they are easier to disinfect before using them. However, these practices might lead to an increase in carbon footprint per student. The implementation of more sustainable management according to the reverse logistics can improve the safety of the packaging, being recyclable, and more facilities to transport. Also, it can mean a significant improvement to regain value from the product or dispose of it, reducing waste.

7.5 High Energy Consumption

There is social distancing in shopping malls, public transportation, restaurants, and others in the new reality; this restriction probably continues in the case of face-to-face education. Because of that, there is student capacity per classroom; with this, the frequency of teaching increases, which means frequent use of university facilities and laboratories, affecting the energy consumption in the buildings. According to Versteijlen et al. [110], virtual learning consumes considerably less energy than campus-based education, which leads to lower carbon emissions in approximately 84%. In this sense, education centers that shift to online or hybrid learning rather than solely face-to-face education achieve energy efficiency and meet their sustainability goals.

7.6 Hybrid Education

After the pandemic, education became entirely virtual in the first months. However, with more people vaccinated and the level of the ones affected by the virus is reducing, many countries are adopting a mix between online and in-person education. It is also known that even though some systems could apply online learning quickly, they struggled in meeting the needs of all the population. For example, in developing countries such as Peru, people in remote areas with little to no Internet access could not study in proper conditions. We can conclude that the pandemic exposed what was lacking in the educational system. Caird and Roy [34] propose a design called blended learning, which is "the design of learning experiences that draw on a combination of face-face, distance, or online delivery methods, learning technologies, delivery multimedia, and pedagogical methodologies to achieve a mix of learning outcomes in educational or training contexts" (p. 4).

One of the objectives of this design is to be able to "support sustainable development, including the social, economic, and environmental dimensions of sustainability and protect global environmental resources to meet the needs of the present and future generations" (p. 4).

Blended learning is widely known as hybrid learning too, and it has become so widespread this year that it may be the new normal. However, an important aspect to consider is that while hybrid education allows a lower environmental impact to be had and contributes to the reduction of the carbon footprint, classes should be planned in such a way that the articulation of the face-to-face and online classes so that they can be part of the same teaching process and not as two parallel educations.

8 Closing Remarks

Education generates a carbon footprint as well as other sectors. In virtual education, to find this footprint, scope 1 refers to the direct emissions produced by the educational center; then scope 2 to the indirect energy emissions; and scope 3 to the other indirect emissions that may be generated, which is a higher percentage of participation unlike scope 1 and scope 2. Also, universities have started talks and activities to raise awareness about climate change to students, but around 44% have not yet received any information on the subject. So, the work is still in process.

Even though there is an approach to hybrid or blended learning, it is yet to have an empirical test to prove its effectiveness in reducing carbon footprint. There need to be further studies that can bring a solution and balance to an effective educational method that includes technology and broader access to education. There would be more accessibility to people who live in rural areas in developing countries with blended learning. However, it would also affect the environment because of the transportation pollution for mobility and energy consumption.

Several practices are not complex and that can be converted into new habits to have a more significant impact on the carbon footprint, such as the proper use of email, managing through labels or folders; the efficient use of laptops, with actions such as lowering the brightness, disconnecting all USB ports that are not used in the device; the efficient use of hybrid education, a point that has demonstrated the large gap in education that exists.

Many countries are preparing for the return to face-to-face classes and with it, a new reality that students face from the moment they leave home until they return. The use of permanent masks, social distancing in public transportation, increased use of water for disinfection, more significant solid waste, higher consumption in university facilities, among others, are changes that students are subjected to do. Therefore, it is a new challenge for universities to determine what actions can be taken to minimize their carbon footprint.

By way of conclusion, it is possible to mention that there are huge innovations, in terms of sustainable education, in both the face-to-face and virtual contexts. Obviously, throughout the present research work, it was noted that most theoretical and practical applications had been carried out in traditional education, that is, face-to-face education. However, many initiatives could be seen for the virtual context, widely expanded due to the COVID-19 conjuncture. In addition, face-to-face activities can serve as a source of inspiration to develop projects that promote a sustainable and focused virtual education, as proposed by the research, in carbon footprint and circularity. In addition, it is exciting that both types of sustainable education mentioned above follow international principles for sustainable development education, which represent a fundamental basis on which great projects can be achieved that benefit student, teachers, communities, and, in general, the public environment and its inhabitants.

References

1. Abdous MH, Yoshimura M (2010) Learner outcomes and satisfaction: a comparison of live video-streamed instruction, satellite broadcast instruction, and face-to-face instruction. Comput Educ 55(2):733–741
2. Adegbeye MJ, Elghandour MMMY, Monroy JC, Abegunde TO, Salem AZM, Barbabosa-Pliego A, Faniyi TO (2019) Potential influence of Yucca extract as feed additive on greenhouse gases emission for a cleaner livestock and aquaculture farming—a review. J Clean Prod 239:118074. https://doi.org/10.1016/j.jclepro.2019.118074
3. Alvarez-Risco A, Del-Aguila-Arcentales S, Delgado-Zegarra J, Yáñez JA, Diaz-Risco S (2019) Doping in sports: findings of the analytical test and its interpretation by the public. Sport Sci Health 15(1):255–257. https://doi.org/10.1007/s11332-018-0484-8
4. Alvarez-Risco A, Del-Aguila-Arcentales S, Diaz-Risco S (2018) Pharmacovigilance as a tool for sustainable development of healthcare in Peru. PharmacoVigilance Rev 10(2):4–6
5. Alvarez-Risco A, Del-Aguila-Arcentales S, Rosen MA, García-Ibarra V, Maycotte-Felkel S, Martínez-Toro GM (2021) Expectations and interests of university students in COVID-19 times about Sustainable Development Goals: evidence from Colombia, Ecuador, Mexico, and Peru. Sustain 13(6):3306. https://doi.org/10.3390/su13063306
7. Alvarez-Risco A, Del-Aguila-Arcentales S, Stevenson JG (2015) Pharmacists and mass communication for implementing pharmaceutical care. Am J Pharm Benefits 7(3):e125–e126
8. Alvarez-Risco A, Del-Aguila-Arcentales S, Yáñez JA (2021) Telemedicine in Peru as a result of the COVID-19 pandemic: perspective from a country with limited internet access. Am J Trop Med Hyg 105(1):6–11. https://doi.org/10.4269/ajtmh.21-0255
6. Alvarez-Risco A, Del-Aguila-Arcentales S, Yáñez JA, Rosen MA, Mejia CR (2021) Influence of technostress on academic performance of university medicine students in Peru during the COVID-19 pandemic. Sustain 13(16):8949. https://doi.org/10.3390/su13168949
9. Alvarez-Risco A, Delgado-Zegarra J, Yáñez JA, Diaz-Risco S, Del-Aguila-Arcentales S (2018) Predation risk by gastronomic boom—case Peru. J Landsc Ecol 11(1):100–103. https://doi.org/10.2478/jlecol-2018-0003
10. Alvarez-Risco A, Estrada-Merino A, Rosen MA, Vargas-Herrera A, Del-Aguila-Arcentales S (2021) Factors for implementation of circular economy in firms in COVID-19 pandemic times: the case of Peru. Environ 8(9):95. https://doi.org/10.3390/environments8090095
11. Alvarez-Risco A, Mejia CR, Delgado-Zegarra J, Del-Aguila-Arcentales S, Arce-Esquivel AA, Valladares-Garrido MJ et al (2020) The Peru approach against the COVID-19 infodemic: insights and strategies. Am J Trop Med Hyg 103(2):583–586. https://doi.org/10.4269/ajtmh.20-0536
12. Alvarez-Risco A, Mil JWFv (2007) Pharmaceutical care in community pharmacies: practice and research in Peru. Ann Pharmacother 41(12):2032–2037. https://doi.org/10.1345/aph.1K117
13. Alvarez-Risco A, Mlodzianowska S, García-Ibarra V, Rosen MA, Del-Aguila-Arcentales S (2021) Factors affecting green entrepreneurship intentions in business university students in COVID-19 pandemic times: case of Ecuador. Sustain 13(11):6447. https://doi.org/10.3390/su13116447
14. Alvarez-Risco A, Quiroz-Delgado D, Del-Aguila-Arcentales S (2016) Pharmaceutical care in hypertension patients in a Peruvian hospital. Indian J Public Health Res Dev 7(3):183–188
15. Alvarez-Risco A, Rosen MA, Del-Aguila-Arcentales S (2020) A new regulation for supporting a circular economy in the plastic industry: the case of Peru (short communication). J Landsc Ecol 13(1):1–3. https://doi.org/10.2478/jlecol-2020-0004
16. Alvarez-Risco A, Zegarra Arellano E, Matos Valerio E, Mejía Acosta N, Solis Tarazona Z (2013) Campaña de atención farmacéutica como estrategia de implementación de los servicios farmacéuticos: Experiencia Perú. Pharm Care España 15(1):35. https://www.pharmcareesp.com/index.php/PharmaCARE/article/view/105

17. Alvarez-Risco A, Estrada-Merino A, Anderson-Seminario MdlM, Mlodzianowska S, García-Ibarra V, Villagomez-Buele C, Carvache-Franco M (2021) Multitasking behavior in online classrooms and academic performance: case of university students in Ecuador during COVID-19 outbreak. Interact Technol Smart Educ 18(3):422–434. https://doi.org/10.1108/ITSE-08-2020-0160
18. Alvarez-Risco A, Mlodzianowska S, Zamora-Ramos U, Del-Aguila S (2021) Green entrepreneurship intention in university students: the case of Peru. Entrep Bus Econ Rev 9(4):85–100. https://doi.org/10.15678/EBER.2021.090406
19. Alvarez-Risco A (2020) Gestión del correo electrónico en teletrabajo: Bandeja cero. Retrieved 01 November 2021 from https://repositorio.ulima.edu.pe/handle/20.500.12724/10830
20. Alvarez-Risco A, Quipuzco-Chicata L, Escudero-Cipriani C (2021) Determinantes de la intención de recompra en línea en tiempos de COVID-19: evidencia de una economía emergente. Lect Econ (96). https://doi.org/10.17533/udea.le.n96a342638
21. Alvarez-Risco A, Turpo-Cama A, Ortiz-Palomino L, Gongora-Amaut N, Del-Aguila-Arcentales S (2016) Barreras para la implementación de la Atención Farmacéutica en establecimientos farmacéuticos de Cusco, Perú. Pharm Care España 18(5):194–205. https://www.pharmcareesp.com/index.php/PharmaCARE/article/view/326
22. Álvarez-Risco A, Dawson J, Johnson W, Conteh-Barrat M, Aslani P, Del-Aguila-Arcentales S, Diaz-Risco S (2021) Ebola virus disease outbreak: a global approach for health systems [Ebola; health professionals; global approach; pharmacist: pharmaceutical care; COVID-19.]. 2021, 53(4). http://www.revfarmacia.sld.cu/index.php/far/article/view/491
23. Alvarez-Risco A, Del-Aguila-Arcentales S (2015) Errores de prescripción como barrera para la Atención Farmacéutica en establecimientos de salud públicos: Experiencia Perú. Pharm Care España 17(6):725–731. https://www.pharmcareesp.com/index.php/PharmaCARE/article/view/24
24. Alvarez-Risco A, Del-Aguila-Arcentales S (2021a) A note on changing regulation in international business: the World Intellectual Property Organization (WIPO) and artificial intelligence. In: Verbeke A, van Tulder R, Rose EL, Wei Y (eds) The multiple dimensions of institutional complexity in international business research, vol. 15. Emerald Publishing Limited, pp 363–371. https://doi.org/10.1108/S1745-886220210000015020
25. Alvarez-Risco A, Del-Aguila-Arcentales S (2021b) Public policies and private efforts to increase women entrepreneurship based on STEM background. In: Mari M, Poggesi S, Foss L (eds) Women's entrepreneurship in STEM disciplines: issues and perspectives. Springer, pp 75–87. https://doi.org/10.1007/978-3-030-83792-1_5
26. Amey L, Plummer R, Pickering G (2020) Website communications for campus sustainability: an analysis of Canadian universities. Int J Sustain High Educ 21(3):531–556. https://doi.org/10.1108/IJSHE-04-2019-0137
27. Apcho-Ccencho L-V, Cuya-Velásquez B-B, Alvarado Rodríguez D, Anderson-Seminario MdlM, Alvarez-Risco A, Estrada-Merino A, Mlodzianowska S (2021) The impact of international price on the technological industry in the United States and China during times of crisis: commercial war and COVID-19. In: Lawrence KD, Klimberg, RK (eds) Advances in business and management forecasting, vol. 14. Emerald Publishing Limited, pp 149–160. https://doi.org/10.1108/S1477-407020210000014010
28. Arıoğlu Akan MÖ, Dhavale DG, Sarkis J (2017) Greenhouse gas emissions in the construction industry: an analysis and evaluation of a concrete supply chain. J Clean Prod 167:1195–1207. https://doi.org/10.1016/j.jclepro.2017.07.225
29. Azadi M, Northey SA, Ali SH, Edraki M (2020) Transparency on greenhouse gas emissions from mining to enable climate change mitigation. Nat Geosci 13(2):100–104. https://doi.org/10.1038/s41561-020-0531-3
30. Basri WS, Alandejani JA, Almadani FM (2018) ICT adoption impact on students' academic performance: evidence from Saudi universities. Educ Res Int 2018:1240197. https://doi.org/10.1155/2018/1240197
31. BrightTalk (2021) COP26 and commodities: what to expect? Retrieved 01 November 2021 from https://www.brighttalk.com/webcast/18654/514637?utm_source=brighttalk-portal&utm_medium=web&utm_campaign=topic&utm_content=on-demand

32. Brusca I, Labrador M, Larran M (2018) The challenge of sustainability and integrated reporting at universities: a case study. J Clean Prod 188:347–354. https://doi.org/10.1016/j.jclepro.2018.03.292
33. Burke, AS, Fedorek B (2017) Does "flipping" promote engagement? A comparison of a traditional, online, and flipped class. Act Learn High Educ, 18(1):11–24
34. Caird S, Roy R (2018) Blended learning and sustainable development. In Leal Filho W (ed) *Encyclopedia of sustainability in higher education*. Springer, pp 1–10. https://doi.org/10.1007/978-3-319-63951-2_197-1
35. Carvache-Franco M, Alvarez-Risco A, Carvache-Franco W, Carvache-Franco O, Estrada-Merino A, Rosen MA (2021) Coastal cities seen from loyalty and their tourist motivations: a study in Lima, Peru. Sustain 13(21):11575. https://doi.org/10.3390/su132111575
36. Carvache-Franco M, Carvache-Franco O, Carvache-Franco W, Alvarez-Risco A, Estrada-Merino A (2021) Motivations and segmentation of the demand for coastal cities: a study in Lima, Peru. Int J Tour Res 23(4):517–531. https://doi.org/10.1002/jtr.2423
37. Carvache-Franco M, Alvarez-Risco A, Carvache-Franco O, Carvache-Franco W, Estrada-Merino A, Villalobos-Alvarez D (2021) Perceived value and its influence on satisfaction and loyalty in a coastal city: a study from Lima, Peru. J Policy Res Tour Leis Events:1–16. https://doi.org/10.1080/19407963.2021.1883634
38. Chafloque-Cespedes R, Alvarez-Risco A, Robayo-Acuña P-V, Gamarra-Chavez C-A, Martinez-Toro G-M, Vicente-Ramos W (2021) Effect of sociodemographic factors in entrepreneurial orientation and entrepreneurial intention in university students of Latin American business schools. In Jones P, Apostolopoulos N, Kakouris A, Moon C, Ratten V, Walmsley A (eds) Universities and entrepreneurship: meeting the educational and social challenges, vol. 11. Emerald Publishing Limited, pp 151–165. https://doi.org/10.1108/S2040-724620210000011010
39. Chafloque-Cespedes R, Vara-Horna A, Asencios-Gonzales Z, Lopez-Odar DR, Alvarez-Risco A, Quipuzco-Chicata L, et al (2020) Presentismo académico y violencia contra las mujeres en escuelas de negocios e ingeniería en universidades peruanas. Lect Econ (93):127–153. https://doi.org/10.17533/udea.le.n93a340726
40. Chen X, Zhang SX, Jahanshahi AA, Alvarez-Risco A, Dai H, Li J, Ibarra VG (2020) Belief in a COVID-19 conspiracy theory as a predictor of mental health and well-being of health care workers in Ecuador: cross-sectional survey study. JMIR Public Health Surveill 6(3):e20737. https://doi.org/10.2196/20737
41. Chung SA, Olivera S, Rojas Román B, Alanoca E, Moscoso S, Limpias Terceros B, et al (2021) Temáticas de la producción científica de la Revista Cubana de Farmacia indizada en Scopus (1967–2020) [Revista Cubana de Farmacia; farmacia; Cuba; evidencia]. 2021 54(1). http://www.revfarmacia.sld.cu/index.php/far/article/view/511
42. Coursera (2021) Course of sustainability. Retrieved 01 November 2021 from https://es.coursera.org/courses?query=sustainability
43. Cruz-Torres W, Alvarez-Risco A, Del-Aguila-Arcentales S (2021) Impact of Enterprise Resource Planning (ERP) implementation on performance of an education enterprise: a Structural Equation Modeling (SEM). Stud Bus Econ 16(2):37–52. https://doi.org/10.3390/su132111575
44. De Hooge IE, van Dam YK (2019) Reach out and touch: student training community projects for sustainability-a case study. Int J Sustain High Educ 20(7):1278–1289 https://doi.org/10.1108/IJSHE-11-2018-0222
45. Del-Aguila-Arcentales S, Alvarez-Risco A (2013) Human error or burnout as explanation for mistakes in pharmaceutical laboratories. Accred Qual Assur 18(5):447–448. https://doi.org/10.1007/s00769-013-1000-0
46. Delgado-Zegarra J, Alvarez-Risco A, Yáñez JA (2018) Uso indiscriminado de pesticidas y ausencia de control sanitario para el mercado interno en Perú. Rev Panam Salud Publica 42:e3. https://doi.org/10.26633/RPSP.2018.3
47. edX (2021) Sustainability courses. Retrieved 01 November 2021 from https://www.edx.org/learn/sustainability

48. Eldesouky A, Mesias FJ, Elghannam A, Escribano M (2018) Can extensification compensate livestock greenhouse gas emissions? A study of the carbon footprint in Spanish agroforestry systems. J Clean Prod 200:28–38. https://doi.org/10.1016/j.jclepro.2018.07.279
49. Enciso-Zarate A, Guzmán-Oviedo J, Sánchez-Cardona F, Martínez-Rohenes D, Rodríguez-Palomino JC, Alvarez-Risco A, et al (2016) Evaluación de la contaminación con agentes citotóxicos en hospitales en Colombia. Pharm Care España 18(6):241–250. https://www.pharmcareesp.com/index.php/PharmaCARE/article/view/354
50. Espino-Díaz L, Fernandez-Caminero G, Hernandez-Lloret C-M, Gonzalez-Gonzalez H, Alvarez-Castillo J-L (2020) Analyzing the impact of COVID-19 on education professionals. toward a paradigm shift: ICT and neuroeducation as a binomial of action. Sustain 12(14):5646. https://doi.org/10.3390/su12145646
51. Fernández-Gutiérrez M, Gimenez G, Calero J (2020) Is the use of ICT in education leading to higher student outcomes? Analysis from the Spanish Autonomous Communities. Comput Educ 157:103969. https://doi.org/10.1016/j.compedu.2020.103969
52. Filimonau V, Archer D, Bellamy L, Smith N, Wintrip R (2021) The carbon footprint of a UK University during the COVID-19 lockdown. Sci Total Environ 756:143964. https://doi.org/10.1016/j.scitotenv.2020.143964
53. Gamarra AR, Lago C, Herrera-Orozco I, Lechón Y, Almeida SM, Lage J, Silva F (2021) Low-carbon economy in schools: environmental footprint and associated externalities of five schools in southwestern Europe. Energies 14(19):6238. https://doi.org/10.3390/en14196238
54. Greer K, Zeller D, Woroniak J, Coulter A, Winchester M, Palomares MLD, Pauly D (2019) Global trends in carbon dioxide (CO_2) emissions from fuel combustion in marine fisheries from 1950 to 2016. Mar Policy 107:103382. https://doi.org/10.1016/j.marpol.2018.12.001
55. Guimarães da Silva M, Costa Muniz AR, Hoffmann R, Luz Lisbôa AC (2018) Impact of greenhouse gases on surface coal mining in Brazil. J Clean Prod 193:206–216. https://doi.org/10.1016/j.jclepro.2018.05.076
56. Helmers E, Chang CC, Dauwels J (2021) Carbon footprinting of universities worldwide: part I—objective comparison by standardized metrics. Environ Sci Eur 33(1):30. https://doi.org/10.1186/s12302-021-00454-6
57. Henderson B, Golub A, Pambudi D, Hertel T, Godde C, Herrero M et al (2018) The power and pain of market-based carbon policies: a global application to greenhouse gases from ruminant livestock production. Mitig Adapt Strat Glob Change 23(3):349–369. https://doi.org/10.1007/s11027-017-9737-0
58. Hornborg S, Smith ADM (2020) Fisheries for the future: greenhouse gas emission consequences of different fishery reference points. ICES J Mar Sci 77(5):1666–1671. https://doi.org/10.1093/icesjms/fsaa077
59. Huang, Holotescu (2020) Guidance on Open Educational Practices during School Closures: Utilizing OER under COVID-19 Pandemic in line with UNESCO OER Recommendation. Beijing: Smart Learning Institute of Beijing Normal University. https://iite.unesco.org/wp-content/uploads/2020/05/Guidance-on-Open-Educational-Practices-during-School-Closures-English-Version-V1_0.pdf
60. Kalpokaite N, Radivojevic I (2020) Teaching qualitative data analysis software online: a comparison of face-toface and e-learning ATLAS. ti courses. Int J Res & Method Educ 43(3):296–310
61. Katta AK, Davis M, Kumar A (2020) Development of disaggregated energy use and greenhouse gas emission footprints in Canada's iron, gold, and potash mining sectors. Resour Conserv Recycl 152:104485. https://doi.org/10.1016/j.resconrec.2019.104485
62. Keramidas CG (2012) Are undergraduate students ready for online learning? A comparison of online and face-to-face sections of a course. Rural Spec Educ Q 31(4):25–32
63. Kirchherr J, Piscicelli L (2019) Towards an education for the circular economy (ECE): five teaching principles and a case study. Resour Conserv Recycl 150:104406. https://doi.org/10.1016/j.resconrec.2019.104406
64. Lam SK, Suter H, Davies R, Bai M, Mosier AR, Sun J, Chen D (2018) Direct and indirect greenhouse gas emissions from two intensive vegetable farms applied with a nitrification inhibitor. Soil Biol Biochem 116:48–51. https://doi.org/10.1016/j.soilbio.2017.10.008

65. Lavey WG (2019) Teaching the health impacts of climate change in many American higher education programs. Int J Sustain High Educ 20(1):39–56. https://doi.org/10.1108/IJSHE-04-2018-0062
66. Leal Filho W, Morgan EA, Godoy ES, Azeiteiro UM, Bacelar-Nicolau P, Veiga Ávila L et al (2018) Implementing climate change research at universities: barriers, potential and actions. J Clean Prod 170:269–277. https://doi.org/10.1016/j.jclepro.2017.09.105
67. Leal Filho W, Sima M, Sharifi A, Luetz JM, Salvia AL, Mifsud M et al (2021) Handling climate change education at universities: an overview. Environ Sci Eur 33(1):109. https://doi.org/10.1186/s12302-021-00552-5
68. Leiva-Martinez M-A, Anderson-Seminario MdlM, Alvarez-Risco A, Estrada-Merino A, Mlodzianowska S (2021) Price variation in lower goods as of previous economic crisis and the contrast of the current price situation in the context of COVID-19 in Peru. In: Lawrence KD, Klimberg RK (eds) Advances in business and management forecasting, vol. 14. Emerald Publishing Limited, pp 161–166. https://doi.org/10.1108/S1477-407020210000014011
69. Li YL, Han MY, Liu SY, Chen GQ (2019) Energy consumption and greenhouse gas emissions by buildings: a multi-scale perspective. Build Environ 151:240–250. https://doi.org/10.1016/j.buildenv.2018.11.003
70. Lopez-Odar D, Alvarez-Risco A, Vara-Horna A, Chafloque-Cespedes R, Sekar MC (2020) Validity and reliability of the questionnaire that evaluates factors associated with perceived environmental behavior and perceived ecological purchasing behavior in Peruvian consumers. Soc Responsib J 16(3):403–417. https://doi.org/10.1108/SRJ-08-2018-0201
71. Lucivero F (2020) Big data, big waste? A reflection on the environmental sustainability of big data initiatives. Sci Eng Ethics 26(2):1009–1030. https://doi.org/10.1007/s11948-019-00171-7
72. Maraseni TN, Deo RC, Qu J, Gentle P, Neupane PR (2018) An international comparison of rice consumption behaviours and greenhouse gas emissions from rice production. J Clean Prod 172:2288–2300. https://doi.org/10.1016/j.jclepro.2017.11.182
73. Mejía-Acosta N, Alvarez-Risco A, Solís-Tarazona Z, Matos-Valerio E, Zegarra-Arellano E, Del-Aguila-Arcentales S (2016) Reacciones Adversas a Medicamentos reportadas como resultado de la implementación de Atención Farmacéutica en la Farmacia Institucional DIGEMID - Ministerio de Salud de Perú. Pharm Care España 18(2):67–74. https://www.pharmcareesp.com/index.php/PharmaCARE/article/view/311
74. Miśkiewicz R (2021) The impact of innovation and information technology on greenhouse gas emissions: a case of the Visegrád countries. J Risk Financ Manag 14(2):59. https://doi.org/10.3390/jrfm14020059
75. Molthan-Hill P, Worsfold N, Nagy GJ, Leal Filho W, Mifsud M (2019) Climate change education for universities: a conceptual framework from an international study. J Clean Prod 226:1092–1101. https://doi.org/10.1016/j.jclepro.2019.04.053
76. Mostashari-Rad F, Nabavi-Pelesaraei A, Soheilifard F, Hosseini-Fashami F, Chau K-W (2019) Energy optimization and greenhouse gas emissions mitigation for agricultural and horticultural systems in Northern Iran. Energy 186:115845. https://doi.org/10.1016/j.energy.2019.07.175
77. Nunes BT, Pollard SJT, Burgess PJ, Ellis G, De los Rios IC, Charnley F (2018) University contributions to the circular economy: professing the hidden curriculum. Sustain 10(8):2719. https://doi.org/10.3390/su10082719
78. OECD (2021) Green Talks LIVE. https://www.oecd.org/env/green-talkslive.htm
79. Ollivier QR, Maher DT, Pitfield C, Macreadie PI (2019) Punching above their weight: large release of greenhouse gases from small agricultural dams. Glob Change Biol 25(2):721–732. https://doi.org/10.1111/gcb.14477
80. Owens TL (2017) Higher education in the sustainable development goals framework. Eur J Educ 52(4):414–420
81. Parker RWR, Blanchard JL, Gardner C, Green BS, Hartmann K, Tyedmers PH, Watson RA (2018) Fuel use and greenhouse gas emissions of world fisheries. Nat Clim Chang 8(4):333–337. https://doi.org/10.1038/s41558-018-0117-x

82. Perales Jarillo M, Pedraza L, Moreno Ger P, Bocos E (2019) Challenges of online higher education in the face of the sustainability objectives of the United Nations: carbon footprint, accessibility and social inclusion. Sustain 11(20):5580. https://doi.org/10.3390/su11205580
83. Price L, Richardson JT, Jelfs A (2007) Face-to-face versus online tutoring support in distance education. Stud High Educ 32(1):1–20
84. Quispe-Cañari JF, Fidel-Rosales E, Manrique D, Mascaró-Zan J, Huamán-Castillón KM, Chamorro-Espinoza SE et al (2021) Self-medication practices during the COVID-19 pandemic among the adult population in Peru: a cross-sectional survey. Saudi Pharm J 29(1):1–11. https://doi.org/10.1016/j.jsps.2020.12.001
85. Radonjič G, Tompa S (2018) Carbon footprint calculation in telecommunications companies—the importance and relevance of scope 3 greenhouse gases emissions. Renew Sustain Energy Rev 98:361–375. https://doi.org/10.1016/j.rser.2018.09.018
86. Ramísio PJ, Pinto LMC, Gouveia N, Costa H, Arezes D (2019) Sustainability strategy in higher education institutions: lessons learned from a nine-year case study. J Clean Prod 222:300–309. https://doi.org/10.1016/j.jclepro.2019.02.257
87. Ren F, Zhang X, Liu J, Sun N, Wu L, Li Z, Xu M (2017) A synthetic analysis of greenhouse gas emissions from manure amended agricultural soils in China. Sci Rep 7(1):8123. https://doi.org/10.1038/s41598-017-07793-6
88. Riches D, Porter I, Dingle G, Gendall A, Grover S (2020) Soil greenhouse gas emissions from Australian sports fields. Sci Total Environ 707:134420. https://doi.org/10.1016/j.scitotenv.2019.134420
89. Rittl TF, Butterbach-Bahl K, Basile CM, Pereira LA, Alms V, Dannenmann M et al (2018) Greenhouse gas emissions from soil amended with agricultural residue biochars: effects of feedstock type, production temperature and soil moisture. Biomass Bioenerg 117:1–9. https://doi.org/10.1016/j.biombioe.2018.07.004
90. Roesch E, Amin A, Gupta J, García-Moreno C (2020) Violence against women during covid-19 pandemic restrictions. BMJ 369:m1712. https://doi.org/10.1136/bmj.m1712
91. Rojas-Osorio M, Alvarez-Risco A (2019) Intention to use smartphones among Peruvian university students [intention to use; Peruvian students; technology acceptance model; smartphone]. 2019 13(03):13. https://doi.org/10.3991/ijim.v13i03.9356
92. Rojas Román B, Moscoso S, Chung SA, Limpias Terceros B, Álvarez-Risco A, Yáñez JA (2020) Tratamiento de la COVID-19 en Perú y Bolivia y los riesgos de la automedicación [COVID-19; tratamiento; estrategia; fármacos; automedicación; Perú; Bolivia]. 2020 53(2). http://www.revfarmacia.sld.cu/index.php/far/article/view/435/351
93. Ross H, Rudd JA, Skains RL, Horry R (2021) How big is my carbon footprint? Understanding young people's engagement with climate change education. Sustain 13(4):1961. https://doi.org/10.3390/su13041961
94. Sandanayake M, Gunasekara C, Law D, Zhang G, Setunge S (2018) Greenhouse gas emissions of different fly ash based geopolymer concretes in building construction. J Clean Prod 204:399–408. https://doi.org/10.1016/j.jclepro.2018.08.311
95. Sandanayake M, Lokuge W, Zhang G, Setunge S, Thushar Q (2018) Greenhouse gas emissions during timber and concrete building construction—a scenario based comparative case study. Sustain Cities Soc 38:91–97. https://doi.org/10.1016/j.scs.2017.12.017
96. Selvaranjan K, Navaratnam S, Rajeev P, Ravintherakumaran N (2021) Environmental challenges induced by extensive use of face masks during COVID-19: a review and potential solutions. Environ Chall 3:100039. https://doi.org/10.1016/j.envc.2021.100039
97. Shakoor A, Shakoor S, Rehman A, Ashraf F, Abdullah M, Shahzad SM et al (2021) Effect of animal manure, crop type, climate zone, and soil attributes on greenhouse gas emissions from agricultural soils—a global meta-analysis. J Clean Prod 278:124019. https://doi.org/10.1016/j.jclepro.2020.124019
98. Shi H, Lai E (2013) An alternative university sustainability rating framework with a structured criteria tree. J Clean Prod 61:59–69. https://doi.org/10.1016/j.jclepro.2013.09.006
99. Tilbury D (2011) Education for sustainable development: an expert review of processes and learning. https://unesdoc.unesco.org/ark:/48223/pf0000191442

100. Tongwane MI, Moeletsi ME (2018) A review of greenhouse gas emissions from the agriculture sector in Africa. Agric Syst 166:124–134. https://doi.org/10.1016/j.agsy.2018.08.011
101. Torres Ramos LK, Carbo Bustinza N, López Gonzales JL (2017) Huella de carbono y los conocimientos, actitudes y prácticas de los estudiantes y personal del nivel secundario sobre emisiones de gases de efecto invernadero. Apuntes Universitarios 7(2):54–63. https://doi.org/10.17162/au.v7i2.196
102. UNDP (2015) UNDP's Knowledge Management Strategy. https://www1.undp.org/content/undp/es/home/librarypage/capacity-building/knowledge/undp_sknowledge-management-strategy.html
103. UNEP (2020) Experience your carbon footprint in VR. Retrieved 01 November 2021 from https://www.unep.org/news-and-stories/story/experience-your-carbon-footprint-vr
104. UNESCO (2021) UNESCO's support: educational response to COVID-19. Retrieved 01 November 2021 from https://en.unesco.org/covid19/educationresponse/support
105. UNFCCC (2021a) Climate Neutral Now Events 2021. Retrieved 01 November 2021 from https://unfccc.int/climate-action/climate-neutral-now/climate-neutral-now-events-2021
106. UNFCCC (2021b). Glasgow Climate Change Conference. Retrieved 01 November 2021 from https://unfccc.int/es/node/307746
107. Udemy (2021) Courses of sustainable development. Retrieved 01 November 2021 from https://www.udemy.com/topic/sustainable-development
108. Universidad de Lima (2021) Centro de Sostenibilidad. Retrieved 01 November 2021 from https://www.ulima.edu.pe/sostenibilidad
109. University of Cambridge (2021) Cambridge Institute for Sustainability Leadership. Retrieved 01 November 2021 from https://www.cisl.cam.ac.uk/
110. Versteijlen M, Perez Salgado F, Janssen Groesbeek M, Counotte A (2017) Pros and cons of online education as a measure to reduce carbon emissions in higher education in The Netherlands. Curr Opin Environ Sustain 28:80–89. https://doi.org/10.1016/j.cosust.2017.09.004
111. Vetter SH, Sapkota TB, Hillier J, Stirling CM, Macdiarmid JI, Aleksandrowicz L et al (2017) Greenhouse gas emissions from agricultural food production to supply Indian diets: implications for climate change mitigation. Agr Ecosyst Environ 237:234–241. https://doi.org/10.1016/j.agee.2016.12.024
112. Wei S, Bai ZH, Chadwick D, Hou Y, Qin W, Zhao ZQ et al (2018) Greenhouse gas and ammonia emissions and mitigation options from livestock production in peri-urban agriculture: Beijing—a case study. J Clean Prod 178:515–525. https://doi.org/10.1016/j.jclepro.2017.12.257
113. Whalen KA, Berlin C, Ekberg J, Barletta I, Hammersberg P (2018) 'All they do is win': lessons learned from use of a serious game for circular economy education. Resour Conserv Recycl 135:335–345. https://doi.org/10.1016/j.resconrec.2017.06.021
114. World Bank (2021) Number of countries with closed schools worldwide in 2021, by status. Retrieved 01 November 2021 from https://www.worldbank.org/en/data/interactive/2020/03/24/world-bank-education-and-covid-19
115. Yan J, Kim S, Zhang SX, Foo M-D, Alvarez-Risco A, Del-Aguila-Arcentales S, Yáñez JA (2021) Hospitality workers' COVID-19 risk perception and depression: a contingent model based on transactional theory of stress model. Int J Hosp Manag 95:102935. https://doi.org/10.1016/j.ijhm.2021.102935
116. Yáñez JA, Afshar Jahanshahi A, Alvarez-Risco A, Li J, Zhang SX (2020) Anxiety, distress, and turnover intention of healthcare workers in Peru by their distance to the epicenter during the COVID-19 crisis. Am J Trop Med Hyg 103(4):1614–1620. https://doi.org/10.4269/ajtmh.20-0800
117. Yáñez JA, Alvarez-Risco A, Delgado-Zegarra J (2020) Covid-19 in Peru: from supervised walks for children to the first case of Kawasaki-like syndrome. BMJ 369:m2418. https://doi.org/10.1136/bmj.m2418
118. Zhang SX, Chen J, Afshar Jahanshahi A, Alvarez-Risco A, Dai H, Li J, Patty-Tito RM (2021) Succumbing to the COVID-19 pandemic—healthcare workers not satisfied and intend to leave their jobs. Int J Ment Heal Addict. https://doi.org/10.1007/s11469-020-00418-6

119. Zhang SX, Chen J, Jahanshahi AA, Alvarez-Risco A, Dai H, Li J, Patty-Tito RM (2021) Correction to: succumbing to the COVID-19 pandemic—healthcare workers not satisfied and intend to leave their jobs. Int J Ment Heal Addict. https://doi.org/10.1007/s11469-021-00502-5
120. Zhang SX, Sun S, Afshar Jahanshahi A, Alvarez-Risco A, Ibarra VG, Li J, Patty-Tito RM (2020) Developing and testing a measure of COVID-19 organizational support of healthcare workers—results from Peru, Ecuador, and Bolivia. Psychiatry Res 291:113174. https://doi.org/10.1016/j.psychres.2020.113174
121. Zhuang M, Lu X, Caro D, Gao J, Zhang J, Cullen B, Li Q (2019) Emissions of non-CO_2 greenhouse gases from livestock in China during 2000–2015: magnitude, trends and spatiotemporal patterns. J Environ Manage 242:40–45. https://doi.org/10.1016/j.jenvman.2019.04.079

GPSR Compliance
The European Union's (EU) General Product Safety Regulation (GPSR) is a set of rules that requires consumer products to be safe and our obligations to ensure this.

If you have any concerns about our products, you can contact us on

ProductSafety@springernature.com

In case Publisher is established outside the EU, the EU authorized representative is:

Springer Nature Customer Service Center GmbH
Europaplatz 3
69115 Heidelberg, Germany

www.ingramcontent.com/pod-product-compliance
Ingram Content Group UK Ltd.
Pitfield, Milton Keynes, MK11 3LW, UK
UKHW022213051026
15624UKWH00001B/142
9789811905483

Breaching the Bronze Wall

Mediterranean Reconfigurations

INTERCULTURAL TRADE, COMMERCIAL LITIGATION, AND LEGAL PLURALISM

Series Editors

Wolfgang Kaiser (*Université Paris I, Panthéon-Sorbonne*)
Guillaume Calafat (*Université Paris I, Panthéon-Sorbonne*)

VOLUME 2

The titles published in this series are listed at *brill.com/cmed*

Breaching the Bronze Wall

Franks at Mamluk and Ottoman Courts and Markets

By

Francisco Apellániz

BRILL

LEIDEN | BOSTON

The publication of this book has been funded by the European Research Council (ERC) [Advanced Grant n° 295868 "Mediterranean Reconfigurations: Intercultural Trade, Commercial Litigation, and Legal Pluralism in historical perspective"]

Cover illustration: Vittore Carpaccio, *Group of Soldiers and Men in Oriental Costume*, Tempera on wood, 68 × 42 cm, Galleria degli Uffizi, Florence.

Library of Congress Cataloging-in-Publication Data

Names: Apellániz Ruiz de Galarreta, Francisco Javier, 1974- author.
Title: Breaching the bronze wall : Franks at Mamluk and Ottoman courts and markets / Francisco Apellániz.
Description: Boston ; Leiden : Brill, 2020. | Series: Mediterranean reconfigurations, intercultural trade, commercial litigation, and legal pluralism, 2494-8772 ; volume 2 | Includes bibliographical references and index.
Identifiers: LCCN 2020018154 (print) | LCCN 2020018155 (ebook) | ISBN 9789004382749 (hardback) | ISBN 9789004431737 (ebook)
Subjects: LCSH: Dhimmis (Islamic law)–Middle East–History. | Witnesses (Islamic law) – Middle East–History. | Authentication–Middle East–History. | Merchants–Legal status, laws, etc.–Middle East–History. | Egypt–History –1250-1517. | Syria–History–1260-1516. | Turkey–History –Ottoman Empire, 1288-1918.
Classification: LCC KMC565 .A64 2020 (print) | LCC KMC565 (ebook) | DDC 347.56/06609023–dc23
LC record available at https://lccn.loc.gov/2020018154
LC ebook record available at https://lccn.loc.gov/2020018155

Typeface for the Latin, Greek, and Cyrillic scripts: "Brill". See and download: brill.com/brill-typeface.

ISSN 2494-8772
ISBN 978-90-04-38274-9 (hardback)
ISBN 978-90-04-43173-7 (e-book)

This book is printed on acid-free paper and produced in a sustainable manner.

Contents

Acknowledgements

This book represents the final step of a long journey that took me through Florence, Rome, Marseilles and Cairo and ended up at the baroque setting of the Orientale University in Naples. This long wandering protracted for more than a decade, involving several historical endeavors and an uncertain quest for an academic safe harbor. Somewhere in the middle of this process, the moral and legal implications of Islamic governance captured my scholarly attention and diverted my efforts away from a previous focus on Mediterranean networks, economic institutions and mutable identities. My teachings of Islamic history at Aix-en-Provence in the years 2009–2012, and the long epilogue of academic wanderings that followed are largely responsible for this turn. The book's moral and legal concern probably begun under the influence of the large share of Muslim students populating my courses at that time, within the contrasting, secular scene of a French University. If only, my exploration of the *bronze wall* helped mitigate the journey's numerous dark moments. The project's instigator, though, is to be searched among the fold of Mediterranean Historians. In 2014 Wolfgang Kaiser invited me to join his European Research Council-funded project Mediterranean Reconfigurations and asked me to venture, together with a bunch of early-modern historians, into the unknown territory of legal relations across confessions, reaching as far as the hitherto (for me) unexplored sixteenth century. At the time, he barely knew me, and was not short of alternatives, what makes my debt of gratitude to his intellectual trust hardly able to repay. I would not be writing these lines as an active researcher were it not for his intransigent attitude to prioritize intellectual concerns. This research is what I render in partial return of his generosity, and I am glad to hear that, at the end, "vous avez très bien joué le jeu."

It is very unlikely that this book would have ever seen the light without the long-standing support to my career by Luca Molà and Wolgang Kaiser himself, together with that of Giovanni Levi, who forcefully added public demonstration of appreciation for my work when I needed it the most. Thanks to their mentoring and to the EU-funded Marie Skłodowska-Curie Actions, I became depositary of a two-year Intra European fellowship at the Department of History and Civilization at the European University Institute. It is in the fertile intellectual environment of my European alma mater where I developed the core research on the Siyāsa courts, published it in Comparative Studies in Society and History, and where I had the chance to benefit from the fleeting, yet profound influence of a seminal lecture by Anver Emon. Emon's talk instilled me the necessity to look at minorities and more generally to 'others' to understand Islamic governance and rule of law. My essay on siyāsah was discussed in several formats and venues by Tijana Krstić, Tamar Herzog, Stéphane Van Damme, Jorge Flores, Mercedes García-Arenal, Maribel Fierro, Yavuz Aykan and Guillaume Calafat. I am particularly indebted to Ersilia Francesca as well as to the legal historian that anonymously reviewed my manuscript at Comparative Studies, and who contributed a great deal to sharpen the expression of my ideas, later crystalized in this book.

Together with Luca Molà, John F. Meloy counts among the number of my supporters. Both acted as mentors for my candidature for the Harvard Center for Renaissance Studies in Florence in 2017–2018, hence deserving a special share of my gratitude, together with the committee that granted me a Deborah Loeb-Brice yearly fellowship at Villa I Tatti, under whose auspices most of the writing was carried out. At I Tatti I benefitted from the surprising views on the moral implications of law and from the erudition of Justin Steinberg. Daily conversations over vermouth with Sanjay Subrahmanyam extended the horizons of law and governance into the wider setting of the Indian Ocean and its talkative Arab Chroniclers. I would like to thank Roberto Tottoli and Michele Bernardini for putting an end to my wanderings and welcoming me, in November 2018, in the Department of Asian African and Mediterranean Studies at the Orientale of Naples, where they witnessed the closing stages of this project.

Research for this book extended for over two years and took me on many stays through the archival collections of Genoa and particularly, at the Archivio di Stato in Venice, in whose gothic refectory I could eavesdrop, almost undisturbed, on many crucial conversations held at Istanbul's Imperial Dīvān. Valentina Ruzzin put her vast knowledge of Genoa's archives at my disposal, serving me with two very significant documents for this research. I am very lucky to have met Davide Gambino and witnessed to his overwhelming knowledge of the Genoese and their archives; without his acquaintance, Jorge Luis Borges' characters would have remained confined to the only sphere of fiction. I enjoyed his valueless help with transcriptions and findings both of Genoese and Florentine collections. I owe to Marga Castells – several times translator of the Nights – the story opening the book, and to Cédric Quertier to share with me his finding of the Florentine Judicial inquiry on Alexandria, that he unearthed in the course of a meticulous scrutiny of the Mercanzia court records. I took advantage of the voluminous knowledge of old-regime Venice by Mauro Bondioli, who put me on the trail of an important issue concerning Ottoman-Venetian relations. Daniel Duran i Duelt helped me with difficult transcriptions and more generally, with his contribution to my research project on Venetian outremer notarial records. Separate mention deserve here my younger colleagues, Alejandro García Montón, Cloe Cavero and José Miguel

Escribano, marching in the same phalanx line during my long post-doctoral journey. During the writing of this book I spent some time in the realm of the jinn but, thanks to my family, came back to see it finished.

Special gratitude is addressed to Maurits H. van den Boogert for housing our publishing project at Brill Publishers and to an anonymous reviewer for his insight and comments. Andromeda Tait reviewed my English and untied the knots of my style; the remaining mistakes are all mine. As a historian without any formal training in Ottoman history, I am grateful to Güneş Işiksel for handing me his painstaking work on reproducing the Bailo a Costantinopoli series, for his help with transcriptions and for having my manuscript read and commented upon. This does not make me immune from misinterpretations in Ottoman history which are entirely my fault.

Francisco Apellániz
Naples, January 2020

Abbreviations

BM	Venice, Biblioteca Marciana
BC	Venice, Biblioteca Correr
BNF, DO	Paris, Bibliothèque Nationale de France, Division Orientale
BNCR	Rome, Biblioteca Nazionale Centrale
ASVe	Venice, Archivio di Stato di Venezia
ASVe, CI, N	Archivio di Stato di Venezia, Cancelleria Inferiore, Notai
ASG	Genoa, Archivio di Stato di Genova
ASG SG	Genoa, Archivio di Stato di Genova, San Giorgio
ASF	Florence, Archivio di Stato di Firenze
ASPo	Prato, Archivio di Stato di Prato

CHAPTER 1

Introduction

> And I have made you this day a fortified city, an iron pillar, and bronze walls against the whole land … They will fight against you but will not prevail
>
> JEREMIAH 1:18

∵

In the three-hundred and forty-third Arabian Night the Mamluk sultan al-Mālik al-Nāṣir sent for the chiefs of police of each the three urban agglomerations of late medieval Cairo: al-Qāhira, the new city within the walls, Būlāq, its riverine port district on the Nile bank, and Fusṭāṭ, Coptic old Cairo. Al-Mālik al-Nāṣir desired each inspector to recount the most astounding story they had encountered in the exercise of their duties. The chief inspector of Būlāq's narrative dealt with counterfeiters and deceivers, while the chief of Fusṭāṭ told a gruesome story about thieves and their executioners. Most significantly for the present work, al-Qāhira, the new city built by the Fatimids and the epicenter of Mamluk life, served as the backdrop for a tale about false witnesses. The inspector narrated the story of two "professional witnesses" (ar. *shuhūd ʿudūl*, or simply *ʿudūl*); upright Muslims of good reputation and sound of mind permitted by the qadi to give testimony about people and facts under his jurisdiction. It was revealed that both witnesses had been secretly leading a life of dissolute ways, indulging in the company of low women and the consumption of wine. The chief of police planned to trap the witnesses, with the complicity of the tavern- and brothel-keepers. Informed by the latter that an episode of debauchery involving the two men was taking place, the chief of police came to the brothel in disguise. With no apparent sign of alarm, the two men, together with the landlord, welcomed the police officer in and proceeded to bribe him for three hundred dinars. Tempted by the money, the chief of police accepted to cover for the two corrupt men, but only that one time. To the police inspector's horror, the following day the local qadi summoned him to appear in court, to answer to a debt of three hundred dinars claimed by the brothelkeeper. The plaintiff exhibited a written deed whereby the chief of police acknowledged the debt, a document duly certified by the two legal witnesses, who attested to the validity of the transaction. Unable to counter this burdensome evidence, the chief of police paid the sum, and left vowing to have his revenge against the two witnesses. The story raises many questions for a contemporary reader: first of all, whether these men really did enjoy such a high reputation in their community, and why were they endowed with the privilege to testify in court. Since the paper exhibited to the qadi clearly bore no signature, why was it accepted by the court? Was the record valid just because it was supported by two "legitimate" witnesses?

This book deals with the Islamic idea that the word of honorable Muslims constitutes the proof par excellence, and that written documents and the testimony given by non-Muslims are of inferior value. According to this view, a merchant hailing from Christian Europe was at great pains to prove even the smallest claim in court, as neither his contracts nor his word were of any value if countered by Muslims. By the same token, a Frank made captive by corsairs or in a borderland attack in the Balkans was expected to prove his free condition by means of Muslim witnesses, failing which he was delivered to whosoever claimed to be his legitimate master. How, if at all, did Franks and Muslims manage to breach or circumvent what the late nineteenth-century jurist Francesco Contuzzi (1855–1925) called the "bronze wall"? This question lies at the heart of this book. In a nutshell, the present work addresses how the so-called 'biases against non-Muslims' were dealt with in medieval and early modern commercial litigation, in diplomatic talks and, more broadly, how discrimination based on religious affiliation could play out in the legal system. Together with the mistrust that Islamic law exhibited towards the word of non-Muslims, this work deals extensively with the chief inspector's dilemma concerning the written deed produced against him and its legal value.

In recent years, historians have focused on the effects of these biases on Islamic societies, asking how they might have impacted Islamic economic institutions, or whether they might have fostered a shift towards a more European style of laws and courts.[1] My research deals instead with these precepts as a historical object, by attempting to understand how, in a typical Islamic

1 Important debates on the capitulatory regime, its economic significance, and issues of proof can be found in Kuran, Timur: *The Long Divergence: How Islamic Law Held Back the Middle East*, Princeton: Princeton Univ. Press, 2010, Van den Boogert, Maurits H.: *The Capitulations and the Ottoman Legal System Qadis, Consuls, and Beratlıs in the 18th Century*, Leiden; Boston: Brill, 2005, Ergene, Boğaç A., "Evidence in Ottoman Courts: Oral and Written Documentation in Early-Modern Courts of Islamic Law", Journal of the American Oriental Society 124 3 (2004), 471–491. Scholarly debates on the impact of European law and institutions revolve around the the "Jurisprudential Shift Hypothesis": Van den Boogert, Maurits H., "Legal Reflections on the 'Jurisprudential Shift Hypothesis' ", Turcica 41 (2009), 373–382, Kuran, Timur

context of legal and religious pluralism, people sought to overcome the challenges posed by these sharīʿa-based principles. In order to answer this fundamental question, I aim to locate, describe and historicize a change in attitude regarding diversity in Middle Eastern marketplaces and courtrooms. It should be noted, however, that my research runs parallel to two important trends in Islamic studies. The first focuses on majority-minority relations, i.e. those binding Muslims to the local Christian and Jewish populations, or dhimmīs. The second looks to Islamic jurisprudence, or *fiqh*; foreigners, in contrast with dhimmīs and Islamic sectarians, did not belong to the Islamic community and, according to jurists, had no interest in the common good, and therefore their role in the legal system was inferior. Unlike dhimmīs, they received little legal attention by Islamic Law. Rather than involve itself in these two dominant research subjects in Islamic studies, this book turns instead to a practical, historical reality that has largely passed under the radar in the scholarly debate. The restriction of proof to the testimony of Muslim witnesses, as well as the denial of validity for written records has been attested since the first century of the ḥijra. Caliph Yazīd b. ʿAbd al-Mālik (ruled 720–723), is credited for banning unbelievers to testify for or against Muslims, a measure qualified by a Syrian chronicler as the *starting point of these perverted laws*.[2]

These attitudes have been presented by researchers as characteristic of the Islamic legal system, surfacing in legal discourse around the second half of the second century,[3] even though the Qurʾān itself does not express any reserves about the drafting of documents and the use, if necessary, of unbelievers as witnesses. These biases have accompanied Islamic societies through the ages and, although some claim they weakened in late modern times, descriptions of Muslim testimony superseding that of Christians and Jews can be found

---

and Scott Lustig, "Structural Inefficiencies of Islamic Courts: Ottoman Justice and Its Implications for Modern Economic Life", ERID Working Paper Number 52 (2010), 1–35, Çizakça, Murat and Macit Kenanoğlu: "Ottoman Merchants and the Jurisprudential Shift Hypothesis", in: *Merchants in the Ottoman Empire*, Edited by S. Faroqhi and Gilles Veinstein, 195–213. Paris-Louvain-Dudley, MA: Peeters, 2008, Çizakça, Murat, "Was Shariʿah indeed the culprit?", MPRA Paper 22865, University Library of Munich, Germany (2010).

2 "Il prescrivit aussi qu'on ne récut point le témoignage d'un Syrien contre un Arabe. Il fixa le prix [du sang] d'un Arabe a douze mille [dinars] et celui d'un Syrien a six mille. C'est là l'origine de ces lois perverses", *Chronique de Denys de Tell-Mahré: quatrième partie*, edited by Jean Baptiste Chabot. Vol. 1, Paris: É. Bouillon, 1895, Tillier, Mathieu, "Califes, émirs et cadis: le droit califal et l'articulation de l'autorité judiciaire à l'époque umayyade", Bulletin d'Etudes Orientales, 63 (2014), 147–190.

3 Hurvitz, Nimrod: "Legal Doctrines, Historical Contexts And Moral Visions: The Case Of Sectarians in The Courts Of Law", in: *The Islamic Scholarly Tradition*, 239–264: Brill, 2011.

even in late-Ottoman, nineteenth-century court practice. Indeed, the impact of these prejudices was not uniquely confined to courts of law. It soon extended to the notarial practice and in customs houses, and had become a cumbersome presence at the Ottoman Imperial Council, where diplomatic talks were held. These legal features were instrumental in shaping specific institutions, artifacts and procedures, and in deepening the legal divide vis-à-vis their Latin Christian counterparts—a divide that became its most pronounced in the late Middle Ages. Islamic archives, if they existed at all, may have looked very different from European ones, as did notaries, their registers, and the way proof was advanced in court. At the end of this period, the divide generated artifacts such as the notarial register in the West, and scrolls up to ten meters long in the Islamic world.

Not all scholars focus solely on jurisprudence, and indeed recent works have looked at how the intricacies of witnessing, proof and evidence have affected the daily practice of justice and the drafting of documents. Despite their undisputed erudition, however, they depart from the chaotic legal world of the Mediterranean port cities addressed in the present work, and, one could argue, describe societies that seem to be inhabited exclusively by qadis. My aim here is to undertake a historical survey of how cross-confessional relations could be affected by these two important features of the Islamic legal system.

Although historical in nature, this work is concerned with the handling of legal principles and ideas by the users of the legal system, often lower-rank users, rather than presenting salient facts in the political history. Yet it aims at contributing to a better understanding of the transition from medieval ways to deal with diversity and unbelief to those common in the Early-Modern Mediterranean. The core facts discussed in the present research extend from the decade of 1350s, when Arab authors became more talkative about the legal affairs of foreigners, and when Venetians felt the need to dispatch their own Latin notaries to Syria and Egypt, hence welcoming the region into the Latin archival Ecumene. The main narrative becomes thicker between the years 1390s and 1440s, due to the impulse for governance changes by sultans such as al-Ẓāhir Barqūq (1382–1389 and 1390–1399), al-Mu'ayyad Shaykh (1412–21) and al-Ashraf Barsbāy (1422–1438), as well as to heftier collections of notarial deeds in Venice's archives than those preserved for the second half of the 15th century. The core narrative ends with the scrutiny of the correspondences sent by Venetian diplomats in Istanbul between the 1480s and the 1550s. The Mamluk sultanate was embedded in both the spice and silk routes and hosted an unprecedented number of foreign merchants, often non-Muslims, ranging from Malacca and Ceylon to France and the Iberian peninsula in the

West.[4] The Arab regions I scrutinize in this work hosted their own scholarly circles and legal institutions, and a tradition of dealing with Franks going back in time to Crusader and Byzantine periods (this precedent, eventually, will justify some breaches in the chronology). The Circassian Sultans (a branch of the Mamluk dynasty that ruled Egypt and Syria between 1382 and 1517) and their forerunners in the mid-fourteenth century, such as al-Nāṣir Muḥammad Ibn Qalāwūn (d. 1341) and his household, modelled a system by which communities were granted full legal coverage, starting from their very presence in the Islamic realm to the legal nature of their activities and the resolution of their eventual disputes. As we will see, one of the features that emerged in the mid-fourteenth century was that mixed disputes, this is, those involving at least a Muslim party, needed to be heard by an Islamic judge and not by a consul or by any other Frankish magistrate. Under the great sultans al-Ashraf Qāytbāy (1468–1496) and Qānṣūh al-Ghawrī (1501–1516), as well as under 16th-century Ottoman rulers, the supremacy of Islamic law in governing cross-confessional disputes was never discussed. Yet in 1625 a Venetian consul in Alexandria sent a report to the Doge on the current state of affairs during his tenure in office. The consul noted how, after centuries of Venetian presence in the country, the spice trade had almost abandoned Egypt, to the benefit of the new transoceanic routes. Most striking is a passage devoted to justice, where the consul assures " to your Most Serene prince that since these peoples hold the consul's justice in such high esteem, often the Turks and Moors of the country, in mixed cases (*nelle cause miste*), while being able to benefit from the justice of the cadis, bring their disputes before my justice, with the other Frankish nations, as they do with the French, whose consul enjoys great reputation".[5] By depicting a scenario where the supremacy of sharīʿa is no longer observed, the *relazione* sets a *terminus ante quem*. Whether the Consul is exaggerating the Muslims' confidence in Frankish justice we do not know, but it is fair to say that by that time the Mediterranean addressed by the present work was long gone.

Needless to say, sketchy, oversimplified view of the biases' role in Islamic history have been challenged on several fronts, starting with the work of the

4 Apellániz, F. (2016). "News on the Bulaq: a Mamluk-Venetian Memorandum on Asian Trade, AD 1503." EUI Working Paper HEC 2016/01.

5 "e prometto a Vostra Serenità esser appresso quelle genti tanto in stima la giustitia del console, che molte volte li turchi et mori del paese, nelle cause miste, che potevano valersi della giustitia delli giudici, et cadi, venivano alla mia giustitia, et volevano da me esser giudicati con le altre nation franche, et in particolare con la francese, che ha console di molta reputatione", Venice, Biblioteca Correr (hereinafter BC), Manoscritti Provenienza Diversa, C 306.

great legal historian Emile Tyan, who long argued that some schools of jurisprudence, such as the Mālikī, were more open to the use of written documents, and that in places where the school was hegemonic this attitude could have modeled court practices. Papyrologists and other specialists in Islamic documents have difficulty accepting the idea that, since documents had no *per se* legal value, Islamic societies did not rely on records as much as their Western counterparts did. Since archives hosting judicial and legal records do not seem to have survived for the Middle Ages, a growing number of researchers have been brandishing extant, fragmentary collections of written artifacts so as to refute the alleged inferiority of Muslims as regards writing and documentation. Chapter Two deals with this highly controversial issue, which has taken center-stage in Islamic studies, and weighs in by arguing that although a proper archival logic nearly emerged under the Ottomans, it certainly did not appear before then.

A privileged locus in the discussion is the judicial archive—the qadi's dīwān—since major scholars such as Wael B. Hallaq argue it has always existed in Islamic societies. Research on medieval judicial documents is now claiming that documents were far more important than previously believed, and that they were systematically preserved by Muslims. Specialists of legal practice and of the functioning of qadi courts have sought to challenge a long-standing assumption: that medieval Islamic law was characterized by the strain placed by the demands of practical life on doctrine. Traditionally, in the absence of judicial archives for the Middle Ages, scholars have presented Islamic medieval law as purely theoretical, and fundamentally divergent from actual practice, a divide only narrowed under the Ottomans, with their pragmatic approach to lawmaking.[6] In the case of both the Mamluk and Ottoman legal systems, authors have felt the need to dispute the established idea that sharīʿa was a formalistic, idealistic and skeptical system that mistrusted the human capacity to attain the truth, and that therefore its precepts were not followed in practice. Recently, the work carried out on a judicial collection found in Jerusalem has helped to challenge this view, dwelling on the sophisticated procedures used by qadis to comply with the demands of sharīʿa, particularly as regards the complex validation procedures found in Mamluk documents. However, these works' high degree of technical sophistication and their restriction to court procedure does not help historians to make sense of daily transactions with unbelievers. Similarly, researchers

6 Schacht, Joseph, An introduction to Islamic law, Oxford-New York, Clarendon Press, 1982, 76–85.

dealing with Ottoman judicial archives tend to conflate actual practice with works of jurisprudence, hence building the image of a legal empire whose jurists dedicated great effort to making the sultan's decrees compatible with sharīʿa. My book does not concern itself with the technical solutions found by the qadis to make documents valid as bearers of truth, nor with the vexing question of whether or not practice followed sharīʿa doctrine. However, it does dwell on these works' findings, which point for example to the Ottoman invention of the register—or *sicil*—a landmark in Islamic history, since without it historians would have no archives to investigate. I am much indebted to the contributions of Reem Meshal, in which she suggests there was an important change in attitude towards writing and documentation in the 16th century, which brought about the invention of the archive, and heralded a totally new approach by common people to documents and courts of justice. A great deal of attention will be paid to the scribes and notaries of the various confessions, to the artifacts they produced, and to the problems encountered by witnesses, either before the judges or simply when attesting to the validity of transactions in the marketplace.

This book is concerned with the actors and practices that have often passed under the radar of traditional scholarship on Islamic law. Much attention will be dedicated to the lowest layer of the legal profession: the notaries. Unlike their European counterparts, Islamic notaries have garnered little interest. Lisān al-Dīn Ibn al-Khaṭīb (1313–74) a secretary from Islamic Granada, described Islamic witnesses as materialistic men of ill-repute who could be found hanging around the markets.[7] According to him, they passed themselves off as versed in law and in the drafting of contracts, often posing at the door of their workshop in the market surrounded by law books, in a calculated display of legal mastery. As in the case of the witnesses' story from the Nights, it takes some effort for the contemporary reader to understand the logic behind some of these episodes reported by Arab chroniclers. Ibn al-Khaṭīb describes notaries showing up at lunchtime at their clients' houses, and the latter feeling obliged to make room at their table for the scribes. Yet if they did not trust a particular clerk, due to his ignorance of the law or a reputation of dishonesty, why did clients not simply go to the next notary workshop in the street? As in any other aspect of daily life, sharīʿa law is entrusted to implement a normativity stemming from Islam as a religion. Notaries were nominated by a qadi, who acknowledged his qualities as a probative and trustworthy witness; thus there

7 Turki, Abdelmagid, "Lisan al-Din Ibn al-Khatib (713–76/1313–74), juriste d'après son oeuvre inédite: *Muthla l-Tariqa fi dhamm al-wathiqa*", Arabica **16** (1969), 155–211 and 279–312.

was no official investiture or any requirement to follow a professional curriculum. The tension between the Islamic notaries and their clients resided in the fact that the former's appointment depended upon local qadis, which meant that the chances of actually choosing a notary were in fact fairly limited. Unlike Muslim ʿudūl, Latin notaries were to be blamed for lying and committing perjury at court, but overall for their capacity to forge notary deeds. This was the main accusation against Ser Cepparello da Prato, the famous character by Ibn al Khaṭīb's contemporary Giovanni Boccaccio (1313–75) *perhaps the worst man that ever was born*, and who tried to cheat God almighty in his deathbed confession.[8]

The kind of legal issues that this book looks into was rarely commented upon by Arab authors; they did not deem it necessary to detail the risks associated with notaries and documents, and as in the story of the Cairo chief of police, it is often difficult to grasp the underlying mechanisms behind some of these episodes. Alongside harsh criticism, favorable portrayals of notaries abound, and indeed descriptions of virtue are often the best way to understand, by contrast, censurable attitudes. For example, Arab authors describe honorable notaries who refused women or dhimmīs the right to enter their workshops as clients. Considered to be socially weak, and whose word was legally inferior in court, it was implied that the latter could be blackmailed by corrupt ʿudūl, and therefore that accepting their legal business could be harmful to a scribe's reputation. Ṭūlūn (1485?–1546) tells the story of Khitāb Ibn ʿUmar al-Shuwaykī, a honorable man who for reasons initially unknown fell progressively ill, ending up in a *maristān*, or hospital, in the Ṣāliḥiyya neighborhood of Damascus. Upon leaving the hospital he stopped at a shop and bought a length of rope for half a dirham, with which he subsequently hung himself. Ibn Ṭūlūn briefly comments on the motivations behind al-Shuwaykī's desperate act: he saw himself in the position of having to handle his wife's estate during her absence for a pilgrimage to Mecca. Indeed, after his suicide, it was found that a deposit (*wadīʿa*) of four hundred and thirty gold dinars had been made to a religious institution. In light of the abundant criticism against notaries by chroniclers, one would expect Ibn Ṭūlūn to conclude that al- Shuwaykī had been a victim of dishonest notaries. Instead, he informs us that one of the madrasa's staff refused to keep the money without the acknowledgement of witnesses. Ibn Ṭūlūn is implying that new, supplementary testimony of the deposit needed to be given, and that a new deed must have been drawn up. This situation would have never occurred in Latin Europe,

8 The Decameron, by Giovanni Boccaccio, translated by J. M. Rigg, Day 1, Tale 1.

where a valid, original deed would have been kept by a notary at the disposal of the right holders.[9]

Another fundamental problem is the immaterial product of the notary's work. Unlike in the West, in the Islamic world, and particularly from the tenth century, witnessing services were divided into two steps. In the first step, testimony of a given fact was taken by a notary-witness (*taḥammul al-shahāda*). The second step implied the performance of witnesses before a judge (*adāʾ al-shahāda*). In general, in this second step the notary and other witnesses (usually two) gave oral testimony for the transaction. Some people were enabled by the judge to take on these services on behalf of the community, and came to be known as professional notaries, 'reputable witnesses' (*shuhūd ʿudūl*, hereinafter simply ʿudūl), or notary-witnesses (who Ibn al-Khaṭīb sarcastically describes for the case of Granada). Although the qadi, as a result of this second step, did sometimes produce some paperwork for his own use, the task of the ʿudūl was in substance limited to witnessing and subsequently giving oral testimony. The ʿudūl therefore, although they delivered a written deed to the parties as an aide-memoire, were not clerks charged with the keeping of original—or 'authentic,' as they are called in Western notarial jargon—documents in case one of the parties lost the copy. The fact that they did not file 'originals' in a ledger or casebook puzzled Latin merchants. Ultimately, clients resorted to the ʿudūl because the latter were authorized by the local judge to act as depositaries of the truth, which implies that it was difficult to ensure the validity of a legal business, ranging from property to family and probate issues, without the notary's consensus, and eventually that of the witnesses accompanying him. To put it plainly, in late medieval times, people needed notaries in order to enjoy and maintain their actual rights, and it was for this reason that from time to time they welcomed the notary at their table.

European notaries have been celebrated as a unique feature of Western society, thanks to their capacity to confer legal validity to peoples' agreements and transactions. They were able to bestow public faith on the deeds they drew up, by virtue of their investiture by legitimate powers, such as the Emperor or the Pope. Islamic notaries, on the other hand, do not benefit from this triumphal view of institutional history, and the ʿudūl's capacity to produce documents endowed with legal validity was far more limited. It is these conflicting views about the notarial institution that I will be discussing in this book. I will be targeting the assumption that the Islamic approach to notarization

9 Ibn Ṭūlūn, Shams al-Dīn Muḥammad Ibn ʿAlī (1485?-1546): *Mufākahat al-Khillān fī Ḥawādith al-Zamān*, Edited by Khalīl al-Manṣūr, Beirut: Dār al-Kutub al-ʿIlmīyah, 1998, 193.

was fundamentally different from the Western one, cast from classical models and Roman law. According to this assumption, the features characterizing the labor of the ʿudūl stem from a religious normativity hailing from Islam as a religion, such as its formalistic principle that only good Muslims can be bearers of truth. Western notaries would have dwelt, instead, upon the old Roman practice of granting public faith to notarial deeds.

Siyāsa justice is another practice that has gone fundamentally unnoticed. Sharīʿa law can be described as a series of divine commandments to the community of believers, an abstract and general notion that historians of Islamic law increasingly refer to as the 'rule of law.' Sharīʿa as a legal ideal differs then from positive law; i.e. the trove of commentaries found in works of jurisprudence (*fiqh*). It is generally agreed that sharīʿa was applied by the Mamluks, together with most medieval sultanates, in two parallel, well-defined areas of jurisdiction. The first court level was run by qadis under the supervision of the four chief justices of the *madhhabs*, or legal schools. Their decisions could be appealed at a second, superior jurisdiction, which came in the form of 'royal courts' presided over by sultans and officials. Under the late Mamluks, these special courts gained in importance, leading to the development of a parallel judiciary that was perceived by many as competing with the traditional qadi courts. Several Arabic terms were used to denote the different manifestations of royal jurisdiction, such as *maẓālim* and *Siyāsa*, however all were grounded in the same legal doctrines. Scholars have traditionally sought to understand these special jurisdictions and in which ways they corresponded with Islamic law. Since the time of the Abbasid Caliphate, jurists have composed works of jurisprudence dealing with governors' responsibilities in the application of justice; these theories are known as *maẓālim* (plural of *maẓlima*, "oppression" and shortened form for *al-naẓar fi-l-maẓālim*, the investigation of "injustices") and Siyāsa, an abbreviation for *al-Siyāsa al-Sharʿiyya*, denoting *governance in accordance with Islamic law*. *Maẓālim* indeed evokes the idea of the 'wrongdoings' committed by administrators, while the etymology of *Siyāsa* carries notions of behavior, public or self-conduct deriving from a primary meaning of the *tending or training* (of beasts).

As a branch of jurisprudence, some eleventh-century authors began to address the role of officials in criminal investigations and the loose procedural methods (in contrast with those adopted by the qadis) that they were supposed to follow, such as torture, or drawing conclusions from circumstantial, non-testimonial, evidence. In the late Middle Ages some mālikīs composed works that sought to furnish the qadi with the efficacy of police methods, and later, Siyāsa became a topic of debate for ḥanafī and ḥanbalī authors dealing more broadly with the "political" jurisdiction of judges. Some of these works are over a thousand pages long, developing Siyāsa into an exercise in legal reasoning that covered the entire sphere of public policies and regulations. Large sections of the work *al-Ṭuruq al-ḥukmīya fī al-Siyāsa al-shar'iyya* by Ibn Qayyim al-Jawzīya (1292–1350) are devoted to proof and procedure, although it also covers a broad range of other topics, from gender to minority issues and market behavior.[10] It should be noted that the aim of early jurists was to provide qadis with the more efficient resources that governors and officials already disposed of. However, the division of labor that eventually emerged saw Mamluk and Ottoman qadis confined to the traditional regime of proof, and the arbitration of disputes brought to them by the parties, while officials were in charge of conducting investigations and preventing crime in a more expedite manner. Siyāsa will be one of the primary concerns of this book, since over a century prior to the Ottoman conquest, Siyāsa judges had begun passing verdict over foreign merchants. As a forum for mixed cases (that is, cases involving Muslims and foreigners hailing from outside the abode of Islam, which therefore could not be resolved autonomously by the community in question), Siyāsa trials have gone largely unnoticed because they did not correspond to either of the two approaches mentioned earlier, focusing, respectively, on dhimmīs and traditional jurisprudence. One the one hand, Siyāsa justice was not delivered by the qadis, but by Mamluk officials and secretaries. On the other, maẓālim has traditionally been understood as justice delivered by the sultans or caliphs at tribunals located in their palaces. Siyāsa as a forum where Frankish merchants were judged has left no material traces on Islamic sources, let alone a series of proper court proceedings. Although in principle, it is agreed that Mamluk Siyāsa was inspired by a specific branch of jurisprudence, and that it respected the general rule of law, legal scholars have often approached Siyāsa with suspicion, seeing it as a chaotic and contingent discipline deprived of the legal rectitude applied by qadis to their court decisions.

The Siyāsa judges share their murkiness with other important actors of the legal system, the Islamic and Latin notaries, the latter subject to Frankish jurisdictions but nonetheless involved in the Middle Eastern cities of commerce. This book seeks not only to reveal the role of actors who have been largely overlooked until now, but also to focus on the forgotten subjects of the legal system; namely, foreigners. Foreignness was equally cast in Islamic legal categories. The foreigners I am addressing in the present work were mostly referred to by the historical sources as *mustā'minūn* (hereinafter *mustā'mins*, or *müstemins*

10 Ibn Qayyim al-Jawzīyah, Muḥammad Ibn Abī Bakr (1292–1350), *al-Ṭuruq al-ḥukmīyah fī al-siyāsah al-shar'īyah* (Mecca, 1428 h).

in the Turkish spelling). A mustāʾmin was a legal foreigner, in contrast with a political enemy (*ifrānj, franjī*), disbelievers hailing from beyond the borders of Islam (*ḥarbīs*) the juridical unbeliever (*kāfir*) and Muslims hailing from distant countries (*gharīb*). Foreigners have been overshadowed by local minorities, or dhimmīs, in the cultural and historical narrative of the Middle East, to a point that is indeed embarrassing, when we consider that Islamic jurists devoted so few lines to the legal status of foreigners. Mustāʾmin is a notion derived from legal theories of hospitality. According to sharīʿa, unbelievers hailing from outside the realm of Islam should be fought by means of jihād. However, there are exceptions to this general rule, including foreigners belonging to diplomatic missions, passing pilgrims belonging to the scriptural religions, or merchants residing in the Islamic polity for a limited period. Ideally, these three groups were granted permission to remain in the realm of Islam upon concession of a safe-conduct, or amān (whence mustāʾmins, or amān-grantees). I will refer in many places to these treaties, and to equivalents such as the Ottoman ahdnames (from Arabic *ʿahd*, a covenant between Muslims and unbelievers). The Sienese traveler, merchant and biographer of Tamerlane, Beltramo Mignanelli, in a letter to Francesco di Marco Datini, described his surprise on seeing such large numbers of foreign merchants in 1390s Damascus, and on observing their inclination towards illicit behavior.[11] And yet in spite of this strong presence, Mamluk history, a discipline mainly concerned with the Islamic elites as represented by the religious learned—the *ulama*—has relegated them to the field of economic history. In contrast with later periods, little has been written about foreigners and their legal and cultural vicissitudes in the pre-modern Middle East.

We know today that dhimmīs frequently appealed to Islamic courts and institutions in their quest for fair justice. Studies ranging from the Geniza period to late-Ottoman times show that dhimmī communities did not represent a legal challenge to the Islamic enterprise of governance. Communal courts and laws were granted broad jurisdiction over internal affairs and family law, yet in spite of occasional unwritten bans by their clergy, dhimmīs showed up at the qadi court and generally managed to cope with the Islamic 'witness system.' Foreigners, and more precisely European foreigners, differed from dhimmīs because their presence in the Muslim polity posed a challenge to Islamic law and governance. Mamluk authors such as Taqī al-Dīn al-Subkī (1284–1355), well

11 "qui sono poche spezie con saraini e tute in grande inchiesta e carestia salvo pepe e ogni di per montare ogni cosa fino aldiciembre pero che ora qui trovara tra catelani et genovesi e con veneziani piu cristiani ci fusono 25 anni ...", ASPo, Datini, *Fondaco di Barcellona*, Damascus, August 2th, 1395, received September 30th, 1395.

known for his interest in dhimmīs, also concerned themselves with the presence of foreigners, questioning whether they belonged to the regular jurisdiction of the qadis or if, on the contrary, they fell within that of the sultan and his amān and trading policies. Franks entered the Islamic polity for trading purposes, but according to most jurists they could neither give testimony against Muslims nor against local Christians and Jews. Foreign and local unbelievers did not belong to the same community and therefore, it was argued, would naturally target each other in court. The very presence of unbelievers impinged on governance, as well as on legal and procedural issues, and contributed to the implementation of Siyāsa courts, an exclusively Mamluk feature.

Although the historiography has little to say about the normative dimension of foreignness, amān theory and practice has garnered a great deal of attention. Amān theory most often materialized in legal acts generally referred to as 'treaties of commerce,' which stipulated the definitions and norms concerning these foreign merchants and communities. However, it has long been a sticking-point for researchers, seen as an erratic discipline in which sharīʿa norms were more often than not disregarded.[12] Attempts by rulers to accommodate for the presence of unbelievers in the Islamic polity, it is argued, were generally at odds with a strict application of *fiqh*. In the present work, I take an approach closer to legal anthropology than to that of jurists; I will be discussing some of these inconsistencies, such as the acceptance of written documents, the Europeans' right to circumvent the jurisdiction of the qadis or, later, to use contracts against the word of Muslim witnesses. As Michael A. Köhler has put it, treaties were signed by common people and not by jurists, and in the same manner not everyone in society was a qadi concerned with the principles of Islamic law.[13] Doctrinal restrictions impinged on the status of foreigners, yet they never managed to hamper diplomatic cooperation; for this reason, I will not attempt to answer the vexing question of whether amān and similar practices were actually compliant with sharīʿa or not.

---

12 Gourdin, Philippe: "Les marchands étrangers ont-ils un statut de dhimmi?", in: *Migrations et diasporas méditerranéennes (Xe-XVIe)*, Edited by M. Balard and A. Ducellier, 435–446. Paris: Publications de la Sorbonne, 2002, Frantz-Murphy, Gladys, "Identity and Security in the Mediterranean World ca. AD 640—ca. 1517", *Proceedings of the Twenty-Fifth International Congress of Papyrology, Ann Arbor 2007* (2010): 253–264, Wansbrough, John, "The Safe-Conduct in Muslim Chancery Practice", Bulletin of the School of Oriental and African Studies 34 1 (1971), 20–35.

13 Kohler, Michael A., *Alliances and treaties between Frankish and Muslim rulers in the Middle East: cross-cultural diplomacy in the period of the Crusades*, edited by P. M. Holt and Konrad Hirschler, Cross-cultural diplomacy in the period of the Crusades, Leiden-Boston: Brill, 2013, 293–4.

Historians have long been fascinated by these sophisticated documents, but they have approached them more as diplomatic artifacts than as historical sources, a fascination that has extended to treating diplomacy as an elite practice. In this work, I aim to put these artifacts in context and historicize the norms they expressed about adjudication, notarization and cross-confessional relations, which, among other things, refer to Siyāsa as a jurisdiction. Specialists of Mamluk studies often claim that Siyāsa justice was to all intents and purposes equal to maẓālim practice, however Mamluk treaties insisted on the right to use local Siyāsa forums for mixed cases, and discouraged recourse to the traditional maẓālim hearings in Cairo. Through a historical approach to amān, I aim to identify the piecemeal, yet significant changes in the ways in which foreigners were dealt with in legal relations, towards the end of the Middle Ages. Medieval amān treaties, I argue, tended to limit the effects of any bias against non-Muslims, setting most issues within a technical-legal framework where a common solution could be found. In the 16th century, the Ottomans introduced rules and discourses regarding the relations between dhimmīs and mustāʾmins in their amāns and decrees, as well as exceptions and amendments to sharīʿa procedures that were not to the jurists' taste. Contrary to medieval practice, which granted foreigners with some procedural privileges, such the right to bring their case to specific courts, the Ottoman ahdnames concentrated all jurisdiction back into the hands of the qadis, and sponsored a purified version of the maẓālim, royal justice. After the Ottoman conquest of the Arab lands in 1516–1517, these biases against non-Muslim witnesses began to take a heavy toll, not only in commercial litigation, but in all sorts of cross-confessional relations.

By looking at the way notaries, archivists, judges and diplomats dealt with unbelief, my initial concern has been to seek explanations for the divergent practices that were used in the production and preservation of proof, as well as the mechanisms involved in governing cross-confessional relations and exchanges, ranging from the taking of oaths to the choice of a forum for mixed trials. Throughout the book, I will examine how court and market practices were affected, one way or another, by the application of the religious biases. Some of these practices—such as archival artifacts or notarial work—changed over the timespan covered here. Different practices echoed divergent notions of proof, something that often puzzled visiting merchants—as was the case for a Genoese merchant who found himself in the uncomfortable position of having to explain to a Latin notary that his Muslim colleagues had not deigned to keep the originals of the deeds proving a transaction. Ultimately, though, I am less interested in seeking explanations than in discovering the meaning behind such practices, and in particular in understanding the significance they might have had for Islamic governance. Behind issues of notarization, archives and court practice, I identify an important change in the way in which cross-confessional relations were handled by Muslim polities. To this aim, I first describe a process whereby Islamic societies refashioned their approach to written documents, progressively adopting what I define as new attitudes towards the written. To be sure, by the end of this process, written documents were being used and produced more freely than in previous centuries, and inspired more confidence than they had previously, for both administrators and common people. Legal records in particular, such as notarized deeds, granted many hitherto-silent actors the right to prove, enjoy and preserve specific rights over the long-term. Together with former slaves, women and dhimmīs, Franks benefitted from this trend. Parallel to this tendency leading up to the development of proper judicial archives in Ottoman times, I will be streamlining a broader process whereby rulers and their secretaries began embracing documents—and in particular archived documents—as part and parcel of the language of governance.

Major trends in adjudication, obligation, and changes in attitude towards written evidence had a direct impact on the conduct of trade between Latins and other confessions. For centuries, several scribal institutions had coexisted in the Middle Eastern marketplace. This is epitomized by the practice, probably in place since the 1340s, of dispatching a Venetian scribe to Alexandria and Damascus; based on the descriptions of legal practice provided by these Venetian clerks, I argue that Latin notaries did not only serve the interests of their consuls and their fellow nationals, but that their deeds were also used by or against Muslims, and that they could, on occasion, put themselves at the disposal of Islamic judges. Italian notaries and Islamic ʿudūl crossed paths in the marketplace, and their deeds mutually acknowledged the binding character of those issued by their colleagues. The Medieval scenario of legal relations and collaboration I describe in this book faded towards the 16th-century Mediterranean when, paradoxically, people involved in mixed dealings increasingly relied on writing and documentation—although by that time such documents were for the most part contracts notarized at the qadi courts.

One could argue that consulates continued to exist in early modern times, particularly under the Ottoman aegis, with their panoply of commercial courts and chanceries. Although I acknowledge the importance of Latin institutions such as the Venetian Bailo of Constantinople in the closing section of this book, the present work is more concerned with depicting the end of a medieval state of affairs, where everyone in the marketplace was aware of each other's institutions and acted accordingly, using all available devices to resolve and prevent cross-confessional disputes. In practice, the solution often implied ample recourse to technicalities. Under the Mamluks, the marketplace witnessed the labor of bilingual courtiers, customs officers acting as judges, and one could encounter a Venetian clerk intervening in the taking of oaths by Franks before a Muslim qadi. Mamluk emirs passed verdicts technically compliant with sharīʿa, however in practice they took a loose approach to the religious standards of evidence and unbelief. The Ottomans, in contrast, endowed the qadi with a jurisdictional monopoly over foreigners in their daily exchanges, making it compulsory to notarize contracts at the courthouse. Unlike in medieval times, the Ottoman enterprise of governance found a direct expression in the proliferation of Islamic contracts and records binding Europeans and Ottoman subjects together.

Together with the benches and courtrooms where Islamic judges sat in justice, the Middle Eastern archive is a key locus for the transformation I analyze here. As a point of departure, I argue that late medieval rulers do not seem to have attributed a specific role to the archive in the enterprise of governance. In this, such research goes against the grain of current scholarship on Islamic documentary studies, which claim that both caliphs and sultans made efforts to preserve the state and court records used in daily life. Scholars have taken recent discoveries of document collections as cues for the existence of Islamic judicial archives, adopting an apologetic tone when trying to explain the non-survival of these archives and collections. If the birth of the judicial archive under the Ottomans is accepted by at least some scholars, much more controversial is the idea that medieval sultans had recourse to documents and archives to a lesser degree than their Western counterparts. In this book I will be discussing current approaches to this debate, arguing that it was not until the Ottoman era that Islamic chanceries and secretaries began to adopt a conservative approach to written records.

In a recent article, a leading scholar in the field of Islamic legal practice, Christian Müller, has suggested that the Middle Ages saw an important shift in attitudes towards witnessing, the basic notarial activity. He argues that, while at first the nature of witnessing remained fundamentally oral, and papers acted simply as aide-memoirs, towards the 10th century a new, 'two-step notarization' pattern emerged.[14] Similarly, although he does not provide the smoking gun, he suggests that the burdensome procedures adopted to keep documents alive were slightly, although significantly, altered during the Mamluk period. This meant that after witnesses had given testimony in court, written

14 Müller, Christian: "The Power of the Pen: Cadis and their Archives in Medieval Islam", in: *Manuscripts and Archives: Comparative Views on Record-Keeping*, edited by Alessandro Bausi, De Gruyter, 2018, 361–385.

aide-memoires were kept by the qadi, who, by applying special certification procedures (tasjīl), could keep these documents alive, at least during his own tenure. Although in practice, mistrust towards writing pervaded legal theory, Müller points towards a wider acceptation of the archived document in the late Mamluk period. Together with storage places and practices such as the 'two-step notarization,' artifacts played an important role in this transformation. In Chapter Two, I sketch out the subtle mutation of scribal artifacts, from the medieval scroll to the Ottoman ledger, a feature that proved to be fundamental to the early modern approach to preservation.

Historians dealing with the Mamluk-Ottoman transition have provided a similar, essential contribution to our understanding of the change in attitude towards written proof. An inquiry into the dispersion of the Mamluk archives by Nicolas Michel has been instrumental in showing a significant change in attitude towards the validity of written documents.[15] In medieval sultanates, the possibility for a government document to be accepted as authoritative was contingent on the person who had been entrusted with its production and safekeeping: the secretary, in whose house these collections were kept. In contrast, Ottoman rulers began to view the archive as a form and manifestation of their own governance and jurisdiction. The Ottoman sultans, therefore—under compelling circumstances related to conquest and taxation, needless to say—took records out of the hands of secretaries and handed them over to archivists instead. Michel's detailed investigation of the whereabouts of Mamluk archives during the turmoil of Ottoman conquest is a striking blow to the theory that archives did not survive due to fortuitous contingencies. Perhaps because the Arab provinces represent a privileged vantage-point for observing legal change, it is historians of early Ottoman Egypt who have most clearly identified this transition to a more impersonal framework in the field of notarization. Reem Meshal has pointed to a fundamental shift that engendered the "mass-production of ḥujjas" for the very first time in Islamic societies. As a result, not only were government records now stored and put at the disposal of rulers, but notarial records (Ar. *ḥujja*, Tr. *hüccet*), now produced mostly by judges, found their way into court archives and were no longer the oral prerogative of the ʿudūl.

While I elaborate on the findings of authors that have furthered the debate on the Islamic judicial archive, I do not vouch for the idea of a teleological race for the Western-like use of documents. The increasing value attributed to

15 Michel, Nicolas: "Les Circassiens avaient brûlé les registres", in: *La Conquête ottomane de l'Égypte en 1517*, Edited by Benjamin Lellouch, 225–268: Brill, 2012.

archived documents in the Middle East did not materialize as a bureaucratic tendency to control foreign presence and to deal with religious difference. Indeed, I will be addressing how these changes ran parallel to a rigid adherence to basic sharīʿa principles, such as the discrimination of some witnesses on religious grounds. The Ottomans relied on writing and archives to a hitherto unknown extent, but at the same time exhibited an adherence to Islamic normativity to an equally unprecedented degree. Ottoman history is often viewed through the prism of ecumenism and social change; however, as far as the handling of cross-confessional relations is concerned, it was the quest for orthodoxy that dominated the broader legal reforms undertaken by the dynasty.

Explaining new attitudes towards archived documents, rather than attempting to explain the documents themselves—the object of my research—is essential to understanding the transformation endured by the Islamic enterprise of governance at the turn of the early modern period. Some researchers see these changes as signifying a move towards building a more centralized, Western-like state. This new proliferation of archived documents has been interpreted as a marker of the Ottoman state's tendency to form citizens, or *proto-citizens*, out of Muslim subjects. For others, efforts to render the law more uniform according to ḥanafī standards, and on top of previous layers of legal traditions, responded to broader objectives of empire-building.[16] According to this interpretation, practices such as reliance on certain ḥanafī compendia of jurisprudence led to the eventual modernization of Islamic law during the Tanzimat period. There could be equal grounds to interpret the transformation of Ottoman governance in light of an increasing confessionalization of society. Indeed, Meshal has gone so far as to suggest that 16th-century judicial practice tended to level differences between free Muslims, including women or former slaves, as part of the Ottoman project of sponsoring a proto-citizenship. According to these views, legal practice drew clear distinctions between Muslim proto-citizens and non-residents or itinerant Muslims, and racially connoted slaves, as well as plaintiffs convicted by previous sentences. Although I do not necessarily propose to embrace the confessionalization or centralization paradigms, it is clear from a closer look at cross-confessional legal relations that not only Muslims, but also dhimmī communities appropriated some elements of Ottoman legal discourse, and sought legal privileges over foreign Christians by demanding the right to be considered a separate community.

16 Ferguson, Heather L., *The proper order of things: language, power, and law in Ottoman administrative discourses*, Stanford University Press, 2018, Meshal, *Sharia and the Making of the Modern Egyptian*, Aykan, Yavuz, *Rendre la justice à Amid: procedures, acteurs et doctrines dans le contexte ottoman du XVIIIème siècle*, Leiden-Boston, Brill, 2016.

Another angle for understanding changes in Islamic governance is encapsulated in the notion of textuality, whereby the state, in the formulation by Guy Burak, is considered as a phenomenon anchored in the authority of a wide variety of texts, ranging from state registers and judicial records to jurisprudence and history, and associated institutions such as the archive. These texts labored to instill confidence in archived documents in the heart of a Muslim society for whom, until that time, trust had been dependent upon the word of trustworthy members of the community. This line of interpretation draws on Brinkley Messick's analysis of the late modern Yemeni state, understood not only as a polity but as a discursive entity. Similarly, for Svetlana Buzov and other authors, the Ottomans sought to build a legal orthodoxy that minimized ethnic, cultural and language differences. This was achieved through the sponsorship of a ḥanafī legal guild, and the parallel implementation of an Ottoman corpus of "secular"—also referred to as "public" or "customary"—law, the *kanun*, which sanctioned the incorporation of previous custom. I will attempt to show how an offshoot of this project to build a legal orthodoxy impacted foreigners, who were the object of legal definitions and acts of power—one such example being the well-known 1613 'Carazo affair,' when an attempt to rigidly apply amān rules threatened Istanbul's entire merchant community.[17] The Ottomans, incidentally, claimed jurisdiction over foreign Muslims, such as those expelled by the Habsburgs of Spain.

Throughout the book, and as a comparative *basso continuo*, the transformation of the Islamic enterprise of governance is understood by looking at the way in which commercial litigation across confessions was dealt with by both dynasties. Relations in the marketplace had received special attention from previous rulers in the region, such as the Byzantines and the Crusaders. As pointed out by the Greek-American scholar of Byzantium Angeliki E. Laiou, foreign merchants were often judged according to specific laws and norms. These relations were handled through legal categories implying access to rights, such as the imperial status of the *burgesioi*, but also through special jurisdictions for mixed cases, and by resorting to technical arrangements governing oaths and witnessing.[18] This book presents evidence to support the theory that under the Mamluks a special jurisdiction—the Siyāsa courts—emerged as a competent site for judging mixed, commercial cases involving foreigners. The idea that a

17 Krstić, Tijana, "Contesting Subjecthood and Sovereignty in Ottoman Galata in the Age of Confessionalization: The Carazo Affair, 1613–1617", *Oriente Moderno*, 93, 2013, 422–453.

18 Laiou-Thomadakis, Angeliki E.: "Institutional Mechanisms of Integration", in Studies on the Internal Diaspora of the Byzantine Empire, Edited by H. Ahrweiler and A. E. Laiou, Washington, D.C.: Dumbarton Oaks Research Library and Collection, 1998, 161–81.

parallel judiciary existed is one that has generated heated debate since its beginnings. In the last decades, historians of Islamic law, critical with the ideas of the legal historian Joseph Schacht (1902–1969) and interested in harmonizing theory with practice, would object that this was no novelty, pointing to the maẓālim tradition of hearing cases at the sultan's court, according to his own judgment. Rulers traditionally held maẓālim justice sessions and in particular heard grievances against the arbitrary (*ẓulm*) decisions meted out by administrators. Sessions of royal justice are documented for the Abbasid period, presided over by the caliph and, on occasion, it involved the ruler's appointees. These practices are well known for Mamluk times, and, although the name and the format changed, similar hearings continued to be held by Ottoman sultans at the Imperial Dīvān. The right for foreigners, and particularly Frankish merchants, to appeal to such courts was an important point in Ottoman ahdnames. With the rise of Istanbul as the epicenter of East-West relations, the Imperial Dīvān, as the site for maẓālim sessions, became the fulcrum of conflict resolution across confessions. Preserved in Venice's archives, decades of detailed reports by the Venetian ambassador (It. *bailo*, pl. *baili*) account for the Dīvān's importance, not only in diplomacy, but also as a court of justice. If we adopt this conservative vision of royal justice, therefore, over time little appears to have really changed in the way Islamic rulers dealt with the affairs of foreigners.

This view is supported by the common conception of maẓālim not as an exceptional feature, but as part and parcel of Islamic jurisprudence. As far as late medieval rulers were concerned, they were implementing some of the doctrines that had been developed centuries before by authors such as al-Māwardī (972–1058). Mamluk maẓālim has been described as the act of making policy-based decisions by the sultans, who issued decrees and held justice sessions at their court in Cairo. Similarly, some historical attention has been paid to the exercise of these functions by appointees, such as the chamberlain (*ḥājib*) or the market inspector (*muḥtasib*). Some recent works have a tendency to present Mamluk Siyāsa as a problematic case, stressing the harsh criticisms advanced by some Mamluk chroniclers such as Ibn Ṭūlūn (1485?-1546), Tāj al-Dīn al-Subkī (1327–1370) or Abū Ḥāmid al-Qudsī (d. 1483), as well as its alleged inconsistencies with sharīʿa.[19] I would argue, however, that if we observe the *longue durée* of cross-confessional relations in the region—ranging from the Crusader period to the late 16th-century situation described in Venetian dispatches—the Mamluk Siyāsa emerges out of this timespan as a distinctive

19 Rapoport, Yossef, "Royal Justice and Religious Law: Siyāsah and Shariʿah under the Mamluks", Mamluk Studies Review XVI (2012), 71–102, 93–4.

historical phenomenon, notwithstanding its legal background. Moreover, Siyāsa and related terms such as *yasa* and *yasaq* were of common use during Ottoman times. Although some aspects of Siyāsa as a doctrine inspired actual practice under the Ottomans, such as for instance, the ruler's prerogative to mete out non-Quranic, physical punishments (taʿzīr), it is generally agreed that Mamluk Siyāsa did not find continuity in Ottoman legal practice. Under the Ottomans, the term Siyāsa came to denote these physical punishments, non-traditional taxes and other features of Ottoman governance, but never a branch of legal doctrine that reflected actual court practice.

Traditional maẓālim hearings, then, without the specific Siyāsa layout, characterized Ottoman adjudication. Franks fell under the regular jurisdiction of the local qadi courts, and could appeal the qadi's decisions at the Imperial Dīvān, as they would have in previous centuries. The sticking point that remained—that is, mistrust towards written documents—had by then been largely overcome by the new Ottoman policy of sponsoring contract registration. Authors such as Timur Kuran see the tendency towards a more rational, "modern" framework to have materialized in this acceptance of written contracts in the cases where Franks were involved, a specifically Ottoman feature that I address in Chapter Four.[20] As Ira M. Lapidus has recently pointed out, when reading recent works on cross-cultural trade, one has the impression that despite cultural barriers, trade was not difficult to arrange.

By approaching the legal devices put in place to govern the Frankish presence as a historical, contingent phenomenon, this work departs from the mainstream line of interpretation of royal justice and posits a more problematic sequence of events. Without necessarily levelling a direct challenge to any of the approaches mentioned above, the broader picture of a world of legal relations characterized by the progressive adoption of Western-inspired values and procedures will not be adopted in this book. Instead, I understand Venetian and Arabic sources to have depicted a scenario in which Siyāsa evolved into a specifically Mamluk mixed court, that was distinct from maẓālim court sessions held in faraway Cairo. Applied to cross-confessional litigation, Siyāsa emerged as a forum widely accepted by Frankish litigants even for cases between themselves. However, instead of being taken as a medieval step towards a pragmatic Islamic version of the 'law merchant,' these courts were dismantled as soon as the troops of Selim I entered Egypt, and were replaced by a traditional framework of adjudication, concentrating jurisdiction in the qadi courts and then, in appeal, in the Imperial Council, the Ottoman version of

20 Kuran, *The Long Divergence*, 253.

maẓālim. The early modern Middle East witnessed widespread recourse to archived documents by judges and secretaries, but when handling the governance of its foreign communities, adhering to the orthodox approach to the religious, skeptical principles of sharīʿa was considered to be just as important as favoring trade with unbelievers.

To understand the important hiatus represented by the Ottoman approach to justice, I discuss the adoption of new attitudes towards the production, use and storage of written evidence. The Mamluks held the same fundamentally traditional attitudes and prejudices about written evidence as those that had characterized their predecessors. The most conspicuous marker of this attitude is the lack of Mamluk archives, and the fact that the only collections that have survived were kept neither by secretaries, nor by qadis, but rather owe their existence to exogenous practices and priorities (such as the series of Mamluk decrees preserved in Christian monasteries). Similar attitudes governed the conduct of affairs within the administration. The mid-15th-century secretary Shams al-Dīn Muḥammad al-Saḥmāwī (d. 868/1464), when describing the polities of the Indian subcontinent, confessed that he was unable to think of an example of correspondence between the Mamluks and some of its principalities since, he admitted, the Chancery had not recently dispatched any letters; thus implying an absence of long-term archive preservation.[21] Yet the Ottoman period has yielded both judiciary and state archives, and as I argue, the logic behind the preservation of both kinds of artifact have much in common.

The present work is thus concerned with the long-standing legal reform sponsored by the Ottomans that was particularly visible in the newly-conquered Arab lands. 16th-century Ottomans were preoccupied by the desire to ensure the governance of their empire through the "secular" and customary laws of the individual territories—the kanun—and the need to preserve and apply these laws contributed to the development of archival practices. Qadis were part and parcel of this process, since they counted among the agents entrusted with applying these secular laws. So too were they expected to preserve Kanunnames and decrees in their courthouses for future reference. In addition, Ottoman legal reform concentrated in the hands of qadis notarizing activities previously entrusted to medieval notary-witnesses, the ʿudūl. Within a few decades, judges saw themselves mass-producing and archiving notarial documents that allowed people to enjoy rights, and in contrast with their

21 al-Saḥmāwī, Shams al-Dīn Muḥammad (d. 868/1464): *al-Thaghr al-bāsim fī ṣināʿat al-kātib wa-al-kātim: al-maʿrūf bi-ism al-Maqṣad al-rafīʿ al-manshā al-hādī li-dīwān al-inshā lil-Khālidī*, edited by Ashraf Muḥammad Anas and Ḥusayn Naṣṣār, 2 vol, Cairo: Maṭbaʿat Dār al-Kutub wa-al-Wathāʾiq al-Qawmīyah bi-al-Qāhirah, 2009, II, 781.

medieval predecessors, these documents were not perishable, and nor did they rely on the word of living witnesses, such as the Granadan blackmailers described by Ibn al-Khaṭīb. The early Ottomans heralded an end to the monopoly of the medieval ʿudūl as bearers of truth.

Throughout this book I stress that new, more receptive attitudes towards the use and preservation of documents were not the result of modernization, so much as they emerged due to legal reform, a different approach to notarization and from a distinctive Ottoman interpretation of ḥanafī doctrines. The new rulers disliked the notarial institution that had proliferated under the Mamluks, and soon the medieval notary witness became obsolete. Since traditional, oral witnesses were being replaced by the notarizing powers of the qadi, the first of the two biases—mistrust for circumstantial evidence such as written records—was profoundly affected and downplayed by Ottoman governance. These new features were clearly beneficial to the conduct of legal affairs involving foreigners, since parallel agreements rapidly allowed Frankish defendants to exhibit their contracts as proofs. Far from adhering to the discourse on modernization still prevalent in the historiography, my work delves into the emergence of a paradox; for, if on the one hand Muslims relied more than ever on documents and archives, their rulers stubbornly upheld the legal biases against minority witnesses. While new attitudes towards the written were being adopted, the enterprise of governance amidst diversity was being redefined around the creed that the word of Muslims was of superior value and could not be overcome. The requirement to advance Muslim witnesses, traditionally confined to court disputes, now extended to multiple areas of governance, and took a heavy toll on the activities of Frankish merchants, diplomats, and captives. The loose approach to the biases adopted by Mamluk officers in Siyāsa hearings gave way to a strict adoption of them at the Imperial Council in Istanbul. This work therefore embraces Ira M. Lapidus' criticism that "the biggest problem in [approaches to] cross-cultural trade was not negotiating among merchants of different cultures *but overcoming non-merchant prejudices, religious laws, and the political interests of rulers*."[22]

Islamic governance has a long tradition of coping with unbelief, going back to the time of the Prophet in Medina and to the Omayyad and Abbasid Caliphates, in dealings with their Greek and Turkic neighbors. Late medieval and early modern jurists relied on available texts by Persian and central Asian authors such as al-Sarakhsī (d. 1096) or al-Marghīnānī (1135–1197) about

22 Lapidus, Ira M., "Religion and Trade: Cross-Cultural Exchanges in World History, 1000–1900 ed. by Francesca Trivellato, Leor Halevi, and Cátia Antunes (review)", Journal of Interdisciplinary History 46, 2 (2015), 267–269.

how to deal with unbelievers at the marketplace and in court. Mamluk jurists did not passively assimilate these much older doctrines either; they also called for an opener interpretation of the tradition. Whether Mamluk rulers made their decisions based on their awareness of such theories, it is difficult to know; on occasion in any case, the sultans seem to have referred to specific solutions described in Siyāsa works when it went in their own favor; to disgrace a political enemy, the shrewd al-Ẓāhir Barqūq had him accused by cheated foreign merchants. The present work aims to demonstrate, however, that Mamluk governance of foreign unbelievers was guided by maẓālim and similar doctrines on governance. The Siyāsa courts, evidence of which can be found in the notarial records left by Frankish merchants, while departing from it in practice, stemmed from the same line of legal reasoning as maẓālim. Thus, in dealing with those controversial areas where traditional sharīʿa proved insufficient, the practical needs of Mamluk governance were left in the hands of the legal system as a whole. In contrast, the fact that Ottoman rulers took cues from the doctrines of the ḥanafī tradition is well known. Defined by A. L. Udovitch as "the law merchant of the middle ages," ḥanafī law was generally favorable to the resolution of disputes in daily relations with local Christians and Jews, although it drew a clear line of demarcation between foreign unbelievers, or mustāʾmins, and local minorities, or dhimmīs. Again, this study challenges the conventional view that jurists tended to align with the rulers, hence sanctioning Ottoman governance. It was in the field of cross-confessional relations that the adherence by Ottoman governance to ḥanafī ideas was put to the test, because the sultans departed from ḥanafī doctrine in this specific, yet important area. One of this book's main arguments is that Ottoman rulers pursued an orthodox position in the way they expected their subjects to treat foreign unbelievers in courts and markets, but that contrary to medieval sultans, they intervened in these matters and imposed their own legal decisions. Thus, even though their choices were expressed in the language of the legally learned, they did not necessarily limit themselves to following the injunctions of the ḥanafī thinkers. Considered as a whole, the array of practical decisions, policies and rulings that legally addressed mustāʾmins certainly constituted an Ottoman orthodoxy and generally complied to ḥanafī thought. However, their decisions sometimes departed from this and, on occasion, the solutions sponsored by Ottoman governance could look puzzling from a Muslim standpoint. The book addresses these variations, against the backdrop of a previous medieval practice where legal issues were fundamentally left to the best judgment of the religious learned, and the rulers limited themselves to promoting the rule of law in the daily life of the market.

Historians have a tendency to present Islamic justice as stemming from one of two different sources. Legal historians have traditionally privileged the normative aspect of the law and the jurists' activity, and recent interest in surviving court records usually underlines the conformity between procedure and jurists' doctrines. On the other hand, a more recent trend interested not in 'the law' per se, but in justice, considers the latter to have ultimately fallen under the ruler's sphere of action, and to include not only maẓālim but also the *ḥisba*, or market inspection, and *shurṭa*, or police activities, as well as Siyāsa.[23] To be sure, Mamluk rulers exerted an undeniable influence over justice, but it was mainly through the promotion of the legally learned, their schools and jurisdictions, and the activities of legal thinkers. In this, they followed the example of most medieval dynasties that came to rule over the vast lands of what once constituted the Caliphate. These sultans were not vested with any charismatic power, nor did they make any hereditary claims to the Prophet's lineage, in contrast with the Omayyad and Abbasid caliphs. Unlike caliphs, although they were certainly concerned with 'the law,' they carefully avoided presenting themselves as lawmakers. For this reason, when it came to their judicial duties, they considered that their role was more to redress grievances than to don the judge's attire and deliver justice.[24] They embraced the Islamic ideal that anyone in the community was capable of delivering justice;[25] as a result, Franks enjoyed the possibility to sue Muslims, without needing to head straight to the local ruler's palace and seek redress from him in person. Instead, in the Siyāsa courts, the sultans' officials passed judgment in provincial towns, treating the affairs of foreigners as an integral aspect of local governance, and some of them even began to study law. As one influential discussion on Mamluk Siyāsa has underlined, the new, parallel jurisdiction started to deal with issues ranging from divorce to bankruptcy and, throughout the long fifteenth century, the sultans made efforts to deploy it in harmony with the traditional courts of the qadis.[26] The present book is not interested in deepening an

---

23 Müller, Christian: "Mamluk Law: a reassessment", in: *Ubi sumus? Quo vademus?*, V&R Unipress, 2013, 263–284, see his criticisms on Stilt, Kristen: *Islamic law in action: authority, discretion, and everyday experiences in Mamluk Egypt*, Oxford-New York: Oxford University Press, 2012.

24 Garcin, Jean-Claude: "The Regime of the Circassian Mamluks", in: *The Cambridge History of Egypt. Vol.1: Islamic Egypt, 640–1517*, Edited by Carl F. Petry, Cambridge University Press, 1998, 290–317.

25 Johansen, Baber: *A Perfect Law in an Imperfect Society. Ibn Taymiyya's Concept of 'Governance in the Name of the Sacred Law'*, in: *The law Applied: contextualizing the islamic shari'a*, edited by Peri Bearman, Bernard G. Weiss and Wolfhart Heinrichs, IB Tauris, 2008, 259–94, 269.

26 Rapoport, "Royal Justice".

artificial divide between these two registers of the Mamluk legal system, or in siding with one of the contending views in this model: either a legal historian's world—inhabited by judges and jurists—or one of unbridled arbitrariness. Frankish merchants never ceased to visit either the stalls of the ʿudūl, or the qadi's court. As far as judicial practice is concerned, the respective perimeters of Siyāsa and sharīʿa were well-defined.

Medieval/Mamluk rule marked an additional step in the post-caliphate trend of legitimizing the sultan's rule by sponsoring the religious learned in general, the four schools of jurisprudence and the different manifestations of Islamic identity, ranging from poetry to Sufism.[27] Mamluk rulers enhanced the display of royal justice by conducting maẓālim sessions in Cairo, by building 'Halls of Justice' in Aleppo and Damascus as well as convents and four-iwān madrasas—religious schools that hosted each of the four rites—and through public trials conducted by Mamluk officers sitting in benches in the street. To demonstrate their legitimacy, officials often formed judicial panels with qadis, as can be seen in the painting held at the Louvre, *The Reception of the Ambassadors*.[28] When dealing with foreigners and diversity, the Mamluks reconciled the doctrines of multiple *madhhabs*, even adopting the teachings of troublesome ḥanbalīs such as those of Ibn Taymīyah (1263–1328). Although medieval ḥanafīs sought to empower the qadis with the same procedural privileges as those of police officers and rulers, the Mamluks never did this. They limited the affairs of foreigners to the jurisdiction of the *ḥājib*, a military function usually rendered as 'chamberlain,' and the latter's justice became popular for mixed cases. Contrary to the situation depicted by the author of the 1625 *relazione*, Siyāsa courts were often the preferred forum even for cases among two ḥarbīs. As I argued in a recent article I extensively rely upon in Chapter Three, in Siyāsa trials, two major procedural obstacles—the bronze wall that inspired the title of this book—were circumvented: the impossibility for non-Muslims to testify against a Muslim,

27 I am much indebted to Ira M. Lapidus' insights into the transition from caliphate to post-caliphate Islamic societies, and its consequent effect on the relationship between religion and the state. Lapidus, Ira M.: *A History of Islamic Societies, 2nd. Edition*, Cambridge: Cambridge University Press, 1988, Rapoport, Yossef, "Legal Diversity in the Age of Taqlid: the Four Chief Qadis Under the Mamluks", Islamic Law and Society 10 2 (2003), 210–28, 210–28; Burak, Guy: *The second formation of Islamic law: the Hanafi school in the early modern Ottoman empire*: Cambridge University Press 2015, *Introduction.*

28 Campbell, Caroline: "The 'Reception of the Venetian Ambassadors in Damascus': Dating, Meaning and Attribution", in: *The Renaissance and the Ottoman World*, Edited by Anna Contadini, Routledge, 2016, 125–138.

and the non-acceptance of written documents. Not surprisingly, therefore, it is difficult to find complaints about the Mamluk system of justice, or even references to the issue of witnessing against a Muslim business partner. Yet with the Ottoman conquest of Egypt and Syria in 1517, and after a long silence, the ban against non-Muslims witnesses reemerged in the records as a major shortcoming of Islamic justice.

In 1517 the Ottomans took over Syria, Egypt and the Red Sea protectorate of Mecca, from whence Indian spices flowed into the Middle East. As Reem Meshal has forcefully demonstrated, military conquest was accompanied by a thorough reform of the local judiciary, by empowering a Turkish ḥanafī legal guild in the local judiciary, to the detriment of non-ḥanafī schools of law. The twofold jurisdictional structure mentioned above, although it did not totally disappear from legal theory, became much foggier in practice. Although Ottomans are all too often assumed to have been political pragmatists, they abandoned the innovative Mamluk approaches to mixed conflict resolution. They did away with the Siyāsa courts, and restored the qadi's authority upon mixed disputes at the basic court level. They returned to the classical maẓālim system, this is, claims could only be judged in appeal by the sultan himself in Istanbul. Since that date, frequent grievances against the biases reemerge in the sources. Chapter Four therefore addresses how mixed and interfaith conflict was dealt with after 1517, after efforts to establish some kind of Islamic merchant court were abandoned. Due to the setback experienced by Siyāsa courts, decisions by local qadis could not be brought in appeal to the ḥājibs in trading places such as Damascus, Tripoli or Aleppo; instead, a case's transfer to the Dīvān in Istanbul, which in many senses involved more hurdles, became mandatory.

Under the Ottomans, in keeping with traditional sharīʿa, witnessing by Franks against Muslims was accepted neither before the local qadi, nor in appeal at the Dīvān, at least in theory. A case study in the closing section sets this issue at the heart of Mamluk-Ottoman transition: it presents the difficulties encountered after a Venetian merchant firm in Syria fell bankrupt in the 1530s, when the local qadi in Aleppo refused to hear Christian witnesses to support its claims. Like many litigants, the Venetians stumbled over similar biases, such as the general principle that facts stemming from Muslim-produced proofs took precedence over facts witnessed by unbelievers. The Priuli bankruptcy, brought to appeal before the Dīvān in Istanbul and pleaded by the experienced Bailo Pietro Zen (1453–1539), provides a unique overview of how the biases were dealt with in early modern times. Solutions were no longer to be found in sharīʿa-based technicalities, but rather on the terrain of diplomacy and negotiation.

The shifting approaches to the problem of proof and unbelief cannot simply be understood by either looking to 'the law,' intended as legal theory and jurisprudence, nor to justice, as a practical matter left in the hands of the ruler, qadis and officials. Recent analyses based on sharīʿa court records put the qadi at the center stage both of doctrine and practice. By applying sophisticated procedures, such as *isjāl*, the qadi had the power to keep everyone's legal rights alive by recertifying written, and therefore perishable, documents, hence safeguarding sharīʿa theoretical concerns for truth-bearing. My aim here is to address the legal system, comprising both legal norms, or laws, and institutions. Throughout this book I will be dealing with the normative texts governing the presence of foreigners, the amān treaties, broadly known as ahdnames in Ottoman times, and, on occasion, legal opinions, Siyāsa treaties and other fiqh works ranging from the 14th to 16th centuries. My interest in institutions mostly centers around scribal culture, but also courtiers and officials operating in the Middle Eastern markets and courts. I refer to judges, and while distinguishing between different jurisdictions, my focus goes to how disputes were actually pleaded and, if at all, resolved, and to what extent the biases played a decisive role in their outcome. I go to the courthouse, where evidence and proof were advanced by the parties, but I also explore in the same measure the archive, since its disputed existence reflects notions and ideas about proof. Rather than attempting to define a hierarchy of institutions, I consider all these devices as constituents of the same legal system, equally affecting the ways in which cross-confessional relations were established, and disputes arbitrated. For this reason, and as a practical device, I often adopt a terminology that opposes medieval with Ottoman/early modern, as well as Latin with Islamic institutions. The legal system this book explores lies at the intersection between vernacular, 'Islamic' institutions and the legal, scribal and consular devices exported by the Franks to the Middle Eastern cities of commerce. Cross-confessional relations were defined by the intersection of both traditions and institutions, and were affected by Western and Islamic legal systems. More often than not—and in contrast with what is often assumed by historians of the Mediterranean—these legal systems interacted with each other far beyond mere mutual acknowledgement, and this was especially the case before the subtle crossing of a 1517 red line.

1.1 Structure of the Book

This book opens with an in-depth analysis of a question I define as the 'archival divide.' In it, I engage in an ongoing debate about why Islamic societies have not yielded judicial archives, while Latin Christian ones did.[29] I discern between historians offering historicist explanations for a lack of Islamic archives, headed by the papyrologist Frédéric Bauden, and those who have gone a step further by suggesting that records did not survive due to deeper choices rooted in culture, path-dependence or the law. Twenty years ago, Michael Chamberlain claimed that Middle Eastern societies entrusted the biographical dictionaries composed by scholars with their strategies of social reproduction and perpetuation, rather than archived documents collected in notary ledgers. Today, Chamberlain's claim continues to provoke strong reactions among scholars. Recent works tend to depart from the views of classic authors such as Claude Cahen, Jean Sauvaget, Samuel Stern or Michael Chamberlain himself, who saw in the non-survival of archives deeper institutional motivations such as, to cite one example, the egalitarian nature of Islam, and the lack of status groups leading to the eventual birth of seigniorial institutions and their archives. By shifting the focus from the legal system to more contingent factors, researchers have gone so far as to claim that Islamic documents were not preserved simply because they were recycled, used as fuel or ritually erased. I then turn from the literature to an analysis of what constituted proof for both Latin Christians and Muslims by tracing the genealogy of concepts such as notarization, testimony, certification, artifacts like the Islamic *sijill* (Tr. *sicil*) or devices serving to store and preserve records, starting with the early medieval *qimaṭr*, the proto-archive where early qadis kept notarized deeds. This diachronic survey helps to explain the emergence of the Ottoman *sicil*, which radically changed patterns of record preservation.

For centuries, the norms for arbitrating cross-confessional conflicts and commercial disputes with Franks were encoded in amān treaties and executive decrees; however, a lack of perspective has led researchers to interpret these normative texts as dictated by pragmatic considerations and balances of power,[30] or as part of a general trend towards modernization, 'europeanization'

29 See, among recent contributions: Hirschler, Konrad, "From Archive to Archival Practices: Rethinking the Preservation of Mamluk Administrative Documents", Journal of the American Oriental Society 136.1 (2016), Hirschler, Konrad, "Document Reuse in Medieval Arabic Manuscripts", COMSt Bulletin 3/1 (2017) (2016), Burak, Guy, "Evidentiary truth claims, imperial registers, and the Ottoman archive: contending legal views of archival and record-keeping practices in Ottoman Greater Syria (seventeenth–nineteenth centuries)", Bulletin of the School of Oriental and African Studies 79.2 (2016), 233–254, Burak, Guy, ""In Compliance with the Old Register": On Ottoman Documentary Depositories and Archival Consciousness", Journal of Economic and Social History of the Orient 62 (2019), 799–823.

30 Valérian, Dominique, "La résolution des conflits dans les communautés européennes dans les ports du Maghreb médiéval, entre métropoles et pouvoir local", Mélanges de

and the western-like codification of Islamic law. A great deal of attention has been directed towards the so-called 'capitulations' signed with the Ottoman Empire, where such pragmatic features were often underlined. I argue instead that medieval agreements comprehensively covered most of the problems and solutions arising from cross-confessional relations. Far from being dictated by relations of power between unequal partners, treaties were based on the respect of technicalities that made it possible to overcome the obstacles presented by religious biases against unbelievers. Heirs of a long tradition of governing a heterogeneous marketplace, the Mamluk sultans were most successful in the diplomatic field, and signed numerous and important treaties with the European powers. A good deal of this success relied on the sultans' ability to enforce sharīʿa-based norms in harmony with the jurists, and not by disputing them. In Chapter Three I therefore explore what this normative framework governing exchanges and the arbitration of mixed disputes actually consisted of. I guide the reader through several centuries of trade agreements and identify the principal trends and variations, as part of a wider process in an ever-changing context. I examine clauses giving instruction on the notarization of transactions, the striking of deals, adjudication and issues of proof, and how to proceed when deals were not honored. Again, religion was not adopted as a point of departure; French Crusaders or Castilian conquerors, for example, faced much the same problems as their Middle-Eastern Muslim counterparts. Paradoxically enough, the Crusader *Cour de la Fonde* was a privileged forum for cross-confessional conflict, where complex rules of procedure secured fair justice across confessions. By looking at precedents of Mamluk approaches to dealing with mixed conflicts, I identify a series of issues that straddled Byzantine, Latin and Islamic societies. In this section I target issues such as whether impartial justice required recourse to the defendant's jurisdiction (as was the case for some Maghrebi rulers and the early Mamluks), the extent to which minority witnessing was accepted, and whether or not the burden of proof was placed on the defendant's witnesses, irrespective of their faith (as was the case for late Crusaders). I also ask whether the law distinguished between people on the basis of their religion or confessional group, and interrogate the texts that stated whether merchants—and thereby foreigners—should fall under a broad application of local laws, be dealt with by different laws or courts, or

l'École française de Rome, Moyen Âge 115/1 (2003), 543–564, De Groot, Alexander H., "The Historical Development of the Capitulatory Regime in the Ottoman Middle East from the Fifteenth to the Nineteenth Centuries", Oriente Moderno 22 (83) 3 (2003), 575–604, Veinstein, Gilles, "Les Fondements Juridiques de la Diplomatie Ottomane en Europe", Oriente Moderno, 88 2 (2008), 509–522.

even by special mixed courts (like the *Cour de la Fonde*, and to some extent, the Mamluk Siyāsa tribunals). Another issue I address is the need for notarization, this is, whether agreements could be concluded and guaranteed by common people, by community witnesses or only by state-appointed officials—as was the case under the Ottomans.[31] I also look to diplomatic agreements to answer questions about the acceptance of either oral or written notarization, and whether the legal system potentially accepted written evidence without the oral support of its authors. Lastly, I approach the judicial oath, an issue related to the validity of witnessing; I focus on whether minority oath was accepted, and if so, in what procedural circumstances it was taken. The recourse to specific places of worship or the choice of sacred books counted among the shared technicalities that legal actors drew on to justify this kind of oath-taking.

Chapter Three also refers extensively to the treaties signed between the Mamluk state and Aragon, Florence, Venice and to some extent Genoa, some of them preserved in their Arabic version. Other sources I use are the drafts and decrees that complemented the capitulations, as long as notarial deeds reflecting the handling of cross-confessional issues. Attention is given to treaties drawn up by powers located elsewhere, as was the case with the Ḥafṣids, and this extends to a survey of sources from Christian polities that were in direct contact with infidels and their legal systems. And for comparison, I observe Castilian legal codes and *fueros* drafted while the Kingdom was in the process of incorporating non-Christian minorities from the conquered lands. Particular attention is paid to the legal codes in French produced in Cyprus, which drew upon previously-recorded law collections from the East. Crusader society produced a most sophisticated set of rules for cross-confessional witnessing, and mixed courts with a commercial bias such as the *Cour de la Fonde*, or courts privative to specific minorities such as the *Cour des Syriens*.

My aim is to determine a series of problems inherent to mixed exchanges, and encoded in the late medieval legal system—problems that I will streamline throughout the final chapter of this book, which deals with actual practice in the Ottoman, sixteenth-century Mediterranean context. I observe legal practice as it took place in the two main cities of commerce under Mamluk rule, Damascus and Alexandria up to 1517, drawing upon the extant collections of Venetian notarial records produced in those cities. The presence of, and exchanges with infidels in Mamluk Syria became a subject of debate for Islamic legal theorists in the Mamluk period, such as Taqī al-Dīn al-Subkī or Badr al-Dīn

31 Faroqhi, Suraiya, "Before 1600: Ottoman attitudes towards merchants from Latin Christendom", Turcica 34 (2002), 69–104, Van den Boogert, *The Capitulations and the Ottoman Legal System.*

al-ʿAynī (1361–1451). More specifically, I provide an account of the emergence of the legal theory on governance under the late Mamluks. As a theory, *al-Siyāsa al-sharʿiyya* aimed to provide the ruler with the means to mete out punishment and serve judgement in special situations upon which sharīʿa was silent, and it eventually led the Mamluks to sponsor Siyāsa courts, where commercial mixes cases, with their trove of procedural issues, were heard. I address works on Siyāsa written by theorists such as Ibn Qayyim al-Jawziyya, allowing me to understand the relationship between legal theory and judicial practice as it happened in a medieval society. If writings on legal theory clearly had an effect on the implementation of Siyāsa justice from above, so too did the subjects of the legal system—in this case Frankish plaintiffs and defendants—contribute to a bottom-up definition of a forum for cross-confessional relations. Mamluk Siyāsa courts, I argue, encapsulate the features that made possible to circumvent the 'bronze wall' that the religious biases represented, since judges were necessarily confronted with the need to examine Frankish testimonies and documents. A last section is thus devoted to describing the actions and the actual hearings of these courts, presided over by the ḥājibs in Damascus and usually by the viceroy (*nāʾib*) in Alexandria. Although verdicts stemmed out from sharīʿa norms, Siyāsa constituted a special jurisdiction where Mamluk officials, rather than qadis, acted as judges. Considered to be a technicality that should be left in the hands of jurists, the legal issues generated by the presence of European merchants were transferred to these special courts as the jurists prescribed. Most of my findings on Mamluk Siyāsa can be found in this section, which extensively develops an article I published in the second 2016 issue of *Comparative Studies in Society and History*, "Judging the Franks."[32] In it, I look at descriptions of trials found in the Venetian notarial casebooks of Alexandria and Damascus. Again, I am interested here in the biases against unbelievers; that is, how notarized documents, private acts and minority witnessing were handled in the procedural practice of these courts. The fact that Siyāsa courts were promoted over previous modalities of royal justice shows that medieval societies had their own ways of responding to diversity and mixed cases, and handled legal norms in ways that were compatible with the specific demands of conflict resolution. The Mamluks, I argue, did this without circumventing due legal processes, and without really challenging the sharīʿa system of norms, but instead by connecting with the subjects concerned by the provisions of Siyāsa.

32 Apellániz, Francisco, "Judging the Franks: Proof, Justice, and Diversity in Late Medieval Alexandria and Damascus", Comparative Studies in Society and History 58 02 (2016), 350–378.

The book's closing section presents an analysis of the transition from Mamluk to Ottoman times; what I interpret to be a significant turning-point around 1517, after the Ottomans took over the majority of the Arab regions. Ottoman rulers abandoned Mamluk innovative approaches to mixed conflict resolution. They did away with the Siyāsa courts and empowered a ḥanafī legal guild and its qadis, often at the expense of local judges and their retinue of court witnesses. From that moment, and quite abruptly, the biases on witnessing reemerged in the European records as a major shortcoming of Islamic justice. In Chapter Four, I shift the focus to new actors and scenarios, such as the Venetian Bailo in Constantinople and his chancery, who played an important role in underpinning mixed dealings, and the Imperial Council—the Dīvān-ı Hümāyūn. The Dīvān, apart from hearing commercial cases, became the principal forum for cross-confessional interactions; together with plaintiffs, consuls and ambassadors negotiated with the pashas on all sorts of matters concerning trade, captives and borderland issues, and found themselves constantly stumbling over the sharīʿa regime of proof when trying to substantiate their claims.

By focusing on the new attitudes adopted by early Ottoman rulers as regards proof and evidence, my research explains why these biases became so prevalent in Christian-Muslim relations from the sixteenth century. It argues that the biases, which medieval sultans had treated as a legal technicality (and thus the object of specific provisions by legal specialists) now formed the foundation of the Ottoman enterprise of governance. Problems stemming from the fact that Franks could no longer act as bearers of truth soon extended beyond the courthouse and the marketplace, and began to impinge on ample areas of governance. This concerned, for example, the management of human property (i.e. captives), because the biases impinged on how servitude and freedom should be defined, and that of territories, as they affected the outcome of borderland disputes; people were enslaved, sold and sent to the Anatolian or Egyptian inland simply because they could not prove their freedom by means of Muslim witnesses. The biases, moreover, were invoked not only by rulers, but also by other social actors and groups—such as local minorities—who found it useful to mobilize similar confessional categories.

in Chapter Four, I develop my research on the Dīvān as a forum for mixed commercial litigation, by providing a detailed description of cases that extended over multiple Dīvān sessions and that required interaction between the Venetian bailo, the pashas, witnesses and legal experts, and entailed the production and use of documentary artifacts. The chapter presents a number of cases brought before the Dīvān that involved Franks, which often revolved around the differences, real or perceived, between Frankish and Ottoman law; such as whether one could sue the dead, whether non-Muslims could stand surety for others, the risk of false witnessing, or whether court proceedings ought to be recorded.

By virtue of the ruler's monopoly on maẓālim, in the sixteenth century the boundary between sharīʿa courts and royal justice, previously so distinct, became less and less clear. A parallel, similar ambiguity played out in the Dīvān: although the Imperial Council is traditionally presented as a diplomatic forum, where the pashas and viziers met European ambassadors, its role as a high court is well known. My aim here is not to discuss whether one of these features predominated over the other, but rather to take this ambiguity as an integral part of Ottoman governance amidst diversity. In the Dīvān, diplomatic or 'political' issues arising from interaction with Franks manifested themselves as legal issues, and were therefore subject to methods and procedures based on sharīʿa. As we shall see, Venetian diplomats were aware of the importance of the law and particularly, of the biases against non-Muslims, and expected their correspondents at the Senate to fully acknowledge the legal intricacies of their mission there.

The Ottoman approach to the legal system is a well-known territory, and 16th-century Ottoman judicial registers are extant for most Arab cities, covering almost all angles of it. In Chapter Two I sketch out the changes the Ottomans sponsored in the legal system and how they paved the way for a more open attitude towards writing and documentation. These policies led to the birth of an Islamic judicial archive, hitherto inexistent, and a parallel crisis in traditional notarization. Against this institutional background, Chapter Four explores further normative alterations to documentary practice contained in kanun and amān treaties, which permitted actors to continue to circumvent the legal biases against non-Muslims. The Ottomans granted privileges stipulating that Franks could defend themselves in court by exhibiting written deeds, more specifically notarized records (Ar. *ḥujja*, Tr. *hüccet*) produced by a qadi. Parallel to this measure, and more importantly, Muslim witnesses could not be heard against such deeds. To be sure, the Ottomans took the injunctions of sharīʿa very seriously; however, I argue that rulers were ready to depart from ḥanafī orthodoxy when it came to certain crucial issues. The work of Mario Grignaschi, which focuses solely on jurisprudence through the study of collections of fatwās on issues of testimony and diversity, provides a fundamental contribution to my argument.[33] In the practice of Ottoman governance, rulers contradicted the opinions of ḥanafī jurists in that the latter distinguished

33 Grignaschi, Mario: "La valeur du témoignage des sujets non musulmans (dhimmi) dans l'Empire ottoman", in: *Recueils de la Société Jean Bodin, La Preuve, 3, Civilisations archaïques, asiatiques et islamiques*, 211–323. Bruxelles Editions de la Librairie encyclopédique, 1963.

between local and foreign Christians and Jews, who they considered to be "two different peoples" and who therefore could not testify for or against one another in court. As regards commercial litigation, however, Ottoman rulers saw all unbelievers as constituting a single people. Similarly, in 1557 Muslim merchants in Aleppo lobbied for, and obtained, a decree disposing new and special conditions for the use of Muslim witnesses in mixed cases and which only applied to that city.

Adopting a broad chronological overview is key to observe an institutional dynamic. The longue durée allows me to capture the link between legal norms and social processes. I aim to demonstrate that the advent of Ottoman rule saw the biases go beyond the marketplace/court medieval binomial to become a decisive tool for governance. Drawing upon venetian dispatches (It. *dispacci*) and more formal consular reports (*relazioni*), I show how the new attitudes towards proof and diversity affected Ottoman management of both physical space and human beings. The case of Marco Priuli, a Venetian who fell bankrupt in Syria in the 1520s, closes this section. His case provides us with a closing analysis of how his default was dealt with—passing first before a Syrian qadi, then to the Dīvān, and finally before Venetian commercial judges. The case is based on a hefty corpus of documents from the *Biblioteca Correr* in Venice, which appears to be the file used by the family attorney, including correspondences and Priuli's testament drawn up in Damascus. A trader in diamonds and luxury items, Marco Priuli's heirs and associates proved to be unfamiliar with the legal practices promoted by the Ottomans, and with the court's requirements for proof and evidence. Priuli's case provides a unique opportunity to observe how, shortly after the Ottoman conquest of Syria and Egypt, judges and officials began to deviate from previous legal practice regarding non-Muslims and mixed cases, and to promote their own practices for producing evidence. This key case of commercial litigation helps us to gauge the meaning of the legal changes the Empire was undergoing. In gross, biases and bans relating to proof and evidence were now invoked not only by judges and officials, but by dhimmīs and merchant groups. For the Ottoman dynasty, the orthodox approach to proof was a means to differentiate the political community it represented, and to mark out its specific vision of the Ottoman sovereign as guarantor of the rule of law. In this sense, the discussion of Priuli's case raises important questions on *confessionalization* as a major vector in Ottoman history.[34]

---

34 Krstić, Tijana, "Contesting Subjecthood and Sovereignty in Ottoman Galata", 422–453, Krstić, Tijana: *Contested Conversions to Islam: Narratives of Religious Change in the Early Modern Ottoman Empire*: Stanford University Press, 2011, Buzov, Snjezana: *The lawgiver*

The thesis of this book is that, although the biases against non-Muslims constituted a veritable divide in the field of legal norms, during the late Middle Ages a trove of customs and institutions nuanced their effects, and maintained a subtle balance that remained in place until early Ottoman times. While medieval sultans accepted that commercial litigation—and more generally, relations with the Franks—should be loosely handled by sharīʿa-based notions of governance, as expressed in Siyāsa treaties, under the Ottomans practices began to adhere much more strictly to the norm. This is not to say that Ottoman sultans were respectful of legal norms, while their predecessors were lenient, but that they approached the problem of governance amidst diversity according to an official orthodoxy. Medieval sultans used all of the legal devices at their disposal to govern their relations with the Venetians, Catalans, French and other Franks, without interfering with the judges' prerogatives to use legal reasoning, issue opinions, hire their own notaries and deputies, and to organize the different schools of law. To tackle the presence of foreign unbelievers, they relied on available doctrines such as the *Siyāsa sharʿīyah* developed by jurists from different madhhabs, but also on dragomans, courtiers (Ar. simsārs) and customs officials in their legal capacities. Where necessary, the Mamluks gave space to critical thinkers—such as the ḥanbalī Ibn Qayyim al-Jawziyah and his master Ibn Taymīyah—who reflected extensively on governance, and who had a more open attitude towards proof and religious diversity, at least in some respects. Mamluk rulers were well aware that Islamic law prevented unbelievers from being actors in the legal system, but they handled the issue as a legal technicality, open to interpretation by jurists and, in cases upon which sharīʿa was silent, to their own intervention as heads of the community. Up until Mamluk times, there was a large consensus on the procedural limitations that had to be observed by the qadis in their quest for truth, such as mistrust for archived records, documents and other artifacts considered to be suspicious, as well as for the word of unbelievers. Since early Islamic times, however, legal doctrines have allowed judges to make procedural exceptions as acts of grace (*istiḥsān*),

---

and his lawmakers: the role of legal discourse in the change of Ottoman imperial culture, PhD. Thesis: University of Chicago, 2005, Rothman, E. Natalie: *Brokering empire: trans-imperial subjects between Venice and Istanbul*, Ithaca, N.Y.: Cornell University Press, 2012, Meshal, Reem A., "Antagonistic Sharīʿas and the Construction of Orthodoxy in Sixteenth-Century Ottoman Cairo", Journal of Islamic Studies 21 (2) (2010), 183–212, Meshal, *Sharia and the Making of the Modern Egyptian*, Ozil, Ayşe: *Orthodox Christians in the late Ottoman Empire: a Study of Communal Relations in Anatolia*, London; New York: Routledge, 2013, Burak, Guy, "Faith, law and empire in the Ottoman 'age of confessionalization' (fifteenth–seventeenth centuries): the case of 'renewal of faith' ", Mediterranean Historical Review 28: 1 (2013), 1–23.

for reasons of necessity (*maṣlaḥa*, *ḍārūra*) and opportunity, and have given the ruler permission to inflict exceptional punishments, and to the judge to fulfill his duties as a governor. Until 1517, this could be done because the Franks were commercial partners who entered the realm of Islam for trade (the ideal situation foreseen by Islamic law). Under the Ottomans, however, this codified, patterned exchange entrusted to jurists and qadis transformed into a diplomatic operation with next-door neighbors. In 16th- and 17th-century Ottoman Istanbul or Cairo, commercial litigation often became tangled up in issues of residence and Islamic jurisdiction, the ransom and ownership of slaves, and border relations. While in Mamluk times contact with unbelievers was circumscribed to specific contexts such as the marketplace, the Ottomans had to deal with a porous and ever-growing inner boundary. This was not only a physical border, but a social contact zone in the tributary territories, or even in the heart of Istanbul. Challenged by the necessities of enforcing the rule of law, the Ottomans brandished a revered ḥanafī doctrine on legal dealings with minorities and unbelievers. They adhered to the letter in matters concerning the legal rights of local minorities, such as their right to testify in court in some cases. However, as I intend to demonstrate, Ottoman rulers also tailored the ḥanafī doctrine as regards foreigners, in order to comply to the same imperative of dealing with diversity under an Islamic system of governance. To all appearances, and probably in all earnestness, safeguarding the highest level of respect for the traditional rule of law, the choices the Ottomans made profoundly altered the way foreigners and Muslims related to each other.

CHAPTER 2

Producing, Handling and Archiving Evidence in Mediterranean Societies

Camios fazen los omes unos con otros de que an meester cartas
ESPÉCULO, LEY XXXVI

Why is it that so few Islamic archives have survived, while Western ones did? Why is it that Ottoman societies have yielded archival collections dating back to the 16th century, whilst their immediate predecessors have not? Is there a reason why archives appear to be confined to the Ottoman era, and why they seem to have been absent under other dynasties? Did medieval Muslims rely on writing and documentation to engage in social competition and reproduction? If they did not, is this because they were less literate, less eager to engage in "impersonal exchange," or less prone to develop institutional participation? If orality was a dominant vector in Islamic society, sanctioned by law, how was it dealt with in practice, for example to keep track of legal, administrative or commercial transactions?

The story behind the whereabouts of Islamic legal and administrative collections is a troubled one, and debates about the lack of archives, often adopting conflicting methods and goals, attest to the different disciplines' fraught relationship with the Islamic past and its records. In this chapter, I analyze how these important issues have been approached in the past and are increasingly being discussed today. In their quest to discover a historiographical locus for the treasures of Islamic chanceries and tribunals, researchers have speculated on the nature of hoards found in Jerusalem, Damascus, Qayrawān, al-Quṣayr, the monastery of St. Catherine at Mount Sinai or in Old Cairo. Were these document collections archives? If so, it is argued, the debate should revolve not so much around differences in sophistication between Eastern and Western institutions, as around the more-or-less accidental causes of their non-survival. At a time in which the Islamic historical narrative, based on literature produced by the piety minded, has superseded the labor of social and economic history, research studies drawing upon Ottoman judiciary and government

DOI:10.1163/9789004431737_003

collections, in contrast, have long outnumbered cultural studies. This is all the more astonishing for a field, such as Islamic history, in which the bias towards the religious-minded, the ulama, and their sources, is so strong that it has led Roy Mottahedeh to label most Islamic social history as mere "ulamology."[1] The great majority of contributions prefer to simply sidestep the issue by acknowledging the lack of archives for the medieval period and focusing instead on processes of document destruction, recycling or hoarding in unusual places—usually concluding with a call for a fresh appraisal of the surviving remains of medieval records.

I open my discussion on the handling of evidence by tackling the current debate on the non-survival of medieval Islamic records and their permanence in the West, the 'archival divide'; a debate that I argue is mostly based on type-cast attitudes and assumptions about Islamic history. Yet it is all the more fruitful because it begs a discussion and a clarification of a series of misconceptions about practices, artifacts, legal institutions on Islamic law, trade and diplomacy—such as those that foreign merchants, diplomats and travelers had to deal with in everyday situations at the marketplace and in court. Thus, important as it is, the debate masks a much broader discussion, revolving around changes in notions of justice, proof and evidence that were becoming apparent towards the turn of the 16th century, in the context of a much more polarized Mediterranean, and for which commercial litigation constitutes a privileged vantage point.

I adopt the view that, by and large, for the medieval period all extant documents are loose items, and a critical mass of artifacts is reached only in the 14th century for the Islamic core regions of Syria, Iran and Anatolia.[2] Document troves deposited in some mosques, together with the so-called *Genizas* and similar trash heaps have turned out to be miscellaneous aggregations rather than archives. Although he nuances the catastrophic vision of a medieval world bereft of archives, Konrad Hirschler, an important voice in this debate, states that there was only "a limited institutional logic of document preservation and that documents were discarded when they ceased to be of relevance for the individual." Indeed, for the period prior to 1517, most extant documents have only survived thanks to private collections and 'counter-archival' practices (such as recycling). More important for my argument is acknowledging the sudden wealth of documents represented by the maturation of the Ottoman Empire: or, in the words of R. Stephen Humphreys, "The 10th/16th century ...

1 Humphreys, R. Stephen: *Islamic history: a framework for inquiry*, Princeton, N.J.: Princeton University Press, 1991, 187.

2 Humphreys, *Islamic history*, 40.

marks a revolution, for from this point on the vast resources of the Ottoman archives lie before us." Judicial archives have yielded collections for Cairo (1522), Alexandria, Damascus (1583), Hama (1536), Jerusalem and Aleppo (1547) starting as early as the 16th century, while Sarajevo, several locations in Albania, Bulgaria, Cyprus, followed by Macedonia, Salonika and Hungary start in the 17th century.[3] The birth of the judicial (and to a great extent also administrative) archive was an Ottoman phenomenon, and accordingly is confined to the lands under the aegis of the Ottoman sultans. However soft the enforcement of Ottoman sovereignty in faraway regions, it is all the more meaningful that the trend towards archive preservation was absent in, say, competing early modern polities such as the Mughal Empire or Morocco. Moreover, the birth of the archive in this region marks a general change of attitude towards what constituted proof, from the signing of contracts to the taking of testimony and oaths.

In this first section, I guide the reader through the current debate on the existence or absence of Islamic medieval archives. The literature on the archival divide is extensive, and the discussion on writing and documentation has reached such a high degree of technicality that it would benefit from reappraisal in a more historical light. That is, by asking how artifacts were produced, by whom, and to what extent this production was subject to change over time. Focusing on Islamic archives, this discussion spans the medieval and early modern periods, and splits into several threads which, in order to be understood, require proficiency in areas of expertise such as comparative law, papyrology and archeology. The debate suffers from a marked division of labor between experts on Mamluk and Ottoman studies and their relative linguistic biases, thus making a diachronic appraisal of the issue, spanning both periods, more challenging. However, the need for one is all the more pressing when we consider that many authors do not appear to have taken account of the stark differences between the medieval sultanates and the Ottoman Empire, in terms of the relationship between the state and religion—something that ultimately affected the functioning of Islamic justice. Ottoman qadis were heavily involved in the administration of the provinces, and were also expected to apply regulations issued by the sultans on the grounds of public, customary law

---

3 Faroqhi, Suraiya: "Sidjill", Encyclopaedia of Islam, Second Edition, IX, 538–545, 1996, Fitzgerald, Timothy Jude: *Ottoman methods of conquest: Legal imperialism and the city of Aleppo, 1480–1570, PhD dissertation*, Harvard University, Vol. 3365261, Ann Arbor, 2009, 222–3. Fizgerald mentions that the earliest registers for Aleppo include a few late Mamluk documents. A good discussion on starting dates for the Arab countries is provided by Alsabagh, Munther H., *Before Banks: Credit, Society, and Law in Sixteenth-Century Palestine and Syria, PhD dissertation*, University of California, Santa Barbara, 2018, 38–9.

(kanun).[4] The need for them to refer to precedent, therefore, meant that their attitude towards archived documents differed greatly from that of medieval judges, whose principal task consisted in arbitrating private disputes according to the mostly oral procedure of sharīʿa.

Since all voices in the choir do not necessarily read from the same sheet music, I find it useful to identify three main axes within the discussion, all revolving around the production and use of proof in society. The first deals with justice, and includes works addressing the Ottoman legal reform and the role of legal proof in society. A second thread addresses the notion of textuality, discussing the changing uses of writing and documentation by the Islamic state. Lastly, a third group identifies major shifts in the production of written proof, or notarization, and its impact on the very "fabric of trust" in Mediterranean societies. My aim is to discuss what I define as historicist, or materialist approaches to the archival divide. A good deal of recent contributions, by attempting to unearth more artifacts in order to prove that larger collections did actually exist, have missed the opportunity to explain the divergent approaches to record preservation, and the very different meanings of proof in Latin and Islamicate societies.

Following this critical overview of the current discussion, I pay particular attention to several artifacts, spaces and institutions related to the presence of foreigners and their commercial disputes, and that will be crucial to the arguments advanced in subsequent chapters. With an emphasis on transformation over time, I tackle the issue of how judicial archives were passed on from one judge to another, the changes undergone by artifacts supporting written proof—such as scrolls and ledgers—and finally, proof-producing, scribal institutions at work in Middle Eastern markets. In the field of law, the production of proof had very direct and relevant consequences. For this reason, and following the views of authors such as Baber Johansen, I look at doctrines on proof and evidence that ultimately determined what responsibility the community had for its storage. I argue that medieval jurisprudence was crystal clear in defining the value of archived documents: after a judge had been dismissed from office, archived documents lost their value as proof, and jurists themselves describe the trimming procedure the successor qadi was expected to adopt. Concepts, but also tools and practices related to the production of evidence changed with Ottoman legal reform; words such as *dīwān*, formerly evoking the idea of court decisions, came to signify a physical place or an "archive", as did the term

4 Aykan, Yavuz: *Rendre la justice à Amid: procedures, acteurs et doctrines dans le contexte ottoman du XVIIIème siècle*, Leiden-Boston: Brill, 2016, 145, İnalcık, Halil: *Suleiman the Lawgiver and Ottoman law*: Mounton, 1969.

sijill, which from "scroll" came to mean the notarized, codex-shaped collection of proceedings in a given courthouse.

In this chapter I aim to convey the dynamism behind these newly-adopted patterns in notarizing legal acts, to capture the changing roles of notaries and scribes, and the enhanced value that written texts had for the Ottomans—what I define here as new attitudes towards the written word. A caveat needs to be advanced here, however: it would be tempting to think of this dynamism as a process of progressive modernization and rationalization, a natural tendency to enhance the written word and the archived document above orality and witnessing.[5] However, as illustrated in the final chapters of this book, new attitudes towards written proof and documentation, which might be associated with modernity, coexisted with much more conservative ones—such as the return to a strict use of oral witnessing in many fields of governance. In the last section of the present chapter, I evoke, by way of example, the problems that could arise at the imperial Dīvān due to a refusal to keep written proceedings. The Dīvān was an institution that combined governance and judicial functions, and was the focal point for legal relations with the Franks. The latter, however, were often puzzled by how little value the pashas gave to written records, and their insistence on the need for Muslim witnesses in order to accept agreements as valid. Mistrust for written records, therefore, did not completely disappear under the Ottomans. Reflecting on this apparent contradiction, Reem Meshal has described the "grafting" of witnessing onto the notarial deed, in an attempt to account for the Ottoman qadis' new role, which included notarizing and archiving what had traditionally been the lot of oral, private witnesses operating in streets and markets. She insists on a more harmonious vision of the relationship between jurisprudence—hostile to written artifacts—and a legal practice keen to use them. Yet despite its innovation, and a more open attitude towards records and archives, after 1517 the complex and sometimes contradictory handling of oral/written proof played out in peculiar ways, and began to pervade Muslim-Christian relations more than ever.

2.1 The 'Archival Divide'

The debate on the non-survival of Islamic archives dates back to early 20th-century investigations into the social and economic history of the Mediterranean. The question of the existence or absence of sources was crucial to the

5 Aykan, *Rendre la justice à Amid*, 228–29.

work of Jean Sauvaget who, in his tireless enthusiasm for unearthing Ottoman archives, contracted a severe illness that eventually led to his death in 1950, aged forty-nine.[6] For Sauvaget, the non-survival of archives could be pinned on the egalitarian character of Islam, and the consequent lack of status groups that might have led to the birth of, for example, the seigniorial institutions and collections characteristic of the European trajectory. A second, legal explanation focused on the divine nature of the law and the fact that the sources of Islamic jurisprudence were rooted in the sunna of the Prophet and the Qur'ān—that is, outside the sphere of temporal powers ("*la volonté du souverain ne suffit pas à créer le droit*").[7] Together with these new contributions to the debate, Sauvaget was the first to advance the argument that will be developed here: that the prevalence of oral over written proof as legal evidence hindered the development of judicial archives.[8]

More than seventy years after the first edition of Sauvaget's book, in which he explained these theories, a solid piece of research on Mamluk documentary production appeared in 2016. In it, Konrad Hirschler addresses again the *vexata quaestio* of the absence of Islamic archives.[9] Like many of his fellow-researchers, Hirschler returns to a hotly-debated passage written by Michael Chamberlain in 1994, where the latter addresses the issue by proposing a "social logic" to the absence of collections. According to Google Scholar, Chamberlain's book has been cited on no less than 344 occasions, and many point to this notorious passage on the lack of archives. For this reason, Chamberlain's text well deserves to be quoted here extensively:

> Is accidental loss the reason that historians have so few original document collections from the high medieval Middle East? ... In the Latin West documents were unmistakable proofs of privilege, exemption, competence, precedent, honor, or possession. As nations, classes, corporations, religious bodies, families, status groups, and factions fought out their struggles with documents, they took measures to preserve them. This accounts in part for the survival of a large number of collections of original documents from high medieval Europe compared to the high medieval Middle East. The critical position of the document within

6 Robert, Louis, "Jean Sauvaget", Revue Historique 207 1 (1952), 173–184, Sauvaget, Jean, "Comment étudier l'histoire du monde arabe, 5" , Revue Africaine 90 (1946), 5.

7 Sauvaget, Jean: *Introduction to the history of the Muslim East: a bibliographical guide*. Berkeley: University of California Press, 1965.

8 Sauvaget, *Introduction to the history of the Muslim East*, 19–21.

9 Hirschler, "From Archive to Archival Practices".

> European social and political competition also shaped the development of modern European historiography. When European historians began to exploit original documents, it was often to examine such symbolic charters to subvert or assert inherited rights, autonomy, sovereignty, titles, and ownership. Collections of documents therefore survived in greater numbers not by accident, but because elite groups exerted themselves to preserve them. And the crucial role of documents in European historiography is in like manner grounded in European practices of social and political competition.
>
> In the high medieval Middle East, however, rulers maintained patrimonial if not absolutist claims, considered most of the wealth of their subjects their own, and permitted other social bodies none of the formal autonomies they had in Europe. Individuals, households, religious bodies, and groups did not brandish documents as proofs of hereditary status, privilege, or property to the extent they did in the Latin West. Nor were their strategies of social reproduction recorded, sanctified, or fought out through documents to the extent they were in Europe.[10]

In his preface to *Knowledge and Social Practice in Medieval Damascus*, Chamberlain unpacks some of his more thought-provoking theses on archival absence. Aside from framing the problem in terms of social competition, he suggests that the alternate forum for social antagonism and perpetuation was not the archive, but the biographical dictionary, a genre that reached its maturity in the late Mamluk period (and in light of these works' comprehensive treatment of society, and the extensive recourse to them by researchers, it seems that a good deal of historians would agree). However, my intention here is not to discuss Sauvaget or Chamberlain's respective theses, even if the latter's analysis is noteworthy because it places the question of the absence of archives beyond the hazards of archaeology, stressing the lack of any defined logic of document safeguarding.

Many historical works attempt to capture the reader's interest by claiming the timeliness and relevance of their topic. It goes beyond mere rhetoric to say that not only in recent years, but even in recent months, attempts to account for the lack of pre-modern Islamic archives have been gaining momentum. Most of these recent works on missing archives attempt to nuance, counter and even mock Chamberlain's arguments on why collections have

10 Chamberlain, Michael: *Knowledge and social practice in medieval Damascus, 1190–1350*, Cambridge; New York: Cambridge University Press, 1994, 13–4.

not survived. As previously noted, however, Hirschler is alone in pointing to a narrow institutional logic for the preservation of records, that "were discarded when they ceased to be of relevance for the individual," although he claims that collections were nonetheless kept through alternative archival practices, different from the building of central state archives. Whether Chamberlain's Weberian polarization between a multi-layered and institutionalized Western society, and an egalitarian/patrimonial Islamic one is justified or not, the truth is that the prestige of the archival record and its weight in Western historiography constitutes a threat to the self-esteem of specialists of Islamic societies. Current reappraisals of the archival divide by Frédéric Bauden, Petra Sijpesteijn, Marina Rustow, Tamer El-Leithy or Hirschler himself are being carried out in parallel with research projects that target alternative archival practices inherent to documentary troves in Damascus, Cairo or miscellaneous Mamluk collections of reused documents.[11]

Although present-day historians are increasingly turning against the formal elements of Chamberlain's argument, they have less often criticized one crucial aspect of it: that what is missing is not so much the archive itself, as specific patterns of document preservation. Differences between Eastern and Western attitudes towards the production and preservation of documents did exist. But differences are visible, too, between the several Islamic powers, and these attitudes also changed over time. A privileged locus for such differences is commercial litigation, along with other forms of exchange, such as cross-confessional diplomacy—that is, by observing the work of agents such as consuls, pashas and chancery clerks. In these places, litigants and diplomatic actors often stumbled over differences in existing conceptions of proof. Rather than approaching the issue, either from a static binomial of two opposing systems, or from a flat paradigm where Islamic and Western trajectories were

11 Bauden, Frédéric: "Du destin des archives en Islam. Analyse des données et éléments de réponse", in: *La correspondance entre souverains, princes et cités-États. Approches croisées entre l'Orient musulman, l'Occident latin et Byzance (XIIIe-début XVIe s.)*, edited by Denise Aigle: Turnhout, 2013, 9–30, Sijpesteijn, Petra M.: *The Archival Mind In Early Islamic Egypt: Two Arabic Papyri*, in: *From al-Andalus to Khurasan: Documents from the Medieval Muslim World*, edited by P. Sijpesteijn and L. Sundelin, 2007, 163–186, Rustow, Marina, "A petition to a woman at the Fatimid court (413–414a.h./1022–23 c.e.)", Bulletin of the School of Oriental and African Studies 73 01 (2010), 1—27, Hirschler, "Document Reuse in Medieval Arabic Manuscripts", Humfress, Caroline: "Institutionalisation between Theory and Practice: Comparative Approaches to Medieval Islamic and Late Roman Law", in: *Diverging paths?: the shapes of power and institutions in medieval Christendom and Islam*, edited by John Hudson and Ana Rodríguez López, Leiden-Boston: Brill, 2014, 16–29, El-Leithy, Tamer, "Living Documents, Dying Archives: Towards a Historical Anthropology of Medieval Arabic Archives", al-Qantara 32 (2011), 389–434.

equal, I argue that each actor's awareness of the other's peculiarities made them more careful in the handling and production of proof. These divergences in approaches to proof and evidence did not at all lessen in Ottoman times, and awareness of them affected the actions and discourses of Frankish consuls and representatives. The debate would have much to gain, therefore, from taking into account the divergent attitudes between Franks and Muslims, between medieval sultans and their Ottoman successors, and the different rationales behind judiciary and government/chancery archives.

As Reem Meshal puts it, legal practices were traditionally dictated by a trove of customary solutions that entered into conflict with Ottoman legal reform. Under the Ottomans, foreigners increasingly felt the need to procure themselves documents backed by qadis to secure their claims, and more generally, to conform to Islamic legal practice for their businesses. Moreover, in the 16th century a new archival locus emerged: the courthouse (*maḥkama*) archive, a development that makes it all the more challenging to distinguish between Eastern and Western approaches to proof. To be sure, in the following pages it is not my intention to delve into the technicalities of how judges and officials certified, used and archived records—a task whose complexity has reached its peak in the analysis of the Ḥaram al-Sharīf documents found in Jerusalem in 1974 and 1976, and of which we will hear more. Instead, I proceed by stressing the comparative dimension of archive survival, and by carrying out a *tour d'horizon* of the relevant literature on the archival divide. My aim is to suggest that, by considering the history of Islamic societies to be fundamentally different from the labor of 'Occidentalists,' all too often researchers have been tempted to look for materialist explanations to justify the lack of Islamic archives. Instead, I argue that exploring different notions and ideas about proof and evidence can help us to understand this lack, and that commercial litigation—in its multiple configurations—can tell us much about the practical ways in which this archival divide was dealt with.

2.1.1 *A Threefold Problem*

Arguments that dwell on the *longue durée* of Islamic law and institutions, such as that advanced by Sauvaget, are long gone. Today, the debate revolves more around the extensive recourse that Islamic societies had to writing, as proof for the existence of archives. Paradoxically, and as a result of this apologetic tendency, the very question of why archives did not survive has been sidelined. The controversy about archives has given way to a parochial, increasingly technical discussion that might benefit from a more historical approach. In order to untangle the archival controversy, I identify three overlapping, yet distinct threads in the historical literature.

a) Justice

The first strand of research concerns Islamic justice, and more specifically judicial practice. According to this line of enquiry, a dramatic shift changed the way in which Middle Eastern societies, non-Muslims and foreign merchants interacted with Islamic justice. This transformation is marked by the appearance of the courthouse—the *maḥkama*—and its archives of court proceedings—the *sicil*—a process that crystalized in Ottoman lands over the course of the 16th century, including in the newly-conquered Arab provinces and even the Balkans, where similar phenomena have been noted.[12] Reem Meshal has also stressed the standardization of judicial procedure, extending from Aleppo to Mecca; indeed, local law schools began to see their traditional dominance challenged by the now superior authority of the ḥanafī judges.[13] In the Arab lands under the aegis of the Mamluks, people chose different courts depending on the nature of the transaction, as some schools had different attitudes towards, say, marriage or debt contracts.[14] Under the Ottomans, qadis were public notaries with ample prerogative for registering most kinds of contracts. Therefore early modern Muslims' daily interactions were now generally framed by a valid legal documentary framework—mostly in the form of notarial deeds, or ḥujjas—and by more consistent access to justice guaranteed by the qadi's court, now enshrined in a permanent institution whose records could be consulted and used to build precedent.

The judicial ledgers included, together with court proceedings, the deeds drawn up by the qadis as notaries, reflecting the different kinds of contractual relationships within society.[15] Reem Meshal has found a remarkable statement in a ḥanafī fatwā that epitomizes these new attitudes to legal practice: "the sijill is a ḥujja", or in other words, "this register is a *de facto* ḥujja", meaning that Ottoman court proceedings, beyond their mere role of physically preserving judicial decisions, provided legal coverage for people's actions, rights, and

12 Peirce, Leslie P.: *Morality tales: law and gender in the Ottoman court of Aintab*, Berkeley: University Of California Press, 2003, Meshal, *Sharia and the Making of the Modern Egyptian*, Al-Qattan, Najwa, "Dhimmis in the Muslim Court: Legal Autonomy and Religious Discrimination", International Journal of Middle East Studies 31 3 (1999), 429–44, Canbakal, Hülya: *Society and politics in an Ottoman town: 'Ayntab in the 17th century*, Leiden; Boston: Brill, 2007.

13 Meshal, Reem A, *The State, the Community and the Individual; Local Custom and the Construction of Orthodoxy in the Sijills of Ottoman-Cairo, 1558–1646, PhD dissertation*, Institute of Islamic Studies, McGill University, 2006, 115, Aykan, *Rendre la justice à Amid*, 227–230.

14 Rapoport, "Legal Diversity in the Age of Taqlid".

15 Aykan, *Rendre la justice à Amid*, 228.

agreements. Ḥujjas were issued on all legal matters by Ottoman qadis, meant, if needed, to be used elsewhere and not only at the maḥkama where it was produced. Ḥujjas concluded with the formula: "What happened was written down and delivered to the applicant, so that he may produce it as proof whenever there is need".[16] The recourse to judges to notarize legal deeds provided right holders with permanent legal coverage, not just for recovering debts, but any situation that could potentially be subject to dispute, such as being a freed former slave or a divorced woman.[17] These actions, undertaken either in or out of court, could be recognized as valid by virtue of the court's capacity to notarize and archive the deeds once and for all. The sijill was a ḥujja for—since most notarized deeds were now kept in the qadi's ledgers- right holders, including European merchants, could now turn to these repositories to validate their rights and claims, and not only in the event of a trial.[18] This was in contrast with the medieval practice whereby, although the notary delivered a copy to the right holder, the client needed to summon both the notary and his witnesses in court in order to give oral testimony of the transaction.

b) Textuality

A second line of inquiry deals with the Islamic state, and the tendency towards textuality exhibited by the Ottoman style of governance; defined by Guy Burak as the "calligraphic language of power that supplemented the shared vocabulary of sovereignty." Brinkley Morris Messick has looked at how the authority of scholars, traditions and oral witnessing in traditional Yemen was replaced by the authority of texts, ranging from university diplomas to legal and administrative regulations. A series of contributions by Burak very forcefully demonstrate new attitudes towards texts that took shape under Ottoman governance. Needless to say, the birth not only of the judicial archive, but also of state central registers contributed much to this new expression of sovereignty and governance. Burak argues, however, that not only archived documents were endowed with the charisma of governance; records such as judicial deeds were accompanied by a wide array of texts, ranging from chronicles to jurisprudence.[19] These researchers have provided a fresh interpretation of

16 Imber, Colin: *Ebu's-su'ud: the Islamic legal tradition*, Edinburgh: Edinburgh University Press, 1997, 52–3, Ottoman ḥanafī jurists started to include the delivering of ḥujjas among the current duties of the qadis, Ibn Nujaym, Zayn al-Dīn Ibn Ibrāhīm (1520–63): *al-Ashbāh wa-al-naẓā'ir 'alá madhhab Abī Ḥanīfah al-Nu'mān,* Damascus: Dar al-Fikr, 2005, 293.

17 Gradeva, Rossitsa, "Orthodox Christians in the Kadi Courts: The Practice of the Sofia Sheriat Court, Seventeenth Century", Islamic Law and Society 4 1 (1997), 37–69.

18 Ergene, Boğaç, "Document Use in Ottoman Courts of Law", Turcica 37 (2005), 83–111.

19 Burak, "Evidentiary truth claims", Burak, "In Compliance with the Old Register", 800.

the changing attitudes towards legal documents—and writing and documentation more generally. For Meshal, textuality made a more impersonal rule possible, and spurred the transition from subject to citizen. For Burak, on the other hand, in the early Ottoman period textuality encouraged both the rulers and the ruled, Muslims and non-Muslims, to increasingly put their trust in documentary artifacts; incorporated into their daily routines it ultimately altered power relationships at all levels of society, in the family and between communities. To Meshal, textuality was used as a motor for social change, whilst Burak considers all nature of texts to have been part and parcel of the Ottoman approach to statecraft and sovereignty.[20] Ottoman governance, not as a polity, but as a discourse, was anchored in a "spectrum of writings and associated institutions," such as archives.[21]

Unfortunately, debates revolving around governance, textuality and state archives often turn into a discussion about the state of medieval Islamic archives. In recent years, scholarship has often adopted a lachrymose, apologetic view of Islamic history, and partly as a result of Western misconceptions about the non-preservation of documents and the lack of archives, it has been victim to Western methodologies and research agendas. Papyrologists, experts in diplomacy and Geniza studies, among others, have touted the popular argument that a virtual archive does exist, but that we historians simply have not realized it yet. And indeed, these authors have been successful in unearthing more or less significant collections of documentary artifacts, whose richness they equate with the archival traditions of medieval Europe. Apart from being overly politically correct, this line of reasoning exhibits some major pitfalls; to mention just one, it still remains to be seen if there is any clear relationship between the production and use of the documents found in these troves, and the existence of any actual logic of preservation. As the epicenter of medieval Islam, and the motherland of scribal culture, land and tax registers and papyrology, it is the fate of Egypt's medieval administrative legacy that has been the most anxiously debated. Although in the past the issue has attracted the interest of social and economic historians, the debate has been largely monopolized by specialists in Islamic documents. At the time of writing this chapter, several research projects were underway to enlarge the trove of medieval Islamic documentary collections, and the idea that these archives have simply not survived, rather than never been kept at all, is gaining an alarming amount of momentum. Most contributions point to the fact that Islamic societies were

20 Meshal, *Sharia and the Making of the Modern Egyptian*, 11–12.

21 Messick, Brinkley Morris: *The Calligraphic State: Textual Domination and History in a Muslim Society*, Berkeley: University of California Press, 1993, 252-257.

highly literate, and acquainted with the use and production of documents. Other voices, however, suggest shifting the focus away from the quest for archival loci such as state archives, and looking instead to the Islamic "archival mind," to "archival practices," "archivalities" and more interestingly and as regards preservation, to "documentary life-cycles" and the reuse of documents. While historians such as Wael B. Hallaq, Frédéric Bauden, Petra Sijpesteijn, Marina Rustow, Maaike Van Berkel or Tamer el-Leithy insist that everyone in Islamic societies, from judges to widows and *dhimmīs*, had their part to play in the proliferation of documents, the fundamental question as to why these documents were not the object of any long-standing logic of preservation remains essentially unanswered. This is mainly because these contributions have failed to distinguish between production and use, on the one hand, and preservation, on the other.

In recent years, authors have claimed that proof for the archival tradition can be found in the diplomas preserved in the private archives of their recipients, such as the collection of medieval government decrees at Saint Catherine's of Sinai. The corpus of decrees preserved at the monastery give examples of medieval maẓālim practice, or policy-based regulations issued by the reigning sultans, together with decisions made as a result of a legal complaint.[22] The Cairo Geniza hosts Fatimid decrees and petitions, and includes a ruling issued to a woman that has been used by Rustow and El-Leithy as proof of an alleged Islamic tradition of archiving petitions. A more popular argument, however, is that secretaries and judges did in fact use and store drafts and copies of original deeds in the exercise of their functions. However, descriptions of such practices point more to temporary storage than to true archiving, and when copies were made, more often than not the motivation for it greatly differed from archival processes. Diplomatic artifacts, for instance, have survived in encyclopedias or vade mecums, leading Bauden and others to claim that copying a chancery document in a handbook served in itself the purposes of archiving.[23]

Maaike Van Berkel's work on the Abbasids similarly fails in its purpose to vindicate an early Islamic archival culture, showing, nonetheless, interesting instances of document use and temporary storage by Abbasid clerks, as

22 Nielsen, Jørgen S.: *Secular Justice in an Islamic State: Maẓālim under the Baḥrī Mamlūks, 662/1264–789/1387*, Leiden: Nederlands Historisch-Archaeologisch Instituut te Istanbul, 1985.

23 In this regard, an interesting discussion on literacy and the use of documentary evidence can be found in Robinson, Chase F.: *Islamic historiography*, Cambridge-New York: Cambridge University Press, 2003, 145-7. To Robinson these practices were widespread in Islamic historiography, yet they cannot be equated with archival practices.

epitomized in the description of vizier ʿAlī Ibn ʿĪsā's chaotic trove of records. Other known practices were the copying of correspondences, the record keeping of agricultural revenue and taxation and those associated with petition justice (*dīwān al-maẓālim*). It has been noted that the Abbasid empire was grounded on a powerful administration, recruiting empire-wide and assuring centralized taxation. In the late Abbasid period, the caliphate eventually yielded to a new, decentralized administrative culture, organized around the idea of the autonomous management of fiefs (iqṭāʿ) and taxation, at the service of a nomadic-minded military. While equating scattered, provincial troves with a proper archival culture, Van Berkel admits a dramatic lack of information on the locales (actual repositories, and physical places) for these practices, the patrimonial notions whereby documents were considered personal property of secretaries, most notably of the viziers, and the dislocation of collections vis-à-vis the political powers, distant from the Caliphal headquarters. A more sophisticated archival culture associated with the Abbasid caliphate—and caliphates in general, vis-à-vis the sultanian model of governance remains a suggestive hypothesis, yet more research efforts will be needed in order to prove it.[24]

The temporary storage of documents to facilitate the drafting of records was, as descriptions of the late Mamluk chancery have made clear, a different practice from the archiving of state papers, and it seldom led to the actual preservation of collections. The survival of records in unexpected places most often implies that a document was not preserved due to the positive value of the information it contained, but to its role as a specimen. This is clearly the case for chancery compendia and manuals of legal formularies, often invoked as meta-archives—such as those of al-Qalqashandī (1355–1418) or Ibn Ḥijjah al-Ḥamāwī (d.1434).[25] To be sure, since the Neolithic no sophisticated polity in the old world has ever managed to survive without attaining a certain degree of sophistication in the production and storage of information; however, the kind of archival practices described by Mamluk secretaries such as al-Qalqashandī can hardly be adduced as proof for the existence of archives. Although it is true that chancery manuals describe the drafting of documents such as petitions, and their storage for future reference, these descriptions present clerks keeping their documents in the back office or in their personal collections at best.

---

24 Van Berkel, Maaike: "Reconstructing Archival Practices in Abbasid Baghdad", Journal of Abbasid Studies, 1, 1 2014: 7–22.

25 Ibn Ḥijjah al-Ḥamawī, Taqī al-Dīn Abū Bakr Ibn ʿAlī (d.1434): *Qahwat al-Inshāʾ*, edited by Rudolf Veselý, Beirut: Klaus Schwarz Verlag, 2005.

The mid fifteenth-century text by Shams al-Dīn Muḥammad al-Saḥmāwī is probably the only authoritative text that deals in some detail with late Mamluk archival practices. These practices are presented as being strictly connected to the drafting of documents, together with others such as the procedure for sealing, using and storing pens. The "archive," or rather, the Dīwān's annex (muta'alliq al-dīwān), al-Saḥmāwī goes on, should be close to the scribes' workshop, and much stress is placed on procedures for controlling access to the chancery and preventing fraud by scribes, particularly in the form of interpolations and other unlawful additions to the register, or *daftar*. The record-keeping procedure puts emphasis on linking entries in the current daftar to a single clerk. This annex guards not only blank decrees but also those already signed by the sultan, such as answers to petitions and other records returned from the palace back to the Chancery. In it is preserved a particular pouch, the *muzarrah*, which appears to have been crucial to late Mamluk chancery practices and linked to the early archival artifacts used by judges (such as the *qimaṭr* and *qimaṭr*-like practices described below). Al-Saḥmāwī was therefore less concerned with preservation than with storage, and gave details on how to guard against the damages of excessive ventilation, humidity or mice.[26]

The terminology used by al-Saḥmāwī to describe the dīwān and its staff differs from that in use in Fatimid times, a period for which authors have allegedly found evidence of more consistent archival practices. Indeed, the description of the Fatimid chancery by Ibn al-Ṣayrafī (1071–1147) has attracted a good deal of attention, since it mentions storage procedures, as well as the presence of a clerk (khāzin) charged with archiving and drafting inventories for official correspondence.[27] Meïr Max Bravmann has also optimistically conjectured the existence of a site dedicated to the storage of archives under the Orthodox Caliph 'Uthmān (644–656), based on references to a place called the *Bayt al-Qarāṭīs* (*House of Documents*) by early authors such as al-Balādhurī and al-Ṭabarī.[28] By the same token, allusions to similar procedures and collections can be found

---

26 al-Saḥmāwī, Shams al-Dīn Muḥammad (d. 868/1464): *al-Thaghr al-bāsim fī ṣināʿat al-kātib wa-al-kātim: al-maʿrūf bi-ism al-Maqṣad al-rafīʿ al-manshā al-hādī li-dīwān al-inshā lil-Khālidī*, edited by Ashraf Muḥammad Anas and Ḥusayn Naṣṣār, 2 vol, Cairo: Maṭbaʿat Dār al-Kutub wa-al-Wathāʾiq al-Qawmīyah bi-al-Qāhirah, 2009, 372–374.

27 Ibn al-Ṣayrafī, 'Alī Ibn Munjib (1071—1147): *al-Qānūn fī dīwān al-rasāʾil*, edited by Ayman Fuʾād Sayyid, Cairo: al-Dār al-Miṣrīyah al-Lubnānīyah, 1990 34-41. al-Qalqashandī, Aḥmad Ibn 'Alī (1355 or 6-1418): *Ṣubḥ al-aʿša fī Kitābāt al-inšā*, 14 vol, Cairo: 1914, 1913 I 133, 135–6. el-Shayyal, Gamal el-Din: "Ibn al-Sayrafi", in: *Encyclopaedia of Islam, Second Edition*, III: 932a.

28 Bravmann, M. M.: "The State Archives In The Early Islamic Era", in: *The spiritual background of early Islam: studies in ancient Arab concepts* 311–314. Leiden; Boston: Brill, 2009.

for the Abbasid Caliphate.[29] Interestingly enough, Bravmann has contested the views of Claude Cahen, who in several of his works has addressed the murky status of the written documents and archives characteristic of the late medieval sultanates, as we shall see in the next section. However, the late medieval authors of chancery handbooks often cited in support of the existence of archives, such as al-Qalqashandī, were well aware that in their own time it was no longer standard practice to draft inventories and file originals (and this is an issue to which I will return).[30] However promising it may seem, the hypothesis of an early archival tradition in Islamic lands would require further empirical evidence, and wherever this tradition did exist, it hardly found continuity after the fall of the Fatimid Caliphate.

To return to late Mamluk practices, the *muzarrah*, or silken bag covering the government decrees for their safe-keeping, is not mentioned by al-Qalqashandī—who, incidentally, does not refer to the sijill as a scroll. While al-Qalqashandī and previous authors repeat the mantra that scribes should keep track (shāhid) of their documentary production, this point seems to be less relevant for al-Saḥmāwī, who places much more emphasis on guarding the daftar against manipulation. The daftar described by al-Saḥmāwī id not contain actual copies of the documents issued by the Chancery, but only brief summaries (mulakhkhaṣ), and nothing is said about a policy to collect these daftars over time.[31] Al-Saḥmāwī, moreover, clearly states that documents did not leave the chancery to join larger state collections, and served only as reference documents for the chancery staff.

The muzarrah is linked to the peculiar spatial layout of Medieval governance, a point of crucial importance to which I will return: blank decrees needed to be protected in its way from the chancery premises to the citadel (ilā al-qaṣr), the center of military power where they would receive the sultan's signature (ʿalāma). The muzarrah aimed at grouping together all the different kinds of decrees and records needing subscription that were previously sent to the Citadel for signature one by one, or in small numbers, 'throughout the whole day' (fī ṭūl al-nahār). It is described as a folded piece of silken cloth, with a bag-like receptacle inside and a closing cord on one of its ends. Rather than a sophisticated archival artifact, the muzarrah appears as the center of a ceremony of document transportation presided over by the dawādār kātib al-sirr,

---

29 Delsalle, Paul: *Une histoire de l'archivistique*, Sainte-Foy, 1998, 48, Van Berkel, "Reconstructing Archival Practices in Abbasid Baghdad".

30 al-Qalqashandī, *Ṣubḥ al-aʿšā fī Kitābāt al-inšā*, I, 139, al-Saḥmāwī, *al-Thaghr al-bāsim*, I, 375.

31 al-Saḥmāwī, *al-Thaghr al-bāsim, I*, 373–4.

an attorney of the Head of the Chancery, and entrusted to a ḥāmil al-muzarrah, or porter, a Mamluk clerk responsible for the integrity of the documents inside the pouch.[32]

One telling admission of a lack of archival references can be found in al-Saḥmāwī's discussion on how to address the different sovereigns; in his digression on India he makes it clear that, if an archive existed at all, clerks in the Mamluk chancery clearly had no access to it. Six Indian polities (mamlaka) are mentioned. First, Delhi, "second only to Cairo among the cities of Islam", whose dynastic politics depend both on investiture by the Timurids and by the Abbasid Caliph in Cairo. The sultanate is followed by Cambay (Kanbāya), a well-known trading partner of the Mamluks in Gujarat, described as a large territorial power. The third mamlaka is Bengal (Bankālā), under the rule of Jalaluddin Muhammad Shah, and Saḥmāwī mentions a letter sent to the Mamluk sultan Barsbāy in 840H/1436.[33] Gulbarga (kālbarkā) is mentioned in fourth place, certainly due to the momentum attained by bilateral relations in the 15th century, with the foundation of the *Ghulbarghiyya* madrasa in Mecca and an embassy dispatched by the Bahmanid Sultan Aḥmad Shāh in 1427. Two less distinct polities close this outline of Indian geography: Ṣanbūb, ruled today by a certain sultan Ibrāhīm and, lastly, Mā Dūkn, a name that probably refers to Mandugarh, capital of the Sultanate of Mālwā. This last polity, represented by a large city lying "at the foot of the mountain" had a king bearing a Muslim name, Maḥmūd (or Sultan Maḥmūd Khaljī, 1436–69), although in diplomatic letters al-Saḥmāwī advised that he be addressed in the same fashion "as the other kings of unbelievers."[34] What is striking in this brief description is the absence of the Mamluks' principal partner, Calicut, and of polities that had sent embassies in the past and that should have been known to the chancery, such as Ceylon—at least, they are not referred to by the usual Arabic placenames (Sīlān). More importantly for our discussion, al-Saḥmāwī admits that, since no letter had been dispatched to Cambay in his own time (fī zamāninā),

32 The muzarrah is a late Ayyubid innovation, introduced by the qadi and vizier Tāj al-Dīn ʿAbd al-Wahāb Ibn Bint al-Aʿaz, al-Saḥmāwī, *al-Thaghr al-bāsim, I,* 375.

33 Behrens-Abouseif, Doris: *Practising Diplomacy in the Mamluk Sultanate: Gifts and Material Culture in the Medieval Islamic World,* London and New York: IB Tauris, 2014, 46.

34 I owe the identification of Saḥmāwī's placename of Mā Dūkn with Mālwā to Prof. Sanjay Subrahmanyam, who considers the inclusion of the inland sultanate as sign of the completeness of Saḥmāwī's description of the Indian subcontinent. Mālwā's direct contacts with the Mamluks are attested for 1467, with an official exchange of letters, Meloy, John L.: "Mecca Entangled" in: *The Mamluk Sultanate from the Perspective of Regional and World History,* edited by Reuven Amitai and Stephan Conermann, Vandenhoeck & Ruprecht; 2019, 453–79, 470–1.

he could not provide specific instructions on how to address its ruler in diplomatic correspondence. This admission is important in that it points to a lack of continuity between the documents held during his tenure, and that of his predecessor; or in other words, the absence of archives extending beyond a single period of office.

In light of Saḥmāwī's admissions, it is no wonder that those who advocate the existence of medieval archives have neglected this important text in favor of other sources, such al-Qalqashandī, of an accumulative nature and insensitive to change, and that often refer to the chancery in Fatimid times. Al-Saḥmāwī presents the daftar an individual decrees as highly sensitive documents, at risk of being manipulated for the issuance of fake rulings, as began to happen under 16th-century Ottoman administration, when scribes organized a forgery network on the basis of reused decrees. Indeed, as I will develop in the following section, al-Qalqashandī admits that such archiving practices were discontinued by Mamluk secretaries. Even more intriguingly, he also reports that the very practice of keeping a daftar tracking the chancery's activities was abandoned for years under sultan al-Ẓāhir Barqūq (1382–1389 and 1390–1399) and his son al-Malik al-Nāṣir Faraj (1405–1412). In the best-case scenario, these practices suggest that preservation was limited to the span of a single generation, and can hardly support the claim that medieval Muslims did not lag behind the West in the field of archival and record preservation. They did, but this does not have to imply any cultural superiority, if we remember to adopt the necessary cultural relativism.

At a time in which the apologetic genre is gaining momentum, I find it necessary to turn instead to a major work that has gone almost unnoticed: Nicolas Michel's essay entitled "The Circassians had burned the registers." In it, Michel focuses on the history of the Medieval archives of Egypt and their gradual disappearance by the time of the Ottoman conquest.[35] Like Hirschler, Michel broaches the question of the register's materiality, but this he considers a secondary issue, as he judges the disappearance of Egypt's archives to be a historical, rather than a contingent phenomenon linked to the way records were handled. He addresses the waning of medieval administrative practices, not by looking to material explanations for the disappearance of paper, but by framing archival practices within the context of processes of Ottomanization, and the new masters' recourse to textuality as the basis for new models of governance. He also looks at the microhistory of the Mamluk administration, and

35 Michel, Nicolas: "Les Circassiens avaient brûlé les registres", in: *La Conquête ottomane de l'Égypte en 1517*, Edited by Benjamin Lellouch, Brill, 2012, 225–268.

the role played by individual secretaries in the potential transmission—or lack thereof—of Egypt's tax and land registers from Mamluk to Ottoman hands. Late-Ottoman clerks explained to the French colonists that the Mamluk archives had been burnt by Egyptian secretaries to prevent them falling into Ottoman hands; however, if the available chronicle material is anything to go by, it seems clear that the medieval Mamluks never saw the possession of archives as an attribute of sovereignty in the first place. Mamluk archives were kept 'downtown' in the secretaries' houses, and the very idea of keeping archives in the citadel—the epicenter of military power—was only an Ottoman innovation.[36] More importantly, while papyrologists still look to the hazards of document destruction to explain the lack of archives, from a historian's perspective this explanation can be found in the fact that the Ottomans, wherever necessary, incorporated the contents of registers and the skill of Mamluk secretaries into the newly-founded administration.

As an important footnote to Michel's work, Wakako Kumakura has recently shed new light on the incorporation of secretarial dynasties into the new administration.[37] In brief, Michel explains that towards 1550 the Ottomans came to the realization that their own cadastral survey, completed in 1528, and upon which the Ottoman taxation system depended, was not trustworthy. After the conquest, the Ottomans set about taking measurements of agricultural land, and to a large extent the survey was informed by depositions by local witnesses and notaries. Due mostly to the latter's intervention in the drafting of the cadaster, tax exemptions and other privileges had proliferated, and were impossible to ascertain on the basis of the 1528 registers. In order to make sense of the old, cryptic Mamluk surveys, and identify the true status of estates prior to 1517, which were now claimed as exempted, the Ottoman rulers also had recourse to Mamluk clerks from the Ibn al-Jīʿān and al-Malakī families, who had been monopolizing secretarial functions under the charge of the *mustawfī dīwān al-jaysh*, the official in charge of the army's accounts. Paradoxically enough, considering their disdain for the previous administrative regime, the Ottomans also issued a decree stipulating that in case of conflict, priority would be given to the old, Mamluk records. Former Mamluk secretaries took great pains to collect the old registers, and coordinated the transfer of the data they contained to the Ottoman tax administration. In the process, the original registers themselves were not preserved, as there had been no logic of record

36 Ibid. 234–5, 254.

37 Kumakura, Wakako, "Who Handed over Mamluk Land Registers to the Ottomans? A Study on the Administrators of Land Records in the Late Mamluk Period", *Mamluk Studies Review* XVIII (2015), 279–298.

preservation from previous dynasties to protect them, and they did not comply with Ottoman standards on authorship and authoritativeness. By virtue of this process, Mamluk scribal knowledge was grafted into the Ottoman archival mind.

The anonymous, unpublished manuscript *Orientale 98*, from the National library in Rome, neatly embodies the subsequent stage in this process of transition between two styles of administrative governance. The manuscript contains five texts concerned with the governance and administration of Egypt. The first is an anonymous treatise entitled *Kitāb al-qawānīn al-sulṭāniyya wa-l-fawā'id al-dīwāniyya* ("Rules of Government and Notes on the Dīwān").[38] This neglected work, most likely produced in the late 1580s, is flanked by two well-known texts: the *Kanunname-i Mısır* and the *Kitāb al-tuḥfah al-sanīyah bi-asmā' al-bilād al-miṣrīyah* by Sharaf al-Dīn Ibn al-Jī'ān (d.1480), as well as two lesser-known, shorter treatises. One is a Mamluk introduction to the combined use of the Coptic, solar calendar and the Arab one, based on the lunar cycles for administrative purposes, while the other is a brief work on weights and measures. The *Kitāb al-Qawānīn* deals with the handling of a variety of state affairs by administrators, in the broad sense of the term, and therefore pays a great deal of attention not only to civil secretaries, but to the administrative role of the Ottoman qadis. It shares some of the same characteristics as a chancery manual, since it deals with the issuing of edicts and court rulings by military and judicial administrators, however it departs from inshā' literature in that it does not provide potential apprentices with actual formularies. Rather than provide templates of decrees and letters *verbatim*, the author gives some context for each case, together with the document's most relevant sections. The *Kitāb al-Qawānīn* was something of a legal treatise in Siyāsa, dwelling on the lawfulness of some administrative practices, taxes and rights associated with the privy council of the sultan.

The *Kitāb al-Qawānīn* is accompanied by the Ottoman corpus of regulations issued by Süleyman for Egypt in 1525 (the *Kanunname*), together with the highlight of Mamluk civil administration, the *Tuḥfah*, a long, updated version of a 14th-century cadastral survey of Egypt. In gross, the *Tuḥfah* is an exhaustive list of agricultural fiefs subjected to taxation, the land-based financial system upon which the Mamluk military institution was based, known as the *iqṭā' system*. The text allowed Mamluk clerks to monitor the attribution of fiefs (iqṭā'āt) between the time of al-Ashraf Sha'bān (1363–1377) and the late fifteenth century. As a compendium, the *Orientale 98* manuscript seeks to illustrate the handover

38 BNCR, Ms. Orientali 98, 1v.

of Egypt's administration to her new rulers after 1517, dwelling on the many problematic aspects that might be of interest for trainee secretaries. As for the *Kitāb al-Qawānīn*, the compendium's only original work, it covers a wide range of topics, including some unexpected ones, such as how to deal with exotic birds. The work's major interest, however, are the miscellaneous aspects of the sultan's finances, such as the ruler's interests in Mecca, fluvial shipping over the Euphrates, taxes on silk imports or the revenues raised from the Khān al-Khalīlī market in Cairo. The author even provides the reader with some valuable figures on topics extending from sugar production to the amounts of spices Egypt provides to the Imperial kitchen in Istanbul.

Often adopting the language of the legally learned, the author of the *Kitāb al-Qawānīn* endeavors, in the first instance, to make a clear distinction between what it identifies as old Mamluk practices (ʿāda qadīma), and the fairer Ottoman ones. At several points in the text, the author presents a more rational and fair fiscal treatment of Indian and *ḥajj* ships arriving mostly at al-Tor in the Sinai peninsula, but also at al-Qusayr and Suez. Discussing a decree issued in 968H/1560, the author claims that the Ottoman administration, contrary to "old practice," taxed pilgrims from Mecca only with half the tithe (ʿushr), while under the Mamluks a full two-thirds was collected.[39] Similarly, the legal one-tenth was levied on Indian spices, while previous illicit taxes had now been abolished. The Ottomans, it was thus argued, conformed to sharīʿa in matters of taxation. To enforce this policy, the regulations were registered "in the sijills and in the Dīwān's daftars." Copies of this decision (ḥukm) were issued, delivered into the hands of three merchants, Khawājā Sāliḥ, Khawājā Ibn al-Jamāl and Khawājā Ibn Tuʿayma. The new fiscal regime is presented not as having been introduced at the initiative of the ruler, but as the result of intercession by a local qadi; an explanation that thus grounds the decision's motives in sharīʿa. Together with the sultan's finances, the work turns much of its attention to the handling of agricultural revenue by the Egyptian administration, focusing on the issues of measuring, registering and collecting revenue for the land plots that used to constitute the Mamluk fiefs. The author deals extensively with the issuing of executive decrees (marsūms) and judicial rulings (ḥukm, arḍ), and how they ought to be filed and archived in daftars and sijills, as well as the procedure for delivering copies as an integral part of Ottoman governance.

The *Kitāb al-Qawānīn* differs from similar Mamluk-era treatises in that it does not clearly differentiate between the judiciary and the chancery spheres. Similarly, al-Qalqashandī and al-Saḥmāwī seldom mention the terms sijill

39 BNCR, Ms. Orientali 98, 15v.

(pl. sijillāt), and when they do they refer to documents issued in the context of diplomatic exchange. Al-Qalqashandī includes some sections on agricultural taxes, or kharāj, and also uses sijill to mean *registration documents* for land plots (sijillāt al-taḥḍīr), a use that can also be found in the *Kitāb al-Qawānīn*. None of these texts, however, describes the sijill as a major byproduct of the qadi's functions, and for the case of al-Qalqashandī it is unclear whether the author even understands sijill as an artifact taking the shape of a scroll. In none of these works do secretaries appear to handle court rulings and proceedings as part of their daily activities. For the two most authoritative Mamluk authors on the subject, then, work in the chancery had almost no connection with the judiciary.

In the *Kitāb al-Qawānīn*, the term sijill refers to a variety of records produced by the qadis, who acted as administrators. Furthermore, references to a double registration in both the sijill and the daftar present decisions as pertaining to both the judicial and executive spheres.[40] Incidentally, Mamluk treatises never refer very clearly to actual archiving and to the issuing of copies. The work they describe is instead presented as a succession of *ṣūras*, or copies of documents issued by civil authorities. Indeed, when comparing Mamluk and Ottoman inshāʾ literature, we realize that in the latter period the use of judicial records extended to ample areas of governance and administration. The *Kitāb al-Qawānīn* refers, for example, to the importance of the sijills in the governance of the Red Sea. The ʿushr regime was imposed on Indian ships hailing from "known places" duly registered in a sijill, implying that unidentified foreigners were subjected to different forms of taxation.[41] There exists, therefore, a sijill containing the names of the countries covered by amān or similar legal devices. The practice of drafting notarial deeds is a manifestation of Ottoman governance: tax bureaus (qalams) were in the habit of levying a number of illicit taxes, on paper, bread, highway use and many other activities. The Ottoman system reduced these taxes to five types, two of which are related, all the more significantly, to the drafting of registered testimonies (shahāda) and sijills.[42]

A second feature of the *Kitāb al-Qawānīn* is that it adopts a specifically Ottoman attitude that censured Mamluk, and more broadly pre-Ottoman customary legal practice. As I argue in this book, the traditional notarial framework was not abolished, but was considered insufficient, and much attention was paid to the registration of acknowledgements, decisions and notarial deeds,

40 BNCR, Ms. Orientali 98, 15v, 19v.
41 BNCR, Ms. Orientali 98, 5v.
42 BNCR, Ms. Orientali 98, 11r.

both in administrative inventories and judicial registers. The author addresses the problems that could arise when a Muslim produced a witness' certificate (shahāda) without further supporting evidence, or literally, "carries a shahāda in his hands" (bi-anna fulān bi-yadihi shahāda nawāḥī kadha wa innahu mustaqīm). He reports a ruling by a local judge, who argued that if the rightholder was an honest man, and recognized as such by his peers, traditional practices should be respected and the shahāda accepted, and that there was no injustice in it because such decisions respected the general right of trustworthy Muslims to give testimony for facts. In response to the judge's ruling, an executive order was issued, warning local magistrates and clerks not to accept any witnesses that did not enjoy a formal appointment (taqrīr).[43] According to Ibn Iyās (1448–ca. 1524), Ottoman policies regarding witnesses sought to limit their number and to control their activities. From the outset, under Ottoman rule only registered notaries were allowed to act in the Ṣāliḥiyya madrasa, where the chief ḥanafī justice held court.[44] Ibn Iyās mentions the imprisonment of Shams al-Dīn Muḥammad al-Munāwī, a ḥanafī deputy judge, for bearing witness outside the Ṣālihiyya in a dispute over debt.[45] For the *Kitāb al-Qawānīn*, customary notions of truth-bearing were upheld, however stress was now placed on its control by the state-appointed judiciary.

Under the Ottomans then, according to the *Kitāb al-Qawānīn,* the notarial activities carried out by the ʿudūl ought mostly to be confined to the rural countryside. In these areas, the judge relied mainly on inspectors (kāshifs, umanāʾ) and notary-witnesses (shāhids) to draw up all manner of deeds and to bear witness to facts. The estimation of crops for tax purposes and other fiscal issues are one of the text's principal concerns, because oppression (ẓulm) and bad governance often stemmed from a judge's poor management of his own notaries and inspectors. These clerks should to be appointed "by virtue of their religiosity, ethical behavior and knowledge." Notaries and assistant judges needed to be monitored, and should only act in the judge's knowledge, and under his authorization. The Ottoman rotation in tenure is well known; judges were often unhappy with the places they had been assigned to, and could resort to appointing "worthless deputies [nāʾibs] with no concern for the local community," leading the court's shuhūd to begin "embezzling the people's

43 BNCR, Ms. Orientali 98, 12v.

44 Ibn Iyās, Muḥammad Ibn Aḥmad (1448-ca.1524): *Badāʾiʿ al-Zuhūr fī Waqāʾiʿ al-Duhūr*, edited by Muḥammad Muṣṭafá, 5 vol, Wiesbaden: Franz Steiner Verlag, 1960–1975, V, 165–6, 466.

45 Baldwin, James E.: *Islamic Law and Empire in Ottoman Cairo*: Edinburgh University Press, 2017, 86.

properties unjustly, and embroiling every legal business that comes into their hands, so that they take what they want from it … and they just start issuing ḥujjas and registering notifications (sijillāt)." Clearly, for the *Kitāb*'s author, notarization played a crucial role in the governance of Egypt's rural areas. According to Ottoman Kanun, it was up to the villagers of local communities (qariā) to appoint reputable shāhids, failing which registers (sijillāt) rife with untruths could proliferate.[46]

The *Kitāb* reports that in many places instructions were given for the assignment, registration and taxation of land plots. On occasion, inspectors were required to produce legal notifications (sijill sharʿī) to be filed in the notaries' daftar.[47] The author alludes here to the fact that, at least in cases of notarial deeds for which taxes had to be levied, such as marriage deeds, Ottoman notaries did keep registers, hence departing from medieval practice. A good deal of attention is thus paid to the fees required for registering deeds at the courthouse (maḥkama). Disputes over land plots were often arbitrated on the basis of information contained in the "dīwān's daftars." The judge was advised not to content himself with the mere acknowledgement and summoning of witnesses (išhād) but to verify the contents, accompany its drafting and eventual registration in the inventories or daftars.[48]

The administrative technicalities in the *Kitāb al-Qawānīn* all point to a new approach to the written proof by the Egyptian administration, and indeed more examples could be unearthed in this text. Written a few decades after the collapse of the Mamluk state, it echoes the crisis, identified by Michel, that Ottoman governance was undergoing in the 1550s. The administrative practice it describes reflects the new limits now being set for traditional notarization, the sudden proliferation of registering activities, and the enhanced role of qadis as administrators, all expressed in a new language of governance.

c) Notarization

The third thread in the historical literature that can help us more clearly understand the archival divide revolves around the idea of notarization. Late-medieval, post-caliphate societies, and in particular polities ruled by sultans—such as the Mamluk state—witnessed the rise of the notarial profession. Under the Mamluks, the recourse to ʿudūl transcended the traditional notarial framework to which it had previously been limited—that is, as witnesses to private

46 BNCR, Ms. Orientali 98, 16r-v.
47 BNCR, Ms. Orientali 98, 19v.
48 BNCR, Ms. Orientali 98, 12r, 19v.

transactions—and was extended to multiple aspects of administration.[49] As was the case for Latin Europe, notarial clerks were attached to many administrations, and in their daily activities Frankish merchants dealt with several categories of scribes, including Muslim ʿudūl, in places such as the customs house and the port in Alexandria.[50] The Franks invariably distinguished between *testimoni* (witnesses) and common scribes, hence underlining the importance placed on sight and hearing over writing. It seems that most Mamluk administrative departments, such as the Khizāna,[51] the Bayt al-Māl and the Citadel[52] relied on their own notarial staff. This proliferation of ʿudūl in the administration responded to the need to guarantee the validity of legal acts, such as transactions and sales contracts notarized at the customs house. Their very presence in these departments could only have been a hindrance to the development of these administrations' archives, since transactions were guaranteed by the traditional reliance on eyewitnesses. Orality had its limits, however, and we have indirect evidence that, when working for these administrations, records were systematically kept by the clerks. In the following section I discuss a Frankish inquiry conducted in Alexandria in 1387, in which it emerges that no clear distinction could be drawn between the notarial clerk's own records and the archives proper to each bureau. These administrations' records were not placed in an institutional repository, but rather kept by the ʿudūl who notarized them as personal archives.

Mamluk ʿudūl enjoyed the capacity of certifiers of truth, and did so on a local, neighborhood basis. Mamluk chronicles depict the ʿudūl's activities as embedded in the daily life of markets, mosques and neighborhoods. Particularly in Syria, mentions of the ʿudūl and their workplaces (maḥallāt, marākiz, often located near the city gates) give the impression that there existed a market for truth-bearing that was specific to a given territory, each one controlled by its

49 "the function of the notary was traditionally private, i.e. commissioned outside of court and performed in the name of the notary rather than the judge", Meshal, "The State, the Community and the Individual", 118. Veselý, Rudolf, "Die Hauptprobleme der Diplomatik arabischer Privaturkurden aus dem spätmittelalterlichen Ägypten", Archiv Orientali XL 40 (1972), 312–43, 324.

50 Ibn Taghrībirdī, Abū al-Maḥāsin Yūsuf (1411–1470): *al-Nujūm al-Zāhirah fī Mulūk Miṣr wa-al-Qāhirah*, edited by Muḥammad Ḥusayn Shams al-Dīn, Beirut: Dār al-Kutub al-ʿIlmīyah, 1992, VII, 313.

51 al-ʿAynī, Badr al-Dīn Maḥmūd Ibn Aḥmad (1361–1451): *ʿIqd al-jumān fī tārīkh ahl al-zamān: ʿaṣr salāṭīn al-Mamālīk*, edited by Muḥammad Muḥammad Amīn, Cairo: al-Hayʾah al-Miṣrīyah al-ʿĀmmah lil-Kitāb, 1987, III, 29, Ibn Taghrībirdī, *al-Nujūm al-Zāhirah*, X, 217.

52 al-ʿAynī, *ʿIqd al-jumān fī tārīkh ahl al-zamān: ʿaṣr salāṭīn al-Mamālīk*, III, 32, 173, 175–6.

respective team of ʿudūl.[53] Even if we should not make generalizations based on criticisms against corrupt ʿudūl, one has to acknowledge that notaries were linked to sources of power such as qadi courts and emir households, and belonged to socially dominant families.[54] These criticisms do however abound, pointing to the fact that very often notarial services were used for illegitimate purposes throughout Mamluk society. The fact that it was not the written notarial deed, but the notary and his assistants' testimony itself that ultimately mattered, was a formidable source of power, and in part at the service of Mamluk governance. For this reason, Ottoman rule aimed to control the ʿudūl, as well as to dismiss large branches of the notarial profession as a major step in the judicial reform program, an important point raised by Meshal. Other authors such as Najwa Al-Qattan and Hülya Canbakal have noted that Ottoman reform, which enhanced the notarial functions of judges, rendered the medieval ʿudūl obsolete outside of Egypt.[55] Not surprisingly, they counted among the first victims of the Ottoman legal reforms, and some were even deported to Istanbul shortly after the occupation.[56] Considered to be corrupt, large-scale dismissals by Ottoman reformers began in 1517, and then again in 1520. The frequency with which they seem to have been reappointed is however some indication of how crucial the notarial institution was to Egyptian society, and more broadly, of the divergent interests of central and local judiciaries.[57]

The influential work of Ronald C. Jennings further elaborates on the new role of notaries in the Ottoman era by addressing the functions of the *shuhūd al-ḥāl* in the 17th-century Ottoman city of Kayseri, and argues that a closed milieu of medieval ʿudūl—exerting a monopoly over truth-bearing—transformed into to an opener system in Ottoman times. Jennings shifted the focus to the shuhūd el-ḥāl, the 'instrumental witnesses,' or trustworthy Muslims attached to the courthouse who, rather than notarizing acts themselves, monitored the notarial and judicial functions of the qadi, hence acting to curb the latter's

53 Mandaville, Jon Elliot: *The Muslim Judiciary of Damascus in the Late Mamluk Period, Princeton University, Ph.D.*, 1969, 9–10, 121–2, Ḥamzah, ʿĀdil ʿAbd al-Ḥāfiz: *Niyābat Ḥalab fī ʿAṣr Salāṭīn al-Mamālīk (1250–1517 M/648–923 H)*. Vol. 2: al-Hay'ah al-Miṣrīyah al-ʿĀmmah lil-Kitāb, 2000, 254–5.

54 Kosei, Morimoto, "What Ibn Khaldun Saw: The Judiciary of Mamluk Egypt", *Mamluk Studies Review* 6 (2002), 109–131, 112.

55 Canbakal, *Society and politics in an Ottoman town: ʿAyntab in the 17th century*, 129–30, Al-Qattan, "Dhimmis in the Muslim Court: Legal Autonomy and Religious Discrimination".

56 Ibn Iyās, *Badāʾiʿ al-Zuhūr*, V, 178–9.

57 Meshal, "The State, the Community and the Individual", 121, Meshal, "Antagonistic Sharīʿas", 183, 199–200, Meshal, *Sharia and the Making of the Modern Egyptian*, 89–93, Michel: "Les Circassiens avaient brûlé les registres", 253.

power.[58] Historians of Ottoman law generally agree that these 'court' witnesses ought to be distinguished from notary-witnesses—that is, the ʿudūl drawing up deeds in the markets and other public places. Although the actual functions of the *shuhūd el-ḥāl* differed from one city to another, their general role was to attest to the validity of court proceedings, and sometimes to carry out scribal functions at court. While mentions to the ʿudūl are rare after 1500, instrumental witnesses were central to Ottoman legal procedure. They were appointed by judges, and seem to have received some kind of legal training.[59]

Together with the notary's status, the object that they were most often solicited to produce, the ḥujja, underwent similar changes. Changes in the use, production and preservation of the ḥujja were closely tied to the spread of the maḥkama/sijill binomial in Ottoman society. While formally the legal definition and value of the ḥujja remained the same, it is the way in which it was used in society that was radically different. It is difficult to know the extent to which the notarized deeds produced by scribes in their stalls continued to circulate as the fundamental probative artifact in Muslim societies, and if so, they certainly were accompanied by an increasing number of notarized artifacts generated by Ottoman judges.[60] The Ottomans empowered the ḥanafī judges over all other notarial actors, put them in charge of notarizing the most important contracts, and made it compulsory to register certain agreements before them. Thus, if the Ottoman takeover was bad news for the Arab notaries that staffed administrative departments, even the ʿudūl working in the markets had their functions as bearers of truth increasingly curtailed by the judge-sanctioned ḥujja. Indeed, the very oral essence of medieval notarization had made it unnecessary to preserve records. In medieval notarial practice,

58 Jennings, R., "Kadi, court, and legal procedure in 17th C. Ottoman Kayseri", *Studia Islamica* 48 (1978), 133, Imber, Colin *The Ottoman Empire, 1300–1650: The Structure of Power*: Palgrave, 2002, 233.

59 Baldwin, *Islamic Law and Empire in Ottoman Cairo,* 36–7. Müller, Christian: "Ecrire pour établir Ia preuve orale en Islam: la pratique d'un tribunal à Jérusalem au XIVe siècle", in: *Les outils de la pensée: Etude historique et comparative des textes*, edited by Akira Saito and Yusuke Nakamura, 63–99: Editions de la Maison des sciences de I'homme, 2010, 75, Müller, Christian: "The Power of the Pen: Cadis and their Archives in Medieval Islam", in: *Manuscripts and Archives: Comparative Views on Record-Keeping*, edited by Alessandro Bausi, 361–385: Walter de Gruyter, 2018, 374–5, Sonbol, Amira, "Women in Shariʿah courts: a historical and methodological discussion", Fordham International Law Journal 27 1 (2003), 225–253, 243.

60 ʿAlī Ibrāhīm Mīlād, Salwā, "Registres judiciaires du tribunal de la Salihiyya Nagmiyya—Etude des archives", *Annales Islamologiques* 12 (1974), 161, 182–88, provides a detailed description of how documents notarized at a law court, such as property deeds, were eventually produced by the parties as the basis for new transactions.

the trustworthiness of a given document resided in the capacity of two original witnesses to verify it. After the death of these witnesses, the record's validity, if it had any at all, was always contested. With the qadi's enhanced role as an agent of Ottoman governance, together with his activities as a regular notary, the medieval 'qadi archive,' soon became obsolete. Despite efforts by senior scholars such as Wael B. Hallaq to prove that a medieval archival tradition existed, these works have been more successful in unveiling the technicalities used by medieval judges to cope with the fragile status of paper, than to prove any continuity with early modern practice.[61] Before Ottoman times, there was absolutely no logic of long-term preservation for judicial records. Rather, we know that only a few sensitive documents survived when a new judge took up his functions, and that the qadi was not expected to rely on the papers produced by his predecessor. In gross, the medieval 'qadi archive' comprised a series of documents whose validity was kept alive by the judge, who validated them through the complex procedure of isjāl. Baber Johansen has written of the "ephemeral archive" and the peculiar rules for its transmission, based on his research on handbooks for qadis. Hallaq's views on the continuity between the medieval qadi archive and Ottoman practice have certainly been challenged from the perspective of court procedure, scrutinized by Müller in several studies of the Ḥaram al-Sharīf's papers.[62]

Amalia Zomeño has devoted many efforts to analyze the role of writing and documentation in Spain's Islamic communities on the eve of the conquest. In fifteenth-century Granada notarial deeds were subjected to the same evidentiary rules as everywhere else in the Islamic world, so that "if a Muslim was challenged on the use of a piece of land that he considered to be his property, he had to go to court with his sale contract and show it to the judge. However, the judge himself would only accept the facts narrated in this contract if the professional witnesses, who wrote the deed, confirmed in his presence that the signatures in the document were theirs."[63] Therefore it is no surprise if no

61 Hallaq, Wael B., "The qādī's dīwān (sijill) before the Ottomans", *BSOAS* 61 (1998), 415–436.

62 Johansen, Baber, "Formes de langage et fonctions publiques: Stéreotypes, témoins et offices dans la preuve par l'écrit en droit musulman", *Arabica* 44 3 (1997), 333–376, Müller, Christian, "The Ḥaram al-Sharīf collection of Arabic legal documents in Jerusalem: a Mamlūk court archive?", *al-Qantara* XXXII 2 (2011), 435–459, Burak, "Evidentiary truth claims",235, 252, Unlike Müller, the work of Burak has challenged Hallaq's conclusions not by studying the materiality of records but from the viewpoint of the role played by texts in Ottoman governance.

63 Carro, Sergio and Amalia Zomeño: "Identifying the 'udūl in Fifteenth-Century Granada", in: *Legal Documents as Sources for the History of Muslim Societies: Studies in Honour of Rudolph Peters*, Edited by Léon Buskens and Petra M. Sijpesteijn, 109–129. Leiden; Boston: Brill, 2017, Zomeño, Amalia, "From Private Collections to Archives; How Christians

significant collections of judicial records have survived for the centuries prior to the conquest. In the *mudéjar* period (1492–1500), when Muslims were still tolerated, and willing to prove and transfer property as quickly as possible, they started brandishing notarial deeds issued according to Islamic law. Yet only after the decrees of expulsion were enacted Islamic records started to be translated, quoted and archived no longer in private collections, but by Christian institutions. Members of the new society deemed those records useful to sanction their newly acquired properties and status. Records were incorporated to court proceedings "according to the uses of the Muslims" and validated by converts. Arabic notarial deeds were advanced as proof in lawsuits over property opposing two Christian parties at a much later date. Paradoxically enough, it was only as a result of the Christian conquest that a significant trove of Arab documents have been preserved.

In this book, I refer to the medieval practice of record-keeping as the '*qimaṭr archive*.' The *qimaṭr* was the transportable wooden box containing time-sensitive records that were considered worth preserving, and that early judges used to carry with them. In medieval times, the practice adopted by judges upon assuming their functions in a given locality was to re-certify a selection of records that might be needed in the future and, at least under the early caliphate, these documents were deposited and carried in the qimaṭr, as sessions were itinerant. The artifact adopted different forms over time, such as bags, and these ephemeral archives—mobile qimaṭr-archives—together with the trimmings of deeds drawn up under previous qadis, changed under the Ottomans. If only a minority of authors attribute the invention of the judicial archive *tout court* to the Ottomans, most specialists agree that they at least provided the qadi's papers with a stable locale.

Beyond the causality behind this historical watershed and the debate it sets off, one thing is for certain: with their approach to keeping registers (*sicils*), and the status of legal proof attached to them, the Ottomans brought with them major changes in the way Islamic societies handled writing and documentation. In a nutshell, Ottoman judges and secretaries were pressured to provide continuity for their practice. In the case of judges, the revolution came in the shape of a register, and the transformation of what until then

---

Kept Arabic Legal Documents in Granada", Al-Qantara: Revista de Estudios Arabes 32 2 (2011), 461–479, Zomeño, Amalia: "Notaries and Their Formulas: The Legacies from the University Library of Granada". dans *From al-Andalus to Khurasan. Documents from the Medieval Islamic World*, edited by P. Sijpesteijn and L. Sundelin, Leiden—Boston: E. J. Brill, 2007, 59–80.

had been scattered records—or, at best, scrolls comprising a series of appendixes collated to an original chart—into bound sets of legal documents. The legal innovation at the basis of the ledger was that the whole documentary trove pertaining to a single lawsuit—that is, anything that transited through the qadi's hands—was now entered into the sicil, or ledger, along with court proceedings, and was therefore certified by the judge. In medieval practice, witnesses would have been summoned to renew the deed's validity, a practice that was now abandoned as long as notarized documents included in the sicil did not require further validation. Under the Ottomans, all judicial papers included in the sicil had permanent validity, and were sewn together in a codex that was itself the object of archival procedure. Starting from the sixteenth century, Cairo registers were completed with a closing sheet and archived in the Dīvān's court for further reference, since copies of their contents could be delivered upon request.[64] As for the ḥujja, the basic unit of the legal system in and out of court, research conducted on the basis of judicial archives is now showing that everyone in society, and lower-rank individuals more forcefully, now had at their disposal a tool to claim, keep and protect their rights from violation.

The major transformation experienced by notarization in the sixteenth century was intimately tied to changes introduced in the administration of justice on the one hand, and on the other, to a new tendency towards textuality in Ottoman governance. New patterns of notarization, the rise of judicial sicil-keeping, and the tendency for Ottoman governance to rely on authoritative texts were all part of the same phenomenon, and all had an impact, albeit unequally, on the way ordinary people dealt with writing and documentation. While medieval ʿudūl saw virtue in keeping mustāʾmins and women away from their workshops, scholars remark an unprecedented rise in women, foreigners and other excluded categories gaining access to the legal system.[65] Taken together, these three threads constitute a *fil rouge* straddling the periods and topics covered by this research. After this survey of the historical literature, I will return in the next section to the narrative on Islamic archives between medieval and early modern times.

---

64 ʿAlī Ibrāhīm Mīlād, "Registres judiciaires du tribunal de la Salihiyya Nagmiyya—Etude des archives", 180.

65 Meshal, *Sharia and the Making of the Modern Egyptian,* 133–8, Peirce, *Morality tales: law and gender in the Ottoman court of Aintab,* 389, Sonbol, "Women in Shariʿah courts", Gradeva, "Orthodox Christians in the Kadi Courts", discusses access to justice by dhimmīs and their approach to evidence.

2.1.2 *Materialist Explanations*

Since the twentieth century, explanations for the limited development of legally organized professional and corporate bodies have alternated with more materialist views. To the arguments advanced by Sauvaget, Claude Cahen (1909–1991) added political turmoil, or the hazards of the Ottoman takeover of the Medieval Arab administration. Together with these motives, he suggested that Islamic judicial systems were characterized by a rapid rotation in appointments, and that Islamic foundations had no legal personality, which implied that documents—such as waqf documents—were kept by qadis and authorities, hence hampering the development of institutional archives. Elaborating on a well-known theory according to which medieval Islam did not develop corporations, Samuel M. Stern suggested that the dearth of surviving archives was rooted in certain features of Islamic institutions, such as the lack of stable professional bodies.[66] Despite this, for the early modern period—when such guilds actually came into being—Ottomanists have turned instead to judicial and government collections when dealing with artisan organizations.[67]

A good deal of recent literature, often critical of Chamberlain's thesis, have shifted the focus entirely away from record-keeping and preservation, and turned instead to the issue of whether documents were produced or not, used or not, and if they have survived in sufficient numbers. Launched in 2011, this approach to the debate is best epitomized by an article by Tamer El-Leithy, for whom Chamberlain's thesis, that "documents don't survive," is a "a totalizing argument" that "forecloses the space for any historical investigation of the social uses of medieval documents." El-Leithy goes on to claim that "*legal documents were routinely produced by notaries and courts; they were assiduously preserved by individuals and families, who later consulted and brandished these written forms of evidence in disputes and conflicts.*"[68] Thus late medieval Islam,

---

66 Cahen, Claude: "Y a-t-il eu des corporations professionnelles dans le monde musulman classique? quelques notes et réflexions", in: *The Islamic City*, edited by S. M. Stern and A. Hourani, 51–63. Oxford, 1970, Cahen, Claude: "Du Moyen Âge aux Temps modernes", in: *Les Arabes par leurs archives: XVIe-XXe siècles, Colloques internationaux du Centre national de la recherche scientifique n. 555*, edited by Jacques Berque and Dominique Chevallier, 9–15. Paris: CNRS, 1976, Cahen, Claude: "Considérations sur l'utilisation des ouvrages de Droit musulman par l'historien", in: *Les peuples musulmans dans l'histoire médiévale*, 71–76. Cairo: Presses de l'Ifpo, 2014, Stern, S.M.: *Fātimid Decrees: Original Documents from the Fātimid Chancery*, London: Faber and Faber, 1964.

67 Faroqhi, Suraiya: *Artisans of Empire: Crafts and Craftspeople Under the Ottomans*, London and New York: I. B. Tauris, 2011, xix-xx.

68 El-Leithy, "Living Documents, Dying Archives", Rustow, "A petition to a woman at the Fatimid court".

as epitomized by the Mamluk sultanate, witnessed the rise of sophisticated and literate societies, which naturally engendered archival institutions. To be sure, some authors have pushed the issue even further than El-Leithy, reducing it to a mere question of miscalculation. For Frédéric Bauden—papyrologist, diplomatist and expert on al-Maqrīzī—the very existence of documents for the medieval period proves that Islamic societies never lagged behind the West in the consumption and safe-keeping of records; it is simply that we have not yet realized the extent to which documents have been preserved. Indeed, Bauden brandishes papyruses and other surviving artifacts as rebuttals of Chamberlain's thesis, and the "calamitous" assumption that the Islamic premodern archive never existed.[69] In the same vein, in a subsequent text Bauden counters Chamberlain's argument that Islamic societies were less prone to record preservation by claiming that secretaries were advised to keep personal copies of some charts.[70] This particular line of reasoning is based on a vexing passage by al-Qalqashandī describing scribal practices at the Mamluk chancery, in which he reminds the clerks to keep trace (shāhid) of the documents they have drawn up. Bauden similarly challenges Chase F. Robinson's view that Arab societies placed more importance on historiography, and therefore on biography and prosopography, than on the preservation of records; a point that responds to Chamberlain's idea that it was the biographical dictionary that Muslim societies vested with an archival role.[71]

Bauden addresses three main arguments in the existing literature; that we should look further afield than state archives, that Islamic society had a skeptical attitude towards written artifacts, and that Islamic judicial records received a specific treatment.[72] Bauden seems to accept that there was an

69 Bauden, F., "Mamluk Era Documentary Studies: The State of the Art", *Mamluk Studies Review* 9 (2005), 15–60, 16., Bauden: "Du destin des archives en Islam", 31–33.

70 This argument is based on a passage by al-Qalqashandī, *Ṣubḥ al-aʿsha fī Kitābāt al-inshā*, VI, 198.

71 Robinson, *Islamic historiography*, 146. 66–79, 187–89.

72 "One of the most repeated [explanations] is that, unlike what happened in Europe, Islam had no legally organized social bodies which could have preserved archives, the unique exception being the waqf documents, as shown by Carl Petry. This means that only state archives could exist. Secondly, it has been argued that written documents do not establish the law (kitābun yushbih kitāban: one writing looks like another writing and can be exchanged with it), but if so why would non-Muslim communities have held for centuries documents that had no legal value and that were referred to in case of necessity? Thirdly, it has been alleged that in Europe most of the documents are of a judicial nature, while in Islam, on the other hand, these kinds of documents were kept by the qadis. When they became useless or obsolete, they were discarded", Bauden, "Mamluk Era Documentary Studies: The State of the Art", 17.

absence of professional and religious bodies that might have produced archives, and seems aware that, as we shall see, Islamic judges were not expected to preserve their records after they left office. Similarly, the Islamic suspicion for the written word (*kitābun yushbih kitāban*) has been noted; however, Bauden echoes a popular argument used to support the existence of the Islamic medieval archive, when he asks whether the preservation of documents in monasteries and by other recipients contradicts these views. Unfortunately, he misses the opportunity to reflect on why issuing authorities did not preserve decrees, while their non-Muslim addressees did. Unlike the trove of factual, materialist explanations for the non-existence of archives, these arguments are phrased in an elusive manner and do not receive sufficiently thorough treatment.

A second attempt to defend the existence of the Islamic medieval archive came in 2013, when Bauden took a firmer stance against the theses of Chamberlain and Chase F. Robinson. Countering Chamberlain's "naïve" idea that Islamic societies had a more limited recourse to documents, Bauden refers extensively to the drafting of circumstantial copies, the reuse of documents, temporary storage, and archive destruction, to support the idea that collections did exist, but that due to ideological motivations they were either neglected or simply destroyed—the presumed destruction by Saladin of the Fatimid archives being one such example. However, according to a recent reappraisal of this idea by Fozia Bora, the destruction by new Islamic rulers of their predecessors' archives and libraries has become an Orientalist, often unfounded, historical commonplace.[73] All too often, in Islamic diplomatics voices have spoken out against positions like Chamberlain's, claiming that they are making arguments from silence. Yet the mere existence of fragmentary groups of documents or collections in unexpected places can hardly account for the intentionality of preservation.

Although Chamberlain acknowledges that tax and land records were kept, and decrees issued, his thesis holds that state, corporate, and household documents held little power in social competition interactions. For example, he argues that commercial contracts were considered to be separate from the political sphere, and that the nature of political relations was fundamentally non-contractual. Engin D. Akarlı's analysis of legal relations in the marketplace quotes a legal case brought to the imperial council in 1814, stressing this fundamental difference with *old régime* European societies. On that occasion,

73 Bauden: "Du destin des archives en Islam", 35, a view contested by Bora, Fozia, "Did Ṣalāḥ al-Dīn Destroy the Fatimids' Books? An Historiographical Enquiry", *Journal of the Royal Asiatic Society* 25 1 (2015), 21–39.

Sultan Mahmud II argued that his decisions, after consultation with legal experts, were legally binding, irrespective of the existence of previous rulings on the same matter. In response to a legal dispute among three different groups of cloth dyers, the sultan had issued a decree that levelled customary statutory distinctions between different dyers and raised the rents they paid to a waqf. Some refused to accept the sultan's decision, claiming precedence for old legal records that confirmed their former rights. The sultan reprimanded them, asking "What does it mean to act in defiance of my decree? ... Destroy the documents in their hands and issue them new ones ... If any of the dyers dares to resist let me know! The new order will surely hold, once a few of them are hanged!"[74] To return to Chamberlain's argument, Islamic societies are characterized as lacking in hereditary status; precedent was therefore far less relevant both to sharīʿa and in administrative practice. In response, Bauden cannot help but "smile" at the "naïve" idea that Islamic societies were less "drowned in paperwork" than their Western counterparts, and responds to the Chamberlain dilemma by reminding us that, whenever paper presented some value to the state, it was far more valuable for junk dealers. Although his argument that document reuse may have hindered the preservation of Islamic archives might well be worthy of consideration—provided it was supported with some documentary evidence—what is more difficult to accept is the claim by Bauden and others that reused documents stand as proof in themselves of the existence of archives.[75]

Occam's razor would have it that materialist explanations cannot be discarded merely because they are simpler, and therefore less satisfactory to intellectual taste. In the current debate, factual claims appear to suffice to counter the social and political constituents behind the lack of archives. This kind of reasoning frees authors from the need to substantiate their claims from an empirical standpoint, and allows them to speculate that archives always existed, and even enjoyed continuity across dynasties, even if they may have suffered the hazards of episodic destruction. It is useful here to return to the passage by Nicolas Michel mentioned above, which contradicts the more detailed historical work on this issue. One of the holes in Bauden's argument is his claim that something similar to a diplomatic Fatimid archive had existed, but that it did not find continuity in successive dynasties; however, he does not provide much evidence, either for its existence or destruction, beyond mere hypothesis.

74 Akarlı, Engin Deniz: "Law in the Marketplace: Istanbul, 1730–1840", in: *Dispensing Justice in Islam: Qadis and Their Judgments*, edited by Muhammad Khalid Masud, Rudolph Peters, and David S. Powers, 245–70. Leiden: Brill, 2006, 264–5.

75 Bauden: "Du destin des archives en Islam", 33.

According to Bauden "Le déménagement des archives fatimides, qui devaient être conservées dans les ministères situés à proximité des palais califaux, dans leur nouvel emplacement, la citadelle construite par Saladin, a dû constituer de ce point de vue un moment particulièrement destructeur". Such uncorroborated assumptions about the transfer of the collection to the new center of power, the citadel, can only further contribute to building a house of cards. Fatimid archives did exist, and they were kept close to the seat of power as a symbolic attribute of governance. And when Saladin modeled himself as a post-caliphate, sultanian—and therefore military—ruler, he did take the records with him up to the castle. However, where papyrologists simply content themselves with the assumption that the empirical evidence for is out there somewhere—and the documents do, after all, exist—it takes painstaking historical work to demonstrate the existence, or lack thereof, of all and every one of the elements in Bauden's hasty assumptions. Michel turns his attention to the jargon used by Mamluk secretaries when referring to where, and by whom, records were kept, and convincingly demonstrates that Mamluk records were held "downtown" in the secretaries' houses, and moved up not by Saladin, but by the Ottomans about four hundred years later. Records were first and foremost left in the hands of high-ranking secretaries, as the documents were not considered to represent either a form, or a manifestation of the sultan's governance.[76] As we have seen, the muzarrah, an Ayyubid innovation, served the purposes of transporting decrees back and forth from the scribal workshops and the Citadel.

In the same fashion, Marina Rustow combines two of the arguments advanced so far in the debate on the archival divide: that many papyri have survived, and that written artifacts were used in many aspects of daily life in the medieval Middle East. It is worth noting that in her case study of a Fatimid petition from a woman, found in the Cairo Geniza collection, she concludes that the petition was probably preserved not so much as positive information, as to provide a model for future applications. This is an important point, as one might argue that many of the documents put forward by researchers to demonstrate that archives did exist were actually preserved to serve as models

76 "In addition to the records of the chancery, the wazir of the ʿAbbasids kept copies of the most important documents in an archives office of his own.", Posner, Ernst, "Archives in Medieval Islam", *The American Archivist* 35 3-4 (1972), 291-316, 297, an argument confirmed by Van Berkel, "Reconstructing Archival Practices in Abbasid Baghdad", 13–4, Messick, *The Calligraphic State,*referring to the 1950s: adds that "The handling of administrative registers, which were always retained at home as the personal effects of officials or secretaries, also changed", 247. Hirschler, "From Archive to Archival Practices", 18–9.

and formularies for clerks and scribes, hence implying a very different logic behind such practices:

> In fact, the evidence of pre-Ottoman archives and archival practices is abundant, but it is also hardly investigated. Comparing the tens of thousands of surviving original Arabic papyri with early medieval Latin documents copied into cartularies suggests that the shopworn comparison with medieval Europe requires some rethinking. Frédéric Bauden has rightly called the notion that few documents have survived from the medieval Near East "calamitous," and made every effort to correct it for the Mamluk period in particular[77][…] Writing and documentation, in short, pervaded the medieval Near East, even if those fully competent in their use and production were few. To deny this and assert instead a preference for perpetuating social hierarchies through biographical dictionaries is to make a virtue of a false necessity: there were documents, and more survived than is commonly understood.[78]

To be sure, the assumption that the document makes the archive exhibits several weak spots. By flattening out the archival divide, historians have contributed to equating collections of extant documents with actual archives. As we shall see, this idea has been overthrown by recent scholarship, which has in particular sought to take a closer look at the Islamic judicial archive found in the Ḥaram al-Sharīf, a trove of documents that belonged to a 14th-century qāḍī in Jerusalem. The finding was initially saluted by scholars such as Bauden and Hallaq as material proof that Islamic archives did exist, and was first studied by Donald P. Little, who saw in it a qadi archive, although he later adopted a more cautious view.[79] Although it is indeed larger than any other Arabic contemporary remains found to date, did the Ḥaram papers really constitute an archive, or were they simply an accidental trove of judicial records? A lively debate has arisen around the nature of the collection; for reasons that will be addressed in the following pages, the verdict by authorities has been negative.

Together with his numerous studies of individual documents, Bauden sponsors a number of archeological, factual and materialist explanations for the

77 Rustow, "A petition to a woman at the Fatimid court", 3.

78 Ibid. 23.

79 Little, Donald P., "The Significance of the Ḥaram Documents for the Study of Medieval Islamic History", *Der Islam* 57 2 (1980), 189–219, Little, Donald P.: "Sidjill", Encyclopaedia of Islam, Second Edition, IX, 538–545, 1996.

absence of Islamic archives. Parallel to the alleged destruction of the Fatimid archives, he speculated on the end of the Egyptian medieval administrative collections. He argued that Mamluk archives were destroyed when the Circassian sultans acceded to power, an explanation based on a famous passage from al-Maqrīzī:

> During the period between the end of al-Ẓāhir Barqūq's reign and before it was re-established [i.e., between 784/1382 and 791/1389] many affairs came into disorder, among them the matters of the chancery's room (qāʿat al-inshāʾ) in the citadel. It was abandoned, all the papers (awrāq) in it were taken, sold by weight, and the information contained in them was forgotten (wa-nusiya rasmuhā).[80]

The explanation cited above has been discussed by Hirschler, who convincingly demonstrates the limited role played by the chancery in the preservation of state records, and who argues that the decentralized nature of the Mamluk archives makes the hypothesis of an episodic loss unlikely. In 2013, Bauden further developed his views that archival documents vanished because they were recycled.[81] Elaborating on the findings of Joseph Sadan, he mentions some interesting jurisprudence that reveals Islamic attitudes towards the ritual disposal of records potentially bearing the name of god (purification/text-canceling by water, burying, etc.).[82] However, Bauden's findings refer to the best ways to destroy documents bearing god's name, rather than patterns of record preservation, and interesting as they may be—and as Bauden himself agrees—they more support the idea that medieval Muslims did not have archives. If old legal papers lost their value for the living, documents could equally be compromising for the rulers. Ibn Faḍl Allāh al-ʿUmarī (m. 749/1349) narrates what happened with the sealed boxes left by a deceased man in Aleppo. They contained letters form the Zaydī imams of Yemen, hence threatening Mamluk sunni rule, and, accordingly, the governor ordered to have the letters ritually washed.[83] Instead on targeting patterns of record preservation, these findings refer to archive-dismantling procedures. By the same token, the idea that paper had material value, and that families were therefore incentivized

80 Hirschler, "From Archive to Archival Practices", 9.

81 Bauden: "Du destin des archives en Islam", 42.

82 Sadan, Joseph, "Genizah and genizah-like practices in Islamic and Jewish traditions. Customs concerning the disposal of worn-out Sacred Books in the Middle Ages, according to an Ottoman source", *Bibliotheca orientalis* 43 1-2 (1986), 36–58.

83 Tillier, "Califes, émirs et cadis", 160.

to sell it, goes more in favor of Chamberlains' argument, than they support Bauden's own stance.

The only procedural practice that might potentially have led to the non-subsistence of records is the abandonment of the Fatimid archival practices mentioned by Bauden, which sought to preserve diplomatic correspondence. This idea is based on a statement by al-Qalqashandī that copies were kept in ad hoc registers, and that this practice ceased under the Mamluk administration.[84] Indeed, other Mamluk authors confirm that hiatuses in archival and scribal practices were common. If al-Qalqashandī mentions that the collection of diplomatic missives was not continued under the Mamluks, our last and most important fifteenth-century informant from inside the Chancery, al-Saḥmāwī, alludes to another significant hiatus in the preservation of chancery material. Mamluk clerks were expected to report the drafts (*musawwadāt*) they prepared in a more formal register, the *daftar*—which, as we have seen above, was an artifact that kept track of scribal production, and for which various procedures had been put in place to secure it against forgery. According to al-Saḥmāwī, under the sultanate of al-Ẓāhir Barqūq, the drafting of copies in the daftars was abandoned for several years, and clerks "limited themselves to the production of drafts."[85] The practice of keeping updated daftars containing all important documents was reinstated by Qadi Nāṣir al-Dīn al-Bārizī, when he joined the administration of sultan al-Mu'ayyad Shaykh (1412–21), and of whom he was an active supporter. Al-Bārizī is remembered as a leading expert in chancery matters and was appointed archivist (khāzin al-kutub); significantly, not of any state department, where such a charge never existed under the Mamluks, but of the Mu'ayyadiyya madrasa's library.[86]

It is by collating several case studies that El-Leithy supports his claim that Muslims produced and preserved records with the same intensity as their Western counterparts. As mentioned previously, his aim is to dismantle Chamberlain's thesis that the social norms and cultural attitudes popular among the Arabs made records less important than in the West.[87] He first does this by

84 Bauden: "Du destin des archives en Islam", 34–5.

85 "fī ayām al-Ẓāhir Barqūq tarakahu wa iqtaṣara 'alā al-musawwadāt", al-Saḥmāwī, *al-Thaghr al-bāsim*, Vol 1, 375.

86 Martel-Thoumian, Bernadette: *Les civils et l'administration dans l'État militaire mamluk* (*IXe/XVe siècle*), Damas: Institut Français de Damas, 1992, 250–1, 258. The function of the khāzin al-kutub is described by Abū Ḥāmid al-Qudsī (d. 1483): *Badhl al-nasā'iḥ al-shar'iyya fī mā 'ala l-sulṭān wa-wulāt al-umūr wa-sā'ir al-rā'iyya, MA thesis, Riyadh*, edited by Sālim al-Shamarī, 2 vols. Vol. 1, 1996, 226.

87 El-Leithy, "Living Documents, Dying Archives: Towards a Historical Anthropology of Medieval Arabic Archives".

arguing that actors may have conspired to burn the evidence, as was the case for a notarial document where a religious conversion was attested. Secondly, he presents the same surviving petition from a Jewish woman advanced by Rustow; however, we know that this was a model document to guide scribes in the composition of similar requests. Thirdly, he dwells on an example, presented by Bauden, of a reused piece of paper that al-Maqrīzī retrieved from the ashes of the Mamluk archives. And indeed, Bauden has demonstrated that al-Maqrīzī systematically contributed to the destruction of chancery collections, with no less than 509 pages of 14th- and 15th-century originals reused in his own manuscripts.[88] As a result, the reader is left with documents destroyed on purpose, documents that owe their survival to their value as formularies rather than for their actual contents, and finally, with recycled records that are only extant thanks to the physical medium they were written on. Although less blatantly than in Bauden's case, the arguments advanced by El-Leithy to justify the lack of archives do not seem to support the idea that clerical, serial procedures leading to the preservation of collections did actually exist—at best, they explain why collections were short-lived.

Although accidents did happen, and fortuitous destruction, just like random preservation, does account for the disappearance of some document troves, we cannot assume from there that they go any way to proving that record preservation was systematically put into practice. One of the key documents for this research tells the story of several Frankish debtors in medieval Béjaïa in Algeria, and how they dealt with the problem of archived documents. They owed money to a resident Genoese merchant, and both parties had their contracts notarized before the local, Islamic ʿudūl. Presumably, the Arab notaries delivered copies to the parties, however they did not keep a personal copy or original deed, since their validity resided in the presence of witnesses who could potentially be summoned to court. The debtors managed to destroy the documents in the creditor's house, which left the latter with no evidence to exhibit back in Europe. The story has come down to us through the preservation of Genoese notarial archives; the victim, indeed, reported his story to a local notary sometime after, whose protocol has obviously been preserved in the state archives. To be sure, the episode raises as many questions as it answers, however it is the contrasting attitudes towards preservation that are most problematic and difficult to ignore.[89]

88 Bauden: "Du destin des archives en Islam", 37–9.

89 See the discussion later in this section, Archivio di Stato di Genova (hereinafter ASG), Notai Antichi, 871, doc. 295, 296.

However important forgery and episodic loss might be, such phenomena can hardly account for the complete absence of medieval archives. It is significant that many of the notarial documents edited by papyrologists and other researchers either belonged to vade mecums or notarial handbooks, or were found in the recipient's archives, ending up in faraway locations such as Venice. What sources suggest is that, at best, notaries and clerks kept records on an ad hoc basis—as was the case for the extensive diplomas they included in the encyclopedias and administrative handbooks they produced—or for the personal archives of scribes and judges. None of the practices described above—voluntary destruction, the use of formularies, and recycling practices—are sufficient in themselves to identify a preservation logic that might actually support the existence of an archival culture.

A passage by Ibn Khaldūn, in his description of Egypt's judicial system in 1400, has often been used to support the idea of a medieval archival tradition. In it, he claims that notaries "serve as witnesses when testimony is to be taken, testify during a lawsuit, and fill in the registers (sijillāt) containing the rights, possessions, and debts of people and other (legal) transactions."[90] Apart from the fact that prior to Ottoman domination such registers have not survived, the sijill in this context should not be understood as an archived artifact, but instead as a single-document court decision comprising a series of court proceedings. A more telling narrative can be found in the attempts, more than a century later, by the Ottoman administration to gather together the records produced by local 'udūl. Ibn Iyās describes the large-scale judicial reform undertaken in Egypt during the years 1521–1522, which included a pyramidal refashioning of the judiciary—with the ḥanafī qadi at the summit of the hierarchy of legal schools—the transfer of notarial activities to the courthouses, and the requirement for local judges to incorporate their deeds into the sijills, or court registers, now subjected to archival procedures. Extant Cairo sijills begin in August 1522,[91] but more interestingly, the documents deposited by the few notaries that complied with Ottoman injunctions were produced in the last years of Mamluk domination, and in some cases date back to 1505.[92] These deeds constitute, to my best knowledge, the only Mamluk notarial material transmitted through the bias of an archival institution and not as the right-holder's personal

90 Ibn Khaldūn, 'Abd al-Raḥmān (1332–1406), *Muqaddimat Ibn Khaldûn. Prolégomènes d'Ebn-Khaldoun, edited by* Etienne Quatremère, 3 vol, Paris,: Institut impérial de France, 1858, 405.

91 Michel: "Les Circassiens avaient brûlé les registres", 253–4, n. 94.

92 Meshal, *Sharia and the Making of the Modern Egyptian,* 112–3.

belongings. The kind of deeds that both Ibn Khaldūn and Ibn Iyās refer to were those notarized in court, rather than those drawn up in the streets and markets by private ʿudūl, and for which, as made clear by the Genoese merchant in Béjaïa, notaries did not keep the originals. However, when acting as attachés to courts of justice and administrations, these notaries left traces of the records they produced, and we have reason to believe that registers, if they were preserved at all, were kept by qadis and notaries in their private residences for personal reference.

The most illuminating example of the how official administrative documents were stored and preserved in their lifetimes is a visit undertaken in 1387 by a Venetian consul to several administrative departments in Mamluk Alexandria. The consul's report offers precious insight into the actual functioning of the Mamluk customs administration, and sheds light on notarization practices and the handling of records. The consul, Luigi Morosini, acting at the request of Florence's principal commercial tribunal, the *Giudici di Mercanzia*, was assisting in an investigation into the activities of a Florentine merchant, Michele di Francesco, in Egypt. Wishing to know about Michele's purchases of diamonds and pearls, the court asked the Venetian consul to dig through the archives in search of tax and commercial records. The Florentine document describing his findings is unique, as it depicts actual scribal practices, and because the right-holders in this case were Frankish merchants, it does so from a cross-confessional perspective. The consul first went through the customs books, mostly to find out whether the Florentine had concealed evidence of these purchases. It is unclear, however, who actually owned the records. It may be that a first group of accounts was held by an official, called the mustawfī (*per libros dogane mostafi alexandrie*). Within the same department, the consul checked the books where Islamic notary-witnesses recorded transactions—in the weighing office, or *qabbān* (*pro libros dogane testium gabani in quibus mercata scribuntur*).[93] Then the Venetian moved on to check the private accounts of a Muslim merchant, with all probability Burhān al-Dīn Ibrāhīm Ibn ʿUmar al-Maḥallī (d.1403) (*in libris Brandin Elmaeli pro barata perlarum*). A different group of ʿudūl operated elsewhere in the customs house, in an area known as the *Duchela*, and their books were also checked by the consul. In passing, we learn that al-Maḥallī, an important businessman and diplomat in the service of the sultan, had Syrian Christians in his staff, and indeed we know that Oriental Christians were traditionally employed by the Mamluk customs (*feci*

93 Archivio di Stato di Firenze (hereinafter ASF), Mercanzia 212, deliberazioni, 1387, 72r.

perquiri libros testium duchele et libros ac scripturas brandim elmaeli quas habuit quidam christianus della cintura scriba ac factor supradicti brandim el maeli).[94] More records belonging to al-Maḥallī, apparently not included in his books, were handed over by a scribe called Mina, as well as excerpts from the books of a Christian clerk working for the fiscal administration (*Copia delle scripture extracte deli libri de Mene Cristiano che riceve lentrata del soldano in quanto spetta alle ragioni de michele de Francesco da Firençe*).[95] Finally, we learn that sworn testimony was taken from both Christians and Jews, "according to their laws."[96]

At several points in the Florentine document, it is mentioned that both tax officers and notaries kept books (*Copia delle scripture extracte di quaderni de Mene cristian scrivan aricevere lentrade de soldan; Copia delle scripture extracte de quaderni di testimoni dal gaban che scrive li mercadi*), excerpts of which could be handed over to the right-holders upon request. However, the document does not mention a specific physical place where these records were deposited, and nor do other Venetian deeds depicting similar inquiries at the customs house. Instead, they suggest that books and records were kept by the same scribes that produced them; and indeed the consul had to return twice to check the deeds drawn up at the *Duchela* and, again, twice to check on the tax records, as if they were not held in the same place or could not be consulted all at once. As Nicolas Michel has demonstrated for the Mamluks, treating records as personal effects was established practice, and this has been confirmed for other Islamic polities. More importantly, the 1387 document confirms Ibn Khaldūn's assessment that, although the ʿudūl did preserve some kinds of records, what we are dealing with here is notaries acting as attached to an administration and not as regular witnesses, and there is no trace of an institutional repository. Secondly, the document mentions that mere scribes, mostly Christian, worked alongside proper Islamic ʿudūl, hence it is clear that two different levels of truth-bearing were at work in the customs house. Third, the taking of oaths and widespread recourse to the ʿudūl, instead of relying on

---

94 al-Ashqar, Muḥammad ʿAbd al-Ghanī: *Tujjār al-Tawābil fī Miṣr fī al-ʿAṣr al-Mamlūkī*, Cairo: al-Hay'ah al-Miṣrīyah al-ʿĀmmah lil-Kitāb, 1999, 115, Fischel, Walter J., "The Spice Trade in Mamluk Egypt: A Contribution to the Economic History of Medieval Islam", Journal of the Economic and Social History of the Orient 1 (1958), 157–174, 169, Behrens-Abouseif, *Practising Diplomacy in the Mamluk Sultanate: Gifts and Material Culture in the Medieval Islamic World*, 43–4.

95 ASF Mercanzia 212, deliberazioni, 1387, 72v.

96 "Item muse çudio dito borgno dise avere conprado da Michel de Francesco panni ... li qual lui dise avere dadi a bet melo per lo dreto del soldan e questo çuralo per la soa fe", ASF Mercanzia 212, deliberazioni, 1387, 77r.

the services of rank-and-file scribes, point to the fact that many administrative issues were recorded 'judicially,' this is, in ways compliant with Islamic law and in a manner compatible with judicial evidence.

When dealing with the materiality of records, the debate on the archival divide exhibits three serious biases. First, there is an ideological willingness to redeem the Muslims of the anathema of being a people without archives. It is clear, however, that medieval administrations and law courts relied on their own production of records that were kept for reference, and that this was entirely possible without necessarily leading to long-term storage, or to the adoption of archive-building policies. In part, this was due to the fact that some transactions, such as business deals between Franks and Muslims, had to be notarized and therefore fell under the sphere of competence of the ʿudūl, who were reluctant to surrender their records, if they had any. This proved to be true in 1521, when the Ottoman rulers called on Cairo's notaries to deposit any records, such as marriage deeds, that were subject to taxation, and almost none complied. Islamic deeds could not be endowed with public faith without the oral support of their authors, and for notary-witnesses depositing their deeds in an archive may have meant losing their personal monopoly over truth-bearing in society, now transferred to perishable objects. There was therefore no motivation to build archives of notarial protocols, as was the case for Latin Europe. In many cases, unfinished notarial business was followed up by a notary's sons, which meant that the ʿudūl class was often made up of notarial dynasties—a fact that echoed the widespread assumption that trustworthiness was a family feature and resided in lineage. Jār Allah Ibn Fahd (1486–1547), who indulges in long descriptions of the ʿudūl milieu in Mecca, suggests that notaries were more concerned with taking over their predecessors' affairs than in inheriting potential document collections. He reports the story of Khawājā Ibn Qawān, a merchant charged by the sultan of Gujerat with the endowment of a religious foundation on his behalf. When Ibn Qawān's notary died in the process, he had recourse to the scribe's son, since he had already paid the fee to the deceased father. The son took advantage of this to request a supplementary fare (*ujra*), claiming that *whenever his father owed the Indians money, they should file a complaint against his last will.* The resulting deeds (*mustanadāt*) were sanctioned by the Ottoman-sponsored, ḥanafī qadi, and it is clear from Ibn Fahd's narrative that clients, and the handling of sensitive transactions were, all too often, left in the hands of legal agents and their dynasties. In the case in question, Khawājā Ibn Qawān had already invested money in some buildings and needed the notarial agreement to have them converted to a tax-exempted, religious foundation, in order to avoid being left with a bunch of houses in Mecca, a circumstance used by the notary's son to blackmail his client.[97] Be that as it may, a number of factors conspired to keep notarized deeds in the realm of memory or, at best, in the personal collections of notaries.

A second bias in the debate is the confusion between, on the one hand, the production and use of written artifacts and, on the other, their preservation. Apart from arguments such as those advanced by Sauvaget on the poor legal value of the written word, the debate has quickly shifted away from the logic of how people preserved information. Readers of European history, for instance, would find the assertions by Hallaq and Hirschler puzzling, which take it for granted that documents "lost their legal value," and that there was therefore no need to preserve them. Just as astonishing might be the repeated mantra that collections ultimately ended up in the private hands of qadis and scribes, rather than in institutional repositories. It is clear at this point that, whether the materialist explanations are convincing or not, the idea of preservation that is conveyed in these studies greatly differs from that which underlay the creation of European judicial and state archives.

Lastly, the debate has little to gain from the way in which the nature of documents is considered, taking several different kinds of records from the same standpoint, when in reality they looked very different from each other in the eyes of early modern Muslims. Particularly risky in this regard is the equal treatment given to notarized deeds and other documents that could be recorded by mere scribes, and even by unbelievers not endowed with sharīʿa notions of trustworthiness.

2.1.3 *Non-Materialist Explanations*

A more productive line of investigation is one that runs parallel to the quest for material explanations. The pioneering 1998 article by Wael Hallaq[98] shows that since early Islam some kind of documentation was being produced and temporarily preserved by the qadis. Less convincingly, Hallaq goes on to question the very idea that archives started with the Ottomans, and stresses that there was a continuity between modern and medieval practices. As regards Islamic patterns and ideas of record preservation, his denial of any Ottoman innovation is not backed by any of his own evidence, even though he acknowledges that, with some exceptions, the Ottomans began to systematically bind

97 Ibn Fahd, Jār Allāh Muḥammad Ibn ʿAbd al-ʿAzīz al-Hāshimī (1486–1547): *Kitāb Nayl al-Muná bi-Dhayl Bulūgh al-Qurá li-Takmilat Itḥāf al-Wará* (*Tārīkh Makkah al-Mukarramah min Sanat 922 H ilá 946 H*), edited by Muḥammad al-Ḥabīb al-Hīlah, 2 vols., Beirut: Mu'assasat al-Furqān lil-Turāth al-Islāmī, 2000, 258–9.

98 Hallaq, "The qādī's dīwān (sijill) before the Ottomans".

the records, and certainly provided official storage for the qadi's dīwān.[99] Be that as it may, Hallaq's article represents a landmark in our understanding of the origins and evolution of the judicial collections, mainly because it has provided a thorough analysis of the transmission of judicial archives from one qadi to his successor. By so doing, Hallaq acknowledged that the qadi archives did exist, but were of a temporary nature, and were not fully transmitted to the new judge, and that this transfer did not correspond to any long-term preservation pattern.

The most fruitful approaches to non-materialist explanations began as a reaction to Hallaq's work. Although he could probably have made this argument more explicit, Konrad Hirschler has deviated from the traditional attempt to justify a lack of archives, to delve instead into the complexities of Mamluk administrations. Documents, he argues, were used in many aspects of daily life and preserved by institutions, according to different logics. Quite forcefully, his article engages in a description of the polycentric, decentralized, and even transportable nature of the Mamluk archives. As mentioned before, the end of the Abbasid caliphate witnessed the advent of feudal-like military states, funded trough land allotments, called iqṭāʿ, and based on a different, non centralized administrative logic. Mamluk officers enjoyed iqṭāʿ grants attributed by the sovereign; however, as these allotments were not fixed the fief-holder could potentially be transferred to a different location in Egypt, and hence an officer would bring his personal archives with him. The army administration produced and stored a great many state documents, as did, I might add, the Sultan's department (Dīwān al-Khāṣṣ). According to a recently discovered memorandum containing information on Mamluk links with India and Southeast Asia, it was the Dīwān al-Khāṣṣ, the sultan's privy department in Cairo supported by its own postal service, that handled data collected by custom officers as far-flung as Jedda.[100]

Much as Hallaq has demonstrated the complexities of the qadi archive, Hirschler has turned his attention to the way documents were produced and temporarily preserved by different state actors. Apart from illustrating documentary practices, as regards the archival divide Hirschler sees the production

99 "There is no suggestion that the Ottomans introduced any changes to the institution, save perhaps for providing a crucial public space for storing the documents.", Ibid. 436, Peirce, *Morality tales: law and gender in the Ottoman court of Aintab,* 98–102, Faroqhi: "Sidjill", Encyclopaedia of Islam, Second Edition, IX", Meshal, "The State, the Community and the Individual", 129–30.

100 Apellániz, Francisco, "News on the Bulaq: a Mamluk-Venetian Memorandum on Asian Trade, AD 1503", EUI Working Paper HEC 2016/01 (2016).

of documents in high numbers as justification for discarding Chamberlain's thesis of a 'social' logic underpinning the non-survival of archives.[101] One could well conclude that at best there were archives, but that they were short-lived, not only due to the hazards of preservation, but because they were never conceived as permanent structures and therefore not considered bound to last. In Hallaq's own terms, in pre-Ottoman times, "there is no hint whatsoever that the qadis, upon dismissal or death, were required to deposit their diwans, in any form or manner, in a state-owned building or other public space."[102] Indeed, Hirschler admits that "there was only a limited institutional logic of document preservation and that documents were discarded when they ceased to be of relevance for the individual." Hirschler presents multiple patterns for temporary storage, and different logics of short-term storage. Yet as in the case for the qadi archives, he admits, it is difficult to draw a line between what constituted official state papers, and documents belonging to the personal collection of a clerk, whose interest in preservation was limited to his own work—hence diverging from a purported state concern for the safe-keeping of positive administrative knowledge.

Together with the quest for non-state archives and practices, the discovery of documentary collections in unexpected places has been summoned in support of a medieval archival tradition. This line of reasoning assumes that if recipients archived records, the authorities that issued them might have done so as well. Government decrees addressed to Christian monasteries such as those at Mount Sinai, scriptures buried in synagogues, text deposits entrusted to mosques as pious endowments, together with the remains of company records, all escaped the hazards faced by state archives and allegedly confirm that preservation was the norm. Apart from the obvious interest these collections present to historians, when it comes to defining the logic behind their preservation, specialists most often discard archival intentions. This is clearly Christian Müller's interpretation for the Jerusalem papers, Miklos Muranyi's conclusions on the library of the Great Mosque in Qayrawān, or Li Guo's understanding of the letters found in the Shaikh's house in al-Quṣayr. For the case of the Cairo Geniza, the debate revolves around whether these records are simply sacred trash, or whether they survived due to institutional care and continuity. Discussing Geniza-like practices among Jews and Muslims, Joseph Sadan has admitted that when historians refer to these repositories as archives, it is more a figure of speech. In fact, he argues that "it is by no means clear that

101 Hirschler, "From Archive to Archival Practices", 27.
102 Hallaq, "The qādī's dīwān (sijill) before the Ottomans", 435.

putting material in a small storeroom at the Ibn ʿEzra synagogue in Old Cairo through an opening in the women's balcony on the second floor, or burying it at Basatin, implies any particular intention of depositing material, bequeathing it to posterity, or preserving it."[103]

The survival of the Cairo waqfiyya stands uncomfortably as the elephant in the room in the discussion about whether random deposits can really prove the existence of an established practice. The very existence of this archive, which brings together the endowment deeds of Egypt's pious foundations, forces us to acknowledge how important any logic of preservation is to ensuring archive survival. Citing the waqfiyya as a surviving prototype of the Islamic archive, many authors forget that waqf documents, unlike state papers, survived because Islamic welfare and the livelihood of most ulama depended on their preservation, as they described the resources that were to be attributed to the payment of clerks and religious functions. Religious foundations were intended to last long after the death of their founder—theoretically forever, indeed—and waqf records regulated the allocation of resources for generations of clerks and staff members to come. For this reason, waqf deeds played a crucial role in social reproduction and competition. They needed to be preserved, transmitted and validated through the procedure of isjāl, whereby new witnesses certified the authenticity of old authentications by previous witnesses. In gross, these documents took the shape of a scroll with additional sheets of paper collated to an original document, mostly subsequent isjāls, like the Saladin waqf deed (certified time and again for over five centuries), or the Damascene document published by Donald S. Richard that covers a period of two hundred and fifty-five years.[104] Zomeño mentions an addition, dated 1488, to a 1432 document describing some properties in the Andalusian city of Baza. In it, two new witnesses identified the original witnesses' writing, and acknowledged they had been ʿudūl in Baza and had maintained their honorability until they died.[105] Endowment deeds, together with control of the foundation's resources, guaranteed the founder's offspring with the right

103 Müller, "The Ḥaram al-Sharīf collection", Muranyi, Miklos, "Geniza or hubus: some observations on the library of the great mosque in Qayrawan", Jerusalem Studies in Arabic and Islam 42 (2015), 183–201, Sadan, "Genizah and genizah-like practices in Islamic and Jewish traditions", Guo, Li: *Commerce, Culture, and Community in a Red Sea Port in the Thirteenth Century: The Arabic Documents of Quseir*. Leiden: Brill, 2004, 90–1.

104 Richards, D. S., "A Damascus Scroll Relating to a Waqf for the Yūnusiyya", *Journal of the Royal Asiatic Society of Great Britain and Ireland* 2 (1990), 267-281, 279, Salati, M., "Un documento di epoca mamelucca sul waqf di ʿIzz al-Dîn Abû l-Makârim Hamza b. Zuhra al-Husaynî al-Ishâqî al-Halabî (ca. 707/1307)", *Annali di Ca' Foscari* XXXIII, 3 (1994), 97–137.

105 Zomeño, "From Private Collections to Archives", 469.

to maintain patronage relations with clerks and ulama. Thus, in contrast with more short-lived transactions, the Islamic judiciary took great care to preserve and transmit these records across generations and dynasties.

In contrast with what has become a commonplace in the genre, it is not my aim to engage here in a thorough analysis of the archival exceptions to Chamberlain's hypothesis. Rather, I propose returning to an argument by Sauvaget that runs parallel to Chamberlain's: that judicial documents, together with other notarized administrative documents were not legally valid beyond the life of their authors, and were therefore neither useful to any logic of governance, nor for social competition. In gross—and I will be returning to this issue—proof in a medieval Islamic context needed to be backed by personal knowledge and trustworthy witnessing, and therefore judicial and state papers only lasted as long as the human capacity to attest to their contents. In the remaining section of this chapter I will explore the works that have allowed us to better understand the logic of proof production in Islamic societies before Ottoman times. Notions and doctrines of proof did have an impact on the way documents were preserved; instead of attempting inductive reasoning by looking for explanations—materialist of not—that might account for an archival culture on the basis of actual remains, there is much to be gained in looking at the way such notions evolved over time. A thorough analysis of the epistemological grounds for proof has been provided, in the last decades, by Baber Johansen, who explored the norm, but also identified hiatuses and transformations in doctrines on proof and evidence. After sketching out a summary of this transformation, I will turn to the artifacts that embodied documents, such as scrolls, ledgers and bookcases, and to certain practices, such as archive transmission, that reflected these changing notions and ideas.

2.2 Islamic Notions and Doctrines on Proof and Evidence

In a series of articles dealing with late medieval attitudes towards proof and notarial evidence, the legal historian Baber Johansen coined a crucial notion: the "epistemological skepticism" characterizing Islamic law. Only god can know the truth, then the best that a qadi can do is to avoid sinful action by giving credence to doubtful evidence. The principal idea being that, due to the divine nature of the Law, records and other physical media can hardly be bearers of truth. Such physical media—i.e., three-dimensional objects—have a volume and are susceptible to trick the senses, and therefore are not reliable. In particular, a judge who chooses to put his trust in such fallible objects when deliberating on a verdict risks committing sin. It is only the word of a trustworthy Muslim that deserves to be taken into consideration in the quest for truth. Arguments against the written emphasized the defense of the illiterate, and discussed whether familiar handwriting could ultimately be recognized and trusted, even going so far as to question whether a judge should rely on his ability to correctly recognize his own calligraphy. Islamic law thus adopts a cautious approach: witnesses are interviewed, notarial deeds are backed by oral testimony, and even the judge's recourse to his own archives may require certification procedures. This was the mainstream, commonly adopted approach by medieval and early-modern jurists, against which innovative voices raised in specific times and places such as tenth-century central Asia or by some Mālikīs in the West. Whilst the skeptical approach of the "classic" Iraqi ḥanafī school was generally upheld, more innovative thinkers were inclined to exceptionally accept the records advanced by merchants, foreigners and governors.[106] The written word was only resorted to with great caution, and throughout Islamic history, up to late Ottoman times, it was generally upheld that oral testimony was the proof par excellence. In daily practice, it was this epistemological skepticism that underpinned Islamic courts and their approach to jurisprudence, up until the 19th-century Ottoman legal reforms, when testimonial rights became a mainstay issue.[107]

Jessica Marglin has recently pointed out that, in the context of the 19th-century Maghreb, and under the influence of Mālikī doctrine, legal practice took a different direction, and notarized deeds began to be accepted more freely than in previous periods. This argument is based on specific Mālikī attitudes, such as their opener approach to the forensic examination of handwriting in order to determine the authorship of a document.[108] Marglin demonstrates that, in Morocco, notaries testified through their signatures, and that judges generally considered documents to be written embodiments of oral testimony. Curiously enough, however, this "triumph of the written" did not lead to the development of judicial archives for legal papers, and in the Moroccan case such documents seem to have ended up in family collections. Although Marglin's research proves persuasive in illustrating Moroccan court practice, these opener attitudes to the written appear to have been confined to a very recent period. Instead of accounting for a new interpretation of the sharīʿa approach

---

106 Johansen, "Formes de langage", 369–371. Turki, Abdelmagid "Lisān al-dīn Ibn al-Ḫaṭīb", 195.

107 Canbakal, *Society and politics in an Ottoman town: ʿAyntab in the 17th century*, 148.

108 Tyan, Émile: *Le notariat et le régime de la preuve par écrit dans la pratique di droit musulman*, Harissa: Imp. St. Paul, 1945, 76–7, Marglin, Jessica M., "Written and Oral in Islamic Law: Documentary Evidence and Non-Muslims in Moroccan Shariʿa Courts", *Comparative Studies in Society and History*, 59 4 (2017), 884–911, 907.

to truth, this example provides more a snapshot of a late 19th-century phenomenon, and Marglin does not attempt to explain why and when such attitudes led to a deviation from mainstream practice in the legal system. The extent to which these practices might account for Mālikī practice at other times and in other places is not free of contention either. Mālikī-dominated al-Andalus, for instance, never conceded to a similar triumph of the written, and case of the specific 19th-century Maghrebi attitude cited above stands more as an exception; thus such cases point more towards a general principle of mistrust for written evidence, than they do negate it.[109] All in all, the exceptional character of late early modern Moroccan court practice is nevertheless intriguing, and more research is needed on the action of Mālikī judges appointed in Cairo, Damascus and Jerusalem.

Léon Buskens has recently contributed to the discussion with a colorful description of his experience as a document collector in Morocco. It is clear from his narrative that, even in the presence of a burgeoning antiquarian market, old notarial documents were unanimously considered as useless and as the most difficult item to place for sellers. Buskens uncovers vernacular attitudes according to which deeds entered *a kind of limbo state* when their owners lost interest in them. Even if, in late modern Morocco, notarized documents had acquired independent value as legal proof—a question not addressed by Buskens- the standard procedure to make them invalid was cutting the ʿudūl's signatures off. Making witnesses to the deed untraceable killed the truth contained in them, that dwelt entirely on the latter's endorsement.[110]

Yossef Rapoport has noted that in Mamluk times written documents could sometimes be accepted as valid elements of proof. In his seminal article on Mamlūk Siyāsa, he mentions in passing a clause contained in Mamluk appointment decrees, whereby "the Mālikī chief qadi is enjoined to apply his school doctrine so as to"—among other prerogatives—"permit the use of documentary evidence."[111] This may mean that, at least in specific circumstances, Mālikī judges were allowed to base their judgement on this form of evidence, which was otherwise considered circumstantial. Legal historians have built on

109 Oulddali, Ahmed: "Recevabilité du témoignage du ḏimmī d'après les juristes mālikites d'Afrique du Nord", in: *The legal status of ḏimmī-s in the Islamic West* (*second/eighth-ninth/fifteenth centuries*), Turnhout: Brepols, 2013, 275–292.

110 Buskens, Léon: "From Trash to Treasure: Ethnographic Notes on Collecting Legal Documents in Morocco" in: *Legal Documents as Sources for the History of Muslim Societies*, edited by P. Sijpesteijn, Maaike Van Berkel, et Léon Buskens, Leiden—Boston: Brill, 2017, 180–208.

111 Rapoport, "Royal Justice", 78.

Emile Tyan's 1945 monographic work on notaries to point to a Mālikī tradition of openness towards written proof, as exhibited by authors such as Ibn Farḥūn (1358–1397). Although he is best known for his work in Egypt, the Mālikī jurist of Andalusian descent was also active in Medina in the Arabian peninsula, and was probably responsible for introducing notions of Mālikī doctrine in Egypt—a school that was hegemonic in al-Andalus and the Maghreb, but fundamentally marginal to Mamluk legal practice. An interesting testimony, although certainly an indirect one, comes to us from Francesco Suriano (1445–1481?), an Italian friar who had twice been appointed head of the Franciscans in Palestine, and who, in his *Trattato di Terra Santa e dell'Oriente*, exhibits a profound knowledge of the Mamluk regime, administration, and judiciary. Suriano describes a personal acquaintance of his, the Mālikī qadi in Jerusalem, who was of Maghrebi origin and shared the same legal background as the western Mālikī chief justices that were traditionally appointed to Damascus.[112] Suriano's narrative presents the Mālikī judge on his deathbed, rather unsurprisingly using his last breath to renounce Islam in favor of Christianity. In his deathbed confession, the qadi repeatedly claims that he is all the more guilty because he had been abusing his office by forging written deeds.[113]

Many authors have argued that sharīʿa adopted the same legal dialectic for dealing with the affairs of unbelievers, as it did for Muslim subjects. In other words, it was not Islamic law that worked differently for unbelievers, but instead the formalist, religious-grounded bias against dhimmī witnesses that preceded legal reasoning, and therefore affected its outcome. Mario Grignaschi presents Ottoman legal attitudes as being marked by mistrust for the capacity of unbelievers to bear truth, rather than any unwillingness to grant them rights.[114] And indeed, the literature describing complex cross-confessional

112 Zaʿrūr, Ibrāhīm, "al-Quḍāh al-Andalusīyūn wa-al-Maghāribah fī Bilād al-Shām fī ʿAṣr al-Mamālīk", Dirāsāt Tārīkhīyah 53–54 (1995), 59–79, Ben Muʿammar, Muḥammad "al-maghāriba wa manṣab qāḍī al-quḍḍat al-mālikī fī dimashq al-mamlūkiyya", Majalla al-ādāb wa al-ʿulūm al-insāniya—Jāmiʿa Dimashq (2008), 145–162.

113 "Infirmandose el Chadi Melechi, stete in transito da la matina sino a completa. Ritornato che lui fo in sè, disse asstante molta gente: Tristi nui Saraceni che tuti ne damnamo et io *étiam* sum damnato per le scripture che ho falsificato, e solum li Christiani se salvano. E chridando tuti che fosse arso, fo risposto propriamente, como habiamo dicto de quella dona de Damasco; et io lo cognobi et pocco avanti ch'el morisse, gli parlai. Tandem dicte che l'hebe quelle parole, immediate morite; e confessò lui dover haver mazor pene de l'altri, per l'instrumenti e carte che lui havia falsificato." Suriano, Francesco: *Il trattato di Terra Santa e dell'Oriente*, edited by Girolamo Golubovich, Milano, Tipografia editrice Artigianelli 1900, 216.

114 Müller, Christian: "Non-Muslims as part of Islamic law: Juridical casuistry in a fifth/eleventh-century law manual," in: *The Legal Status of Dhimmis in the Islamic West*, edited

debt cases, involving dhimmīs, not mustā'mins, is abundant. For example, we know from this literature that the Muslim creditor of a deceased infidel has precedence over a Christian one, if both rely solely on Christian witnesses, however this right of precedence can be superseded by the Christian creditor if this latter is able to produce Muslim witnesses.

It seems that for most late medieval and early modern polities, and particularly in the territories under the Ottoman umbrella, priority was given to safeguarding the formal standards of the judge's decisions and their compliance with religious principles. What happened, then, if the testimony advanced in court was false? Johansen argues that lying made a witness a sinner who would be punished in hell. Yet even when a witness admitted he had lied during a trial, if the decision by the judge had been taken on appropriate formal grounds it needed to be upheld, although the lying witness was considered deserving of punishment.[115] More importantly for the present study, formal preconceived notions heavily affected the witness system, thus impacting the way diversity was dealt with. Collections of fatwās dating from the early modern period offer many examples of what ought to be done in cases where witnesses exist, but none comply to Islamic standards. This problem is epitomized in a hypothetical case whereby the only Muslim inhabitant of a Christian village murders a Christian. In such an eventuality, the Muslim could never be convicted on the basis of Christian witnessing, because one of the basic prerequisites for the judge to trust a witness resides in his being Muslim. If, as a system of legal reasoning based on rules and principles, Islamic law did not discriminate, the biases against non-Muslims preceded and impinged on the application of these very principles, hence conditioning the system's outcome.

The Islamic formalism that underpins Johansen's analysis of biases against non-Muslims had an obvious impact on the artifacts used to bear witness. The nonexistence of archives was indeed intimately tied to the issue of preserving and transmitting truth. Since in practice it was people, and not objects, that were considered to be the depositaries of earthly truth, the only means for transmitting it over time was through an uninterrupted chain (ittiṣāl) of witnesses, as in the isjāl procedure that provided for the certification of endowment deeds, as described above.[116] In order to be valid, documents underwritten by long-dead

---

by Maribel Fierro and John Tolan, 2013, 21–65, Grignaschi: "témoignage des sujets non musulmans".

115 Atçil, Abdurrahman: *Procedure in the Ottoman court and the duties of kadis, M. A. Thesis, Bilkent University*: the Institute of Economics and social sciences of Bİlkent University, 2002, 76, Johansen, Baber, "Le jugement comme preuve. Preuve juridique et vérité religieuse dans le Droit Islamique Hanéfite", *Studia Islamica* 72 (1990), 5–17. 14–5.

116 Johansen, "Formes de langage", 349.

witnesses had to include verification of the original signatures by currently living witnesses. When transmitted to a newly-appointed judge, the medieval, *qimaṭr*-like, 'qadi archive' only contained the few pieces that needed to be submitted to this recertifying procedure. For this reason, endowment deeds, which needed to last over generations, were recertified time and again by new cohorts of witnesses in scroll-shaped documents.[117] It is no coincidence that the only proper archive that has survived in Egypt is a waqfiyya, or repository of deeds relating to the functioning of pious foundations, on which successive generations based their decisions regarding the appointment and funding of charges. Endowment deeds are a kind of document bearing numerous isjāls, examples of which are well known. The recertifying procedure has been documented until late modern times by anthropologists, although some authors have raised doubts about whether the procedure was seriously implemented, as it seems that the chain of transmitters was sometimes broken. Even if, on occasion, incomplete chains of certifications were accepted, accepting these documents was always problematic for jurists. "You permit the testimony of a man after his death, if you found his signature on a document," argued a Mālikī jurist in 15th-century Iberia.[118] In this sense, Ottoman judicial practice represented a groundbreaking deviation from this medieval procedure, as all pieces included in the sicil were considered as certified once and for all, and therefore no further confirmation by witnesses was needed.[119]

Ultimately, as Johansen has developed elsewhere, the formalistic approach to proof was rooted in the epistemological skepticism of Islamic law. Proof, including trustworthy witnessing, was a power put into play by either party in order to alter the conditions of the world of appearances. Truth ultimately belonged to God, while the judge was fundamentally incapable of attaining the inner (bāṭin) truth of existence; he was thus obliged to rely on fallible utterances in order to safeguard justice and other earthly aims.[120] Johansen's work, therefore, delves into the moral dimension of proof-production, and in so doing yields a better understanding of its abiding framework. The regime of proof, which ultimately determined the fate of written documents and the

117 Müller, Christian, "A Legal Instrument in the Service of People and Institutions: Endowments in Mamluk Jerusalem as Mirrored in the Ḥaram Documents", Mamluk Studies Review 12 1 (2008), 173–191, 184, Richards, "A Damascus Scroll Relating to a Waqf for the Yūnusiyya".

118 Marglin, "Written and Oral in Islamic Law: Documentary Evidence and Non-Muslims in Moroccan Shari'a Courts", 108.

119 Müller, "The Ḥaram al-Sharīf collection", 458.

120 Johansen, "Le jugement comme preuve. preuve juridique et vérité religieuse dans le Droit Islamique Hanéfite".

biases against minority witnesses, was anchored in Islamic epistemology, and not simply the result of practical choices.

Legal practice partly developed out of a highly formal and theoretical context. In a pioneering work published in 2002, Johansen also addressed the daily difficulties encountered by rulers and their delegates in the application of justice. He explored the works of theorists dealing with the so-called *secular, political* dimension of justice, or Siyāsa, which will be addressed in Chapter Three. Siyāsa has existed since the time of the caliphs, as a branch of jurisprudence devoted to empowering the ruler with legal solutions for the conduct of governance within the rule of law, particularly in areas upon which sharīʿa remains silent. Siyāsa was all the more important in that—beyond its spectacular development in Mamluk times by the ḥanbalī jurists of Damascus—it played out in real court practice and "Siyāsa" judges heard, among other areas of interest, cases involving the Franks. Commenting on the ideas of Ibn Farḥūn and Ibn Qayyim al-Jawziyya, Johansen has painted a dynamic picture of Islamic jurisprudence in the "post-classical" period (according to his own periodization), notably through the study of doctrines of governance, arguing that sharīʿa did not only work under the ideal framework of epistemological skepticism, but could also operate in a context of 'altered normativity', or, according to a later formulation "a perverted law" (al-sharʿ al-mubaddal). In plainer words, daily governance was at odds with the narrow dictates of sharīʿa described above, and accordingly, jurists labored to provide the ruler with a more operational framework for investigating the truth and inflicting punishments, in situations such as one in which witnesses deliberately lie in order to obtain an unfair ruling.[121]

Just as he has demonstrated that jurists sought to adapt sharīʿa to practical governance, Johansen has also explored the limits of the formalist approach to proof, observing the problems ḥanafīs encountered when attempting to maintain the doctrinal unity of their school. He focuses in particular on the central Asian doctrines dating from the 10th-12th centuries, which addressed the legal value of the use of direct speech in legal writing, and the status given, for example, to private documents written by merchants. This "Transoxiana school," epitomized by legal thinkers such as al-Sarakhsī, emerged in a context marked by long-distance trade and a dominant written culture. As we shall see, these thinkers set precedence on regulating dealings with unbelievers, extending their reasoning not only to dhimmīs but also to foreign unbelievers,

121 Johansen, "A Perfect Law in an Imperfect Society", 265. Johansen, Baber, "Signs as Evidence: The Doctrine of Ibn Taymiyya (1263–1328) and Ibn Qayyim al-Jawziyya (d. 1351) on Proof", Islamic Law and Society 9 2 (2002), 168–193, 182.

or ḥarbīs. They fashioned a more open attitude towards written evidence, and were inclined to consider not only notarized deeds but also authoritative documents as proof, such as those emitted by rulers and administrators, even in the absence of oral certification.[122] By the fourteenth century this had become common practice, and examples of uncertified documents issued by authorities have been found among the Jerusalem papers, as elements used in support of the judge's rulings.[123] Indeed, although Johansen focused on some developments within the ḥanafī school, from Abbasid to late medieval times, some views of innovative ḥanafīs, such as the attitude towards archived documents, were shared by other jurists, including the ḥanbalīs, who were influential under the Mamluks. Ibn Qayyim agreed with the ḥanafīs in that testaments ought to be accepted, as well as the qadi's right to rely on his own personal archives.[124] Denying value to last wills, Ibn Qayyim argued, was in overt contradiction with the reliance on writing and copies for matters as important as God's Revelation, of which only the first generations of Muslims had had direct experience.

Johansen traces the genealogy, extending out from central Asia, and going as far back as Abbasid times, of important issues on proof and procedure in Islamic law. He has showed that the ḥanafī school fostered an opener attitude towards the use of the written, without actually challenging sharīʿa evidentiary standards and its reliance on the oral performance of trustworthy Muslims. Through his attention to chronology and changes over time, Johansen's works has given us a better understanding of the norms governing proof and evidence and helped uncover their epistemological and religious roots, but also set a line of demarcation between the normative sphere and the priorities of Islamic governance. These priorities materialized in theories such as the Siyāsa doctrine, which challenged the normative-centered view of Islamic law and acquired relevance under the Mamluks. Johansen also explores the evolution that theories of governance underwent over time, shaped first by mālikīs and later by ḥanafīs, and hence provides us with a better understanding of specific late medieval attitudes towards Siyāsa. Finally, Johansen's overview points to the meager value of the archived document. Significantly, it is in this context that late medieval Siyāsa thinkers, such as the disciples of Ibn Taymīyah, raised

---

122 Johansen, "Formes de langage", 365–374, Johansen, Baber, "Le contrat salam: Droit et formation du capital dans l'Empire abbasside (XIe-XIIe siècle)", Annales. Histoire, Sciences Sociales 61 4 (2006), 863–899 Johansen, "Signs as Evidence".

123 Müller: "Ecrire pour établir Ia preuve orale en Islam: Ia pratique d'un tribunal à Jérusalem au XIVe siècle", 77.

124 Ibn Qayyim al-Jawzīyah, Muḥammad Ibn Abī Bakr (1292–1350): *al-Ṭuruq al-ḥukmīyah fī al-siyāsah al-sharʿīyah*, Mecca: Dar al-ʿAlim, 1428 h, 544–568.

objections against the generally-accepted framework and called for an opener attitude towards testaments and other written records.[125] Such claims are all the more important when we consider that Siyāsa materialized in Mamluk times, in the courts that were in the habit of judging issues concerning foreign merchants, and which necessarily had to deal with evidence produced by unbelievers. As far as sharīʿa-based norms were concerned, however, the status of records and archives remained unvaried throughout medieval times. Although materialist explanations for the lack of collections suggest that qadi archives existed but have been lost, a closer look at the issue of archive transmission demonstrates that medieval legal theory never supported such a view.

2.2.1 *Archive Transmission*

As mentioned previously, in a context where orality was widely accepted, a legal act created with the intention of surviving over several generations needed to be transmitted by an uninterrupted chain of witnesses. In this context, two detailed investigations into archival transmission by Johansen and Hallaq, published in the same year, have proven crucial; both make it clear that, even though some archives were transmitted from a judge to his successor, this did not imply that all records were still valid. Indeed, Hallaq has demonstrated that the qadi's archive was not systematically copied and certified in its totality by his successor. We know from descriptions in handbooks for judges, by the jurists al-Khaṣṣāf (d. 874) and Ibn Abī l-Dam (d. 1244), how qadis transmitted their records to their successors. To put it in very rough terms, the qadi's archives could not be handed over to the new judge without him guaranteeing the chain of transmission for every single document.[126] Limitations to relying on personal archives could also affect a qadi if he was temporarily dismissed from his office and subsequently reinstated.

---

125 Ibid., 232.

126 Johansen, "Formes de langage", 350–1, "We should recall here that not all cases were copied down in the incoming qadi's diwān. Those which have become inconsequential because they are considered, for example, old (where all the concerned parties have died), would expectedly be left out. This practice has the important implication that the bulk of the diwan's material did not grow cumulatively but was constantly subjected, at the stage of copying, to a measure of trimming", Hallaq, "The qāḍī's dīwān (sijill) before the Ottomans", 435, n.16, "The Ayyubi judge Ibn Abī l-Dam (d. 642/1244) says that the dīwān al-ḥukm containing maḥāḍir and siğillāt was handed over to the successor who had to scrutinize (taṣaffaḥa) whether the witnesses to these documents were still alive. In that case, there was no need to renew authentication (iṯbāt). However, if most witnesses were dead and only two of them alive, the judge had to summon them for authentication at his court, Müller, "The Ḥaram al-Sharīf collection", 456.

Previous witnesses bore testimony for both the document and its contents to the new witnesses that succeeded them. The new judge "who wished to take over his predecessor's papers" needed to send two trustworthy Muslims who could certify every single record, in the presence of the former judge. What is more, they needed to ascertain whether the witnesses to these legal deeds were still alive; if there were only two surviving witnesses, they needed to be summoned to court to authenticate a claim. The crowning example of this procedure is the founding act for the sufi convent established in Jerusalem by Saladin, whose certification was renewed from 1187 to 1614.[127] In the middle ages, unlike in more recent times, the probity of witnesses was a concern of the judge, and specific procedures had to be respected to ascertain their trustworthiness.[128]

The preservation of judicial records, even if they constituted real archives, was seriously compromised by the supremacy of orality, and the fact that the judge needed to have firsthand knowledge of each document's validity. As it could not be otherwise, only a minority of records was considered of future interest and therefore submitted to the re-authentication procedure, and the rest were discarded. With this analysis of the procedure by al-Khaṣṣāf, Johansen never inferred that judicial archives actually existed; rather, he underlined the fact that most notarized deeds lost their validity when the qadi ended his tenure, and found their way, at best, into the private collections of legal practitioners and litigants—where they have been found today—but not into archives. Specialists in legal practice and procedure have loosely assumed that the qadi archive did actually exist for an undetermined timespan, to have been a temporary repository, and to have followed a logic of preservation that greatly differed from Western practice—which, as we shall see, was based on the late medieval legal fiction that notarized records were vested with *publica fides*. For his part, Hallaq has interpreted this contingent storage of judicial papers as a real institutional repository, and levelled differences between the temporary collections held by the medieval qadis and the Ottoman sicils systematically kept in courthouses. While I must stress that he provides the reader with some particularly innovative insights, Hallaq's vindication of the medieval Arab archive is unpersuasive, because he provides the reader with no proof of such continuity; he denies the Ottoman practice of collecting court proceedings in bound ledgers and, paradoxically, pushes

127 Müller, "A Legal Instrument in the Service of People and Institutions", 184. Müller, "The Ḥaram al-Sharīf collection", 458, n. 148.

128 Tyan, Émile: *Histoire de l'organisation judiciaire en pays d'Islam,* 2 vol, Paris: Librairie du Recueil Sirey, 1938, I, 354–7.

the invention of the register back to late Mamluk times, a claim not backed by any evidence.[129]

2.2.2 *Was the* qimaṭr *an Archive?*

The tendency to equate documents with archives has found its clearest embodiment in the archival artifact called the qimaṭr. The qimaṭr is documented as having been the mobile archive of early judges, who held sessions in changing locations, as no such thing as a courthouse existed at that time. It came in different shapes and forms—a simple bag, a basket or a wooden box—and contained the certified documents described above, validated by living witnesses known to the court and to the judge in charge, and therefore ready for use in open cases. The surviving inventories, reports and depositions identified as part of court proceedings were not actually included in the qimaṭr, which was more a movable repository than it was an archive. In gross, as a mobile repository containing a selection of 'living' certified deeds ready for use by the qadi in charge, the qimaṭr is evocative of the distinction between the Islamic approach to document use, and a proper judicial archive *à l'Européenne*; and this difference resided, ultimately, in diverging notions and ideas about validity.[130]

Another archival locus that deserves to be mentioned here is the Ḥaram documents, a collection that was identified soon after its discovery as the archives of a judge active in Jerusalem in the fourteenth century. The discovery of a coherent series of judicial papers pushed the initial research team, led by Donald P. Little, to interpret the collection as proof that medieval Arabs regularly kept and stored their judicial archives. However, a successive generation of researchers of the Ḥaram al-Sharīf records has questioned the very idea that they represent a qadi archive. A convincing inquiry by Müller into the nature of the Jerusalem collection dealt a definitive blow to the idea that the survival of documents can be conflated with the existence of an Islamic archival tradition, and confirmed that the Ḥaram collection cannot be defined as an archive.[131] Very forcefully, Müller

129 Hallaq, "The qāḍī's dīwān (sijill) before the Ottomans", 434.

130 Tillier, Mathieu: "Le statut et la conservation des archives judiciaires dans l'Orient abbasside (IIe/VIIIe-IVe/Xe siècle): un réexamen", in: *L'autorité de l'écrit au moyen âge: orient-occident: XXXIXe congrès de la SHMESP, Le Caire, 30 avril-5 mai 2008*, Edited by Publications de la Sorbonne,263–276, 2009, 5, Müller: "The Power of the Pen: Cadis and their Archives in Medieval Islam", 372.

131 Little, "The Significance of the Ḥaram Documents", Müller, "The Ḥaram al-Sharīf collection", 459.

demonstrated that the Jerusalem papers were not the result of a systematic effort to safeguard a series of records. Instead, they consist of an array of papers kept for the sole purpose of serving an ongoing investigation. The Ḥaram collection was merely evidence in a trial against the judge in charge, Ibn Ghānim, for corruption, and included documents related to some cases passed before him during his tenure. This material related to old suits would have been discarded if it had not been for the investigation against Ibn Ghānim. Müller's demonstration is all the more important in that it undermines the views, most notably of Hallaq, that medieval judicial archives were kept and that Ottoman practice simply stemmed from Mamluk precedent. Despite this, most writers dealing with this issue continue to unquestioningly adopt the view that the Ḥaram papers prove that judicial archives, comprising whole court proceedings, were kept. This tendency has been epitomized by a contribution by Bauden in 2013, in which he acknowledges Müller's conclusions, but pushes them down into the footnotes, despite the fact that the latter's analysis invalidates Bauden's own views that medieval archives actually existed.[132]

2.2.3 *The Sijill: from Scroll to Codex*

Another clue as to the evidentiary divide can be found in the new meanings that the Arabic word *sijill* acquired over time. It has traditionally been assumed to be of Latin origin (*sigillum*, or seal), and to have come to Arabic via Greek and Aramean variations. During the early middle ages, it was the word originally used for the physical attribute of Byzantine officials—the seal—that acquired the meaning of the physical support it was impressed upon—the scroll—and as a result began to signify a legal effect, such as a ruling or certification.[133] It has been noted that Byzantine seals served the purpose of securing documents, as well as validating archival copies. The seal, moreover, was in itself an object of archival practices, and collections of seals have been discovered in Constantinople and Preslav. If we stick to this formal, artifact-based viewpoint, the medieval Islamic sijill was not perceived as a seal, as it was in classical antiquity, but as a document certified by that instrument. Ibn Iyās bitterly complaints that Sultan Qānṣūh al-Ghawrī (ruled 1501–1516) neglected the

132 Bauden: "Du destin des archives en Islam", 29, n.9.

133 Little: "Sidjill", Encyclopaedia of Islam, Second Edition, IX: 538b, Meshal, "The State, the Community and the Individual", 105–6. The etymology of *sijill* is further complicated by its proximity to the Aramaic word for 'clay', and therefore to the notion of seal, Robinson, Neal: "Clay", in: *Encyclopaedia of the Qur'ān*, edited by Jane Dammen McAuliffe, Brill, 2003, 1:339a, and Troupeau, Gérard: "Metals and Minerals", op. cit., 3:383a.

affairs of the state and tended to procrastinate the signing of administrative edicts, to the point that Mamluk secretaries found themselves forced to buy "old sealings" to stick to the new decrees.[134] Here, the word for sealing is *'alāma*, not *sijill*, signifying 'sealing' or the 'well-known motto' that was stamped by the judge onto legal documents to certify them.[135] Rendered as *sigilletto*, or less frequently as *cozetto* (from Tr. hüccet, or notarial deed), in the jargon of Italian-speaking merchants in Ottoman lands, the term is often used in post-1517 European sources.

In the debate on the existence of Islamic archives, one front has devoted its energy to questioning whether the Arabic term sijill refers to a body of documents, to the judge's archive, or to a physical register where his deeds were drawn up. Authors dealing with the medieval judiciary call the judge's collection the dīwān al-qāḍī, however Ottomanists prefer instead the Turkish word *sicill*, to refer to the qadi archives. If we hold uniquely to its formal definition, we learn that the Qur'ān clearly understands sijill to mean a scroll-shaped record. It should be noted that both the Qur'ān and early traditions adopted a positive attitude towards scrolls as proof, as does the Revelation as regards written contracts.[136] It may seem that medieval Muslims understood sijill to mean a certified, most probably scroll-shaped, legal document. Researchers dealing with Abbasid sources such as Aḥmad Ibn 'Umar al-Khaṣṣāf (d. 874) read it to mean 'note of verdict' (*enregistrement du jugement*), a term that captures the legal nature of the concept, rather than its physical dimension, of which a copy was delivered to the plaintiff.[137] Similarly and for Mamluk times, Müller arrives at the conclusion that the word signified 'certificate scroll,' and referred to the legal act that became the heading of a scroll-shaped document, followed by leaves containing successive certifications. Appending additional certifications to a given initial document is,

---

134 Ibn Iyās, *Badā'i' al-Zuhūr*, V, 92.

135 Müller: "The Power of the Pen: Cadis and their Archives in Medieval Islam", 374, 376, Johansen, "Signs as Evidence", 187, mentions the meaning acquired by the term in the semantic field of proof. al-Saḥmāwī, *al-Thaghr al-bāsim*, 375.

136 Heck, Paul L.: "Scrolls", in: *Encyclopaedia of the Qur'ān*, 4:569b: Brill, 2003.

137 Tillier, Mathieu: "Le statut et la conservation des archives judiciaires", Müller: "The Power of the Pen: Cadis and their Archives in Medieval Islam", According to Müller "Early on, a siǧill was a notification of a court decision that required the original witnesses to be questioned again by the new judge ... From the tenth century onwards, cadis' certificates (siǧillāt) combined the use of notarial documents with the attestation of court procedure", 381. Khaṣṣāf, Aḥmad Ibn 'Umar (d. 874): *Ādāb al-Qāḍī = Islamic legal and judicial system*, edited by 'Umar Ibn 'Abd al-'Azīz Ṣadr al-Shahīd and Munir Ahmad Mughal, New Delhi, Adam Publishers, 2017.

incidentally, the most obvious way of adding such a (potentially unlimited) series of attachments.

Insisting on the continuity between medieval and modern documentary practices, Hallaq has maintained that before and during Ottoman times the sijill was a large ledger; a claim that has been contradicted by thorough examination of the Jerusalem collection by Müller, who showed that the Mamluk sijill was in fact a scroll. Because Hallaq did not attempt to provide any elements to support his views, he overlooked the term's changing meaning between medieval and modern times. The Ottoman sijill (sicil) took the form of a codex—that is, a collection of paper leaves bound together and forming a single unit. Indeed, upon closer inspection a clear transition can be identified between the 14th and the 16th centuries: in the words of Müller, in contrast with Mamluk practice the Ottoman court registers "were organized differently and taken together functioned as a siğill (certificate): written mostly in chronological order in a codex." In an article published in 2018, Müller explicitly sanctions the definitive transformation of the scribal artifact: "Mamluk archives with their scrolls stored in boxes were replaced by Ottoman sharīʿa court records in the form of registered books."[138] The material transition from scroll to codex implies a willingness to bind together groups of documents at a single time, hence creating the notion of court proceedings. Indeed, Müller notes that in Ottoman judicial ledgers the documents included under each entry of the register "were considered evidence" and did not need to be resubmitted for further certification. Cases could continue to be investigated under newly-appointed judges without the need for prior certification or additional isjāl procedures. The Ottoman court register, in sum, rendered obsolete the Mamluk sijill as a certification procedure and as a scribal artifact. Sicil, therefore, became equated with the judge's archives, under the form of ledgers; a physical artifact safeguarding validated court proceedings, but which also included the judge's activities as a notary.

It should be clear by now that a strong relationship existed between the idea of legal certification, and that of legal preservation, and that judicial documents functioned as simple aide-memoires. In the Ḥaram al-Sharīf collection, different versions of the same act (such as inventories) can be found, containing small but important lexical differences; textual differences were of little importance, because the contents of these documents needed anyway to be orally confirmed by the witnesses, when needed, before the judge.[139]

---

138 Müller: "The Power of the Pen: Cadis and their Archives in Medieval Islam", 380.

139 Müller: "Ecrire pour établir Ia preuve orale en Islam: Ia pratique d'un tribunal à Jérusalem au XIVe siècle".

Truth resided in the utterances pronounced by trustworthy Muslims, irrespective of written variations present in the aide-memoires that supported their testimonies. It is for this reason that Monica Gronke has rightly described the style of notarial deeds as cursory and often incomplete.[140] She forgets, however, that in order for them to have been legally valid—before a judge, for example—the ritual of notarization required a reconstruction of the conditions in which the word had been transformed into writing. Significantly enough, a recent study of 150 notarial documents from Granada makes it clear that the deeds in the collection were simple aide-memoires to be corroborated in court, where the judge validated testimonies by writing *shahida* ("he testified") over their names. Signatures were often absent or illegible, and we can extend this same generalization concerning oral validation procedures to the case of 14th-century Jerusalem, as studied by Müller.[141] Even in Latin Europe, notaries sought publicity in the drafting of their deeds. In his discussion of the performative and ceremonial aspects of the medieval notaries of Marseilles, Daniel L. Smail notes that *the ceremony of notarization appears to have played a role in fixing events in memory: notarization, paradoxically, served the interests of memory*.[142] In medieval Islamic societies, records that for some reason needed to be kept valid, and that were therefore certified time and again, were preserved (first in the early qimaṭr, or later in the middle ages, in the qadi's perishable dīwān), while regular documents tended to lose validity over time and were destroyed.[143] The perishability of documents was to change radically at some point in the early modern period, and it is in the practice of ledger record-keeping that this transformation found its most complete expression.

140 Gronke, Monika, "La rédaction des actes privés dans le monde musulman médiéval: théorie et pratique", Studia Islamica 59 (1984), 159–174.

141 Carro and Zomeño: "Identifying the 'udūl in Fifteenth-Century Granada".

142 Smail, Daniel Lord: "Notaries, Courts and the legal Culture of Late Medieval Marseille", in: *Urban and Rural Communities in Medieval France, Provence and Languedoc, 1000–1500*, Edited by Kathryn Reyerson and John Victor Drendel, 23–51. Leiden—Boston: Brill, 1998, 49.

143 For some authors acknowledging the preservation of archives and records for the early-modern period, Eddé, Anne-Marie: "Documents et archives d'Orient: conclusions provisoires et tendances de la recherche actuelle", in: *L'autorité de l'écrit au Moyen Âge (Orient-Occident): XXXIXe Congrès de la SHMESP (Le Caire, 30 avril-5 mai 2008)*, Éditions de la Sorbonne, 385–400. Paris, 2009, Fadel, Mohammad: "al-Qaḍi", in: *The Oxford Handbook of Islamic Law*, edited by Rumee Ahmed and Anver M. Emon, 317–8.

2.3 Notaries in the Cross-Confessional Middle Ages

Documents and the material artifacts that embodied them need to be understood not only in light of specific archival practices and surviving collections, but through a complex kaleidoscope of notions and beliefs about truth and evidence. Moreover, the logics of document production, use and preservation not only changed over time, but also differed from those followed in Latin Europe. In this section I turn to notarization, as the debate on archives has focused more often on the circumstances determining whether a document was preserved or not, than on its actual inception. I first underline the differences between the southern European, 'Latin,' notaries and their Islamic counterparts, as regards sources of legitimacy, notarial skills, and their respective institutional environments. To illustrate actual practice, I dedicate special attention to the Mamluk commercial cities before 1517, where both types of notarial culture coexisted, and to a certain degree, interacted. The cities of commerce such as Damascus, Aleppo or Mecca abound with references to local ʿudūl operating in the streets and markets. Port-cities harboring Frankish vessels such as Alexandria held a special status (thughūr) as frontier zones,[144] and hosted many notaries attached to the administration, such as those we encounter in the 1387 Florentine document discussed in section 2.1.2.

While Mamluk sources can be overly greedy in technical details, and often express themselves in an allusive tone, they contain abundant references to notaries. Harsh criticisms abounded, and this has led contemporary historians to hasten to describe the Mamluk judiciary as fundamentally corrupt. Ibn Ṭūlūn and al-Qudsī report sayings such as "all men are honest (ʿudūl) except the ʿudūl," and popular verses targeting the notaries as "a tribe brandishing the spears of false witnessing," labeling them as "sultans in the realm of sijills" and property deeds, who "shed blood through the nibs of their pens."[145] While authors such as al-Maqrīzī or Ibn Iyās (particularly the latter) considered the Mamluk judiciary to be far more versed in sharīʿa than the Ottoman conquerors, they also seem to admit that there was always a shāhid ready to testify when needed. However, it would be unfair to pretend that references to honest notaries did not also abound, most often in obituaries praising their religious qualities; indeed, one sign of their piety that was praised, and which might

144 Catlos, Brian A.: *The victors and the vanquished: Christians and Muslims of Catalonia and Aragon, 1050–1300*, New York: Cambridge University Press, 2004, 23–71.

145 Ibn Ṭūlūn, Shams al-Dīn Muḥammad Ibn ʿAlī (1485?-1546): *Naqd al-ṭālib li-zaghal al-manāṣib*, edited by Muhammad Ahmad Dahman and Khalid Muhammad Dahman, Beyrouth, 1992, 88., al-Qudsī, *Badhl al-nasāʾiḥ al-sharʿiyya*, 212.

astonish the contemporary reader, was their insistence on keeping women and dhimmīs away from their stalls. It is difficult, though, not to note that the Muslim notary, precisely because of his prerogative as bearer of truth, found himself in a position of superiority, and often abused it. Despite this potential abuse of power, as regards their labor in bridging cross-confessional relations, notarial activity was fostered by Mamluk governance. By supporting the notarization of deals between Muslims and Franks, the sultan emphasized his protection over unbelievers engaged in trade and diplomacy, and thus fulfilled one of his religious duties as the head of the community. The arrival of the Ottomans and the general reorganization of notarial activities had, therefore, a twofold effect: on the one hand, ordinary subjects saw their access to rights facilitated through the drafting of ḥujjas and the availability of qadi courts, while foreign merchants' activities were now tightly governed by sharīʿa courts and regulations. On the other, mixed transactions were now registered by the qadis, and consequently the traditional actors of notarization withdrew from the marketplace.

Acknowledging differences between Western and Islamic notarial systems has often led historians to assume that they constituted two opposite, conflicting legal systems which, rather than interact with each other, simply coexisted in a divided Mediterranean. In the previous section, I sketched out the transformation that artifacts and storage facilities underwent after 1517, even though actors of notarization have traditionally been described as representing a certain continuity. Latin notaries in particular have been portrayed as the natural heirs of Roman law, and the notarial deed as finding its universal, public validity on the same classic grounds. In contrast, Islamic notaries are presented as a separate genealogy, one that did not share the Roman confidence in trustworthy documents, and whose legal system, therefore, missed out on the "triumph of the written" experienced in Latin Europe. Although divergent notarial systems engendered different preservation logics, and are therefore at the very root of the archival divide, it would be misleading to assume that the genealogies for Latin and Muslim notaries evolved on entirely distinct, separate lines—an assumption that is often shared by historians of Islamic law. On the contrary, the idea that notarized documents were endowed with public faith, and therefore valid without the oral support of their authors, is a legal contingency that only appeared at a later stage. Indeed, historians dealing with late Roman, Byzantine law depict a notarial system that in many respects exhibits striking similarities with the drafting of documents by Islamic ʿudūl.

When interpreting the uses of writing and documentation in cross-confessional environments, historians have stressed that a clear boundary existed between Western and Islamic traditions. According to a commonly-held interpretation, in legal or diplomatic contexts the two normative systems simply coexisted, and the primary locus for interaction was the occasional acceptance of each other's proofs. Extraneous proofs, it has been argued, were only accepted on condition that the counterpart advanced proof that had been validated according to his own evidentiary rules. This line of reasoning conveys a cross-cultural approach, assuming fundamentally separated notions of cultures and identities and, as we shall see, largely obscures the legal relations and interactions that actually took place. I argue instead that late medieval Mediterranean markets and courts witnessed the coexistence of two complementary notarial systems within a shared legal space. Multiple scribal institutions were known to merchants and other legal actors, who possessed a finely-honed understanding of these differences, and acted accordingly. While under the Mamluks mixed transactions were underpinned by different scribal institutions, a fundamental transformation was brought about by Ottoman legal reform, which meant that the registration of legal transactions suddenly looked very different after the Middle ages.

2.3.1 *Mediterranean Notarial Traditions*

The word *shāhid* signifies both a professional notary, and any "righteous witness" (*shāhid ʿadl*) in a given community. Ideally, any believer of good reputation could notarize, provided he opened shop at the market and was recognized by a local judge.[146] According to classical Islamic jurisprudence, only upstanding male Muslims were allowed to give testimony, and in time some of these witnesses came to perform professional notarial services. Since early Islamic times, the need to rely on a list of available trustworthy Muslims who could attest at court sessions favored registration at the maḥkama, in order to differentiate them from occasional witnesses to facts. Claude Cahen has acknowledged, however, that this distinction between ordinary and professional witnesses was of a purely practical nature, and had little significance in the early days of Islam.[147]

In Ottoman times, instrumental witnesses (designated as *shuhūd al-ḥāl*) worked as court staff, where they attested to the authenticity of all legal documents and their incorporation into the sijill. They were also scribes, and often

146 The classic account is still Tyan, *Le notariat et le régime de la preuve par écrit dans la pratique du droit musulman,* 18–21, Tyan, *Histoire de l'organisation judiciaire en pays d'Islam,* I, 349–372, Wakin, Jeanette A.: *The function of documents in Islamic law: the chapters on sales from Ṭaḥāwī's Kitāb al-shurūṭ al-kabīr,* Albany: State University of New York Press, 1972, 7.

147 Cahen, Claude, "A propos des Shuhud", *Studia Islamica* 31 (1970), 71–79.

drafted the contracts that were later to be notarized by the qadi.[148] As has been mentioned, the recognition that the Ottomans gave to the instrumental witnesses came at the expense of the ʿudūl, and by the 16th century the most notarial functions of the medieval ʿudūl were being transferred to the Ottoman judge and his court staff. This fact has moved authors such as Reem Meshal to regard the functions of court notaries as "no longer having a private function," and to note a general shift "from local notarial class to Ottoman-trained clerks."[149] In medieval times, in contrast, it was sufficient to obtain recognition from a qadi and to display competence in Islamic law in order to be counted as a reliable witness. However, in practice social conventions biased access to the notarial profession; we know that in cities such as medieval Aleppo, for example, the ʿudūl were often chosen from among a pool of families considered to be endowed with a collective reputation of trustworthiness.[150] Although respected notaries counted among professional religious scholars, Mamluk authors make it clear that other members of the religious learned community, often including merchants, could also practice the notarial profession on a temporary basis.

While any upstanding member of the community could act as a notary in Islamic society, in southern Europe notaries were instead public officers invested by the legitimate powers. These were the Pope, the Emperor, or his delegates. Only they could create the law, and therefore only they could delegate this creative power to the *tabelliones*—the Roman clerks charged with drafting private deeds. This stands in marked contrast with the realm of Islam, where God enjoyed the monopoly on lawmaking, and therefore the notary's legitimacy resided in his righteousness, and not on any given status.[151] In practice, in the lands under theoretical imperial jurisdiction, feudal lords such as the

148 Fitzgerald, *Ottoman methods of conquest,* 109–13. Wilkins, Charles L.: "Witnesses and Testimony in the Courts of Seventeenth-Century Ottoman Aleppo", in: *Lire et écrire l'histoire ottomane*, Edited by Vanessa Guéno and Stefan Knost, Beirut and Damascus: IFPO—Orient-Institut Beirut, 2015, 107–129. Wilkins describes them as signatory or "notarial witnesses," in contrast with eyewitnesses and informants summoned before the judge. A most useful description of the *shuhūd al-ḥāl* is provided by Baldwin, *Islamic Law and Empire in Ottoman Cairo,* 36–37.

149 Meshal, "The State, the Community and the Individual", 118.

150 Ḥamzah, *Niyābat Ḥalab fī ʿAṣr Salāṭīn al-Mamālīk (1250–1517 M/648–923 H)*, 254–5.

151 Cook, Michael: "Early Medieval Christian and Muslim Attitudes to Pagan Law: A Comparison", in: *Islam and Its Past: Jahiliyya, Late Antiquity, and the Qur'an,* edited by Carol Bakhos and Michael Cook. Oxford: Oxford University Press, 2017. Medieval issues revolving around notarization, the value of the written and witnessing are discussed by Madero, Marta, "Façons de croire. Les témoins et le juge dans l'œuvre juridique d'Alphonse X le Sage, roi de Castille", Annales. Histoire, Sciences Sociales (1999), 197–218.

Count palatine of medieval Genoa took over their social superiors' right to appoint notaries.[152] Southern European city-states sought to impose controls on a candidate's knowledge of jurisprudence, and the latter were required to pass exams. In places such as Venice, measures to control access to the profession increased throughout the Middle Ages, and culminated in the early sixteenth century, when quotas were established.[153] Islamic legal practice tended instead to keep a close watch on witnesses' morality, rather than on their effective legal knowledge, and judges were charged with monitoring the ʿudūl's rectitude. In exchange, the notaries of a given qadi were permitted to sit around him at court hearings, according to precedence, and wearing a particular kerchief.[154] Al-Maqrīzī describes a large-scale inspection of notaries and their workshops by judicial authorities, in order to ascertain the reputation of every single scribe.[155] This is also reflected in the role of the purifier (muzakkī) described by legal historians, a court clerk entrusted with the task of verifying the moral credentials of witnesses and who, on occasion, could conduct investigations in local neighborhoods to ascertain the probity of individuals. Emile Tyan has described the efforts by eighth and ninth-century Egyptian qadis to cope with the problem of witnessing, first by drafting a list of authorized witnesses, then by implementing the institution of the muzakkī, measures often contested by the population. Eventually, in the late Middle Ages the muzakkī was replaced by a mechanism whereby candidates were elected into the body of ʿudūl by their peers.[156] Sharīʿa-based notarization, apart from the fact that it played a role in the conduct of administrative business, was required in ample areas of daily life. Witnessing was needed in matters ranging from the grinding of harvest crops to the announcement of the official start of the lunar months. Religious festivities, for example, had to be determined by eye-witnessing of the phases of the moon. Ibn Ṭūlūn reports an episode in which a group of ulama refused to take part in the festivities at the end of Ramaḍān because, they argued, the witnesses attesting to the beginning of the month "were not upright."[157]

152 Airaldi, Gabriella: *Studi e documenti su Genova e l'oltremare*, Genova: Università di Genova, 1974, 197–241.

153 Pedani, Maria: *Veneta auctoritate notarius: storia del notariato veneziano: 1514–1797*, Milano: Giuffrè, 1996, Airaldi, *Studi e documenti su Genova e l'oltremare*, 227.

154 Ibn al-Ḥanbalī, Muḥammad Ibn Ibrāhīm al-Ḥalabī (1502 or 3-1563): *Durr al-habab fi tarikh aʿyan Halab*, edited by Mahmud Fakhuri and Yahya Abbarah, Damascus: Wizārat al-Thaqāfah, 1972–1974, I, 30 and n. 3.

155 al-Maqrīzī, Aḥmad Ibn ʿAlī (1364–1442): *Kitāb al-sulūk li-maʿrifat duwal al-mulūk*, edited by Said A. F. ʿAshour, Cairo: Matbaʿat Dar-al-Kutub, 1970–1973 III/2, 933.

156 Tyan, *Histoire de l'organisation judiciaire en pays d'Islam*, I, 357.

157 Ibn Ṭūlūn, *Mufākahat al-Khillān*, 412.

There were not only formal differences between the Venetian clerks and their Islamic counterparts; Latin notaries were able to produce written documents endowed with legal value in court (empowered with public *fides*), while Islamic notarial deeds were not valid without the oral support of their authors.[158] An Islamic document was not considered primary evidence until the witnesses and the notary who had drafted it had appeared in the courtroom and certified both the document and its contents. The Franks involved in trade in Mamluk lands describe the *ʿudūl*, therefore, as "witnesses who wrote."[159] They invariably used the term *testimoni* to refer to the *ʿudūl* and to notarial deeds, but never for the regular scribes, often of Christian origin, who were employed in the customs houses and in the personal dīwāns of prominent merchants.

In the late Middle Ages, southern European cities witnessed the emergence of the professional notary. By this period, most Italian city-states had come to rely heavily on a new category of notary empowered with official authority, who drafted "public" notarial deeds, and not just private acts, and it is generally agreed that it is around this time that the notarial act acquired juridical value.[160] The genuine nature of a contract came to depend on the officer's signature, rather than on the presence of witnesses. Notaries were incorporated into chanceries, courts of justice and communal administrations, such as the consulates in Alexandria and Damascus, where Venetian clerks became regular figures. They offered their services to the whole Frankish community, which, particularly for matters of debt, could benefit from the validity of notarial acts in their own city's courts.[161] As a result, deeds drawn up in Alexandria

---

158 For the use of deeds in other mixed, southern European contexts, see Burns, Robert Ignatius: *Jews in the notarial culture: Latinate wills in Mediterranean Spain, 1250–1350*, Berkeley: University of California Press, 1996, 32–51, McKee, Sally: *Uncommon dominion: Venetian Crete and the Myth of Ethnic Purity*, Philadelphia: University of Pennsylvania Press, 2000, 19–57, Smail, Daniel Lord: *The consumption of justice: emotions, publicity, and legal culture in Marseille, 1264–1423*, Ithaca: Cornell University Press, 2003 and his influential chapter: "Notaries, Courts and the legal Culture of Late Medieval Marseille".

159 ASVe, CI, N, B. 222, Notary A. Vactaciis, f. 43r-v, undated (March 5–9, 1401): "Item soluit testibus saracenis a gabano qui scripserunt, f.108r-v, Dec. 22, 1405: in testificatione testium saracenorum." Examples of this can be multiplied, with numerous mentions in the accounts of the Foscari firm in Alexandria, whose staff records the sums paid for notarial sevices at the customs: "per testimonio de doana per so chortexia", ASVe, Procuratori di San Marco, Misti, XXIX, B. 44A, Accounts Antonio Coppo 1481, f.2r.

160 Costamagna, Giorgio: *Il notaio a Genova tra prestigio e potere*, Roma: Consiglio nazionale del notariato, 1970, 7–32.

161 Most noteworthy among the profuse Italian scholarship on notaries, are the monographs on Genoa (ed. G. Costamagna) and Venice (M. P. Pedani) in the *Studi storici sul notariato italiano* series, together with several conference proceedings edited by V. Piergiovanni,

and Damascus found their way into trial proceedings held elsewhere. Before a Genoese court of justice in Famagusta, for example, a Damascene deed dated 1447 by the Venetian clerk Andrea Michiel was produced as proof by one of the litigants.[162] Many of the notarial deeds drawn up by these *outremer* notaries describe issues that ended up before Islamic judges. When they were brought before the qadis, Frankish merchants generally asked for an official record of proceedings, in order to be able to report back to Europe, and this is why they described their experiences with local justice to their Latin notaries.

The most striking divergences between Western and Islamic notarial traditions can be observed in the probative status of the written deed, and the very different approaches to its preservation. Venetian notaries kept a detailed copy of the deeds that needed to be preserved in state archives. Indeed, the Venetian authorities put great emphasis on the preservation of registers in the state archives, and issued regulations to that effect. According to Venetian regulations, *outremer* notaries were expected to draft their deeds in codex-shaped, paper protocols.[163] Venetians were equally invested in the safeguarding of the notarial and administrative collections of lost colonies, such as Crete after it fell into Ottoman hands in 1669.[164] In contrast, *ʿudūl* records were valid only during the lifetime of their authors, and in most cases there was no need to preserve them afterwards. Somewhat like Byzantine *tabelliones*, Islamic notaries do not seem to have preserved notarial acts *en minute*, a fact that Frankish traders found abhorrent. These differences demonstrate that both notarial systems, if they were not necessarily in conflict with each other, exhibited alternative and complementary features, which ultimately accounted for the medieval presence of Latin notaries in the Islamic cities of commerce—a presence that, not surprisingly, ended with the Ottoman takeover of the Arab provinces.

---

such as a recent one devoted to notarial public faith: Piergiovanni, Vito.: *Hinc publica fides. Il notaio e l'amministrazione della giustizia*, Milan: A. Giuffrè, 2006. On the Roman period, see Amelotti, Mario: "Fides, fides publica in età romana", in: *Hinc publica fides. Il notaio e l'amministrazione della giustizia*, Edited by V. Piergiovanni, 9–20. Milan: A. Giuffrè, 2006. On the validity of notarial documents as juridical items "in court and outside," see Burns, *Jews in the notarial culture*, 38–43, Pratesi, Alessandro: *Genesi e forme del documento medievale*, Roma: Jouvence, 1979, 47–55, and, for Genoa and Venice, Bartoli Langeli, Attilio: *Notai: scrivere documenti nell'Italia medievale*, Roma: Viella, 2006, 59–87, where an exhaustive bibliography is provided.

162 ASG, SG 590/1289, f.106v.

163 Bigaglia, Marco. A.: *Capitulare legum notariis publicis Venetiarum*, Venice: Apud Andream Poleti, 1689, 16, 24–27, Wheeler, Joseph Russell, "The sestiere of San Polo: a cross section of Venetian society in the second half of the fifteenth century. PhD thesis, University of Warwick", (1995), 12–14.

164 McKee, *Uncommon dominion: Venetian Crete and the Myth of Ethnic Purity*, viii.

More importantly for my argument, these differences between Western and Islamic notarial systems motivated a promiscuous recourse to notarial institutions by merchants. A series of documents from the notarial collections of Genoa are most illuminating in this regard. In two deeds drawn up in Genoa in 1475 and 1479 by the notary Emanuele Granello, a merchant summoned witnesses to prove the losses he incurred while he was a resident in Algeria. According to the witnesses, this Genoese merchant, Urbano de Dernisio, was living in Béjaïa in 1475, where, as reflected by both Islamic and Latin deeds, the Catalans Benet Spital and Joan Sala owed him money. Together with an accomplice named Venturino, himself a debtor, Spital broke into the house of his Genoese creditor and stole a few golden objects, some merchandise, together with Dernisio's book of accounts, as well as "diversas alias scripturas, tam in arabico quam in latino scriptas." Spital tore and burnt the Islamic deeds mentioning his debts for a total amount of 672 golden dinars. As a result, the Genoese merchant was left with no means to prove his debts. As Dernisio himself bitterly argued, "Muslim notaries do not preserve authentic copies of the deeds they draw up," and in a second deed related to the same facts he added "they do not keep registers."[165] In other words, according to Dernisio the ʿudūl neither kept originals, nor did they preserve them in codex-shaped artifacts for future reference. This may have been the norm for private contracts, even though, as suggested by the Florentine enquiry, some legal transactions subject to taxation by the Mamluk courtiers (Ar. simsār), such as sales contracts, necessarily left some documentary trace behind them. Destroying Dernisio's contracts may well have been useless under Islamic jurisdiction, since jurisprudence allowed witnesses to give testimony for debts and similar legal actions in case papers went missing, because they entailed effective rights.[166] The

165 "dictus Venturinus simul cum Benedicto Spitale barchinoniensi furtive acceperunt de scriptorio domus habitacionis dicti urbani in dicto loco buzee diversa cartularia rationis dicti urbani et diversas alias scripturas scriptas tam in arabico quam in latino inter quas erant dicta dua instrumenta scripta in arabico unum videlicet per quod constabat Benedictum Spital et Johannem Sala barchinionienses dare debere dicto urbano duplas 600 buzee et aliud sicut dictus benedictus erat debitor dicti urbani de duplas 72 et uno quarto buzee. Que instrumenta dictus venturinus tradidit dicto benedicto qui ea laceravit et combusit. Quas duplas 672 et quartum unum dictus urbanus exigere non potuit a dictis benedicto et johani. eo quia non habebat dicta duo instrumenta ut supra capta furtive per dictos Benedictum et Venturinum, et quia notarii barbari non tenent registrum instrumentorum per ipsos conpositorum", ASG, Notai Antichi, 871, doc. 274. With small lexical variations, the episode is also described in doc. 296: "mauri non tenent autenticum instrumentorum per ipsos conpositorum".

166 Johansen, "Formes de langage", 350, notes that this was not the case for documents not engendering effective rights, such as court proceedings.

creditors were much more likely, however, to have been targeting Dernisio's options for recourse back in Latin Europe, where he would not be able to substantiate his claims. In these documents, Dernisio was instructing the notary about the two divergent notarial systems, and this legal argument was indeed crucial to his case, because it helped to explain his inability to prove his Spital's debts with documentary proof. Dernisio's argument clearly refers to the Islamic notarial system in general (here referred to as *notari mauri, notari barbari*), and goes far beyond a more narrow reading that "the keeping of minutes … did not exist in Béjaïa."[167]

Abd al-Raḥmān Ibn ʿUmar al-Jawbarī (died after 1222) counts notaries among the Damascene low life he describes in his *al-mukhtār fī kashf al-asrār*, a work devoted to the adventures of swindlers, alchemists, beggars and thieves. According to al-Jawbarī, the ʿudūl cooked the deeds by introducing ambiguous wording and formal errors, allowing one of the parties to have the transaction invalidated. He suggests that clients often repented and that, with the help of the ʿudūl, the parties knew that notarized transactions could be contested and businesses, eventually, overturned. More interestingly, the murky status of Islamic documents made witnesses to the deed easy to manipulate. A ḥujja sanctioning either unimportant or well-known facts, such as small debts or transactions related to deceased clients, could be erased and a new deed forged. In the new ḥujja, the original witnesses' signatures were reused, who, in that case, would never be able to deny the validity of the new document's contents (lā yankarūn al-shuhūd khuṭūṭahum).[168] If uncorroborated Islamic deeds could hardly prove anything before a qadi, paradoxically, Dernisio or al-Jawbarī's stories suggest that corrupted of absent ḥujjas seriously compromised the witnesses' commitment to the truth, and led to the loss of rights.

2.3.2 *Legal Fiction and Roman Ancestry*

The traditional narrative on the genealogy of the notarial system sees late medieval notaries as heirs of the Roman past. Often associated with chanceries and administrations, southern European communal cities witnessed the

---

167 "la pratique du minutier […], n'existait donc pas à Bougie", Valérian, Dominique: "Le recours à l'écrit dans les pratiques marchandes en contexte interculturel. Les contrats de commerce entre chrétiens et musulmans en Méditerranée", in: *L'autorité de l'écrit au moyen âge: orient-occident: XXXIXe congrès de la SHMESP, Le Caire, 30 avril-5 mai 2008*, Publications de la Sorbonne, 2009, 59–73.

168 al-Jawbarī, ʿAbd al-Raḥmān Ibn ʿUmar: *Al-Ǧawbarī und sein Kašf al-asrār: ein Sittenbild des Gauners im arabisch-islamischen Mittelalter (7./13. Jahrhundert): Einführung, Edition und Kommentar*, edited by Manuela Höglmeier, Berlin: Klaus Schwarz, 2006, 306.

ascension of notaries able to draft deeds that were vested with universal validity. Unlike the ʿudūl, whose authority relied on religious legitimization, Latin notaries claimed it for themselves thanks to investiture by spiritual or temporal powers. Even in late medieval times, when the Holy Roman Empire was a political fiction, notaries were appointed by feudal lords claiming some imperial legitimacy.[169] In practical terms, this implied that their deeds' legitimacy extended throughout the Holy Roman Empire, and included the portion of the Mediterranean now under Muslim rule.[170] As a universal ruler, the Emperor guaranteed universal validity for their deeds, which were valid *ubique terrarum*, anywhere on earth.[171] Historians often assume that a direct link exists between the Roman notaries, or *tabelliones*, and the medieval Latin notarial institution, seeing it as part of a revival of classical Roman Law, a legal system sensitive to written proof. Under this system, it was natural that such documents, issued by clerks vested with legitimate authority, be valid without necessitating the support of oral witnessing. Public faith, therefore, went hand-in-hand with the notary's imperial investiture. This narrative naturally leads us to the idea that Western notaries were 'Roman' in origin, and Muslim notaries 'Islamic.' In my view, however, such an assumption is problematic for several reasons.

Roman documents were endowed with public faith and were actually kept in some kind of state archive, but this was the prerogative of public documents produced by a government official. However, the Roman *tabellio* limited himself to drafting private deeds; although they bound both parties, such deeds did not constitute firsthand and definitive proof in court, even if they were drawn up in the presence of witnesses (*publice confectum*), and included some additional assurances (such as the notary's *completio*). Indeed, there are descriptions of Roman and Byzantine notaries receiving clients in

169 Airaldi, *Studi e documenti su Genova e l'oltremare*, 199.

170 "Mais l'institution notariale avait un caractère international; l'acte public fait autorité dans toute la Chrétienté", Richard, Jean: "Aspects du Notariat public à Chypre sous les Lusignan", in: *Diplomatics in the Eastern Mediterranean 1000–1500: Aspects of Cross-Cultural Communication*, edited by Alexander D. Beihammer, Maria G. Parani, *et al.*: Brill, 2008, 207–222, 220. Airaldi, *Studi e documenti su Genova e l'oltremare*, 209, Airaldi, elaborating on the investiture of an outremer notary dating from 1454, notes that jurisdiction extends "per omnes terras et loca que Romanum profitentur Imperium". On universal validity, "la facoltà di nomina dei notai è valida in qualunque sito essi si trovino, anche nel Levante, purché in esso viga il diritto del Sacrum Romanum Imperium", 228–29. See also Murray, James M.: *Notarial instruments in Flanders between 1280 and 1452*, Bruxelles Académie royale de Belgique, 1995, 11.

171 This formula is often used by the imperial notary in Cyprus; Nicola de Boateriis *Nicola de Boateriis notaio in Famagosta e Venezia (1355–1365)*, edited by Antonio Lombardo, Venezia: Comitato per la pubblicazione delle fonti relative alla storia di Venezia, 1973.

their workshops (*stationes*), and this offers a striking parallel with the stalls (marākiz, ḥawānīt) that hosted Islamic notaries in the streets of medieval Syria and Egypt. As concerns public faith, in a recent monograph on early modern Roman notaries, Laurie Nussdorfer suggests that endowing notaries with *publica fides* was a medieval invention by Italian commentators, something that came in handy in the construction of the Ius Commune.[172] Studies on early notarial practice show that at some point scribes started to reclaim imperial/universal authority and validity for their deeds, with no apparent precedent. Marta Madero has contributed to further elaborating this picture by reminding us that medieval jurists were perfectly aware that believing documents over witnessing was a legal fiction, and a necessary 'mirage.' Innocent IV, in a famous passage of the *Apparatus in quinque libros Decretalium* (*II, 22, 15*, ca. 1245) establishes a necessary link between the problem of ensuring legitimate investiture for the notary, and the validity of the written document. Notaries could only be instated by Emperors and Popes, as the latter were the only valid sources of law who could delegate such power to a third party. Delegating this power implied acknowledging the capacity of notaries to legitimize a legal fiction, by giving validity to signs "written on a dead animal's skin."[173]

The probative nature of artifacts, the role of medieval notaries, and issues of jurisdiction were indeed interrelated. Much can be gained, therefore, in seeing the attribution of public faith to notarial deeds, not as a direct appropriation from the classical past, but rather as the result of a process that was evolving throughout medieval times. By disrupting the alleged linearity of notarial genealogies, historians of Roman law have detected a willingness by successive emperors to endow written documents with probative value, until the reigns of Valentinian and Valens (ruled 364–375).[174] The opposite tendency, to mistrust written proof and to enhance the value of oral witnessing, surfaced shortly after this period, and reached its peak under Justinian (ruled 527–565). A similar hiatus can be found in Justinian codes, forbidding Jews from testifying against Christians, hence diverging from classical precedent, when religion had not been an issue, and indeed converging with sharīʿa practice.[175] It is possible that

172 Nussdorfer, Laurie: *Brokers of Public Trust: Notaries in Early Modern Rome*, Baltimore: Johns Hopkins University Press, 2009, 9.

173 Madero, "Façons de croire. Les témoins et le juge dans l'œuvre juridique d'Alphonse X le Sage, roi de Castille", 202.

174 Ankum, Hans A., "Les tabellions romains, ancêtres directs des notaires modernes", Atlas du Notariat. Le Notariat dans le Monde. Huit Siècles de Notariat Latin, *Ars notariatus*, 42, Amsterdam: Kluwer, 1989, 5–44, 15–6.

175 Fattal, Antoine: *Le statut légal des non-musulmans en pays d'Islam*, Beyrouth: Imprimerie Catholique, 1958, 361-64.

late Roman law adopted a different attitude, that might well be connected with the Islamic biases against the written and against unbelief; this is all the more striking when we consider that, as noted earlier, the Qur'ān has a positive attitude towards contracts and writing, and does not expressly ban non-Muslim witnesses. The biases, it has been noted, stemmed from Middle Eastern legal practice in Umayyad times, not from scriptural injunctions. The "triumph of the written," and of public faith in the West, was therefore a late medieval, contingent development. In Genoa, use of the private deed without recourse to witnesses emerged in parallel with the emergence of notaries endowed with a "stronger fides" towards the end of the 12th century, who signed not just as "notarius," but also claimed imperial nomination (*notarius sacri imperii, sacri palacii*).[176] In his research into Ravenna's notaries, Mark Steinhoff sees this coincidence between universal authority and public faith in notarial texts as coming about shortly earlier, around 1050. He quotes a text, dated 1033, equating *publica* with *imperial* authority, as the earliest manifestation of what would become the norm in subsequent centuries.[177]

Research on Justinian's law offers some additional parallels between Byzantine and Islamic notaries. According to descriptions in Justinian's *Novels 44* and *73*, the logic of production, use and validation of documents followed by Byzantine notaries was identical to that of their Islamic successors. As was the case for Muslim notaries, *minutes* were not kept by the former, and both notaries and witnesses had to be summoned to court to prove the authenticity of a given document. Witnesses to private deeds were expected to attest to the veracity of transactions, and in the eventuality that they passed away, judges proceeded by comparing samples of their handwriting with that in the deed (*comparatio litterarum*)—although Justinian's law mistrusted this procedure.[178] Both Justinian's law and sharīʿa mistrusted the holographic will—an issue that was later contested in Siyāsa works of the Mamluk period, which called for freer recourse to written evidence, first by Ibn Taymīyah, and later by his disciple Ibn Qayyim al-Jawziyya. Similarly to the Muslim ʿudūl,

176 Rovere, Antonella, "I "publici testes" e la prassi documentale genovese (secc. XII-XIII)", *Serta antiqua et mediaevalia* 1 (1997), 291–332, 325–6, Costamagna, *Il notaio a Genova tra prestigio e potere*, 55–6.

177 "This moment in history, circa 1050, is a critical point in the evolution of the authentic act; notarial acts acquired *publica fides* at the same time as the idea that notaries should be appointed by some universal authority, either imperial or papal, took hold". *Steinhoff, Mark Wayne: Origins and Development of the Notariate at Ravenna* (*Sixth through Thirteenth Centuries*), Thesis (Ph. D.), New York University: Ann Arbor, Mich.: University Microfilms International, 1976, 74.

178 Lévy, Jean-Philippe: *Autour de la preuve dans les droits de l'antiquité*, Naples: Jovene, 1992, 155–75, Amelotti: "Fides, fides publica in età romana".

the Byzantine *tabellio* conferred in the *statio* with his client, acquiring first-hand knowledge of the business at hand, so that he could later testify to its veracity in court, even when the effective drafting of the deed was entrusted to an assistant. It is clear from this parallel that what both notaries were producing were private deeds serving as *aide-mémoire*, and that such documents required additional certification by those who participated in their redaction. At least under Islamic law, authentic documents were neither kept nor collected. In contrast, the use of *protocols* in Byzantine notarial practice has been noted; however, according to *Novella 44* the Byzantine protocol should be understood, not as the Ottoman bound ledger guaranteeing that originals were collected, but as a kind of official stamped paper covering the initial part of the deed. This first sheet of the document contained the name of the Count of the Imperial Exchequer, and hence served the purposes of validating the document's existence, rather than ensuring its preservation as an archival artifact.[179] Nothing similar to an actual notarial protocol has survived from Islamic lands. Although we do have compilations of notarial formularies, they were most probably used as handbooks for notarial trainees, and owe their survival to their role as specimens, rather than to their legal validity, or even as aide-mémoires for future reference. Lastly, as can be inferred from the Nessana papyruses produced by hellenized Arabs shortly after the rise of Islam, it was the client, not the notary, who was credited with the deed's *absolutio*. This "oriental custom" held that it was the client who actually delivered the notarized document to the other party, while the notary simply limited himself to its drafting and to bearing witness to its validity.[180]

Apart from highlighting the continuity between late Roman and Islamic notaries, comparing the two traditions' understanding of the notarial function also helps us to question a deeper assumption: that Romans and Byzantines had public notaries at all. As we have seen, the *tabelliones* emerged first and foremost as clerks who drafted private deeds with no real public character, they were not initially public servants appointed by state authorities,

179 Amelotti: "Fides, fides publica in età romana", 17. Ankum, "Les tabellions romains, ancêtres directs des notaires modernes", 42, "We add hereto, that notaries shall not make the final draft of a document on any paper other than that which has in front what is called a protocoll, containing the name of the officiating glorious Count of the Imperial Exchequer, the time when the document was drawn and other things usually contained thereon; nor shall the protocoll be cut off, but bound with the paper; for we know that many documents written on such papers have in the past been and are now being proven to have been forged," See Novella 44, c. 2, *The Codex of Justinian: a new annotated translation, with parallel Latin and Greek text based on a translation by Justice Fred H. Blume*, 3 vols., Cambridge: Cambridge University Press, 2016.

180 Ankum, "Les tabellions romains, ancêtres directs des notaires modernes", 25.

and they kept no minutes/originals.[181] What would arise later as a distinctly Western Christian, or Latin attribute, is the public character of documents, not the Roman ancestry of the notarial institution. In addition, mistrust for written evidence is neither a necessary nor a specifically 'Islamic' approach to proof, and indeed this idea was most probably imported by early Islamic jurisprudence from their Byzantine imperial predecessors. Holding a black-and-white, essentialist vision of two opposing notarial systems is therefore to forget that both were directly, if unequally, rooted in antiquity, and moreover leads us into the error of conceiving of the 'Western' notariate as a direct heir of classical Roman law.

2.4 The Case of the Outremer Notaries

Towards the mid-fourteenth century, the Venetian state began to send notaries to Alexandria and Damascus, most probably as a result of the restoration of the galley convoys to the Levant in 1344, following the end of papal bans on trade with the infidels.[182] The latter were imperial notaries, enjoying imperial jurisdiction, and hence should be distinguished from the common public notaries acting under Venetian Law (*more veneto*). The imperial notaries were recruited and trained differently, and were assigned to Venetian mainland possessions, which unlike the city of Venice were under imperial jurisdiction. Venice's state archives host a healthy collection of deeds drawn up in Alexandria or in Damascus by imperial notaries for the fist half of the 15th century, and even some from the second half of the 14th century. There is however a dramatic decrease in deeds for the subsequent century; due to a fire in 1577, most documents for the period 1450–1517 have been lost, and it would appear that after that date, the Venetian government ceased to send notaries to the Levant for the rest of the 16th century. No extant protocols, even fragmentary, have survived for Syria or Egypt under Ottoman domination; I have come across two isolated references to notaries active for that period, but they seem to have worked exclusively as consular chancellors, and not as public notaries, most probably because the

181 "trois différences entre les tabelliones du droit de Justinien et les notaires modernes sont frappantes. Ces tabelliones n'étaient pas comme les notaires modernes des fonctionnaires publics nommés par les autorités de l'Etat; leurs actes n'etaient pas des actes authentiques (ils n'avaient pas de fides publica) et, enfin, ils ne conservaient pas de minutes.," *ibid.*, *44*.

182 Lane, Frederic Chapin: "The Venetian Galleys to Alexandria, 1344", in: *Wirtschaftskräfte und Wirtschaftswege. I: Mittelmeer und Kontinent: Festschrift für Hermann Kellenbenz*, edited by Jürgen Schneider, Stuttgart, 1978, 431–440.

task had by then fallen under the qadi's prerogative.[183] Genoese and Venetian notaries traditionally operated in the Black Sea, in Lesser Armenia and in the Byzantine territories, and in the Arabic-speaking *échelles du Levant* they met the ʿudūl for almost two centuries.

For the period between 1350 and 1500, Damascus and Alexandria saw Venetian notaries and ʿudūl crossing paths in the souks in search of clients, and thus the encounter between the two notarial traditions. In contrast, no Frankish notaries seem to have been dispatched to Mamluk Aleppo, a major city of commerce, despite the presence of consular institutions. By the same token, but for a vague isolated reference to them, no Genoese notaries appear to have operated in the Mamluk territories. This unique configuration, it seems to me, points to the conventional nature of notarial markets, with cities marked by the presence of a Venetian or a Genoese notary, but rarely both at the same time. In the Maghreb, Genoese clerks predominated, whilst Syria and Egypt were covered by the Venetian notaries. Indeed, Mamluk markets and courts looked different from those in Ottoman times because, after 1517, the Latin notary is missing from the picture, a fact that has received strikingly little attention. We have good coverage for Alexandria and Damascus, with notaries resident for two-year periods spanning from the 1390s to the 1450s, and then scattered deeds and testament collections for the rest of the Mamluk period; for the Ottoman period, however, the coexistence of notarial institutions, if it existed at all, has left no Latin traces.

Parallel to the Italian republics, in Mamluk times the notarial profession grew considerably, and commercial cities are in particular described as burgeoning with workshops in the markets, at the city gates and in the mosques. Indeed, the Mamluks saw the adoption of written contracts between Muslims and Franks as a first step towards building a framework for interfaith conflict resolution, and by the end of the Middle Ages Mamluk treaties signed with Franks began including new provisions that the ʿudūl systematically notarize all mixed transactions. While 13th-century treaties accepted transactions either in or out of customs (*in dogana … extra dogana*), notarized by witnesses or not (*cum testibus … sine testibus*), the tendency in the fifteenth century was to require that all mixed transactions be notarized before Islamic ʿudūl

183 ASVe Cancelleria Inferiore, Micellanea 2, Notary Claudio Giovanni, with a single deed in favor of the abbot of Saint Catherine, dated September 1567, and Iacopo Vigulo who drafted Marco Priuli's testament discussed in Chapter Four. The Bailo's chancery has yielded extensive series of notarial records preserved in ASVe *Bailo di Costantinopoli* series.

and, where available, that sales be concluded within the customs facilities.[184] By 1400, notarial witnesses were attached to several of the different customs departments—such as the *Duchela* and the Dīwān al-Qabbān, which are mentioned in the Florentine report of 1387—and this was part of a general trend extending to other administrative and judiciary offices.[185]

How, then, did merchants decide which of two potential legal institutions to use—Venetian or Mamluk? Although it may seem fairly obvious that Latin clients would have chosen Venetian notaries—where they could find legal, linguistic and personal affinities—in the following pages, I aim to demonstrate that such a strict confessional divide did not exist; Latin notaries coexisted alongside Muslim ones, and both offered their services to a mixed clientele. Different clients sought the support of the *ʿudūl* to guarantee the validity of their transactions in local markets and courts, and they did so irrespective of their own religion. Among the ʿudūl's clients were Franks who sought locally-produced evidence of their transactions; however such evidence only had legal value in local courts if it could be supported by those Muslims who had witnessed the transaction. Similarly, 15th-century Venetian notaries sold their services to merchants of various religious and ethnic backgrounds, particularly when recognition of their transactions was sought before Frankish associates and legal institutions. Thus, two Latins could have recourse to the ʿudūl to "seal their engagements," and conversely, there are many documented instances of Muslims requesting the services of Venetian clerks for their transactions with Frankish merchants. Finally, although this was much rarer, two Muslim parties could underwrite a Latin contract before a Venetian notary.[186]

184 *Diplomatarium veneto-levantinum sive acta et diplomata res venetas graecas atque levantis illustrantia*, edited by G.M. Thomas and R. Predelli. Vol. 2, Venezia: Deputazione veneta di storia patria, 1880–1899, I 295 (art. 22), For fifteenth-century treaties signed with Barcelona, Florence and Venice, Ruiz-Orsatti, Reginaldo, "Tratado de Paz entre Alfonso V de Aragón y el sultán de Egipto, al-Malik al-Ashraf Barsbay", *Al-Andalus* 4 (1939), 333–389, 343, 345, 361 (art. 11, 15, 26), Wansbrough, John: "A Mamluk Commercial Treaty Concluded with the Republic of Florence, 894/1489", in: *Documents from Islamic Chanceries*, edited by S. M. Stern, Oxford: Bruno Cassirer 1965, 39–79, Wansbrough, John, "Venice and Florence in the Mamluk Commercial Privileges", Bulletin of the School of Oriental and African Studies 28 3 (1965), 483–523, 488 (art. IV), 512 (art.V), 498:35 (art. II).

185 Popper, William: *Egypt and Syria under the Circassian sultans, 1382–1468 AD; systematic notes to Ibn Taghrî Birdî's chronicles of Egypt*, Berkeley: University of California Press, 1955, 1957, 1963, Part I, 97–103, 107.

186 ASVe, CI, N, B. 222, Notary A. Vactaciis, f. 17r, 2 Mar. 1400: "procuratore ... pro quadam carta moresca." Sometimes deeds were drawn up to deal with third-party Muslim associates: ASVe, CI, N, B. 83II, Notary C. Del Fiore, 2 May 1461; ASVe, CI, N, B. 148, Notary P. Pellacan, 7 Oct. 1444: "una charta moresca mi rechiedete chio fazi a chonfirmatione de uno chompromesso fatto tra ser fra Antonio Mozzo e me"; ASVe, Notarile Testamenti,

Aside from giving us insight into cross-confessional recourse to notaries, the Venetian *outremer* clerks can also help us understand two major areas of interaction: first, the promiscuous use of legal artifacts, and specifically Islamic records produced by the parties to support claims to Latin contracts, and second, the action of legal agents in contexts that fell outside their own legal system. References to the mixed use of documents are relatively frequent. In 1435, for example, a certain Abdrexach submitted to the Venetian notary of Alexandria an Arabic document written by a long-term Florentine resident in Egypt, Francesco Mannelli, where the latter acknowledged his debts to the former. Abdrexach had the private document notarized and later requested that the notary draft a power of attorney. With this second document, Abdrexach appointed a Genoese agent to recover his debts on his behalf back in Christian lands, with particular mention to the courts of Rhodes.[187] In another similar case from 1400, a Jew from Crete gave an Islamic power of attorney to a Rhodes consul, empowering the latter to retrieve debts traceable in Latin deeds.[188]

Interactions across legal systems are obviously easier to trace in the protocols of the Levantine *outremer* notaries than in those drafted in the metropolis. It has been noted that there are no discussions in Latin jurisprudence of how to deal with proof produced in the language of infidels, or under exogenous legal systems.[189] Such issues were instead dealt with in practice, and it seems that similar procedures were also implemented by Latin notaries when handling cross-confessional business deals. One example of this practical approach to cross-confessional cases can be found in that of the Genoese agent Giovanni Italiano, who found himself in Almería in 1483, in the very last years of Islamic rule. Italiano was acting on behalf of a Venetian merchant called Andrea Mocenigo, and reclaimed the debts owed to the latter by a Muslim merchant named Macomet Ubecher. Mocenigo had agreed to sell spices to Ubecher, but had for some reason preferred to stay aboard a Venetian galley and appoint Italiano as his attorney on land. For his part, the Muslim argued that the entire arrangement was not compliant with Islamic law, refused to

B. 215, Notary S. Peccator, 14 Oct. 1448; ASVe, CI, N, B. 211, Notary N. Turiano, f. 6r-v, 21 May 1455.

187 ASVe, CI, N, B. 211, Notary N. Turiano, f.36v, and f.38 r-v, February 8, 1435. "quadam carta debiti in lingua arabicha scripta nobili et egregio viro Francisco Mannelli quondam Rinaldi [...] super quam cartam sunt anotate due subscriptionis manibus christianorum franchorum".

188 ASVe, CI, N, B. 222, Notary A. Vactaciis, f. 17r, March 2, 1400: "procuratore ... pro quadam carta moresca."

189 Airaldi, Gabriella, "Genovesi nel mondo islamico: « carta sarracenica » e « carta in arabico »", Critica Storica IX/1 (1972), 106–121.

recognize Italiano as Mocenigo's attorney, and declined to pay. Italiano responded by demanding that Macomet Ubecher put his refusal in writing, which he accepted. A group of five Genoese merchants in Almería undersigned the Arabic document, giving testimony that Ubecher had effectively written the deed with his own hand. Four of them signed in Latin, and one in vernacular Genoese. Indeed, this exact same procedure was also followed by the Florentine Francesco Mannelli just mentioned, who had his due bill undersigned by fellow Christian witnesses.

Later that year, in December, a certain Giovanni Doria showed up at a notary's office in Genoa in the company of some Venetian representatives. Doria asked the notary to have Ubecher's Arabic document translated, and have a notarial deed based on it drawn up. The notary found himself obliged to call upon two Muslims settled in Genoa—Sadat Abeyn Aven, son of Muḥammad, from Granada, and Ucoy son of Saad Ariz from Tunis—fluent in both the Spanish and Genoese vernacular languages.[190] Both attested that the document was an original and, apparently, translated the record by reading it aloud. This would not suffice to the notary who called on the additional services of two Genoese merchants knowledgeable in Arabic, *vernacular Genoese and Spanish*.[191] These Genoese, in their turn, attested to the correctness of the translation provided by the two Muslims. The document is silent as to the reasons for this complicated procedure, although in addition to validating the version provided by the Muslims, I conjecture that the two Genoese had knowledge of spoken Arabic but were unable to read it. By drawing up these *testificationes*, then the *instrumentum registrationis et traductionis*, and by incorporating them into his protocol, the notaries Testa and Foglietta were able to give public faith to an Arabic private deed.

Two observations can be made about the exceptional procedure described above. First, that an Arabic record was converted into a Latin notarial deed

190 "Fuit suprascripta scriptura traducta de lingua arabicha in linguam vulgarem ianuensem et ispanam per interpetres, videlicet Sadat Abeyn Aven quondam Macometi, mercatorem regni Granate et Ucoy filium Saad Ariz, mercatorem Tunetis, mauros existentes in presenti civitate Ianue, afirmantes eorum iuramento quod dictam scripturam in lingua arabicha scriptam fore scriptam manu propria dicti Machometi Urbeche.", ASG, Notai Antichi 1144bis, not. Pellegro Testa, n. 316.

191 "et expertos et in vulgari ydiomate et sermoni Ianuensium, vulgari Ianuensium et yspani, presentibusque in hac eadem traductione nobilibus viris Guillelmo Embrono et Gotifredo Spinula, civibus Ianue, ambobus expertis dicte lingue arabice et yspane ac lattine afirmantibus sub eorum iuramento tactis corporaliter scripturis sese intelexisse mediantibus dictis mauris legentibus dictam scripturam, ipsam traductionem fuisse vere factam.", ASG, Notai Antichi 1144bis, notary Pellegro Testa, n. 316.

thanks to its undersigning by several Christian witnesses on the beach of Almería, and as a result was subject to a certain degree of hybridization. Secondly, the statement given by Macomet Ubecher was not even notarized by a shāhid and proper witnesses, making it all the more risky for the notary to endow it with executive legal validity (severe penalties could be imposed in case of forgery, such the notary's hand being cut off—a punishment that was equally reserved for Ottoman notaries).[192] Overall, however, the view that proof established by unbelievers was accepted if fully compliant with evidentiary standards does not apply here. It is perhaps for this reason that Giovanni Doria returned several times the notary workshop, in an attempt to definitively wrap up the Almería issue. In November, he requested that two Genoese citizens certify the trustworthiness of the Almería document (referred to as "litteris mauritaneis in appapiro scriptis"). They witnessed to the fact that the Genoese signatures in the Arabic document ("sub quibusdam litteris mauritaneis") were original; these witnesses attested knowing these merchants' handwriting, since they had received letters from them in the past. As if the public faith of a single notary did not suffice, the deed was transferred to a second notary, Oberto Foglietta, with Pellegro Testa now acting as scribe and witness to the deed. In addition, the chapel of Genoa's notarial guild was chosen as the place of signature for the deed, in an attempt to lend further credibility to its registration.[193]

Grounded as they were in very different juridical traditions, Latin and Muslim notaries had very different approaches to the fundamental issue of evidence and proof; however, in practice they complemented each other. The validity of acts drawn up by the ʿudūl was tied to the local courts, while Latin deeds had to secure rights back in Europe, where they had probative value. The two institutions together produced evidentiary artifacts to legalize private transactions, to aid arbitration panels, and to guarantee the validity of private contractual arrangements before the parties entered into a lawsuit. In court, only professional witnessing by the ʿudūl was acceptable to the qadis. However, the Mamluk officers acting as judges in Siyāsa trials were an exception to this rule, in that they seem to have reached decisions not only on the basis of notarial deeds, but also correspondence, written accounts, and other kinds

192 Carosi, Carlo: "II tradimento della fides: il falso", in: *Hinc publica fides. Il notaio e l'amministrazione della giustizia*, edited by V. Piergiovanni, 127–151. Milan: A. Giuffrè, 2006, 141, Fodor, P., "How to forge documents? (a Case of Corruption within the Ottoman Bureaucracy around 1590)", *Acta Orientalia Academiae Scientiarum Hungaricae* 48 3 (1995), 383–389, 385.

193 ASG, Notai antichi, 787, Oberto Foglietta doc. 53, 25 November 1483.

of written evidence produced by unbelievers. Describing Ibn Qayyim's reflections on the regime of proof, Baber Johansen aptly stresses that daily governance inevitably had to deal with an "altered normativity" (*sharʿ mubaddal*).[194] Dealings with the Franks fell within this category—a point that I will develop later in this chapter—and, accordingly, during the 14th century commercial cases dealing with all kinds of foreigners, ranging from Persians to Franks, were transferred to officials who served in the capacity of judges. The Islamic notarial system, moreover, interacted with that of the Latin scribes to a greater extent than has been traditionally been held, as did notarial agents. Descriptions of the Siyāsa trials held in Damascus tell us that, on occasion, judges relied on the collaboration of Venetian notaries. The consulate clerks could be called before the Siyāsa judges, for example, to hold the scriptures upon which Frankish defendants took their oaths. Conversely, at least in Damascus, some scribes were specialized in recording the transactions contracted between Latins and Muslims, apparently on the basis of their linguistic expertise.

2.4.1 *Merchants and Notaries in the Eastern Mediterranean*

The earliest surviving collection of a substantial number of Latin deeds produced in the Levant are in the papers of the notary Giovanni Campione, who stayed in Alexandria for twenty-three months between 1361 and 1363. Of the 165 people mentioned in his notebook, only one identified himself as an Eastern Christian, and one other is labeled as a Muslim.[195] As Egypt and Syria became the center of networks dealing in the spice trade, the notary's task became ever more complex. In extant fourteenth-century casebooks, notarial deeds almost exclusively concern Latins. However, in the fifteenth century, it was not only local Muslims, but characters from a variety of backgrounds that found their way into the protocols. In order to mark a clear distinction between the religious minorities under the protection of the Western powers, and those who were subjects of the local Islamic state, the Latin notaries enlarged their formulaic, largely fossilized Latin terminology. Hence, the dhimmī Jews were sometimes referred to as *judeus ebraicus*, as a way to distinguish them from the Jews of Venice.[196] On occasion, the notaries adapted their terminology to clearly identify Muslims (for instance, as "Saracen Moors"), as opposed to the

194 Johansen, "Signs as Evidence", 182.

195 ASVe, CI, N, B. 36, Notary G. Campione, Oct. 27, 1362 : *christianus a centura,* Oct. 28, 1362, Oct 30, 1362: *saraxino.*

196 ASVe, CI, N, B. 230, Notary N. Venier, f. 13r, 3 Apr. 1419: "cuidam Ellie, judeo ebraicho, illo tunc existenti in Damasco et ad presens habitatori dicte civitatis Nichosie."

more common Arabic-speaking Christians.[197] Muslims could also be identified as "foreign," like the Maghrebis, or vassals of Christian rulers such as the Iberian Muslims. Consequently, in their Latin jargon the notaries conveyed individuals' jurisdictional location as best they could, as in the complex case of *Abdella, judeus ebraicus magrabi de Tunisio, habitator in Damasco*.[198] The major challenge, however, was labeling the different kinds of Christians. Most often, Oriental Christians were indistinctly categorized as "Christians of the Girdle"; however the notary also had to deal with members of Oriental churches living in Cyprus, or elsewhere in the former Byzantine territories, who, to avoid confusion, were labeled differently.

The complexity of the notarial taxonomies is particularly visible in deeds where Oriental Christians from Islamic lands engaged in business transactions with their coreligionists from places such as Rhodes or Cyprus.[199] In such circumstances, complex juridical situations that are difficult for us to reconstruct could hide behind the vague religious and ethnic categories used by notaries. We have two examples of Latin clerks who used the word *fazolati* to describe the Arabic-speaking Christian minority operating mainly from Cyprus.[200] Indeed, the increasing complexity of notarial terminology, which reached its peak by the fifteenth century, suggests that Venetian clerks were operating at the intersection of two spheres of jurisdiction, which transcended a simplistic model of Christian-Muslim dichotomy. The matter was of crucial importance, as the Christians and Jews of Islam were charged only half the amount of taxes as their coreligionists subject to Christian powers. This meant that both the notaries and the Mamluk secretaries were increasingly faced with the legal issue of determining whose jurisdiction these Christians and Jews fell under. While the Venetian notaries' response was to adopt an increasingly complex terminology, the chancery manual by al-Saḥmāwī suggests that discussions revolved more around identifying the different Christian sects, in order to clarify which were led by Oriental patriarchs, and therefore which individuals

197 ASVe, CI, N, B. 230, Notary N. Venier, f. 9v, 18 Oct. 1418: "aliquibus mercatoribus saracenis moris."

198 ASVe, CI, N, B. 230, Notary N. Venier, f. 13r, 4 May 1419.

199 For references to mixed Christian networks: ASVe, CI, N, B. 222, Notary A. Vactaciis, f. 74v, Oct. 20th, 1404, f. 183v, July 29th, 1405, ASVe, CI, N, B. 211, Notary N. Turiano, f. 59v, Oct. 4th, 1455.

200 ASVe, CI, N, B. 222, Notary A. Vactaciis, f. 80r, 3 Dec. 1404: "Salem façolato habitatori nicosie presenti et intelligenti per Nessinum interpretem venetorum lingua Arabica"; ASVe, CI, N, B. 230, Notary N. Venier, f. 19v, 29 May 1419: "in su laqual nave I era haver de mori e fazolati."

should be considered subjects of the sultan, irrespective of where their members lived.[201]

During the fifteenth century, Muslims and Christians who were subjects of Islamic states became frequent clients of Venetian notaries. They often gave power of attorney to Franks, for matters ranging from the capture of a slave to the recovery of debt.[202] Providing services to these Mamluk subjects implied the *de facto* acceptance that the validity of notarial deeds was universal (*ubique terrarum*), at least in the territories under the theoretical jurisdiction of the Roman Empire, which included the Mediterranean Levant.[203] Together with the very idea of public faith, this second legal fiction permitted notaries to present themselves as a universal institution to which any merchant could appeal. Moreover, by extending their services to individuals under Islamic jurisdiction, notaries were implicitly granting public faith to their legal documents. The notary often quoted documents in Arabic, generally referred to as *Moorish letters*, but also in Hebrew, presented by clients as evidence of previous business relations. The Venetian clerks had account books and customs records translated, and, on occasion, they made references to private Arabic acts and contracts.[204] Among these documents, we can find Islamic notarial deeds drawn up by the ʿudūl and validated by witnesses. In a much-disputed case, two Muslims were called to witness an agreement between a Florentine consul and a Venetian jeweler in Alexandria ("ad conficiendum saltem pro duos testes mauros unam cartam morescam"). These *Moorish letters* were later used to resolve disputes before Christian courts in Alexandria and Rhodes.[205] Similarly,

---

201 al-Saḥmāwī, *al-Thaghr al-bāsim*, 424–8. As for the classic system of taxation (ʿushr) applied to dhimmīs and ḥarbīs, see Ibn Taymīyah, Aḥmad Ibn ʿAbd al-Ḥalīm (1263–1328): *al-Siyāsah al-sharʿīyah fī iṣlāḥ al-rāʿī wa-al-raʿīyah*, edited by ʿAlī Ibn Muḥammad ʿUmrān, Mecca: Dār ʿĀlam al-Fawāʾid, 1429 H, 55–6.

202 ASVe, CI, N, B. 222, Notary A. Vactaciis, f. 41r, 25 Feb. 1401; f. 119v, 16 Aug. 1406; ASVe, CI, N, B. 22, Notary V. Bonfantin, 28 June 1419; ASVe, CI, N, B. 230, Notary N. Venier, f. 5v–6r,18 May 1418; f. 6r–v, 16 May 1418; ASVe, CI, N, B. 148, Notary P. Pellacan, 9 Nov. 1444; ASVe, Notarile Testamenti, B. 215, Notary S. Peccator, 10 Oct. 1448; ASVe, CI, N, B. 83II, Notary C. Del Fiore, f. 15v, 14 June 1426; f. 24r, 30 May 1426; ASVe, CI, N, B. 211, Notary N. Turiano, f. 38, 8 Feb. 1435.

203 Airaldi, *Studi e documenti su Genova e l'oltremare*, 209, Murray, *Notarial instruments in Flanders between 1280 and 1452*, 11.

204 ASVe, CI, N, B. 222, Notary A. Vactaciis, f. 17r, 2 Mar. 1400, mentions an Arabic contract drawn up in Cyprus; ASVe, CI, N, B. 83II, Notary C. Del Fiore, 2 May 1461: vigore certe carte arabice, se constituerit plezium. As for contracts in Hebrew: vigore unius scripti anotati in ydiomate ebreo, ASVe, CI, N, B. 211, Notary N. Turiano, f. 66r, 26 July 1428.

205 ASVe, CI, N, B. 148, Notary P. Pellacan, 7 Oct. 1444; ASVe, CI, N, B. 211, Notary N. Turiano, f. 38, 8 Feb. 1435, an Arabic contract is produced as evidence to be used before Rhodian courts.

Franks who went into partnerships with Muslims often made use of Arabic contracts ("cartas, instrumenta et scripturas lingua Arabica scriptas").[206] It has been argued that legal relations between Franks and Muslims were limited to the occasional validation of either party's probative artifacts;[207] according to this vision, evidence produced under the Islamic system was accepted by Franks only if compliant with Islamic rules of proof, and vice versa. However, the combined, mutually enforced use of legal instruments that we have seen here suggests that legal relations went beyond a simple Islamic/Frankish polarity, and in reality had a truly multi-faceted configuration. As Kate Fleet has suggested for judicial cooperation in pre-Ottoman Turkey, notarial culture contributed to the full and mutual integration of Frankish and Mamluk legal devices.[208]

Depositions and sworn testimonies by Muslims are a recurring feature in the ledgers of Latin notaries. They accepted statements in Arabic and in Turkish, whose contents could be asserted either by official dragomans (interpreters) or by simple merchants. Hence, someone might speak "in a translated voice" (*dixit et testificatum fuit suo sacramento iurando per vocem turcimatam*) or be "understood" (*intelligenti pro interprete lingua Arabica*) through the intermediary of a third party.[209] One such example is the a disagreement that arose in 1404 over a shipping contract signed between two Cypriots. The contract had been drawn up in Arabic by one of the merchants, Salem, who spoke Arabic and who subsequently sought the mediation of a Venetian dragoman. The arbitration, accordingly, was notarized on the basis of the linguistic mediation and the validity of the Arabic document. Beyond this cultural sophistication delivered by notarization, mutual recognition required a certain legal accommodation, particularly with witnesses. The notary was compelled to accept oaths and testimonies from Muslim courtiers and dragomans even when, embarrassingly, they were Christian renegades. Their condition as apostates is generally made explicit (*olim christianus fidelis ad presens saracenus*), and indeed, although the Venetian authorities recommended that oaths by non-Christians be sworn on "their old texts", when renegades were called to swear

206 ASVe, CI, N, B. 211, Notary N. Turiano, f. 46v, 11 Sept. 1455.

207 Valérian: "Le recours à l'écrit", 68.

208 Fleet, Kate, "Turkish-Latin Diplomatic Relations in the Fourteenth Century: The Case of the Consul", Oriente Moderno 22 (83) 3 (2003), 605–611.

209 Examples are two sworn testimonies by Muslims: ASVe, CI, N, B. 211, Notary N. Turiano, f. 30v, 4 Mar. 1435; and again f. 48r, 27 Apr. 1435: per vocem turcimatam cuidam vocati Acmar saraceni.

on "the scriptures," we are not informed which scriptures these might have been.[210]

Although there was widespread recourse to Venetian notaries by the fifteenth century, could everyone be a witness? Siyāsa thinkers such as Ibn Qayyim, and the Venetian colonial authorities were equally concerned by this issue. Gathering testimony from unbelievers represented a challenge to the accepted norms in Islamic jurisprudence because, either in court or before the notary, witnessing implied becoming an actor in the legal system, not simply a passive subject. Again, a comparative description of how norms were put into practice suggests that common attitudes were adopted in spite of doctrinal differences. Apart from the ḥanafīs, who expressed reserves for intra-communal cases, Islamic jurists denied the right of non-Muslims to guarantee the intentions or claims of others as witnesses. As a result, the rare extant Islamic notarial acts where non-Muslim witnesses are involved always regard other non-Muslims.[211] In contrast with sharīʿa, nothing in Venetian law prevented non-Christians from acting as witnesses to deeds, though in practice this was only the case for contracts where their coreligionists were involved. In Alexandria and Damascus, notaries stuck uniformly to this practice.[212] Similarly, specialists of Genoese history underline that, even in the absence of an explicit prohibition, in practice the Jews of Genoa only acted as witnesses for deeds underwritten by other Jews.[213]

By shifting the focus from theoretical differences to actual practice, my intention is not to pass over the real differences between sharīʿa and Western law. I aim rather to stress that, even in matters where Latins and Muslims relied on different legal doctrines, striking coincidences emerge in the practical ways in

210 ASVe, CI, N, B. 222, Notary A. Vactaciis, f. 80r, 3 Dec. 1404; ASVe, CI, N, B. 211, Notary N. Turiano, f. 58v–59r, 9 Aug. 1435; f. 61r, 19 Aug. 1435; f. 30r, 8 Dec. 1434; ASVe, Notai di Venezia, 14832, Notary I. Dalla Torre, f. 2 (n. 2), mentions a Genoese dragoman, "olim cristiano," 31 May 1412.

211 Ragib, Yusuf: *Actes de vente d'esclaves et d'animaux d'Egypte médiévale*. Vol. 2, Le Caire: Institut Français d'Archéologie Orientale, 2006, 107.

212 Non-Christian witnesses invariably appear in deeds related to other Non-Christians: ASVe, CI, N, B. 22, Notary V. Bonfantin, 17 Jan. 1393; 28 June 1419; ASVe, Notarile Testamenti, B.215, Notary S. Peccator, 2 May 1448; 5 Oct. 1448; 14 Oct. 1448; ASVe, CI, N, B. 211, Notary N. Turiano, f. 6v, 21 May 1455.

213 On the Islamic biases against non-Muslim witnesses, Fattal, *Le statut légal*, 361–4. Emon, Anver M.: *Religious Pluralism and Islamic Law: "Dhimmīs" and Others in the Empire of Law*, Oxford: Oxford University Press, 2012, 136–141, Ragib, *Actes de vente d'esclaves et d'animaux d'Egypte médiévale*, 105–15. For minority witnessing in the Genoese colonies, Argenti, Philip P.: *The religious minorities of Chios: Jews and Roman Catholics*, Cambridge Eng.: Cambridge University Press, 1970, 100–146.

which norms were implemented. By the same token, both legal systems dealt with the judicial oath of minorities in a very similar way. In Venetian legal practice, the taking of judicial oaths by non-Christians was a familiar procedure. In Venetian Crete, a special procedure was established in 1340 to validate pledges pronounced by Jews. As *infideles*, they could not swear by the cross, so members of the Jewish community were subjected to a special oath-taking procedure that required their presence at the synagogue. In the same fashion, when the Inquisition started to gather depositions from Venetian Jews, they were allowed to swear to the truth of their testimony with a Jewish formula, a solution that coincides with that resorted to in sharīʿa.[214] What these apparently surprising coincidences reveal is that Islamic concepts of difference had actually been incorporated into legal practice by the Christian powers; it was the doctrine of the mālikīs, shāfiʿīs and some ḥanbalīs that recommended that Jews and Christians take their oaths in their respective houses of prayer.[215] This late medieval transfer was not specific to Venice, but found its way into the legal codes of other expanding Christian powers, such as Castile, and it is a subject that I will look into in more detail in Chapter Three.[216] Similarly, the biases against minority witnessing were not specifically Islamic, but had precedent in Byzantine legal concepts. The Justinian Code I.V.25 issued in September 531 set the standard for later bans on cross-confessional witnessing. It not only limited the ability for heretics to give testimony in cases concerning Christians, but also anticipated exceptions allowing them to testify for or against each other in inter-communal disputes—the Byzantine equivalent of the so-called 'ḥanafī exception' which, as we shall see, would become a mainstay of Ottoman governance.[217] The

214 Santschi, Elisabeth, "Contribution à l'étude de la communauté juive en Crète vénitienne au XIVe siècle, d' après des sources administratives et juridiques", *Studi Veneziani* XV (1973), 177–211, 207–8, Head, R. C., "Religious Boundaries and the Inquisition in Venice: Trials of Jews and Judaizers, 1548–1580", *The Journal of medieval and Renaissance Studies*, 20 2 (1990), 175–204.

215 Bechor, Guy: *God in the Courtroom: The Transformation of Courtroom Oath and Perjury Between Islamic and Franco-Egyptian Law*, Leiden; Boston: Brill, 2011, 122–127.

216 Jecker, Mélanie: "Jurer selon sa religion. La figure de l'autre dans le droit médiéval castillan", in: *La culture judiciaire: Discours, représentations et usages de la justice du Moyen Age à nos jours*, edited by Lucien Faggion, Christophe Regina, *et al.*, Dijon, Glick, 2014, 241–269, Thomas F.: *Islamic and Christian Spain in the Early Middle Ages*, Leiden Netherlands; Boston: Brill, 2005 reprint, 190.

217 "We ordain that no heretic and those who cherish the Jewish superstition shall give testimony against orthodox (Christians) whether one of them is orthodox or the other. 1. But if heretics or Jews want to litigate among themselves, we permit promiscuous agreement and witnesses worthy of them to be introduced." *The Codex of Justinian: a new annotated translation, with parallel Latin and Greek text based on a translation by Justice*

prejudices against female witnesses are perhaps the best illustration of this continuous borrowing of solutions for fundamental legal questions. Qur'ān 2:282 prescribes the need for two female witnesses to replace one man, which is often counted among one of the major legal biases of sharīʿa, and was generally upheld in legal practice. When the Great Council of Venice reinforced the role of witnesses for final wills and testaments in 1475, two women were required to take the place of one man.[218]

It is generally agreed that fifteenth-century Mamluk governance witnessed the blossoming of merchant nations involved in the spice trade, along with their consular institutions. The interaction of scribal institutions, among a number of other factors, played a significant role in consolidating these cross-confessional dealings, and in the prevention and resolution of conflicts. Significantly enough, in the context of the general reorganization of trade routes in the early Ottoman era, this Mediterranean configuration of notarial activity ceased to exist in both Syria and Egypt. The only similar instance of an active, unbiased notarial workshop existed under the aegis of the Venetian Bailo of Constantinople, to whom a notarial clerk was attached. The Bailo's archives, which have recently been opened for consultation, suggest that similar practices were continued in early modern times, but in the far more restricted context of the Imperial capital.[219] As will become clear in the following chapters, the Ottomans empowered local qadis to act as notaries and to hear mixed cases. Together with Venetian notaries, other legal actors such as the Siyāsa judges ceased to exist in the early modern era, after the advent of Ottoman rule in Syria and Egypt in 1516–7. The ways in which proof was produced in cross-confessional settings saw a subsequent, very significant turn, and in consequence the trove of solutions allowing actors to overcome the biases had to be refashioned.

2.5 New Attitudes towards the Written

The changes brought about by the Ottomans in the field of commercial litigation neither altered the probative status of written documents, nor that of oral witnessing. As regards procedure, descriptions of court hearings suggest that

---

Fred H. Blume, Cambridge: Cambridge University Press, 2016, See also Fattal, *Le statut légal*, 361–4.

218 Bigaglia, *Capitulare*, 28–9.

219 ASVe, Bailo a Costantinopoli, see the recent *Ordinamento ed inventario, Bailo a Costantinopoli*, a cura di Giustiniana Migliardi O'Riordan.

cases continued to be heard in the same way as they had been in previous periods, and that proof continued to be produced by litigants according to precedent.[220] Yet, in spite of normative continuity or, at best, in the context of slow legal change, early modern times witnessed the emergence of new attitudes towards the written. These attitudes were the result of technical adjustments and changes in the organization of justice, rather than due to a general shift in jurisprudence. Although the notarized deed, the ḥujja, retained its traditional legal recognition in sharīʿa, and oral witnessing continued to constitute the proof par excellence, these technical changes brought about the progressive demise of the medieval ʿudūl, more extensive recourse to the qadi's court and its capacity to certify and archive documents, and, lastly, a reorganization of the 'witness system.' These changes had an enormous impact on the field of commercial litigation; the Ottoman 'invention' of the courthouse affected the practical way in which foreigners established contracts and records were certified, bringing about the disappearance of the notarial pluralism that had been built up previously under the Mamluks. Ultimately, these new attitudes towards writing and documentation culminated, to cite one major change, in a prescription against using Muslim witnesses against Franks, who could not be convicted in the absence of notarized contracts. Jurists did not necessarily agree with the validity of such measures, which were at odds with sharīʿa, and the jurisprudence contained in collections of fatwās continued to adhere to traditional approaches to proof and procedure. The following section is an attempt to identify the piecemeal modification of some aspects of the law in the Ottoman era, and in particular the regime of proof that meant that cross-confessional relations looked very different in comparison with medieval times.

2.5.1 *New Attitudes and Archives*

Probably due to researchers' reluctance to admit to the absence of medieval archives, specialists of Ottoman legal practice have often assumed that some kind of archive should have existed for medieval judges, and have refrained from claiming that the Ottomans invented the judicial archive.[221] However, as Reem Meshal, among other authors, has acknowledged, the *maḥkama* brought about crucial changes in the consumption and preservation of records. For instance, it is generally admitted that it was the Ottomans who first provided a stable storage facility for records—i.e., an archive—particularly in large

220 Canbakal, *Society and politics in an Ottoman town: ʿAyntab in the 17th century*, 149.
221 Peirce, *Morality tales: law and gender in the Ottoman court of Aintab*, 98–102.

administrative centers.[222] Notarial personnel were also appointed to register cases at courts in Damascus.[223]

To account for this transformation, Halil İnalcık has highlighted the specific needs of Ottoman judges, derived from the application of administrative, customary law in the Ottoman provinces. The increasing production of state or public law, the kanun, played an important role in the development of archives. İnalcık has pointed out that the need to refer to precedent to inform new cases encouraged Ottoman qadis to archive their decisions, together with copies of the customary law codes—the *kanunnames*—and related decrees, or *kanun hukms*.[224] The kanun encoded the legacy of legal practice, complementing the provisions of the jurists, issued and sponsored by the dynasty, and it was often referred to as 'customary' (Tr. örfi) law. The written nature of kanun is all the more important in that it stood in marked contrast with medieval practice. Early Ottoman governance crystalized around this specific feature, the promulgation of the kanunnames, while, it is generally agreed that such a corpus of public law never existed for the medieval sultanates.[225]

To cite Nicolas Michel's blunt evaluation, "the Mamluks did not have written laws." This rather bold statement stresses that in medieval times there was a tendency to consider legal acts as being bound solely to the sultan that had issued them, and only during his lifetime. Indeed, this notion appears to have impregnated Mamluk governance, and doubtless hampered the development of archives. In recent years, however, an interesting debate has revolved around the Mamluk precedents of kanun. Ottoman rulers alluded in legal texts to the "Kanun of Qāytbāy" (ruled 1468–1496), assuming that citing similar legislation issued by previous respected sultans would endow Ottoman regulations with additional legitimacy. Evidence of these legal texts has not been found, although the Franks also referred to the "old laws of Qaytbay" being enforced as late as 1560.[226] Of course, the Cairo chancery emitted decrees (*marāsīm*)

222 Faroqhi: "Sidjill", Encyclopaedia of Islam, Second Edition, IX:538b.

223 Bakhit, Muhammad A. S., "The Ottoman Province of Damascus in the sixteenth century", Ph.D. Thesis, University of London, School of Oriental and African Studies, 1972, 138.

224 İnalcık, *Suleiman the Lawgiver and Ottoman law*, Aykan, *Rendre la justice à Amid: procedures, acteurs et doctrines dans le contexte ottoman du XVIIIème siècle*, 145, Fitzgerald, *Ottoman methods of conquest*, 222.

225 Burak, Guy, "Between the Ḳānūn of Qāytbāy and Ottoman Yasaq: A Note on the Ottomans' Dynastic Law", *Journal of Islamic Studies* 26 1 (2015), 1–23.

226 ASVe, Senato, Dispacci Consoli, Egitto B1, 23, Mar. 8, 1559, Consul Giambattista Querini complains about an alleged decree by Qaytbāy granting universal protection granted to Frankish traders arriving in Alexandria and upheld until his own time, abused by corsairs, "Che io potro non dando un quatrino al moro del navilio, son per ridurmi in Allexandria dove saro libero di questi periculi come sono li sudiiti del re di Spagna per fino ancho

to implement Mamluk governance, which covered a diverse array of themes; however, they were executed by *ad hoc* deputies, called khaṣṣakīs, through military officials and civil secretaries, and such decrees largely remained outside of the remit of the judicial sphere. All this conspired to hinder the development of a Mamluk qadi archive, while in contrast Suleiman's kanunname exhorted judges to archive and transmit the sijill containing, among others, their decisions in 'örfi cases.' "Thus," İnalcık concludes, "for the history of Ottoman law, every copy, especially the annotated ones at the courts, had the merit of an original."[227] Similarly, another process fundamental to archive-making was the tendency to centralize the attribution of timar revenues; designating timar-holders in Istanbul and not in the provinces meant that a central register developed, with a growing staff of clerks to manage it.[228]

There existed an Ottoman chancery archive, since the Imperial Dīvān has left traces of its activities in two series preserved at the Başbakanlık Osmanlı Arşivi. I believe that this groundbreaking development owed as much to the role of the Imperial Dīvān as a court of justice, as to its activities as a diplomatic forum, a point that I will elaborate below.[229] Descriptions of the role of the Nişancı—the head scribe of the Ottoman court—and, later, that of the Re'īs ül-Küttāb, however, suggest that these officials were more responsible for transmitting drafts to the sultan for approval, than they were for keeping diplomatic decisions and records.[230] Emphasis was laid, not on preserving diplomatic records in a central archive, but on the signing of documents with the tughra, the sultan's monogram, without which they lacked any legal value. On their side, Venetian envoys insisted on the importance not of drafting the final, physical document, but on its authentication by Muslim witnesses.[231] It is in this sense revealing that early in the 16th century a whole industry for the forgery of fake decrees grew up, involving scribal staff, mainly through the reuse of old decrees signed with the official tughra, a technique similar to the recycling

Maltesi et li ladri medesimi che vano in corso sotto habito de mercanti per privilegio di una antigua leze del condam Cait Bej fu soltan del cairo che sicura hogni nacione che vien per mercanzia in Allexandria."

227 İnalcık, *Suleiman the Lawgiver and Ottoman law*, 117.

228 Howard, D., "The Historical Development of the Ottoman Imperial Registry "(Defter-i Hakanî)": Mid-Fifteenth to Mid-Seventeenth Centuries", *Archivum Ottomanicum* 11 (1986), 213.

229 Faroqhi, Suraiya, "The Venetian presence in the Ottoman Empire (1600—1630)", *The journal of European economic history* 15 2 (1986), 345–384, 353.

230 Deny, J.: "Re'īs ül-Küttāb", in: *Encyclopaedia of Islam, Second Edition*, VIII: 481b: Brill.

231 Işiksel, Güneş: "Hierarchy and Friendship: Ottoman Practices of Diplomatic Culture and Communication (1290s-1600)", *The Medieval History Journal*, 2019: 1–20, 5.

of 'black market' clay seals mentioned by Ibn Iyās (cf. 2.2.3) or to the rewriting of discarded deeds described by al-Jawbarī.[232] Again, assuming that the preservation of all official records was a necessity for all literate societies may be misleading inasmuch as, for early modern Muslims, preserving documents ran the risk of their being used out of their original context.

It would be inaccurate to assume, however, that new attitudes towards writing and documentation were solely limited to or defined by the practice of registering deeds, and even less an alleged 'triumph of the written' in sharīʿa courts. Researchers have revealed many elements of continuity with the medieval past in terms of judicial practice. According to 16th-century ḥanafīs such as Ibn Nujaym (1520–67), judges were expected to grant all subjects access to their sijills and to deliver copies to the right holders.[233] If, out of court, this meant that actual rights were now recognized for many subjects of the legal system, this was not in antagonism with the traditional ceremony of justice. The weight of oral and written proof in legal procedure has been thoroughly studied by Boğaç Ergene, who, by examining the qadi courts of small communities, has concluded that Ottoman litigants and judges tended to respect the preeminence of oral witnessing.[234] Written documents were advanced as proof, but litigants had almost no chance of success if these contracts could not be backed by oral witnessing. Although depositions by witnesses were now included in the sicil—therefore giving historians access to these voices—we should keep in mind that these depositions and the witnessing that accompanied it were transcribed very summarily in court proceedings. Even when transcripts were established, they continued in practice to serve simply as aide memoires for actual witnessing.[235]

2.5.2 *Legal Reform and the Written*

Alongside the continuity of judicial practice between the Mamluk and Ottoman eras, current research has also identified elements of change: although in principle the oral nature of proof was not fundamentally compromised, the

232 Fodor, "How to forge documents? (A Case of Corruption within the Ottoman Bureaucracy around 1590)".

233 Ibn Nujaym, *al-Ashbāh wa-al-naẓāʾir ʿalá madhhab Abī Ḥanīfah al-Nuʿmān*, 293.

234 Ergene, "Evidence in Ottoman Courts: Oral and Written Documentation in Early-Modern Courts of Islamic Law", Wilkins: "Witnesses and Testimony in the Courts of Seventeenth-Century Ottoman Aleppo", Cohen, Amnon: *Jewish life under Islam: Jerusalem in the sixteenth century*, Cambridge, Mass.: Harvard University Press, 1984, 110–140.

235 Masters, Bruce Alan: *Christians and Jews in the Ottoman Arab world: the roots of sectarianism*, New York: Cambridge University Press, 2001, 31–33, Imber, *Ebu's-suʿud: the Islamic legal tradition*, 53.

Ottomans' changing attitude towards proof was not without consequences. From the beginning, Ottoman reform concentrated the power to notarize and register most contracts in the hands of Ottoman, ḥanafī judges, to the discontent of the local judiciary. From Syria to Mecca, in the former Mamluk territories this usually meant a foreign, Turkish magistrate, at best associated with the Ottoman regime and legal traditions (referred to as *yasaq*) and in any case belonging to a *madhhab* different from the shāfiʿī, this latter having traditionally been dominant under the Mamluks. Ibn Fahd describes the fraught case of a pious foundation endowed by a merchant named Khawājā al-Ḥamawī in 1519. The institution of a waqf usually meant that the heirs of a founder had to kiss goodbye to a juicy estate, and consequently attempts at fraud by the latter were a temptation.[236] The merchant's daughter tried to circumvent the testament with the assistance of a corrupt Mālikī judge, who pressured a local notary to steal the receipt (*mustanad*) associated with it. Both parties tried to build their case on the basis of notarized testimonies, however because depositions had to be registered (tasjīl) by the ḥanafī judge and included in his sijill, he had the final word in most quarrels. The notary was investigated and questioned, and finally admitted to his role in the disappearance of the receipt. A dispute eventually arose between judges and witnesses from different madhhabs, and during discussions an Ottoman official, the shāhbandar of Jedda, stated that the truth was that the notary had lied and had to be dismissed, and that "this was the law in his own time, and so it should be under the Ottoman yasaq" (*al-yasaq al-ʿuthmānī*).[237]

As portrayed by Ibn Fahd, early Ottoman notaries often appear involved in similar quarrels revolving around the estates of the deceased. Rather than notarizing private contracts in their stalls, they are mentioned as legalizing through their testimonies the legal acts associated to inheritances. In a much embroiled case, two factions in the judiciary litigated over the estate of a deceased qadi named Jamāl al-Dīn al-Murshidī. Notaries suffered pressures from both parties in order to modify their statements. They were instrumental in having al-Murshidī's initial will legally disqualified, for which the other spent

---

236 Gökbilgin, Tayyib: "La preuve et le témoignage dans la jurisprudence des Fetva d'Ebussud et quelques exemples d'application aux tribunaux ottomans du XVI siècle", in: *Recueils de la Société Jean Bodin, La Preuve, 3, Civilisations archaïques, asiatiques et islamiques*, Bruxelles, Editions de la Librairie encyclopédique, 1964, 205–9, Gökbilgin mentions a fatwā related to the endowment of a house as waqf that the heirs, presumably, managed to frustrate by hampering its notarial registration.

237 Ibn Fahd, *Nayl al-Muná bi-Dhayl Bulūgh al-Qurá*, 184–5. For the Yasaq; Bakhit, "The Ottoman Province of Damascus in the sixteenth century" 139.

money "on the witnesses and the judge". On one occasion, one of the notaries was framed when asked details about the time when he supposedly gave testimony; in another the customers denied having ever met the notary. The episode ended up with the clerk punished in front of the qadi's house, paraded through the sūqs with the head uncovered, while criers announced "this is a crooked notary who had brought discord among the judges".[238]

It has been noted that under the Ottomans, notary-witnesses were integrated into the *maḥkama*, and their status reduced to that of clerks and scribes, serving as instrumental witnesses and tasked with the collection of fees. Judges charged a fee for drawing up ḥujjas, but also for registering them in the sicils, as well as for delivering copies. The collection of fees on marriage contracts, for example, has been identified as one of the breaches with Egyptian judiciary and, more broadly, with the local population. Muhammad Bakhit remarks that, if under the Mamluks the ʿudūl made a living out of the notarization of private transactions, and judges received a state salary, the Ottoman qadis' income was dependent upon judicial fees.[239] Ottoman qadis limited other scribes' access to the notarial offices as a means to secure the collection of levies, often referred to as *yasaq*—a derogatory term that loosely meant, in a post-conquest context, 'Ottoman custom.' And similarly, in Syria too the Ottomans adopted a policy of limiting the number of active notaries and controlling their activities. Thus in January 1517, in the aftermath of the Ottoman conquest, a 'Rūmī' judge, Zayn al-ʿĀbidīn Ibn al-Fanarī, decreed that each chief qadi should be limited to two notary-witnesses, and forbade the remaining freelance notaries from drawing up deeds—hence, according to Ibn Ṭūlūn, "severely harming the shuhūd of the country." This decision meant that all notarial activities became concentrated in the Jawziyya Madrasa, now known as the 'Hall of Law,' and that fees had to be collected for each receipt (mustanad). In particular, al-Fanarī targeted the ʿudūl practice of drafting marriage contracts, now subject to judicial permission upon the collection of a Ottoman, "yasaq" tax.[240]

The way in which "Selim removed the notaries from their stalls" and subsequently replaced them with contending powers in their own workshops

238 Ibn Fahd, *Nayl al-Muná bi-Dhayl Bulūgh al-Qurá*, 308–10.

239 Bakhit, "The Ottoman Province of Damascus in the sixteenth century", 137–8.

240 Ibn Ṭūlūn, Shams al-Dīn Muḥammad Ibn ʿAlī (1485?-1546): *Iʿlām al-Wará bi-Man Wulliya Nāʾiban min al-Atrāk bi-Dimashq al-Shām al-Kubrá*, edited by Muḥammad Aḥmad Dahmān, Damascus: Wizārat al-Thaqāfah wa-al-Irshād al-Qawmī, 1964, 316–17, The same episode is addressed in his chronicle *Mufākahat al-Khillān*, Ibn Ṭūlūn, *Mufākahat al-Khillān*, 338, 346–7.

is in itself significant; it is symbolic of a very real struggle over writing and notarization, and the different meanings attached to them by Islamic governance. In February 1518, Governor Jānbirdī al-Ghazālī, at odds with the aggressive centralist policies seeking to curb the local judiciary, reinstated the system "as it was under the Circassian Mamluks," granting the ʿudūl liberty to draw up deeds wherever they chose, and setting exemptions from notarization for certain types of legal acts.[241] The intrigues of the judiciary fascinated the Syrian chroniclers of the day; although their interpretations of these facts differ slightly, according to al-Ghazzī, Governor Jānbirdī clashed with the local ḥanafī qadi Walī al-Dīn al-Farfūr. The leading clan of the Mamluk judiciary under the reigns of Qāytbāy (1468–1496) and Qānṣūh al-Ghawrī (1501–1516), The Farfūr family had long taken advantage of their responsibility to handle nominations to judicial charges, largely to their own benefit. As the overlord of the Syrian administration of justice, Walī al-Dīn al-Farfūr had succeed in becoming integrated into the Ottoman system as chief qadi, and by converting to the now-dominant ḥanafī rite. Ottoman reforms "pointed towards the adoption of yasaq," and al-Farfūr himself "persisted in the kanun of the Turks." Governor Jānbirdī al-Ghazālī, who sought to present himself as defender of the traditional order, forced al-Farfūr into exile. Between September 1520 and February 1521, the province of Damascus witnessed Jānbirdī's short-lived revolt against his Ottoman overlords, a major episode in the history of the Ottoman conquest, and in which the new attitudes towards writing and documentation had their part to play.[242]

Ottoman reforms aimed to keep three vital functions in the hands of the ḥanafī, Ottoman judges. In newly-conquered Egypt, judicial policies targeted the local, non-ḥanafī judges' right to pass verdict on important matters, their

241 Ibn Ṭūlūn, *Mufākahat al-Khillān,* 348, 386–7.

242 al-Ghazzī, Najm al-Dīn Muḥammad Ibn Muḥammad (1570–1651): *al-Kawākib al-sā'irah bi-aʿyān al-mi'ah al-ʿāshirah,* 3 vol, Beirut: Dār al-Kutub al-ʿIlmīyah, 1997, I, 170–1. Ayalon, David, "The End of the Mamluk Sultanate (Why did the Ottomans spare the Mamluks of Egypt and wipe out the mamluks of Syria?)", *Studia Islamica* 65 (1987), 125–148, 136. Alsabagh, "Before Banks: Credit, Society, and Law in Sixteenth-Century Palestine and Syria", 55–63, Meshal, *Sharia and the Making of the Modern Egyptian,* 87–9. Alsabagh's reading of Ibn Ṭūlūn stresses a continuity with the previous regime in al-Farfūr's behavior, while Meshal presents him as a victim of Ottoman changes in the judiciary. In general, on Ibn Farfūr and the Syrian judiciary, see Mandaville, *The Muslim Judiciary Of Damascus In The Late Mamluk Period, Princeton University, Ph.D.,* 12–66, Martel-Thoumian, *Les civils et l'administration dans l'État militaire mamluk* (*IXe/XVe siècle*), Martel-Thoumian, Bernadette: "Al-Diyā' al-Mawfūr fi Aʿyān Banī Farfūr: un exemple de littérature prosopographique", in: *Actas XVI Congreso UEAI,* Edited by C. Vázquez and Miguel Ángel Rodríguez, 323–335, Salamanca: Agencia Espanola de Cooperación Internacional, 1995.

right to engage in independent legal reasoning, and their grasp of witnessing and notarizing. Although the judges in Arab provinces were less affected, these reforms were bad news for the medieval ʿudūl, whose weight in courts and administrative offices was curtailed by the Ottoman reorganization. Part of the struggle for the judiciary was played out in the field of the production of proof, and particularly, written proof, at a time when a general increase in the use of writing and documentation has been noted. Insofar as concerns qadi local courts, Yavuz Aykan has argued that the recourse to written documents cannot be reduced to the issue of validity in court, since the qadi provided notarial services for a wider range of social relations and transactions. The research of Leslie Peirce, Gilles Veinstein or Nicolas Michel, as well as that of Reem Meshal for Egypt, or Rosistza Gradeva for the Balkans, all concur that the recourse to notarization, now provided by the judge, became generalized in early modern Middle Eastern communities, and people began making unprecedented use of documents, although not necessarily as evidence to be used in a trial. There is therefore no contradiction between the Ottomans' attempt to undercut the influence of the notary-witness, and the parallel tendency towards writing and documentation becoming a more present feature in society. Ottoman policies sponsoring notarization extended to foreigners and minorities, since they could not testify against Muslims in court and would have benefited from a greater reliance on notarized deeds compliant with sharīʿa.[243]

In some respects, Muslim women experienced similar limitations to minorities; they are frequently mentioned as benefiting from the new role of documents in society, together with other disadvantaged groups, such as former slaves.[244] This situation was in stark contrast with women's access to the notarial services under the Mamluks; indeed according to Arab chroniclers, trustworthy notaries took great care to avoid contact with female and, sometimes, minority clients. A witness to the Mamluk-Ottoman transition, Ibn Ṭūlūn was much preoccupied with the notaries' misbehavior, and addressed repeated accusations against some notarial workshops, such as those in the Bāb al-Ṣaghīr district in Damascus. As a result, some of the notaries of this district ended up in prison, and the *markaz* was involved in the embezzlement of estates

243 Faroqhi, "Before 1600: Ottoman attitudes towards merchants from Latin Christendom", 77, Faroqhi: Sidjill, Encyclopaedia of Islam, Second Edition, IX:538b.

244 Meshal, *Sharia and the Making of the Modern Egyptian*, 134–9, Peirce, *Morality tales: law and gender in the Ottoman court of Aintab*, 371–4, 388–9, Aykan, *Rendre la justice à Amid: procedures, acteurs et doctrines dans le contexte ottoman du XVIIIème siècle*, 130–5, Alsabagh, "Before Banks: Credit, Society, and Law in Sixteenth-Century Palestine and Syria", 318.

endowed as waqf. The setting of most *marākiz* near Damascus' city gates, close to the suburbs where uprooted immigrants from the countryside settled, did not certainly get the notary's stall much respectability. Other workshops were located in the financial districts, where foreigners resided, such as one in the Khān al-Sulṭān, an inn hosting Frankish merchants and other travellers.[245] Ibn Ṭūlūn recounts that in 1491 the governor of Damascus had the notaries removed from the Omayyad Mosque and transferred to one of the city gates due to their poor reputation. As proof of this he claims that the ʿudūl "were said" to have had relations with dhimmīs and women. Denoting a similar attitude, a notary of Aleppo conspicuously refused to notarize deeds for female clients, which Ibn al-Ḥanbalī (1502–1563) judged to be an "excess of piety."[246] We might conjecture then that female or minority clients, considered to be socially weak, were more likely to be exposed to potential abuse by corrupt ʿudūl, and that consequently their presence at the notaries' stall could be regarded as suspicious, or at best undesirable, and in any case unsuitable for the Omayyad Mosque.

New attitudes towards writing and notarization encompassed the disciplining of the notarial practice, and its confinement to the physical space of the courthouse, hence making it more available to all subjects of the legal system. These individuals now came to the court to register, in the form of ḥujjas, their family and marital status, or to attest to their being freed from slavery. The criticisms to medieval notaries by Arab authors suggest that right holders needed to keep alive their claims, in the same manner documents were kept valid through tasjīl procedure. For the Ottoman period, Reem Meshal has found evidence of the recipients' capacity to request copies of ḥujjas confirming their rights, whose originals were kept in official archives, and these documents could be exhibited in other places in the Empire without recourse to oral validation by witnesses. Irrespective of the actual scale of this phenomenon, it suggests a major break with medieval attitudes towards the production and archiving of proof.

2.5.3 Judicializing *the Written, Writing* Judicially*, and Handling Orality*

New approaches to the legal validity of the written were not only limited to administrative practice, but also extended to the field of jurisprudence. In his

245 Mandaville, *The Muslim Judiciary Of Damascus In The Late Mamluk Period*, 121–2, Apellániz, Francisco, "The Funduqs of Damascus seen by Frankish Notaries and Merchants" in: Travel in the Middle Ages, edited by M. Sureda, Barcelona, IEMed, 2015, 65–71, 266–270.

246 Ibn Ṭūlūn, *Mufākahat al-Khillān*, 114, Rabiʿ II 896H, Ibn al-Ḥanbalī, *Durr al-habab fi tarikh aʾyan Halab*, II, 633.

highly innovative work, Guy Burak has demonstrated that some seventeenth-century jurists sought to justify recourse to the Imperial registers as legal proof, even when it was uncorroborated and did not come accompanied with further probative support.[247] As discussed earlier, a ḥanafī tradition already existed, which could consider official documents as proof by virtue of their authoritativeness in some exceptional cases, and this practice followed by Mamluk judges is documented in the Ḥaram records.[248] Moreover, Guy Burak has recently observed that some forms of literary knowledge, such as chronicles, could be certified and used as legal proof in order to validate facts of daily life (which, incidentally, indirectly supports Chamberlains' claims as to the social function of literary works, such as the biographical dictionary). Burak has called this process *defterization*—the tendency to equate trustworthy information with government-produced records, hence suggesting an Ottoman propensity to "judicialize the written."[249] Together with *defterization*, this research attempts to highlight the parallel—though different—Islamic tendency to "write judicially"; that is, the practice of drafting non-judicial documents in a legally valid form, properly notarized by witnesses. Such artifacts were perceived to be legal valid acts reflecting a truth supported by living, male Muslims, rather than merely recordings of administrative facts.

The oral legal culture of Islam pervaded many aspects of cross-confessional relations, such as diplomatic exchanges, and was indeed a constant source of misunderstanding. Most diplomatic issues were dealt with at the Imperial Council, the Dīvān-ı Hümāyūn, which also functioned as a supreme court of justice. The Venetian Bailo and his delegates attended the Dīvān meetings daily; in these sessions, they met the Ottoman dignitaries, the pashas, who acted as representatives of the sultan and would hear and refer to the ruler. As the fulcrum of Ottoman decision-making and governance, the Dīvān hardly suits the distinction between 'political' and purely legal areas of activity, since all decisions were subjected to legal validation—like the taking of testimony or sanctioning by the chief judges. Historians of cross-confessional relations often neglect that it was not only judicial issues that were treated as judicial acts, but also other administrative and diplomatic decisions, and therefore, to be valid, all needed to be "sworn" by witnesses. While government-produced and other authoritative documents were given probative value—most probably the result of an old ḥanafī central-Asian

247 Burak, "Evidentiary truth claims".
248 Ayoub, Samy, ""The Sulṭān Says": State Authority in the Late Ḥanafī Tradition", Islamic Law and Society 23 3 (2016), 239–278, Johansen, "Formes de langage", 375.
249 Burak, "Evidentiary truth claims", 3.

tradition—conversely many ordinary documents—or at least they were so for the Franks- were for the Muslims mere artifacts that evoked legally binding, oral agreements.

While it is tempting to interpret the new attitudes towards writing and documentation as a sign of modernity, research on Ottoman judicial sources underlines that, if not in practice, traditional notions and ideas about truth and notarization prevailed. As sanctioned by sharīʿa, the oral nature of truth needed to be taken into account and, particularly during diplomatic and judicial negotiations, Venetian representatives and their Ottoman counterparts invested a great deal of energy into this task. To cite an obvious example, even after they had been written down according to the agreement reached between the two parties, the ahdnames were only considered valid if they were certified by Muslim witnesses in a different time and place. As the Venetians often mention, these documents needed 'to be sworn'.[250] During the last stages of the Veneto-Ottoman negotiations in 1502, the notions and practices involved in the drafting of the decisions provoked a great deal of misunderstanding. For the Ottoman officials, the final document was considered simply to be an aide-mémoire of previous diplomatic talks, rather than as a binding contract. While Venetians stuck to the letter of the document, the final signature was protracted time and again, on the basis that it did not reflect all and every single element agreed upon orally during the negotiation process. Indeed, before proceeding to the final draft, the pashas felt the need to interrogate the dragomans as to whether or not they remembered certain specific conversations that might have arisen during negotiations, on the Venetians' right to use the harbor of Alessio (Lezhë) in Albania. This point was brought up several times at the Dīvān, meaning that omitting parts of a fulfilled agreement was considered to be disrespectful of the truth.[251] When they were discussed at the Dīvān, the Capitulations needed first to be 'accepted' by the sultan, then 'solemnly sworn,' and finally submitted to the procedure of the 'golden seal'- the tughra or sultan's monogram. The long impasse incurred by this final drafting opened new opportunities to twist the Venetians' arms; arguing that the Capitulations were ready, the Ottomans allegedly demanded additional concessions regarding

250 Venice, Biblioteca Nazionale Marciana (hereinafter BM), Cod. It. VII, 878, 8652, Dispacci Gritti (1502), f.16r., "i capitoli siano sta giurati".

251 BM, Cod. It. VII, 878, 8652, Dispacci Gritti (1502), f. 32r, 37r-v. "un giorno per loro vedermi dicto una cosa, et per el signor de sua bocha propia quella confirmada, et laltro I suo bassa Josarla cum nove pensate, che son cosse da fare I omini vacillare."

the possession of Cephalonia, under the implicit threat of delaying the final issuance of the treaty.[252]

During the entire negotiation process for the Capitulations, the idea was upheld that it was the sultan who was the ultimate author for the legal act, together with that of guaranteeing oral transmission between himself and the acting negotiators: the *capitoli* could only be confirmed as valid if written by the sultan's hand or bearing his personal seal (*over sigillata cum el annello*).[253] Until the final draft had been solemnly sworn, points could be added, but not after the witnesses had taken the final oath on the document's contents. Even treaties of the utmost importance were, at least in a way, considered to be merely the trace of a verbal agreement concluded with the head of the protected community, memory of which ultimately lived in the heart of trustworthy Muslims. It is generally agreed that the ahdnames were presented as unilateral acts by which the sovereign bestowed concessions upon foreigners, but it also needs to be stressed that the final diplomatic documents were shaped in a totally Muslim context, and were subjected to the sharīʿa regime of proof. It is perhaps for this reason that the pashas complained of the way Venetians brandished papers under their noses. Over the course of a highly sensitive debt case—the *Abdellatif affair*—a consul in Alexandria was imprisoned for someone else's debts, in violation of the ahdname. However, the letter of the treaty clashed here with other legitimate proof advanced by the Muslim plaintiff, Abdellatif, who furnished both contracts and witnesses (allegedly forged) attesting that the consul had personally agreed to repay these debts. "Questi toi venetiani spazzano sempre le cose de nostri con un pezzo di carta" ("these Venetians of yours bump off every issue on our merchants with a piece of paper") the pasha argued to the Bailo, and later complained that the Franks were "always putting these clauses under my eyes" (*mi mettete negli occhi questi nostri capitoli sempre*).[254]

The pashas, moreover, seemed to be conscious of the perils of oral agreements and—quite frustratingly for the Bailo—avoided giving their opinion during negotiations. In the Imperial Council, where a vast array of dealings were discussed on a daily basis, the pashas presented themselves publicly as mere intermediaries of the sultan, often refraining from expressing their

252 BM, Cod. It. VII, 878, 8652, Dispacci Gritti (1502), f. 26v, referring to the Ahdname draft, "I capitoli […] esser acceptadi et solemniter jurati, et de quelli presentada la copia con el sigillo d'oro".

253 BM, Cod. It. VII, 878, 8652, Dispacci Gritti (1502), f.30r.

254 ASVe, Senato, Dispacci degli ambasciatori e residenti, Costantinopoli, 1A, Reg. 1552–57, f.365, 374–6.

personal views on the issues discussed. A Bailo complained that "the pashas did not properly answer any of my questions, as they are not used to actually stating their arguments." Here, he is referring to the pashas' tendency, at the Dīvān, to limit themselves to a theoretical intermediary role, hence refusing to elaborate "for their duty is to hear and refer, but they said they would make everything known to His Excellency the Sultan, together with the other requests."[255]

Although the Ottoman certification system signaled a giant step towards the generalization of written artifacts as legal evidence, it arose within a context still largely dominated by orality, and more generally by sharīʿa-based restrictions. If sixteenth-century Ottoman judges and officials preserved more legal documents than their medieval predecessors, early modern Italians did not hide their astonishment at the continued importance of the oral in qadi courts in the Ottoman Empire. In particular, the Venetians remained puzzled by the fact that, during the discussions, no minutes were jotted down by the Dīvān. In principle, when acting as a court of justice the Imperial Council functioned according to the same procedural rules as those in a sharīʿa court, and it was not easy for foreigners to discern between the discussion of a given commercial issue that was part of a diplomatic, bilateral agenda, and its hearing as a legal case involving Franks. This is why, in Ottoman times, the prevalence of orality in the production of proof provoked both cultural misunderstandings and practical inconveniences. One of the most striking examples of the divergent practices between these two notarial cultures is the lack of transcription of what was discussed during the Dīvān's sessions, something that was common practice for Venice's council meetings. At the Dīvān, most controversial issues were protracted over several days, until a final decision was reached. These decisions were notarized by the Nişancı, present at the hearings, and constitute today the bulk of the mühimme registers, beginning in the 1550s. But, again, only verdicts and final decisions were notarized in the daftars or issued as fermans. Court proceedings were not kept *in extenso*, thus making it impossible to recreate the legal arguments and doctrines that had been mobilized in order to reach these decisions.

The contradictions between the Latin and Ottoman systems crystallized in a much-disputed hearing at the Dīvān reported by Bailo Antonio Barbarigo in September 1552. Lengthy discussions had revolved around a series of debts,

255 "non me resposeno ordinatamente ad cosse alcuna, per non essere loro costume formar parole: et poichè etiam l'officio loro era de udir et referir: ma dixeno che fariano entender a la exma. del signor el tuto insieme con le altre dimande, si del bailo come da I altro marcadanti.", BM, Cod. It. VII, 878, 8652, Dispacci Gritti (1502), f. 22r.

incurred by both Ottoman Muslims and a Venetian merchant who had fallen bankrupt. To counter the Bailo's arguments, the pasha brought up an old case concerning a Christian pirate released upon the intervention of a former ambassador, claiming the Venetians should now return the favor by cancelling the debts owed by the Muslim merchants. The pirate had been freed upon a legal decision made by a Muslim judge, and the Bailo claimed that the former must have been proven innocent or have reached an agreement with the plaintiffs, and suggested that the pasha seek out the qadi's decision (the *sigilletto*) in the archives. The pasha then explained that the trial had not taken place at a local qadi court but had, instead, been transferred to the Imperial Council. It was known, the pasha went on, that in the Dīvān "they do not make notes" and therefore "my word should be given more credit than a qadi's sijill", and *I say to you the pirate was found guilty* and released only as a concession to the former Bailo.[256] The pasha clearly highlights here the tension between the oral performance of trusted men, and the value given to notarized deeds. However, he also acknowledges the weight given to the authoritativeness of Muslims and to oral agreements. Indeed, the pirate had been released on the Bailo's word (*sopra la parola del Bailo*), since—as I will argue in Chapter Four—particular weight was given to it as coming from the head of a protected community. When the Bailo asked to be given a memorandum—"in order to know what to write"- the pasha answered that he should speak personally with the former Bailo and his dragoman, a man called Zanesino. The former Bailo was dead, and due to his age and the time that had passed since the events, his dragoman did not remember; however the pasha insisted that, in order to shed light on the matter, the information should come from this source, in spite of the dragoman's old age. Indeed, the dragomans' capacity to correctly remember facts was a requirement for their appointment, and not only due to the need to memorize vocabulary. Dragomans were required to be present in court for mixed trials, and descriptions of these trials strongly suggest that they took an active role in their clients' defense. This was the case in similar cross-confessional contexts;[257] in a description of the old Dragoman Franco de Negro the Bailo noted

256 "in divano.. dove che non si fa nota … la mia parola è creduta piu che sigilletto, et io vi dico che fu provato addosso di lui ogni cosa, le risposi che non contradicevo alla parola della magnificenzia sua, ma che sendo stà rilassato attrovandosi gli adversarii presenti bisognava credere che fusse stà rilassato di volontà loro o per l'innocentia sua o per qualche accordo seguito tra essi.", ASVe, Senato, Dispacci degli ambasciatori e residenti, Costantinopoli, 1A, Reg. 1552–57, f.432.

257 Zecevic, Selma, "Translating Ottoman Justice: Ragusan Dragomans As Interpreters of Ottoman Law", Islamic Law and Society 21 4 (2014), 388–418.

that he had become old and was losing his "very necessary memory," so that the merchants preferred not to be represented by him before the qadi.[258]

In the extensive literature on the—mainly commercial—exchanges with the Ottomans, one comes across complaints time and again about the disregard for the clauses in the ahdnames. Uncertainty often abounded about the duration of the ahdnames, or the very meaning of the word 'truce.'[259] As we have seen, clarifications on previous discussions that might have had an influence on the drafting of clauses were often required, making it necessary to interview former participants. On their own side, historians often complain about the mysterious absence of originals, while copies circulated freely, and often translations can be found in diplomatic European collections.[260] As we shall see, crucial clauses were sometimes absent from the letters of the treaties, such as the privilege for Franks to be convicted only on the basis of notarized documents. My contention is that ahdnames belonged to the realm of the judicially written, and were therefore governed by the logic of memory and orality (and this was strictly connected, incidentally, to the necessity for them to be renewed by every new sultan). Yet as the discussion between Antonio Barbarigo and the pasha makes clear, in spite of divergent attitudes towards the written document, even at the epicenter of cross-confessional justice—the Imperial Dīvān—archiving practices were of little significance. The very act of exchanging correspondence with the metropolis was regarded with suspicion: during the talks of 1502 the ambassador's correspondence was often intercepted, and he was sometimes prevented from sending letters to the Doge. This happened because, ten years before, Bayezid II had felt offended by the Bailo's choice to dispatch encrypted messages to Venice.[261] From the point of view of sharīʿa, diplomacy fitted into the laws of obligation, and therefore followed the logic of orality and trustworthiness governing legal relations. In such an ideal framework of majority-minority contacts, there was no room for official secrets.

Some participants in the debate on archives have argued that it is only by moving away from the state as the principal actor, that a coherent picture of the Mamluk archival mind emerges. In a similar fashion, this research suggests

---

258 ASVe, Senato, Dispacci degli ambasciatori e residenti, Costantinopoli, 1A, f. 200v.

259 "gli avisi al magnifico bassa letti che sua magnifitentia gli hebbe, per che questi nel tradurgli in turco dicono la parola tregua sotto nome di pace conditionata le feci da Zanesino per ogni buon rispetto dichiarire a bocca che detta pace conditionata se intendeva la suspensione d'arme delli 40 giorni che gli haveno gia' l'altra fiale detto.", ASVe, Senato, Dispacci degli ambasciatori e residenti, Costantinopoli, 1A, f. 377–384.

260 Steensgaard, Niels, "Consuls and nations in the Levant from 1570 to 1650", *Scandinavian Economic History Review* 15, 1–2 (1967), 13–55, 15.

261 BM, Cod. It. VII, 878, 8652, Dispacci Gritti (1502), f. 34, 79.

that the traditional distinctions between judiciary and state, or administrative archives are equally problematic. The Mamluk archive appears, at best, to be a very contingent construct: when needed, records and referencing were left in the hands of the men who produced them, and this was never understood as a challenge to Islamic administrative sophistication. Again, I argue that it is the conscious act of preservation, rather than the actors who preserved them, that deserves more attention in future research. As for actual archiving, the Mamluk-Ottoman transition is an important crossing-point that has thus far received remarkably little attention; if there is little to gain in flattening out the differences between the two, the transition from medieval sultanates to Ottoman governance provides us with a fruitful token of comparison for many areas of Islamic history. Notions of governance, such as the using of the Caliphal title, the emergence of a ḥanafī legal guild, and the ascendency of kanun law are just a few well-known traits of Ottoman rule that made it look different from its medieval predecessors. Similarly, more historical work is needed to scrutinize early medieval Muslims' attitudes towards documents. Allusions by Ibn al-Ṣayrafī, head of the Fatimid Chancery for a forty-year timespan, and by al-Qalqashandī suggest to us that, for the pre-Mamluk periods ranging from early Islamic times to the Abbasid and Fatimid caliphates, a positive attitude towards archiving may have surfaced.[262] Whenever Caliphate archives actually existed, they succumbed in late medieval times to the sharīʿa-based regime of proof endorsed by the clerical, legally learned, for which the literary and intellectual spheres of these societies were tailored. This explains why descriptions of notarial practices in late medieval Granada and Jerusalem look surprisingly similar, despite being based on the opposite shores of the Mediterranean.

On the opposite end of the chronology under study, Ottoman governance imposed administrative practices that led to the development of archival devices such as the register, and a long-standing legal reform pushed towards the preservation of judicial collections. There was a logic underpinning the preservation of administrative law; it allowed judges to draw upon a trove of customary rules that could serve as precedent. On the other hand, Ottoman chancery practice departed from precedent and generated its own archive. This contrasts with the tendency of the Mamluks to not preserve copies of the *murabbaʿāt*, or minor decrees, which constituted the most common tool of Cairo's governance.[263] This was particularly true as concerns diplomatic dealings, and more generally relations with the Franks, for which such decrees

262 Sijpesteijn, *The Archival Mind In Early Islamic Egypt: Two Arabic Papyri*, 164.

263 Popper, *Egypt and Syria under the Circassian sultans, 1382–1468 AD; systematic notes to Ibn Taghrî Birdî's chronicles of Egypt*, part II, 24., Hirschler, "From Archive to Archival

were constantly being issued.[264] In sum, rethinking the debate on the archival divide and highlighting the different needs, attitudes and logics across time and space makes it possible to draw relevant conclusions on governance and the management of cross-confessional relations. For this reason, in the following chapter I concentrate on excavating a deeper layer in these relations: the realm of doctrines, norms and competent courts. I will be addressing, from a medieval, pre-Ottoman viewpoint, the ways in which cases were adjudicated, and to which judges and courts they were assigned, but also what treaties and normative texts advised for the conclusion and notarization of business deals. More importantly, I look to Venetian notarial evidence for the activity of the Siyāsa courts, run by Mamluk officials, with their characteristic approach to who and what could provide trustworthy testimony.

Practices", 14, notes that, again, examples have survived in the collections of the recipient, such as the Ḥaram papers.

264 They were referred to in the Venetian sources as *commandamenti maraba*: Pedani, Maria, "Gli ultimi accordi tra i sultani mamelucchi d'Egitto e la repubblica di Venezia", *Quaderni di Studi Arabi* 12 (1994), 49–64, 52, Horii, Yutaka, "The Venetian Consul and Residents in Egypt under the Ottoman Conquest", *Quaderni di Studi Arabi* 15 (1997), 121–132, 123, *Ambasciata straordinaria al sultano d'Egitto (1489–1490)*, edited by Franco Rossi, Venice: Comitato per la pubblicazione delle fonti relative alla storia di Venezia, 1988, is a hefty corpus of documentation produced during a diplomatic mission to Egypt, where the term is frequently mentioned. See, among others, 218, 221, 232.

CHAPTER 3

'*Men Like the Franks*': Dealing with Diversity in Medieval Norms and Courts

> Car encores seient il Suriens et Grifons ou Judes ou Samaritans ou Nestourins ou Sarasins, si sont il auci homes come les Frans de paier et de rendre ce que iuge sera, tout auci come est etabli en la cort des borgeis
>
> Livre des Assises de la Cour des Bourgeois, Assise CCXXXVI

∴

Until recently, an enquiry into cross-confessional exchanges would have naturally started with the amān charts and similar documents, the diplomatic formulation of the so-called 'treaties of commerce.' And indeed, this is despite the fact that historians of Islam have long since pointed out the problematic nature of such artifacts, contesting the idea that amān treaties and similar artifacts were real, bilateral agreements and even questioning the very idea of Islamic diplomacy. Recent research insists on the heterogeneous nature of these documents, which grouped together a diverse array of institutions, such as the truce and the safe-conduct, and also highlights the perils of understanding medieval treaties in light of the later, uneven Ottoman 'capitulations.'[1] One is struck by the importance attached to such diplomatic artifacts by Western historiography, an interest that does not seem to have declined in recent years. One argument that is often cited is that these texts depart from the basic postulates of legal theory, such as the imposition of taxes considered illegal by sharīʿa, the reliance on written documents as proof, or the unbeliever's status in Islamic lands and the duration of his stay.

1 Theunissen, Hans: *Ottoman-Venetian Diplomatics: The 'Ahd-Names: the historical background and the development of a category of political-commercial instruments together with an annotated edition of a corpus of relevant documents*. Vol. I, 1998, Frantz-Murphy, "Identity and Security in the Mediterranean World ca. AD 640—ca. 1517", 253–264.

DOI:10.1163/9789004431737_004

Researchers of cross-confessional diplomacy have often pointed out inconsistencies between chancery practice and juridical theory.[2] In his reading of Max Weber's writings on Islamic law, Abraham Udovitch has described the problems faced by qadis who had to comply with the fixed rules of a rational, formal law, in a chaotic context of commerce and minorities pervaded by forms of legal pluralism and particularism.[3] In this domain, Islamic law, while invoking its direct affiliation with the Revelation, opened the door to the silent, piecemeal incorporation of merchant customs or 'popular practice.' To describe the changing status of foreign merchants, Angeliki Laiou has introduced the concept of accommodation—*oikonomia*—to explain the abandonment by Byzantine judges of the abiding principle that the same law should be applicable to all subjects.[4] In his recently translated monograph, Michael Köhler has insisted on the strain that Islamic legal theory imposed on cross-cultural diplomacy. However, echoing Udovitch's view that local practice complemented the general principles of divine law, he acknowledges that the biases against Muslims inherent to Crusader law were not fully enforced in the daily practice of diplomacy. Indeed, Köhler argues that such limitations did not hinder contenders from reaching durable agreements, which owed much of their flexibility to their imprecise nature. The main device used to circumvent major legal biases was to resort to *technicalities*: agreements were not drawn up by jurists, but relied upon the parties' respect for technical instruments, such as duration and suspension clauses, bans on fortification or formulas of shared sovereignty, and respect for procedural patterns such as oaths and validation techniques.[5] To be sure, amān theory imposed upon Muslims the duty of pursuing jihād against the ḥarbīs, except for those involved in trade, pilgrimage and diplomacy, yet it is most striking that a clause in the 1283 treaty between the Mamluks

2 Wansbrough, "The Safe-Conduct in Muslim Chancery Practice", Brunschvig, Robert: *La Berbérie orientale sous les Hafsides*, Paris: Adrien-Maisonneuve, 1940, I, 431–40, Moukarzel, Pierre: "La législation des autorités religieuses et politiques sur les marchands Européens dans le sultanat mamelouk (1250–1517)", in: *Religious cohabitation in European towns (10th-15th centuries)*, Turnhout, Brepols Publishers, 2014, 121–139.

3 Udovitch, A. L., "Les échanges de marché dans l'Islam médiéval: théorie du droit et savoir local", *Studia Islamica* 65 (1987), 5–30, Gourdin: "Les marchands étrangers ont-ils un statut de dhimmi?", 437.

4 Laiou-Thomadakis, "Institutional Mechanisms of Integration".

5 Köhler, Michael A.: *Alliances and treaties between Frankish and Muslim rulers in the Middle East: cross-cultural diplomacy in the period of the Crusades*, edited by P. M. Holt and Konrad Hirschler, Leiden-Boston: Brill, 2013, Riley-Smith, Jonathan: "Government and the indigenous in the Latin Kingdom of Jerusalem", in: *Medieval frontiers: concepts and practices*, edited by David Abulafia and Nora Berend, Aldershot, Burlington, VT: Ashgate, 2002, 121–131.

and the kingdom of Jerusalem foresaw that Christians should give the Muslim contracting party two months' notice before the arrival of a new Crusade.[6]

When it comes to examining the multifaceted artifact of the diplomatic agreement, with its heavy burden of political and economic connotations (*privileges, capitulations, traités de commerce*, etc.), it is surprising that it has received so little attention from a comparative perspective. As trophies hanging on the walls of the offices of papyrologists and diplomatists, they are most often treated in isolation, and at best considered in light of similar treaties underwritten by a single polity. As a result, we still know relatively little about the actual origins of many decisions, particularly regarding the choice of jurisdiction for mixed cases, and other practicalities concerning the management of disputes. In spite of the significance of the treaties—through their attempt to set a normative framework for relating to foreigners—not all of the norms governing contact between Muslims and Franks were included in them. In the previous section, I elaborated on the idea that it was not only the ruler who set legal norms, to then be followed by actors; rather, they stemmed from different spheres of normativity, such as government reforms, policy-based regulations, the discourse of jurists and notarial practice, and in addition they changed a great deal over time. As for treaties, they were not necessarily fully implemented, and nor were they exhaustive, but were increasingly complemented by a series of *ad hoc* decrees that could even depart from the letter of the treaty itself and contradict its terms. As suggested earlier, Muslims saw the treaty as an artifact that merely pointed to more detailed oral discussions and agreements, and for which proper, Islamic witnessing needed to be provided. Niels Steensgaard mentions an episode in which the Ottoman authorities asked the European consuls to hand over the texts of the ahdnames in their possession, so that they could be modified.[7] The new versions were handed to the Nations only after repeated demonstration.

In this section, I turn to a subject that was often covered in treaties, particularly from late medieval times: adjudication. Treaties and legal codes tackled the conditions under which exchanges should be actually made; transactions

6 Köhler, *Alliances and treaties between Frankish and Muslim rulers*, 305.

7 Steensgaard, "Consuls and nations in the Levant from 1570 to 1650", 18. Işiksel stresses the conflicting views and misunderstandings around the diplomatic agreements, "Resident ambassadors' complaints sprang from their belief that the capitulations were bilateral agreements, enacted according to the norms of *ius gentium*. However, this was not quite true because the sultan could alter and even remove articles previously granted, whenever he deemed such a policy appropriate—at times without even notifying the other party", Işiksel, Güneş: "Hierarchy and Friendship: Ottoman Practices of Diplomatic Culture", 16.

needed to be concluded under fair conditions, but these conditions also needed to remain valid over time. For this reason, clauses not only described how business deals should be struck, but also how the handling of evidence should be dealt with, in case of dispute further down the line. In the eastern Mediterranean, early treaties such as those underwritten with the founding Mamluk sultans al-Ẓāhir Baybars (ruled 1260–1277) and al-Manṣūr Qalāwūn (ruled 1279–1290) mostly dealt with issues of non-aggression and were political in nature, while in the late Middle Ages and up to late Ottoman times clauses dealing both with the closing of deals and with issues of witnessing and registering took precedence over other topics. It is my contention that these clauses dealing with proving, transacting and bringing deals to court encoded the many changing attitudes towards the fundamental problem of dealing with diversity.

Traditional approaches to the diplomatic artifact are of little interest to this research; rather than attempting to define an Islamic orthodoxy on the grounds of which Europeans and Muslims were supposed to have interacted, my aim here is instead to capture the reconfiguration of cross-confessional relations at the turn of the early modern era. Encoded in the letter of diplomatic negotiations, this reconfiguration is located at the intersection between several sources of normativity, including legal thought, amān provisions and bottom-up legal practices, which defined relations at the marketplace and in court. In this chapter, I deal with treaties both diachronically and from a comparative perspective; more importantly, however, I broaden my analysis to the actual sphere of adjudication, as in the courts and judgeships entrusted by the treaties to hear disputes between Muslims and Franks. I refer to a series of problems inherent to mixed exchanges, which I further develop throughout Chapters Three and Four, and which straddled late medieval and early modern times.

The first issue with mixed disputes in Islamic polities was determining whether, in order to ensure impartial justice, the defendant should be allowed to have recourse to their home jurisdiction. This principle, commonly referred to with the Latin phrase *actor sequitur forum rei*, could intervene in conflict resolution by granting, for example, a non-Muslim defendant the right to bring a Muslim plaintiff to his own consular court, hence limiting the prevalence of sharīʿa jurisdiction over particular legal devices. A second, major issue was defining the extent to which minority witnessing could be accepted. Crusader law can give us valuable insight into this question, as it set an important precedent for the handling of minority witnessing in cross-confessional environments. Another aspect of interfaith litigation was whether or not the burden of proof lay on the defendant's witnesses. Although Crusader law stemmed from the feudal legal system exported to the Kingdom of Jerusalem, paradoxically, it denied Latin Christian witnesses the right to give testimony against Syrian, Jewish or other indigenous defendants. A third problem was determining whether the law should distinguish between people on the basis of their religion, or of their confessional group. Closely related to this, a fourth issue concerns the existence or not of mixed courts. Merchants—and thereby foreigners- were often dealt with through a broad application of local laws, or through different laws or courts. Although they were based upon different theological and legal concepts, both Crusader and Islamic law incorporated biases against minority confessions. Making people different before the law was a fundamental challenge to the safekeeping of cross-cultural relations, and similarities emerge in the ways in which both legal systems coped with this. The *Cour de la Fonde* emerged as a privileged setting to arbitrate inter-communal and market disputes in Crusader lands, and it found a parallel in the Siyāsa courts sponsored by the Mamluks and presided over by the ḥājib, a military officer.

A fifth issue related to cross-confessional arbitration concerns the need for notarization; that is, whether agreements could be concluded between and guaranteed by ordinary people, by community witnesses, or by state-appointed officials. A sixth issue questions whether notarization was oral or written, and ultimately, if a given legal system accepted written evidence without the oral support of its authors. Again, similarities can be traced between the ways in which Crusader and early Mamluk normative texts addressed the problem. Although my aim is not necessarily to level the differences between distinct traditions, some striking resemblances can be observed between the jurors in Crusader courts and the Islamic ʿudūl and instrumental witnesses. Lastly, the taking of oaths played an important role in the handling of cross-confessional relations, in and outside the courts of justice. This is an issue related to the validity of witnessing, although Mediterranean societies with an opener attitude to minority witnessing were not necessarily more eager to accept the oaths of unbelievers, and vice-versa. Legal texts, but also practice-oriented chancery handbooks described the procedural circumstances and conditions under which minority oaths should be accepted.

This seven problems articulated in practice much of the intricacies of cross-confessional dealings. Diplomatic treaties and the branch of Crusader law known as *Burgess law* paid a great deal of attention to oaths, special courts and jurisdictions and, more generally, to dealings with natives and the evidence they produced. When dealing with this heterogeneous array of issues, lawmakers often departed from a strictly juridical approach; for example, nowhere in Venetian and Genoese legal codes is it explicitly stated that a Jew cannot testify, however Mediterranean practice had it that they could only appear as witnesses for cases involving two Jewish parties. Examples abound, and in the same vein the recourse to shared technicalities has been noted as the practical grounds for the signature of truces, also in cross-confessional relations at the marketplace a trove of procedural, practical solutions to actual problems prevailed over the application of legal biases against unbelievers. This spirit also pervaded many of the efforts by Mamluk jurists to grant the Islamic ruler with legal solutions to cope with the presence of foreign merchants and colonies of Franks in Islamic lands. One of these solutions was found in the so called 'royal' or Siyāsa courts (in contrast with the 'qādī' or 'sharīʿa' courts), which late Mamluk treaties increasingly designated as the places where the ruler and his delegates should hear the grievances of foreign merchants.

3.1 An Introduction to Siyāsa

The new role played by royal justice under the Mamluks has been identified by historians as an important shift in the history of Islamic Law. Since early Islamic times, the courts of the qadis, where sharīʿa was applied according to traditional jurisprudence with complex rules of procedure, had been supplemented by more expedient courts. The most well-known among these royal jurisdictions was the *maẓālim*, in whose courts the ruler theatrically displayed his justice and gave verdicts according to his own judgment. In this chapter I will be dealing with the *Siyāsa*, a similar royal jurisdiction derived from the maẓālim, which frequently overlapped with it, and where justice was instead administered by the chamberlain (*ḥājib*) and other military officers, such as the head emir in Alexandria.[8]

For centuries, we have erroneously believed that Siyāsa referred to a lost legal code imported from Asia. Contemporaries believed that the word Siyāsa was etymologically connected to the Mongol word *yasa*, and with the Turkish *yasaq*, which, as we have seen, was a term used by Arab subjects of the Ottomans to refer to Ottoman customary law—principally taxes considered unlawful and imposed by the Turkish government. We know now that it has its origins in a lost Arabic root meaning 'the tending and training' (of beasts). *Siyāsa* is commonly translated as 'politics,' and sometimes rendered as 'governance,' or 'statecraft.' The conceptual world of the jurists was ideally meant

8 For a discussion of siyāsah under the Mamluks, Nielsen, *Secular Justice*, Rapoport, "Royal Justice", Irwin, Robert, "The Privatization of "Justice" under the Circassian Mamluks", Mamluk Studies Review 5, (2002), 63–70, Fuess, Albert, "Zulm by Mazālim? The Political Implications of the Use of Mazālim Jurisdiction by the Mamluk Sultans", Mamluk Studies Review 13 (2009), 121–147.

to harmonize *Siyāsa Sharʿīyah*, a legal theory of governance, with the general rule of law, or sharīʿa.[9] This was a more troublesome task for Mamluk judges, sultans and officials, who, as Mamluks chronicles show, daily entered into conflict with qādīs while administering justice.[10] Regarded with suspicion by both contemporaries and modern historians, Mamluk Siyāsa is now stirring up a lively scholarly debate, perhaps due to growing recognition of its importance in the later modernization of Islamic law.[11] The debate on Siyāsa has revolved around the question of whether it was compliant with religious law, was a new invention by the Mamluks or predated them, was secular or not, or whether it was simply a tool of political legitimacy for the sultans.[12] As mentioned in the Introduction, important research by Baber Johansen on the disciples of Ibn Taymīyah (d. 1328) has underlined the efforts of Siyāsa thinkers to overcome the limitations of the justice administered by the qadis, particularly in the field of proof and evidence, precisely the question at issue here.[13] In the first section of this chapter, I use an asymmetrical comparison between Islamic and Byzantine and other Christian societies to describe Crusader and early Mamluk approaches to the recording and closing of deals, and how transactions could be challenged if they went wrong. I subsequently turn this line of inquiry to the royal courts, by exploring the role of Mamluk Siyāsa as a judicial practice, and

---

9 On late medieval siyāsah, Lambton, Ann K. S.: *State and Government in Medieval Islam: an Introduction to the Study of Islamic Political Theory: the Jurists*, Oxford; New York: Oxford University Press, 1981, 138–152, Hallaq, Wael B.: *The origins and evolution of Islamic law*, Cambridge, UK; New York: Cambridge University Press, 2005, 99–101, Black, Antony: *The history of Islamic political thought: from the Prophet to the present*, Edinburgh: Edinburgh University Press, 2001, 158–164.

10 Rapoport, "Royal Justice", 100–1, Irwin, "Privatization", 66.

11 For the origins, meaning, and different perceptions of the term, Vogel, Frank E.: "Siyasa", in: *Encyclopaedia of Islam, Second Edition,* IX: 693b, Lewis, Bernard: "Siyasa", in: *In Quest of an Islamic Humanism: Arabic and Islamic studies in memory of Mohamed al Nowaihi*, edited by A.H. Green, Cairo: American University in Cairo Press, 1984, 3–14, Najjar, Fauzi M.: "Siyasa in Islamic Political Philosophy", in: *Islamic theology and philosophy: studies in honor of George F. Hourani*, edited by George F. Hourani and Michael E. Marmura, 92–110, Albany: State University of New York Press, 1984, Masud, Muhammad Khalid, "The Doctrine of Siyāsah in Islamic Law", Recht van de Islam 18 (2001), 1–29.

12 Fuess, "Zulm by Mazālim?", 132, 141. I do not think it necessary to further elaborate here on this debate, which is thoroughly addressed by Youssef Rapoport in his "Royal Justice", 73–80. Neither is it my intention to create an artificial divide with scholarship on Mamluk *maẓālim*; instead, I aim to shift the focus away from the well-known maẓālim court in Cairo, to a broader legal and geographical setting, Nielsen, *Secular Justice,* 32.

13 Johansen, "Signs as Evidence", Johansen, Baber: "Vérité et torture. Ius commune et droit musulman entre le Xe et le XIIIe siècle", in: *De la violence*, Edited by Françoise Héritier, Paris: Odile Jacob, 1996, 123–168.

focusing on a hitherto unknown aspect of it; namely, its role in settling mixed commercial disputes.

Capturing the emergence of Siyāsa as a judicial practice is difficult for two reasons. First, although normative texts can provide us with some snapshots of a given society's attitudes towards diversity, a more dynamic picture of what actually happened in practice requires a series of judicial records that simply do not exist. Descriptions of Siyāsa trials suggest that an oral approach to procedure was adopted, and rarely mention verdicts or legal acts being put into writing. The second issue is that islamologists concerned with the history of Islamic justice often attempt to reconcile their findings with legal theory, hence privileging the quest for precedent over social change. Since the times of the Caliphate, Islamic rulers had a long tradition of sitting in justice at the palace to hear petitions and grievances about the unjust rulings of secretaries and officials.[14] The very existence of this practice, called maẓālim, appears to support the idea that the Mamluk Siyāsa courts were not in themselves an innovative feature.[15] In this chapter, I adopt the view that, although both developments drew upon the same kind of legal reasoning, Siyāsa constituted a separate concern for jurists, in the same way that it had in 12th-century Transoxiana, while in the Mamluk context the maẓālim was increasingly understood as a court of appeal.[16] The closing section of this chapter dwells extensively on Mamluk Siyāsa courts as a forum for mixed commercial conflict; a picture that, significantly enough, emerges out of the Venetian notarial records drawn up in Alexandria, which reveal a dimension of Islamic law invisible to Arabic sources. Mamluk Siyāsa was clearly a late byproduct of the doctrines dealing with the legal attributions of the Islamic sovereign, such as *maẓālim* and *taʿzīr*, the sovereign's right to mete out punishments. However, it also emerged in the

---

14 Tillier, Mathieu: "Qadis and the political use of the mazalim jurisdiction under the Abbasids", in: *Public Violence in Islamic Societies: Power, Discipline, and the Construction of the Public Sphere, 7th-19th centuries C.E*, edited by Christian Robert Lange and Maribel Fierro, Edinburgh: Edinburgh University Press, 2009, 42–67, Tillier, Mathieu: "The Mazalim in Historiography", in: *The Oxford Handbook of Islamic Law*, Edited by Anver M. Emon and Rumee Ahmed, Oxford: Oxford University Press, 2015, 356–380.

15 Moukarzel: "La législation des autorités religieuses et politiques sur les marchands Européens dans le sultanat mamelouk (1250–1517)", Christ, Georg: *Trading Conflicts: Venetian Merchants and Mamluk Officials in Late Medieval Alexandria*, Leiden-Boston: Brill, 2012, Fuess, "Zulm by Mazālim?" , Winter, Michael, "The judiciary of late Mamluk and early Ottoman Damascus: The administrative, social and cultural transformation of the system", in: History and Society During the Mamluk Period (1250–1517) 5, Göttingen 2014, 193.

16 Johansen: "Vérité et torture. Ius commune et droit musulman entre le Xe et le XIIIe siècle", 132.

context of previous efforts undertaken in Palestine or in the Byzantine Empire to define a technical framework for resolving cross-confessional conflict. To be sure, Siyāsa drew on available legal theory, but as regards its competence for foreigners, it materialized in the hitherto unexplored institution of Islamic commercial mixed courts, and was endowed with an open approach to proof, procedure and unbelief. Together with legal theory, the rise of Siyāsa is better understood in light of these available precedents.

Similarly to diplomatic treaties, Crusader law addressed most of the crucial issues mentioned earlier on proof and adjudication, and dealt with cross-confessional relations with an unprecedented intensity. Their legal codes, or Assises, anticipated much of the concerns surrounding the amān treaties underwritten with Muslims. If the Crusaders exported Latin-Christian and particularly feudal law to the Kingdom of Jerusalem, the resulting legal system cross-pollinated with sharīʿa and with Byzantine precedent, approaching their solutions on matters of witnessing, in the opening to judicial autonomy for minorities and in the introduction of some religious biases. Feudal law governed relations among the nobles, and was complemented by a second legal layer known as Burgess law. In principle, Burgess law was meant to be applied to the second principal group in Frankish lay society: non-nobles, liegemen, or *burgenses*.[17]

The keystone of the Crusader adjudication system was the High Court, which applied feudal law among the nobles, and was complemented both by an eminently civil court known as Cour des Bourgeois and—more interesting for our study—a commercial and mixed court called the Cour de la Fonde.[18] I will leave the High Court out of this discussion, as it is not strictly relevant to cross-confessional relations, and focus on the last two tribunals, which applied the Assises des Bourgeois, or Burgess law. Christians native to the Crusader States, who mostly belonged to the confessional group referred to as 'Syrians,' seem to have enjoyed some legal autonomy on the basis of their own courts and officials. We have little knowledge about the nature of the rulings passed by the *Cour des Syriens*, although we do know that the court applied the customary law of the Oriental Christian communities, and sources describe it as being composed of the *raïs* and a mixed panel of jurors. In addition, it cannot be excluded that, in the Crusader states, some disputes between Muslims were

17 Prawer, Joshua: "Social Classes in the Crusader States: The "Minorities"", in: *A History of the Crusades: The Impact of the Crusades on the Near East*, edited by K. M. Setton, N. P. Zacour, *et al.*, vol. V, 59–115: University of Wisconsin Press, 1985.

18 Nader, Marwan: *Burgesses and Burgess law in the Latin Kingdoms of Jerusalem and Cyprus, 1099–1325*: Ashgate, 2006, 156.

arbitrated by Islamic law, however specialists have not found any evidence for Crusaders giving official recognition to this.[19] As concerns cross-confessional relations in the marketplace, in any case, Crusader law set a precedent in creating a complex framework of courts of justice. At the center of the adjudication system was a mixed, commercial court, set up as the principal forum for cross-confessional dispute. In complement to this a sophisticated, highly technical set of rules governing witnessing and the production of evidence in mixed settings was developed.

The Cour de la Fonde, therefore, counted as an immediate forerunner of the Siyāsa tribunals, which were empowered to pass judgment on merchants of different confessions. Moreover, this koinè of solutions in Crusader law incorporated late Byzantine law, upon which the Crusader system must have drawn extensively. Angeliki Laiou draws a link between the Crusader status of the bourgeois and the Byzantine category of *burgesioi*, which was often applied to different categories of foreign merchants and bestowed by Manuel I Komnenos upon Venetians.[20] As long as they kept their political independence, the Byzantines governed cross-confessional relations in the market by transferring mixed cases to special, often imperial courts, by imposing technical solutions on the taking of oaths according to religious affiliation, and by imposing a ban on Jewish witnesses. However, until the Fourth Crusade brought about a loss of political independence, consular justice for issues between foreigners was not allowed, at least officially, in favor of Byzantine, local courts. This legal balance shifted in favor of foreigners after 1204, when jurisdiction over mixed issues was transferred to the foreigners' courts and judges. In any case, if we leave aside the repertoire of specific formulas that were applied to each case, it appears that the set of concepts and devices employed by Crusaders bore some similarities with those used by Middle Eastern societies to solve the issue of cross-confessional relations and disputes.

3.2 The Crusader Marketplace

The fulcrum of commercial arbitration in Crusader lands was the *Cour de la Fonde*, a denomination related to the Islamic notion of *funduq*, or urban caravanserai. Set up in Acre and in other cities, it passed verdicts for both commercial and interfaith cases; that is, it passed judgment for all commercial disputes

19 Mayer, Hans Eberhard, "Latins, Muslims And Greeks In The Latin Kingdom Of Jerusalem", *History* 63, 208 (1978), 175–192.

20 Laiou-Thomadakis: "Institutional Mechanisms of Integration".

and any lawsuit involving native Christians, Jews, Muslims and Samaritans. Although it acted as a mixed commercial court, it is unclear whether it had jurisdiction over Latin traders, who may have been able to resort to the Cour des Bourgeois if they so wished, even if this appears to have been unlikely from the viewpoint of enforcing transactions within a composite merchant community. Most of what we know about Crusader commercial jurisdiction comes to us through late medieval sources, which reflect the situation after the fall of Acre in 1291, when Crusader society had resettled in Cyprus and founded an epigone state on the island. Late legal codes were apparently compiled by the Venetians when they took over the island later in the 1470s.[21] In general, late Palestinian sources describe a situation where the Cour de la Fonde gradually absorbed the Cour des Syriens as a forum for minority disputes.[22] Similarly, in late Crusader Cyprus there is no evidence of the Fonde jurisdiction and the Cour des Bourgeois rose to prominence in most cross-confessional and civil matters, gradually attracting all jurisdiction for mixed and commercial disputes. As mentioned earlier, in the Kingdom of Jerusalem the system of feudal and bourgeois law was perfected with the court of the Syrians, who enjoyed judicial autonomy by virtue of the privileges granted to them by the Crusaders. In the Latin kingdom of Jerusalem, this court passed verdict in Jerusalem, Nablus, Tyre, Bethlehem, Nicosia and Famagusta. Syrians did not obey the Roman church—that is, they deviated from the *loi de Rome*—however they were granted tax exemptions and privileges both in the Holy Land and, later, in Cyprus, thanks to their political alliances. As for the Cour des Syriens, it survived after the fall of Acre as one of the Cypriot jurisdictions; however, intriguing as this may seem, the Cour des Syriens and its *raïs* do not seem to have continued to handle cross-confessional relations after this point, and became limited to a murky, special jurisdiction for some elites in the 15th century, before ultimately vanishing under Venetian legal reform. In 15th-century Cyprus, it has to be noted, the closest example of a mixed court was the tribunal held by the *Capitano di Famagosta*, a Genoese magistrate.[23] The late Crusader legal system foresaw the supremacy of feudal law, as epitomized by the High Court, where

21 A.Beugnot: *Abrégé du Livre des Assises de la Cour des Bourgeois, in Recueil des historiens des croisades, vol. 2: Lois*, Paris, 1843, XXXV.

22 Ibid. XXIV.

23 Otten, Cathérine: "Le registre de la Curia du capitaine Génois de Famagouste au Milieu du XV siècle: Une source pour l'étude d'une société multiculturelle", in: *Diplomatics in the Eastern Mediterranean 1000–1500: Aspects of Cross-Cultural Communication*, Edited by Alexander D. Beihammer, Maria G. Parani, *et al.*, Brill, 2008, 251–274, Fossati Raiteri, Silvana: *Genova e Cipro: l'inchiesta su Pietro de Marco, capitano di Genova in Famagosta, 1448–1449*, Genova: Università di Genova, Istituto di medievistica, 1984, ix-lxxxi.

irrational forms of proof and evidence always existed. Muslim visitors were amazed by the duels, ordeals and trials by fire or water they witnessed, even if such methods were restricted to the Frankish elite and its internal disputes. Exceptionally, duels were accepted for commercial disputes, but only if these suits involved claims for more than a silver marc.

3.2.1 *Muslims and Crusader Courts*

The complex configuration of Crusader Latin courts reflects the importance that Western legal systems attached to the notion of status; as an old-regime legal system, it was status, and not its human, physical vessel that was considered to be the depositary of rights. The Crusader legal system was therefore characterized by a double tier of courts; one for those of high or noble status, and another for the urban class known as the *burgesses*. The abiding principle was that both groups could only be judged by their peers—which, incidentally, became all the more complex as time went on and these groups became ever more heterogeneous, hence opening the door to more and more exceptions and technicalities. This idea of hierarchized and separate spheres of courts clashed with the Islamic legal conceptions, in which status groups were not granted any specific treatment. Sharīʿa was conceived as an egalitarian system dominated by the personality of the law; in principle, the same law applies to all Muslims, irrespective of where they live and of their social standing. Dhimmīs could have recourse to the qādī, and it has been argued that the so-called dhimmī rules and biases against non-Muslims do not correspond to fixed statuses, but rather answer to the dhimmīs' refusal to accept the Revelation, precisely what these rules encourage dhimmīs to do.[24] In the legal world that came into being under the Crusaders, Muslims were granted the lowest status by their social superiors, however this status did not apply in the exceptional framework governing relations in the marketplace.[25] Significantly, insofar as the Cour de la Fonde was concerned, Muslims were accepted as subjects of the legal system.[26] Strikingly, this echoes the observation by Johansen that, for

24 Emon, *Religious Pluralism,* 66–7, Friedmann, Yohanan: "Classification of unbelievers in Sunni Muslim Law and Tradition" , in: *Tolerance and coercion in Islam: interfaith relations in the Muslim tradition*, 54–86. New York: Cambridge University Press, 2003.

25 Edbury, Peter W.: "Latins and Greeks on Crusader Cyprus" , in: *Medieval Frontiers; Concepts and Practices*, Edited by David Abulafia and Nora Berend,133–143. Aldershot: Ashgate, 2002, 137.

26 Nader, Marwan, "Urban Muslims, Latin Laws, and Legal Institutions in the Kingdom of Jerusalem" , *Medieval Encounters* 13 2 (2007), 243—270, 256–7. Nader focuses on courts and procedure as facilitators of cross-confessional relations. On the Crusaders' attitude toward natives, and particularly Muslims, see Riley-Smith, Jonathan: "Government and

Islamic law, religious and gender hierarchies dictated social transactions, but not commercial exchange. As regards debts and contractual obligations, differences between Muslims and dhimmīs were flattened out by jurists.[27]

Muslims are mentioned among the plaintiffs who brought their claims to the Cour de la Fonde, as we learn from a well-known passage by Ibn Jubayr (1145–1217) that has contributed to a—probably undeservedly—positive image of Crusader rule among plural confessions.[28] In addition, borrowings from Islamic practice included obvious examples, such as the institution of the muḥtasib (market inspector), which survived in late medieval Cyprus.[29] The Cours were presided over by the bailiffs of the funduqs, assisted by jurors of different Christian confessions, namely four Syrians and two Franks.[30] It is perhaps for this reason that in early Mamluk treaties we find a marked tendency to consider the head of the Customs and similar officials as potential judges for mixed cases. And indeed, the functioning of the Cours points to a tendency to accept a certain degree of legal pluralism. Contrary to what happened in the High Court, and to a certain extent in the Cour des Bourgeois, in the Cour de la Fonde irrational means of proof were not accepted, and native Christians were not subjected to ordeals or allowed to resolve disputes through duels. Similarly, a crucial feature was the presence of scribes of different confessions; there were Arabic-speaking scribes both at the Cour de la Fonde and at the Cour de la Chaîne: "escrivein sarasinois ou fransois ... a la fonde on a la chaene."[31] In light of the abundant evidence for Arab Christian scribal families working in Cyprus, Jean Richard has assumed that here Saracen/*sarasinois* might mean Syrian Christian, something that has also been confirmed by Ibn Jubayr, who

---

the indigenous in the Latin Kingdom of Jerusalem", in: *Medieval Frontiers; Concepts and Practices*, edited by David Abulafia and Nora Berend, 121–131. Aldershot: Ashgate, 2002, Mayer, "Latins, Muslims And Greeks In The Latin Kingdom Of Jerusalem", Prawer: "Social Classes in the Crusader States: The 'Minorities' ".

27 Johansen, Baber: "Commercial exchange and social order in Hanafite Law", in: *Law and the Islamic World. Past and Present*, edited by Ch. Toll and J. Skovgaard-Petersen, 81–95. Copenhagen: Royal Danish Academy of Sciences and Letters, 1995, Müller, "Non-Muslims as part of Islamic law", 45.

28 As for an intra-communal forum for issues between Muslims, its existence is much debated, Mayer, "Latins, Muslims And Greeks In The Latin Kingdom Of Jerusalem", 181, 185, Bishop, Adam, "Criminal law and the development of the assizes of the crusader Kingdom of Jerusalem in the twelfth century", PhD Dissertation, University of Toronto, 2011, 125.

29 Nader, *Burgesses and Burgess law*, 142.

30 A.Beugnot, *Abrégé du Livre des Assises de la Cour des Bourgeois*, XXV.

31 E. Kausler: *Les Livres des Assises et des Usages dou Reaume de Jerusalem sive Leges et Instituta Regni Hierosolymitani*, Stuttgart: Adolf Krabbe, 1839, 344-5, Assise 284.

mentions these Christian scribes ("kuttāb al-dīwān min al-naṣāra").[32] Another striking parallel can be found in the 13th-century treaties concluded by the Mamluks with Genoa and Venice, which stipulated the right for Franks to dispose of their own Christian scribe at the customs house.[33]

3.2.2 *Jurors and Witnesses*

The presence of jurors in commercial and civil Crusader courts also raises questions about what might have been borrowed from previous Islamic practice, as well as possible continuities with it. As a professional body versed in local law, the jurors were regularly present in court, where they overlooked the drafting of proceedings, provided the parties with legal advice, and could sometimes be appointed as judges to preside over the court. According to the detailed description provided by Marwan Nader, the jurors at the Cour des Bourgeois acted as witnesses and certified as to the legitimacy and permanency of transactions; in this sense, their role might be equated with that of the Islamic instrumental witnesses first described by Claude Cahen, and later by researchers of Ottoman justice.[34] Initially called witnesses by the sources, as in the case of the ʿudūl, lists of eligible witnesses/jurors were drafted for the use of the *cours*. They were appointed upon decision of the *seigneur justicier* and ultimately, the king, and therefore could be dismissed. However, they also worked outside the court, where they were vested with a certain degree of notarizing power, since their presence at the conclusion of a sales contract sufficed to validate it. Like the Muslim ʿudūl, jurors had to prove themselves to be "wise and good men."[35] In addition, only those who were "bourgeois et frans, de la loi de Rome" qualified for the job. Religious biases against minorities, such as those against Monophysite and Oriental Christians—who could not be jurors in the case

32 Richard, Jean: "Le plurilinguisme dans les actes de l'Orient latin", in: *La langue des actes. Actes du XIe Congrès international de diplomatique (Troyes, jeudi 11-samedi 13 septembre 2003)*, Paris, O. Guyot-Jeannin 2004, supported with further evidence by Jacoby, David: "The fonde of Crusader Acre and its tariff: some new considerations", in: *Dei gesta per Francos: études sur les croisades dédiées à Jean Richard, crusader studies in honour of Jean Richard*, edited by B. Z. Kędar, Jonathan Simon, *et al.*, Ashgate, 2001, 277–293, Ibn Jubayr, Muḥammad Ibn Aḥmad (1145—1217): *Riḥlat Ibn Jubayr*, Beirut: Dar wa Maktabat al-Hilal, n.d., 248.

33 Tafel, G.L.F. and G.M. Thomas: *Urkunden zur älteren Handels- und Staatsgeschichte der Republik Venedig, mit besonderer Beziehung auf Byzanz und die Levante: Vom neunten bis zum Ausgang des fünfzehnten Jahrhunderts*: Hof- und Staatsdruckerei, 1856, 488, treaty of 1254, "Capitulum. Item, quando applicuerint, habere debeant unum scribanum Christianum, qui clarificet in doana, et debeat scire suas rationes per totum tempus."

34 Baldwin, *Islamic Law and Empire in Ottoman Cairo,* 35, Cahen, "A propos des Shuhud".

35 Nader, *Burgesses and Burgess law,* 144–9, 152.

of the Crusader states—often presented as specific uniquely to Islamic judicial practice, were in fact shared with other eastern Mediterranean societies, including the Crusaders and Byzantines. Indeed, the importance of jurors in Crusader court procedure has been traced back to Carolingian institutions, even if it seems fair to say that it bore more points in common with its closer, Muslim counterpart.

3.2.3 *Courts and Bans*

Little is known about how the Crusader courts actually functioned in practice, despite the amount of Crusader jurisprudence that has come down to us. When the Venetians took over the Kingdom of Cyprus they made an impressive effort to collect all available legal texts, mostly late ones. It is generally agreed that those addressing procedure in the High Court reflect an earlier stage in Crusader law, notably the *Livre des Assises* by John of Ibelin, while those dealing with the Cour des Bourgeois, principally the *Livre des Assises de la Cour des Bourgeois* and the *Livre contrefais* are considered to be late. Early High Court Assises do not mention Muslims at all, because they reflect the legal world of the elites of Crusader society, however the extant legal codes on bourgeois jurisdiction, which included the Cour de la Fonde, describe a society in which transactions between individuals of different religions and confessional groups were very frequent. In the Cour de la Fonde, bourgeois law was applied rather than Syrian customary law, although it was a version of Crusader law that addressed a mixed and minority public, and excluded, for example, recourse to ordeals.

Franks, and among them, the burgesses, appear as a minority group in bourgeois jurisprudence, along with the many Syrians, Jews, Samaritans, Armenians and Muslims involved in trade; the issue of witnessing and guaranteeing for others is thus a primary focus of the Assises. A striking contrast between law produced in the Kingdom of Jerusalem and its later versions can be observed in the status of witnesses (*garans*). Early legal texts referring to the High Court are clear: only baptized Christians could testify, and only those obeying *la loi de Rome* were allowed to bear witness for important matters.[36] In contrast with this strict approach to witnessing and testimony, the issue of minority witnessing is more amply developed in the *Assises des Bourgeois*, and indeed this text

36 "Et qui viaut prover par garans fié ou chose de fié ou autre chose, il covient qu'il soient ii loiaus garans ou plus de la loy de Rome, se ce n'est de prover aage ou lignage. Que l'on peut prover ses ii dites choses par chascun home ou feme, mais que il soit crestien batié et qu'il soient.ii. ou plus et que il s'acordent bien ensenble a une parole.", Edbury, Peter W.: *John of Ibelin: Le livre des assises*, Leiden; Boston, Brill, 2003, 165–66.

represents a point in time when Crusader law dealing with the legal handling of cross-confessional relations was at its most complex.

Crusader jurisprudence, therefore, was extremely attentive to transactions in which the parties' different confessions might give rise to complications or dispute. It cannot be said that the Crusaders considered everyone to be actors in the legal system, or that they went so far as to apply the same law to Latins and non-Latins alike. However, it is clear that, if feudal and seigniorial courts continued to judge nobles according to traditional jurisprudence, fundamental biases against non-nobles and non-Latins were kept out of the marketplace. One example can be found in the case of the ban on heretic witnessing, in which Crusader law departed from the stance taken by Roman law, developed in the Republican period, that did not impinge on individual creeds. As we saw earlier, Justinian's legal codes insisted on excluding Jews from witnessing, and this found continuity in both Crusader and sharīʿa law. However, in the Cour de la Fonde this was not the case; as we shall see, in this court all witnesses had equal value in mixed cases. To cope with these inconsistencies, the Crusaders created a system of courts and bans that made legal enforcement compatible among confessions, while safeguarding the theoretical supremacy of noble status above the burgesses, Syrians and heretics. Although the early legal text of John of Ibelin does not mention Muslims at all, they emerge in bourgeois legal codes, and it is clear that the late Crusaders considered them to be, if not actors, at least subjects of the legal system. Often quoted by proponents of an alleged Crusader tolerance, a passage in Assise CCXXXVI refers to Muslims' and other non-Latins' right to justice and their involvement in mixed dealings, acknowledging that "si sont il auci homes come les Frans."

3.2.4 *Writing Down Transactions*

Parallel to underlying principles of status and religion, Crusader justice introduced a series of technical solutions for dealing with the way transactions were produced, recorded, and the eventual recourse to dispute resolution in court. According to bourgeois legal codes, sales contracts could be concluded in court, which meant they were notarized in *recounoissance* and were therefore deemed valid; if not, ("se la chose nen estoit faite en la cort") they needed to be underwritten in the presence of witnesses.[37] As regards the Fonde, it appears that important parallels can be drawn between bourgeois law and Mamluk and some Maghrebi contemporary amān documents. Treaties signed with the Mamluks in the thirteenth and fourteenth centuries describe a situation in

37 E. Kausler, *Les Livres des Assises et des Usages dou Reaume de Jerusalem*, Assises LIX, LXIII.

which deals were concluded in a similar setting: at the customs house (*dogana*, ar. Dīwān). Outside of this ideal framework, presumably in urban marketplaces (usually referred to with the expression *in bazaro*) treaties encouraged the conclusion of deals in the presence of witnesses. In the Mamluk *dogana*, these agreements foresaw the assistance of a Christian scribe.[38] In the Fonde and the Chaîne courts, which presented a similar tribunal for maritime issues, it was Latin and Arab scribes who recorded transactions.

Although the Assises and Mamluk treaties underline the usefulness of notarization, the jurist and editor of the Assises Auguste-Arthur Beugnot (1797–1865) notes that one important commercial transaction—the loan—needed in Crusader lands to be proven by means of oral witnessing.[39] This fact is particularly relevant in that it presents a deviation from the general tendency in southern European cities to enhance the role of notarization. As we saw in Chapter Two, the tendency in communal law was to ascribe probative value to the notarial deed; in the Crusader lands, however, due to the influence of French law, seigniorial approval was required to validate a document. This was noted, for instance, by Francesco Balducci Pegolotti (1290–1347) in his *Pratica della mercatura*, who states that, at least for matters of debt, in Cyprus notarial contracts were deprived of legal value. Transactions continued to be registered by royal officials, the "escrivains du comerc," similarly to how they were at the Fonde.[40] The role of witnesses to transactions must therefore have been all the more important in places where this parallel, non-Latin notarial tradition existed.[41] Beugnot claimed that frequent recourse to oral agreements pushed the Crusader legal mind to develop

38 "Item, quando applicuerint, habere debeant unum scribanum Christianum, qui clarificet in doana, et debeat scire suas rationes per totum tempus," 1254 treaty with the Venetians, Tafel and Thomas, *Urkunden zur älteren Handels*, 488, and the 1230 treaty with Tunis, "Et quod valeant(ur) habere scribanum Christianum suum in doana ad eorum voluntatem," 305.

39 A.Beugnot, *Abrégé du Livre des Assises de la Cour des Bourgeois*, xliii.

40 "in Cipro … in nulla parte dell'isola non vale nulla carta de notario se non fosse de testamento o de dota o di schiavo comperato o de navoleggiamento salvo se lo re lo fasse valere per grazia a cui volere mettere avanti per usare sa ragione," Richard: "Aspects du Notariat public à Chypre sous les Lusignan", Otten, Cathérine: "Quelques aspects de la justice à Famagouste pendant la période génoise", in: *Πρακτικά του Τρίτου Διεθνούς Κυπρολογικού Συνεδρίου, Λευκωσία, 16–20 Απριλίου 1996. 2, Μεσαιωνικό Τμήμα*, Nicosia, 2001, 333–351, mentions the existence of a "cour d'enregistrement" in Famagusta where contracts were registered, 338.

41 "Il se peut que nous ayons là le résultat d'une évolution et qu'antérieurement au 15e siècle les notaires investis par l'autorité impériale aient été normalement des Latins, les membres des autres communautés ayant habituellement recours à des notaires ou à des personnes jouissant des mêmes pouvoirs issus de leurs propres rangs," Richard: "Aspects du Notariat public à Chypre sous les Lusignan", 207–9. Richard, Jean, "La diplomatique royale

a highly complex set of rules underpinning cross-confessional witnessing. Determining what religion the witness should be in the case of deals concluded between men of different faiths presented a very real challenge. And indeed, the late Assises present a complex casuistry of cross-confessional exchange involving Latins, Jews, Jacobites, Nestorians, Armenians, Syrians, Muslims and even some obscure Christian sects.

Assise CCXXXVI in the *Livre des Assises de la Cour des Bourgeois* describes the Cour de la Fonde and its procedure. Although most Assises come in the form of ad hoc casuistry for precise legal issues, the text acquires a more normative tenor when describing the court, its staff and issues of procedure. After introducing its role in the marketplace, as well as stressing the need for the bailiff and the jurors to commit to maintaining justice amidst diversity, the text quickly moves on to matters of proof and evidence. Its primary focus is the general ban on cross-confessional witnessing, after which it turns to the question of oaths.[42] The Crusader witness system described both in the Palestinian Assises, and then in later Assises from Cyprus is well-known. In gross, in late Burgess law we find a total ban on cross-confessional witnessing. A Latin Christian witness could not give testimony against a Jacobite defendant, and neither could a Jacobite be used to testify against a Nestorian defendant, and so on, including heretics (that is, Jews). In practice, no one was allowed to bear witness against a person other than those of his own confessional group. In addition, oaths played an important role; Assise CCXXXVI goes on to detail the modalities of oath-taking by Jews and other non-Latins, such as the sacred texts involved in the ceremony of oath-taking (mentioning the Torah, Qur'ān, Pentateuch for Samaritans, etc.). The Assise closes with a discussion of the procedure to follow in the case of a dispute opposing two parties belonging to the same faith, in which case religion of the witnesses did not matter.

Several crucial features regarding the legal framework described in Assise CCXXXVI should be pointed out here. Firstly, the cross-confessional ban, as it was practiced in late Crusader times, was total, and also extended to the social elite—that is, the feudatories and their descendants. Latin Christians were barred, for instance, from testifying against Syrians, or even against Muslims. Secondly, at the Cour de la Fonde social, status, and religion categories had been flattened out, and all the non-Franks now belonged to a generic second rank in society. In contrast with this general tendency, the social elite

dans les royaumes d'Arménie et de Chypre (XIIe-XVe siècles)", Bibliothèque de l'école des chartes (1986), 69–86.

42 See the note on this topic written by Adam Bishop for the RELMIN research project, http://www.cn-telma.fr/relmin/extrait136984/.

maintained a procedural privilege. In cases where they were accused without witnesses, Muslims were obliged to take an oath, whereas a Frankish defendant need not do so.[43] It should be noted that the regime governing the use of oaths by court procedure in the Assises des Bourgeois closely resembles that of sharīʿa: Shāfiʿīs, Mālikīs and ḥanbalīs accept the defendant's oath provided that the plaintiff does not have sound proof (*bayyina*).[44] Unlike Crusader law, however, Islamic law did not impose religious distinctions limiting the validity of oaths by non-Muslims: discrimination applied to witnessing, while everyone was allowed to take oaths.[45] It did, however, provide ample detail on the validation of minority oaths, such as the books considered to be sacred and by the Jews, Muslims or Samaritans. In passing, it should be noted that complex rules on oath-taking also appear in late Byzantine law.[46]

The procedural privilege of the social elite follows the same logic as the *actor sequitur forum rei* principle disccussed later in this section, in the limited sense that it strengthened the position of a Latin Christian defendant, but not if the same Latin is the plaintiff. A debate within Islamic jurisprudence argued that minorities, who could have cause for resentment due to their inferior position in society, did not have a vested interest in safeguarding the public good, and therefore would naturally target their social superiors in court by providing false testimony. The Crusader witness system seems to have been based on a similar principle: if the oath privilege system did not exist, the reasoning went, an ill-meaning non-Latin could have a Frank convicted on the basis of a forged accusation. The privilege allowed the Frankish defendant to win the case without needing to invoke the name of god, and eventually, committing a sinful action such as perjury. It has to be noted however, that, at least theoretically, this privilege was the keystone of Latin legal superiority in commercial litigation; but for this final pledge of a Latin Christian denying his opponent's claims without taking an oath, there appears to be no device at the Fonde that might have granted any advantage to Latins over non-Latins and infidels. It is for this latter reason that the Cour de la Fonde and the Crusader witness system stand

43 "Ici orres la raison dou Franc et dou Sarasin. Se un Franc se clame en la cort dun Sarazin daveir que il li deit, et le Sarasin li nee laveir, et le Franc nen a garens: la raison comande que le Sarasin deit iurer sur sa lei que il rien ne li deit, et atant en deit estre quite. Encement et se un Sarasin se clame dun Franc en la cort daveir que il li deit, et le Franc li nee laveir, et le Sarasin nen a garens: le dreit comande que le Franc ne deit pas faire sairement au Sarasin, se aucune chose nen i avoit de recounoissance.", E.Kausler, *Les Livres des Assises et des Usages dou Reaume de Jerusalem*, 89, Assisse LIX.

44 Bechor, *God in the Courtroom*, 30–34.

45 Ibid., 339–41.

46 Laiou-Thomadakis: "Institutional Mechanisms of Integration", 170.

out as exceptions in the history of cross-confessional conflict resolution, and indeed they found an unexpected parallel in the late Mamluk Siyāsa courts that I will discuss below. The Crusader witness system was not only protective of minorities when their transactions were notarized at the Fonde, but also when they were concluded in the marketplace, because the cross-confessional nature of witnessing was imposed on all plaintiffs, meaning that no actor had the upper hand in terms of production of proof. In contrast, in a sharīʿa court, it was the minority defendant who might look for Muslim witnesses to plead his case. This potential threat to the advantageous legal position of feudatories, I must stress, applied only to commercial litigation, and was limited to bourgeois jurisdiction. At the opposite end of the Crusader system of adjudication, as expressed in the *Livre des Assises* for the High Court, the early, abiding bias against non-baptized witnesses was absolute for cases involving nobles.[47]

As described in bourgeois law books, biases against certain witnesses were considered to be functional to cross-confessional relations. This is epitomized in the lawsuit involving two equal parties; in derogation of the general principle that nobody could bear witness in the High Court against someone of a different confession ("nul ne peut porter garantie en la haute court contre persones qui ne sont de sa nacion"), in a case involving, say, two Christians, and therefore deprived of cross-communal resentment, other unbelievers, such as Muslims, might allegedly be able to testify, since they could not be targeting either party on religious grounds. In terms of its applications in the Fonde at least, the Crusader witness system epitomizes the notion that dealing with diversity was clearly a legal sphere of its own, where superior—and often formalistic—legal principles were not overtly contradicted, but subject to a technical framework of implementation.

3.2.5 *Crusader Cyprus*

The Crusaders' legal system went through still further transformations when it was exported to its offshoot society in Cyprus in 1291. We know that

47 "Et qui viaut prover par garans fié ou chose de fié ou autre chose, il covient qu'il soient ii loiaus garans ou plus de la loy de Rome, se ce n'est de prover aage ou lignage. Que l'on peut prover ses ii dites choses par chascun home ou feme, mais que il soit crestien batié et qu'il soient ii ou plus et que il s'acordent bien ensenble a une parole," Edbury, *John of Ibelin: Le livre des assises,* 165, and again: "Ne gens de nassion qui ne sont obeissans a Rome ne pevent porter garantie en la haute cort, se ce n'est contre celui ou ceaus qui sont de sa nacion que des ii dites choses. Que nul ne peut porter garantie en la haute court contre persones qui ne sont de sa nacion," 167. This point is central to Edbury's view of the Crusader legal system, Edbury: "Latins and Greeks on Crusader Cyprus".

in Palestine the Cour des Syriens appears to have been absorbed by the Cour de la Fonde, perhaps because it comprised Syrian jurors. If the Cour de la Fonde disappeared in Cyprus, we know that the need to deal with cross-confessional conflict did not diminish. The Cour des Syriens, headed by the *raïs*, and comprising at least two jurors and a scribe, persisted. It passed verdicts in civil cases and served the purposes of registering contracts, and had jurisdiction in cases where at least one of the parties could claim some 'oriental,' Greek or Syrian, status; however, apart from constituting a privative legal forum for Syrians, it was not competent on commercial litigation.[48] The disappearance of the Fonde has been attributed to the fact that there were no Muslims living on the Island. Indeed, Marwan Nader has noted that the courts had difficulty providing enough jurors of Latin stock, hence jeopardizing the abiding principle that Latins should only be judged by their peers. Apart from the mutations in the Kingdom's social composition after 1291, in Cyprus Crusader law entered into competition with the legal systems of the Italian communes, and with other commercial courts. The subsequent transfer of ownership of the city of Famagusta from the Lusignan Crusader monarchy to the Commune of Genoa in 1373 was responsible for introducing Genoese status and jurisdiction to the Island. Venice also promoted her interests in Cyprus by granting citizenship privileges or by recognizing the Crusader-Venetian status for some Cypriots.[49] Although in the beginning it appears that some commercial courts are mentioned by the sources to have been under royal jurisdiction, by the fifteenth century the Italian city republics had set up consular tribunals throughout the Mediterranean, and exported their own courts to subjected territories like the city of Famagusta.

48 Richard, Jean: "La cour des Syriens de Famagouste d'après un texte de 1448", in: *Croisades et États latins d'Orient: points de vue et documents*, 383–398: Aldershot, 1992, Nader, *Burgesses and Burgess law*, 138–42, Otten: "Quelques aspects de la justice à Famagouste pendant la période génoise", 339–41.

49 Jacoby, David: "Citoyens, sujets et protégés de Venise et de Gênes en Chypre du XIIIe au XVe siècle", in: *Recherches sur la Méditerranée orientale du XIIe au XVe siécle: peuples, sociétés, économies*, London: Variorum Reprints, 1979, 159–188, Balard, Michel: "La Massaria Génoise de Famagouste", in: *Diplomatics in the Eastern Mediterranean 1000–1500: Aspects of Cross-Cultural Communication*, edited by Alexander D. Beihammer, Maria G. Parani, *et al.*: Brill, 2008, 235–249, Balard, Michel: "Note sull'amministrazione Genovese di Cipro nel Quattrocento", in: *La storia dei Genovesi; Atti del Convegno di studi sui ceti dirigenti nelle istituzioni della Repubblica di Genova, Genova 11–14 Giugno 1991*, Edited by Tipolitografia Sorriso Francescano,XII, parte 1, 83–94. Genova 1994.

On Cyprus, a commercial tribunal in Nicosia was held by the Venetian Bailo, while that in Famagusta was held by the Genoese Capitano.[50] The Venetian consul in Famagusta sat in justice for Venetian residents, for cases worth less than 50 Byzantines, however this threshold did not apply for travelling merchants (*homeni de passazo*).[51] The logic behind this exception was that travelling merchants found themselves in a weaker position, and were therefore in need of additional legal protection—a notion to which I will return, as it played an important role in the development of Mamluk Siyāsa. Like those in action in the Italian cities, a Genoese *Mercanzia* court heard cases under the loggia at Saint Francis' Church in Famagusta (where a Genoese coat-of-arms can still be seen), and similar tribunals such as the *bailli du comerc* are mentioned for the 14th century.[52] After the transfer of royal dominion over Famagusta to the Genoese, the *Podestà/Capitano* was endowed with jurisdiction over the *bourgeois*, but also over the Syrians, Jews and Franks who had traditionally fallen under the jurisdiction of the Fonde. Indeed, the treaty of 1374 foresaw that the Commune should apply the law according to the customs and assises of Cyprus.[53] In terms of actual judicial archives, the Cour of the Capitano is the only one to have yielded a consistent series of proceedings, and if on the one hand these documents offer a window into the legal world of their Crusader predecessors, on the other it departs from the Eastern Mediterranean tradition of counterbalancing religious discrimination through the application of a set of technicalities and notions of legal particularism. The new legal regime brought about after Famagusta was incorporated to Genoese dominion resulted in a freer approach to the rigid Crusader notions of status. Thus, under Genoese rule, we find the Capitano's court recognizing the right for some Jews and Syrians to enjoy the status of *burgesses*.[54] This went a step further under subsequent Italian rules, such as the Venetians who took over

50 Otten, Cathérine: "Les droits du consul des Vénitiens à Famagouste au xve siècle", in: *Mélanges Cécile Morrisson*, 619–631. Paris: Collège de France—CNRS, 2010, Otten: "Le registre de la Curia".

51 "Che'l possa far rason a Venitiani fina ala suma de bisanti 50 de Famagosta et da là in zoso, non astrenzando a questa quantità i homeni de passazo, sì in domandar come in responder, ai quali el possa far raxon de ogni quantitade." *Otten: "Les droits du consul des Vénitiens à Famagouste au xve siècle",* 629.

52 ASG SG 590/1292, 75v, mentions a panel of four arbitrators of the "officium mercantie" in Famagosta.

53 Otten: "Quelques aspects de la justice à Famagouste pendant la période génoise", 337.

54 ASG SG 590/1288, 76, 127, ASG SG 590/1290, 57r., Musso, Giangiacomo, "Gli ebrei nel Levante genovese: ricerche di archivio", La Berio a. 10., n.2 maggio-agosto (1970), 5–27, 21–2, confirms this with additional notarial evidence.

the Island a century later, and who suppressed the remnants of the Crusader courts, including the Cour des Syriens.

Tantamount to the merging of Bourgeois and Fonde jurisdictions, these communal courts should have attracted a great deal of cross-confessional litigation, as in Cyprus acquiring Venetian or Genoese citizenship could be most valuable to anyone involved in trade. Besides Latin migrants accessing citizenship, even Syrian Christians and Jews embraced some lesser Genoese and Venetian status in Famagusta (the so-called 'white' Genoese or 'white' Venetians), and one could be a citizen of these republics while still remaining a vassal of the Lusignan king according to feudal law. Finally, Genoa and Venice fostered the progressive transfer of commercial and cross-confessional cases to the new courts sponsored by the Italian powers. In this context, the Capitano attracted the most cross-confessional civil cases, and his jurisdiction was extended to cover commercial disputes.

Together with the statutes of the Venetian consular court, the Capitano is probably the only tribunal for which proceedings have been preserved. Although Muslims do not appear in the surviving ledgers, it is clear that fair justice was at least delivered by the Capitano in cross-confessional cases. Armenians, Copts, and a group of Arabic-speaking Syrian Christians referred to as *Fazolati* appear in the preserved ledgers, both as defendants and as plaintiffs. Although the court was used by Latins of Crusader lineage, such as the so-called 'white Genoese,' the frequency of instances of *fazolati* and Jewish plaintiffs in the registers makes it clear that this court can be considered the Fonde's true heir.[55] At the Capitano, contracts between Jews and involving Jewish and Christian parties were notarized; in my view, the notarizing role of this court responded to a 1374 injunction to observe the ancient Crusader laws ("gubernare burgenses ... secundum usus et asisias regni Cypri"). Thus we find in the registers a Jew named Azariel who had a loan registered at the Court of the Syrians, and which he produced before the Capitano when he took his debtor to court. Jews sued each other and notarized arbitral agreements between each other at the court, and, although the scribes do not record any of the parties' origins, it seems that it also served as a forum for disputes arising in Nicosia and in other cities. At the Genoese court of Famagusta, for example, debtors could be arrested following denunciation by a Jewish creditor.[56] The Capitano, moreover, enforced decisions made by the *Mercanzia* and the Court of the Syrians. In addition, the court registered the appointment of courtiers charged with the supervision of

55 Jew Salomon Habibi sued a woman named Levantina, ASG SG 590/1291, 95r.

56 ASG SG 590/1290 November 1438, f. 64, 20 March.

various crafts and market activities. Among these *censarii* we find some women, and market brokers exhibited some degree of confessional variety, including Jews, Armenians, Fazolati and White Genoese. Aspirants to the censarii post were sponsored by a patron, who then signed the act of appointment—and indeed, in one case we find a Jew acting as guarantor.[57]

It is obvious that from a late medieval perspective Courts such as those of the Genoese permitted important procedural advantages for parties. The main benefit was that plaintiffs could produce written evidence that would be considered by the court without the intervention of witnesses. Unlike the situation in Crusader courts described by Pegolotti, where notarial deeds had no probative value, they were freely accepted by the judge. And indeed, plaintiffs produced all kinds of notarial documents before the Capitano. Deeds drawn up by Venetian notaries in Damascus turned up during the hearings, hence confirming that the Middle East was included in a Latin, imperial notarial oecumene.[58] Greek notarial documents were just as readily translated and incorporated into the proceedings by the court notary. Hebrew documents were the object of sworn translations and used as a basis for judicial decisions. In contrast, apart from the case of an apostate, Muslims never appeared before the Court, nor are Islamic deeds mentioned, although we know that Muslims had been present on the island since it had become tributary to the Mamluks in 1420.[59] I have come across only one instance of a Muslim involved in a judiciary act, and the way written evidence was handled in this case is very significant. In April 1455, the Capitano had a vessel seized at Famagusta harbor, on the basis of a claim by Giorgio Manson, an individual of Syrian-Crusader origin who had entered into business with a certain Ali Sulumano of Tripoli, who owned the boat jointly with two Latins. Solumano owed a good deal of money to Manson, according to an acknowledgement of debt exhibited to the court, although, meaningfully enough, it does not mention in which language this document was written. Nor did the court, contrary to its own practice with notarial deeds produced elsewhere, appear to find it useful to have the document incorporated into the proceedings. The reader is left none the wiser, therefore, as to whether the court accepted evidence produced in Islamic lands, and if Muslims could be admitted as real actors of the legal system, or simply have their goods seized.[60]

---

57 ASG SG 590/1291, 412–14, ASG SG 590/1292, 187–201.

58 ASG SG 590/1289, f.106v.

59 ASG SG 590/1288, 114v, a certain Petro de Soria *olim saracenus*.

60 ASG SG 590/1291, 123v.

Nowhere in the Genoese legal system are religious-based limitations against minorities expressly stated. In general, metropolitan jurisprudence never refers to how proof advanced by unbelievers or in alien languages should be dealt with, for instance, by notaries. In this, the Genoese legal system matched the Venetian one, and more broadly, those related to the continental tradition of the Ius Commune, and differed from the Islamic, Byzantine and Crusader ones. Yet it has been noted that, as witnesses, the Jews of the Genoese colonies appear with less regularity in the registers than Greeks, and, when they do, it is uniquely as witnesses to the transactions of their coreligionists.[61] The legal historian Elisabeth Santschi has remarked that, as regards minorities under Venetian dominion, the *Statuti* of Venice applied to everyone, although some legal particularisms were observed for family law.[62] Although she does not provide empirical evidence for court practice, she notes that Jews in the colonies could testify either for or against each other, but also against Christians. However, this contrasts with the praxis observed by Venetian notaries in Alexandria and Damascus, where Muslims and Jews only signed documents that included at least one non-Christian party. This can be observed, without exception, in a serial survey of deeds running between 1360 and 1450.[63]

At this point, we can draw several conclusions from the Crusader legal system and its transfer to Cyprus. First, that the legal systems of the Latin city states differed from Middle Eastern ones in that they lacked explicit legal biases. However, despite this difference, it should be noted that in Latin European market and court practice, the role of non-Christians as actors of the legal system was limited, and this practice had features common to the Islamic, Byzantine and Crusader 'witness systems.' Under all three of these Eastern Mediterranean regimes, justice was indiscriminately dispensed to believers and unbelievers alike. In the podestral court of Famagusta, in commercial courts such as those in Florence, and even in Venice, before the *giudici di petizion*, whose *sentenze* have come down to us, this did not happen, since

61 Becker, Brian Nathaniel: *Life and Local Administration in Fifteenth-Century Genoese Chios*, Ph.D. Thesis, Western Michigan University: ProQuest Dissertations and Theses, 2010, 213, Balard, Michel: "Il notaio e l'amministrazione della giustizia nell'oltremare genovese", in: *Hinc publica fides. Il notaio e l'amministrazione della giustizia*, edited by V. Piergiovanni,. Milan: A. Giuffrè, 2006, 355–369, Balard quotes an interesting case in which non-Latins inhibited from appearing in court. Argenti, *The religious minorities of Chios: Jews and Roman Catholics*, 100–41.

62 Santschi, Elisabeth: "Contribution à l'étude de la communauté juive", 177–211.

63 ASVe, CI, N, B. 22, Notary V. Bonfantin, Jan. 17, 1393, June 28, 1419, ASVe, Notarile Testamenti, B. 215, Notary S. Peccator, May 2, 1448, Oct. 5, 1448, Oct. 14, 1448, ASVe, CI, N, B. 211, Notary N. Turiano, f. 6v, May 21, 1455.

Muslims were always absent, even if the city counted with a nurtured community of Ottoman merchants. If the legal systems of Genoa and Venice had opener attitudes towards diversity, paradoxically enough, in practice unbelievers enjoyed even less access to the legal system, and Muslims never appear as actors or subjects of it in the surviving records. In this, the Genoese tribunal in Cyprus sets a definitive line of demarcation with its Crusader past.

3.3 The *actor sequitur forum rei* Principle

If the Cour de la Fonde provided a more neutral forum for commercial litigation, it would not be fair to say that Muslim rulers were exclusively interested in asserting the supremacy of sharīʿa and of the local qadi courts. Indeed, a most interesting development in the field of adjudication is the inhibition of the sharīʿa and the acceptance of alternate systems of mutual enforcement—that is, by defining a procedure involving both foreign and local courts. The general principle by which Roman law defined the geographical jurisdiction under which a given case should be tried is known as *actor sequitur forum rei*, according to which disputes were heard in the court that had jurisdiction over the defendant, rather than the plaintiff. Again, this abiding *actor sequitur* principle echoed Byzantine and Crusader procedure, in whose courts disputes between burgesses from different cities were heard in the court of the defendant.[64]

Actor sequitur was the main principle governing the settlement of mixed disputes in treaties between European and Muslim states, and in particular with the early Mamluks, which stipulated that it was the religious and legal status of the defendant that determined the court in which a dispute should be tried. In cases involving a Muslim party, the *actor sequitur* principle stood at odds with the supremacy of sharīʿa in Islamic societies, since it implied that, if the defendant was a Muslim, the case should be heard by an Islamic judge, but that if a Muslim sued a Frank the issue should be transferred to the latter's consul.

The *actor sequitur* principle was adopted for specific times and places, and was not applied universally. Kate Fleet has found evidence for mixed panels comprising Frankish and indigenous judges for pre-Ottoman Turkey and the Tatar khanates, however the sources are not sufficiently clear on the extent or

64 Nader, *Burgesses and Burgess law*, 154. Macrides, Ruth J.: "The competent court", in: *Kinship and Justice in Byzantium, 11th-15th Centuries*, Ashgate 1999, 117–130.

regularity to which such solutions were resorted to.[65] Use of the *actor sequitur* principle is documented for Islamic Iberia and Christian societies during the early Reconquista, and in a commercial context it was sanctioned in very early Mamluk treaties. According to the Mamluk-Venetian treaty of 1254, signed with the Mamluk sultan Muʿīzz al-dīn Aybak (1250–1257), this principle stood as a major guideline for adjudication:

> Item, quod, si aliquis Sarracenus clamauerit se de aliquo Veneto, diffiniatur causa ante consulem Venetorum. Et si aliquis Venetos proclamauerit se de aliquo Sarraceno, diffiniatur ratio ante illum, qui fuerit loco Soldani; et potestatem habeat consul faciendi rationem inter ipsos.[66]

In the Maghreb the same principle of transferring jurisdiction to the defendant's forum is mentioned in the Ḥafsid treaties with Genoa (1343) and Pisa (1397).[67] This put Frankish judges in the position of potentially passing verdict on Muslims, which in turn represented a challenge to the dominance of sharīʿa within the Dār al-Islām. The Venetians seem to have promoted this cross-communal adjudication system in their treaties with Lesser Armenia (1307), Cyprus (1306) and even in Tabriz (1320).[68] This makes it all the more intriguing that it should be adopted in the first Mamluk treaties with European powers, which may have drawn extensively from Crusader practice. As we saw earlier, it is still up for debate whether the Crusader legal system was truly as pluralistic as some would have it believed, to the extent of recognizing parallel jurisdictions for every community, even for Jews and Muslims. What is clear, however, is that aside from defining a forum for mixed disputes at the Cour de la Fonde, it privileged the defendant's confession by imposing the burden of proof upon his own community witnesses.

The 1290 treaty between al-Manṣūr Qalāwūn and Genoa confirmed the initial Mamluk tendency to rely on the *actor sequitur* principle. In it, the Islamic ruler bestows upon foreign consuls the right to pass judgment in cases where

65 Fleet, Kate: "Turks, Mamluks, and Latin Merchants: Commerce, Conflict, and Cooperation in the Eastern Mediterranean", in: *Byzantines, Latins, and Turks in the Eastern Mediterranean world after 1150*, edited by Jonathan Harris, Catherine J. Holmes, *et al.*, Oxford University Press, 2012, 327–344, 341, Fleet, "Turkish-Latin Relations", 611, Orlando, Ermanno, "Venezia, il diritto pattizio e il commercio mediterraneo nel basso medioevo", *Reti Medievali Rivista* 17 (1) (2016), 3–33, mentions the treaty of 1342 with Zanibech.

66 Tafeland Thomas, *Urkunden zur älteren Handels*, 487.

67 Amari, M.: *I diplomi arabi del R. Archivio Fiorentino*, Florence: Le Monnier, 1863, 320, Valérian, "La résolution des conflits", 557.

68 Orlando, "Venezia, il diritto pattizio e il commercio mediterraneo nel basso medioevo", 17.

the plaintiff is a Muslim; however, if the lawsuit is initiated by a Frank, the adjudication must go in the opposite direction, to the sultans' officers ("deffiniatur ratio ante illum quo fuerit loco soldani").[69] It has to be noted that the Mamluks do not seem to have dwelt much upon the precedent set by their immediate Muslim predecessors, the Ayyubids, who in 1238 concentrated all jurisdiction for mixed cases in the hands of Islamic judges.[70] The *actor sequitur* rule was not only enforced for interfaith cases, but also for disputes opposing two Franks.[71] For the Byzantine context, Angeliki Laiou and others have interpreted the transfer of disputes to foreign courts as a sign of weakness resulting, predictably enough, from the Latin takeover of Constantinople during the Fourth Crusade. In his discussion of the 1323 treaty, or even when addressing maẓālim jurisdiction, Dominique Valérian shares this tendency to explain all decisions made by Muslims as dictated by an unfavorable balance of power.[72] However, the balance of power between Crusaders and Muslims was much more advantageous to the Mamluks than to the Ayyubids, hence explaining the adoption of the *actor sequitur* principle on this basis appears questionable.[73]

3.4 Empowering One Consul over the Others

A technique similar to the *actor sequitur* principle, for ensuring intra-communal dispute resolution amongst non-Muslims, was to empower one consul over the others. This seems only to have been used for cases between Franks of different origins, and the granting of such privileges seems to have concerned mainly the Maghreb. This approach probably has its roots in the Islamic tendency to deal with non-Muslims in an inter-communal framework, and therefore to consider all Franks to be a single community governed by the laws of Islamic

69 Holt, P. M.: *Early Mamluk diplomacy, 1260–1290: treaties of Baybars and Qalāwūn with Christian rulers*, Leiden; New York: E.J. Brill, 1995, 145.

70 "Item, si aliquis Venetus habuerit placitum cum aliquo Christiano, diffiniatur ante Consulem. Et si habuerit placiturr cum Saraceno, diffiniatur ante justitiam terre. Et sic de hoc respondimus.", Tafel and Thomas, *Urkunden zur älteren Handels*, 338.

71 "Item, quod si aliquis Florentinorum injuriatus aliquem fuerit, quod ejus consul teneatur eum punire juxta ipsorum rictum. Et si alius offenderit Florentino, quod consul offendentis teneatur ipsum punire.", Houssaye Michienzi, Ingrid, *Datini, Majorque et le Maghreb (14e-15e siecles): réseaux, espaces Mediterranéens et stratégies marchandes*, Leiden, Brill, 2013, 173–4, 1421 treaty between Florence and Tunis, article 2.

72 Valérian, "La résolution des conflits" , 557, 563–4.

73 Humphreys, R Stephen: "Ayyubids, Mamluks, and the Latin East in the thirteenth century", *Mamluk studies review*, 2, 1998: 1–17, 10.

legal pluralism. Perhaps for this reason, Maghrebi rulers qualified the Pisan consuls to hear any dispute between Franks, irrespective of their political belonging. This clause is stipulated in the 1358 treaty with the Marinids, who gave precedence to the qādīs for mixed cases, but empowered the Pisan consul as judge among all Franks.[74] While the Marinids in the west and the Ayyubids in the east maintained the precedence of sharīʿa courts over cases involving at least one Muslim, the early Mamluks and the Ḥafṣids, in contrast, opened the door to the *actor sequitur forum rei* principle. Moreover, Mamluk and Ḥafṣid treaties do not necessarily refer to qadis, but mention instead the judicial functions of officials, such as those attached to the customs offices.[75]

In any case, after the second half of the fourteenth century all alternative solutions for dealing with mixed cases were progressively abandoned in the Eastern Mediterranean, and no subsequent Mamluk treaty mentions the *actor sequitur* principle. The 1496 agreement with Florence makes it clear that officials should pass judgment irrespective of the identity of the plaintiff or the defendant,[76] and neither the Mamluks nor the Ottomans empowered a single European consul over the others. In other words, while up until the fourteenth century for Islamic rulers and policy-makers it was conceivable that a fellow Muslim might appear before a Frankish judge—normally a consul—this tendency changed during the fifteenth century. In fifteenth-century treaties, the supremacy of sharīʿa is not challenged, and therefore no Muslim could be brought before a Christian judge. A similar tendency has been noted for the Turkish principalities, which progressively discarded the *actor sequitur* principle in favor of Islamic adjudication for mixed disputes, irrespective of who the defendant was.[77] As a minor exception to this principle, the Catalan treaty with

74 "E questo è il capitolo undecimo, lo quale havete domandato. Che se alcuno mercatante pisano havesse quistione con un altro Cristiano d'altra lingua, che sia la quistione dinanzi del vostro consolo; salvo che se la quistione fusse grande che portasse pondo, che vengha a sententiarla al cadi' della terra. E quando nel luogo non havesse consolo e la detta questione fusse, che la veggia tra loro lo aveli (al-walī) de la terra, e sino lo signore del castello. Et habbiamovelo conceduto questo. E quando la quistione fusse dal Saracino al Cristiano, che torni alla ragione de' Saracini e de' loro cadi." Amari, *I diplomi,* 311.

75 Art. 5 of the treaty between Pisa and Tunis: "Et se alcuno Saracino si ramaricherà d'alcuno Pisano, sia tenuto farlo richiedere dinanzi al consolo de' Pisani; e'l consolo debba quegli spedire et fargli ragione; et se questo non facesse, allora et in quel caso il Saracino si possa lamentare al signore della doana. Et se alcuno Pisano, o chi per Pisano sia astretto, vorrà o dovrà adomandare d' alcuno Saracino, o da alcuna altra persona che sia sotto la pace del detto re, allora il Pisano debba adomandare ragione in doana; et la doana sia tenuta di fare a lui ragione, et quello da lui spacciare," ibid., 320.

76 Ibid. 188, "min muslim ʿalā bunduqī aw ʿalā muslim min bunduqī."

77 Fleet, "Turkish-Latin Relations".

Barsbay allowed Muslims to have recourse to a merchants' arbitration panel if they so wished.[78] Technical arrangements for the resolution of conflicts were sometimes at odds with legal theory, since no Muslim jurist would have agreed to hand a case concerning Muslims over a Frankish consular court. The Mamluks had a similar tendency to transfer disputed cases from other nations to the Venetian consul, even if this was never sanctioned in the treaties, and was often the source of bitter complaint from, for example, Genoese merchants.

The transfer of litigation to the defendant's court gradually faded from the adjudication system in the East over the course of the fourteenth century, acknowledging the rule of islamic law over the issues derived from the presence of foreigners. However, this tendency to abandon the *actor sequitur* principle cannot exclusively be interpreted as a bid to renounce legal pluralism. Islamic rule of law did not prevent, but fostered consular jurisdiction, provided European consuls limited themselves to arbitrating intra-communal disputes alone, criminal cases excluded. Although the fourteenth century saw a progressive fixing of cross-confessional litigations in the field of action of Islamic judges, either qadis or officials, out of court many cross-confessional interactions were dealt with by notaries and consular institutions. Imperial investiture granted notaries the title of *judex ordinarius*, who had ample judicial prerogatives and assisted private arbitration panels. The attribution of cases to consular officials had been developing since late medieval times, not by virtue of any exceptional privilege, as has often been assumed, but due to a growing Islamic acceptance of the legal particularism of dhimmīs and mustā'min communities. Moreover, the courts and notaries in the Italian colonies, as well as those under consular jurisdiction, tended to level differences between Latin Christians and aliens, admitting anyone as an actor of the legal system.

Instead of putting an end to legal pluralism, the unprecedented ascendancy of consuls and their notaries went hand in hand with the rise of an Islamic, yet alternative jurisdiction to the qadi courts: the Siyāsa courts. From the mid-14th century, and up to 1517, treaties such as the agreements between the Mamluks and Cyprus started to mention the transfer of mixed cases to Siyāsa officials, rather than to qadis in sharīʿa courts. In the Arab Middle East, beyond informal solutions such as arbitration by peers, commercial litigation was solved either in the framework of communal institutions, or, in all mixed cases, before the officials presiding over Siyāsa courts.

The disappearance of *actor sequitur* clauses for mixed cases coincided, at least under the Mamluks, with the rise of Siyāsa as a legal theory, and with

78 Ruiz-Orsatti, "Tratado de Paz".

its materialization as a mechanism for enforcing court decisions. The *Siyāsa* courts, rather than displacing previous practice, came as an added layer to a widespread tradition of relying on ad hoc, technical solutions to the problem of governing amidst diversity. These technical solutions included, as we have seen, a panoply of devices that encompassed the taking of oaths, the selection of witnesses on religious grounds and the acceptance of cross-confessional instances of adjudication. The precedent that the Mamluks set from this moment on was that they repositioned the problem of coping with diversity in the realm of Islamic law. If the Roman-law *actor sequitur* principle, the ultimate expression of this late medieval solution to mixed litigation, persisted in Castile, in contrast in the East it was abandoned entirely.

3.5 An Iberian Epilogue

Some of the elements that had made cross-confessional relations possible persisted under different conditions in the far Western extremity of the Mediterranean basin, in Castile. From a second-rank polity with narrow territorial claims in northern Christian Iberia, Castile underwent a social transformation accompanying its political expansion, at the expense of both neighbors and infidels. Expanding from a peripheral county to a hegemonic Iberian power from the 11th to the 13th centuries, it incorporated an unprecedented number of Muslim and Jewish communities under the aegis of a Christian king. The Battle at Navas de Tolosa in 1212 witnessed the unlocking of the Almohad stronghold in southern Spain, and led to the fall of the larger cities, such as Seville in 1248. Parallel to this expansion, several legal codes were issued in an effort to extend a uniform legal layer over heterogeneous communities. The laws in Castile, ranging from earlier ones such as the *Espéculo* to the late compilation of *Las Siete Partidas*, devoted a good deal of attention to issues of cross-confessional witnessing. In particular legal realities, as was the case of Seville, Islamic notions and ideas on governing diversity were adopted and rephrased by Christian rulers. As a result, in Reconquista Castile, Muslims found themselves bound by similar laws of obligation as those that had been imposed on dhimmīs under Islamic rule.

In 1251, the Genoese concluded an agreement with the Castilian King that exhibits numerous analogies with amān treaties.[79] Among these, the Genoese

79 Gourdin: "Les marchands étrangers ont-ils un statut de dhimmi?", Gallego, Isidoro González, "El Libro de los privilegios de la nación genovesa", *Historia. Instituciones. Documentos* 1 (1974), 275–358.

were granted permission to enjoy a public bath and quarters, and consuls were considered not as exogenous magistrates but as officials appointed by the King, in the same fashion that Mamluk and Ḥafṣid rulers did. More interestingly for our argument, adjudication was guided by the *actor sequitur forum rei* principle, whereby a Castilian plaintiff could potentially end up before the Genoese Consul if he sued a Genoese citizen. For two centuries, the Castilian monarchs mirrored the dhimmī status, imposing it upon Muslims and Jews, and reserved a similar treatment for both foreign merchants and religious minorities. Forms of legal particularism and private jurisdiction and courts were recognized in Seville to Muslims, Jews and Genoese merchants, then extended in 1282 to Catalan traders. The 1251 treaty distinguished between Genoese residents and those "from abroad," considering the former to be local subjects, just as it happens to mustāʾmins remaining in Islamic land after the expiration of their amān.[80] The 1251 document expresses the King's prerogative to judge criminal offenses, as well as other technicalities of adjudication also present in Ḥafsid treaties; *mutatis mutandis*, the Castilian text mirrors Islamic amān practice and theory.[81]

As for the Islamic biases against witnesses, they were grafted into a specific legal context; that of the early codes such as the *Fuero Real*, which uniquely envisaged the action of Christian witnesses. The process has been documented by Mélanie Jecker, who has examined the codification of local laws under Alfonso X of Castile, although in her interpretation the novelty of it can be found not in the influence of sharīʿa, but in the revival of Roman law. In fact, early Castilian law books do not mention any particular biases, and the first allusions to applying alternate methods of oath-taking to Muslims and Jews appear only in late appendixes to the *Fuero Real*. The *Espéculo* announces its intention to ban non-Christian witnesses for cross-confessional disputes, although it allows exceptions for cases where no Christian witnesses are available. In this, the thirteenth-century Castilian code drew direct inspiration from Qurʾān II: 282, which elaborates on the need to have recourse

80 "E estos consoles que non puedan judgar ningund juyzio de sangre nin puedan judgar a vezino de Ia çibdat de Sevilla mas que iudguen entre los genueses que vinieren de fuera que non fueren vezinos de Sevilla. E si por aventura el ginoes que viniere de fuera oviere querella del vezino de Sevilla quel lieve antel fuero e los alcaldes de Sevilla e si el vezino de Sevilla oviere querella del genues que viniere de fuera quel lieve otro si ante los consules.", Gallego, "El Libro de los privilegios de la nación genovesa", 290.

81 For example, the right for an indigenous subject to appeal to his own institutions if unsatisfied with the foreign consul's verdict, see Amari, *I diplomi*, 320, quoted above.

to women and unbelievers when male, Muslim witnesses are not available.[82] When foreseeing the absence of believers to act as witnesses, the Revelation refers to the historical reality of the early ḥijra, when Muslims were a dominant minority and no real project of mass conversion was even on the agenda of Islamic governance. This situation of isolated communities is with all certainty what the Castilian lawmaker also had in mind, leaving few doubts as to crosspollination between the two texts, and more generally between Castilian law and Islamic *fiqh*. Like in Islamic jurisprudence, emphasis is placed on the trustworthiness of these Islamic witnesses vis-à-vis their own communities, a point brought up by contemporary thinkers such as Ibn Qayyim, in a passage on minority testimony included in his Siyāsa treatise.[83] Lastly, a direct parallel can be traced between Castilian Law and Mālikī doctrines—dominant in al-Andalus—which allowed for torture if a defendant was found guilty of public infamy.[84]

The author of the *Espéculo* seems to mirror, *mutatis mutandis*, the Islamic idea that Revelation, rather than contradictory breaks, has experienced a historical continuity. A crucial notion in Islam, it is reflected in the fact that conversions to faiths other than Islam are not allowed.[85] In the *Espéculo*, in contradiction with the general rule in force for mixed marriages, a Jew who had converted to the last version of the divine message, hence embracing Christianity, should not be allowed to keep his former Jewish wife.[86] The lawmaker stresses the continuity between both versions of the Revelation, the Jewish faith

---

82 *Opúsculos legales del rey don Alfonso el Sabio, Tomo I, El espéculo o espejo de todos los derechos*, edited by Real Academia de la historia, Madrid,: Imprenta Real, 1836, 194, "Testigo non deve seer ome que sea de otra ley, asi como judio, o moro, o herege, o ome que aya otra crencia que non sea de la nuestra. Ca atal como este non puede testiguar contra christiano, sinon si fuer en algún fecho malo que feziese alguno, o quisiese fazer, o fuese en conseio de lo fazer contral rey o contra el regno, o en otro fecho malo que feziese otrosí, en algún logar que non acaesciesen y christianos con que lo podiesen provar."

83 Ibn Qayyim al-Jawzīyah, *al-Ṭuruq al-ḥukmīyah*, 512, "idhā qabalna shahāda baʿḍahum ʿalā baʿḍ iʿtabarnā ʿadālatahum fī dīnihim."

84 *El espéculo o espejo de todos los derechos*, 194, "Mas si aquellos que fuesen acusados desta manera fuesen en ante enramados dotro fecho malo, dezimos que el testimonio destos que dudemos con el enfamamiento, que aquellos acusados avian ante, es ayuda para metello a tormento para saber la verdat de aquel fecho.", Johansen, "Signs as Evidence", 170.

85 Friedmann: "Classification of unbelievers in Sunni Muslim Law and Tradition", 66–9, Emon, *Religious Pluralism*, 66–7.

86 "Otrosí dezimos, que si el marido e la mugier fueren de señas leys, e seyendo en uno denostare el de la otra ley al que fuere christiano, ol conseiase cosa por que pierda su alma, si el christiano se quisier partir del que fuer dotra ley, non deve el otro seer entregado del, maguer le demande." *El espéculo o espejo de todos los derechos*, 382–3, Ley XXXIV.

"being the beginning and the precedent of ours".[87] Following a similar logic as that evoked for ḥarbīs and their lack of interest to preserve the common good, thus it was assumed that a Jewish woman would tend to drag her husband back to the previous version of the message expressed by the Jewish prophets.[88] This principle echoes Qur'ān 60:10, dealing with the problem of women who emigrated with the Prophet to Medina and to whether or not they remained the wives of unbelievers back in Mecca. If marrying a dhimmī wife is lawful, in this case the marriage contract has been severed by conversion, hence the Qur'anic injunction "do not hold on to your marriages with unbelieving women". The Castilians consider Moors and heathens not exposed to the same risk as Jews, since their beliefs could not be proved "by prophets and saints," and who would therefore be supposedly less tempted to return to their original unbelief.

Similarities between Islamic and Castilian law can also be found in discussions about the transfer of proof from one judge to another,[89] and in particular the application of a special procedure for the taking of oaths. A great amount of detail was given on the words that needed to be said, on the sacred character of the places of worship and on the use of sacred texts. In this, Castilian law needs to be read against the backdrop of late medieval developments in Islamic law, more precisely notarial manuals such as *Jawāhir al-ʿuqūd* by Muḥammad Ibn Shihāb al-Dīn al-Suyūṭī (1413/1414–1475) or al-Saḥmāwī's chancery manual.[90] Saḥmāwī was deeply preoccupied with the issue of oaths, and devoted a long section to the topic, identifying specific formulae for the different confessions of Jews, Samaritans, Christians, Zoroastrians, and an intriguing category of "Greek philosophers." As concerns the Franks, he provided the example of an oath taken to sanction a treaty in 1371.[91] The technical handling of cross-confessional relations in both Castilian and Islamic lands reached its peak in this period, with al-Saḥmāwī's list of a long series of non-Muslim sacred artifacts and notions, and similar elaborations on oaths made by Jews

87 "e es probado por muchas profetas e por muchos santos, e es la su ley comienzo e testimonio de la *nuestra*", Ibid., 383, Ley XXXV.

88 "los que se convertiesen a la nuestra ley [...] puñaríen de los engañar, e de los tornar a la su creencia, e sacarlos de la nuestra", Ibid., 384.

89 Ibid., 205, Ley XXII. Hallaq, Wael B., "Qadis Communicating: Legal Change and the Law of Documentary Evidence", *al-Qantara,* 20 2, 1999, 437–66.

90 al-Suyūṭī, Muḥammad Ibn Shihāb al-Dīn (1413 or 14–1475): *Jawāhir al-ʿUqūd wa-Muʿīn al-Quḍāh wa-al-Muwaqqiʿīn wa-al-Shuhūd* Edited by Muḥammad Ḥāmid al-Fiqī, 2 vols. vol, Cairo: Maṭbaʿat al-Sunnah al-Muḥammadīyah, 1992, II, 339–352.

91 al-Saḥmāwī, *al-Thaghr al-bāsim,* 853–95.

and Muslims from the author of the *Espéculo* and the *Partidas*.[92] The text of the *Partidas* recommends oath-taking at the synagogue, echoing Islamic law's ideas about the efficacy of validating an oath in houses of prayer. Elisabeth Santschi has detected the same interest in oath-taking in two rulings on the topic by the Venetian authorities in Crete.[93]

A further evolution can be noted in the later texts produced under Alfonso the Wise. Without falling in major contradiction, the *Partidas* deal with proof and unbelief with much more parsimony, and, for example, do not mention the issue of the availability of majority witnesses. The *Partidas* declare a total ban on cross-confessional witnessing, without explicitly accepting the validity of the ḥanafī exception; that is, that in Castile Jews could not give testimony for or against Muslims, and vice versa.[94] This ḥanafī exception is instead provided by the *Espéculo*, allowing for a technical acceptance of non-Christians as witnesses, provided, like in Islamic law, they were considered to be upstanding by their peers. Castilian law equally foresaw the participation of Muslims in cases of *lèse-majesté*, and where no upright Christians were available to testify.[95]

One is tempted to see an Islamic genealogy in the Castilian jurisprudence on these matters. Baber Johansen realized that, despite a backdrop of fundamental differences regarding truth and the purpose of the legal process, some legal issues developed along parallel lines in the late Middle ages. In several works, Johansen elaborates on the common interest that canon law, *ius commune* and sharīʿa had in allowing judicial torture as a means to get to the truth.[96] This issue, as we shall see in the following pages, was on the agenda of Mamluk rulers and jurists in the fourteenth century. The fact that such genealogies cannot be

92 *El espéculo o espejo de todos los derechos*, 406–9, *Las siete partidas del rey Don Alfonso el Sabio, cotejadas con varios codices antiguos por la Real academia de la historia*, edited by Real Academia de la historia Madrid, 3 vols., Madrid: Imprenta real, 1807, 485–7.

93 Santschi, "Contribution à l'étude de la communauté juive", 207–8. Bechor, *God in the Courtroom*, 122–4.

94 *Las siete partidas*, II, 519, "et aun decimos que home de otra ley asi como judio, o moro o herege, que non puede testiguar contra cristiano, fueras ende en pleyto de traycion que quisiesen facer al rey ó al regno; ca entonce bien puede seer cabido su testimonio, seyendo tal home que los otros de su ley nol podiesen desechar con derecho para non valer lo que testiguase, et seyendo el fecho averiguado por otras pruebas ó presunciones ciertas: mas quando los que fuesen de otra ley hobiesen pleyto entre sí mesmos, bien pueden testiguar unos contra otros en juicio et fuera del."

95 *El espéculo o espejo de todos los derechos*, 194, "Ca en tal manera como esta, tanbien deven yr sus testimonias de omes, que sean de otra ley, seyendo tales, que non los podiesen desechar de testimonio otros omes que fuesen de su ley misma."

96 Johansen, "Signs as Evidence", Johansen: "Vérité et torture. Ius commune et droit musulman entre le Xe et le XIIIe siècle".

definitively proved does not justify the tendency by scholars to interpret every novelty as reminiscent of Roman law. Lastly, repeated mentions of heathens (*gentiles*) in fourteenth-century Castilian legal codes probably echo Islamic jurisprudence, which since very early started nuancing between, on the one hand, Christian and Jews, and non scriptuaries on the other.

Islamic influences surface in another point of doctrine expressed in Castilian law, as regards the deciphering of witnessing. As I mentioned in the previous chapter, Siyāsa theorists were concerned with the problem of enlarging the narrow investigative methods traditionally granted to the qādī. Ibn Qayyim (1292–1350) deals with the issue of witnessing amidst diversity at three different points in his major work on Siyāsa, *al-Ṭuruq al-ḥukmīyah*. Also departing from an orthodox approach to the regime of proof, the judge is obliged to pay attention to signs and clues that might shed light on the witness's motivations. Contrary to traditional jurisprudence, he should not simply rely on the utterances of an *a priori* trustworthy, Muslim witness while neglecting his personal knowledge of the surrounding context in which witnessing is rendered. Confronted with a defendant who claimed to have regularly deposited a certain sum of money with the plaintiff over the last fifteenth years, the judge was advised to examine his purse and inquire about the coinage in order to ascertain whether the coins in the bag were actually in circulation at that time. Incidentally, Ibn Ḥijjah al-Ḥamawī (died 1434), included the same story in an anthology of prose entitled *Thamarāt al-Awrāq*, in which he stressed the dishonesty of the notaries who initially received the money as a deposit and later attempted to fraud the merchant by replacing the dinars with dirhams.[97]

The judge is also required to go beyond simply hearing the witnesses' depositions, and to scrutinize the "faces of the adversaries" (ujūh al-khuṣūm). In this way, the judge might acquire a certain expertise (*durbah*) in interpreting the true intentions of lying witnesses, so as "to avoid making mistakes."[98] Parallels with this approach can be found in Castilian law; Marta Madero has identified a passage in *Partidas 3.16.26* that describes the procedure to be adopted when hearing witnesses, advising judges to scrutinize the facial expressions of the speaker.[99]

97 Ibn Ḥijjah al-Ḥamawī, Taqī al-Dīn Abū Bakr Ibn ʿAlī (d.1434): *Thamarāt al-Awrāq*, edited by Muḥammad Abū al-Faḍl Ibrāhīm, Beirut: Dār al-Kutub al-ʿIlmīyah, 1983, 118, Ibn Qayyim al-Jawzīyah, *al-Ṭuruq al-ḥukmīyah,* 69.

98 Ibn Qayyim al-Jawzīyah, *al-Ṭuruq al-ḥukmīyah,* 71.

99 "et desde quel testigo comenzare á decir debe el judgador oirle mansamente, et callar fasta que haya acabado catandol todavia en la cara," *Las siete partidas,* II, 527.

Madero points out that this particular point draws from the writings of jurists such as Baldo, as well as previous classic authorities. In the same vein as the Castilian lawmaker, Ibn Qayyim insisted that the judge should inquire as to the circumstances in which the witness had acquired their knowledge of the facts.[100] Whatever the origins, coincidental or not, of these comparable procedures for ascertaining the veracity of oral testimony, they attest to a common tendency to attempt to rationalize procedure and the regime of proof, and to entrust the judicial process with the task of unveiling the truth.

Similarly to Castilian lawmakers, Ibn Qayyim dealt extensively with the unbeliever's credibility as a witness. He condones hearing non-Muslims in cases related to last wills and similar circumstances where no Muslims are available, and insists that dhimmīs propend to seek the truth in their community lives, that they are committed to the Islamic enterprise of governance and that Muslims, therefore, should rely on minority community leaders, the *aʿyān*. Unbelievers, thus, should be taken as eyewitnesses in the same manner as the Prophet relied on vernacular, non-Muslim guides during his journey to Medina.[101] The Qur'ān, moreover, recommends relying on Muslims, but does not forbid unbelievers from bearing witness. Internecine quarrels and religious hatred between confessions should not invalidate their witnessing, Ibn Qayyim reasoned, just as it did not for shīʿī and sunnī Muslims, who were often embroiled in similar disputes. The fact that non-Muslims were considered liars in religious matters did not make them liars to the judge, as they disobeyed the Revelation out of ignorance, and therefore could be considered to be acting in good faith. Muslims relied on the unbelievers' community leaders and its upstanding members for daily dealings, hence trusting their judgment—unless these individuals were considered to be notorious perjurers.[102] Together with infidels, Ibn Qayyim's arguments repeatedly mention merchants and debt issues as relevant examples, and arguments in favor of minority witnessing multiply in *al-Ṭuruq al-ḥukmīyah.* It is my contention that the issue of cross-confessional legal relations greatly inspired Mamluk Siyāsa in theory and practice, and, together with the presence of foreign merchants, contributed to new developments in proof and procedure in Islamic law, at least until a new agenda took priority under the Ottomans.

---

100 "et en aquel logar que fallare que dice que sabe el fecho debel preguntar como lo sabe faciendol decir por qué razón lo sabe, si por vista, ó por oida ó por creencia: et la razón que dixiere debela facer escrebir"; ibid. II, 527, Ibn Qayyim al-Jawzīyah, *al-Ṭuruq al-ḥukmīyah,* 65.

101 Ibn Qayyim al-Jawzīyah, *al-Ṭuruq al-ḥukmīyah,* 63.

102 Ibid., 483–5.

3.6 Siyāsa Justice in Theory and Practice

In spite of recent interest in Siyāsa as a special Muslim jurisdiction, its authority over foreigners has gone almost totally unnoticed. We know that mixed disputes were dealt with by some sort of 'secular justice'; however only an imperceptible legal shift from maẓālim to Siyāsa can explain the exact nature of these courts. The way in which Siyāsa emerged in the Mamluk context fits into the model of legal change conceived by Wael Hallaq, whereby jurists proved to be sensitive to issues related to, in this case, the presence of foreigners and unbelievers, a concern which in turn was reflected in their fatwās and writings. Siyāsa found its way into mixed cases through the "piecemeal modification of particular aspects of the law" and the articulation of a new doctrine about fatwās and commentaries, in response to new and "wide ranging circumstances."[103]

To further complicate the study of Siyāsa, court proceedings, if they ever existed, have not survived, and trials can only be reconstructed through descriptions provided by the Franks to their own notaries. The courts of the ḥājibs existed until the mid-fourteenth century as a special jurisdiction administering justice among the military.[104] Before that, the numerous bilateral treaties that regulated the activities of Frankish traders make no mention of the ḥājib, the military official who presided over the courts. Again, it is through a comparative and diachronic reading of the treaties that we are able to understand one of the principal features of Siyāsa: the transfer of mixed cases out of the qādī's hands, and into those of other officials, such as the ḥājib.[105] In early treaties, the ḥājib is never mentioned, while the qādī clearly adjudicates.[106] The 1271 treaty with Genoa foresaw that in some circumstances "the case should be brought before the Muslim judge (*archadi*, i.e. al-qādī)."[107] Similarly, the treaty of 1303 with Venice describes disputes being settled by the qādī: "questio oriretur, debeat diffiniri per cadhy terre,"[108] as is the case for Article 22 of the

103 Hallaq, Wael B.: "Islamic law: history and transformation", in: *The New Cambridge History of Islam, Vol. 4*, 2010, 142–183, 171 and Rapoport, "Royal Justice", 73.

104 al-Saḥmāwī, *al-Thaghr al-bāsim*, 393, describes the changing role of the ḥājib throughout the Mamluk period.

105 al-Qudsī, *Badhl al-nasā'iḥ al-shar'iyya*, 184, states that, in 'our days', the ḥājib has been charged by the Turks (i.e. the Mamluks) with the arbitration of disputes.

106 Holt, *Mamluk diplomacy*, 136, Fleet: "Turks, Mamluks, and Latin Merchants", 340.

107 Holt, *Mamluk diplomacy*, 145–6.

108 Thomas, G. M.and R. Predelli: *Diplomatarium veneto-levantinum sive acta et diplomata res venetas graecas atque levantis illustrantia*, Vol. 1, Venice: Deputazione veneta di storia patria, 1880, 7.

Mamluk-Venetian treaty of 1345: "tunc uenditor et emptor debeant ire ad rationem coram el cadi."[109] I have previously referred to the late medieval tendency to concentrate all mixed jurisdiction in the hands of Muslim judges. Indeed, up until the mid-fourteenth century Middle Eastern qādīs passed verdict on lawsuits involving Frankish and Muslim parties. However, this did not contradict the general right to appeal to the sultan's maẓālim, invariably mentioned in Ayyubid, Mamluk and Ottoman treaties. In principle, early jurists writing about Siyāsa sought to empower qādīs with the same inquisitive power enjoyed by the officials traditionally entrusted with the repression of crime, however in Mamluk times Siyāsa justice was in practice entrusted to the ḥājibs, some emirs and, on occasion, the head customs official.

3.6.1 *Persians in Cairo, Franks in Acre*

Treaties signed before the 1360s allowed the qadis to adjudicate disagreements between Muslims and Christians, while the right to appeal directly to the sultan's maẓālim court in Cairo was always recognized. The first explicit mention of the Siyāsa courts as a competent jurisdiction in mixed trials can be found in the 1368 draft treaty with Cyprus.[110] The treaty deals extensively with issues of justice, repeatedly mentioning that disputes involving Cypriots and Saracens should be heard by judges who are unspecified by the document, but clearly different from qadis. Unfortunately the extant Latin document accounts only for the Cypriots' requests, as it was not ratified by the sultan. This draft is nonetheless important as it sets a line of demarcation with previous practice. From this point on, treaties abandoned principles of equity, such as the defendant's right to have recourse to his own court, leaving sharīʿa as the unique and

109 Ibid., 205, "22. Item, si mercationes Venetorum uenderentur in doana, tunc doanarij teneantur soluere mercatori, cuius essent mercationes; et si uenderentur extra doanam cum truzimano et testibus, simili modo teneantur doamirij far.ere fieri solutionem mercatori; et si uenderentur sine testibus et truzimano, tunc uenditor et emptor debeant ire ad rationem coram el cadi."

110 "Et quando contra Saracenum aliquem aut alium vel alios sibi extraneos, scilicet non Chipriensem, habebunt aliquid faciendum, suam exigendam rationem seu jus, ipsi recursum habebunt ad ilium judicem ordinarium vel ipsos a quibus vel a quo jus sibi faciendum petere voluerint. Et ipse judex, Saracenus vel alter qualis extet, tenebitur eisdem in prorupto jus facere ac etiam rationem. Et si judex predictus vel predicti voluerint vel noluerit dictis Chipriensibus vel alteri eorumdem in prompto facere rationem atque jus, dicti Chiprienses vel alter ipsorum, si casus postulabit, recursum habebunt vel habebit ad alios qui dicuntur vulgariter Cadis, vel ad officiales dicti domini soldani ad quos hoc, causa suorum officiorum, pertinebit et spectabit cum una petitione dicti consulis requirentis ...", De Mas Latrie, René: *Histoire de l'île de Chypre sous le règne des princes de la maison de Lusignan, 3 Vols.*, Paris: Imprimerie impériale, 1852–1861, II, 293.

abiding legal system; this development meant that a Muslim was now guaranteed to be judged by his peers, and in this case a Mamluk official. Few details are given at this stage as to the nature of these judges, who were expected to pass judgement "by virtue of their office," and should they decline to do so, the case could be transferred to the qadi.

According to the Arab historian Taqī al-Dīn al-Maqrīzī, it is an episode in 1352–1353 that triggered the expansion of Siyāsa jurisdiction to the affairs of foreigners. According to this famous passage, in that year some Persian merchants arrived in Egypt fleeing mistreatment by the Mongols. They struck a business deal in Cairo with local merchants that turned bad. The Persians appealed to the ḥanafī qadi, but the defendants found a loophole in the ḥanafī bankruptcy regulations and managed to get away with their debts. The ḥanafī school prescribed the imprisonment of debtors for a limited period before they could be declared bankrupt, and it seems that the Cairo merchants were ready to spend some time in jail waiting to be declared default, so that they would not have to pay. In other words, sharīʿa regulations on bankruptcy prevented the plaintiffs from obtaining a satisfactory verdict. However, the Persians complained to the sultan in the Dār al-ʿAdl, the *Hall of Justice* where maẓālim sessions took place, who for the first time handed the case over to the royal courts. The ḥājib, in turn, "punished" the Cairene merchants (presumably, by torturing them, ar. *ʿāqabahum*) and forced them to pay their debts. The former were acting in accordance with juridical doctrines such as the *Siyāsa Sharʿīyah*—"Governance According to Islamic Law"—of Ibn Taymīyah, which recommended the sultan inflict corporal punishment on defaulters who were, we must imagine, hiding their wealth. After this first intervention on issues of debt backed by the sultan himself, al-Maqrīzī argues, the ḥājib took advantage of the episode to reprimand the ḥanafī chief qādī and forbid him from hearing future cases "concerning merchants and debtors."[111]

If the rise of Siyāsa under the Mamluks is connected with other major judicial developments, it also needs to be placed in the wider context of the Mongol and Timurid invasions. Previous dynasties displaced by nomadic invasions, such as the Zanjīs and the Ayyubids were at odds with the dominance of the shāfiʿī school. They had established four chief judges in Syria, promoted professorships in the Ṣāliḥiyya Madrasa and appointed judgeships from the four

111 Ibn Taymīyah, *al-Siyāsah al-sharʿīyah,* 60. al-Maqrīzī, Aḥmad Ibn ʿAlī (1364–1442): *Kitāb al-Mawāʿiẓ wa-al-iʿtibār fī dhikr al-khiṭaṭ wa-al-āthār,* Cairo: Maktabat al-Thaqafa al-Diniyya, n.d. vol. 2, 220–22, Rapoport, "Royal Justice", 82–3, Irwin, "Privatization", 66, Moukarzel: "La législation des autorités religieuses et politiques sur les marchands Européens dans le sultanat mamelouk (1250–1517)", 132.

madhhabs during the 13th century. Joseph H. Escovitz has devoted a detailed study to the decision by the Mamluk Sultan Baybars to establish full chief judgeships for each of the four madhhabs in 1265. For contemporaries, the institution of the qadi al-quḍāt came about as the result of conflicts between royal and qadi justice, often taking place at the Dār al-ʿAdl, the siege of Royal justice. In several narratives, the shāfiʿī qadi Ibn Bint al-Aʿazz bears the brunt for the Mamluks' wrath. He is repeated blamed by the sultan and the emirs due to his slowness on making decisions, the inadequate character of penalties, his lack of expedience when forcing defendants to refund victims and his arbitrariness to accept some testimonies.[112] Together with this judicial discontent, contemporaries highlighted the great number of people flocking to Cairo, which had now become the siege of sultanian power, thriving with scholars and adherents of the different legal rites. The influx of Persian refugees, and particularly merchants, has been noted in a wider, Mamluk-dominated area extending from Cairo to Jedda and Yemen.[113] Lastly, Siyāsa justice apprears as a post-Crusader phenomenon, at a time when trade across confessional boundaries could not be conducted anymore through the European commercial outlets in Palestine and Lesser Armenia.

A second episode, not hitherto linked with al-Maqrīzī's story of the Persian merchants, can help us to further understand how the problem of foreigners was handled by Mamluk jurists. In April 1353, the shāfiʿī jurist Taqī al-Dīn al-Subkī issued a legal opinion (*fatwā*) on a similar topic regarding the juridical situation of some Frankish merchants in Acre. The merchants, according to the petitioner, a provincial governor, had gone beyond the terms of their agreement when they started to publicly celebrate religious ceremonies that offended local Muslims (apparently, they hired Muslim porters during a procession). In his response to the consultation, al-Subkī placed all jurisdiction over foreign merchants in the sultan's hands, rather than in those of the qadis; it was the ruler and his agents who enjoyed discretionary power to punish offenders in this case, as their offence was not clearly specified by sharīʿa.[114] It was this

112 Escovitz, Joseph H.: *The Office of Qâdî al-Quḍât in Cairo under the Baḥrî Mamlûks*, Berlin: Klaus Schwarz, 1984, 23–31.

113 Power, Timothy C.: "Trade Cycles and Settlement Patterns in the Red Sea Region (c. AD 1050–1250)", in: *Navigated Spaces, Connected Places. Proceedings of Red Sea Project V. Held at the University of Exeter 16–19 September 2010*, edited by D.A. Agius, J. P. Cooper, *et al.*, Oxford: Archaeopress, 2012, 137–45, Vallet, Eric: *L'Arabie marchande: Etat et commerce sous les sultans rasûlides du Yémen (626–858/1229–1454)*, Paris: Publications de la Sorbonne, 2011, 646–7.

114 Atiyya, 'Aziz: "An unpublished XIVth century Fatwā", in: *Studien zur Geschichte und Kultur des Nahen und Fernen Ostens*, Edited by W. Heffening, P. Kahle, *et al.*, Brill, 1935, 55–68,

same concern for preventing the qāḍī from adjudicating in cases concerning merchants and debts, according to al-Maqrīzī, that motivated the Sultan's intervention in the Persians' case. Al-Subkī, who had been appointed as one of the first official legal advisors of the royal courts in Damascus, took inspiration from the theories of governance championed by Ibn Taymīyah and his disciples, who promoted the which regulated the application of discretionary punishment (*taʿzīr*) by the ruler.[115]

In 1370, al-Subkī's son, Abu-l-Barakāt, who followed his father as legal advisor (muftī) at the Dār al-ʿAdl, enriched his father's text with a long commentary on the juridical situation of Frankish merchants.[116] The latter, legally enemies of Islam, could enter the realm of Islam for trading purposes upon acceptance of a pact. The basic legal concept here is that any foreign merchant in Islamic lands could benefit from a safe-conduct (*amān*) protecting his life and property for a limited period. Outside this protection framework (for instance, when it expired or when its terms were broken) the *amān*-holder lost his legal status as a protected foreigner (*mustāʾmin*), and in consequence any tax or extraterritoriality privileges, such as consular jurisdiction, expired. While from a European viewpoint commercial privileges constituted the main scope of these treaties, for the Muslim authorities they were also the instrument that solved the juridical dilemma of the European presence in Islamic lands, providing merchants with a clear legal personality and settling jurisdictional issues. Together with other prerogatives, treaties included recognition of the right for European consulates and consular courts to deal with issues among Franks. Government-sponsored jurists like the Subkīs took the issue of the Franks' safe-conduct very seriously, placing the presence of Frankish merchants in the sphere of public interest (*maṣlaḥat al-Islām*) and stating that officers, not qadis, had jurisdiction over issues concerning their legal status. In so doing, they were opening the door for action by royal courts over these disputes.[117]

---

al-Subkī, Taqī al-Dīn ʿAlī Ibn ʿAbd al-Kāfī (1284–1355): *Fatāwā al-Subkī,* edited by Ḥusām al-Dīn Qudsī, 2 vols., 1355 Vol. 2, 417–21.

115 On Ibn Taymīyah and taʿzīr, see Masud, "The Doctrine of Siyāsah in Islamic Law", 11. On discretionary punishment in general, Schacht, Joseph: *An introduction to Islamic law*, Oxford; New York, Clarendon Press, 1982, 175–187, Heffening, W.: "Ta'zīr", in: *The Encyclopaedia of Islam: Second Edition*, X, 406.

116 Aḥmad Abū al-Barakāt b. ʿĀli b. ʿAbd al-Kāfī al-Subkī, (d. 773 AH), for whom a lengthy biography can be found in Ibn Ḥajar al-ʿAsqalānī, Aḥmad Ibn ʿAlī (1372–1449): *al-Durar al-Kāminah fī Aʿyān al-Miʾah al-Thāminah,* edited by Muḥammad Sayyid Jād al-Ḥaqq, 5 vols., Cairo: Dār al-Kutub al-Ḥadīthah, 1966–1968, I, 210–6., Nielsen, *Secular Justice,* 171, provides a list of muftīs at the Dār al-ʿAdl.

117 Schacht, Joseph: "Amān", in: *The Encyclopaedia of Islam: Second Edition*, I, 429–430. Leiden 1986, Schacht, Joseph: "'Ahd", in: *The Encyclopaedia of Islam: Second Edition*, I, 255. Leiden

In recent times, authors have spotted the episode of the Persian merchants as a relevant landmark in the history of Mamluk justice. It connects the important presence of foreigners and their fragile legal position, on the one hand, with the shortcomings of sharīʿa in terms of enforcement and the rulers' renewed interest for Siyāsa. Rarely, however, have these episodes been associated with contemporary works on Siyāsa, such as that in which Ibn Taymīyah avouched for torturing dishonest defendants in debt cases. One might object that it is after all an exaggeration to assume that isolated snapshots such as that of the merchants in Acre could account for the expansion of Siyāsa under the Mamluks.[118] It is true, however, that Taqī al-Dīn Ibn al-Subkī was concerned with issues of governance and mustāʾmins beyond his fatwā on Acre's merchants. Najm al-Dīn al-Ṭarsūsī (1320-ca. 1356), one of al-Subkī's acquaintances, reports that the latter intervened in a dispute between jurists on whether rulers and officials should accept gifts from Frankish kings.[119] Al-Ṭarsūsī and al-Maqrīzī traced a common genealogy of royal justice, the ḥājibs and the Hall of Justice in their writings, and both point to the reigns of al-Nāṣir Muḥammad and al-Ẓāhir Barqūq as the peaks of their development.[120] It has been suggested that the rise of Siyāsa might also be attributed to Timur's conquests in mainland Persia, which pushed local merchants out of their homeland and towards the Mamluk area of influence. The increasing presence of traders from the former Ilkhanid lands, and the borrowing of the Persian term *khawājā* to designate them may well come in support of these views. Indeed, a third *vignette*, discussed below, may be connected with both Persian presence and the rise of royal justice under the Mamluks.

The specific legal argument advanced by Ibn Taymīyah on debt issues was again invoked, this time by Sultan Barqūq during his second reign in 792/1390, against the shāfiʿī judge Shihāb al-Dīn al-Qurshī al-Malḥī. It has to be noted that Barqūq's rise to the sultanate is equated with a coup d'état against the

---

1986, Wansbrough, "The Safe-Conduct in Muslim Chancery Practice", On Mamluk jurists and the idea of public interest, Cook, M.A.: *Commanding right and forbidding wrong in Islamic thought*, Cambridge, Cambridge University Press, 2004, 151–6, Masud, "The Doctrine of Siyāsah in Islamic Law", Najjar: "Siyasa in Islamic Political Philosophy".

118 Alsabagh, "Before Banks: Credit, Society, and Law in Sixteenth-Century Palestine and Syria", 7–8.

119 al-Ṭarsūsī, Najm al-Dīn Ibrāhīm Ibn ʿAlī (1320-ca.1356), *Tuḥfat al-Turk fī-mā Yajib an Yuʿmal fī al-Mulk*, edited by Riḍwān al-Sayyid, Ibn al-Azraq Center for Political Heritage Studies, Beirut, 2012, 125–8., and a similar discussion on inheritance and the public treasury involving al-Subkī, ibid. 43–5.

120 Ibid. 48–9, al-Maqrīzī, *al-khiṭaṭ*, 207–8.

previous Mamluk regime, which comprised an aristocracy of emirs and a narrow circle of families amassing appointments in the judiciary. After almost a decade in power, Barqūq endured a rebellion that ended up with his own imprisonment in the fortress of al-Karak, awaiting a death sentence. What came about during his reinstatement to power is telling of the nature of the relationship between sultans and judges before Ottoman times. As Michael Winter has argued, despite the fact that they are often presented as autocratic Oriental despots, and in spite of their frequent disputes with the ulama, for the Mamluk sultans getting rid of the former by violent means was seldom an option, since it challenged the sultan's legitimacy as promoter of the religious learned. The Ottomans, instead, had three şeyhülislams executed in 1634, 1656 and 1703.[121] In Cairo, Barqūq's rivals labored to obtain a condemnatory fatwā securing a death penalty for him, underwritten by most of the chief judges, including Ibn Khaldūn, Barqūq's former protégé. Barqūq eventually managed to escape death and to put down the revolt—and the fatwā episode must indeed have been a bitter disappointment for him, as he had supported Ibn Khaldūn since his arrival from the Maghreb, granting him offices and defending him from his numerous rivals.[122] Having marked himself out as one of Barqūq's opponents during the revolt, al-Qurshī was summoned to Cairo. Probably for the very same reason he reinstated some treacherous emirs, Barqūq was forced to look for a legal loophole to deliver al-Qurshī to the executioner. The sultan accused the qādī of belonging to a sectarian group, probably seeking to obtain a harsh verdict or even a death sentence against him, as Mālikī qadis often ruled in similar cases.

Although, significantly, many chroniclers mention al-Qurshī's gruesome fate, only Ibn Ṭūlūn and Ibn Ḥajar al-ʿAsqalānī (1372–1449) report that a second legal strategy was actually adopted to that end. A "Persian merchant" showed up during a hearing, Ibn Ṭūlūn reports, and "sued al-Qurshī on the basis that he was keeping money and fabrics from him." Al-Qurshī denied this, although it did not save him from being whipped fifty times, then delivered to the judicial officers to look for the plaintiff's money by means, first of the stick, then again

121 Winter, Michael: "The judiciary of late Mamluk and early Ottoman Damascus: The administrative, social and cultural transformation of the system", in: *History and society during the Mamluk Period: (1250—1517)*, V&R Unipress, Göttingen, 2014, 193–220. Clayer, Nathalie, "L'Autorité religieuse dans l'islam ottoman sous le contrôle de l'État?", *Archives de sciences sociales des religions* [*En ligne*] 125, janvier—mars 2004, 45–62.

122 Fischel, Walter J.: *Ibn Khaldun in Egypt: His Public Functions and His Historical Research (1382–1406); A Study in Islamic Historiography*, Berkeley and Los Angeles: University of California, 1967.

the whip.[123] Al-Qurshī died at the Khizāna prison shortly afterwards. Ibn Ḥajar insists on the summary nature of al-Qurshī's trial, and on the brutality in the administration of the taʿzīr. Historians have rightly pointed to more spectacular developments in late medieval legal history, such as the appearance of the hall of justice under the Ayyubids and the appointment of four chief qadis by Sultan Baybars. However impressionistic and anecdotal the abovementioned stories might be, they point to the endorsement of Siyāsa by late Mamluk rulers, to the borrowing of concepts from jurists to that end, and attest to the importance of foreign merchants' legal needs in this process.

On doctrinal grounds, it would be a mistake to see a strict separation between Siyāsa and preceding versions of royal justice, such as the maẓālim sessions delivered in Cairo by the sultans. As a theory, Siyāsa took precedent from the much older doctrine of maẓālim, and the former's development in late medieval times by ḥanbalī and mālikī thinkers is therefore uncontroversial.[124] However, if we focus on legal change instead of looking for precedent, the ḥājib's court appears as an "expansion of royal jurisdiction," "parallel to the sharīʿah courts of the qādīs."[125] Although late Mamluk treaties signed with Florence, Genoa, Venice and Aragon have attracted a great deal of attention, the mid-fourteenth-century shift towards Siyāsa has not been fully understood. The real novelty brought about in Mamluk times happened in the field of judicial practice: with precedents since the fourteenth century, treaties ruled the jurisdiction of the qadis definitely out of mixed issues, even as a court of appeal.

Fifteenth-century treaties started to include clauses according to which mixed trials should be heard by "the viceroy or chamberlain (ḥājib) or officials of the province, and none other than the above-mentioned should adjudicate between them." As early as 1415, the Venetian government instructed ambassadors dispatched to Cairo to plea that disputes be heard in the sultan's presence, or at least that of the viceroy (*vel naibys*, ar. nā'ib), or the ḥājib (*aut agebis*) and not before the qadis (*cadi legis*).[126] The Venetians' requests were met in the amān granted by Sultan al-Mu'ayyad Shaykh (1412–1421), which included this

123 Ibn Ṭūlūn, Shams al-Dīn Muḥammad Ibn ʿAlī (1485?-1546): *Qudāt Dimashq: al-Thaghr al-Bassām fī Dhikr Man Wulliya Qaḍā' al-Shām*, edited by Ṣalāḥ al-Dīn al-Munajjid, Damascus: al-Majmaʿ al-ʿIlmī al-ʿArabī, 1956, 117., Ibn Ḥajar al-ʿAsqalānī, Aḥmad Ibn ʿAlī (1372–1449): *Inbā' al-Ghumr bi-Anbā' al-ʿUmr*, edited by Ḥasan Ḥabashī, 3 vols., Cairo: al-Majlis al-Āʿlá lil-Shu'ūn al-Islāmīyah, 1969–1972, 1/416–7, and his obituary, 423.

124 Johansen: "Vérité et torture. Ius commune et droit musulman entre le Xe et le XIIIe siècle".

125 Rapoport, "Royal Justice", 75, 101.

126 Iorga, Nicolae, "Notes et extraits pour servir à l'histoire des croisades au XVe siècle", Revue de l'Orient Latin 4 (1896), 545–6.

provision. Recourse to the courts of the ḥājibs applied to any situation involving Franks, and therefore determined the abandonment of the *actor sequitur* principle.[127] The same clause was stipulated again in the subsequent treaties of 1422, 1442 and 1497.[128]

There is another aspect that made Siyāsa look different from previous forms of royal justice. The right to appeal to the maẓālim courts, set by the sultans in most cases in the capital city, is as old as political Islam, and is more or less explicitly stated in every amān granted by a respectful sovereign. However, going to Cairo or wherever the sultan delivered his justice represented a burden and indeed Muslim merchants abused this right by suing Franks, but not deigning to turn up in court later in Cairo. Maẓālim was therefore a source of "damage" (*ġarāma*) and "difficulty" (*mashaqqa*), and Frankish governments lobbied to avoid their subjects having to resort to the sultan's court in Cairo in mixed cases.[129] Maẓālim was generally considered to be a court of appeal, a *board for grievances* chaired by officials who made decisions on unjust decisions made by qadis. Mamluk Siyāsa differed, in practical terms, from maẓālim in that it gave Franks the right to be heard *on the spot* by an official applying less stringent procedures than those adopted by the qadis. As for the procedural advantages of Siyāsa, they are well known, comprising permission to accept non-Muslim witnesses, the using of personal and public knowledge by the judge, the reliance on documents considered to be trustworthy even if non-notarized, the power to depute officials to make inquiries, and in general ample latitude in the examination of evidence and the deliverance of punishments.[130]

Ibn Qayyim, for instance, did not share the mistrust of ḥanafī, shāfiʿī and mālikī jurists about the use of written evidence by qadis. He describes how in practice, in order to use his own documents, a judge needed to personally remember, file, register and have records sealed and sworn by witnesses. Ibn Taymīyah, he argues, accepted the validation of testimony on the basis of a written document if the witness himself was no longer available. If we could not trust the written word, this would be the ruin of Islam; since the traditions (sunna) of the Prophet's life are no longer in the hands of ordinary men, we are forced to rely instead on old, revered manuscripts. The same, indeed, was the case for jurisprudence, since copies of legal texts were consulted for reference.

---

127 *Diplomatarium veneto-levantinum, Vol. 1*, 311.

128 Wansbrough, "Venice and Florence", 488 (Mamluk-Venetian treaty of 1442), 512 (Mamluk-Florentine treaty of 1497).

129 Amari, *I diplomi*, Treaty of 1496, ch. 10, 192, negotiations in 1488, 376, 11.

130 Nielsen, *Secular Justice*, 12–26.

The Prophet himself relied on writing and exchanged correspondence with rulers. In deathbed situations involving last wills, for example, where witnessing is problematic, the ḥadīth recommends relying on written documents: "It is not permissible for any Muslim who has something to will, to stay for two nights without having his last will and testament written and kept ready with him" (Bukharī, Ṣaḥīḥ IV, 51, 1). Ibn Qayyim goes on to quote instances in which the judge should proceed by examining handwriting, and by securing written testaments by having them read and certified by witnesses.

For Ibn Qayyim, the act of writing asserted one's willingness to bear testimony; being equated to the oral utterances of Muslims, writing needed to be subjected to forensic examination by the judge, in the same way that a witness looks at and inspects the reality for which he would bear testimony. To counter an excessive reliance on oral proof, such as the need to read documents out loud, writing is a way to enunciate, and enunciating unveils the willingness of the individual. Doubting the written word was not much different from doubting one's own sight and hearing, and indeed it often happened that witnesses to someone's handwriting could express uncertainty.[131] As in the case of minority witnessing mentioned earlier, the thought given by Siyāsa theorists to written evidence tended to undermine the fundamental biases against it, with which this book is concerned. Although jurists thought of maẓālim and Siyāsa as forms of procedure, rather than as separated doctrines and notions of justice, in practice Siyāsa circumvented the traditional forms of procedure in use in sharīʿa courts and made petitioning to the maẓālim in Cairo unnecessary.

The legal change brought about by Siyāsa should not be interpreted as the mere substitution of the qadi courts and their sharīʿa-based norms with new, "secular" ones. Rather, Mamluk jurists were providing rulers with the necessary legal space to manage the political realities of a European presence. Works endorsing *al-siyāsa al-sharʿiyya* justified the existence of civil judges, and not only qadis, in the community, who could administer justice based on state interests, and not just traditional jurisprudence.[132] According to Siyāsa theorists, the ruler and his delegates should sit in judgment and deliver physical punishments, not due to any exceptional power, but as part of their obligation to made decisions for the benefit of the community—an issue that is explicitly mentioned in both of al-Subkī's fatwās on Acre and al-Maqrīzī's story of the Persian merchants. "The Imām," al-Subkī states "can deal with them [the Franks] ... not according to his pleasure, but according to what seems to be for the good

131 Ibn Qayyim al-Jawzīyah, *al-Ṭuruq al-ḥukmīyah,* chapter 23, 544–567.
132 Emon, *Religious Pluralism,* 179–83, Johansen, "Signs as Evidence".

of Muslims."[133] As can be inferred from the spirit of the new *amān* treaties, dealing with Frankish traders fell within the imperatives of governance, and it was up to the royal courts to pass judgment on their affairs. The Mamluk government expanded upon this with the parallel development of royal courts in Damascus, Aleppo, Cairo and Tripoli, with state-appointed *muftīs* and its own hierarchy of ḥājibs, thus expanding Siyāsa jurisdiction over criminal law as well as civil cases, and away from the jurisdiction of the qadis.[134] Its practical implementation and geographic coverage outside Cairo made Mamluk Siyāsa appear fundamentally different from previous versions of royal justice. Moreover, Siyāsa judges were granted jurisdiction over the judiciary.[135] They prosecuted qadis in cases where favoritism led to the appointment of colleagues who were "ignorant of the law," or to embezzlement from charitable trusts. As most Mamluk chroniclers belonged to the same religious establishment as the qadis, straightforward resentment against Siyāsa can be found in many of the sources; in one case, the historian Ibn al-Ḥimṣī was arrested in the course of a ḥājib investigation. Siyāsa judges set up their own detention facilities, and further quarrels emerged regarding the jail in which a detainee should be kept, although prison conditions—at least for the Frankish merchants—seem to have been relatively fair. According to Ibn Ṭūlūn who had no sympathy for the new parallel judiciary, Franks accused of debauchery could encounter arrested judges in the ḥājib prison of Damascus.[136]

Apart from delineating a legal sphere of action for the ruler, Siyāsa theorists and the sultans who sponsored them were launching a critique of the procedural limits of sharīʿa, which, as in the case of the Persian merchants narrated by al-Maqrīzī, could prove harmful to foreign merchants. Siyāsa courts

---

133 The delegates' (*nāʾib*) responsibility, punishment (*taʿzīr*), and public good (*maṣlaḥat al-muslimīn*) are explicitly addressed by al-Subkī, Atiyya: "An unpublished XIVth century Fatwā", 65–66, 60. Maqrīzī also insists on punishment, but instead uses the term *ʿāqabahum*.

134 ASVe, *Giudici di Petizion*, Reg. 98, f. 151v, mentions a trial in Tripoli before the ḥājib: "davanti lazebo el qual el dixe e qual el fexe sentenciar ...".

135 The function is described in the chancery manual by al-Saḥmāwī, *al-Thaghr al-bāsim*, 393.

136 Ibn al-Ḥimṣī, Aḥmad Ibn Muḥammad (1473–1527 or 8): *Ḥawādith al-zamān wa-wafayāt al-shuyūkh wa-al-aqrān*, edited by ʿUmar Tadmurī, 3 vol, Ṣaydā: al-Maktabah al-ʿAṣrīyah, 1999, Vol. 2, 201, 212, 220, 227. Ibn Ṭawq, Aḥmad Ibn Muḥammad (1430 or 31–1509): *al-Taʿlīq: yawmīyāt Shihāb al-Dīn Aḥmad Ibn Ṭawq, 834–915 H/1430–1509 M: mudhakkirāt kutibat bi-Dimashq fī awākhir al-ʿahd al-Mamlūkī, 885–908 H/1480–1502 M*, edited by Jaʿfar Muhājir, Damascus: IFEAD, 2000–2007, I, 119. Ibn Ṭūlūn, *Iʿlām al-Wará*, 117. al-Buṣrawī, ʿAlī Ibn Yūsuf (1439–1499): *Tārīkh al-Buṣrawī: ṣafaḥāt majhūlah min tārīkh Dimashq fī ʿAṣr al-Mamālīk, min sanat 871 H li-ghāyat 904 H*, edited by Akram Ḥasan ʿUlabī, Damascus; Beirut: Dār al-Maʾmūn lil-Turāth, 1988, 119.

expanded their jurisdiction to cover various cases where sharīʿa's "formalistic attitude to proof and evidence prevented the application of justice."[137] For instance, the ḥājib sat in judgment in divorce cases, because the qadi courts required four eyewitnesses to prove adultery. Siyāsa theorists criticized the qadis' formalistic system of proof, and went so far as to legalize judicial torture, a method considered illegitimate in sharīʿa. Indeed, this criticism is implicit to al-Maqrīzī's account; had the cheating merchants not been "punished" by the ḥājib, as explicitly recommended by Ibn Taymīyah, justice would never have been served. By claiming royal jurisdiction for mixed affairs, diplomats, sultans and jurists placed mixed cases in an area of legal practice where the major biases of traditional Islamic justice could be circumvented. Ibn Taymīya's disciple Ibn Qayyim and mālikīs such as Ibn Farḥūn (1358–97), rationalized court procedure by stressing the importance of written and circumstantial evidence and by allowing the judge to rely on signs and indicators, and not only the word of witnesses.[138] Indeed, Ibn Qayyim went so far as to reform his own school's views on the issue, and to claim that nothing in the ḥanbalī tradition prevented Jews, Christians and Zoroastrians from acting as witnesses for mixed cases and, in cases of necessity, even in lawsuits concerning Muslims.[139]

3.7 Conflict Resolution in and out of the Courtroom

Venetian descriptions of Siyāsa lawsuits offer a new perspective on these problems and on Mamluk legal attitudes towards non-Muslims. According to the treaties, Siyāsa courts heard mixed cases, but how did they deal with the proof and testimony provided by Franks? For travelling merchants, proving claims in the courtroom was fundamentally a matter of producing written evidence. To secure proof of their transactions, merchants had both Islamic and Western notaries at their disposal. But was the legal value of their deeds equal? Could Islamic courts accept Latin deeds? Conversely, could a Venetian notary acknowledge the trustworthiness and probity of an Islamic contract?

Thinking about legal pluralism in the Mediterranean as the ability to switch between Islamic and foreign courts, but in which the former was simply a

137 Rapoport, "Royal Justice", 80.

138 Johansen, "Le jugement comme preuve. Preuve juridique et vérité religieuse dans le Droit Islamique Hanéfite", Johansen, "Signs as Evidence", Turki, Abdelmagid, "Lisan al-Din Ibn al-Khatib", 192–4, On Ibn Qayyim's attitude to written documents, Bechor, *God in the Courtroom*, 347.

139 Ibn Qayyim al-Jawzīyah, *al-Ṭuruq al-ḥukmīyah*, 470–82.

second-best option, is to oversimplify the nature of justice. Many mixed conflicts were solved out of court, Siyāsa trials being, as far as the Franks were concerned, the keystone of the judicial system. Arbitral, consular and Islamic courts enforced each other, and a common notarial culture was involved at all levels. Notaries provided evidentiary support to settle and prevent disputes and thus helped breach the fundamental limitations of consular justice. Consuls had no jurisdiction over Franks from outside their own nation, nor over the sultan's subjects. When Muslims failed to "honor their agreements," the Venetian consular court, "not having power over them," had no choice but to boycott the merchant in question so that no member of the Venetian community could engage in business with them.[140] Although, on occasion, foreigners voluntarily submitted to the jurisdiction of other consuls, an extant register of the Venetian consular court of Alexandria suggests that consuls almost exclusively settled internal disputes.[141] The preferred extrajudicial way to solve cross-national conflict was through arbitration, in which consuls, but also trustworthy merchants formed arbitration panels. Though notarized arbitration emerged mainly for issues among Latins, it should be noted that, to issue their verdicts, arbiters inevitably relied on the customs administration; lawsuits revolved around evidence produced by the Christian scribes ("scribani doane, scribas christianos a centura dicte doane") and the Muslim ʿudūl ("testes saracenorum") attached to the customs authorities, and translated by dragomans, mostly Jews.[142]

All too often, historians have attributed the success of arbitration to a desire to avoid 'formal' justice, and this preference for arbitration over litigation has emerged in recent scholarship as the keystone for solving social conflict in late medieval cities. Genoa and its overseas cosmopolitan colonies, for example, witnessed the emergence of mixed arbitration courts similar to those of Alexandria and Damascus.[143] However, the capacity of notarized arbitration to

140 ASVe, CI, N, B. 229, Notary L. de Valle, *Verbali del consiglio*, May 11th, 1402: "cum mulcti mercatori saraceni et aliis forensis faciant mercata cum mercatoribus nostris ... sed quem super ipsos non possit dare ordo necesse est super mercatores nostros providere."

141 ASVe, CI, N, B. 229, Notary L. de Valle, *Verbali del consiglio*, Christ, *Trading Conflicts: Venetian Merchants and Mamluk Officials in Late Medieval Alexandria*, 72.

142 ASVe, CI, N, B. 222, Notary A. Vactaciis, f.108r-v, Dec. 22, 1405, ASVe, CI, N, B. 211, Notary N. Turiano, f. 32v, 34v, March 11, 1435. Mixed arbitration was allowed, if voluntary, by some treaties, Ruiz-Orsatti, "Tratado de Paz", 343, 361. Fleet, "Turkish-Latin Relations", 609–10 mentions episodes of mixed arbitration.

143 Wray, Shona Kelly, "Instruments of Concord: Making Peace and Settling Disputes through a Notary in the City and Contado of Late Medieval Bologna", *Journal of Social History* 42 3 (2009), 733–760, Kuehn, Thomas: "Law and Arbitration in Renaissance Florence", in: *Law, family & women: toward a legal anthropology of Renaissance Italy*, 19–74.

enforce the law in mixed contexts presented its own limitations. Arbitration implied agreement by both parties over the election of judges, but there was often disagreement over the nation they should belong to, and the number of arbitrators. A Catalan refused the decision made by two arbiters, on the basis that neither of them was Catalan.[144] Acrimony could push the parties to enlarge the panel up to eight members. In Alexandria, for contentious cases arbitration courts began resorting to drawing lots to decide the makeup of the panel. Moreover, arbitration was limited to cases in which both parties voluntarily submitted to the court's decision. Needless to say, decisions by the arbitrators were not always respected by the losing party.[145]

It may be tempting to view recourse to Islamic courts as being motivated by a need for coercion, and to see Siyāsa tribunals only as courts of appeal when arbitration failed.[146] This would be a rigid oversimplification, as parties were often not interested in obtaining a satisfactory decision by means that could be considered prejudicial to their reputation. The behavior of a Muslim from Mecca, al-Sharīf Ḥasan, may serve as an example to how solutions stemming from notarial culture could intermingle with the formal authority of Islamic courts. In 1441, two Catalan merchants committed to providing a certain number of goods to Ḥasan, who intended to send them back to Mecca with the seasonal caravan. A Florentine merchant backed the operation as a third-party guarantor by underwriting an Arabic document. The Catalans never honored their agreement, the caravans departed for Mecca and the Florentine was held responsible for the loss. Even though his responsibility as guarantor was clear to all and sundry in the city, both parties agreed to submit the question to a

Chicago: University of Chicago Press, 1991. For Genoa and its Greek colonies, see Epstein, Steven: *Genoa & the Genoese, 958–1528*, Chapel Hill: University of North Carolina Press, 1996 p. 64–5, Becker, *Life and Local Administration in Fifteenth-Century Genoese Chios*, p. 214–5.

144 ASVe, Notai di Venezia, 14832, Notary I. Dalla Torre, f. 3v, Sept. 17, 1412: "vos non habui nec habeo pro meos judices, qui debent esse quator vel quinque … et nichil contra nationem catelanorum non habetis ad iudicandi."

145 For an eight-member panel, partially drawn by lots, ASVe, CI, N, B. 148, Notary P. Pellacan, September 29, 1444. ASVe, CI, N, B. 211, The parties added supplementary members to judge again in case of disagreement with the final decision. ASVe, CI, N, B. 222, Notary A. Vactaciis, f.93r, Sep. 2, 1405. A consul compelled a reluctant party to accept arbitration, yet they were at liberty to choose who would be on the board, ASVe, CI, N, B. 211, Notary N. Turiano, f. 21r-23r, Aug. 16, 1455:"veniatis ad arbitrium mercatorum cuiuscumque nationis quam velitis."

146 As seems to have been the case in Geniza times, Goldberg, Jessica: *Trade and Institutions in the Medieval Mediterranean: the Geniza Merchants and their Business World*, Cambridge: Cambridge University Press, 2012, 161.

Latin arbitration court. As a Muslim, Ḥasan had no need to present himself before such a tribunal, yet this choice served his interest in establishing himself in the public eye as someone who would not bring his Frankish partners to the Islamic courts. Given the Florentine's discontent with the verdict, the case was eventually brought "as usual" before a royal court held by the emir, who had the parties heard again and convicted the Florentine for a second time. Even in that case, the Muslim merchant asked the emir to consult Frankish community about the issue. The Franks gathered at an inn and again had both parties heard, and as a result of this unanimously pronounced a verdict condemning the Florentine for a third time. The episode, which left a long trail of notarized statements and depositions, and where Islamic and Latin documents were used for evidentiary purposes, demonstrates the complexities of administering interfaith justice, and the complementary role of the legal devices involved.[147] Going beyond mere coexistence, courts proved to be complementary in enforcing verdicts, as did both notarial systems in proving claims by the litigants. In 1444, an eight-member panel dealing with a quarrel allowed the winning party to turn to Islamic justice to enforce the panel's decision,[148] and on at least two occasions, Mamluk officials handed a dispute over to the Venetian consuls.[149]

3.8 Merchants at the Islamic Courts: a Lender of Last Resort?

The consolidation of *Siyāsa Sharʿīyah* as a doctrinal legitimation of state authority, with its emphasis on utility and public good, set the conceptual groundwork for transferring jurisdiction over Frankish merchants to the royal courts. However, Siyāsa should not to be understood solely as a normative imposition by the sultans, but rather as a solution developed to respond to mixed conflict cases. In this regard, it is interesting to note that, in the first place, Siyāsa justice did not totally override the jurisdiction of the qadi courts. Muslim plaintiffs continued to bring Franks before the qadis for relatively

147 Arbiters inspected the Arabic deed ("visa quadam carta more saracenorum") and the consulate registers: "carta testificationis ... in libro actium", ASVe, CI, N, B. 211, Notary N. Turiano, f.42r-45v, Sept. 9–10, 1455.

148 ASVe, CI, N, B. 148, Notary P. Pellacan, September 29, 1444, "pro qua executione per partem victorem contra partem tunc victam possit licite peti et implorari ac obtineri iuditium subsidium et favorem maurorum et alterius cuiuscumque generationis."

149 ASVe, CI, N, B. 222, Notary A. Vactaciis, f.108r-v, Dec. 22, 1405: "electo et constituto iudice per magistratus alexandrie." See also the de Negro case discussed below.

simple cases, in which the judge could call upon the testimony of the ʿudūl. As plaintiffs, however, Franks only had recourse to the Siyāsa tribunals, and by the mid-fifteenth century, Latins mentioned the royal courts in their contracts as the local forum where suits should be filed.[150] Second, despite clauses defining the competent courts for different groups in fifteenth-century treaties, Siyāsa did not deal with interfaith cases alone, but frequently intervened in disputes among the Franks themselves. Perhaps most significantly, Siyāsa justice broke the unwritten rule that conflicts among Latins should be solved among themselves, without involving the Muslim authorities. The frequency with which these injunctions were disobeyed suggests that, in many cases, Latins considered Siyāsa to be a suitable and even desirable solution for dispute resolution.

3.9 Mixed Cases at the Qadi Court

Although, as we saw earlier, treaties signed after 1360 took mixed cases out of the hands of the qadis,[151] daily commercial practice could deviate from the letter of the treaties and the doctrine of jurists as Muslim claimants were still in the habit of bringing mixed cases before the qadis. Two Damascene lawsuits, dated from 1418 and 1434, can give us some insight into the hybrid solutions resorted to in qadi courts in mixed cases. These cases revolved around testimony provided by a courtier (*simsār*) about exchanges between Muslims and Franks. The simsār acted here as a professional witness, almost certainly registered as one of the trustworthy ʿūdūl at the court, and the cases were easily solved in favor of the Muslim plaintiffs, as the former was able to certify the transactions previously concluded in his presence before the qadi. In compliance with the procedural norms of sharīʿa, he did this by reading the written records he had previously drawn up. As, according to the new treaties, all mixed transactions had to be notarized, a specialized courtier was called upon ("publicum sansarium inter mercatores cristianos et saracenos"). This peculiar simsār-dragoman acted as a notary—and therefore appeared in court as a professional witness on behalf of the Franks. This institution illustrates perfectly how interaction generated solutions to some fundamental biases of

150 ASVe, CI, N, B. 83II, Notary C. Del Fiore, f. 15v, Nov. 5th, 1463: "comparendi in quocumque iudicio et officio et coram quibuscumque dominis saracenis … et universis officialibus mauris."

151 For the 1271 treaty with Genoa, see Holt, *Mamluk diplomacy*, 145–6.

sharīʿa, such as minority witnessing, without challenging the accepted norms followed by the qadis. In both cases, the simsār was brought forward to testify "more saracenorum" in a separate juridical act, this time before the Venetian notary. In both cases, the defendants were agents of third-party investors and, most probably, had the testimony from the simsār notarized as a disclaimer in future lawsuits.[152] Finally, one single document makes reference to two Frankish litigants appealing to the qāḍī court. The parties 'had recourse to Christian justice' to settle their dispute in Damascus, then turned to the local qāḍī, and eventually to arbitration. The parties litigated over many years and eventually settled the dispute in Cyprus.[153] The Damascene trials in particular suggest that Muslim claimants preferred to address their legal claims to the qadis, especially when they could rely on evidence produced in due Islamic form. Siyāsa judges heard more complex cases than those that went before the qadi courts, requiring the use of circumstantial evidence, such as written Latin deeds, or documents not supported by certified witnesses.

3.10 Mixed Cases before Siyāsa Courts

In light of accusations by learned men that the royal courts made arbitrary decisions and were contrary to the spirit of sharīʿa, many authors have seen the ḥājibs as usurpers of the judicial functions of the qadis.[154] It is doubtful, however, that in transferring mixed cases to the ḥājibs the Mamluks were attempting to promote an arbitrary alternative to sharīʿa. Venetian sources suggest that only slight differences in procedure were adopted, together with more flexible approaches to proof and investigation. Indeed, when we compare the issues brought before the qadis with those brought to Siyāsa courts, the fundamental difference resides more in the nature of the cases heard and the kind of evidence produced by the litigants, than in their actual approach to Islamic law.

152 ASVe, CI, N, B. 230, Notary N. Venier, October 12, 1418, ASVe, CI, N, B. 222, Notary A. Vactaciis, f. 4r-v, Sept. 2, 1435.

153 ASVe, CI, N, B. 222, Notary A. Vactaciis, f. 5v-6r, January 8, 1436: "tam coram iudicio cristiano videlicet coram domino consule veneciis […] quam coram domino er Cadi ipsius civitatis damasci certas lites habuerint et coram etiam quibusdam arbitris et arbitratoribus." The archivists' attribution of this fragmentary ledger to A. Vactaciis is incorrect, and needs to be connected with the material in ASVe, CI, B. 122, int. 25. A Venetian was sentenced by the qāḍī Ḥanbalī of Damascus, ASVe, CI, N, B. 83II, Notary C. Del Fiore, Oct. 25, 1463.

154 For a critique to these positions see Rapoport, "Royal Justice",. Irwin, "Privatization", 64–5, Nielsen, *Secular Justice*, 105.

To the extent that they can be reconstructed through Venetian eyes, the commercial suits heard by Siyāsa judges were of great complexity, often involving forms of evidence and testimony that were difficult to contain within the formalist requirements of the qadis. And indeed, the dozen or so trials reported by Venetian sources suggest that most often, Siyāsa judges did their best to comply, at least externally, with the procedural traditions of sharīʿa.[155]

In one Siyāsa case reported by Venetian sources, Muslim merchants appealed to the ḥājib of Damascus to enact reprisals upon Catalan merchants after a Catalan pirate attacked a ship and seized merchandise belonging to both Muslims and Arabic-speaking Christians. The ḥājib opened a trial that was far from arbitrary in its methods, as it resorted to the principal forms of traditional procedure. As the Catalans were operating mainly through intermediaries, the ḥājib focused on whether the merchandise could be considered Catalan, and therefore be seized. Needless to say, the Venetians and other Franks who were apparently handling the Catalan goods did their best to embroil the judge in a complicated web of transactions. The ḥājib, unlike the qadi, displayed ample executive powers outside of the court, and had an intermediary based in Beirut brought to Damascus to testify. After collecting circumstantial evidence, the ḥājib used coercion to gain a confession. When the merchants did not provide a satisfactory explanation, the ḥājib had everyone sent to jail until they could produce a statement accusing other merchants for the losses incurred by the plaintiffs.[156] However, the way in which this new evidence was produced suggests a different approach to procedure by the ḥājib: the imprisoned merchants took an oath by swearing on the Gospels held by the Venetian notary-priest. The oath, taken outside of court and presumably handed in written form to the ḥājib, was doubtless accepted by the court, as it succeeded in improving the situation of the defendants. In situations such as these, the interaction between the Siyāsa courts and the Venetian notary did not end with a pronouncement by the judges in mixed cases. To enforce the court's decisions, Muslim litigants or the ḥājib himself went before the Latin clerk to publicize the decisions made in the courtroom. For example, one Muḥammad Ibn Mūsā notarized a receipt for the money his Frankish opponent was sentenced to pay.[157]

155 For the greater liberty of royal courts to examine documentary evidence: Nielsen, *Secular Justice*, 25–8.

156 ASVe, CI, N, B. 230, Notary N. Venier, f. 15r-16r, May 18, 1419: "dimandandole mori alazebo chostoro abia de le robe de catellani," ibid., 24.

157 ASVe, CI, N, B. 83II, Notary C. Del Fiore, f. 24r, May 31, 1426, f.15v, June 14, 1426: "Mahomet ebne Muse morus … recepisse per sententiam Admirati Alexandrie."

Complex rules of procedure were also followed in a mixed suit brought before the emir of Alexandria by two Muslims in 1401. A ship flying a Genoese flag had just docked in the port of Alexandria loaded with Frankish merchants and their cargoes. Unexpectedly, these two Mamluk subjects claimed to be the owners of most of the ship's freight and demanded that the wages be paid by the merchants on board.[158] A judicial panel deliberated over the lawsuit that followed—a format known in other forms of royal justice as the maẓālim—and included the emir and two qadis. The defendant, a Dalmatian merchant, appeared in court advancing written evidence (the original freight contract notarized in Senj), and he paid for the services of both a translator and an unspecified "attorney" (*machademus*, ar. muqaddam). The Mamluk judicial machinery involved other actors; the Muslim claimants did not immediately turn to royal justice, but first had judicial officers sent to interview the defendant over the course of several days ("mittentis in zimis per plures dies"). One of the major accusations against Siyāsa––the judges' habit of selling verdicts for money––is mentioned in this trial; the Frank reported having bribed one of the qadis in exchange for pronouncing a less severe sentence.[159]

3.11 Siyāsa among the Franks

The expanding role of Siyāsa as a commercial jurisdiction soon overstepped the spirit of the treaties, as we saw above; in particular, royal courts took to hearing cases where both parties were Franks, and not just interfaith cases involving Mamluk subjects *and* protected merchants. Siyāsa trials were frequent in Damascus, perhaps because consular institutions were less developed than in Alexandria. The first mention of such a trial dates back to 1397; Andrea di Sinibaldo, agent of the Portinari Bank, and the Venetian Bartolomeo Lombardo had set up a partnership in Damascus, however when Bartolomeo died owing money to his partner, the former's family in Venice rejected the Florentine's claims. Pressed by his Arab creditors, Sinibaldo brought the issue to the ḥājib, who forced reprisals on the Venetian merchant community.[160]

Appealing to the Siyāsa courts impinged on unwritten customs regarding dispute resolution among Franks. Many complaints by defendants mention

158 ASVe, CI, N, B. 222, Notary A. Vactaciis, f. 38v-39r, Jan. 18th, 1401: "asserunt se esse parcionabiles dicte coche ferazium pro medietate et melechi pro 1 tertium."

159 ASVe, CI, N, B. 222, Notary A. Vactaciis, f. 38v-39r, Jan. 18th, 1401, f. 43r-v, undated (however, drawn up between March 5–9, 1401).

160 ASVe, *Senato*, Deliberazioni, Misti, 44, f. 56r.

a tacit agreement not to appeal to local justice to resolve disputes among Franks, most particularly when the two parties belonged to the same nation. Siyāsa, however, undermined this agreement; indeed, these courts intervened so frequently that, as the Franks themselves admitted, they became the only possible solution for disputes between subjects under the jurisdiction of different consulates. This was the case for a merchant from Montpellier compelled by the Genoese consul to pay some taxes. Genoa was temporarily under French protection, and this argument was used by the Genoese consul to present himself as a representative of the French king. The French merchant protested that the consul had applied to the ḥājib, "who holds *the justice of the sultan* in Alexandria" and that he should instead have advanced his claims before the French representative, as, he argued, "my consul has power over those in his funduq. … and knows better the facts between Frank and Frank than the justice of the Moors does." The Genoese consul then attempted to "prove before the ḥājib" that the Frenchman was handling Genoese goods, and that he had gone into partnership with Genoese merchants, something the consul could hardly do without the help of Latin records and witnesses.[161]

By the same token, in October 1460 a Venetian in Damascus appeared before the Muslim authorities accusing a fellow national of several misdeeds, including silk smuggling and illegally trading slaves. The defendant denied the charges and accused the plaintiff of forging evidence, however his principal defense revolved around the argument that "it is against our laws and customs and against the consul's duties to bring our differences before the Muslim authorities, between Franks and particularly between Venetians." He reserved the right to protest to the consul for having tolerated this anomaly and apologized "before God and the world and before every merchant present here, that litigation before the Muslims has taken place; not by my doing, but because of you and your commissioners, violating our laws and our authorities' dispositions." However theatrical the merchants' prejudice against Islamic justice might seem, it did in any case sound genuine.[162] In 1403, two powerful consulates in Alexandria, those of Venice and Genoa, engaged in a

161 ASVe, CI, N, B. 222, Notary A. Vactaciis, f. 101r-v, Dec. 8, 1405: "davant la jegp que ten la justiçia dels moros … senyos de consols e franch que al present son en alexandria los cals coneixeran mells lo fach de franch afranch que non fara la justiçia dels moros," f. 107v, Dec. 14, 1405: "davant la jegp dalesandria local ten en lo dit loch laiustiçia per lo soldan."

162 ASVe, CI, N, B. 83II, Notary C. Del Fiore, Oct. 21, 1460: "et cum sie chelsia contra leze et consuetudine nostre et contra la commission del consolo a metter davanti segnorie de mori tal gare et defferentie tra francho e francho e maxime tra venezian e venezian."

dispute concerning not a single individual, but a larger group of merchants. The Genoese consul refused to elect an arbitration panel to resolve the dispute, and instead sought the justice of the emir, upon which the Venetian consul complained, calling such practice "against all justice and equity." The Genoese consul reversed this argument by reminding the Venetian consul that the emir and governor "has always been and is the arbitrator and judge between the different Frankish nations, and his decisions and will cannot be disobeyed."[163] After the emir reached his decision, the losing parties complained that being judged by the Islamic court was "against the law and against justice," and that it was their fellow countrymen who should preside over such trials.[164]

Royal justice was also called upon to intervene in complex financial matters. One of these trials revolved around the close examination of written evidence and accounts. The trial was initiated by the Genoese consul, who, in the process of dealing with the consulate's finances, clashed with a merchant, Nicola de Negro, over some debts. To twist de Negro's arm, the consul brought him before the emir of Alexandria, accusing him, in addition to failing to pay his debts, of defrauding the sultan's treasury. He first publicly accused de Negro before the customs officers; then the case was brought to the emir, in whose house the session took place. The strategy consisted in proving the defendant's guilt on the basis of account books, something not technically possible at the qadi courts, who would never have taken into consideration written evidence without the support of righteous witnesses. The defendant, in turn, presented official correspondence from the Genoese authorities exempting him from these debts. According to the account furnished by de Negro, the emir found the consul's claims exaggerated and "not in accordance with the law," although he declined to make a decision and handed the case, surprisingly enough, to the Venetian consular court. As a subject of Genoa, de Negro had no need to come before a Venetian tribunal. Therefore, he voluntarily submitted himself to the judgment of the Venetians, though "only *de iure*," knowing that if the

---

163 ASVe, CI, N, B. 229, Notary L. de Valle, March 10, 1403, "et nec nos domino consul Ianuensis cum omnibus nostris mercatoribus contra quantibus equitatis et iustitiam mihi veneritis coram iuditio moresco ... contra iuditium dicti armiragii de quacumque re sit vel contingerit inter nos et vos dicere non possumus nec ultra voluntatis ipsius armiragii facere non possumus."

164 As was the case for a trial among people from Gaeta, ASVe, CI, N, B. 211, Notary N. Turiano, f. 8v-9v, September 1, 1434: "contra ius et justicia ... secundum mores et consuetudines et legem saracenorum et non secundum ... mores christianorum," ASVe, CI, N, B. 211, Notary N. Turiano, f. 8v-9v, September 1, 1434.

subsequent trial took place before an arbitration court, its decision would not be binding.[165]

That Mamluk Siyāsa was part and parcel of the adjudication process is difficult to deny in light of its recurring presence in the treaties. However, the different attitudes of Franks towards it can tell us much about Frankish merchants' actual acceptance of Siyāsa as a competent court. One exceptional example can be found in the case of a Venetian notary turning up at one the Siyāsa hearings on the September 9th, 1422. There he met, as defendants, the Genoese traders in the city, and as plaintiffs the Rhodian consul Giovanni Lacana and the merchant Matteo de Soris. Presiding over the Siyāsa court was a panel composed of the emir, a military official, and the qadi nāẓir al-thaġr. The latter was the head of the civil administration of the customs, appearing in contemporary documents as linked to the treasury (Bayt al-māl), and involved in taxation as well as complex financial operations. While some secretaries could be referred to as qadis by their contemporaries, it does not seem that the qadi nazir was a standard religious judge, but rather an administrator entrusted with judicial duties. Mamluk and Venetian sources increasingly designate this and other civil charges as qadis, not by virtue of any particular religious training but in harmony with the idea that officials were responsible for delivering justice.[166] Now acting in collaboration with the emir, his role as arbitrator was mentioned in early Mamluk treaties.

Four days before turning up before the Siyāsa court, de Soris had appeared before the Venetian notary to file a complaint against the Genoese. In gross, according to Soris, the Knights of Rhodes had seized the goods (saffron) from an unspecified Turkish merchant in the Island. For this reason, the Genoese consul Gabriele Cattaneo was himself held responsible in Cairo and forced to pay the losses. Back in Alexandria, Cattaneo had de Soris stopped at the port's gates by judicial officers,[167] claiming that the saffron was in de Soris'

165 ASG, Governo, *Archivio Segreto*, Materie politiche, F. 18B- 2737B, n. 72: "offerendosse voler provar questo cum li libri de la massaria."

166 "It seems evident that at least some Muslim jurists of the Mamluk period consider as judges (ḥākim or qāḍī) any major administrative and political officials who impose sanctions or obligations on persons subject to their authority," Johansen, "A Perfect Law in an Imperfect Society", 269.

167 "Me habia fato astrenzer davanti larmiraio digando voler che io ge reffaza zerta quantità de denari i qual i dixe esser sta astrecti al chaiero a pagar per certa quantità de zaffaran che el gran maestro de rodo tosse da I turcho per lo passsado metandome in zime per farme retegnir ale porte, voiandome astrenzer che io meta fuora le dite merze. E per che dal dito mio maestro mai non havi simel hordene ...", ASVe, CI, N, B. 230, Notary N. Venier, 5 Sept. 5, 1422, f.54v-55r.

possession, which he denied ('my master never gave me such an order', he claimed). Probably fearing a condemnatory verdict, on September 5th de Soris sued the Genoese by having a notarial letter of protest drawn up. Indeed, four days later he summoned the same notary again, this time to the Siyāsa hearing, to have the sentence notarized as well. Probably as de Soris himself expected, he was ordered to release 150 loads of raisins belonging to a Rhodian from one of the city's funduqs. Though the documents are not talkative, we can surmise that he may have called upon the judges to seek redress for the initial losses in Cairo. It may well be that Cattaneo, in turning to the Siyāsa court, was trying to avoid passing through consular or diplomatic channels. In any case for de Soris, the theatrical display of the Siyāsa court issuing a sentence was an event he considered worth notarizing, obviously in support of future legal action elsewhere.

A final example of procedural cooperation between several different instances comes from a deed dated February 12th, 1418. The evidence presented thus far in this section has concerned justice administered by the qadis, ḥājibs and emirs, and referred to trials which occurred at different times and places. In contrast, this Venetian notarial deed was drawn up during one of the Siyāsa sessions in Damascus. The deed is dated "Damasci in domo residentie prefati magnifici domini Azebi prope banchum juris", that is, at the ḥājib's house, and it mentions the platform (*dikkah*) from which the Mamluk officials gave their verdicts. While mosques were the preferred places for qadis to pass verdicts, muslim jurists indeed advised to perform hearings in private houses so that non-Muslims may attend.[168] A siyāsah trial has just finished, and the ḥājib has made a decision. A Genoese and a Venetian merchant have applied to the ḥājib and he has found their claims to be just (decernens atque considerans petitione ipsorum [...] justas fore). In consequence, the ḥājib has seized some merchandise held by another Genoese merchant. Circumstantial evidence is mentioned, in the form of correspondence setting out the ownership of the merchandise. As a result, the defendant has been asked to take an oath as to the veracity of the testimony by the plaintiffs, and supported by the correspondence. The ḥājib sticks here to traditional sharīʿa procedure, which allowed anyone to take oaths, not only Muslims. For that purpose, he calles upon the Venetian notary to witness the pledge, who holds up the Gospels up while the Genoese merchant swears upon them. This time, the defendant makes it easy for the judge, acknowledging the validity of the plaintiffs' claims, after which the ḥājib makes a decision "by virtue of his office." Although formally couched

168 Müller, "Non-Muslims as part of Islamic law", 39.

in the procedural rules of sharīʿa, the nature and scope of the justice dispensed in this case clearly evokes the spirit of Siyāsa. Like in the qadi courts, the burden of proof was laid on the oath, rather than on the written document. However, the ḥājib allowed a Latin notary to directly collaborate in the production of the proof to be used in the Siyāsa session. Proceedings were recorded and translated by the Muslim dragoman in the presence of both parties and the notary. On his side, the Venetian clerk drew up his own Latin deed before the same dragoman and two Venetians. In this way, the outcome of a single juridical act could be conveyed to both legal systems.[169]

The participation of Venetian scribes in Siyāsa trials, together with the testimonial role of bilingual simsār, the borrowing of legal concepts, or the way courts enforced each other's decisions and were accessed across confessional boundaries, all appear to have been responses to the specific problems of dealing with diversity. The instances of legal cooperation examined so far demonstrate that legal relations and collaboration went far beyond mere tolerance and coexistence, as has often been suggested; rather, it required adjustments in procedural matters, and implied a common notarial culture for facilitating transactions between strangers. All too often, legal systems are believed to have been kept fundamentally separate until the eve of European colonization, and that Islamic law maintained biases against non-Muslims, as well as its main formalist traits. However, the kind of legal relations that the late Mamluks maintained with their Latin neighbors breaks with the teleological vision of a stagnant legal system that was only 'modernized' in the nineteenth century—with the inevitable adoption of Western legal principles and codification, such as Article 1736 of the Ottoman *Majalla*, which gave recognition to written documents. The imperceptible, yet significant shift represented by the transition from maẓālim to Siyāsa shows that medieval societies had their own way of managing diversity and mixed cases, and handled legal norms in ways that were compatible with the necessities of conflict resolution. The Mamluks did this without allowing anyone to bypass Islamic courts and without really challenging the sharīʿa system of norms.

As close examination of the treaties has shown, the Mamluks inherited some of their predecessors' solutions for dealing with the affairs of foreigners, and belonged to a much larger koinè of polities facing the same difficulties. It is difficult to imagine why the victorious Mamluks might have otherwise deigned to adopt the *actor sequitur* principle and accept that Muslims be judged by Frankish consuls. If we can only speculate as to the reasons for the rise of Siyāsa

169 ASVe, CI, N, B. 230, Notary N. Venier, f. 10v-11r, Feb. 12th, 1419.

in Mamluk society, we can say with some certainty that the problems associated with foreign merchants played an important part in it. In Mamluk markets and courts, legal change can be observed in the way in which biases against written evidence and minority witnessing were dealt with. New technicalities were introduced, such as a higher standard of notarization, and new patterns of adjudication. To be sure, Siyāsa stirred up a great deal of polemic among the legally learned, for which the intellectual framework was tailored in Mamluk and similar societies. Many voices were raised against Siyāsa, and among them Shams al-Dīn Ibn Ṭūlūn and Tāj al-Dīn al-Subkī, and contemporary historians were all but tender when speaking of the Sultan's meddling with justice. And indeed, these complaints account for the unprecedented nature of Mamluk Siyāsa. Legal change did not however follow a gradual, predictable pattern, but was contingent upon and circumscribed by medieval exchange in a specific historical context. Indeed, Siyāsa courts were not enhanced, but dismantled by the Ottomans, who instead reinstated a more traditional version of royal justice and placed the biases at the center of their relationship with the Franks.

Notarial casebooks record the remarkable, yet unexpected, emergence of the Islamic Siyāsa courts as an institution capable of enforcing justice not only for mixed cases, but also among Latin Christians. By turning from legal theory to take a closer look at the mixed trials described in this chapter, we can observe how an institution that issued from an Islamic legal background consolidated to settle disputes between strangers. Siyāsa justice, it should be noted, emerged even where such an institution was undesired, and it grew out of an unfavorable juridical tradition. There is perhaps no better proof for the effectiveness of Siyāsa than the fact that we have only come across a single diplomatic misunderstanding regarding the Islamic witness system. During the negotiations leading up to the Florentine-Mamluk treaty of 1481, the Florentine government requested that the word of a Florentine be acceptable as valid in Islamic trials: "should the Florentines be in need of producing witnesses in the market or at court, be the Florentines allowed to present and produce witnesses from all nations, both Christians and Moors, to whom credit should be given."[170] One can imagine the embarrassment of the Mamluk diplomats upon reading the Florentine request; the Mamluks had sponsored Siyāsa judges and their pragmatic trials to cope with the problem, and the request of an official

170 "Nel capitolo secondo de' Vinitiani, ove si fa mentione di testimoni, si agiunga et dichiari conceduto a Fiorentini questo, cioè: Che avendo Fiorentini a porre o produrre testimoni in mercati o iudicii, si possa per i Fiorentini porre e produrre et usare testimoni di ogni natione, et così Mori come Christiani; a' quali si habbi a prestare fede." Amari, *I diplomi*, 361.

admission of both witnesses' equal dignity could only come from inexperienced diplomats—indeed, the request was simply ignored in the final draft. While in Mamluk Syria and Egypt, and for more than a century, there appear to have been no complaints about the witness system, they quickly surfaced in the sources when the Ottoman judges began sitting in justice in the former Mamluk provinces.

The normative framework governing mixed conflict did not answer to a given "rapport de forces," nor to mere pragmatism. The early Mamluks, in spite of a favorable balance of power—since they had managed to expel the Crusaders and prevent new invasions—indulged in the adoption of some principles of equity governing the adjudication of commercial and mixed cases. And indeed, had clauses on jurisdiction been exclusively dependent upon, allegedly, unfavorable balances of power, Muslim authorities would have handed over commercial or even criminal cases to consulates (a claim that was sometimes advanced). However the Mamluks and later, the Early Ottomans concentrated all jurisdiction on Islamic judges. If in Mamluk times different scribal institutions existed, under the Ottomans they left room for the notarizing qadi alone, and in a very similar manner Siyāsa judges were replaced by Rūmī, ḥanafī qadis as the main judicial instance empowered to pass verdict. An orthodox approach to sharīʿa and the biases against non-Muslims, therefore, was progressively enhanced by the Ottomans, despite the fact that they are often depicted as mere pragmatists who yielded to the granting of privileges and capitulations demanded by Western polities.

Under the Mamluks, the moving forces dictating which forum or procedure should be followed drew, not from political opportunism, but from a repertoire of legal principles, such as the postulates of Islamic governance/Siyāsa and the doctrines of obligation, such as ʿahd and amān. Particularly after the mid-fourteenth century, the technical elaboration by jurists of these legal principles and their application by judges came to dominate the governance of cross-confessional issues. While the Mamluks set the discussion in the field of legal, technical reasoning handled by jurists, Ottoman rulers dismantled the Siyāsa courts and set their own standards concerning the biases against unbelievers. Soon after 1517, issues of witnessing and unbelief not only reemerged, but took center stage in legal and diplomatic interaction.

CHAPTER 4

Ottoman Legal Attitudes towards Diversity

> nui non haueuamo fatto paxe cum li soi tallasumani ne loro cum li nostri preti
>
> Ambassador GIOVANNI DARIO (1484)

4.0 The 'Witness System': a Bronze Wall?

In 1484, the Senate of Venice dispatched one of its citizens, Giovanni Dario, as ambassador to the Court of Sultan Bayezid II. An experienced diplomat, he had been appointed as ambassador to the sultan's predecessor Mehmed II five years before; on that occasion, Dario had to deal with the humiliating losses of a number of Venetian outposts, such as Scutari in Albania, and the rich colony of Negroponte in the Aegean, which had belonged for centuries to the *Serenissima*. After dealing with painful borderland issues during his first mission in Istanbul, Dario labored to prevent Ottoman involvement in Italian affairs pursuant to the war of Ferrara, which opposed the Papacy and his allies against Venice. During this episode of the Italian internecine wars, Pope Sixtus IV (1471–84) went so far as to invoke the Ottoman assistance against Venice, and proceeded to excommunicate the Venetians. The Turks, it has to be noted, declined to receive the Pope's ambassador to Istanbul on that occasion. In the context of these diplomatic intrigues, Dario attended numerous meetings with Ottoman dignitaries, the pashas, and as a result sent twenty-two dispatches reporting on his progress to the Doge and the Senate. In his missives, Dario addresses, among other subjects, a fair number of borderland issues, ranging from the enslavement of former Venetian subjects in regions recently lost to the Ottomans to the building of bridges across borderland watersides. It is in this context too that both the Venetian envoy and the pashas exchanged jokes about the excommunication imposed by Sixtus IV, and even about the news of the Pope's death. On several occasions, Dario confesses having found amusement in these irreverent exchanges with the pashas: "the other day the pashas said to me that the new pope had not yet absolved them of his excommunication, and I answered that this was not necessary, since the former pope had died, carrying the excommunication with himself, so they roared with laughter before they retired to meet the sultan, and in all sincerity I was unable to answer them otherwise."[1] On another occasion, jokes about Venice's Italian enemies could hardly hide Ottoman concern for the fate of the Republic: "They questioned me about the Pope's recent death, of which I replied I had no news; they informed me of a letter sent to his excellency the sultan regarding this, confirming the news. Mehmet Pasha then started joking, we indulged in revelry-making and His Excellency rolled over with laughter."[2] And indeed, we know that Venice's salvation in the war of the League of Cambrai was the object of public celebrations in the streets of Damascus in 1509.[3]

During one of these meetings, Dario advanced Venice's complaints regarding Christians living on the Ottoman-Venetian borders, who were being made prisoner during peacetime. In such cases, in order to ascertain the circumstances under which a given person had been delivered to Muslim hands as a slave, the Turks required the Venetians to advance two Turkish witnesses fitting Islamic standards for trustworthiness. According to these standards, witnesses needed to know the person whose slave status was being disputed, "should know the slave's mother and father by their names," and had to "be reputable Muslims, able to say their prayers." These "difficult and impossible conditions are so rigid," Dario informs the Doge, "that it actually means the Ottomans do not wish to return any prisoners, hence we were left in a worse condition in times of peace that we were during hostilities," since, he argued, at war "one takes precautions in order to defend oneself—but now, trusting the peace agreement, we are exposed and suffer attacks both from land and sea, since the former is packed with corsairs, damaging every day our towns and fleet."[4]

1 "L'altro giorno i pascià mi dissero che il nuovo papa non li aveva ancora assolti dalla scomunica ed io risposi che non era necessario perché il papa precedente era morto, ed aveva portato con se la scomunica; hanno riso a più non posso e ridendo entrarono dal Signore e io, a dire il vero, non potevo dar loro altra risposta.", Dario, Giovanni: *22 dispacci da Costantinopoli al doge Giovanni Mocenigo*, edited by Giuseppe Calò, Venezia, Corbo e Fiore, 1992, 227.

2 "me domando de la morte del papa, li respusi che mi non haueua alchuno auiso me disse che la Signoria soa haueua habudo lettera et che la iera certa. Mehemet Bassa comenzo a dir mille piaxeuelleze et fessemo un tanferuzo la dentro. et el Signor rideua quanto podeua," ibid., 119.

3 Apellániz, Francisco: *Pouvoir et finance en Méditerranée pré-moderne: le deuxième état mamelouk et le commerce des épices* (*1382–1517*), Barcelona: CSIC, 2009, 245, n.25.

4 "me agreuai poi cum lui de li captiui nostri fati in tempo de paxe li quali voleuano che mostrasse do testimonij turchi che cognossesse per nome el pare et anche la mare et che fosse persone da ben et che sauesse dir le soe oracion et che questa condicion cusi stretta et impossibille me significaua che non haueria voia de restituirne alchuno captiuo et aquesto

The Ottoman *subaşı* and officials, Dario went on, "go bird hunting, although their quarry are men, *and* our peasants' sons are apprehended as soon as they reach their farms." Dario mentions that these issues were brought before the Ottoman judicial authorities: "we have recourse to their voivodas and subaşıs for help, but it is those same officials who are responsible for these misdeeds, and who actually mock the reasons of our plaintiffs."

Just like the cases dealt with by the Venetian ambassador Dario, we often stumble upon descriptions of Ottoman officials hearing disputes and leading criminal investigations, and it is clear that traditional approaches to proof and procedure, such as those adopted by the qadis, were being followed in these cases.[5] "These extreme conditions we cannot stand, and," Dario said to the pashas, "since you are respectful of the law when acquiring slaves with your own money, even more discretion would be expected when dealing with us, your friends,"[6] to which the pashas answered that "requiring Muslim witnesses of good standing is an article of faith." In his missive, Dario further noted that "such is the injunction of their gospels, that they could not counter their own laws. I replied by saying that we Venetians have not concluded a truce with your muezzins and neither have you with our priests and friars ... should we heed them, we would never be at peace, and never conduct any commerce, nor any dialogue. Instead, the rulers who seek to preserve their state leave the priests to sing their hymns in their houses of prayer, they sign peace treaties with whomever they wish and so they keep their promises."[7] The pasha, surmising a Venetian connection to the unexpected death of Sixtus IV, then retorted that "you Christians can do this openly, since you have killed your pope and therefore are free of honoring any obligation, but we cannot do that."

modo eramo in pezor condicion in tempo de paxe, che non erimo in tempo de guerra", Dario, *22 dispacci*, 91.

5 Baldwin, *Islamic Law and Empire in Ottoman Cairo*, 41. Hosainy, Hadi: "Ottoman legal practice and non-judicial actors in Seventeenth-century Istanbul", in: *Law and Legality in the Ottoman Empire and Republic of Turkey*, edited by Kent F. Schull, Bloomington: Indiana University Press, 2016, 23–33, 28.

6 "che queste erano estreme condicion le qual non podeuamo pui suportar et che le Signorie soe usauano descrecion cum li soi schiaui compradi cum li soi dinari tanto pui doueuano far cum li soi amici," Dario, *22 dispacci*, 91–2.

7 "me respose chal fato de voler testimonii turchi de bona condicion era de iure diuino et cusi comandaua el so euanzelio et che elli non podeuano contrauegnire a le soe leze. Li respusi che nui non haueuamo fatto paxe cum li soi tallasumani ne loro cum li nostri preti et frati per che se vardassamo alor non stessamo mai in paxe ne hauessano insieme alchuno comerchio ne pur loquella ma li Signori che volleno la conseruation di soi stadi lassano li eclesiastici ne le soe giesie a cantar li soi officij et fano paxe cum chi che li piaxe et attendeno a quello che prometteno," ibid., 92.

To judge by the hilarious tone of the conversation, the pashas were not lacking in a sense of humor, however what is more important for our argument is that the Turks were attempting to stress the moral obligations carried by governance, including respecting the biases against minority witnesses. Dario went on to claim, decidedly much less convincingly, that the Pope "might well have died, but the Church still lives and governs among us Christians."[8] At this point, Dario introduced the problematic issue of captives from the Ottoman-Venetian borderlands: "his excellency laughed, but I cried, for I recall having seen in Constantinople, in chains, people known and close to me lost during the truce, begging for my help but, not being able to advance Turkish witnesses knowledgeable of the law, able to say their prayers and knowing their father and mother, I was unable to help them, so that they ended up being carried away and shipped to Anatolia with utmost cruelty, and as a result I clearly see it is impossible to keep the peace between us."[9] As Dario was losing faith in the conversation, other officials present at the meeting began conversing, seemingly agreeing with Dario's argument—that is, that it was simply impossible to recover any of the Venetian captives by appealing to the qadi courts.

A delicate diplomatic mission similar to that encountered by Dario was entrusted to Andrea Gritti (1455–1538) in 1502, after the Ottoman–Venetian war.[10] Lengthy negotiations led by Gritti and his secretary revolved around the Florentines' attempt to gain access to Ottoman lands through the Adriatic, and the proposal to create a foothold for Ottoman traders in Venetian-dominated Ravenna.[11] Gritti faced the usual controversies that emerge during the drafting of treaties, such the handling of the estates of Muslims dead in Venice (and conversely, the seizure by the Islamic treasury of those belonging to Venetians) or the return of a number of slaves who were Muslims of Corfu and therefore Venetian subjects. The ambassador pushed the Ottomans to accept Venetian

8 "me respuse che vui podeti ben far questo ad abaldeza perche haueti morto el vostro Papa et seti asolti da ogni hobligo. ma nui non podemo far questo. Li respusi che sel Papa era morto e uiueua la giesia la qual era imortale et gouernaua nui altri cristiani." Ibid., 92.

9 "et che la Signoria soa rideua ma io pianzeua quando me arecordai huer vista in Constantinopoli in cadene persone ami notte et familiar persi in tempo de paxe li quali implorauano lo mio adiuto et per non hauer testimonij turchi et anche dottori che sapia la oracion che cognosesse el pare e la mare non li podeua souenire et veniuano strasinadi et tragettadi in la Natolia cum grande crudelta et che vedeua chiaramente che a questo partido non era possibile che fosse bona paxe entra de nui," ibid., 92.

10 Davis, James Cushman: *Shipping and Spying in the Early Career of a Venetian Doge, 1496–1502*, Florence, Leo S. Olschki, 1974 (Estr. da: Studi veneziani, vol. 16).

11 Discussions revolved around the Ottoman request to gain free access for Muslims to Ottoman lands through the Adriatic, a claim mentioned in successive treaties, De Groot, "The Historical Development of the Capitulatory Regime", 591.

testimony for issues concerning the status and belongings of these Franks, and for which Muslim testimony was not available or undesired. In his own *dispacci*, Gritti received the same moral objections mentioned by his colleague Giovanni Dario: "Davut Pasha replied to this by saying that, according to their law, they could not accept witnesses other than Muslim, since otherwise they would carry that burden until Judgment Day; I feigned not to understand."[12]

The apparently isolated issues mentioned above in the diplomatic narratives of Dario or Gritti actually revolved around the same legal problem; that is, the biases of Islamic law concerning proof and unbelief. Since the rise of the Ottoman Empire as a military power, if not earlier, the biases had been brought to the forefront of governance and diplomacy. In this, the Ottomans marked a definitive break with a long-established Mediterranean tradition of weaving trade and diplomacy relations and of legally dealing with foreigners in the marketplace. To be sure, voices were most frequently raised in the bazar, particularly in the Arab provinces, where the legal change brought about by Ottoman domination most strikingly countered previous practice. In the 1530s, a Venetian merchant wandered through the streets of Tripoli and Aleppo trying to ascertain certain sums owed by Muslims to a dead relative. He inquired after the former partners of the deceased, but hesitated to actually seek redress before the qādīs, since "no one is allowed to give testimony of their knowledge but Muslims."[13] The Venetian consul in Egypt Lunardo Emo, discussing in July 1560 some businesses gone bad, mentioned that of a certain Pandolfo Contarini, whose ship had been seized. Emo deliberated about presenting Contarini's case to the Dīvān, as other Venetian residents had discouraged him from doing so: None of them, Emo argued, is able to testify, *if only the Ottomans accepted the testimony of our merchants;* [*but*] *they are not used to this, since it is their custom and law, as Your Most Serene prince knows, to require Muslim witnesses*.[14] It is in any case hard to ignore that the biases had become

12 "Daut rispose etiam a questa ultima parte cum dir, che per la leze sua non se poteva aceptar altri testimoni che musulmani, che facendosse altramente ne haveriano gran carico fina al giorno del iuditio. Mostrai de non lo entender ...", Venice, Biblioteca Marciana (hereinafter BM), It. VII, 878/8652, Dispacci Gritti (1502), f. 22r.

13 "saper quella niuno satrova in aleppo apresente non so si debbo farla meter de qui e per non aver niuno possa testimoniar di saper salvo Musulmani," Venice, Biblioteca Correr (hereinafter BC), *Manoscritti Provenienza Diversa* c. 508c/6.

14 "Non sentono che'l se appresenti, non vi essendo manco alcuno di loro che potesse giustificar la carata della preditta nave ... chel magnifico meser pandolfo contarini non ne abbia a far niente, come vuol il commandamento, posto chel [ciphered text] si contentasse delle giustification de nostri, il che non usano, essendo suo costume et legge come sa Vostra Serenità di voler testimonii musulmani.", ASVe Senato, Dispacci Consoli, Egitto B1, 32, July 7th, 1560.

commonplace in diplomatic conversations, so that awareness of their importance was expected of the Doge and the members of the Senate to whom the dispatches were addressed.

As I argued in Chapter Two, in Ottoman times people increasingly had recourse to ḥujjas and similar deeds registered in court, but this was most often done to protect their rights from potential threat, as has been shown by Rossistza Gradeva for the case of former slaves seeking to guarantee their freed status.[15] However, in sharīʿa courts hearings continued to prioritize oral testimony, and almost no litigant adopted a line of defense based solely on documentary evidence. Closely related to this phenomenon, complaints also appeared about the unclear status of written documents. When he attempted to resolve some debt issues in the 1540s, the Venetian Bailo stumbled over the Islamic mistrust for written artifacts as proof. He described what happened when he approached the Grand Vizier Rustem Pasha (d. 1561) with some Venetian letters, accompanied by an official Turkish translation: "I was required to prove that this money belonged to us, I showed the aforementioned letters, but he asked me whether I could advance further testimony other than those letters, and I exposed all my arguments to his excellency to no avail, so that the money has been seized. I wonder whether the pasha, under the pretext of a lack of oral witnesses, is plotting to appropriate the money for his treasury."[16] Gianfrancesco Morosini, Bailo for thirty-five months in the 1580s, complained that "as for their justice, the legal procedure used by the Turks to implement their governance is summary," and that criminal cases "are not heard in proper trials, but verdicts are reached through the hearing of oral witnesses." Incidentally, the Turks did not hesitate to torture criminal defendants, it was argued in several *relazioni*. "In civil cases, they do not take written documents into consideration," and it was noted that trials were started by the plaintiff before the qādīs or the *kadıasker*, who heard the parties and delivered a sentence on the basis of the witnesses' utterances. For family issues, instead, "they rely on the

15 Gradeva, Rossitsa, "On the Judicial Functions of Kadi Courts: Glimpses from Sofia in the Seventeenth Century," Wiener Zeitschrift zur Geschichte der Neuzeit, 5/2 (2005), 15–43. Gradeva has advanced similar considerations regarding the motivations for minorities or women seeking to have documents notarized, Gradeva, "Orthodox Christians in the Kadi Courts: The Practice of the Sofia Sheriat Court, Seventeenth Century".

16 "questi danari, dovesse probar, che fossero de nostri et io mostrandoli le lettere predicte diceva, che non poteva darli maggior testimonio che dette lettere, nientedimeno non mi ha valso alcuna ragion che habbia detto in questo proposito a sua magnificentia di modo che li danari stano cosi in deposito. Forse chel detto bassà con questa difficultà mossa da probar con testimonii, ha pensato di appropriarli alla casenda.", ASVe, Archivi Propri Constantinopoli, 1, fasc. IV, f. 116, (154-47).

Koran." For the 16th-century author of the *Viaje de Turquía* "the justice of the Turks treats everyone as equals: Christians, Jews and Turks". He describes qadis as having "a crufifix on their desks to take the oaths of Christians and a Bible for the Jews". Second only to the sultan, The kadıasker is the chief justice, who proceeds by hearing witnesses, while criminal matters are left in the hands of governors and officials such as the subaşı. The latter are familiar with physical punishments: they kill murderers, hang thieves and impale traitors. Ottoman governance is particularly ruthless with debtors hiding their wealth. Qadis, instead, apply punishments to minor offenses by means of the stick. Great attention is paid to false witnesses: they are paraded on the back of donkeys, their faces smeared with ink, they are finally marked with an iron and invalidated to give testimony, a point of particular importance to which I will return.[17] This thumbnail sketch of Ottoman justice therefore stresses a civil procedure that was marked by the two biases under consideration in this book, by the importance of testimonial witnessing in a cross-confessional setting, by the qadis' inability to independently start trials,[18] the role of officials, the widespread adoption of taʿzīr or discretionary punishment in criminal cases, as well as a family law characterized by a religious imprint. For Bailo Giovanni Moro, "judges seek promptness, neglecting long speeches by attorneys," reached quick decisions and paid little heed to the reading of documents, "quite in contrast with the procedure in other countries."[19] The requirement to produce Muslim witnesses heavily impinged upon cross-confessional relations, but it also went hand-in-hand with the summary nature of Islamic justice and its mistrust of writing and documentation.

When dealing with Ottoman judges and judicial authorities, not only Venetian merchants and diplomats, but also craftsmen, seamen and even slaves encountered unexpected legal difficulties. Indeed, as the opening quotations of this chapter underline, the principal obstacle these foreigners came up against was the Ottoman 'witness system,' which privileged Islamic notions of proof and evidence.[20] In a 16th-century context, the difficulties posed by the biases were not confined to the domain of jurists and litigants, but pervaded political

17 De Villalón, Cristóbal: *Viaje de Turquía*, edited by Antonio G. Solalinde. 2 vols., *Colección Austral; 246*. Madrid: Espasa-Calpe, 1965, 213.

18 Tyan, *Histoire de l'organisation judiciaire en pays d'Islam,* II, 23.

19 "Attendono i giudici alla brevità e, tralasciando i lunghi sermoni d'avvocati, risolvono prestamente le cause con poca lettura di scritture, molto diversamente dall'uso degli altri paesi, quando non siano comprobate col detto di testimoni," *Relazioni degli ambasciatori veneti al Senato, Serie III, Volume III*, edited by E. Albèri. Vol. IX, Florence: Società editrice fiorentina, 1855, 376.

20 I borrow this expression from: Ozil, *Orthodox Christians in the late Ottoman Empire.*

and diplomatic encounters. Even though Gritti—significantly—pretended not to have understood the implications of the biases, he and his colleagues were of course well aware of these legal inconveniences, and they expected the same knowledge of their metropolitan correspondents. In this final chapter, I look at the legal attitudes adopted by the Ottomans, which made relations at the market and in court look so different than they had previously in the medieval Mediterranean. Whatever their origins, these attitudes and ideas about sovereignty and legal order have been connected with those adopted by other central Asian, post-Mongol states. By 1517 the Ottoman Empire was an old polity, and indeed these attitudes formed part of a broader tendency to regulate the ḥanafī school, reaching back to the mid-fifteenth century.[21] Giovanni Dario explicitly formulated the problem some thirty-five years before the Ottoman conquest of the Arab provinces. In this chapter, I focus on the fulcrum of cross-confessional diplomatic and legal exchange, the Dīvān-ı Hümāyūn in Istanbul, although on occasion I will turn to Egypt and to the Syrian provinces, where Ottoman choices had a greater impact on the local judiciary and in the markets. The concluding section of this chapter, which looks at a case of bankruptcy, makes it clear that, shortly after the conquest of the Arab provinces, Frankish litigants were disoriented by the stubborn insistence of authorities on having transactions registered at the courthouse, and by their adherence to the biases against minority witnessing.

Whatever importance the Ottoman witness system exhibited in the field of norms, and despite recurring complaints by Venetians and other users of the justice system, it was not, as the Bailos would have it, a mere arbitrary device, but rather a tool for expressing a specific version of the legal order. To be sure, the reliance on local witnesses led to abuses that required the attention of the Ottoman authorities, such as the proliferation of false witnesses selling their services to the highest bidder. However, it cannot be said that this strict adherence to the biases actually hindered trade, or overly favored local Muslims; rather, the biases played out in the complex social reality of the Empire, where they were manipulated, and sometimes contested, by local minorities, by the Franks themselves and, on occasion, benefitted foreign Muslims. By the same token, implementation of such discriminatory measures in the daily life of the markets entailed the adoption of a series of exceptions and amendments to the general rule that, on occasion, came at odds with the very same ḥanafī legal order the Ottomans sought to embrace. Indeed, the biases took center stage

21 Burak, *The second formation of Islamic law*, 207–23, Burak, Guy: *The Abū Ḥanīfah of his Time: Islamic Law, Jurisprudential Authority and Empire in the Ottoman Domains (16th-17th Centuries), PhD thesis, New York University*, 2012, 25–6.

in all kinds of exchanges and relations across confessional boundaries, as the Bailos experienced, but also generated many controversies and often provoked contradictory and chaotic legal situations. Venetian negotiators were of course open to accommodation with the pashas and actually kept an open diplomatic channel in the Dīvān, far beyond the occasional exchange of embassies. Yet in light of the passages examined at the beginning of this chapter, one can hardly present the pashas as pragmatic diplomats ready to put aside Islamic notions and ideas in order to reach a profitable agreement. Rather, the Venetian ambassadors took great care to inform the Doge and the Senate of the difficulties posed by sharīʿa when dealing with issues of proof and testimony. This strict adherence to skeptic sharīʿa principles only further complicated the already chaotic sphere of legal definitions and relations with unbelievers, which as we have seen manifested itself in the divergences between amān theory and practice, and in the difficulties arising from its actual application. The adoption of the biases as one of the founding principles of Ottoman governance certainly drew a line of demarcation between Muslims and non-Muslims in cross-confessional disputes, but it ended up generating more problems than it actually solved. Indeed, this very central role given to the biases made them so cumbersome as to generate a series of contradictory discourses and practices, epitomized in the 'Aleppo ferman' issued in 1557.

I open this chapter by taking a juristic approach to the question, and specifically, the two vectors governing, according to ḥanafī doctrine, the ways in which unbelievers could participate in the legal system. Yet it is my contention that a study of the Ottoman framework of governance amidst diversity needs to take account of actual practice, including some of these contradictory aspects and discourses. In this chapter, my aim is to historicize these specific attitudes and notions about proof, justice and belonging, as described by Frankish litigants. I will be dealing with some of these anomalies, not as uninteresting exceptions, but rather as peculiarities that encapsulate the very essence of the Ottoman distinctive attitude to justice. I will be addressing the objections raised by dhimmī minorities—the '*carazari* clause'—and exceptional measures such as those sponsored for Aleppo, where foreign Muslims were actually prejudiced by the specifically Ottoman approach to proof and evidence. In connection with this, I look into the proliferation of false witnessing, a phenomenon that began to develop in the sixteenth century, in a similar manner to the forgery of documents evoked in Chapter One. The most significant among all these alterations to the general normative framework of justice, however, was the privilege granted to Franks according to which Muslim witnesses were not accepted in mixed lawsuits. In all of these contentious cases, the strife between juristic discourse, government-sponsored practice, and actors' reactions from below pushed the system into a state of tension.

I conclude this chapter with the story of a Venetian merchant who fell bankrupt in Syria in the late 1520s, and by describing how his default was dealt with by Ottoman judges and judicial authorities.[22] A trader in diamonds, rubies, pearls and luxury armors, Marco Priuli found his customers among the Syrian upper class, including judges and officials, as well as their wives and daughters. Muslims, but also Jews and Christians were all equally affected by his default; as his creditors, his debtors, and as his associates. The lawsuits that ensued after Priuli's death in the Ottoman prisons are representative of many of the obstacles commonly met by people dealing with cross-confessional affairs in the Early Modern Mediterranean. If there was little doubt about Priuli's responsibility for his own bankruptcy, the judges concerned themselves more with determining the liabilities of Priuli's partners, as well as the claims of his debtors. The several courts involved faced many pressures when attempting to decide whose credits should be paid, by whom, and what witnesses should be heard first. The legal discussion revolved certainly around facts, but surprisingly plaintiffs, judges and attorneys spent a great deal of energy and time dealing with issues of proof and procedure, and with their application in the context of religious diversity.

Marco Priuli's Venetian heirs and associates proved to be unfamiliar with the legal practices sponsored by the Ottomans, and with the court's requirements for proof and evidence. This seems particularly surprising for members of a veteran trading nation that proudly claimed five centuries of presence in the area. Indeed, they seem to have clashed over particular issues whose implementation under the Ottomans had started to deviate from previous legal practice. As an early Ottoman example of commercial litigation, the Priuli lawsuit echoes the discomfort generated by the Ottoman stance vis-à-vis Frankish merchants. I have referred to the fact that, for more than a century, there had apparently been no complaints about the witness system. Yet once Ottoman judges began sitting in justice in the former Mamluk provinces, these complaints began surfacing in the documentation. The abiding principle that an unbeliever's word should not be used to harm the interests of a Muslim was invoked first by a Syrian qadi entrusted with the affair, and later again by Muslim plaintiffs at the Imperial Council—the Dīvān-ı Hümāyūn—where the Priuli lawsuit was heard in appeal. Apart from reflecting well-known Ottoman legal attitudes, the Priuli lawsuit also offers instances of accommodation, as well as

22 BC, *Manoscritti Provenienza Diversa* c. 508c.

some puzzling anomalies, such as the repeated claim that a Frank could not bring forward a Muslim witnesses.

We know that the Ottomans disapproved of the way in which legal principles had been implemented under the Mamluks. It was not so much the principles themselves that they took issue with, as their formalistic approach to them. The medieval ʿudūl, recognized by local judges and registered in court, had become in Mamluk times numerous in tribunals and administrative offices.[23] Ibn Iyās describes the measures taken against Cairo's ʿudūl, targeted by Ottoman legal reforms, soon after the conquest, in 1517 and in 1520, and Ibn al-Ḥanbalī puts it in very bold terms, stating that in Aleppo, "under Ottoman rule, the notarial stalls were closed," reporting personal accounts of notaries who were forced to seek a different profession "after the Circassians."[24] Under the Ottomans, as became evident in the Priuli lawsuit, traditional recourse to the notary-witness began to be replaced by the practice of having transactions notarized at the qādī court.[25] The lawsuit sheds light on the problematic adaptations that had to be resorted to in response to this new situation, provides us with an example of a case being transferred from a qādī court to the Imperial Dīvān, and offers a glimpse into both the Ottoman version of the witness system and its loopholes.

By doing away with some of the practices prevalent under the Mamluks, such as Siyāsa trials for foreigners, the Ottomans set a line of demarcation with the conquest of Mamluk Syria and Egypt in 1517. This, again, is viewed in a problematic light by researchers dealing with the late Ottoman court registers. Indeed, the strong biases against unbelievers relaxed in the late Ottoman period, thus creating conditions more conducive to mixed justice in the contemporary era. The seminal research conducted by authors such as Najwa al-Qattan has shown that the judges of nineteenth-century Damascus or Aleppo considered it justifiable, if needed, to rely on the word of a Christian against that of a Muslim.[26] The Priuli lawsuit epitomizes an early modern scenario where Franks had to deal with qadis and sharīʿa as sponsored by a rising Ottoman power, thus enhancing the historical contrast with medieval practice. In this early modern scenario, the legalistic approach to cross-confessional relations was taken very seriously, sometimes definitely at the Ottomans' own advantage but also when the ruthlessness of such rules made agents fronting

23 Meshal, "Antagonistic Sharīʿas", p. 183–212.

24 Ibn al-Ḥanbalī, *Durr al-habab fi tarikh aʿyan Halab,* 921–2, 980.

25 Fitzgerald, *Ottoman methods of conquest,* 229.

26 Masters, *Christians and Jews in the Ottoman Arab world,* 31–33; Al-Qattan, "Dhimmis in the Muslim Court: Legal Autonomy and Religious Discrimination".

the new legal regime scratch their heads. By the same token, the conditions under the Priuli lawsuit stand in marked contrast with the late-Ottoman context, where Capitulations, privileges and berats favored, allegedly, the emergence of a pathway of European legal institutions alternative to sharīʿa.

4.0.1 *The Early Ottoman Adjudication System*

The Ottomans gave the ḥanafī qāḍī a much more central role in the judicial stage, at the expense of other potential judges, such as the officials who used to pass verdicts in Siyāsa courts under the Mamluks. As was the case for the Priuli bankruptcy, in the Ottoman provinces all mixed affairs were first adjudicated by the local qāḍī. Moreover, both Muslims and unbelievers were expected to register their debts, contracts and depositions at the qāḍī court, who would stamp his seal on the resulting deed, and in the presence of honorable Muslims.[27] This constituted a break with the Mamluk past as regards mixed transactions, when the Franks had their contracts notarized by Latin notaries, Muslim notary-witnesses, by courtiers or even concluded oral agreements.[28] Having become compulsory under Ottoman rule, this certifying function of the Ottoman judge rendered obsolete previous forms of notarization, not only for the Islamic notary-witness, but also for the Latin notaries who, during the late Middle Ages, were usually sent to Islamic countries to draw up Latin deeds.[29] Under the new Ottoman system, as long as an act had been certified before a judge, one's religion didn't really matter. The problem, however, was that many transactions, agreements, debts and oral contracts were concluded outside of this ideal framework, and when businesses turned bad the parties could not always prove that a given agreement had been reached, or that a particular debt had been contracted. By the time Priuli fell bankrupt, proving and notarizing had become synonymous to a hitherto unknown extent.

The Ottoman legal discourse put emphasis on the Islamic communal vision of governance and justice. Although, in practice, foreigners of all sorts were in great abundance in places such as Istanbul, in principle unbelievers sojourned by virtue of a communal agreement underwritten with the ruler, and they did

---

27 Van den Boogert, *The Capitulations and the Ottoman Legal System*, 44–5; Wittmann, Richard: *Before Qadi and Grand Vizier: Intra-communal Dispute Resolution and Legal Transactions Among Christians and Jews in the Plural Society of Seventeenth Century Istanbul, Ph.D. Dissertation*: Harvard University, 2008, 72–3.

28 On the widespread recourse and compulsory character of notarization, Faroqhi, "Before 1600: Ottoman attitudes towards merchants from Latin Christendom", 77 and n. 31.

29 Canbakal, *Society and politics in an Ottoman town: ʿAyntab in the 17th century*, 129–30, argues that "whether or not the Ottoman *ʿudul* continued to act as notaries, could not be established with any certainty."

so on a temporary basis.[30] According to this ideal principle, the Islamic ruler was disposed to hear his guests' complaints against oppressive officials and qādīs,[31] and a foreigner dissatisfied with a qādī sentence could therefore seek redress from the sultan at the Dīvān-ı Hümāyūn. As I have been arguing, this jurisdiction corresponded with medieval maẓālim although, under the Ottomans, it was not referred to by this name. Consequently, the Dīvān passed verdicts in appeal for many mixed disputes, and at least during the sixteenth century these trials exhibited a strong communal bias.[32] Contrary to ordinary Muslim petitioners, European Christians were represented by the head of their community. In practice, ambassadors such as the Venetian Bailo brought their subjects' claims to the Imperial Council, where Ottoman pashas would hear the case and refer it to the ruler. In practice, the Ottomans discouraged recourse to the Dīvān, limiting it to claims exceeding a certain amount. While before 1517, any Frank could sue a Muslim before the ḥājib and other officials in any Mamluk town, after this date it was only the Ottoman sultan in Istanbul who dealt with mixed issues, and treated only with community representatives. Bringing grievances before the sultan in Cairo—the maẓālim—although the option always existed, was discouraged under the Mamluks. Under the Ottomans, instead, a trial at the Dīvān was in fact the only possible option in appeal after a local qadi had heard the case.

Historians of Ottoman treaties underline the fact that the Ottomans refused collective responsibility for individual debts, and treated the Bailo simply as a diplomatic representative, rather than as the head of a confessional group.[33] Yet as we shall see, this was subject to some limitations. The Bailo Antonio Rizzo admitted in his dispatches to the Doge that, unlike in cases judged by a local qādī, the Bailo was not totally exempt of communal responsibility at the Dīvān. Historians have a tendency to ignore differences with later periods, presenting the European ambassadors with their later incarnation in mind, as professional diplomats devoted to the granting of privileges and capitulations,

30 Krstić, "Contesting Subjecthood and Sovereignty in Ottoman Galata", 1613–1617.

31 Guy Gerber, Haim: *State, Society, and Law in Islam: Ottoman Law in Comparative Perspective*, Albany: State University of New York Press, 1994, 58–79, Ursinus, Michael: *Grievance Administration (şikayet) in an Ottoman Province: the Kaymakam of Rumelia's 'Record book of complaints' of 1781–1783*, New York: Routledge, 2004.

32 On the Dīvān's role as a court of appeal, Ginio, Eyal: "Coping with the State's Agents 'from Below': Petitions, Legal Appeal and the Sultan's Justice in Ottoman Legal Practice", in *Popular Protest and Political Participation in the Ottoman Empire—Studies in Honor of Suraiya Faroqhi*, edited by Eleni Gara, M. Erdem Kabadayı and Christoph K. Neumann, İstanbul Bilgi University Press, 2011, 41–57, 45.

33 Faroqhi, "Before 1600: Ottoman attitudes towards merchants from Latin Christendom", 77.

who intervened in the Dīvān court as part of their professional duties, or simply as personal favors.[34] The role played by the experienced Venetian Bailo Pietro Zen suggests instead that Bailos were considered more as the grantees of the amān and as heads of a protected community, and therefore expected to be responsible for their subjects in court.[35]

Although the Ottoman certification system signified a giant's step towards the generalization of written artifacts as legal evidence, it arose within a context still largely dominated by orality. During a conversation between a Bailo and a Pasha quoted by the former in 1557, the ottoman official admitted that "in the Dīvān, where they do not make notes, my word is trusted more than a qadi's sijill".[36] Bailo Barbarigo, in his report to the Senate, expressed his irritation with the Gran Vizier Rustem Pasha, who undermined the role of the Re'īs ül-Küttāb and the Nişancı and monopolized the Dīvān's decision-making process for himself.[37] Although sixteenth-century Ottoman judges preserved more legal documents than their medieval predecessors, the Venetians remained puzzled by the fact that no court records were kept by the Dīvān. The Priuli affair was heard by both Ottoman qadis and pashas, before a third trial was filed before the Venetian commercial courts. The extant Venetian documents describe the litigants' recourse to multiple forums back in the Ottoman Empire, and comprise vivid descriptions of the hearings held both by the qādīs and pashas, with a particular interest for issues of proof and procedure.

4.1 The Legal Grounds of the Ottoman Witness System

Half a century ago, as part of the proceedings published by the erudite academy Société Jean Bodin, the Italian historian Mario Grignaschi devoted a study to the treatment of witnessing and religious diversity by Ottoman jurists. To be sure, Grignaschi did not seek to compare his findings with actual courtroom practice; however, his results did nonetheless shed light on the link between practice and jurisprudence. Grignaschi searched Istanbul's libraries for

34 Dumas, Juliette "Müste'min Dealing with the Ottoman Justice: Role and Strategy of the Ambassador", *Oriente Moderno* 93 (2013), 477–494.

35 As Işiksel rightly argues, consuls and Bailos were considered both as heads of a community and as part of the Ottoman administration, and differently from the ambassadors leading occasional diplomatic missions, Işiksel, Güneş: "Hierarchy and Friendship: Ottoman Practices of Diplomatic Culture", 15–6.

36 "in divano, dove che non si fa nota … la mia parola è creduta più che sigilleto," ASVe, Senato, Dispacci degli ambasciatori e residenti, Costantinopoli, B. 1A, f. 432.

37 *Relazioni degli ambasciatori veneti al Senato,* Serie III, Volume III, 155–6.

manuscripts containing fatwās collected by Ottoman jurists, but also by Arabs and scholars based in the Mughal Empire. His work offered a new angle on the transformations experienced by the regime of proof around 1500. Grignaschi argued that special attention was devoted to infidels when dealing with issues of witnessing and testimony. Arab—i.e., medieval- jurisprudence was instead more focused on issues of conversion, while a real interest for witnessing amidst diversity only arrived with the Ottomans. Ottoman scholars, at the time, did not yet have much of a taste for novelty, and Grignaschi established very strong ties between Ottoman views on the problem and available precedent, in particular with ḥanafī medieval thinkers such al-Sarakhsī. Incidentally, al-Sarakhsī (d.1106), flagged as the reference for later Ottoman views on diversity, was originally from Transoxiana—a central Asian commercial crossroads—and therefore belonged to the regional legal school responsible for a more dynamic approach to exchange with non-Muslims, to the use of documentary evidence, and which coined expressions such as "the ḥanafī exception."[38]

Half a century after the publication of Grignaschi's seminal work, my focus moves away from a jurists' approach to law, to concentrate more on the practical handling of these burdensome sharīʿa principles in forums such as the Dīvān. For this purpose, I make extensive recourse to the colorful narratives provided by Venetian representatives (*dispacci, relazioni*), particularly since documentary production at the Dīvān was limited to the drafting of petitions and the resulting decrees, leaving most of the decision-making process undocumented. An examination of cross-confessional issues of proof and evidence can give us a better understanding of how Middle Eastern societies made use of their jurists, and shed light on the relationship between scholars, rulers and the people involved in the consumption of justice.

4.1.1 *Vector One: the 'Ḥanafī Exception'*

One phenomenon observed by Grignaschi is particularly worth examining, as it emerged as an important vector in legally relations between the Ottomans and Frankish consuls, merchants, and others: since an early date in Islamic law, ḥanafī scholars such as al-Sarakhsī endorsed the view that all dhimmīs constituted a single people under the aegis of their Muslim social superiors. The idea of a single dhimmī nation was based on the assumption that the ties binding different religions together should be considered stronger than their

38 Johansen, "Formes de langage", 365–9, Johansen, Baber: "Entre révélation et tyrannie: le droit des non-Musulmans d'après les juristes musulmans", in: *Contingency in a Sacred Law*, 219–37. Leiden: Brill, 1999, 138.

communal differences, by virtue of their shared, subordinated socio-political status, and their common enmity for Muslims. By claiming that unbelievers all belonged to the same people, and that dhimmīs could bear witness against each other even if they did not share the same beliefs, al-Sarakhsī set himself in opposition to most shāfiʿī, mālikī and ḥanbalī scholars.[39] Mālik, Ibn Ḥanbal, Ibn Abī Laylā, al-Awzāʿī and to some extent al-Shāfiʿī expressed reserves to the fundamental unity in infidelity, distinguishing between scriptuaries and not, stressing the plurality of beliefs and setting limits to inheritances across confessional boundaries.[40] Such open attitudes were probably not the norm in Islamic history, and the ḥanafīs backed their views with the precedent established by the Prophet, who acted as judge in Medina, as well on the cosmopolitan garrison-cities of early Islam, where judges were often compelled to sit in justice over non-Muslims. Although many jurists simply refused unbelievers any right to testify, Abū Yūsuf (d. 729) and Ibn Abī Laylā (d. circa 702) accepted dhimmī witnessing, although only for claims within their own communities.[41] Most ḥanafīs argued that Muslims were neither present at the dhimmīs' deathbed, nor did they attend their weddings, and therefore their own witnesses should be permitted, as otherwise this could lead to a "loss of rights" for the dhimmīs.[42] Jurists argued that women's testimony, for this same reason, was accepted for events that occurred inside the ḥammām, or in other places were upright male Muslims were absent. This argument was usually extended to encompass the acceptance of other 'diverse' witnesses, such as a child brought to trial for having torn another child's clothes.[43] This legal reconfiguration, by virtue of which Christians and Jews could testify either for or against each other, but not against Muslims, is therefore known as the ḥanafī exception. The ḥanafī exception was therefore a first vector towards forming legal attitudes about Franks and other foreigners, not only in court, but also in more political scenarios, such as at the Imperial and provincial Dīvān.

---

39 Yusuf Ragib, *Actes de vente d'esclaves et d'animaux d'Egypte médiévale*, vol. 2, Le Caire: Institut Français d'Archéologie Orientale, 2006, 105.

40 Friedmann, "Classification of unbelievers in Sunni Muslim Law and Tradition", 57–8.

41 Anver M. Emon, *Religious Pluralism*, 136–41; Ragib, *Actes de vente d'esclaves et d'animaux*, 105–15.

42 Sarakhsī, Muḥammad Ibn Aḥmad: *Kitāb al-mabsūṭ*, edited by Muḥammad Rādī, 30 vols., Cairo, Maṭbaʿat al-Saʿadah, 1324-31H, 1906–13, XVI, 135. Ibn Nujaym, Zayn al-Dīn Ibn Ibrāhīm (1520–63): *al-Ashbāh wa-al-naẓāʾir ʿalá madhhab Abī Ḥanīfah al-Nuʿmān*, Damascus: Dar al-Fikr, 2005, III, 283.

43 Sarakhsī, *Kitāb al-mabsūṭ*, XVI, 136.

4.1.2 *Vector Two: Mustāʾmins Cannot Testify*

Muslim jurists took great care to differentiate between the people of the book—and more precisely local Christians and Jews—and apostates, heathens and other categories of people from outside the Dār al-Islām. In contrast with local minorities, the latter did not belong to the abode of Islam and therefore did not share the same motivations as the dhimmīs. The crucial term underpinning this argument is *walāya*, evoking notions of friendship and sanctity, but most often those of obedience and authority. Contrary to local dhimmīs, mistrust for the mustāʾmins as bearers of truth was justified by the latter's alleged lack of *walāya* vis-à-vis the Muslims. Frankish merchants protected by ahdnames constituted an intermediate and, in early modern Istanbul, numerous category (20.000 people, rumors had it, in the 1610s).[44] According to the ḥanafīs, mustāʾmins were outsiders and therefore not supposed to be interested in safeguarding the common good if summoned to court as witnesses.[45] Actual relations between Franks and local minorities in Istanbul confirm the views of medieval ḥanafī jurists on the inferior legal status of Franks vis-à-vis local minorities. In the light of the legal claims they raised, Ottoman local minorites had good reason to mistrust foreign Christians. Indeed, for more than two centuries ḥanafī jurists advanced the claim that cross-confessional witnessing might be harmful, since Frankish witnesses might be tempted to target rival, local minorities in the course of litigation. Since local communities were subject to the payment of the poll tax, or *carazo* in Italian-Turkish jargon, I will refer to this legal claim as the 'carazari clause,' often discussed in diplomatic exchanges. Conversely, and as we shall see, the Ottoman official position was to claim that all unbelievers present in Ottoman territory constituted *a single people*, and were all included in the Muslim enterprise of governance and therefore equal as subjects of the sharīʿa-based legal system. In other words, both dhimmīs and mustāʾmins were expected to testify for and against each other. By so doing, the Ottomans were not just embracing the ḥanafī exception, but twisting the arm of standard doctrine so as to include foreigners in it. In this matter, the Ottomans departed from the letter of the jurists, hence tracing a thin red line between legal theory and actual governance.

To be sure, it was not the original intention of the ḥanafī jurists to include ḥarbīs (disbelievers hailing from outside the lands of Islam) in their argumentation. As we have seen, in principle the ḥanafīs mistrusted the mustāʾmins, and the ḥanafī exception did not offer legal coverage for them. In plain words,

44 Krstić, "Contesting Subjecthood and Sovereignty in Ottoman Galata", 438.

45 Emon, *Religious Pluralism*, 31.

mustā'mins—Frankish merchants in our case, were not allowed by ḥanafī jurists to testify against local Christians and Jews. High medieval ḥanafīs such as al-Sarakhsī,[46] al-Marghīnānī and later ones such as Badr al-Dīn al-ʿAynī (1361–1451) explicitly denied mustā'mins the right to act as witnesses, since they were not residents in the Muslim polity and could not be trusted to act in others' best interests.[47] Dhimmīs were placed in a higher position within the Islamic polity, and belonged to the abode of Islam ('our household' ar. *daruna*), which was not the case for foreign Christians, or *ḥarbīs*, hailing from the abode of war (*ḥarb*), where social relations were based on violence (*ġaṣb*). Although the founding fathers of jurisprudence such as al-Shāfiʿī and Mālik denied dhimmīs any right to act as witnesses, al-ʿAynī argued that the Prophet accepted their testimony when judging their intra-communal affairs. Most importantly, and unlike local minorities, foreign unbelievers were only allowed to testify in cases were both parties in the case belonged to the same nation (ar. *dār*).[48] Ḥanafī texts often quote a passage by al-Sarakhsī explaining that outsiders such as "a Byzantine and a Turk," lack common belonging (*walāya*) and unlike dhimmīs, who are under "our *dār*," would naturally tend to target each other.[49] These attitudes were crystalized in a fatwā in which the ḥanafī Ottoman jurist Ebu's-suʿud (circa 1490–1574) questioned what would happen if Amer, a hypothetical Venetian Bailo (in spite of his Turkish name), was sued by a dhimmī creditor. After initially denying his debts, the Bailo confesses that the debt actually did exist, but claims it has already been honored. He produces two Venetian witnesses who give evidence that counter the claims advanced by the dhimmī. Should these testimonies then be accepted? Ebu's-Suʿud's answer is clearly in the negative.[50]

Burhān al-dīn Ibrāhīm Ibn Muḥammad al-Ḥalabī (1460–1549) shares these views in his Multaqā al-Abḥur, that became an authoritative handbook of the ḥanafī law under the Ottomans. al-Ḥalabī received legal training in the Mamluk sultanate and moved to Istanbul towards 1500. In the Ottoman Empire, he built a successful career, receiving appointments in elite teaching institutions, something that would soon become unusual for a Syrian. The Multaqā

46 Sarakhsī, *Kitāb al-mabsūṭ*, XVI, 139.

47 al-ʿAynī, Badr al-Dīn Maḥmūd Ibn Aḥmad (1361–1451): *al-Bināya Sharḥ al-Hidāya*, edited by Ayman Sālih Sha'bān, Beirut: Dār al-Kutub al-ʿIlmiyya, 2000 IX, 154.

48 Heyd, Uriel: *Studies in old Ottoman criminal law*, edited by V. L. Ménage, Oxford, 1973, 245.

49 al-ʿAynī, *al-Bināya Sharḥ al-Hidāya*, IX, 152–155.

50 Grignaschi: "témoignage des sujets non musulmans", 253–5. On Ebu's-suʿud's general mistrust towards dhimmī witnesses, Kermeli, Eugenia: "Ebussuud Efendi", in: *Christian Muslim relations. Volume 7, Central and Eastern Europe, Asia, Africa and South America (1500–1600): a bibliographical history*, Brill, Leiden; Boston 2015, 715–23.

al-Abḥur was completed in 1517. In the passage devoted to minority witnessing, he sticks to the ḥanafī orthodoxy as expressed by al-ʿAynī, in that dhimmīs can testify for or against mustāʾmins "although not vice-versa" (dūn ʿaksihi). If al-Ḥalabī does not mention walāya, he seems to allude to the fact that everyone in society, even the most disgraceful, ranging from dhimmīs to mustāʾmins and from sons of adulterers to the effeminate, can give testimony provided unequal power relations do not destroy the truth contained in it. A slave (mamlūk), therefore, cannot testify, although a manumitted servant (muʿtaq) can be trusted as witness for or against his master (muʿtiq). Indeed and as we will see, Bailo Pietro Bembo complained in 1484 about a Venetian made prisoner on the basis of an affranchised slave's testimony in favor of his former master. Neither al-Ḥalabī considers religious enmity as a necessary obstacle to fair witnessing, a point previously raised by Ibn Qayyim. Inherent vices and corrupted morality, instead, disqualify potential witnesses, and al-Ḥalabī lists a long list mentioning from the most flagrant cases to pigeon raisers as well as music, chess and backgammon players.[51]

Even though the ḥanafī exception is generally accepted to have been adopted as a distinctive feature of Ottoman governance, this linear path between theory and practice was not applied in the case of the Franks. Although jurists did not accept Frankish testimony against local dhimmīs, rulers overthrew this claim by affirming that the mustāʾmin formed a single community with the Ottoman subjects of Christian and Jewish stock.[52] Placing the Franks on the same legal footing as that enjoyed by the dhimmīs constituted an appalling disregard for jurisprudence and particularly for ḥanafī doctrine. The Ottoman official position is made clear in the Italian and French versions of

51 Pigeon raisers were often excluded from witnessing, probably due to the knowledge on their neighbors' lives they unlawfully acquire over rooftops, al-Ḥalabī, Ibrāhīm Ibn Muḥammad (1460–1549): *Kitāb Multaqá al-abḥur,* edited by Wahbī Sulaymān Ghāwijī al-Bānī: Dar al-Bayrouti, 2005, 459–60. The interdiction, however, lay its roots in Jewish law: Malka, Orit: "Disqualified Witnesses between Tannaitic Halakha and Roman Law: The Archeology of a Legal Institution", *Law and History Review*: 1–34. As for backgammon and chess players, they are usually targeted by the ḥanbalīs, Cook, *Commanding right and forbidding wrong in Islamic thought,* 93–96, 146.

52 "Il est intéressant de noter que le fiqh refusait catégoriquement aux étrangers (musta'minin) le droit de témoigner contre les dhimmi. Leur témoignage n'était accepté que dans les procès mettant en présence des parties appartenant au même pays (dar). La raison de ces limitations nous est indiquée par Sarakhsi, quand il insiste sur l'opposition entre le daru-l-islam, constituant une unité spirituelle et juridique si forte, en dépit de ses divisions politiques, qu'elle établissait un lien parmi les dhimmi eux-mêmes, et le monde étranger, dans lequel les Etats n'étaient basés que sur la violence (Sarakhsī Kitāb al mabsut).", Grignaschi: "témoignage des sujets non musulmans", 219–20.

commercial treaties, but also in Turkish firmāns ranging from 1520 to as late as 1698.[53] When asked about the Bailo's right to summon Venetian witnesses to counter the claims of a hypothetical dhimmī, Ebu's-suʿud answered that no school of thought validated such testimony, adding that "in this regard the ahdnames convey traditional mistakes" and that the qādīs should ignore the clauses allowing mustāʾmins to bear witness for or against dhimmīs, failing which they deserved to be removed from office.[54] If it cannot be said that the Ottomans were in strict adherence to ḥanafī doctrine, nor can they be cast as mere pragmatists—an angle often adopted by commentators of the ahdnames and treaties with Muslims. Indeed, the Ottomans' decision to put the Franks on the same legal footing as their local coreligionists was applied to all merchant nations, even small ones, and upheld for over two centuries. By the same token, the tendency to invariably present the Ottomans as innovative and dynamic in their attitude towards institutions should be adopted with caution. Indeed, Grignaschi's assumption that the considerations mentioned above about ḥarbīs and issues of evidence was a distinctively Ottoman trait is no longer tenable. We have seen in previous chapters that Mamluk thinkers such as al-Subkī were concerned with similar issues; the position of the ḥanafīs and other schools were discussed in the 14th-century treatise on Siyāsa by Ibn Qayyim, and later thoroughly addressed by the ḥanafīs al-ʿAynī and al-Ḥalabī. The former, a native of ʿAyntab in southern Anatolia, was a Turkish-speaker who translated his own works from Arabic into Turkish and taught jurisprudence to several Turkish-speaking Mamluk sultans, whilst the latter pursued a career in Istanbul based on his prestige as a Syrian jurist.[55] Naturally, Syria, to which European merchants flocked in great numbers throughout the long 15th century, provided scholars with material for these fatwās and fiqh works, in the same manner that the region was host to major developments in the application of Mamluk Siyāsa. Conversely, and probably due to the quarrel between jurists and rulers on that point, not all ḥanafīs stuck to the same specific framework when dealing with diversity. Şeyhülislams such as Zakariyazade Yahya Efendi (1634–1643) and Menteshizade ʿAbdu-r-Rahim Efendi (in office 1715–6) allowed cross-confessional dhimmī witnessing while excluding mustāʾmins, whereas Zayn al-Dīn Ibn Ibrāhīm Ibn Nujaym (1520–1563), writing at a time when such controversial issues were only just emerging, and who thoroughly explored the

53 Ibid., 235, quotes a *ferman* adressed to the Lords of Samos, and the Treaty of Karlowitz, respectively.

54 Ibid., 253.

55 Marçais, W. (2012), al-ʿAynī, *Encyclopaedia of Islam, Second Edition*, 1:790b.

relationship between writing and testifying in his chapter on proof, prudently preferred to leave this question aside.[56]

4.1.3 *Foreign Unbelievers: Between Mistrust and Accommodation*

Probably due to the current emphasis on describing cross-confessional practices, often stemming from non-Muslim sources, we are getting used to a flexible and pragmatist interpretation of Ottoman governance. This was not the norm half a century ago, and Grignaschi, by contrast, held the assumption that ḥanafī doctrine constituted a major vector in understanding Ottoman legal attitudes. Most schools of law, including some ḥanafīs, mistrusted dhimmī testimony, however jurists were open to accepting it for the sake of convenience (istiḥsān, or juristic discretion). Accordingly, so too the Ottoman legal order must have sought a balance between these two principles. Grignaschi initially speculated on the apparently specifically Ottoman tendency to allow dhimmīs the right to produce proof for reasons of necessity, focusing on legal situations where Islamic laws on evidence were strictly applied, even if it conspired to curtail the spread of Islam.[57] The demonstration of this claim, however, relied on extreme examples, such as the fatwā by Çatalcalı Ali Efendi (1631–1692) reasoning that, if a Christian freed his Muslim slave before two Christian witnesses, the latter could not be proved to be free and should remain therefore in slavery. Initially Grignaschi's had stressed how Ottoman law extended to dhimmīs the same legal reasoning adopted for issues concerning Muslims, a viewpoint shared by contemporary legal historians.[58] Not necessarily in contradiction with this, Grignaschi then adopts a more critical tone, claiming that Ottoman jurisprudence was dominated by a second principle: mistrust for the dhimmīs' ability to testify the truth. As regards regulations on matters of debt, Grignaschi focused on the principle that Muslim creditors enjoyed precedence over their non-Muslim counterpart, although he acknowledged that such a privilege was not relevant when the witnesses were Muslim, irrespective of the parties' religion. Yet as we shall see in the case of bankruptcy studied later in this chapter, in claims for debt the biases against non-Muslims provoked attitudes that cannot be easily explained by jurisprudence alone. Just as dhimmīs' rights were extended to mustā'mins, in disregard for the rules of fiqh, during

56 Ibn Nujaym, *al-Ashbāh wa-al-naẓā'ir*, 257–293, Grignaschi: "témoignage des sujets non musulmans", 234.

57 Heyd, *Studies in old Ottoman criminal law*, 186, on istiḥsān, Kermeli: "Ebussuud Efendi", elaborates on Ebu's-su'ud's accommodation for neighborly relations, and for family law. Imber, *Ebu's-su'ud: the Islamic legal tradition*, 108–9.

58 Müller: "Non-Muslims as part of Islamic law," *28*.

the Priuli lawsuit Franks seem to have been denied the right to have recourse to their own Muslim witnesses.

Legal historians debate over the justice that was delivered at the qādī courts. For some, decisions were dictated by the need to maintain the rule of law, with no biases against the litigants' social condition. For others, gender and race prejudices were strong, and judges tended to validate testimony from witnesses of a privileged social background.[59] My analysis of Ottoman intervention in cross-confessional issues aims to distance itself both from claims of mere pragmatism, and an excessive reliance on jurists' law. Rather, with all of its inconsistencies, attitudes towards unbelievers, maintained over time, signal the existence of a conscious policy to attempt to deal with the affairs of the mustā'mins.

Born in Cairo under Ottoman rule, and the author of an important work on ḥanafī jurisprudence, Zayn al-Dīn Ibn Nujaym (1520–1563) combined both the new attitudes towards written evidence addressed in section 2.5, and suspicion towards minorities as actors of the legal system. Ibn Nujaym mistrusted minority witnessing, while maintaining an open attitude towards the widespread use of ḥujjas.[60] To justify reliance on written artifacts, he pointed to the use of safe-conducts by ḥarbīs: deeds of appointment (berats), granted to ḥarbīs accounted for the legal status of the holder (ḥāmil), and, since these documents referred in the last instance to the head of the community, they were endowed with the trust granted to official documents, and hence should be equated with the persons of officials.[61] Similarly to the Mamluk case examined in the previous chapter, where courtier-witnesses bore testimony for cross-confessional transactions, merchants and bankers -Ibn Nujaym goes- rely on the registering of transactions by these courtiers, and hence their *daftars* should be admitted as proof. Ibn Nujaym, it has been noted, counts among the earliest scholars exhibiting an open attitude towards custom as an important vector in Islamic law, hence anticipating nineteenth-century developments in this much-disputed matter.[62] Written evidence, Ibn Nujaym argued, was always more trustworthy than a witness who claimed to no longer remember

59 Hosainy: "Ottoman legal practice and non-judicial actors in Seventeenth-century Istanbul", Wilkins: "Witnesses and Testimony in the Courts of Seventeenth-Century Ottoman Aleppo".

60 Johansen, "Formes de langage", 369, Ayoub, Samy, ""The Sulṭān Says": State Authority in the Late Ḥanafī Tradition", *Islamic Law and Society* 23 (2016), 239–278.

61 Ibn Nujaym, *al-Ashbāh wa-al-naẓā'ir 'alá madhhab Abī Ḥanīfah al-Nu'mān*, 257.

62 Francesca, Ersilia: "L'applicazione del diritto islamico nell'impero ottomano (1500–1800)", in: *Storia dell'Europa e del Mediterraneo: Culture, religioni, saperi, vol. XI (II della sezione Età moderna)*, edited by Roberto Bizzocchi, 475–523, Roma: Salerno editrice, 2011.

the facts, since good faith should be assumed for those who bore testimony, and not for those who admitted not knowing. Moreover, Ibn Nujaym seems to have mistrusted the well-known principle *kitābun yushbih kitāban*—"one writing looks like another," and therefore cannot be trusted. "How, then, shall we proceed?" he asked: by taking responsibility for the writing of only one's own deeds. The ḥujja could thus be counted among the standard types of sound proof, he reasoned, and—according to Ottoman practice examined in Chapter Two—what had been accepted by a judge during the hearings and had been included in his sijill should in theory allow the ḥujja-holder to enjoy their rights, in and out of court.[63]

Ibn Nujaym's tone, favorable towards written evidence, changed however when addressing the issue of who could give testimony. The unbeliever's testimony could be accepted for cases where mixed dealings had led one of the parties to appoint a commission agent of another confession, however as was the case for matters of debt, minority witnessing was only accepted if it favored Muslim partners and creditors. Several times, Ibn Nujaym equated unbelievers with the deaf and children, and insisted that witnessing should not be accepted when it came from an individual unable to claim a clear religious status.[64] As Giovanni Dario was reminded by the pashas, witnesses needed to be known to the community, together with their families. Ibn Qayyim and a number of other theorists held that the ḥanafī exception was generally applicable, except for cases of conversion. Generally, the probative effectiveness of a statement was dependent upon the religion of the witness, and not on that of either party; in support of which Ibn Nujaym cited the usual case of cross-confessional ownership of slaves, deathbed conversion to Islam, etc., accepting the principle that Muslim witnesses needed to be trusted, even in cases in which their testimony proved to be prejudicial to the interests of Islam. This was the case, for example, for a Christian who had had three sons, two of which were Muslims and who could witness to the fact that the man had died a Christian. If a third, Christian son swore that he died a Muslim, he was not to be trusted.[65] In *al-Ashbāh wa-al-Naẓā'ir*, Ibn Nujaym adopts an open-minded attitude towards the use of writing for probative purposes, however he is reluctant to regard non-Muslims as actors in the legal system. He also prefers not to comment on

63 Ibn Nujaym, *al-Ashbāh wa-al-naẓā'ir 'alá madhhab Abī Ḥanīfah al-Nu'mān*, 293.

64 Heyd, *Studies in old Ottoman criminal law*, 245, Ibn Nujaym's opinion in this regard is shared by Ebu's-Suud, who ruled that an infidel who did not frequent his house of prayer was not competent to testify, Ibn Nujaym, *al-Ashbāh wa-al-naẓā'ir 'alá madhhab Abī Ḥanīfah al-Nu'mān*, 280.

65 Ibn Nujaym, *al-Ashbāh wa-al-naẓā'ir 'alá madhhab Abī Ḥanīfah al-Nu'mān*, 473.

the status of the mustāʾmins. Ibn Nujaym's work is representative of Ottoman intervention in the legal system, but also suggests that government-sponsored practice may have influenced legal reasoning, and not the other way around.

Judging by its longevity, the rigid application of the biases, and particularly against Christian witnesses, could be labeled an Ottoman orthodoxy. However, this official doctrine needs to be interpreted against the backdrop of actual choices and policies, and not solely against the writings of jurists. As argued by Snjezana Buzov, such orthodoxy was tautological when applied to the decisions made by a Muslim ruler, who asserted his commitment to implementing the rule of law. In the following pages I will refrain from adopting either a purely pragmatic, or a strictly juristic view of Ottoman governance amidst diversity. Although fiqh and practice were obviously linked to each other, the field of legal relations with foreigners was rife with conflict, and Ottoman rulers made their own choices, which were not necessarily always in harmony with the advice of jurists. In the case of Ebu's-suʿud, usually described as a pragmatist, and considered to be the jurist who brought kanun in agreement with sharīʿa, it is precisely on the issue of Frankish witnessing that he departs from government-sponsored practices.[66]

Ottoman jurists dealing with unbelievers were actually used to tackling issues of inheritance and debt. Muslims were not often present at a Christian's deathbed, nor did they attend their weddings, and for this reason dhimmī witnessing was actually necessary when it came to claims of inheritance, succession, or determining relations of kinship. Unbelievers were heard for reasons of equity or as an act of grace (*istiḥsān*), or due to necessity (*ḍārūra*).[67] A classic example was the inheritance of a deceased dhimmī; in the absence of righteous heirs, these estates were reclaimed by the treasury, the Islamic *Bayt al-māl*; thus, dhimmī witnessing was indispensable for proving intra-communal matters, such as those pertaining to family law. Hence sharīʿa biases against dhimmī witnessing were mitigated by the principles of *istiḥsān*, as they had been under the Mamluks in texts invoking the *maṣlaḥat al-Islām*. A case concerning the estate of a merchant from Dubrovnik is illuminating as regards commercial litigation. The so-called 'Dubro-venedik' case, tackled in Ebu's-Suʿud's legal writings, marks the limits to the application of *istiḥsān* exceptions, and to the Ottoman authorities' willingness to accommodate for non-Muslim testimony. Based on broad consensus by scholars, Ottoman

66 Buzov, *The lawgiver and his lawmakers*, 16. Significantly, Kermeli: "Ebussuud Efendi", hesitates between characterizing Ebusuud as a pragmatic thinker or as a conservative as regards inter-faith relations.

67 Ibn Nujaym, *al-Ashbāh wa-al-naẓāʾir ʿalá madhhab Abī Ḥanīfah al-Nuʿmān*, 268.

authorities had traditionally allowed dhimmīs to testify for the property left by a deceased coreligionist. However, a decree issued in 1543 apparently contradicted this right, and triggered a famous fatwā by Ebu'ʾs-Suʿud. In that year, after the death of a merchant from Dubrovnik, the case was brought to the attention of the Porte, which issued a request to look for Muslim witnesses for the merchant's last will. The Dīvān's request, which was in contradiction with both scholarly consensus and actual court practice, alarmed the qadis, who consulted Ebuʾs-Suʿud on whether the qadis were expected to continue applying the *istiḥsān* procedure—commonly accepted for family and probate issues—or whether they should ask for permission from the Porte. Ebuʾs-suʿud answered that the decree was an *ad hoc* measure, due to suspicion of fraud in the witnesses' depositions, and therefore should not be applied to future cases. Contemporary commentators suggest that the merchant was a mustāʾmin who died in the Ottoman Empire; the fact that he was not a dhimmī but a foreigner enjoying a lower legal status may have motivated the suspension of the *istiḥsān* procedure.[68] Be that as it may, the episode is telling in that it reveals the constraints Ottoman rulers could impose on the labor of jurists.

The principle of istiḥsān was in particular frequently resorted to in the European areas under Ottoman dominion, for issues concerning payments to timar-holders, and more generally, those revolving around the property of land. In claims over debt, last wills, and other fields, there was room for an accommodating interpretation of evidentiary rules. However, this cannot simply be attributed to the fact that the Empire extended over regions largely inhabited by non-Muslims; indeed, it was the norm to give priority to Muslim witnesses in the late Ottoman Rumelia courts, such as those of seventeenth-century Sofia.[69] Moreover, it has to be noted that the decision to accommodate for dhimmī witnessing occurred at the height of Ottoman military strength, so it can hardly be viewed as a concession granted according to balances of power. In other European principalities, accommodation affected other aspects of procedural law: In Dubrovnik, Bulgaria or Serbia, Ottoman qadis accepted to consider vernacular forms of contracts and even collaborated with dragomans knowledgeable in Islamic law.[70] However, to

68 Imber, *Ebuʾs-suʿud: the Islamic legal tradition,* 108–9, Van den Boogert, *The Capitulations and the Ottoman Legal System,* 174–5, Grignaschi: "témoignage des sujets non musulmans", 241–2, Gökbilgin: "La preuve et le témoignage dans la jurisprudence des Fetva d'Ebussud".

69 Gradeva, "Orthodox Christians in the Kadi Courts: The Practice of the Sofia Sheriat Court, Seventeenth Century", Speros Vryonis, Jr., "The Byzantine Legacy and Ottoman Forms", Dumbarton Oaks Papers Vol. 23/24 (1969/1970), 251–308.

70 Fotic, Aleksandar: "Non-Ottoman Documents in the Kadis,' Courts (Moloviya, Medieval Charters): Examples from the Archive of the Hilandar Monastery (15th- 18th C.)",

the difference of dhimmīs, accommodation never governed the relations with mustā'mins, who had nevertheless to deal with many obstacles to the validity of their testimony. Minority testimony in all other matters, as at the qadi courts, was completely valueless, and in the context of large communities of dhimmīs and mustā'mins, for Frankish merchants the 'bronze wall' rose higher and higher, with no apparent mechanism to balance out its effects, as was the case for dhimmīs with the istiḥsān procedure. This is at least a late-nineteenth century reading of the problem, such as that advanced by the Neapolitan jurist Francesco Paolo Contuzzi. In his book, which is significantly enough devoted to the consular institutions, he notes that: "the mustā'mins can be witnesses only for their internal disputes, and when they belong to the same nation. The testimony of a mustā'min against a Muslim is not accepted. Here emerges a bronze wall, raised between Muslims and Christians."[71]

To be sure, the cumbersome nature of legal bans took a heavy toll on the interactions between Muslims and foreign unbelievers, be they borderland issues or simply market transactions. Yet in spite of these strong prejudices, mechanisms did exist to unblock the legalistic catch-22 imposed upon the Franks. Indeed, the Ottomans adopted their own, opener attitudes towards written evidence and, whatever the motivations behind the adoption of these practical measures, they served to mitigate the effects of the legal biases against minority witnessing. Most notably, I will be addressing a policy that has largely passed under the radar, and that allowed Franks to rely entirely upon documents for their legal defense, and which I refer to hereinafter as the "ban on Muslim witnesses." Although it marked a clear division between Islam and unbelief, combined together, Ottoman legal attitudes created a legal framework that was able to guarantee the striking of deals, supported by the notarizing functions of the qadi, and made it possible to have recourse to cross-confessional witnessing to support agreements between Europeans and local Jews and Christians. Ottoman attitudes did not simply adapt to the standard solutions provided by ḥanafī jurists, and even when they departed from jurisprudence Ottoman policies were implemented without necessarily challenging fiqh. While tolerating business and daily exchanges, the Ottoman

in: *Frontiers of Ottoman studies* (*Vol. 2*), edited by Colin Imber, Keiko Kiyotaki, *et al.*, London, 2005, 63–73, Zecevic, "Translating Ottoman Justice: Ragusan Dragomans As Interpreters of Ottoman Law".

71 Contuzzi, Francesco: *La istituzione dei consolati ed il diritto internazionale europeo nella sua applicabilità in Oriente*, Naples: E. Anfossi, 1885, 145.

sultan was able to present himself as a guarantor of the supremacy of Muslims as legal actors, and as an enforcer of the rule of law.

Three months after his return to Venice, in April 1522, the ambassador Marco Minio had his final *relazione* read before the Senate. Venice and the Porte were by that time engaged in a diplomatic effort to define the conditions of the mustā'min presence in post-conquest Egypt and Syria. To the two agreements drafted in barely five years, Minio added the clauses contained in the new treaty issued in 1521. The Venetians had placed a demand on the negotiating table, asking that, for disputes concerning the return of Frankish captives "Christian witnesses should be accepted."[72] This request represented a challenge to one of the founding principles of the law on evidence, and one can hardly believe the Venetians' request was advanced out of pure ignorance of this. To include this issue in the agenda, Minio adds, was the doge in person. Needless to say, the Ottomans avoided giving a straight answer, suggesting that such disputes needed to be discussed at the Dīvān, where the Bailo's word would be taken into consideration—an issue to which I will return. As I have mentioned earlier, such a naive request was never raised by Venice during the Mamluk period, although, significantly enough, the less-experienced Florentines did so in 1481. It is noteworthy, then, that the Venetians felt the need to bring minority witnessing up so soon after the conquest of Egypt and Syria. Without giving much hope to the Venetian negotiators, the Ottomans made clear that, although minority witnessing was non-negotiable, further accommodation at the Dīvān could be reached, particularly if the Bailo himself personally vouched for the Venetian party. In light of the conversations held between Giovanni Dario and the pashas thirty years previously, there is little doubt that the allegedly orthodox approach to non-Muslim witnessing had a directed influence on, if not entirely governed, relations with the Franks.

The strict adherence of Ottoman governance to the doctrines of the ḥanafīs is patent as regards the adoption of the ḥanafī exception. As eyewitnesses to this legalistic approach to governance, the Bailos and the ambassadors noticed the official sponsoring of ḥanafī views on proof and procedure, and the empowerment of ḥanafī 'Turkish' qadis over preexisting local judiciaries. They

72 "Il terzo, che se uno mercadante veneziano fosse convenuto davanti il cadi per qualche differenzia. non possa il cadi giudicarlo non essendo il suo dragoman li presente. Un altro capitolo è delli schiavi, circa la sua ricuperazione, il qual non ho potuto far che fosse posto secondo la intenzione di Vostra Serenità, cioè che fosse creduto a' testimonii cristiani," *Relazioni degli ambasciatori veneti al Senato, Serie III, Volume III,* 88. Theunissen, *Ottoman-Venetian Diplomatics,* 415–36.

commented on the summary nature of justice, the lack of interest paid to written deeds, but above all on the complex nature of the Dīvān itself, where there was no clear division between state and purely legal affairs. As Bailo Barbarigo put it in the 1550s, "they deal with all requests (arz); that is, lawsuits, petitions, along with all sorts of private disputes and state affairs."[73] Venetian diplomats, whose talkative dispatches provide us with a good deal of insider information, were firsthand witnesses to this ḥanafī orthodoxy, as well as being directly involved in accommodations, negotiations, and the quest for loopholes in the bronze wall of sharīʿa norms.[74]

However, what the Bailos perceived as an Ottoman orthodoxy needs to be read against the backdrop of these apparent contradictions and exceptions to the general rules. Mirroring the two vectors governing, in theory, the handling of proof and unbelief, two distinctive features characterized, in practice, early Ottoman attitudes towards the Franks in the court and at the marketplace. The first is the departure from ḥanafī legal texts in one important respect; in their disregard for the ḥanafī principle that mustā'mins could not testify against dhimmīs. This was an important footnote to the ḥanafī masters, and as pointed out by Grignaschi, Ebu'ʾs-Suʿud, the champion of Ottoman legal orthodoxy, explicitly raised objections against the sultans' actions on this point.[75] As expressed in the jurists' jargon, the official position held by Ottoman rulers was that all unbelievers constituted "a single nation." However, by affirming this, they departed from the revered opinions of the ḥanafīs. On similar grounds, a second feature that marked all kinds of exchanges in practice is the unexpected ban on Muslim witnesses against Franks. Ad hoc rearrangements of the legal system were found for important locations such as Aleppo, where a specific ferman inhibited the general legal framework enforced in other cities of commerce.

---

73 "dove giunti, subito seduti li pascià, si appresentano in mano di Rusten tutti li artz, cioè querele, suppliche ed ogni altra sorte di scritture appartenenti cosi a materie e cause private come a, cose di Stato", *Relazioni degli ambasciatori veneti al Senato, Serie III, Volume III*, 155–6.

74 Graf, Tobias P.: *The Sultan's renegades: Christian-European converts to Islam and the making of the Ottoman elite, 1575–1610*, Oxford-New York, 2017, 6.

75 Fatwās issued by Ebu'ʾs-Suʿud published in Friedrich Selle, *Prozessrecht des 16. Jahrhunderts im Osmanischen Reich. Auf Grund von Fetwas der Scheichülislame Ebüssuud und anderer unter der Regierung des Sultans Süleiman des Prächtigen*, Wiesbaden, Harrassowitz, 1962, 41 (91), §§ 10–11, and Paul Horster, *Zur Anwendung des islamischen Rechts im 16. Jahrhundert: die "juristischen Darlegungen" (Maʿrūżāt) des Schejch ül-Islam Ebū Suʿūd (gest. 1574)*, Stuttgart, Kohlhammer, 1935, 53 (91), cited by Heyd, *Studies in old Ottoman criminal law*, 245, and n. 7.

4.2 The Ban on Muslim Witnesses

At some point in the early sixteenth century, the Ottoman chancery started granting a privilege that allowed Franks to defend themselves in court by exhibiting written deeds, more specifically a ḥujja (Tr. hüccet) produced by a qadi. Parallel to this measure, and more importantly, Muslim witnesses could not be heard to counter the contents of these legal certificates. As pointed out by Maurits van den Boogert, the very idea of preventing the oral performance of Muslims at court is, in principle, at odds with Islamic laws of evidence. Evidently, the argument could be turned the other way around, as a hüccet was by definition an agreement supported by Muslim witnesses. And indeed, this counter-argument was not unknown to late medieval commentators; according to Ibn Qayyim and his master Ibn Taymīya, if a document was known to be authentic, a judge could rely upon it for a ruling, since "writing is the expression of utterings and ultimately, of willingness and determination."[76] In choosing to issue such a ruling, the Ottomans probably had in mind the claims of contemporary jurists such as Ibn Nujaym, to whom the ḥujja counted among "sound forms" of proof (*bayyina ʿādila*), together with confession (*iqrār*), the defendant's refusal to take an oath, the oath itself, collective oaths (*qasāma*), the judge's personal knowledge of facts, or reasonable assumptions (qarīna qāṭiʿa).[77] For European commentators and scholars focusing on the external manifestations of this practice, such a privilege departed from sharīʿa principles. From a juristic perspective, however, jurists such as Ibn Qayyim or, in Ottoman times, Ibn Nujaym refashioned the argument in favor of written evidence, on the grounds that words such as *bayyina* and *ḥujja*, although usually understood to mean, respectively, 'oral' and 'written' witnessing, actually carried the more abstract notions of 'proof' of 'arguments of authority', and that the Qur'ān did not intend the term *bayyina* to refer exclusively to testimony.[78]

Whatever its origins, the move to grant Franks the right to present ḥujjas as proof should be taken against the backdrop of a more generalized Ottoman-sponsored practice; that is, that any business deal—although it continued to be an oral agreement freely reached by two parties, needed to be supported by a written deed, a hüccet registered before a ḥanafī qadi and subscribed by witnesses.[79] Of course, this was nothing entirely new: we know that some

76 Ibn Qayyim al-Jawzīyah, *al-Ṭuruq al-ḥukmīyah*, 549.

77 Ibn Nujaym, *al-Ashbāh wa-al-naẓā'ir ʿalá madhhab Abī Ḥanīfah al-Nuʿmān*, 293, Johansen, "Signs as Evidence", 336.

78 Bechor, *God in the Courtroom*, 348.

79 Baldwin, *Islamic Law and Empire in Ottoman Cairo*, 35, and n. 16 illustrates the process of contract registration, on the basis of the Cairo court registers.

contracts were registered in court under the Mamluks, presumably in a very similar manner to that now rendered compulsory by the Ottomans.[80] We also know that under the Mamluks, Franks and Muslims had recourse to the notary-witnesses and, when available, to the Latin notary, although we have no references to mixed commercial contracts registered at the courthouse.

The new registering policy promoted by the Ottomans had the effect of empowering the ḥanafī, Ottoman-appointed qadis, as they were endowed with the monopoly of certification (*tasjīl*). In addition, the measure was applied to the entire territory of the Empire, and affected faraway commercial hubs such as 16th-century Jedda and Mecca. In these cities, although the parties had recourse in the first place to their own witnesses and qadis, they were later compelled to have their acts registered before the ḥanafī—sometimes referred to as 'Turkish'—judge. This meant that the ḥanafī judge had the last word in much-disputed cases, particularly for issues related to the endowment of religious foundations or last wills. And indeed, the descriptions of Mecca given by Jār Allāh Ibn Fahd (1486–1547) place the ḥanafī chief qadi Badīʿ al-Zamān Ibn al-Ḍayāʾ at the center of many quarrels, and portrays him manipulating their legal outcome by imposing his own, corrupt witnesses for the final drafting of hüccets.[81]

The treaty signed between the Ottomans and Florence in 1488 does not make any mention of the interdiction on testimonial proof.[82] Similarly, the 'Egyptian' treaty underwritten with Venice in 1517, drafted in Arabic, adheres to traditional methods of notarization—although, it could be argued, it is a treaty that should be interpreted in the context of Egyptian judicial custom. According to Nicolas Vatin, the latter was a provisional ahdname issued when the effective incorporation of Egypt into the Ottoman Empire had not yet been decided.[83] Gladys Frantz-Murphy asserts that the text is "Mamluk," meaning probably that it was drafted by the Mamluk chancery, or that it followed conventional Mamluk patterns of treaty-writing.[84] Be that as it may, the 'Egyptian'

---

80 Rapoport, "Legal Diversity in the Age of Taqlid: the Four Chief Qadis Under the Mamluks", 222.

81 Ibn Fahd, *Nayl al-Muná bi-Dhayl Bulūgh al-Qurá, 184–5, 208–9, 308–10.*

82 Müller, Joseph: *Documenti sulle relazioni delle città toscane coll' Oriente cristiano e coi Turchi fino all'anno MDXXXI*, Firenze: M. Cellini e c., 1879.

83 Moritz, Bernhard: "Ein Firman des Sultans Selim für die Venzianer vom Jahre 1517", in: *Festschrift Eduard Sachau*, Edited by G. Weil, 422–443. Berlin: Georg Reiner, 1915, Theunissen, Hans, "Cairo Revisited (I): Four Documents Pertinent to the Ottoman-Venetian Treaty of 1517", Electronic Journal of Oriental Studies 2 (1999).

84 Frantz-Murphy, "Identity and Security in the Mediterranean World ca. AD 640—ca. 1517", 256, n.16.

ahdname, whose terms seem to have been limited to stipulating Venetian trade in Egypt, reflects the most pressing priorities of the Ottoman reform agenda. The 1517 text makes it clear that the Siyāsa courts were no longer part of the adjudication system, although it does not yet concern itself with the other procedural aspects of mixed trials.

Early Ottoman treaties—that is, those signed before the 1530s—clearly evoke the continuity of traditional patterns of notarization, with contracts underwritten before traditional notary-witnesses. In the same fashion, nothing is said about notarizing at the qadi's court in the 1478 ahdname signed with Venice.[85] The Egyptian treaty stresses continuity as regards how business deals should be concluded, and on several occasions mentions the port's notary-witnesses (shuhūd), thus confirming that notary-witnesses attached to the customs administration, were still an option. However, the document quite insistently makes a clear distinction between Franks, Christians (*naṣrānī*), Jews and Maghrebis. Moreover, it marks a growing division of labor between the agents of notarization and those entrusted with linguistic mediation, the dragomans, two tasks whose contours had been growing more and more blurred in the late Mamluk years. The ahdname states that if a dragoman has recourse to the shāhid to draw up a deed, the latter should not be prevented from doing so. This probably means that deals were often concluded solely in the presence of a translator, but that in some cases one of the parties insisted on having the deal notarized by a proper shāhid. In contrast with the Egyptian, Arabic treaty issued immediately after the conquest, neither the second, more official Turkish treaty signed by ambassador Alvise Mocenigo in 1517, nor the 1521 ahdname mention issues of notarization, and seldom do they refer to legal forums for disputes.[86] In this, the Ottoman chancery marked a break with the Mamluk past, increasingly concerned as it was throughout the 15th century with adjudication norms and with the securing of deals.

The ban on Islamic witnesses for mixed trials is mentioned in the treaty with François I in 1535, but not in the treaty concluded a few years before, in 1528.[87]

85 Wright, Diana Gillilandand Pierre A. MacKay, "When the Serenissima and the Gran Turco Made Love: The Peace Treaty of 1478", Studi Veneziani 53 (2007), 261–277.

86 Theunissen, *Ottoman-Venetian Diplomatics*, 400–36.

87 Saint-Priest, François-Emmanuel Guignardand Charles Henri Auguste Schefer: *Mémoires sur l'ambassade de France en Turquie et sur le commerce des Français dans le Levant*, Paris,: E. Leroux, 1877 345-353, Charrière, Ernest: *Négociations de la France dans le Levant, ou Correspondances, mémoires et actes diplomatiques des ambassadeurs de France à Constantinople, etc.* Vol. 1, Paris: Imprimerie Imperiale, 1848, 287, Jensen, De Lamar, "The Ottoman Turks in Sixteenth Century French Diplomacy", The Sixteenth Century Journal 16 4 (1985), 451–470, 453–5 and n. 18.

Thus, if this chronological speculation is correct, the ban on Muslim witnesses against Frankish merchants crystallized in chancery artifacts sometime in the early 1530s. This would account for the general climate of confusion surrounding the Priuli bankruptcy dealt with at the end of the present chapter, and for the hesitancy of Western merchants to register contracts in and out of court. When Marco Priuli died in prison in Damascus, the executors of his will found both Arabic contracts and deeds drawn up by the Venetian notary in Syria, however his heirs had difficulty backing their claims due to the absence of contracts properly registered at the qadi courts. Encoded in the much-disputed treaty with France in 1535, is the first formulation of a ban on Muslim witnesses for mixed cases:

> Item. Que en cause civille contre les Turcqs, carrachiers ou autres subgets du Grand Seigneur les marchans et subjectz du Roy ne puyssent estre demandez, molestez ne jugez si lesdicts Turcqs, carrachiers et subgetz du G. S. ne monstrent escritures de la main de l'adversaire ou coget (c'est instrument) du caddi, baille ou consul, hors de laquelle escriture ou coget, ne sera vallable ne receu aucun tesmoignage de Turcq, carrachiers ne autre en quelque part que ce soit de testât et seigneurie dudict G. S. et les caddi, sousbassy ne aultres ne pourront ouyir ne juger lesdicts subgetz du Roy sans la présence de leur dragoman.[88]

We can only speculate as to why the lawmakers felt it necessary to make the norm explicit in treaties signed with inexperienced trading nations, such as Tuscany and France, but not with historic partners such as Venice. France, although it was an important trading nation by that time, was a newcomer to Middle Eastern diplomacy. In any case, the right to prove acts with registered contracts was rendered explicit in all following 16th-century French capitulations, and in those underwritten with another rising commercial power, the Grand Duchy of Tuscany. Some authors have cast doubt on the 1535 treaty's authenticity, an argument based on stylistic grounds, however De Lamar Jensen believes that the text reflects the momentum attained by Franco-Ottoman relations at the time of the alliance between Süleyman I and François I.[89] Generally considered to be a concession to France, the early formulation of the ban on Muslim witnesses could indeed support this idea, together with the fact

88 Saint-Priest and Schefer, *Mémoires sur l'ambassade de France en Turquie et sur le commerce des Français dans le Levant,* 356.

89 Jensen, "The Ottoman Turks in Sixteenth Century French Diplomacy", 455.

that the treaty also mentions the Ottoman position that the mustā'min's word could be used against that of a dhimmī.[90]

Whatever its origins, the ban on Muslim witnesses against Frankish litigants slowly crystallized in Ottoman chancery documents, and invariably appears in the extant translations of the treaties. The French version of the 1535 treaty stipulates that only contracts registered by the "caddi, baille ou consul" should be counted as valid, although as time went on, the phrasing did away with references to any notarial artifact other than the qadi-registered hüccet, as is the case and in 1569, where the text appears to refer to the validity only of written deeds registered at the courthouse.[91] Incidentally, the Turkish version of the 1569 treaty just assumes that in case where no documents are available, the parties will naturally tend to produce false witnesses (*şāhid-i zūr*), an important point to which I will return.[92] More explicit provisions excluding Muslim witnesses were laid down in a draft treaty between the Porte and Tuscany in 1578, or in the French capitulations of 1597:

> Che nessuna persona possa astringier per debito subditi o mercanti delli Stati del Granduca di Toscana a nessun tribunale, se gia non fussi scrittura apparente, o sigillo del Cadi, non occorrendo in tal caso testimonianza di persona.[93]
>
> Que les François, leurs consulz et interprettes ou ceux des lieux qui deppendent d'eux ayent en leurs ventes et achaptz, pleigeries et tous autres pointz d'en faire acte devant le cady, au deffaut de quoy, ceux qui

90 "Item. Que en causes criminelles, lesdits marchans el autres subgetz du roy ne puyssent estre appelles des Turcqs, carrachiers ne autres devant les caddis ne autres officiers du Grand Seigneur, et que lesdits caddis ne officiers ne les puyssent juger; ains, sur l'heure, les doyvent mander a l'excelse Porte, et en l'absence d'icelle Porte au principal lieutenant du grant seigneur, Ia où vaudra le tesmoignage du subget du roy el du carrachas du Grand Seigneur l'un contre l'aultre." Charrière, *Négociations de la France dans le Levant,* 287–88.

91 Saint-Priest and Schefer, *Mémoires sur l'ambassade de France en Turquie et sur le commerce des Français dans le Levant*, 370, n.6.

92 "Frānça ve aña tābiʿ olan yerlerüñ tācirleri ve tercümānları ve ḳonsōlōsları memālik-i maḥrūsemde beyʿ ü şirā ve ticāret ü kefālet ḫuṣūṣunda ve sā'ir umūr-ı şer'iyyede ḳāḍīye varub šebt-i sicill étdürüb veyā ḫüccet alalar. Ṣoñra nizāʿ olursa sicille veyā ḫüccete nażar olınub mūcibi ile ʿamel olına. Bu ikiden biri olmayub mücerred şāhid-i zūr iḳāmet étmekile ḫilāf-ı şerʿ nesne daʿvā éderlerse mā-dām ki ḳāḍīlerden ḫüccetleri olmayub ve yāḫūd sicillde muḳayyed bulınmaya anuñ gibilere tezvīr étdürilmeyüb ḫilāf-ı şerʿ olan davāları istimāʿ olınmaya," Paris, BNF, DO, Ms. Turc 130, f. 2r–8r, transcription courtesy of Güneş Işiksel, 4r.

93 Camerani, Sergio, "Contributo alla storia dei trattati commerciali fra la Toscana e i Turchi", Archivio Storico Italiano 47 (1939), 83–101, C. 28, 100.

> auront quelque prétention contre eux ne le faisant apparoir par constract publicq enregistré au lieu de nos juges voullans prendre tesmoings, voulions et commandons qu'ils ne soient escoutez, ains soit donné foy aux contracts passés devant noz juges ou, n'y en ayant d'enregistré, que les demandes ne soient adjugées; et se tienne la main qu'il n'arrive chose contre la sacrée justice.[94]

The ban on Muslim witnesses in absence of documentary evidence gradually consolidated into a general principle governing relations with all Frankish nations—even, as we shall see, with Venice, in whose treaties such stipulations were previously omitted. The ban found its way into Ottoman-Polish diplomatic practice for the first time in 1553, and then again in four successive treaties underwritten during the 16th century.[95] As a basic mechanism that underpinned the enforcement of transactions, it is mentioned in French capitulations as late as 1740.[96] Timur Kuran reports the case of the English ambassador Finch (1660–1669) vs. the Iraqi merchant Mehmet, in which the former brandished the Ottoman-English capitulations, stating that only hüccet-supported claims could be advanced against Franks. Mehmet lacked valid documentation, and as a result the case was thrown out on procedural grounds.[97] Similarly, in order to back his claims against an Ottoman Jew, an English subject produced a decree by Sultan Mehmet IV, and supporting fatwās by two şeyhülislams; ultimately, the Ottoman subject was unable to produce valid hüccets to defend his case.[98] Kuran has found similar cases spanning several decades in the late 17th century. It should be noted, however, that the ban clause was conversely neither mentioned in the Venetian treaty of 1575, nor in that underwritten in 1619. Similarly, the ahdname of Osman II (ruled 1618–1622) deals extensively with the role of witnesses and cross-confessional relations, but is silent on the ban on Muslim witnesses, and makes no reference to the need to register transactions, the very pillar of mixed deals under the Ottomans.[99] In

---

94 Saint-Priest and Schefer, *Mémoires sur l'ambassade de France en Turquie et sur le commerce des Français dans le Levant*, 406–7.

95 Kołodziejczyk, Dariusz: *Ottoman-Polish diplomatic relations (15th-18th century): an annotated edition of 'ahdnames and other documents*, Leiden; Boston: Brill, 2000 241, 258, 277, 292, 311, 322.

96 Contuzzi, *La istituzione dei consolati ed il diritto internazionale europeo nella sua applicabilità in Oriente*, 205.

97 Kuran, *The Long Divergence*, 228–9, 339, n.1.

98 Ibid. 339, n.2.

99 Mumcu, Serap, "1619 il trattato di pace fra l'Impero Ottomano e la Serenissima Reppubblica Ahidname di Osman II (electronic article)", https://www.academia.edu/26041394/LAhidname_di_Osman_II, last accessed June 2019.

the Treaty of Passarowitz (1718) the ban is implicit, suggesting that disputes are to be resolved through the examination of documents, and only secondarily through the hearing of witnesses.[100] Whether the ban on Muslim witnesses was considered a detail to be stipulated in the ferman s and other sultan's decrees complementing the treaties, or Venice and the Porte found it useful to leave the question open to discussion deserves further investigation. However, the very fact that the ban is repeatedly mentioned not only in in Polish and Tuscan treaties, but also in those with increasingly prominent partners such as the French and the English, makes all the more intriguing its absence in Ottoman-Venetian ahdnames. Venice was the Ottomans' most favored trading nation, and the missing clause certainly cannot be pinned on her diplomats' inexperience.

Be that as it may, the 1718 Ottoman-Venetian agreement explicitly mentions both the Carazari clause, and the probative value of documents, and this late treaty provides us with a more sophisticated version of documentary practices. The agreement states that for mixed deals contracts needed to be registered and the subsequent hüccet delivered to both parties; for potential claims, the judges were obliged to compare the contract itself with the hüccet and the notarial protocol—that is, the qadi's sicil.[101] In the absence of any of these articles of proof, the judge should proceed by hearing righteous witnesses, taking special care to ensure their probity. The 1718 treaty makes it clear that contracts, presumably written, could be concluded outside of a court of law; for example, contracts involving one or two Muslim parties and signed before the Bailo's notary were not unknown. In March 1602 a certain Osman Re'is appointed Cussem Re'is as his attorney for a maritime business, and so did two janissaries in the following days, having these deeds notarized by the chancellor of Bailo Girolamo Cappello.[102] However, when it came to the resolution of mixed conflicts, ideally contracts, a proper hüccet delivered on request to the parties, and the original record bound in the qadi's sicil had to be inspected,

100 Miltitz, Alexandre: *Manuel des consuls*, Londres & Berlin: A. Asher, 1837–1841, II, 1489.

101 "ils devront faire enregistrer leurs contrats et recevoir le hodjet ou autre acte juridique; et s'il s'élève un différend, on devra comparer le contrat, le registre ou Protocole et le hodjet, et juger d'après la Conformité qui existera entre ces trois pièces; et lorsqu'on ne pourra produire aucune de ces trois pièces, et que l'équité exigera néanmoins que le différend soit jugé, les juges, en vertu de leurs pouvoirs judiciaires, devront accueillir les demandes en justice complètement et équitablement: ils devront peser, avec l'attention convenable, les témoignages … ," ibid. II, 1489.

102 ASVe, Bailo a Costantinopoli, Atti e Sentenze. Protocolli 271/ 67v-68r, 4 March 1602, 68v, 7 March 1602.

and only in cases in which none of these documents could be produced were witnesses summoned.

The obligation to register contracts before the qadi, and their capacity to overrule witness testimony has been signaled as a deviation from the spirit of sharīʿa or, for Timur Kuran, as a sign of the yielding of a sharīʿa-based legal system to the economic realities of the time—a process that would be completed in the nineteenth century. However, we should take into consideration the possibility that it was the rigid application of the witness system itself that caused this development. The compulsory nature of the hüccet might also have come about due to the very attitudes towards diversity adopted by the Ottomans. To be sure, apart from intra-communal matters, the ḥanafī school was particularly rigid in its refusal to accept dhimmīs as eligible witnesses; compared to under the Mamluks, we find no trace of bilingual simsārs, the Christian and Jewish scribes working for the customs, nor is there mention of the ḥājibs' courts and their flexible approach to proof and procedure. Similarly, the Ottomans did away with the tendency to blur the boundaries between linguistic mediation and notarization, as was the case under the Mamluks. Dragomans were only recognized in their capacity to provide translation and interpreting services, and their presence was made compulsory in court, whereas it had not been deemed necessary in medieval times.

In the context that emerged in the early decades of the 16th century, actors discovered that legal disputes were necessarily heard by the religious learned, who had no sympathy for procedural deviations. As Giovanni Dario put it during discussions about the building of a bridge near Ulcinj in Montenegro, "we especially have to deal with bigots, and with people who observe the Holy Scriptures and who are not willing to be contradicted, notably the pashas who are religious learned men."[103] Accordingly, in a given trial where the parties needed to prove their positions by means of testimonial evidence, the temptation to forge fake depositions was strong. As I claim in the section devoted to false witnesses, the danger was real enough to provoke a strong reaction from the Ottoman authorities against the proliferation of false witnesses, by publicly summoning and punishing a notorious liar in Istanbul. When a lawsuit revolved around two contradictory statements, by witnesses of different religions, judges were necessarily inclined to give the Muslim party precedence, or to be more precise, to the party defended by Muslim witnesses—a position epitomized by Ebu's-Suʿud's opinion on the village inhabited exclusively by

103 "et maxime habiando afar cum pizochari che e prima el Signor che man ti en la lege Santa che non vol aldir parola contraria. poi sono do Bassa che sono homeni de la fede," Dario, *22 dispacci*, 220.

infidels, where one Muslim murders a dhimmī, and where the Muslim cannot be found guilty on the basis of dhimmī testimony.

It is indeed possible that the supremacy of oral witnessing may have caused contracts exhibited by Franks be overturned by false witnesses, as was often the case in disputes involving Christian prisoners enslaved by Muslims. As an obvious solution for this problem, the ahdnames began to insist on the compulsory nature of ḥujjas, and to introduce explicit limitations to testimonial evidence. This may go some way to explaining a crucial question in the chronology: the ban on Muslim witnesses took a few decades to consolidate—for example in the realm of norms governing cross-confessional exchanges—while the promiscuity between dhimmīs and mustā'mins in court—the Carazari clause- had been sponsored since 1500, if not earlier. Grignaschi showed that even jurists adopting a stricter approach to ḥanafī doctrine on matters of witnessing, such as Ebu's-Suʿud, ended up making similar concessions to the Ottomans, and eventually accepting dhimmī witnessing to prove issues related to the ownership of land. As we saw earlier, the *maṣlaha* had been invoked by medieval Arab jurists dealing with Siyāsa, and in the same manner in the Ottoman context many debates revolved around the notion of *istiḥsān* to allow judges to rely on dhimmī witnesses for specific cases. While it is tempting to interpret such changes as pragmatic steps towards a more secular version of the rule of law, in promoting such measures the Ottomans were more motivated by their own quest for orthodoxy, and by their desire to distinguish their own approach from previous Mamluk practice, and from the corruptness of the Mamluk ʿudūl.

4.3 Dhimmī Claims on Communal Exclusivity: the *Carazari* Clause

Cross-confessional relations constituted an arena in which convoluted interactions played out between jurists' ideas, the Ottoman-sponsored version of these principles, and bottom-up reactions from legal actors. Although unbelievers could not be heard against Muslims, their testimony was accepted for cases involving other unbelievers. An area rife with contention, the ḥanafī exception began in those years to deviate into very peculiar ideas and practices about proof that, I argue here, have gone unnoticed and open new angles for understanding early Ottoman legal attitudes. For example, for more than a century local Christians and Jews brandished the ḥanafī exception so as to clearly distinguish between themselves and the mustā'mins in court. Dhimmīs claimed that they formed a legally separate community, and that therefore only the word of other Ottoman Christians and Jews could be used against them in court. Consequently, they would request the exclusion of Franks as witnesses against them, a measure that, if implemented, could compromise the resolution of disputes between these two prominent groups in Ottoman trade. These privileges were never fully embraced by the Ottoman government, but the idea that witnessing and proving claims should be associated with one's confessional group, rather than one's religion, permeated legally relations between the Ottomans and their neighbors. Probably from an early stage in the history of diplomatic agreements, and certainly since 1500, dhimmīs had been raising this specific point that, if successful, would have invalidated the word of Franks in mixed trials. Extant texts mentioning this request, such as the 1575 Ottoman-Venetian treaty quoted below, make it clear that in mixed cases dhimmīs wished for the opposing party to prove claims by means of local, dhimmī witnesses:

> Che essendo la prima differentia tra quelli della natione venetiana et alcuni christiani carazari, et vegando li veneziani produceano testimoni della loro natione, onde li adversari, con dire bisogna testimoni christiani di questo loco, davano travagli alli detti, et non accettavano la testimonianza delli christiani veneziani. Pero per essere christiani tutti d'una fede ho commandato che se li veneziani haveranno differentia con li christiani, et che gli faccia bisogno produr testimonii, producendo christiani, siano di che nazione si voglia, posano testificar, et secondo l'honesta legge del profeta sia accettata la testimonianza loro.[104]

In the present work, I refer to this legal claim as the *carazari* clause. *Carazari* is a term that became popular in the context of Ottoman relations with the Franks; originally deriving from *carazo* (ar. kharāj), referring to the poll tax paid by the dhimmī population, Christian and Jewish Ottoman subjects came to be known as the *carazari* in commercial-diplomatic jargon. The carazari clause was often invoked by Jewish and Christian subjects of the Ottoman Empire, advanced in contexts of commercial litigation and in the drafting of ahdnames and agreements. The mainstream ḥanafī approach to the question, as epitomized in the Multaqā of Ibn al-Ḥalabī, tended to lend support to this claim raised by Ottoman dhimmīs.[105]

104 ASVe, *Libri commemoriali*, Registri, Reg. n. 24, chapter 26 of the 1575 Venetian-Ottoman Treaty. Published in Theunissen, *Ottoman-Venetian Diplomatics*, 530–46 ("*carazari* clause", 539), from a different source: ASVe, Documenti Turchi 828.

105 Heyd, Uriel and V. L. Ménage: *Studies in old Ottoman criminal law*, Oxford: Clarendon Press, 1973, 245.

To the best of my knowledge, the first reference to the carazari clause can be found in the 1500 treaty with Florence. The text points an accusing finger at, in the first place, Jews, followed by "any other nation" except the Muslims. The setting for this early mention is, most probably, the intense trade Florentine merchants were conducting in Bursa and Constantinople.

> Whenever Florentines have any difference with Jews or any other nation, except with Turks, the former often refuse to accept the testimony of Florentines. And I command that they should accept the witnessing by a Florentine against any nation, except against Muslims, and the qadi must sentence in their favor.[106]

As for the Turkish versions, early treaties render the carazari clause in full, as is the case for the agreement drafted by the Porte in 1502 and ratified by Venice in May 1503, followed by a subsequent renewal by Selim I in 1513. According to these treaties, in legal conflicts with non-Muslim tributaries, Venetian, unbeliever witnesses were to be admissible in court. The 1502–1503 rendering is succinct, simply mentioning the validity of cross-confessional witnessing, while the 1513 version adds new discursive elements to the clause's formulation. Ottoman rulers claimed that, in this regard, they were following the example of the Prophet Muḥammad and that all Christians were to be considered as a single nation (*millet-i-vahıde*). These elements would crystallize in later documents, as in that underwritten in 1575. In all probability, both arguments for and against cross-confessional witnessing centered around the ḥadīths related to the activities of Muḥammad in Medina and his role delivering justice among the city's different communities. Whatever justifications they used, the sultans made their legal choices not by virtue of any arbitrary power, but by underlining their conformity with specific acts described in the Prophet's sunna.[107]

One would expect to encounter the carazari clause in all successive treaties after that of 1503, however this is not necessarily the case; the French-Catalan treaty signed with Süleyman I in 1528, for example, does not mention

106 "Et havendo detti Fiorentini differentia con Giudei o con qualche altra natione, da' Turchi in fuora, piu volte non vogliono acceptar testimoni fiorentini. Et io ho comandato che contro ogni natione, eccetto Turchi, debbino acceptar testimonianza di detti Fiorentini, et il Cadi debbe sententiare in favor di essi." Camerani, "Contributo alla storia dei trattati commerciali fra la Toscana e i Turchi", 94.

107 De Groot, "The Historical Development of the Capitulatory Regime", 591, Theunissen, *Ottoman-Venetian Diplomatics,* for the treaty of 1503, 386, art. 35. For that of 1513, 398, arts. 41, 42, 43.

the carazari clause.[108] The text is considered to be a renewal of the Mamluk privileges enjoyed under Sultan Qānṣūh al-Ghawrī, which is probably the reason why the issue is not mentioned. As we can observe in late Mamluk treaties with the Italian cities, the Mamluk chancery was not concerned with the problem, and in fact it is absent from late agreements such as that granted to Florence in 1489, although issues of adjudication and procedure are dealt with at great length.[109] The 1528 text refers to the conditions necessary for the conclusion of deals, making it clear that traditional patterns of notarization were still in use. Even though the dhimmīs are not mentioned, the text raises the issue of witnessing amidst diversity. In the first instance, a murky passage in the French translation refers to the fact that Muslims have full witnessing capacity, particularly in cross-confessional disputes, while Franks do not. Secondly, the text depicts a situation where business deals could apparently still be concluded without the presence of witnesses, but in the event that witnesses were summoned, Franks could only bear witness on behalf of their coreligionists.[110]

Unlike the French treaties, the Venetian ones invoke the carazari clause starting with those issued in 1503, 1517 and 1575, and up to the eighteenth century.[111] In 1575, explicit mention was made of the idea that Frankish witnesses should be admissible for or against dhimmīs, because all Christians practice the same religion, and that differences in residence should not be used as an argument against their capacity to testify.[112] Sixty years after the 1513 forerunner, the ahdname evokes the language used by the religious learned in legal reasoning, mentioning the example of the Prophet Muḥammad, or the Ottoman-sponsored idea that dhimmīs and mustāʾmins belonged to the same *dār*. It is clear at this point that ahdnames issued to the Venetians were addressing a particular concern encoded in the legal debate, and for this purpose echoed the language used by ḥanafī authors such as al-ʿAynī a few decades previously. In contrast, Ottoman ulama such as Ebuʾs-Suʿud were equally aware of

108 Charrière, *Négociations de la France dans le Levant*, 121–9.

109 Wansbrough: "A Mamluk Commercial Treaty", such as clause XX and XXVI for adjudication and XXVII for notarization.

110 "S'il se conclurra marché en la présence de tesmoins, que les tesmoins soyent escritz le Franc avec le Franc, comme il s'escrit, et le More avec le More et avec le Franc, et s'ilz voudront que les tesmoins se soubzcrivent, qu'ilz ne le pujssent refuser et ne leur soit empêché comme aussi de la police de recevoir." *Saint-Priest and Schefer, Mémoires sur l'ambassade de France en Turquie et sur le commerce des Français dans le Levant,* 350, Charrière, *Négociations de la France dans le Levant*, 124, 128.

111 Theunissen, *Ottoman-Venetian Diplomatics*, 392, 539, 411.

112 Ibid., ch. XXVI, 539., Miltitz, *Manuel des consuls*, II, 1486.

the ruler's approach to witnessing and unbelief, and expressed disapproval for the "traditional mistakes" conveyed by the treaties.[113]

One could ask: if priority is given to historical context over legal theory, was the ban on Muslim witnesses invoked at all times, and for all the trading nations? Although an unequivocal answer to this question would require a more thorough scrutiny of Ottoman diplomatic production, a preliminary comparison of treaties between different nations does demonstrate that the dhimmī biases against Frankish witnesses were particularly intense in the case of the Italians. Ottoman-Polish treaties began invoking the need to notarize by means of *hüccet* and *sicillat* in 1553, and then reiterated the compulsory nature of this measure in 1564, 1557, 1591, 1597, and 1598. Incidentally, these treaties only implicitly mention the ban on Muslim witnesses, as they argue that in the absence of valid documents witnesses will be heard, putting a strong emphasis on the need to rely on honorable ones. No mention is made however of the carazari clause in 16th-century Polish treaties, which can only mean that the merchants from the Italian nations, and first and foremost the Venetians, were much more prone to litigate in court than their Polish counterparts, addressed in these agreements. After all, in the negotiations for the 1575 agreement, it was the Venetians themselves who included the argument in the diplomatic agenda, which later crystalized in the ahdname. In other words, the specific nature of Veneto-Ottoman relations differed from relations with other powers, since they went far beyond the quest for legal security and commercial privileges. Such relations implied delicate diplomatic exchanges, in which not only specific claims, but Islamic legal doctrines and positions were invoked by all actors, and opened the arena of negotiation to include just as equally the Porte, the Doge, the Ottoman tax-paying subjects, and Frankish merchants. The carazari claimed that local and foreign Christians belonged to different sects, and that their testimony naturally tended to be biased against one another. Jurists such as Ebu'ʾs-Suʿud adhered to ḥanafī doctrine in that he insisted that dhimmī witnessing should be accepted across confessions, but that this should not include the Franks. The Ottomans ran counter to this, repeatedly asserted that it was "the Law of the Prophet" that conveyed the idea of a single community of unbelievers, and that this community included Frankish merchants, hence deviating from the mainstream ḥanafī interpretation and distancing themselves from their own flagship jurist.

113 "Non. Le temoignage des « musta'man » contre les « dhimmi» n'est guère admis par aucune des écoles juridiques. Dans les traités qui ont été accordés aux « musta'man », il y a des erreurs traditionnelles. On ne leur donnera pas exécution." Grignaschi: "témoignage des sujets non musulmans", 255.

Several remarks can be made at this juncture. Venetian diplomacy was early to underline the importance of the carazari clause, however it would take her over two hundred years to claim the right for their subjects to refuse Muslim witnesses in the absence of written documents. The carazari clause seems to have predated rulings on the compulsory nature of notarization, or at least was prioritized in the drafting of treaties. Moreover, it was entirely absent from treaties with *a priori* minor nations such as Poland.

The absence of the ban on Muslim witnesses in Venetian-Ottoman ahdnames suggests that Ottoman action was more concerned with governing amidst diversity, than with the opportunistic granting of privileges to meet economic imperatives. It also speaks of the political nature of Venice's relationship with the Ottoman Empire. Venice refrained from requesting a privilege known to all and sundry—the ban on Muslim witnesses—that would have put her subjects in the same position as those of inexperienced nations. Unlike Venice, latecomers lacked sufficient resources to prevent conflicts with local dhimmīs, and they could not deter the actions of cheating plaintiffs. By this, I am not implying that Venetian merchants simply renounced commonly-accepted procedural advantages; indeed, when necessary they actually brandished the decrees relative to their rights. However, it is clear that as a nation marked by an unrestrained proximity with the dhimmī community—with many of its members even enjoying ambiguous statuses—the Venetians were eager to avoid having cases thrown out on procedural grounds. Similarly, I conjecture, we should not exclude the possibility that the Ottomans refrained from granting exceptional privileges to a nation whose members often had poorly-defined statuses, and many of whom appeared in poll-tax registers.

Issues constantly arose due to the Venetians' closeness with the carazari community, and their contested statuses often came to the fore when the estates of ambiguous subjects were at stake. For example, a certain Piero Ginardi, "our Venetian ... passed himself off as a dhimmī (carazaro) for his businesses with Bursa, since he figured in the poll-tax books, although under a not so clear name," noted the Bailo, before describing his efforts to reclaim Ginardi's estate.[114] Eric Dursteler has devoted a long section in his 2006 study to

114 "Piero Ginardi nostro veneziano, si spazzava per carazaro nelle facende che facea per bursa, ... et era notato sopra li libri delli carazi, se bene con uno nome non tanto chiaro, subito, che fu morto, andorono li beltramaggi che sono come li cattaveri di vostra serenita et bollorono tutte le robbe sue in casa, et nelli magazeni. Perche di tutti li carazari che muoreno in queste parti senza haver testato alla turchesca, et senza figlioli, è herede questo serenissimo signor, ho trovato una fattica estrema, et anco con qualche spesa, non già di vostra serenita per far liberar dette robbe, sopra le quali ho fatto metter il sigillo per

a description of the many Venetians in Constantinople of ambiguous status, and which included a large group of exiles.[115] Under these circumstances, indulging in legal definitions was deemed less useful than handling claims as they came in the daily practice of courts. Since nothing of the like was ever at stake in Mamluk times, these two issues can be described as characteristic of the Ottoman agenda in the field of cross-confessional relations. However, they can also tell us something about the Ottomans' priorities and methods; although contested, the two anomalies mentioned above referred to sharīʿa as justification, and were expressed by the lawgiver in the language of the jurists. Combined together, this piecemeal innovation in amān treaties underlines the priority given to the ḥanafī qadi in the daily life of mixed relations, and sets a line of demarcation between the thriving dhimmī community and their Muslim social superiors, rather than between dhimmīs and the undisciplined mustaʾmin community. Closing the circle of enforcement, Muslims, by virtue of the ban on Muslim witnesses against Franks, were reminded not to abuse the legal superiority granted them by sharīʿa. Such a warning was all the more relevant in light of the perilous proliferation of false witnessing.

4.4 False Witnessing

Just as reliance on written documents favored the circulation of forgeries, the Ottoman position on testimonial evidence fostered the action of false witnesses. The criminal code of Süleyman the Magnificent devotes special attention to the forgery of witnessing and ḥujjas, together with official decrees (ḥukms).[116] In my research, I have not come across any references to the issue in Mamluk commercial litigation, nor does the recourse to false witnesses seem to have concerned diplomatic negotiators.[117] Under Ottoman rule, however, complaints had become frequent by the 1530s. "The fact that justice depends on the witnesses' depositions leads to much abuse in Turkey, since by means of money there will always be witnesses for whoever needs one," Bailo Gianfrancesco

la man cancellaria," ASVe, Senato, Dispacci degli ambasciatori e residenti, Costantinopoli, 1A, 326–8.

115 Dursteler, Eric: *Venetians in Constantinople: nation, identity, and coexistence in the early modern Mediterranean*, Baltimore: Johns Hopkins University Press, 2006, 61–103.

116 Heyd and Ménage, *Studies in old Ottoman criminal law*, 121.

117 In *Ambasciata straordinaria al sultano d'Egitto (1489–1490)*, Franco Rossi has edited the only extant, complete documentary trove pertaining to a Venetian embassy to tihe Mamluks. Justice and procedural issues were never mentioned in the diplomatic agenda.

Morosini wrote in his *relazione*.[118] According to Giovanni Moro, writing in the 1580s, if the Turks relied on witnesses, it was due to the summary handling of justice and their disregard for documents, which "are only valid if corroborated by witnesses." For these reasons "witnesses abound, who for a little money can confirm whatever the parties wish."[119]

False witnesses came in handy when proper documentation of deals was absent. In parallel to the proliferation of false witnesses, it soon became customary to forbid, in mixed cases, the issuing of verdicts based on the depositions of forged testimonies alone, without any supporting documentation. This tautological clause found its way into the drafting of ahdnames, and is mentioned, for example, in the six treaties signed with Poland throughout the sixteenth century. The problem was brought up by French diplomats in 1540 and 1569: when deals turned bad, false witnesses showed up in support of forged accusations of insults to Ottoman subjects.[120] French merchants, the diplomats complain, fall bankrupt, and in order to recover their losses, Muslim merchants sue other Frenchmen. For this purpose, Muslim witnesses are forged, hence the necessity to notarize.

> Bien souvent, il arrive que quelques marchands François font banqueroute de grandes sommes aux sujects du Grand Seigneur, qui pour se récompenser de telles pertes, s'en prennent aux autres marchands de la nation, et leur veulent faire payer leur perte, comme s'ils estoient obligez de ce faire, se servants, pour c'est effet, de faux tesmoins. Pour à quoy remédier, il est dit en l'article trente quatre, que s'il n'apparoist que ces marchands ainsi poursuivis, soient cautions par contracts authentiques, qu'ils ne soient molestez, ny tenus des debtes des fuïards.[121]

118 "Il far dipendere la giustizia dal detto de' testimoni causa le molte avanie che si usano in Turchia, perché con denari mai mancano testimoni falsi a chi ne avere. E per concludere, tutta questa giustizia, cosi civile come criminale, si può chiamar grandissima ingiustizia, poiché ella depende da testimoni che si comprano, e viene fatta da giudici che per il più la vendono," *Relazioni degli ambasciatori veneti al Senato, Serie III, Volume III,* 273.

119 "Attendono i giudici alla brevità e, tralasciando i lunghi sermoni d'avvocati, risolvono prestamente le cause con poca lettura di scritture, molto diversamente dall'uso degli altri paesi, quando non siano comprobate col detto di testimonj; dei quali là più che altrove ne abbondano de' falsi che con poco danaro confermano quello che è desiderio della parte," ibid., 376.

120 Saint-Priest and Schefer, *Mémoires sur l'ambassade de France en Turquie et sur le commerce des Français dans le Levant,* 370–1, 388 (1581), 407 (treaty of 1597). For the treaty of 1569, BNF, DO, Ms. Turc 130, f.4r.

121 Saint-Priest and Schefer, *Mémoires sur l'ambassade de France en Turquie et sur le commerce des Français dans le Levant,* 427.

French consuls seem to have been deeply concerned by this issue. Numerous times, they insisted on the tendency to charge foreign subjects by means of false witnesses, which could even include serious accusations such as blasphemy, hence psychologically blackmailing judges on religious grounds.[122] Unlike the French, the Venetians do not seem to have included this issue in the letter of their treaties, although Venetian representatives were well aware of the problem in their daily activities. The consul in Egypt complained of being wrongly prosecuted by a Sanjakbey, who had fooled the qadi of Alexandria by advancing false witnesses.[123] So too did Giovanni Dario stumble upon a Greek renegade who, allegedly, bore false witness in an important smuggling affair. After his version was dismantled by the ambassador, the Greek witness was delivered to the executioner.[124]

The Abdellatif affair was a highly disputed case that took place in the mid-1550s, and epitomizes the threat posed by false witnessing to cross-confessional relations in general. Abdellatif's alleged fraud, like in the novel by Carlo Emilio Gadda, acquired the contours of an "awful mess," reaching far across the Ottoman territories and involving several courts and consulates. The affair was triggered when, after the seizure of his cargo by the Venetians of Cyprus, Abdellatif brought a complaint before the qadi in Cairo. Although we do not know the grounds for this decision, the case was transferred to the Syrian courts. Eventually, the Venetian Bailo had wind of it, who brought the dispute to the attention of the imperial Dīvān in Istanbul. In gross, Abdellatif requested that the Cairo consul be held liable for his losses. Allegedly, he had in his possession a declaration by the consul in which the latter admitted his responsibility for the seizure of the cargo and committed to reimburse the Muslim. Armed with these ḥujjas, he moved to Tripoli where similar commitments were obtained, backed by the local qadi. Abdellatif eventually showed up before the beylerbey in Aleppo in 1555. The latter's intervention led to the detention of the Venetian consul in Aleppo, forcing resident Venetians to urgently dispatch a messenger to Istanbul. At this point, the Abdellatif affair had acquired cumbersome

122 "Qu'estant dressé quelque embusche contre les François pour les accuser d'avoir injurié et blasphémé contre nostre Sainte Religion et produisant des tesmoings faux pour trouver moien de les travailler," Ibid., 407.

123 ASVe, Senato, Dispacci Consoli, Egitto B1, 33, f.2, "avendo detto sanzacco voluto con falsi testimoni dare ad intendere che'l fu battalado ingiustamente, con haver fin contentado il cadi vecchio d'Alessandria perchè dicesse ben di lui".

124 "poi comenci a parlar io et breuemente et saldamente ho reprouado tuto quello che haueuaho ditto: per modo che li Bassa intexe le justificacion mie fulmirono contra de luj et contra quel rebaldo Schiauo che haueua fato false relacion ala porta per guadagnar questi dinari. et subito lo feceno metere in man di boia," Dario, *22 dispacci*, 212.

proportions, and the bailo hastened to the imperial Dīvān to demand the beylerbey's head. The legal arguments on Abdellatif's side resided in the ḥujjas he had presented as proof, whereby the consul (it is unclear to what extent the consul in Cairo was also involved) accepted his own liability. The consuls denied the very existence of these records, and refused to pay, on the grounds that well-known clauses in the treaties protected representatives from being held liable for their subjects' losses. The consul in Aleppo cited this argument in his refusal to produce witnesses in his favor, since such a decision would have been "prejudicial to the capitulations." Upon this, the beylerbey was summoned to Istanbul, where he exhibited a memorandum drawn up by the qadi in Aleppo, in which the latter acknowledged the validity of the ḥujjas incriminating the consul.[125] The bailo's secretaries examined the records and concluded that they had not been effectively notarized, and that it was not logical that the consul would have chosen to become involved in this affair.

Probably due to the involvement of officials, for whom alternative forums were envisioned, the dispute was brought before the kadıasker, the competent high authority empowered to judge the military class. However, the Venetians refused the kadıasker as the competent court, and maneuvered to have the dispute placed in the hands of the pashas at the Dīvān.[126] For the bailo, Abdellatif had followed different legal strategies in Cairo and in Istanbul, most notably by playing out evidence that had been produced elsewhere. As we have seen in the Chapter Two, medieval jurisprudence tended to severely curtail the transfer of probative artifacts from one court to another, whilst, as shown by Reem Meshal, under the Ottomans even copies of ḥujjas began travelling with their right-holders across the Empire. In the end, aside from his cargo, Abdellatif managed to recover the loss of one of his slaves, who had escaped in Cyprus and been forced to convert to Christianity—a highly sensitive issue at the Dīvān.[127] For the bailo, Abdellatif had been unable to find his slave through legal channels, so he had deemed it more useful to return to Egypt and procure himself forged evidence to back his claims, and then sue the Venetians in Syria with the help of these documents. The bailo narrates a previous episode in

125 "Il quale dapoi aver detto quanto gli parve sopra cio, presento un'arz dal cadi di Aleppo, per il qual pare, che quel cadi faccia fede, che sara provato per sigilletto che dal magnifico consolo sia stata fatta la sopradetta promissione al turco.", ASVe, Senato, Dispacci degli ambasciatori e residenti, Costantinopoli B. 1A, 228c.

126 "che dimane devessero ritornare avanti li cadilaschieri, alquale fu risposto per essi secretari, che le cose di vostra serenità non haveano daesser giudicate, se non dal Serenissimo Signor.", ASVe, Senato, Dispacci degli ambasciatori e residenti, Costantinopoli, B. 1A, 228c.

127 ASVe, Senato, Dispacci degli ambasciatori e residenti, Costantinopoli, B. 1A, 394.

1550, when a Christian slave of his had successfully sued Abdellatif, arguing that the latter had experience manipulating the judges, fabricating evidence and abusing religious pretexts in court. More importantly for our argument, Abdellatif frequented Christian chanceries such as that of the bailo in Constantinople, where contracts were registered by the consular notary. In the bailo's narrative, after having contracts drawn up by Latin notaries, when needed Abdellatif was prone to contest such transactions before Islamic judges with the help of false witnesses.

It is worth noting here an interesting point regarding both parties' performances in court. Abdellatif adopted dilatory tactics, often failing to show up at the Dīvān. The consulate, on the other hand, put the story into writing and had this read by the pashas. The pasha then learned that the Cypriot magistrates "had proceeded to open an inquiry, and have passed a verdict in the Christian way." In his self-congratulatory narrative, the bailo took up cudgels for Venice's justice: "and I answered him that justice is universal, and that irrespective of the place it is done, it is always fairly done, and most particularly that emanating from Your Most Serene Grace ... he then roared with laughter, and said 'write to the doge that I have ascertained the validity of the deed, and will issue an arz for the Sultan.' "[128]

Antonio Barbarigo suspected that not only Abdellatif, but probably also the pashas had an interest in protracting the dispute indefinitely. By that time the word *dīvān,* which before had been used uniquely to talk about the ceremony of diplomatic and legal exchange, had taken on a temporal meaning, with expressions such as "during these last dīvāns," etc. In that period too, the bailo attended some sessions and paid private visits to the pashas, but for minor hearings relied on a team of secretaries directed by a most experienced dragoman called Giannesino. Abdellatif, who seemed to be monitoring his opponents' appearances in court, showed up unexpectedly at the Dīvān at one of these sessions, in a calculated performance by means of which he obtained the transfer of his lawsuit from the pasha to the kadıaskers. As had happened months earlier, this transfer was deemed prejudicial to Venetian interests, and Barbarigo and his secretary labored to bring Abdellatif to court, with

128 "la ho veduta, quelli giudici di cipro hanno fatto teftis, et giustitia alla cristianesca, ond'io rispondendogli che la giustitia al mondo era una sola, et pero' in ogni parte quando era fatta essa giustitia era sempre bona, et giustamente fatta, et in particolare da vostra serenità, et sui rapatante, ella si mise a ridere, et disse scrivete alla Signoria ch'io ho accertata la scrittura, et che ne faro' arz al gran Signor, ne manchero' di aver questa cosa à cuore, come lei desidera, la ringratiai.", ASVe, Senato, Dispacci degli ambasciatori e residenti, B. 1A, f. 400–402.

the intention of having the Venetian plea read in his presence. By so doing, they expected to force an acknowledgement from the plaintiff's side. Again, the kadıaskers admitted they were granting Abdellatif a further opportunity to substantiate his claims by means of qadi-produced evidence. The Muslim merchant was indeed laboring to obtain a favorable sentence from a qadi in Istanbul, a decision the kadıaskers intended to examine before transferring the case again to the pashas.[129]

As we saw from the Venetians' point of view, the Ottoman version of the witness system, as well as their increasing reliance on qadi-registered instruments, was becoming an obstacle to mixed justice. According to the bailos, fraud could be committed by qadis acting as notaries, and, as we shall see, claims from captives who had been forced to subscribe false debt contracts before the qadis were often reported.[130] Although post-conquest reform sought to do away with traditional notary-witnesses drafting deeds from their stalls, the new praxis ended up favoring the presentation of serial witnesses at hearings, among whom hid serial false witnesses. Neither did the new procedure prevent the action of ill-meaning cadis, such as those allegedly operating for Abdellatif's cause. Hence the bitterness of Ibn Nujaym, who pointed out that in the case of unbelievers, children, and witnesses afflicted by temporary loss of hearing, in some cases (not against Muslims), whenever these limitations were overcome, their testimony was accepted. Although repentance was accepted in a person who had once given false testimony, it was not accepted in the case of the professional notary-witnesses.[131]

A story confirming the proliferation of false witnesses is reported in July 1547 in the following terms. Zuan Burletto of Chios, probably a tax farmer for the Ottomans, wished to reclaim money from a certain Venetian called Francesco Michiel. Michiel was a merchant in Constantinople where he represented the interests, among others, of a certain Zaccaria Morosini. Both men belonged, in all probability, to the Venetian ruling elite, the nobility. Morosini was indebted to Burletto for a much larger sum, of at least 23,000 ducats. In all probability, Burletto, with the help of false witnesses, expected to recover part of his sizeable credit not directly from Morosini but from Michiel, his commercial agent in Istanbul. Burletto had his own motivations, as he owed

129 ASVe, Senato, Dispacci degli ambasciatori e residenti, Costantinopoli, B. 1A, 400–2, 431.

130 ASVe, Senato, Dispacci degli ambasciatori e residenti, Costantinopoli, B. 1A, 429, mentions a friar forced to sign 'a sijill' committing to ransom Muslim captives, ASVe Senato, Dispacci Consoli, Egitto B1, 42, describes how captives who may have been the object of Frankish interest were forced to sign bills of exchange and obligations.

131 Ibn Nujaym, *al-Ashbāh wa-al-naẓā'ir 'alá madhhab Abī Ḥanīfah al-Nu'mān*, 270, 275.

three thousand ducats himself to the *hazine*, or Imperial Treasury, and for this reason he counted on the support of some secretaries ("the defterdars") in his plot against Michiel. Endorsed with the hazine's support, Burletto managed to have the judicial officers (*chiaus*) sent after Michiel and have him brought before the defterdar. Since the case involved a protected mustā'min, this official had the trial transferred to the Dīvān, where the bailo took charge of Michiel's defense in several hearings. At the Imperial Council, the bailo encountered the kadıasker, who was sitting in justice, and both parties were heard. The bailo stuck to the argument that Burletto was targeting Michiel on behalf of Zaccaria Morosini, hence violating the principle of individual responsibility stipulated in the treaties. In addition, the bailo cited an executive order by the sultan (*commandamento*, most probably a *ferman*) affirming that in cases over five thousand aspre, witnesses should not be heard in the absence of written documents, properly notarized by the qadi and incorporated in the court's sicil.[132] Incidentally, this confirms that the Venetians made use of the ban on Muslim witnesses without having it explicitly mentioned in the treaties. Yet in spite of the bailo's sound defense, the kadıasker decided to hear Burletto's witnesses. At this point, the reader readies himself to encounter the usual narrative of bitter grievances about unfair Muslim justice. Instead, the bailo reports to the Doge what constituted a conscious and calculated Ottoman reaction, not only to Burletto's plot, but to the larger threat posed by false witnessing in the conduct of cross-confessional justice. The kadıasker, the bailo argues, was planning, with the help of Ottoman officials, to have Burletto and his witnesses punished. If we are to believe the bailo, some Muslims in Istanbul were making a living out of bearing false testimony, and that Burletto's witnesses were false was known to all and sundry. The vizier Rüstem Pasha had them arrested and ordered the usual penalty inflicted upon corrupt judges and witnesses; they were paraded on the backs of donkeys for three days throughout the streets of Istanbul, bearing a cow's intestine around their necks, after which Burletto's witnesses were put to the oar in the galleys. Ibn al-Ḥimṣī describes the fall of a mālikī qadi in Mamluk Damascus a few years previously, to whom the same treatment was delivered, and we have mentioned similar descriptions by Ibn Fahd for Mecca.[133]

132 "uno commandamento del serenissimo gran Signor che vuole che testimoni delle cause da @ 5000 in su non siano admessi, ne ascoltati, se non vi è scritto in sigilleto del cadi. Non di meno parse al predicto magnifico Caddilaschier di voler admetter, et esaminar doi testimoni produtti per il predicto Burletto," ASVe Archivi Propri Constantinopoli, 1, fasc. IV, f.181–2, 17 July 1547.

133 Ibn al-Ḥimṣī, Aḥmad Ibn Muḥammad (1473–1527 or 8): *Ḥawādith al-zamān wa-wafayāt al-shuyūkh wa-al-aqrān*, edited by ʿAbd al-ʿAziz Fayyad Harfush. Vol. 1–3, Beyrouth: Dār

As for Burletto, the dispatch from the bailo adds that he was threatened with being *bollato* or *sigillato*, meaning "marked by the iron in three different places", in addition to being sentenced to the oar.[134] Burletto was more frightened by the mark than by the gallows. Brought before the pasha at the Dīvān, he converted to Islam, hoping to exploit his new religious status to avoid "such a shameful marking".[135] However, he still remained under some kind of judiciary detainment (he was still put in the hands of the *chiaus*). The bailo went to the Dīvān to congratulate the authorities "for this good justice", which would serve as a reminder to the "many false witnesses" who earned a living in this manner. Such "demonstration against false witnesses" had not been seen, the bailo added, in the last twenty years. Nevertheless, the episode did cost Michiel a good deal of money, as the pasha made it known that he desired a gift for freeing Michiel from the plot. The gift, considered to be part of the community's contribution to 'protection costs' was charged to the consulate, and the consulate's ruling "council of twelve" recommended Michiel spend "as little as possible" on it.

The Burletto episode described above is a reminder of the existence of executive decrees accompanying and complementing the ahdnames. A practice that was already in place in Mamluk-Venetian diplomacy, it was not confined to the labor of contingent embassies, but characterized by an open, uninterrupted diplomatic dialogue, with chanceries constantly issuing documents with amendments and updates. Ambassador Marco Minio reported, for example, a delicate decision concerning captives that the Ottomans had excluded from the formal treaty, but accepted instead to draw up in a separate decree.[136] In Burletto's case, it was one of these documents that set the threshold for accepting witness testimony without documentary evidence at five thousand aspre. This means that adjudication was affected by two thresholds; one setting a minimum sum under contest for a case to be considered for a Dīvān hearing, and another setting a minimum amount before a judge should consider inhibiting standard sharīʿa procedure on the hearing of witnesses.

al-Nafāʾis, 1996 261. Muʿammar, "al-maghāriba wa manṣab qāḍī al-quḍḍat al-mālikī fī dimashq al-mamlūkiyya", 159, Ibn Fahd, Nayl al-Muná bi-Dhayl Bulūgh al-Qurá, 308–10.

134 De Villalón, Cristóbal: *Viaje de Turquía*, 213.

135 ASVe Archivi Propri Constantinopoli, 1, fasc. IV, f.182, "per fuggire questa vituperosa sigillatura, senza altra difficultà si fece musulmano".

136 "lo volea etiam che fosse posto in detti capitoli che li schiavi, quali si avessero francato per qualunque modo si voglia, si potessero partir senza alcuna spesa né impedimento; questo non hanno voluto poner nelli capitoli, ma mi hanno ben dato un comandamento della detta contenenzia," *Relazioni degli ambasciatori veneti al Senato, Serie III, Volume III*, 89.

Another interesting aspect of the Burletto case is the role played by the kadıaskers at the Imperial Council; permanent members of the Dīvān, their presence was—according to some authors—provided for when lawsuits were too numerous to be handled by the pashas alone.[137] Two important diplomatic affairs that were transferred from the pashas' hands to the kadıaskers, were also sensitive issues from a religious standpoint. The 'Tirabosco affair' involved the debts left by Venetians who had either become Ottoman tributaries (*carazari*) or married dhimmī women.[138] A vague, yet important article in diplomatic treaties dealt with this specific status, for which the Ottomans made an exception to the principle of individual responsibility. They also considered the consuls liable for the debts of Venetians who, having become official residents, had fled to their homeland. Bailo Domenico Trevisan warns the doge that in such cases, if Venetian merchants of carazari status died, their estates *would be lost*, since they would go either to their dhimmī sons or to the Bayt al-Māl.[139] This was a delicate issue involving legal notions of amān and Islamic laws of obligation, and a certain Ahmet Pasha labored to have it transferred to the kadıasker.[140] Lastly, the Venetians managed to unblock Dīvān negotiations by obtaining from a kadıasker the abrogation of an executive order blocking a ship.[141]

---

137 Heyd and Ménage, *Studies in old Ottoman criminal law*, 225, Atçil, *Procedure in the Ottoman court and the duties of kadis, M. A. Thesis, Bilkent University*, 75.

138 "Nella causa del tirabosco, et in alcune altre occorse al tempo di q. magnifico Acmat bassà, non le havendo posute finire con sua magnificentia volendo ella andar molto riservata si come le ho scritto altre fíate, fui forzato mandare il secretario et zanesino al divano, li quali sebene furono rimessi alli caddi leschieri, che cosi sono qui giudici di giustitia come sono in quella sudita citta li giudici delle corti ordinarie di vostra Serenità", ASVe, Senato, Dispacci degli ambasciatori e residenti, Costantinopoli, B. 1A, f. 184–5.

139 "Di questi mercadanti, alcuni nel tempo della guerra passata, per fuggire il pericolo della prigione, e per salvar la roba, si sono fatti carazzari di volontà, oltre quelli che avendo moglie e figli sono fatti carazzari per legge; dal che avviene che occorrendo la morte di quelli, sarebbe pericolo che fosse per seguire che il bailo di vostra serenità avesse travaglio, massime se quel tale restasse debitore di Turchi, e che li maestri che avessero mandate le faccende fossero quasi certi di perder tutto il lor capitale, perchè li figliuoli, se ve ne fossero, ovvero li deputati sui beni dei morti senza eredi, leveriano il tutto dalle man di quelli, e volendo li maestri ricuperare il loro sarebbero sforzati richiederlo alla giustizia, con le leggi e testimonianze a modo loro," *Relazioni degli ambasciatori veneti al Senato, Serie III, Volume I*, edited by E. Albèri, Florence: Società Editrice Fiorentina, 1840, 185–6.

140 "Ben le diro riverentemente, che succederano molte fiate di necesita delle dimande simili a questa contra il tirabosco, fino cheli sudditi sui, che vengono in queste parti, si fanno carazari di questo serenissimo signore overo si maritano qui, perche se quelli venirano in quella città, restando debitori de qui, saranno dimandati alla serenità vostra secondo li capitoli, et se li maritati lassarano figlioli sarà dimandato il conto della robba sua," ASVe, Senato, Dispacci degli ambasciatori e residenti, Costantinopoli, B. 1A, 184–5.

141 ASVe Senato, Dispacci Ambasciatori e Residenti, Constantinopoli. B. 1A, 374–6.

The Antonio Priuli affair, not to be mistaken with the Marco Priuli bankruptcy discussed at the end of this chapter, concerned instead Muslim creditors against Venetian debtors.[142] Himself the victim of unpaid debts of other Venetians, Antonio Priuli ended up owing 42.000 ducats to the treasury. Pursuant to the huge debts incurred by Priuli and his associates, a bitter quarrel arose at the Dīvān on whether a certain Jew could stand surety for Priuli, and to what extent the Venetian consul could vouch for his own merchants. During the discussions, the kadıaskers provided the parties with legal advice, 'based on their law'. The Jew, Salomon Alaschar, was not eligible as guarantor, since he was himself debtor to the treasury, hence the Ottoman interest in obtaining a guarantee from the Venetian consul in Cairo, Zuan Battista Querini. Some of these principles irritated the Venetian consul; he was requested to certify that at least some of the Venetian subjects concerned were both trustworthy men and merchants. The first, Querini argued, was an attempt to use this legal certificate as a personal commitment as guarantor. As for their being merchants, and unwilling to commit to signing any paper, the consul argued that it sufficed to check the customs registers, but he was answered that those records were kept by Jews, and therefore they could not be considered as proof (*che sono libri de ebrei, et che non hano credencia*).[143] Needless to say, this clashed with Querini's notions of official documents: "this is absurd, since if someone is recorded as debtor in the customs books he is a debtor of the king."[144] An Ottoman official, "a man of sharp intelligence" was looking for a loophole in the Venetian law of agency to force Querini to accept his own liability, using persuasion, some threats, and forcing the consul to fall in contradiction. In this, he eventually succeeded; since he could not commit to a pledge for his own subjects, the pashas drafted a list of five Venetians standing surety for the debtors, including some of the debtors' associates. Querini was required to approve the list of

142 ASVe, Senato, Dispacci degli ambasciatori e residenti, Costantinopoli, B. 1A, 258–60.

143 Horii, Yutaka, "Venetians in Alexandria in the First Half of the Sixteenth Century", AJAMES 20 2 (2005), 131–144, 137–8.Reimer, Michael J., "Ottoman Alexandria: The Paradox of Decline and the Reconfiguration of Power in Eighteenth-Century Arab Provinces", Journal of the Economic and Social History of the Orient 37 2 (1994), 107–146.

144 "Io confesso a vostra serenita che se fuse novo in turchia fácilmente per questa fiata mi hariano ingannato, ma habisi questo per fermo et constante che in materia de piezarie quando uno aproba il piezo esser homo da bene tanto bien dito quanto chel sii sicuro, et de qui nase che loro (per sue leze) poi tanta accione hano verso colui che ha fato tal depositione, quanto verso colui che ha fata la piezaria, ala ragione de libri de doana mi bien dito che sono libri de ebrei, et che non hano credencia, questo e uno absurdo a considerarlo non che a dirlo, pero che se uno si trova debitor in diti libri e tenuto debitor del re senza alcuna contradicione et se dubitamo che diti libri da se non sino certi facino che il cadi laschiero faci la legalita," ASVe Senato, Dispacci Consoli, Egitto B1, 12.

guarantors, something the consul “was unable to decline”. The kadıasker immediately notarized a warranty deed with his personal sign (*sigiletto*) on it.[145]

If discrimination derived from formal sharīʿa principles, Frankish biases against scriptuaries were not better understood by Muslims. In 1557 a Pasha had commissioned to a Jewish Woman named (probably Gracia) Mendes, a piece of silken cloth. The item had been seized in Venice, Barbarigo argued, probably because it was not made clear that to whom it was directed, but that in any case she would not have been believed, since “truth is rarely found among the Jews”. The Pasha, who had been previously trying to convince Barbarigo to procure him red silk for his daughter, replied irritated “cannot then Jews, by any chance, bring fabrics into Venice?[146]

To be sure, the only source of normativity in Ottoman judicial practice was sharīʿa, and the procedure followed at the Dīvān tended to mimic that in use in the qadi courts. Yet the Abdellatif and similar affairs suggest that there existed alternate paths for the handling of sensitive legal issues at the Imperial Council. Transfer from the pashas, more sensitive to the logic of kanun and customary law, to the stricter sharīʿa-based viewpoint of the kadıaskers was undesirable, at least from the point of view of Venetian interests. This, again, was in stark contrast with the widely-accepted Mamluk procedure to leave decisions involving foreigners in the hands of the emirs and officials. In an episode I examine in the next section, the kadıaskers made a sound defense of the necessity to rely only on Muslim witnesses for borderland issues, although Burletto's story proves that they were equally vigilant about the alarming proliferation of false witnesses in the imperial capital.

4.5 Proving Enslavement

The Dīvān was a privileged place for observing the confessional meaning given in Ottoman times to issues of proof and evidence, not only from the government's side but also for other subjects on the legal system. Endless conversations between the Venetian bailo, his secretaries, the pashas and the kadıaskers revolved around procedural and evidentiary matters, and mostly around the

145 “Et dapoi alcuni quesiti fatti per il signor bassa, sua sia. Disse che bisognava che io aprobase questi piezi, li fu risposto che io non podevo impedirmi, il cadi laschiero li notò piezi in sigiletto, et poi fezeno che il magnifico dolfino dette in nota alquante sue robbe.”, ASVe Senato, Dispacci Consoli, Egitto B1, 14.

146 “non possono forse hebrei portar panni in venetia?”, ASVe, Senato, Dispacci degli ambasciatori e residenti, Costantinopoli B. 1A, 473–4.

need to prove claims through the means of a qadi's sijill and before Muslim witnesses. Historians have devoted a great deal of effort to understanding the mechanisms behind the phenomenon of ransoming, a practice that emerged as an expression of early modern cross-confessional coexistence. For the bailos, it was instead the recovery of prisoners that was rife with conflict, and they saw the witness system as the principal device that biased the legal system in favor of Muslims. Reports addressed to the Senate by the ambassadors bitterly complained about the manner in which slave-owning disputes were conducted.

The *relazione*, or conclusive report, by Antonio Rizzo, read before the Senate in 1557, deals extensively with borderland attacks involving the seizure of Venetians, and the difficulties experienced by those attempting to free the prisoners. Rizzo presents three major scenarios of maritime attacks and the subsequent arguments put forward by the Ottomans to support their decisions. Two of these scenarios concern captives, and the third involves the requisition of cargo. According to Rizzo the first line of defense against returning captives held that the offenders were corsairs, acting in isolation and therefore not subject to the sultan's obedience; they therefore could not be punished, and in any case Ottoman officials, as interested parties in these misdeeds, would not enforce corrective decrees addressing the issue. The second scenario regarded damages caused by men who could not deny their Ottoman status; the pashas' strategy was generally to deny any involvement, or the existence of any proof of it. This was a particularly difficult argument to counter, Rizzo argues, because in the borderlands the only witnesses present were those who had inflicted the damages and those who had suffered them. Admittedly, the former were not willing to talk, whilst the latter could hardly be given any credit, both for being the plaintiffs and because of their being Christians. "They [the Turks] call the latter *giauri*, or infidels, whose testimony is not accepted against Muslims." In other words, because there were no unbiased witnesses present to attest to their enslavement, it was impossible to secure the release of such captives.

The third maritime scenario described by Rizzo regards the requisitioning of cargo on Venetian vessels, which, it was claimed, often belonged to Ottoman subjects, and whose seizure was therefore not illicit. Rizzo goes on to mention the recent confiscations by Dragut Bey in Corfu and by the Beylerbeyi of Algiers Salah Re'is (1552–1556). In these cases the witness system, he argues, hampered any redress for the Venetians.[147] In both episodes, according to the

147 On the same episode, copy of a letter by Secretary Giannesino mentioning the investigation (taftīsh, it. teftes) to Dragut, Amasia, 19 Feb. 1554, ASVe Senato, Dispacci Ambasciatori e Residenti, Constantinopoli. B. 1A, 79r.

relazione, the confiscated goods were not returned to their Ottoman owners, who were themselves enslaved. These captives, when reclaimed by the Venetian authorities, needed again to prove their free status by means of Muslim witnesses. At this point, Rizzo's narrative appears somewhat contradictory: who are, one wonders, these Ottoman subjects whose goods and freedom are the bailo's concern? Rizzo here elaborates on an episode that happened during his tenure, narrated in his regular reports (*dispacci*) preserved in the Senato, *Dispacci Costantinopoli* series. These reports refer to a highly disputed case where the Ottomans of Rhodes captured three vessels and conducted them to Constantinople. The ships were transporting irregular troops of Greek stock—it. *levendi*—who had allegedly deserted Turkish ranks. The ships arrived in Istanbul, upon which Rizzo was alerted to the suspicion that the ships might be carrying Venetian subjects; he managed to trigger an official investigation to ascertain the presence of Venetians, first at Constantinople's arsenal, then at the Dīvān before the pasha. Rizzo conducted a delicate negotiation in order to identify, before Muslim officials, a group of twenty-five Venetian subjects from Zante, Corfu and Cephalonia previously captured and put to the oar by the *levendi*. Playing his cards carefully at the Dīvān, Rizzo soon obtained the release of the Venetians on board the ship.[148] Marcantonio Donini, the bailo's secretary, adds in a *relazione* written shortly afterwards that the *levendi* manned their galleys principally with Venetian prisoners, and that before they called at Istanbul they dropped these prisoners off at Mytilene for fear of the bailo having them freed. Ostensibly, the *levendi* preferred to sell captives unfit for the oar in Anatolia, since in the Aegean it was more likely that the Venetian authorities would get wind of their presence and intervene on their behalf.[149]

148 "Sua m. mi dimandò, se io sappea li nomi d'essi, gli risposi, che io non li potea sapere, perche erano stati presi in diversi tempi, et che sono homini di bassa sorte, laqual mi disse, chel bisognava pur veder per giustitia se fusero stati comprati, overo come fussero stati posti alla cathena, io non volsi risponder all'hora, che a niun modo poteano esser schiavi per [207r] metter la cosa in disputa, fino che non fussero levati dalla cathenali schiavi; ma le dissi che non volea altra giustitia che quella delle magnificentia sua; la qual pregava che fusse contenta mandar uno delli homini sui con un mio dragomano per riconoscere li schiavi, accio non fussero strabalzati, perche poi io non temea della giustitia, et gli replicai, che non volea altra giustitia che di sua magnificentia, la quale principio a ridere, et mi disse, che volea servirmi di buon inchiostro, et che volea mandar a chiamare il prothogero del capitano di Rhodi, et intendere questo fatto. Io, sebene non mi piacque molto questo principio, pur non possendo far altro dissi, che era contento," ASVe Senato, Dispacci Ambasciatori e Residenti, Constantinopoli. B. 1A, 206v-207r.

149 "sudditi di Vostra Serenità, delli quali se ne pigliano alcuno che a ciò non sia atto, lo vendono o barattano più presto nella Natolia che in altre parti, perchè in luoghi vicini a quelli della Serenità Vostra non ritrovano cosi facilmente quel che vogliono d'esso, dubitandosi li

Failing this, their abductors arranged for Islamic witnesses who could attest to these captives having debts and bonds of servitude for their masters, or sometimes forced them to acknowledge debts so that their masters could legally reclaim these sums as ransom.[150] In his final report to the Senate, Rizzo uses the episode of the Venetian captives to blame the Ottomans' attachment to the witness system.

> In such cases goods are never returned to these subjects, since they are conducted as slaves, and when your Grace the Prince requests their release they pretend their free status to be proven by Muslim witnesses, which, being difficult enough as regards the goods, I judge is in every respect impossible to obtain as concerns the latter, something aggravated by a legal bias making it impossible to free giauri slaves from the Muslims' hands.[151]

For Donini, the pashas admitted the damages but required Muslim witnesses to assess the losses.[152] Lastly, Rizzo indulges in the cliché of chained Venetians paraded before their own representatives ("sopra la faccia delli rappresentanti suoi"), claiming that in spite of his good offices most slaves were sent to Anatolia, from whence they would never return. The cliché has a characteristically early-Ottoman flavor; seventy years earlier, the scene was depicted by Zuan Dario in almost the same terms, stressing that these captives displayed before the consuls in chains were enslaved during peacetime.[153]

---

compratori che in qualche modo quel tale gli debba esser levato dalli ministri," *Relazioni degli ambasciatori veneti al Senato, Serie III, Volume III*, 194.

150 "quando loro occorre venire a Costantinopoli per alcun negozio della Maestà Sua, lasciano li schiavi, se sono sudditi di Vostra Serenità, o a Metelino o in qualche altro luogo, dubitando che dalli clarissimi baili quelli siano fatti liberare; e se pure ve ne conducono alcuno, dicono e provano per testimonj musulmani che li hanno comprati, o che hanno loro prestati danari con obbligazione ch'essi li abbiano a servire per qualche numero di anni nella fusta o galeotta loro, facendo appresso ch'essi medesimi schiavi confessino l' istesso per forza di bastonate di modo che li detti leventi offeriscono poi essi schiavi per la quantità del danaro ch'hanno detto aver esborsato come di sopra, e a questo modo non v'è rimedio di poter liberare alcuno di loro," ibid., 194–5.

151 Ibid., 143.

152 "volendo che per musulmani sia fatta la giustificazione della qualità e della quantità di essi danni. Il che quanto poi sia difficile a farsi, lo lascio considerare alla molta prudenza e ottimo giudizio della Serenità Vostra; alla quale, per mancamento di tali testimoni, da alcuni anni in qua sono stati fatti tanti danni di gente," ibid. 195–6.

153 "hauaua molto dispiacesto ala v.ra .S.ta et anche ogni zorno li reclami che vano a venexia de li dani fati et de questi nostri captiui fati in tempo de paxe strasinadi in cadene auanti ali nostril ochii.", Dario, *22 dispacci*, 110.

The fact that all decisions of royal justice were referred in the last instance to the sultan contributed to placing the biases at the center of an Ottoman-sponsored legal order. An early occurrence of the problem can be found during the tenure of Bailo Pietro Bembo; in the course of negotiations at the Dīvān concerning Venetian slaves, the pashas eventually withdrew from their initial concessions, on the basis that Muslim witnesses were needed. The pashas approached the Venetian diplomats gently, arguing that the sultan himself "wants no prisoner to be released without Muslim-produced proof." Allegedly, the pashas had interceded on behalf of the Venetians: "they have no Turks to prove it," an argument against which the sultan "stood firmly" with the justification that "he did not intend to contravene their own law."[154] On another occasion, Zuan Dario brought the question up in the course of a conversation with Mehmet Pasha, a deputy of Grand Vizier Koca Davud Pasha (1482–1497). Both men were discussing the effects of borderland strife in Albania and Morea, which had provoked a worrying increase of the seizure of captives. Dario took advantage of the friendly tone to mention the shortcomings of requiring "Turkish witnesses, that cannot be found in our lands," however Mehmet Pasha only "shrugged his shoulders, saying he could not contravene his people's faith."[155] Dario closes his narrative by declaring himself "an instrument of two powerful states diverse in language, costume, and law."

Bailo Bembo was not the only representative to mention the flow of miserable slaves aboard ships from Corfu and "lost places" such as Valona. After an initial stop at Gallipoli, in the absence of sharīʿa-compliant testimonies, the

154 ASVe Senato, Dispacci Ambasciatori e Residenti, Constantinopoli. B. 1A, 22 april 1484, "Innocupato chome eramo e maxime circa questi miseri schiavi che se atrova presenti et chi sono per venire sperandoche per le cosse tratate per Ser Zuan Dario se dovesse miorar le condition nostre in quelli alla qual parte venuti con essi bassà con parole dolce et humane resposeno chel signor non voleva ne havessamo alguno senza prova de turchi et che loro bassa lihavea dito nui diseamo non haver turchi da probar el signor fermo digando non voler contravenir alle leze soe e che non sene parlasse piui. Poi [...] molti altri che per comandamento de questi bassà erano stati mandati da la turchia per uno schiavo per loro mandato de la qual alguni garzoni fati turchi et tolti da le man di padri loro dei altri non trovando prove di turchi hali ttornati alleso padroni che questi bassà se defendeno con el signor che non vogi e mi credo che loro siano quelli." Bembo stumbled upon the same difficulties regarding some captives from Corfu shortly afterwards: ASVe Senato, Dispacci Ambasciatori e Residenti, Constantinopoli, 17 June 1484, 10a.

155 "li dissi poi del fatto di schiaui et del torto che ne veniua fato a domandare testimonij turchi che non nasseno nei nostri paexi. el strense le spale digando non poder contrauenire ala lor fede. li respusi che non era possibille che fosse bona paxe intra de nui se non se meteua qualche bon mezo a questa cason [...] ma per tornar al fatto mio Ser.mo principe io me trouo hauer gran cargo su le spale atrouandome instrumento tra do cusi potentissimi stadi diuersi de lengua de leze et de costumi." Dario, *22 dispacci*, 102.

resident bailo could not prevent the captives' definitive departure for Anatolia. Transfer to the hinterland seriously affected captives "of lower stock," a practice about which the ambassadors in Egypt also complained. When Franks were enslaved, the Venetians argued, if the consul invoked the clauses in the treaties that forbade this practice, high-ranking Egyptian owners made these slaves disappear. Indeed, these captives were forced to sign acknowledgements of debt and other written contracts, so that the treaties' dispositions would be invalid. According to several dispatches from Cairo, slave owners took great care to have these agreements notarized so that they could not be abrogated.[156] Contrary to the Maghreb, where a institutionalized 'ransom economy' emerged, in the Levant the rigid application of the Islamic laws of evidence impinged over definitions of foreignness and captivity. According to Suraiya Faroqhi, individuals were often imprisoned as fugitive slaves, then released after proving their free status. Manumitted slaves needed to possess ḥujjas certifying their freedom, since a particular kind of bounty hunters could lawfully capture escaped slaves, and they "might claim that whatever strangers they encountered were escaped slaves."[157]

By bringing up the contentious issues reported by Venetian representatives, it is not my intention to present the Islamic regime of proof as a mere device for covering up all misdeeds. Neither were the Venetians, in spite of some bitter *relazioni*, such as that sent by Rizzo—omitting his successful recovery of twenty-five captives—ignored of the fact that false claims were often overthrown on the very same procedural grounds. Alvise Mocenigo, ambassador to sultan Selim after the conquest of Egypt, admitted a pasha's goodwill, after an attack to Naxos and Mitylene by Kurtoğlu Muslihiddin Reis (1487–1535), in sending orders to qadis and officials in Anatolia to return to the Bailo the eventual prisoners being sold after the event.[158] Bembo mentions in his *dispacci* a group of slaves forcibly converted to Islam who were returned to the Venetians, since no Muslim could attest to the validity of their enslavement. By the same token, a judge's suspicions could be aroused if he found the versions delivered by slave owners unconvincing, as in the case of a young citizen from Bergamo whom the bailo managed to have freed.[159] Cases of captives fleeing to Venetian

156 "li fanno per sigilleto di cadi, che è impossibile tagliarli.", ASVe Senato, Dispacci Consoli, Egitto B1, n.42, n.46.

157 Faroqhi, Suraiya: *A cultural history of the Ottomans. The imperial elite and its artefacts*, I.B. Tauris, London-New York, 2016, 87–8.

158 Sanudo, Marino: *I diarii di Marino Sanuto*, edited by F. Stefani, Nicolò Barozzi and G. Berchet, Venezia, 1889, Vol. XXV, 274, 564.

159 ASVe Senato, Dispacci Ambasciatori e Residenti, Constantinopoli. B. 1A, dispacci Pietro Bembo, 22 April 1484.

lands, such as Crete, were often discussed, and concerned Ottoman officials pleaded their cases at the Dīvān by producing Muslim witnesses, often sending deputies to collect such testimonies from places as far as Naxos. Ottomans advanced grievances concerning dhimmīs who had abandoned the sultan's domains and settled in Venetian territories, as in the case of a large group of *carazari* migrants to Lepanto.[160] The Venetians often had to deal with these claims in exchange for concessions regarding their own subjects. "Do you realize the courtesies we grant to your subjects?" a pasha asked the bailo after a kadıasker had released a Venetian ship to satisfy the complaint of an Ottoman Jew.[161] In their conversations with the pashas, the consuls and ambassadors admitted that the object of their complaints were false witnesses and arbitrariness, rather than to challenge the sultan's justice and its procedures.[162] A pasha explained to Dario that, if Muslim witnesses were sometimes impossible to find, neither could the pashas simply accept the bailo's word on the matter, since claims concerning Christians from many different places, ranging from Germany to Puglia, were heard and examined at the Dīvān, "so that our judgment would not seem biased or suspicious."[163] The examination of witnesses, Italianized as *teftes* (ar. taftīsh) emerges time and again in the *dispacci* literature, and is proof that the Venetians were more familiar with sharīʿa procedure than is often assumed.[164] However discriminatory it may appear, the defense of the witness system cannot be labeled as simply opportunistic—and indeed this is epitomized in the reprisals against Burletto. Not only Muslims, but also the carazari sponsored the discrimination of witnesses on religious grounds. The Venetians, on their side, seem to have accepted the witness system to be part and parcel of the Ottoman version of the legal system, and one of the foundations of Islamic sovereignty. The bias against minority witnessing was not felt simply as an imposition, but perceived as point of doctrine that played out in the daily practice of bilateral relations.

---

160 Dario, *22 dispacci*, 148.

161 "voi vedete quello che noi facemo per li homeni vostri?" ASVe Senato, Dispacci Ambasciatori e Residenti, Constantinopoli. B. 1A, 374–6.

162 See Bailo Antonio Rizzo's speech reported in note 148.

163 "et li patroni di schiaui sempre diriano esser sta inganadi dal Baylo per che luno cristiano aiuta laltro et aliberar schiaui ogni vno e in clinado da la natura ma la via del mezo seria questa che messe da canto le leze nostre cusi strette. nui Bassa insieme cum vui fossamo cognitori de questa facenda et che ne fosse apuntadi li schiaui et che li vedessamo et aldissamo et considerade le circonstantie li zudegassamo et chel nostro judicio seria perfetto et non seria suspetto," Dario, *22 dispacci*, 94.

164 Heyd and Ménage, *Studies in old Ottoman criminal law*, 228–9.

Issues of religion and political identity impinged upon, and often complicated, the outcome of disputes. Christians captured in the Ottoman territories were forcibly converted as a means to prevent their restitution, probably because ḥanafī jurists objected to return Muslim slaves on the basis of dhimmī testimony (since non-Muslims converted after enslavement remained slaves).[165] Conversely, episodes in which Muslims were the object of forced conversion were often heard at the Dīvān, about which the "pashas seemed very angry." Bailo Barbarigo mentions a complaint filed at the Dīvān in 1557, concerning two hundred Muslims put to the oar in Venetian galleys off the coast of Cyprus, according to a written report by a captive sipahi forced to convert.[166] Religious issues arose when a Venetian outlaw—a nurtured community in Constantinople described by Eric Dursteler—laid claim on the property of a certain Niccolò, a Cretan mariner stuck in Gallipoli for over two years. The Venetian dissident filed a complaint before the qadi, allegedly bribing him, and produced two witnesses backing his claims over Niccolò. According to Bembo, who was clearly suggesting that evidentiary standards were not always observed, a closer look revealed that one was a manumitted slave of the plaintiff, and the other an eighty-year old man. Muslim jurists generally sketch a typology of valid witnesses, excluding those who had some kind of power relationship with one of the parties, as well as the disabled, but as we have seen a manumitted slave qualified as witness. As I claimed in the section devoted to false witnessing, the Ottomans' attachment to the biases against minority witnessing favored the proliferation of forged testimonies, an issue that was of great concern to the authorities. Quite theatrically, the mariner was carried away from the court crying out before the qadi and everyone present "I am a Venetian and a free man, and I have never been a slave before." Bembo then complained of the difficulties brought about by the exercise of justice as applied by the Ottomans, which, as we have seen, limited the taking of legal decisions to the sultan's court. "Had the court been there or in Adrianople, I would have rushed to protest …", "If the Porte was here I would have complained so loudly that I would have probably been heard," the bailo argues. Indeed, by the late fifteenth century, the Ottoman court was still itinerant, and as mentioned in the treaties, in case of the absence of the sultan the Dīvān-ı Hümāyūn handed petitions to a chief qadi, before the institution came to be dominated by the grand vizier and his bureaucracy in the seventeenth and eighteenth

165 Ibn Nujaym, *al-Ashbāh wa-al-naẓā'ir*, 284.

166 "uno spachi che aviano fatto far christiano per forza, la qual querella stata letta in divano haveva molto alterati tutti essi bassà," 26 September 1557, ASVe Senato, Dispacci Ambasciatori e Residenti, Constantinopoli. B. 1A, 437–39.

centuries.[167] Decisions invariably appeared to have been taken personally by the sultan after consultation with the pashas, yet in case of his absence, Bembo argued, the verdict issued by a "crazy qadi" led to the definitive enslavement of Niccolò and his eventual transfer to Anatolia.[168]

4.6 Legal Truth and the Governance of Frontier Zones

The pashas were sensitive to similar claims on religion advanced by their Christian counterpart, provided they were compliant with Islamic standards of proof. Bailo Antonio Rizzo vividly describes the stay of a Rumelian bishop in Istanbul, during which he lobbied to free a large contingent of Bosnian Christians. The bishop brandished a "stash of notarial deeds" to support his claims, with statements by both Franks and "honorable Muslims". The bishop had been laboring in some episcopal sees, then in Naples and in Rome to obtain a papal decree (defined as "an excommunication") on the grounds that these slaves' masters were themselves Christians. Eventually, the strategy proved to be successful, thanks to the bishop's perseverance with the pashas whom, just like the bailo, he visited and entertained at their private residences.[169]

167 Atçil, *Procedure in the Ottoman court and the duties of kadis, M. A. Thesis, Bilkent University*, 75–6, Baldwin, *Islamic Law and Empire in Ottoman Cairo*, 39.

168 "aviso la vostra Magnificentia come era de qui uno candioto nominato Nicolo el qual era de qui fin la rota de la nave de Nicolo Caro che do anni e mezo passati et a hora se hatrova uno ribaldo che disea quello esser suo schiavo et halo menado avanti allo cadi apresentado do testimoni. L'uno era so schiavo franchado e laltro vechio de anni 80. Esi testi dicono che lera so schiavo. Lo cadi che havea manzato l'a sentenentiado che lel debi menar via per so schiavo et sta menà in l'Anatoli. El povero gridava e son libero e venetian e mai non fo schiavo in tempo de la mia vita pocho me ha valesto el mio gridare avanti el cadi e de tuti de questo locho; dio lo sa che se fosse sta la porta li o in andrinopoli seria andato a far lamento. Per tanto puol intender la vostra magnificentia la inzuria che ne fa questo mato del cadi [...]. Quello che con questi mezi puol seguir non el voio dechiarir perche per si medemo el se intende. Io non ne saperia dar remedio non so che altri che Dio el possi far. Pur se la porta fosse qui gridaria tanto che forse seria aldido. Idio proveda lui chel po far." ASVe Senato, Dispacci Ambasciatori e Residenti, Constantinopoli. B. 1A, Constantinople, 14 July 1484.

169 "appresso le quali haveva molti sigilletti de sanzachi, cadi, et nasiri, che attestavano a questa porta, che un numero di cinque, et piu mille homeni sudditi di questo serenissimo sr erano ritornati alle case loro (...) Con queste attestationi introdussi esso reverendo episcopo al magnifico bassà, et dissi à sua m. Che vedendo io per molti sigilletti de mussulmani honorati, et anco de nostri, questo essere assai benemerito della eccelsa porta, lo raccomandava assai alla m. Sua, alla quale lui esponerebbe li particolari della causa sua. Sua m. Lo fecce introdure, alla quale lui presentò uno gran fasso di sigilletti che faceano fede ...", ASVe Senato, Dispacci Ambasciatori e Residenti, Constantinopoli. B. 1A, 297.

Language and personal affinity played their part, too, in the eventual success of negotiations; the bishop and the pasha communicated in a Slavic language, in the same fashion Zuan Dario claimed his knowledge of Greek to have granted him the favors of his counterpart at the Dīvān. This Greek-speaking pasha, in passing, was eventually dismissed from the Dīvān due to his excessive proximity to the Venetians.[170] Yet it was ultimately the series of documents drafted "by qadis, sanjakbeys and nāẓirs," advanced by both the bishop and the bailo, that procured the former the leverage needed for his mission. In this, slave-ownership disputes did not differ much from other borderland quarrels. Zuan Dario spent a great deal of time discussing the issue of a bridge that had been built on the fringes of Ottoman Bosnia; the discussion revolved around whether the bridge had been built on Ottoman territory, and on the reasons behind its construction. Dario wanted the bridge to be demolished since, he stated, it facilitated Ottoman raids in Venetian Dulcigno (Ulcinj). The Dīvān sent an inquiry to the qadi of Scutari, who replied that the bridge was a pious foundation built, in accordance with religious law, for welfare purposes. An Ottoman official summoned to the Dīvān argued that the artifact had saved the lives of local peasants, who often drowned crossing the river. The debate soon got bogged down into an inquiry on whether or not the borders had been agreed upon by the Venetians and Ottomans and properly notarized, and on the availability of such documents. The pashas pushed the bridge affair into the terrain of religious legitimacy, insisting that it was a waqf founded on the sultan's lands, and summoning Ottoman qadis to the Dīvān to attest to its status as a pious foundation. Dario, irritated, ended up acknowledging that the pashas' attachment to religion, and the absence of sharīʿa-compliant records, made the issue impossible to resolve.[171]

A similar climate of religious orthodoxy was felt during a discussion between the bailo and Rustem Pasha (d. 1561) regarding frontier disputes near Vrana, in today's Croatia. The Bailo had presented Venice's willingness to discuss borderline disputes in the area and to withdraw the complaints advanced by Venetian subjects. The sultan had agreed that a new *taftīsh*, or witnesses' examination,

170 "missith Bassa era cului che respondeua: compositamente et anche benignamente per esser sta de nobillissima fameia et ben dotado de la natura: et erame molto comodo per la lengua grega che in verita molte fiade me aiutaua dove che li turcimani non afferauano ben le materie" [...] "et el zorno seguente siando zorno de vacacion el pouere zentilhomo ando a spasso a zerto so zerdino vecino ala terra dove li fu subito ala coda un messo de Alma el qual li anuncio per parte soa como laueua casso de Bassa e non voleua che vegnisse pui ala porta", Dario, *22 dispacci*, 170.

171 "dixeuano che quel ponte era fabricado per elemosina perche le vna de le opere pietose comprexe in ne la leze loro" ibid., 214.

should take place in order to ascertain these disputes. Venice's position, the bailo insisted, required "examination of Venetian testimonies upon a foot of equality with Muslims, so that justice can be achieved on both sides." Rustem Pasha objected that he believed this last point to be "forbidden by our laws (*canoni*)," but suggested that the bailo advance his claim at the Dīvān, and committed to personally intervene in the bailo's favor. At the Imperial Council, once the bailo had done as advised, the pasha turned to the kadıaskers and asked them to draft a petition to the sultan for the admission of Venetian subjects as witnesses. The latter replied that such a request did not comply with the law ("non essendo … de iure"), and was "totally forbidden by our canons." The bailo eventually invoked the *carazari* clause for disputes involving Franks and dhimmīs—Christian Morlachs in this case—arguing that in the dispute in question both communities were able to give testimony. "Be patient" Rustem Pasha enjoined the bailo, because "if I addressed such a request to his Majesty he would take offense against me, knowing this goes against the law."[172]

To be sure, the tone of the discussions described above is evocative of the peculiar, political nature of Veneto-Ottoman relations. In the last decades, historians have turned away from the so-called "*ghāzī* paradigm" to focus on the intense borderland mobility found in areas such as the Balkans or North Africa, or on the crossing of community boundaries in places such as Istanbul. It was the Dīvān, however, that became the fulcrum for cross-confessional diplomacy, and in order to understand the political relations that were played out in it, we need to take into account the importance given to Islamic notions of

172 "dicendoli che havendo vostra Serenità intesa la deliberatione […] di voler che da novo sia fatto teftes sopra le querele passate … sopra li ditti confini et che per cio la havesse a mandar uno suo agente sopra il loco, la era contenta di satisfare a quello che sua magnificentia desiderava ancor chel fosse superfluo […] era che le nostre giustificationi fossino tolte, et li nostri testimoni essaminati. qui a voler che la giustizia fosse pare da una parte et dall'altra… il qual mi rispose, io credo che questo sia prohibito dalli nostri canoni nientedimeno ne faro arz al Signore et savendogli io ditto di darli uno Arz sopra di questo nel qual fossero dechiarate le nostre ragioni, mi disse non atendo lassa pur far a me, onde alli 16 del presente andai in Divano per proponergli questa materia, laqual fu da me esposta con li stesse ragioni sopradette, per il che il bassa volto alli magnifici caddi laschier disse, che dovessino farne Arz al Signore li quali risposero che non essendo questo da iure, et del tutto prohibito dalli loro canoni, non volendo far Arz altramente, replicai le ragion nostre piu di una fiata, allegando quel capitulo della pace nostra disponente che nelle differentie, che nascerano tra vostri caraciari, siano admesi testimoni christiani di qualunque sorte ne fu modo che volessino admetterle di modo chel bassa mi disse, non se potendo far tal cosa, habbi pacientia, perche se io ne facesse Arz al Sinore sua Mta. si sdegneria contra di me, sapendo eser contra li canoni," ASVe Archivi Propri Constantinopoli, 1, fasc. IV, 162v, 20 April 1547.

proof and evidence. At the Imperial Council, more than in any other place, the biases against minority witnesses served the purpose of marking an invisible line between the two promiscuous empires and their porous borderlands. This importance is apparent in the numerous discussions over the ownership of both human bodies and physical space, and their outcome did hinge on the religious roots of legal normativity. Red lines were drawn over sensitive issues and seriously respected, such as the jurisdiction over criminal disputes involving Muslims. For example, in 1558 Bailo Barbarigo asked the Porte permission to judge a Turk who had committed serious crimes near Spalato in Venetian dominions, plundering a monastery and having some monks slain. For Barbarigo, territorial notions of law were universal, but they contrasted with the pasha's views on the personal nature of sharīʿa, who saw it as having jurisdiction over all Muslims, irrespective of where they lived, particularly in criminal cases. "The sultan" he argued, "shall never grant you such a ruling, since he does not wish for you to put a Muslim to death."[173]

It would be misleading, however, to see religious discrimination as an overarching category, imposed from above, and conditioning all decisions. If the Porte often justified its position on religious grounds, other actors too, used religion to gain the necessary leverage in commercial litigation. In a most telling account, Hasan Çolak has shown the instrumental role played by Muslim witnessing in changing the status of the City of Istanbul under the patriarchate of Ieremias (1522–1545). Through an expert recourse to Muslim testimony, the Patriarch managed to counter Ottoman attempts to confiscate the churches of Istanbul. These claims, raised by a kadıasker at the Dīvān, were grounded on the legal argument that, if a city is taken by force, then there should be no Christian churches in it. Accordingly, a fatwā had been issued with the injunction of destroying the churches within five days. While the claim that the city had been taken by the sword stemmed from historical accounts found by the Turks "written in their papers", the Patriarch proved able to summon two watchmen who had witnessed the Ottoman takeover in 1453. They attested to the fact that it had been eighty-four years since the time Sultan Mehmed took Constantinople not by force, but upon agreement, and that they were

173 "Nelli luochi nostri doveva anco dalli rapresentanti (di) Vostra Serenità esser castigado cosi per quello e stà osservado sempre come per ogni raggion et leggi del mondo, rispose il bassà questo commandamento non vi concederà mai il gran Signore, per che non vorrà che voi possiate far morir un Mussulmano, ne li capitoli che havete manco lo vogliono, quali dicono che abbiate a consignar gli huomini in man nostra et noi habbiamo conossuto il suo fallo castigargli," ASVe Senato, Dispacci Ambasciatori e Residenti, Constantinopoli. B. 1A, Constantinople, 406–7, 2 May 1558.

eighteen years old at that time. The watchmen orally confirmed the alternative narrative whereby the emperor Constantine handed the keys of the Castle and bowed in front of the sultan. As a result, the sultan issued an imperial order so that the Christians would not be disturbed. The fact that witnesses were aged 102 at the time do not seem to have impinged on formal validity of the decision.[174] By the same token, and notwithstanding the crucial importance of the witnesses' faith, Abdellatif's or Burletto's cases suggest that blackmailing the judges over religious grounds did not always work. Rather than raising a 'bronze wall,' around the Muslim community, in the daily life of diplomatic and legal encounters, the biases played out in relations that were political in nature, rather than in the context marked exclusively by economic interests or military rivalry.

4.7 The Aleppo Ferman

It is clear at this point that diplomatic discussions at the Dīvān often addressed the harm caused by the enhanced role of testimonial evidence. Some of the discussions held between the bailo and the pashas make it clear that not only the mustā'min and the *carazari* advanced claims over the use of witnesses, but also Muslims. Indeed, it seems that the Muslim merchants of Aleppo may have lobbied for, and obtained, a ferman stipulating new special conditions for the use of Muslim witnesses in mixed cases. Taking place in 1557, the episode should be read against the backdrop of Ottoman adjustments and exceptions to general evidentiary law, and in particular, the rule on the inhibition of witnesses when qadi-notarized documents were available. This development is important for several reasons, but first and foremost because the new decision contradicted a superior principle that had become a central issue in the legal order sponsored by the dynasty. In fact, the decision to set different evidentiary standards for Aleppo suggests that, elsewhere, in the main Ottoman cities, the principle invalidating Muslim witnesses against Franks had become customary.

Unfortunately, the story's contours remain, as in previous cases, murky. We know that grievances began to arrive from Aleppo, and that they concerned "foreign" Muslim merchants. Apparently, these foreign merchants struck deals with Franks, or at least with the Venetians. We cannot know from whence these merchants originated, although it is unlikely that they came from other Syrian

174 Çolak, Hasan: *Co-existence and conflict between Muslims and non-Muslims in the 16th century Ottoman Istanbul, MA thesis, Department of History, Bilkent University*, 2008, 38–65.

cities, where Ottoman laws would have been known; in any case, the bailo's report on the Aleppo ferman probably refers to Persians involved in the overland silk trade that had been experiencing a boom in the last few decades. Be that as it may, these business deals often turned bad, and the cheated foreign Muslims sought redress from the qadi courts. The bailo acknowledged having received letters from the consul in Syria, who requested that the former seek the abrogation of an imperial decree granted a few months previously to the beylerbey and the qadi of Aleppo. The ferman in question allowed the qadis to summon Muslim witnesses for or against Frankish defendants, in cases where the other party were Muslims, meaning not local, Aleppine merchants, but "foreign," hailing "from outside the country."[175] The consul argued that once a complaint against Franks had been filed, these foreign Muslims were informed that the Ottoman sultan had forbidden the use of Muslim witnesses in cases against the Franks. Whoever they were, the Muslims in question seem to have seen this measure as completely aberrant, and in any case their complaints soon reached Istanbul. Since they were ignorant of the ban on Muslim witnesses, foreign Muslims would have concluded business deals without requesting the necessary ḥujja from the qadi courts, hence remaining unprotected in case of a lawsuit. The beylerbey and the qadi of Aleppo obtained a decree stating that Muslim witnesses would not be heard for cases concerning the Venetians against local, Aleppine merchants, but that this could be permitted in mixed cases involving Franks and "foreign" Muslims.[176] These foreigners, according to

175 "Lettere dal consolo della Soria [...] che mi ricercano a dimandar revocatione di un commandamento già alcuni mesi ottenuto à questa porta dal magnifico Beglierbei, et Cadi di Alepo, che con li franchi possano esser essaminati testimonii musulmani quando essi habbino differentia con mercanti, che non siano del paese, attento che loro, non sapendo il commandamento che già fece il gran signor che non fossero contra franchi admessi testi musulmani, venivano defraudati da nostri, ne mai potevano aver il suo, andai al magnifico Bassà, et con ogni efficace forma di parole esponendogli li disturbi che ogni dì seriano seguiti alli nostri per questo commandamento del tutto contrario alla prima intentione de sua Magnificentia et all'antiqua usanza lo pregai per giustizia, et quiete delli predicti mercanti, li quali erano di tanto utile alle scalle sue, che volesse darmi la revocazione, et egli mi disse che mandasse il secretario a divano che uderia detti comandamenti, et mi risolveria, et cosi mandatolo, la mattina ebbe in risposta che non si poteva rimovere detto commandamento perche era sta ordinato dal gran signore di bocca propria, et poi era necessario per li predicti mussulmani forestieri, per prohibire che non fussero di continuo ingannati, il secretario a questo secondo che gli aveno commesso le disse che poteano piu facilmente li nostri mercanti anco loro forestieri esser ingannati, che loro ingannar altri, ma che se pure la magnificentia sua aveva questa oppinione." ASVe Senato, Dispacci degli ambasciatori e residenti, B. 1A, 385.

176 "et letta che la hebbe parve che rimanesse sodisfatta, poi gli esposi il desiderio che aveva Vostra Celsitudine che fusse per la Magnificentia sua revocato il commandamento

the beylerbey and the qadi, "are not familiar with the decree Our Lord issued banning the use of Muslim witnesses against the Franks," and, as a result, "they were being cheated by our merchants, and were never able to retrieve their money." The bailo hastened to the Dīvān, arguing to the pashas that such an annulment, apart from being prejudicial to the Venetians, was contrary to the sultan's will.

Far from being a minor issue, the affair of the foreign merchants in Aleppo was included in Venice's diplomatic agenda in the following months, but the Porte upheld its original decision to make an exception for Aleppo.[177] The bailo protested on the grounds that the new decision contradicted superior norms, but the pasha replied that as long as the foreigners were ignorant of the standard ban on Muslim witnesses they would be easily deceived. According to Bailo Barbarigo, it would have been simpler to have had the measure announced by criers at the markets, or notified by the simsārs present to witness business deals. For Barbarigo, the measure threatened the property rights of Venetians, not only in their deals with foreigners, but also with local Muslims "acting in bad faith." Indeed, in the latter instance, the problem with the *commandamento* was that it departed from sharīʿa. By distinguishing between local and foreign Muslims, it implied a territorial aspect that is fundamentally foreign to the personality of Islamic law. Indeed, in that same year the Porte had intervened in favor of some Muslims of Granada imprisoned in Venice at the request of the Spanish ambassador, on the basis that, as Muslims, they were subjects of the sultan and therefore were protected by the ahdnames.[178] Far from representing a mere anecdote, the Aleppo ferman is evocative of the fundamental shift brought about in the 16th century as regards the handling of cross-confessional relations. Whereas medieval sultans took great care to leave legal doctrine in the hands of the religious learned, the Ottomans sponsored their own solutions, most often adhering to the dictates of the ḥanafī tradition, and in other

ottenuto dal beglierbei d'Aleppo che contra franchi fussero essaminati testimoni musulmani però in differenze che abbino con mercanti che non siano del paese. Egli propostosi [...] di tenerlo fermo, mi disse il commandamento non vi viene alterato se non in quella parte de mercanti forestieri, onde non vi dovete dolere, per chè à questo modo sta bene, che altrimenti non potendo loro forestieri saper il vostro commandamento facilmente sariano ingannati." ASVe Senato, Dispacci degli ambasciatori e residenti, B. 1A, 406–8.

177 "Ho mandato le sue lettere per Soria già 6 giorni à quel illustrissimo consolo con quel più di information che havevo circa l'arz per la revocation del commandamento novo, che intervenendo mercanti forestieri siano admessi testimoni mussulmani contra franchi et spero. ..." Following this are two lines in cipher, a sign of the importance given to the issue by Barbarigo, ASVe, Senato, Dispacci degli ambasciatori e residenti, Costantinopoli, B. 1A, 420v, 6 July 1557.

178 ASVe, Senato, Dispacci degli ambasciatori e residenti, Costantinopoli, B. 1A, 422.

cases vouching for their own interpretations, for example when choosing to consider mustā'mins and dhimmīs to be a single people. It is fair to see the ban on Muslim witnesses, at least, as being at odds with sharīʿa. Apart from the fact that, as a corrective measure the Aleppo ferman contravened a superior norm, it also represented a further step in Ottoman intervention in sharīʿa matters, such as the regime of proof. Furthermore, by issuing the ferman, the Ottoman sultan was implicitly admitting that the Ottoman-sponsored legal order might well have looked arbitrary to Muslims from outside the Empire.

4.8 A Death in Damascus

This chapter closes with the analysis of a highly disputed legal issue; the bankruptcy, arrest and subsequent death of a Venetian merchant in Syria, as well as a series of trials ensuing his death, seen through the lens of an exceptional trove of documents. The issue was debated in several legal forums, both in the Ottoman Empire and in Venice. Marco Priuli's bankruptcy was first brought before a qadi in Damascus in 1531, who, as we shall see, handled the affair under a narrow interpretation of Islamic law. A year later, the case was brought, in appeal, before the Imperial Dīvān in Constantinople. Besides its administrative and diplomatic capacities examined so far, the Dīvān also functioned as the supreme court, which included the hearing of mixed cases. The venetian bailos indulged in long descriptions of the Dīvān's procedure, noting the handling both of affairs of state and of lawsuits.[179] To be sure, royal justice aimed to promote the rule of law, or *sharīʿa*, as did ordinary judges; however, for the Priuli case traditional procedure and the witness system were handled with particular tact by the pashas.

Several years after the Priuli case was brought before the Dīvān, a third, final trial was started in Venice by Marco's father, who accused his brother-in-law, Domenico da Molin, of stealing some of the gems left by Marco and of obstructing the investigation. What we know about the bankruptcy, escape and death of Marco Priuli emerges from this last trial, filed before the Venetian *Giudici di Petizion*. Unlike most trials heard by that court, for which only the final verdicts have survived in Venice's state archives, the Priuli vs. Da Molin case has left behind a hefty corpus of documents. This collection appears to

179 "Si riducono la mattina per tempo ed avanti il levar del sole, e vengono in divano; dove giunti, subito seduti li bascià, si appresentano in mano di Rusten tutti li artz, cioè querele, suppliche ed ogni altra sorte di scritture appartenenti cosi a materie e cause private come a cose di stato," *Relazioni degli ambasciatori veneti al Senato, Serie III, Volume III*, 154.

be the file that was used by the plaintiff's attorney, including correspondence between the Priuli family and Marco's associates, the merchant's testament in Damascus, and other judicial documents, such as lists of Muslim and Jewish creditors. It also includes the final sentence given by the court. Marco's father sued Domenico da Molin, Marco's uncle and associate, who resided at the time in the Syrian coastal town of Tripoli, before this Venetian magistracy. According to Domenico, Marco was ready to buy anything, provided it was done by installments, and would even pay ten times its value, and the way he conducted his business in Syria raised murmurings of disapproval. Moreover, Marco cooked his books, cheated his associates, and made little effort to retrieve the credits owed to him. Marco's incompetence and malpractice as a merchant seems to leave little room for doubt, particularly since similar accusations can be found in the letters addressed to him by his own father.[180]

The *Giudici* managed to ascertain some facts regarding the events leading up to Marco Priuli's death. Having accumulated unbearable debts, he had fled Damascus for Tripoli, where he hoped to embark on a Frankish vessel. His uncle, Domenico da Molin, was a merchant involved in organizing the Venetian convoys that regularly arrived in Tripoli's harbor. Marco's creditors, in the meantime, had filed a complaint before the Ottoman pasha of Damascus. The Ottoman authorities dispatched two messengers (*chiausi*, Tr. *çavuş*) to Tripoli, found the uncle in his home, and brought him before the judicial superintendent, the *emin*. Domenico was forced to stand surety for his nephew, in accordance with Muslim custom.[181] When Domenico returned home, he found Marco waiting for him. Fearing his imminent arrest, Marco handed a kerchief containing jewels to his uncle, some of which had been bought in partnership with him. Meanwhile, one of Marco's assistants was tracked down in Tripoli by the authorities, who proceeded to arrest Marco. After a period of detention in Tripoli, Marco was brought to Damascus and imprisoned in the Citadel, the medieval fortified precinct that still hosted the provincial government. Having fallen ill, Marco had his last will notarized in jail on the 5th of October 1530 by the last known Venetian notary in Syria, Iacopo Vigulo. The document illustrates the context of legal hybridity that these merchants operated in; the deed records debts with Jews and Muslims drafted in Islamic contracts, as well as through acknowledgements of debt Marco himself drew up. Priuli admitted debts amounting to more than 5000 ashrafī, and left instructions to have his papers and an inventory of his jewels seized by the notary.

180 BC, Manoscritti Provenienza Diversa c. 508c/2, 30r.

181 BC, Manoscritti Provenienza Diversa c. 508c/3, 2 "dar alle fin piezaria del viso come de lí se usa."

When Marco passed away in the Citadel's dungeons, the judicial authorities had his body searched, and found two compromising letters he had received from his uncle Domenico while in detainment. In the letters, Domenico overtly mentioned the jewels he had received from Marco in Tripoli. These letters were handed to the qadi, who proceeded to notarize them. The discovery of this correspondence seriously complicated Domenico's position, who was finally arrested in Tripoli. However, he managed to delay his transfer to Damascus, as he was busy with the galley season in Tripoli. When he was eventually conducted to the provincial capital, Domenico took the stones, and put into a sealed bag those bought in partnership with his deceased nephew, and which he had tried to save from confiscation on the basis he had a share on them. Domenico also took a "big, long Indian diamond" and a ruby worth 5000 ashrafī—the gems that had precipitated Marco's bankruptcy—and hid them in his clothes. Eventually brought before the pasha of Damascus, who was clearly conducting an investigation, Domenico was questioned about his involvement in his nephew's bankruptcy. According to his first version, Domenico and Marco had purchased some stones together—those he was carrying in the sealed bag—while he denied any share in other purchases made by his nephew. Domenico showed the pasha the jewels both relatives had bought as partners, but the letters were read publicly in his presence, whereupon he admitted having kept the large Indian diamond and the ruby. Though Domenico confessed to the judiciary commission that he had kept the stones, one cannot blame the pasha for not believing Domenico's explanations. Domenico told the pasha that during his journey to Damascus, the two gems fell from his clothes and were lost. As a result, the stones in the sealed bag were confiscated, and Domenico ended up in jail. Domenico's story, as it was reconstructed later by the Venetian judges, presents us with a number of surprises. During his imprisonment, a Muslim woman had found the stones in the streets of Tripoli, and had sold them to two Jewish jewelers, the Cohen brothers. The confused judicial narrative is difficult to reconstruct, but the implausible story of the discovery became known to the authorities, inducing the local governor, the Sanjakbey Süleyman, to confiscate the stones as well as any property he could connect to Marco. To further complicate the story, the governor was subsequently transferred to the Sanjak of Maraş and carried the stones with him.

Another murky chapter in the Priuli affair concerns Marco's account books, which were dispersed. Some of them were sequestered by judicial officers, others ended up in the hands of partners, heirs, and the Cohen brothers.[182]

182 BC, Manoscritti Provenienza Diversa c. 508c/2, 17–18.

Andrea Priuli, a relative dispatched to Syria to square off what was left of Marco's business, managed to obtain a compromising book. It contained evidence of Marco's purchases of pearls and diamonds, and Andrea was scared the creditors would seize it. The Priuli, while accusing Domenico of smuggling Marco's records, were themselves hiding evidence from Ottoman judges. Domenico knew that his nephew's accounts would not be of much legal help. He was a veteran in the trading milieu of Aleppo and was the object of a similar lawsuit in 1540.[183] He recovered the records in the hands of the Cohens and sent them to Marco's brother, Zuan Francesco, in Istanbul. "I do not know what evil frenzy has brought you and your father to accuse me of having hidden these papers," reads the letter accompanying the package.[184] By the end of Marco's financial downfall, Domenico says, his accounts had already been falsified, and some partnerships poisoned with inflated, fake purchases of jewels. Money owed to Muslims had not been properly annotated, and the credits granted by Marco were not always registered by a qadi's ḥujja, which made them impossible to recover.

Overall, however, Marco Priuli's accounts records were of meager evidentiary value for the Ottoman judges. After his death, both Frankish and Ottoman subjects rushed to the qadi to have their debts registered. Andrea himself tried to certify one of Marco's credits left in Aleppo, but found no Muslim still able to bear witness to that transaction.[185] The sultan intervened by sending orders to a qadi of Damascus to have some old credits certified, such as one owed by Marco to a certain Abū Jumʿa al-Maghrebī. Some credits, such as the money owed by the Syrian chief Qadi Ibn Farfūr and his daughter Fāṭima, had been sold to third parties. This "cadi grande Sidi Farfor" was most probably Walī al-Dīn Ibn Farfūr discussed in Chapter Two, who remained in charge until 1530 and whose family is often mentioned as being involved in cases of embezzlement. Moreover, some of Priuli's debtors were dead, a legal problem to which I will return. It may seem that in the early 1530s, the Venetians were not particularly familiar with the certifying procedure, and indeed the consul monitored his subjects' agreements to ensure that they were registered at the courthouse. As Andrea was painfully coming to realize, local Muslims and Jews had taken

183 ASVe Correr di Santa Fosca, 11 / 162, 11 May 1540.

184 BC, Manoscritti Provenienza Diversa c. 508c/3, letter dated 15 April 1531, "le qual scritture non so da qual diabolica frenesia tentato tuo padre, et voi tutti voletti, che io avessi occultato".

185 BC, Manoscritti Provenienza Diversa c. 508c/6, 2v, "niuno s'atrova in Aleppo apresente non so si debbo farla meter de qui e per non aver niuno possa testimoniar di saper salvo Musulmani."

great care to register their debts, which was not the case for the majority of Marco's credits.

The Priulis also came up against the Ottoman custom of privileging the claims of living creditors over those of dead people. "Proof is not accepted against the affairs of the dead," Bailo Antonio Rizzo pointed out in 1556.[186] Rather, it seems it was almost impossible to obtain satisfaction, in practice, for quantities exceeding 30,000 aspre. Large sums owed to dead creditors were often used to repay the smaller debts of the living. The Venetians were puzzled by the sharīʿa approach to this issue, as they were instead used to dealing legally with their deceased ancestors, whose last wills were taken care of by the *Procuratori di San Marco* and other testamentary executors. The legal practice of administering inheritances, or *commissarie*, was deeply rooted in Venetian society. One could sue the dead, who in exchange could collect rents and pay pensions for generations or even centuries after their passing.

4.8.1 *The Priuli Case in Court*

The Priuli affair was first heard by the qadi of Damascus as early as 1531. The judge was unwilling to accept any claim not supported by proper certification. An additional complication emerged: although the Priuli managed to retrieve some of their credit claims with certified documents, the qadi considered this evidence to have been produced by Christians, and, although acceptable, priority was to be given to the satisfaction of Muslims. Fatwā collections and ḥanafī jurisprudence on common practice at the time sponsored this solution for cross-confessional debts.[187] The qadi, therefore, stipulated a "very strict clause" by virtue of which credits held by Muslims had to be repaid first, arguing that otherwise, as Bailo Pietro Zen explained it, "proof produced by us Christians ... could be used against the interests of Muslims."[188] Thirdly, and more intriguingly, the lawsuit echoes the peculiar application of the witness system that was being adopted by the early Ottomans. As the Priuli repeatedly argued, their defense was hampered by a ban on the recourse to Muslim witnesses,

---

186 ASVe Senato, Dispacci Ambasciatori e Residenti, Constantinopoli 1A, 304, 11 July 1556, "che non si admetteano prove contra le cose de morti". BC, Manoscritti Provenienza Diversa c. 508c/6, 2, "l'è consueto fra loro che quando quando una persona mora, cui diebba haver piui di saraphi 1500, che sono aspri 30.000, besogna vadino in iustification [...] coloro che die haver meno de la tal suma che iustificano de qui malmente sono pagati."

187 Grignaschi: "témoignage des sujets non musulmans", 221.

188 BC, Manoscritti Provenienza Diversa c. 508c/1, 15r-v, "che lui chadi vedendo le pruove mie herano di nostri cristiani volse meter una clausula aspera dura e severa che il credito nostro non fusse satisfato salvo da poi satisfar i mori dicendo [15v] che le testificatione de cristiani sariano a preiudicio de mori."

"whose testimony it has been forbidden for us to have."[189] It is possible that this specific ban was due to a cultural misunderstanding, as in mixed cases for non-Muslim plaintiffs the ideal solution was always to seek the support of Muslim witnesses. In any case, the ban on cross-confessional witnessing, as we shall see, was again invoked when the trial was later heard by the Dīvān.

Incidentally, a process heard by the Bailo's consular court in 1545 accounts for this climate prone to cultural misunderstandings and, maybe, for opportunistic uses of religious differences on issues of proof. The defendant, Antonio Calvocoressi invoked a number of legal arguments, supported by his own cross-confessional interpretations, in order to discredit a hostile witness, a courtier named Giacomo Leoncino. Calvocoressi claimed that Leoncino was charged with murder and banished from Venice, where he was legally married, while in Istanbul he had concubines "at the Turkish manner". Calvocoressi brandished quotations from the Old and New testaments against Leoncino's ability to give testimony. Courtiers do not qualify as witnesses, he argued, since they tend naturally to benefit their clients' interests and to "utter a thousand lies". Moreover, courtiers' testimony had been invalidated in some trials in Venice and this was the usual approach by sharīʿa. Whence *such a nonsense* comes from, the judges replied, we do not know, if it is not actually from the *Law of the Turks*. To be sure, Calvocoressi's claims were thrown out by the court, although it is significant that even experienced Levantine traders—Calvocoressi was a well-known family from Chios- attempted, on occasion, to play the card of religion when trying to discard witnesses.[190]

As Marco's relatives viewed things, sharīʿa procedure constituted a serious obstacle to their interests, and the Priuli believed their evidence would be handled differently at the Dīvān-ı Hümāyūn. In order to have their case transferred to the Dīvān, the approval of the local governor was required, who may have collected a large sum for the transfer. In addition, the bailo's "favor and support"

189 "accio non li fusse fato un simel torto, che torto possiamo dirlo, quando per mancamento di testificatione di mori non vi fusse dato il nostro, le qualle testificatione a noi sono prohibite di poter haver.", BC, Manoscritti Provenienza Diversa c. 508c/1, 15v.

190 "Quarta oppostione se le opponne che non é creduto in juditio per far la sansaria si come ne havenno casi seguidi in venetia et manco inquesta terra alla rason turchesca non vengono tolti per buoni testimonij percio che quella sua arte si é fundata quali sensari non ponno far di manco cha non dicano mille busie ni contrattar li mercadi et per far piacer à uno amico" [...] Alla terz oppositione se gli risponde che se ben è compare del predicto messer Piero questo per legge non è possibile, nè manco in osservantia è et si simil corrutella si observasse, seria causa che pochi seriano compari [...] alla quinta opposition oppone per esser sanser non deve esser creduto in Juditio, questa è vana; et non merita risposta alguna non sò dove il Reo se imagina componer di sua immagination tal ambagis, non é manifesto che uno sansaro é creduto nelli soi mercati solum dove bisogna doi testimonij et il sansaro solo é creduto, non so in qual legge lui trova, se non fosse come lui dice alla turchesca", ASVe, Bailo a Costantinopoli, 263, 39r-43r, 17 July 1545.

were required, not because of any consular prerogative, but because of his role as a community representative. This intercessory role raises the question of whether consuls were merely representatives, or were actually held accountable for their subjects' actions. Normative sources such as the ahdnames insist on individual responsibility, but according to Bailo Antonio Rizzo, in 1555 the situation was far more complex. Trials before a local qadi were considered to be "private," but they became "public" when brought for appeal before the Dīvān. In these cases, Rizzo continues, the bailo could be considered responsible for the economic losses incurred by the Venetian subjects under his protection. This seems to have been just as much the case for Venetians who were long-term residents or married to Ottoman subjects, and who had ended up fleeing the Empire to escape debt.[191] In other words, Venetians made a permanent choice when they became dhimmīs, and should avoid the temptation to leave the realm of Islam, since otherwise damages could be pinned on the bailo.

Contrary to what appears to have been common practice with Ottoman petitioners, whose cases were for the most part based on paperwork, there were hearings.[192] The parties were expected to show up in court, although as we saw earlier experienced defendants such as Abdellatif could use delay tactics and calculated absences to defer the trial. Sometimes they decided not to bring their witnesses, or show up just before the end of the session, hence avoiding formal questioning. Sometimes the bailo attended sessions personally, while at other times he simply sent his secretary, depending on his interest in speeding up the matter. Although the traditional approach to evidence was never questioned at the Dīvān, in practice the Priulis' expectation that proof would be handled differently was confirmed. However, even though the Venetians had their own witnesses, Bailo Pietro Zen adopted the strategy of presenting the case as something to be handled by the representatives of the two communities. Zen's witnesses had even been prepared, but insofar as was possible he hoped to avoid their being put upon the witness stand, and he saw the recourse to witnesses as a potential dishonor.[193] This echoes the Syrian consul's choice to allegedly refuse to

191 "questo difficillmente si potrebbe fare, sarebbe voler far publice le cause private, et consequentemente pigliare li debiti nella serenita vostra perche il capitolo dice, che se alcuno averà difficoltà col bailo, la causa sia giudicata alla Porta et questo si intende nelle cause publice solamente perchè uno altro capitolo dice che le cause de privati siano giudicate dal caddi,", ASVe Senato, Dispacci Ambasciatori e Residenti, Constantinopoli 1A, 184–5, 29 October 1555, 204v, 30 November 1555.

192 Baldwin, James E., "Petitioning the Sultan in Ottoman Egypt", *Bulletin of the School of Oriental and African Studies* 75 03 (October 2012), 499–524.

193 BC, Manoscritti Provenienza Diversa c. 508c/4, 47 et ss, "el magnifico imbasator vuol veder ad ogno modo di far di mancho delli testimoni per esser cosa pericolosa di perder poi dillo honor."

produce witnesses against Abdellatif, on the grounds that such a measure would undermine the privileges granted in the treaties banning Muslim witnesses.[194]

As a sign that mixed cases were treated by the Dīvān as a communal matter, rather than issuing from individual petitions, in the Priuli case the bailo's priority soon shifted from defending the Priulis' interests to those of other Venetian associates, such as Girolamo Venier. As the documents report in rather graphic language, if Priuli had already been "put away," some of his partners still had a chance to recover their money. Indeed, Venier was able to prove through registered contracts that he had purchased jewels in partnership with Marco, and was therefore entitled to save his share from confiscation. Zen's strategy was, to everyone's surprise, that of accepting the validity of all Muslim claims against Marco Priuli.[195] In exchange, Zen demanded satisfaction for Venier, whose jewels had been seized and who risked losing everything. This put the pasha in an uncomfortable position; if he accepted Zen's requests on Venier's behalf, giving satisfaction to Christian debtors would thwart the compensation claims of other Muslim debtors (Venier's stones were worth no less than ten thousand ducats). The pasha was displeased with this complication, since the plaintiffs insisted that Venier's claims needed to be confirmed by Muslim witnesses. Yet surprisingly, Venier had produced these Muslim witnesses, and they were ready to testify. This was precisely the kind of situation the qadi of Damascus had labored to avoid. On the other hand, it is clear that the Dīvān could not afford to dismiss every single one of the bailo's claims on the grounds of the witness system. This uncomfortable legal situation materialized during the hearing; both parties had witnesses waiting outside the room, ready to be called to testify. Nonetheless, hearing the witnesses implied permitting an unbeliever to make use of Muslim-produced proof to his own interests. This could threaten the specific Ottoman approach to the witness system, in the same manner that it had earlier in the sharīʿa court. Accordingly, the Ottoman plaintiffs insisted on having their witnesses heard and objected to the Franks that "you cannot have a Moorish witness."[196] Eventually, the Pasha judged it to be more judicious to simply leave the witnesses waiting at the door.

If, for Zen, there was no honor in seeking the testimony of witnesses, the pasha was left with a similar dilemma: either he could validate facts by

194 ASVe Senato, Dispacci Ambasciatori e Residenti, Constantinopoli 1A, 228v, 8 February 1555.

195 "el magnifico imbasator comparse alla porta dove disse haver visto tutte le rason de mori et haver in nota su una poliza li crediti loro secondo vi dico di sopra, et confesso el tal die haver et tal tanti, dove el Bassa vete sta realta che sua [61r] magnificentia trova tutti loro crediti giusti et senza contrastar dove si meraviglio et li piacque molto." BC, Manoscritti Provenienza Diversa c. 508c/4, 60v.

196 BC, Manoscritti Provenienza Diversa c. 508c/4, 60v, "et con molte gran parole dicendo poi di mori voleva contrastar su queste dicendo tu non poi haver testimonianza morescha."

summoning witnesses for Christian plaintiffs, or he could simply trust the bailo's word, held in particular consideration at the Dīvān. This, again, constituted an anomaly in court practice, although there existed precedents in which the bailo's word had been invoked to certify facts in mixed cases, and even used to release a prisoner. In 1557 Antonio Rizzo asked the doge whether, prior to the last treaty that had been concluded, it was customary to accept the bailo's word on the release of captives. Rizzo mentions several occurrences in which he had prisoners freed on his word alone.[197] Similarly, Zuan Dario reports a long conversation about the release of captives, in which the pasha admitted that accepting Christian testimony was out of the question, but reminded the bailo that captives were sometimes set free on his word.[198] Rizzo's successor Antonio Barbarigo had to deal with an episode that had happened twelve years before his tenure, and that the pasha was now laboring to reopen, concerning a Venetian subject, Agostino, accused of having seized the ship of Hacı Kamali, bound for Ancona. The bailo argued that Agostino was not guilty, and that if he had been released then a qadi's ḥujja ("sigilletto del cadi") should exist, attesting to his innocence. The pasha admitted that the issue had been tried at the Dīvān "where they do not make notes," and that Agostino had been found guilty of all charges. When the bailo argued that Agostino must have been released pursuant to an agreement between the parties, the pasha retorted that he had been released on the bailo's word alone, and that now the plaintiffs had become more powerful and were demanding satisfaction.[199] Barbarigo goes on by invoking a number of arguments, such as the existence of a statute of limitations, the fact that the former bailo Stefano Tiepolo had passed away, and that secretary Giannesino had now forgotten the relevant facts. Be that as it may, the Dīvān stuck to sharīʿa forms of procedure, but also hosted notions and practices that reminded the legal autonomy granted to dhimmī communities. Although we have descriptions of the Dīvān's paperwork and the drafting of final ḥujjas, proper proceedings were not kept, and the role of the archive was played by memory and orality. Yet the important function of oral agreements

197 "la qual cosa, per quello che sono informato, si osservava altramente avanti li presenti capitoli, perché era creduto alla parola di Vostra Serenità, ovvero delli rappresentanti suoi; e se ben è successo a me alcuna fiata il medesimo, che sopra la mia parola ne ho avuto alle flate dal magnifico lbraim bassà, al tempo che il Signor era in Persia, e che la Magnificenza Sua era governatore in Costantinopoli, avendoli io detto oh' ella mi credea in molto maggior cose; e ne ho avuto da poi il ritorno del Signor in Costantinopoli con il mezzo del magnifico capitano, il quale mi ha fatto fede al magnifico hassà, ch' erano sudditi di Vostra Serenità", *Relazioni degli ambasciatori veneti al Senato, Serie III, Volume III*, 143.

198 Dario, *22 dispacci*, 92–4.

199 "et lo dimanda à mé, come quello che lo ha rilassato sopra la parola del bailo", ASVe Senato, Dispacci Ambasciatori e Residenti, Constantinopoli 1A, 432r.

did not necessarily play out to the advantage of Muslims alone, as in the case of Agostino's release. In any case, it is clear that, as the head of his community the bailo's word was invested with a particular charisma, and that it contributed to breaching the wall erected by the biases.[200] Accordingly, and to return to the Priuli lawsuit, the pasha then made a highly significant statement: "Bailo, do as it pleases you, I believe you more than I do forty Turks." In other words, regarding the hearing of witnesses, the pasha did not intend to adhere strictly to sharīʿa procedure.[201] In the end, no witnesses were heard, Zen saved his honor and his word was trusted, and Venier recovered his stones, at the expense of other credits owed to Muslims. Everyone was stupefied, Priuli stated, adding that he could not believe it himself.

To be sure, the Burletto, Abdellatif or Priuli affairs are evocative of the complexity of the Veneto-Ottoman "special relationship." Protracted over many Dīvān sessions, the sophisticated debates and concessions from both parties make it clear that this relationship was of a political nature. While this relation is increasingly being addressed by researchers from history of communication, diplomacy or espionage perspectives, my use of these stories and in general, of the very talkative and colored Venetian sources intends rather to illustrate the fact that these door-to-door neighbors were legally different, and became gradually aware of their differences throughout the 16th century. In contrast with the medieval merchant, who seldom concerned himself with the technicalities of the Islamic legal system, I believe that much of the perceived difference between Muslims and Europeans in the early modern era stemmed from the divergent nature of their legal systems, which impacted the conduct of trade, litigation and or the practice of diplomacy. However, this impact has gone mostly unnoticed, probably because specialists in Islamic law and institutions are not concerned with the agenda of world, trans-imperial history and with cross-confessional relations. Similarly, it seems to me, historians of the Ottoman empire are too busy looking for facts and explanations, as for instance when scrutinizing the letter of the ahdnames, and often miss the more important moral meanings people connected to these rules and practices. Studies

200 "mi rispose non vi só dir altro bisogna che si adimandi di là al bailo di quel tempo, et a Zanesino di qui, che saperà informarvi, per il che le dissi che il clarissimo bailo di quel tempo che era il clarissimo messer Stefano Thiepolo [...] era passato a miglior vita, et Zanesino anco non se lo ricordava essendo andati tanti anni da poi", ASVe Senato, Dispacci Ambasciatori e Residenti, Constantinopoli 1A, 432r.

201 "Adunque questo pover'homo de Hieronimo saria ssassinato che sapendo lui Marco dovesse morir, et desputo cosi certe parole su sta cosa, et voleva [61v] far intrar li mori dentro, et pregava el bassa li lassasseno intrar anchor loro, accio dicesseno le sue rason, dove il bassa non volse mai lassarli vegnir et dise a l'ambasator io te credo piui a te che a 40 turchi, fa quel che ti piace, et ordino fosse tolte le buste di mani de el emin de morti." BC, Manoscritti Provenienza Diversa c. 508c/4, 60v-61r.

on dhimmīs abound, yet surprisingly, if we take into account the social landscape of early modern Mediterranean port-cities, there is little written about legal relations and the mustā'mins. In the Dīvān arena for cross-confessional relations, the divine character of the law made issues stemming from religion much more sensitive, as epitomized in the role of witnesses, and accordingly bailos and ambassadors learned to handle these issues with tact. We have seen them coping with technical aspects related to proof and evidence, estate law, or the intricacies of the dhimmī status. Sometimes, a divide opened between actual practice and the letter of the treaties, as was the case for contradictions in the principle of individual responsibility, when the Venetians realized that there was a doctrine on 'public' disputes, for which the whole community was liable. In 1556, Bailo Barbarigo was forced to intervene during a Dīvān hearing because he did not agree on a matter of translation into Turkish of the word 'truce,' demonstrating proof of his knowledge of amān theory.[202]

Whether they were deeply rooted in jurisprudence or not, the ban on cross-confessional witnessing and the dhimmīs' attempts to forbid the use of Frankish witnesses underline the fact that a specific approach to proof was part and parcel of Ottoman governance. Snjezana Buzov has criticized the monolithic vision of historiography of harmony between actual governance and rule of law, according to which the jurists' priority was to provide legitimacy for the sultan's claims.[203] Significantly, Ebu's-Su'ud himself raised objections against Ottoman policies on issues of testimony and unbelief, as in the case of the bailo Amer and his dhimmī witnesses. These peculiar notions and practices were accompanied by other legal ideas of a more anomalous nature. If the ban on Muslim witnesses meant twisting the arm of traditional sharī'a procedure, arbitrary suspensions of it such as that decreed for Aleppo constituted an exception to the exception. The legal grounds upon which Ottomans claimed the existence of a universal dhimmi-mustā'min nation are equally intriguing. Similar questions could be raised about the value given to the word of the bailo at the Dīvān supreme court, and, above all, to the ambiguous ban preventing the Priulis' attorneys from availing themselves of Muslim testimony, a defense universally accepted in Muslim societies. Although it is difficult to know the extent to which these bans, anomalies and adjustments were enforced in practice, there seems to have existed a legal ethical code, based on unwritten rules and shared notions of justice, according to which the production of evidence was held up to a confessional framework.

Guy Burak, Snjezana Buzov and Reem Meshal have demonstrated that Ottoman legal reforms, which were moving towards a more universal and homogeneous law, and away from local custom, were instrumental in fashioning a new

---

202 ASVe, Senato, Dispacci degli ambasciatori e residenti, Costantinopoli, B. 1A, 377–384.

203 Buzov, *The lawgiver and his lawmakers*, 80–1.

imperial sovereignty.[204] It was in fields such as the laws of evidence and obligation that earlier legal customs were more systematically dismissed. To be sure, the Ottomans did not discuss sharīʿa principles governing exchanges with unbelievers, but they did criticize "corrupt" judges' and professional witnesses' formalist attachment to rules and procedures. The unusual approach to minority witnessing, so confusing to Venetian litigants, was part of the Ottoman project to build a religious and legal orthodoxy that stood in contrast with the old Mamluk "orthopraxy."[205] And indeed, the story of the two corrupt ʿudūl in Cairo that opens this book epitomizes the formalistic approach that the Ottomans were contesting. The way in which the different affairs described in this chapter were handled demonstrates that these policies were not just an excuse for dismissing all Frankish claims related to diplomacy, ransoming and borderland relations on religious grounds. The strategies of plaintiffs to gain leverage by playing with conversion were not always successful, and indeed legal principles, such as the superiority of the Muslim word and the ḥanafī exception were not just government injunctions imposed from above, but equally invoked and manipulated by the plaintiffs.

We have seen that the sultans and their representatives invoked time and again before their Venetian counterparts the need to comply with the principles of sacred law. The biases against non-Muslim witnesses characteristic of sharīʿa, treated by the Mamluks as a legal technicality, had important meaning for the Ottomans. These principles served as a means for them to differentiate themselves as a political community, delimited by identity boundaries, and to mark out their vision of the Ottoman sultan as a champion of the rule of law. Perhaps for this reason, claims and bans such as these were heard at the Ottoman courts for at least fifty years, and crystalized around the ahdnames over the following centuries. To be sure, the biases against minority witnessing helped to demarcate social and religious boundaries, but they never constituted the unbreachable 'bronze wall' described by Western jurists.[206] However firm Ottoman adherence was to the principle of Muslims' superiority as legal actors, the potential damages of discrimination were counterbalanced by a new and opener attitude towards written evidence, that did away with the archival divide between Medieval and Modern, East and West. In addition, not only educated diplomats, but all actors, ranging from pirates to merchants and captives, proved to be familiar with each other's legal practices and with some basic sharīʿa principles, making legal divides open to negotiation.

---

204 Ibid., 7.

205 Meshal, "Antagonistic Sharīʿas", 211.

206 Contuzzi, *La istituzione dei consolati ed il diritto internazionale europeo nella sua applicabilità in Oriente*, 145.

CHAPTER 5

Conclusions

> I will erect a strong wall between you and them: "Bring me blocks of iron."
>
> Qur'ān 18: 95-96

∴

If the issue of archival practices is a highly vexed one, the way in which the biases played out in daily business remains a largely neglected research question. Since the outset, I had the chance of gleaning, almost undisturbed, many findings on the legal vicissitudes of Franks living among pre-modern Muslims. Of course, it is far from unexplored territory; a fair amount has been written on the embassies to the Mamluks, ranging from the exchange of diplomatic letters and gifts to the drafting of treaties.[1] The *relazioni* dispatched to the Senate by Venetian diplomats have long been highlighted as a landmark of the European rational approach to cross-cultural diplomacy and information. However, scarce attention has been paid to cross-confessional interactions such as those described in the daily dispatches I have scrutinized in this book.[2] Indeed, the historian's eye has usually turned away from the legal issues foreigners stumbled over, to focus instead on the adjacent artistic, diplomatic and economic aspects of their presence, deemed more important. Some aspects pertaining to the Latin Kingdom of Jerusalem have not passed under the radar, such as the complex witnessing rules in Crusader legal books, yet the Mamluks' major contribution to cross-confessional relations, the siyāsa courts, has been ignored, just as Mamluk institutions are rarely connected with any non-Muslim precedent in the Middle East.

Although in Chapters Three and Four, which concentrate on medieval and Ottoman sultans, facts take precedence over scholarly discussion, this does

1 See, among recent contributions, *Mamluk Cairo, a Crossroads for Embassies*, edited by Frédéric Bauden and Malika Dekkiche, Leiden, Brill, 2019.

2 Benzoni, Gino, "Ranke's favourite source. The Venetian relazioni. Impressions with allusions to later historiography", in: Leopold von Ranke and the Shaping of the Historical Discipline. edited by Georg G. Iggers and James M. Powell. Syracuse University Press, 1990, 45–57.

not mean these facts cannot be read against the strong backdrop of previous scholarship. While I have had the liberty of presenting my findings on siyāsa, Frankish merchants in Medieval Syria and Egypt, or on discussions at the Dīvān in an almost empty field, there is an undeniably hefty mass of recent scholarship on Islamic legal practice, which provides an accurate picture of the changing attitudes of Islamic rulers towards the legal system. Studies on Ottoman justice far outnumber those dealing with the Fatimid, Ayyubid and Mamluk periods, not simply because of the dynasty's larger chronological and geographical venture, but also due to a very simple fact: historians of the Ottoman Empire have archives, whereas those dealing with medieval times do not. On this score, the literature that addresses the development of Ottoman justice agrees on one fundamental point: that the Ottomans sought to reverse the Mamluk approach to legal pluralism, with their flexible handling of judges, notaries and legal schools, what Reem Meshal has defined as the replacement of a pluralistic sharīʿa with an "antagonistic" Ottoman law, based on the predominance of the ḥanafī legal guild.[3] Mamluk rulers, and more generally medieval sultans, neither intervened on matters of doctrine, nor in the functioning of the judiciary and the schools of law. The sultans were regulators of adjudication, and assigned a sphere of action to judges and officials, while reserving for themselves the right to appoint chief justices. Apart from that, they never interfered with the prerogatives of the legally learned, who continued to decide who could exercise legal reasoning or issue legal opinions.[4] When medieval sultans intervened in legal matters, they did so as pious Muslims, according to the Islamic principle that any believer is expected to undertake necessary actions on behalf of the community, rather than by virtue of any exceptional power.[5] The sources mention that several Mamluk sultans and emirs, such as Timurbūghā (ruled 1467–8) and Sayf al-Dīn Ṭāṭār (1421), studied jurisprudence, and portray major rulers such as al-Ashraf Qaytbāy in the act of writing their own fatwās, hence accepting the law of the jurists as an

3 Ibrahim, Ahmed Fekry, "Al-Shaʿrānī's Response to Legal Purism:: A Theory of Legal Pluralism", *Islamic Law and Society* 20 1-2 (2013), 110–140, Burak, *The second formation of Islamic law: the Hanafi school in the early modern Ottoman empire,*. Alsabagh, "Before Banks: Credit, Society, and Law in Sixteenth-Century Palestine and Syria", 52–61.

4 Fitzgerald, *Ottoman methods of conquest,* 114. Meshal, *Sharia and the Making of the Modern Egyptian,* 93–6. Winter: "The judiciary of late Mamluk and early Ottoman Damascus."

5 Hurvitz, Nimrod: "The Contribution of Early Islamic Rulers to Adjudication and Legislation: The Case of the Mazalim Tribunals", in: *Law and empire: ideas, practices, actors,* edited by Jeroen Duindam, Jill Harries, Caroline Humfress and Nimrod Hurvitz, Leiden—Boston: Brill, 2013, 133–156, 136.

unchallenged frame of reference in matters of governance.[6] Most scholars would agree that this drastically changed with the rise of Ottoman kanun and its consolidation as a corpus of public law. Despite their public image in historiography as despotic rulers, Mamluk sultans took care not to appear in the public eye as eager to slay dissident jurists and judges.[7] In sum, legal historians generally agree that the Ottoman dynasty, particularly in the 16th century, grafted not only the law but the entire legal system onto the mechanisms of governance. Just like other post-Mongol dynasties, they actively intervened in the structure of the ḥanafī legal guild and its doctrines, involved themselves in the appointment of muftīs, and developed their own imperial hierarchy of ulama and learning establishments. While this book can only contribute to this general scheme with its findings on the changing attitudes towards proof and unbelief, it may also contribute to the more challenging task of making sense of this grafting.

The rationale for the early modern incorporation of justice into governance is usually pinned on the pragmatic attitude of the Ottoman state, and in the negotiating, accommodating legal culture of the dynasty. Istiḥsān, or juristic preference, intervened in ample areas of Ottoman legal action, such as witnessing, in the same manner that most commentators attribute, one way or another, the many controversial issues related to the capitulations to convenience. An analogous consensus exists regarding the Mamluks' relationship with the legal system, which sees the sultan's patronage over legal scholars and their associations as equitable and symbiotic in nature. Symbiotic, inasmuch as rulers had more to gain from sponsoring scholars, than from challenging them by attempting to dictate legal doctrine, as had happened under the ʿAbbasid caliphs. The Ayyubids and Mamluks drew much of their legitimacy from their sponsorship of the four-madhhab system, which they housed in four-iwān madrasas and mosques built around a central hall, and, similarly, from their support for maẓālim justice displayed in public places and in the Hall of Justice (dār al-ʿadl), also a late-medieval innovation. The legally learned, in exchange for their doctrinal autonomy, rarely attempted to obstruct the sultan's designs. The relationship was equitable in the sense that, while sponsoring all forms of Islamic legal identity, the share of each school was not strictly egalitarian; shāfiʿīs, and sometimes ḥanafīs, were privileged over ḥanbalīs and mālikīs, and some legal agents such as ḥājibs and wakīls were promoted and

6 Irwin, "Privatization", 69–70, Fitzgerald, *Ottoman methods of conquest*, 114.

7 Winter: "The judiciary of late Mamluk and early Ottoman Damascus: The administrative, social and cultural transformation of the system".

their jurisdictions enlarged.[8] The practical expansion of royal justice on the basis of available theories has been, all too often, interpreted in negative terms by historians. Only in recent years have Youssef Rapoport and Christian Müller called for a more articulated vision of the justice dispensed by the qadis, and by those jurisdictions that drew from the executive power of sultans and officials, such as the market inspector or the ḥājib himself. Royal justice, in spite of complaints by many ulama, was rooted in sharīʿa and hence complied to its principal forms and manifestations. Jurists accepted the prerogative granted to Mamluk officials in the vests of judges to mete out punishments and deal with proof and procedure, so as to safeguard the interests of the Muslim community. In turn, Mamluk rulers took good note of the legal solutions issued by jurists—as demonstrated in Sultan Barqūq's opportunistic recourse to a Persian merchant's complaint against the defiant qadi al-Qurshī. Mamluk judges handled procedural issues, such as oaths or written evidence, with the participation of those foreigners to whom the law applied, without overtly challenging Islamic normativity—as would be the case, for example, if a Muslim was imprisoned on the grounds of Frankish testimony alone. Descriptions of siyāsa hearings suggest that judges adopted a juristic view, without necessarily interrogating normative texts (taqlīd).[9] Left in the hands of jurists and judges, the affairs of foreign unbelievers were dealt with fairly within the limits of Islamic traditional normativity. To my knowledge, however, no researcher has thus far sought to offer a positive interpretation of the Mamluk approach to relations with infidels, dhimmīs and mustāʾmins alike.

Researchers have a tendency to present Latin Christians as free from sin, having inherited a classical legal background in which religion was irrelevant in the courtroom and therefore no witness could be prevented from being an actor in the legal system. Yet we have seen that, against this ideal backdrop, in Mediterranean markets no Jew ever dared to present himself as a valid witness for a transaction between their Christian social superiors. As the deeds of the *outremer* notaries demonstrate, minorities bore witness only among themselves. In addition, the claims of Christian actors soon contributed to the erosion of classical Roman law in Byzantium, which asked for the discrimination of unbelievers. This tendency reached its peak in the Crusader courts and markets, with their complex regulations on the taking of oaths and testimony

8 I have dealt extensively on the charge of the wakīl al-sulṭān in Apellániz, Francisco: *Pouvoir et finance en Méditerranée pré-moderne: le deuxième état mamelouk et le commerce des épices (1382–1517)*, Barcelona: CSIC, 2009.

9 Ibrahim, "Al-Shaʿrānī's Response to Legal Purism: A Theory of Legal Pluralism", Rapoport, "Legal Diversity in the Age of Taqlid: the Four Chief Qadis Under the Mamluks."

across confessional boundaries. To cope with the complexities of dealing across confessions, the Crusader marketplace was left in the hands of the jurists, who set up a legalistic system of technicalities, and the vast production of law codes in the period testifies to this. These regulations may well have appeared strange to contemporary rulers on the continent, who lived in increasingly mono-confessional societies where such distinctions had few practical applications. Instead, societies exposed to the presence of unbelievers, such as the ever-expanding Castile, adopted without hesitation elements of Islamic legal pluralism, such as the oath-taking biases and procedures examined here. The myth of a substantive legal system rooted in Roman law as against a formalistic, idealistic and skeptical Islamic system finds its origin-point in the public faith given to notarial deeds. I have attempted to demostrate here that Latin legal systems were far more permeable than is often believed, implying deep-rooted relationships and continuities between different legal cultures, and that the supposedly classical ancestry of some key institutions such as *publica fides* is not convincing. Islamic laws of evidence, in fact, look very similar to those passed under Justinian, while the Western notary is very much a late medieval innovation.

When it came to dealing with foreigners and legal relations, this book's contention is that the Mamluk approach to cross-confessional issues was rooted in Middle Eastern traditions, and that their law courts were conducive to handling mixed conflicts, and even became the preferred forum for issues between Franks. More importantly, this process was achieved without overtly challenging sharīʿa norms, but instead by maintaining a symbiotic relationship with jurists, and their elaboration of available, if dated, doctrines on governance. In addition, the Mamluk enterprise of governance attempted to be inclusive of the judiciary as a whole, since as we have seen, the otherwise suspect ḥanbalīs served the purposes of governors, as in the case of regulations protecting foreign merchants. The rationale for the Mamluk legal handling of foreignness and unbelief was to keep its incumbent, changing presence within the limits of the available legal system. To cope with Franks and their legal needs, the Mamluks' strategy focused on adjudication procedures, laws of obligation towards non-Muslims, and sponsored notarization at all levels. Rather than indulging in an ad hoc, pragmatic policy of grating privileges to European partners, the Mamluks fully exploited the legal resources at their disposal. They mobilized notaries and judges of all sorts—ranging from qadis to officials and from simsārs to customs clerks—and demonstrated a hitherto unknown degree of sophistication in diplomacy, and most particularly in the drafting of amān. Jurists—ranging from the Mamluk-sponsored al-Subkī and al-ʿAynī, to the unruly Ibn Taymīyah and his disciples—played a part in a process that was not only driven from above. In the same vein, bottom-up forces affected the Mamluk model of market/court relations with infidels, as in the case of Franks that called on siyāsa judges to arbitrate their own intra-communal disputes. The jurists summoned sultans and officials to inflict non-Quranic punishments, open investigations and trials, and deal with circumstantial evidence in ways that the qadis were reluctant to do. The sultans, in exchange, drew on the jurists' legitimacy to deal with the businesses and claims of suspect ḥarbīs. Yet at an imprecise time around 1500—to use Ibn Iyās' metaphor describing the Portuguese arrival in the Indian Ocean—the Franks managed to breach the mythical wall erected by Alexander Dhū-l-Qarnayn to keep Gog and Magog away from humankind, and separating the Mediterranean and the China Seas.[10] In a 16th-century Mediterranean marked by the coexistence of empires, such as the Ottoman and the Venetian, a new wall was raised between members of the two principal imperial confessions. The traditional balance maintained by sultans, in their relationship with the legally learned, was profoundly altered, particularly in the post-1517 Arab provinces, in the same manner that Frankish communities and their legal issues soon outgrew their marginal role not only in the courts and markets, but also in political discourse. Just as current views on Mamluk relations with infidels have proven insufficient for understanding the complexity of the sultans' approach, the descriptions presented in this book on the handling of cross-confessional relations by the Ottomans call for a rethinking of the current pragmatist or ecumenic paradigms with which the dynasty has been typecast.

Rather than pointing an accusing finger at Ottoman contradictions, my approach has been instead to acknowledge the emergence of Ottoman practice through two contradictory trends; one whereby the subjects concerned with Ottoman law were accommodated in its practical application, and another, parallel tendency to use the law as a means to arrest social change. To this end, I make use of the conceptual tools proposed by Baki Tezcan for analyzing Ottoman politics, who argues for the existence of two opposing tendencies in medieval and early modern Islamic governance: one moving in an absolutist direction, and another alternative, 'constitutional' one that looked to religious normativity for the limits of sultanian, executive power. In keeping with previous studies on the Ottomans' relationship with the law, with its sponsoring of specific legal actors, courts, archives and written artifacts at the expense of others, it seems clear at this point that Ottoman governance moved in an absolutist direction when dealing with the affairs of foreigners. In their amān clauses,

10 Ibn Iyās, *Badāʾiʿ al-Zuhūr,* IV, 109.

bans and other decrees—such as that issued for Aleppo—the Ottoman sultans began terming their decisions on the conduct of cross-confessional affairs in the first person. However, Tezcan warns that these two tendencies did not dovetail into conservative and liberal political approaches respectively. On the one hand, the sultan publicly expressed his respect for the law of the jurists, by adhering to governing sharīʿa principles as regards proof and evidence. By fixing his positions regarding witnesses, the ruler acknowledged the divine nature of the law as interpreted by jurists, and adopted a conservative attitude by pretending the law to arrest social change. This became all the more evident in the sultans' pretensions over borderland territories and piracy issues, with their resulting loads of captive bodies, where they placed the burden of proof uniquely on Christian plaintiffs and their valueless testimony.

The worldly nature of public, kanun law, however, could on occasion nuance juridical thinking, and even overtly challenge it. Just as important as adhering to the jurists' viewpoint, the Ottoman sultan's own kanun on the rights of mustaʾmins implied the acceptance that the law had to adapt to the realities of social change. The ruler departed from the jurists' viewpoint in that he promoted the thesis that all unbelievers, dhimmīs and foreigners alike, constituted a single community, and some of his positions on amān theory and witnessing challenged the views of alleged supporters of the state such as Ebu's-Suʿud. If the sultan adopted an uncomfortable conservatism as regards the first bias at the center of this book—namely, witnessing—in parallel he issued 'liberal' kanuns protecting Franks from the potential loss of rights, allowing them, for instance, to face trials without the direct intervention of Muslim witnesses. Thus, although the sultan sent an absolutist message in terms of witnessing and unbelief, some elements of that same message were conservative, while others were liberal.

Mamluk amān treaties tend to bear a monotonous resemblance with each other, since juristic solutions necessarily applied to everyone; the Ottomans left instead key issues in the handling of cross-confessional relations to executive decrees, that applied arbitrarily for the different foreign nations.[11] The Mamluk approach to foreigners was, as in the case of their Crusader forerunners, to leave matters in the hands of specialists of sharīʿa, who took care of such issues as a technicality, a matter of *fatwās* and adjudication procedures, and in practice transferred cases to the royal courts and officials. This legalistic, medieval approach to the mustaʾmins was replaced under the Ottomans by a more political logic, of which the Venetian case is the most emblematic. The

11 Wansbrough, "Venice and Florence".

political nature of this relationship, hence not primarily dictated by economic motivations nor by mere convenience, can be appreciated in the absence of any reference in Venetian treaties to the major privilege granted to Europeans in evidentiary matters: the right not to be condemned on the basis of Muslim testimony but on that of written documents. Far from fighting for the acquisition of arbitrary privileges, Venice presented herself at the Porte as declining these procedural concessions, and as a dhimmī-like community whose representatives were heard on an equal footing. The political relation with the Porte manifested itself in the ambiguity of dīvān sessions, where diplomatic agendas were treated as any other legal business. In contrast with their predecessors, who may have discerned more clearly between bilateral and legal issues, when it came to the discrimination of non-Muslims, an orthodox approach to sharīʿa was endorsed by the dynasty. This attitude transcended the sphere of the courthouse to become an integral part of the Ottoman political discourse delivered, time and again, to the European consuls and ambassadors. Against the backdrop of this orthodox approach to divine law, the liberal aspects of the absolutist message were expressed in the language of public law, through executive decrees, including the ban on Muslim witnesses or the Aleppo ferman. The consolidation of public law went hand-in-hand with the inception of a truly archival culture associated with the courthouse, with decrees now included in the qadi's sijills. Conversely, the conservative backbone of this absolutist message adhered to the sharīʿa principles on truth sponsored by early jurists, and it was adopted at the risk of encountering constant hurdles at the borders and at sea. Thus liberal nuances affected, though not exclusively, the new status of written evidence in cross-confessional exchanges.

The Mamluks, instead, saw themselves as unconditional *constitutionalists*. The foundational act marking their rise to power was the delegation of full executive power from a legitimate Abbasid Caliph to the sultan al-Manṣūr Qalāwūn (1279–1290). The leading role of the ulama among the urban population has been sufficiently stressed by classic authors such as Ira M. Lapidus or Ulrich Haarmann. Under the aegis of the Mamluk sultans, spurious elements of the arbitration system were purged, and the moral prerogative of righteous Muslims—as epitomized by the ʿudūl- over the truths of early life extended beyond hitherto known areas. Qadis concentrated jurisdiction over cross-confessional cases and, under the Circassians, they shared it with officials deemed legitimate judges by the revered theories of Siyasah. Relations with unbelievers were increasingly dictated by a repertoire of legal principles such as the postulates of *al-siyāsa al-sharʿiyya* and the doctrines of ʿahd and amān. When paying visits to sufi masters or building welfare complexes in Mecca, when standing the provocations of the ḥanbalīs and refraining from putting away dissident ulema, the Mamluk sultans were acknowledging the role of the sharīʿa-minded in limiting their own executive power. I believe that the prevalence of the piety-minded and their values on areas such as truth-bearing is ultimately responsible for the hampering of a real archival culture in the stronghold of chancery and scribal traditions. Whenever the Abbasid and Fatimid caliphs kept archives, they yielded to the sharīʿa-based regime of truth sponsored by the legally learned. The sophistication of amān and siyāsah theories created the conditions for the flourishing of diplomacy and made Cairo 'a crossroads for embassies'. The legal and piety-minded were now empowered in areas that had traditionally been left in the hands of dragomans, diplomats and secretaries. From the vantage point offered by courts and markets, there emerges a historical time in which governance, chancery and cross-confessional practices were filled with sharīʿa-based notions and ideas. This *age of the ulama* can be located between the rise of siyāsa as a distinct post-Crusade phenomenon and the twilight of the Mamluk commercial empire with the Ottoman conquest of Egypt and Syria.

The pragmatist interpretation of Ottoman attitudes towards the European powers, as opposed to those of the despotic Mamluks, has long identified the Dīvān as a privileged locus for cross-confessional exchange. However, the Istanbul Dīvān and similar forums where privileges were obtained, and where balances of power allegedly led to the adoption of pragmatic decisions—such as the granting of capitulations and berats—are not the only setting in which we can observe cross-confessional diplomacy at work. As demonstrated by discussions revolving around the biases against non-Muslims, Islamic normativity manifested itself in a vast range of exchanges and relationships between Muslim polities and their European counterparts. Against the grain of recent trends in diplomatic history, and in light of the examples explored in this book, it would be fair to note that religious principles were respected and even became guidelines for the actual practice of diplomacy and other cross-confessional interactions. Recent cultural analyses of cross-cultural diplomacy take place in a late-modern, decline scenario where a weakened Ottoman power contravened sharīʿa norms and accepted agreements in order to satisfy imperatives of convenience and balances of power. In the recent analysis by Christian Windler, Islamic normativity was not central to relations at the Dīvān, an interaction that was instead primarily dependent upon the actors' (consuls in this case) experience and skill.[12]

---

12 Windler, Christian, "Diplomatic History as a Field for Cultural Analysis: Muslim-Christian Relations in Tunis, 1700–1840", The Historical Journal 44 1 (2001), 79–106.

Such an analysis sees both parties as being characterized by fundamental differences in terms of their values and vision of the world, and that negotiators attached different meanings to the same gestures and facts. Relations were fueled, in spite of this fundamental divide, by the capacity to reach compromises on specific points of contact. For instance, in the Maghreb regencies, hand-kissing was considered to be most humiliating by the Franks, however once they gained the privilege of kissing the bey's hand more privately, the ceremony was no longer felt to be degrading, while for the bey it kept intact its political meaning of superiority. This cultural interpretation holds that change did actually happen, but that it was circumscribed to variations in the ceremonial, and in the repertoire of norms governing cross-confessional exchanges. In addition, some medievalists have argued that in the field of legal exchange, interaction was limited to the occasional acceptance and exchange of tokens—proofs in this case—produced by the contender, whenever these proofs were compliant with the other's own system of norms.[13] This common approach to Mediterranean history sees agreements as being reached in order to accept a legal or diplomatic item as valid, while a deeper exchange of values never happened. In other words, legal systems were kept fundamentally isolated in the medieval Mediterranean—or, as Michel Balard puts it, "the gates of the funduq separated two worlds that knew nothing of each other."[14] Such a narrow interpretation stands in marked contrast with the legal milieus we see in Alexandria and Damascus in Chapter Three, where actors made promiscuous use of courts and notaries, and where institutions, irrespective of religion, had the capacity to mutually enforce each other's decisions.

Sixteenth-century Istanbul and the imperial dīvān represent a challenging unity of analysis when compared with the unequal setting of the eighteenth- and nineteenth-century regencies. I have not come across any situation in which one of the parties aimed to make their opponent "lose face", the main

13 Valérian: "Le recours à l'écrit", 68.

14 Balard, Michel: "Relations économiques entre l'Occident et le monde islamique à la fin du Moyen Age", in: *Istituto Internazionale di Storia Economica "F. Datini", Atti della xxxviii Settimana di Studi: Relazioni economiche tra Europa e mondo islamico. Secoli XIII–XVIII* t.1, 193–218. Firenze 2007, 218. It was not undue attachment to a given system of norms and values that dictated other aspects of cross-confessional relations, but instead conjectures and balances of power; see, for instance Valérian, "La résolution des conflits" , Valérian, Dominique, "Les marchands latins dans les ports musulmans méditerranéens: une minorité confinée dans des espaces communautaires?", Revue des mondes musulmans et de la Méditerranée 107–110 (2005), 437–458.

device at work in Windler's analysis. At the Porte, hand-kissing, the epitome of the diplomat's struggle for symbolic domination, did not exist. This struggle for symbolic domination stands in marked contrast with the trove of practices I have described, that were functional to the 'political', bilateral relations maintained between Venetians and Turks—exemplified by, as we have seen, the frequent exchange of jokes. Rather than attempting to twist each other's arms with the aim of gaining symbolic trophies for their own system of values, interlocutors at the Dīvān proved not only to be aware of each other's standards, but also—even if they were not necessarily in mutual agreement—the Venetians displayed a strong degree of empathy for the Turks' commitment to divine law. Only twice did Italian diplomats attempt to twist the arm of Ottoman administrators by requesting that unbelievers be accepted as witnesses: once, quickly refused, by the inexperienced Florentines, and a second time in 1522 by a disoriented doge, a request that was quickly censored by the acting ambassador.

Rather than precise agreements whose objective was to 'save the face' of both parties, in a context of conflicting systems of values, one conversation between the Turkish-speaking Giovanni Dario and the pashas can be further added to the exempla offered so far. The two were, once again, discussing the inconveniences encountered when dealing with Christian captives—although, as mentioned earlier, the Ottomans were open to granting value to the bailo's word when dealing with the deliverance of prisoners. The arguments advanced and the tone used by Dario suggest that, beyond seeking to obtain specific goals and privileges, there was no symbolic or linguistic struggle for domination at play, and that the meaning given by both contenders to concepts such as slavery were very much the same. In a discussion dated September 1484, the pasha admitted that "everyone is inclined to free slaves," and that since Frankish captives were arriving in large numbers in Istanbul due to piracy or borderland violence, he suggested that they may well "put our strict laws aside," and proceed "us pashas and you Venetians" together to investigate each individual case and to "draft the captives list," hearing their cases and circumstances before adopting a common decision. In this way "our [the pashas'] judgment" could not be suspected of partiality. Acknowledging the pasha's goodwill, Dario replied that "neither I nor the bailo would ever, in the Doge's name, free a slave not legally belonging to us." "If we did such a thing," he argued, "we would offend first God, since it was no less a sin to unjustly free a legally bought slave than to enslave a free man." So, for Dario, Venetians "simply want assurance that our citizens will no longer be captured, as [Islamic] justice requests, and that those already captured will be freed." Beyond resolving the specific issues on their agenda, Venetian ambassadors rarely clashed over cultural interpretations of the same facts, nor did they seek to challenge Islamic notions and beliefs.[15]

It is not my aim, however, to paint a rosy picture of cross-confessional relations, but rather to reposition our understanding of how foreigners and their legal issues were handled, and away from a scenario insisting upon unequal power relations. My objective has been to present a historical setting—that of the 16th-century Mediterranean—in which new and challenging boundaries were being raised. The barrier was not, like in Qurʾān 18:95–96, cast out of the iron of essentialist definitions of culture, but instead came into being by virtue of divergent normative systems inherited from the past. Adherence to sharīʿa was taken very seriously after 1517, and required a profound reconfiguration of relations across confessions. Commercial litigation, and more generally court-and-market interactions, are a privileged viewpoint for observing a phenomenon that, I believe, has passed by largely unremarked. Rather than constituting a deliberate choice, the adoption of evidentiary standards was the result of parallel, sometimes conflicting forces. The willingness to purge the allegedly corrupt Mamluk judiciary, the empowerment of ḥanafī judges at the expense of ḥājibs, the Ottoman aversion for those ʿudūl blackmailing their clients in the markets, or the rise of a logic of preservation for the written, all conspired for the adoption of a new ethics.

If I am not suggesting that the biases were adopted because they were functional to a given imperial or absolutist project, it is fair to say that the value given to these biases against disbelievers helped to demarcate social and religious boundaries and to mark out the role of the ruler in sanctioning the superiority of Muslims and sharīʿa in an increasingly promiscuous empire. On their side, instead of fostering conflict and disagreements, not only elite mediators but most dhimmīs and Franks involved acknowledged such values and beliefs, hence contributing to the spread of legal knowledge, diplomatic practices and political relations in which merchants, as well as consuls and slaves, were

15 "et aliberar schiaui ogni vno e inclinado da la natura ma la via del mezo seria questa che messe da canto le leze nostre cusi strette. nui Bassa insieme cum vui fossamo cognitori de questa facenda et cbe ne fosse apuntadi li schiaui et che li vedessamo et aldissamo et considerade le circonstantie li zudegassamo et chel nostro judicio seria perfetto et non seria suspetto et che partido che seria mi da la porta misser lo Baylo intrasse in mio luogo et che voleua esser cum el Signor et far che questo modo se obseruasse li resposi che lo aseguraua de questo che ne la M.cia del Baylo ne mi ne persona alchuna che fosse in questa parte per name de la Ex. v.ra franchessamo mai alchuno schiauo che non fosse di nostri per quanta habiamo cara la vita per che offendessamo prima dio per che non era manco peccado a liberar vn schiauo comprado contra rason dezo che era de far schiauo vnomo libero," Dario, *22 dispacci*, 94.

embedded. Beyond setting some crucial lines of demarcation, the 'bronze wall' was not meant to seal all interaction between communities, since it was accompanied by a series of measures and practical interpretations of the biases that favored exchange rather than impeding it. In a similar manner, ransoming oiled commerce with the Ottoman regencies, as has been pointed out by Wolfgang Kaiser, and I believe that the necessary legal knowledge now required of Franks in terms of registering transactions, dealing with judges and the legal principles applied in the qadi courts, was ultimately beneficial to trade. It entailed an unprecedented degree of cultural, and in this case, legal, awareness, and was responsible for the blossoming of Istanbul as a central forum for diplomatic and material exchange.[16]

16 Kaiser, Wolfgang, "L'économie de la rançon en Méditerranée occidentale (xvie-xviie siècle)", Hypothèses 10 1 (2007), 359–368.

Bibliography

a) Manuscripts

Rome, Biblioteca Nazionale Centrale, Ms. Orientali 98 / 1, Anonymous, *Kitāb al-Qawānīn al-sulṭāniyya wa-l-fawā'id al-dīwāniyya* ("Rules of government and notes on the dīwān").

Venice, Biblioteca Marciana, It. VII, 878/8652, *Dispacci Andrea Gritti* (1502).

Venice, Biblioteca Correr, Manoscritti Provenienza Diversa c. 508, *Processo Davanti ai Giudici di Petizion, eredi Marco Priuli Vs. Marin da Molin.*

Venice, Biblioteca Correr, Manoscritti Provenienza Diversa, C 306, *Relazione del Consolato in Alessandria.*

Paris, Bibliothèque Nationale, Division Orientale, Turc 130, *Mémoires de l'ambassade de Monsieur de Brèves en Levant.*

b) Arabic and Islamic Sources

al-'Aynī, Badr al-Dīn Maḥmūd Ibn Aḥmad (1361–1451): *'Iqd al-jumān fī tārīkh ahl al-zamān: 'aṣr salāṭīn al-Mamālīk*, edited by Muḥammad Muḥammad Amīn. Cairo: al-Hay'ah al-Miṣrīyah al-'Āmmah lil-Kitāb, 1987.

al-'Aynī, Badr al-Dīn Maḥmūd Ibn Aḥmad (1361–1451): *al-Bināya Sharḥ al-Hidāya*, edited by Ayman Sălih Sha'bān. Beirut: Dār al-Kutub al-'Ilmiyya, 2000.

al-Buṣrawī, 'Alī Ibn Yūsuf (1439–1499): *Tārīkh al-Buṣrawī: ṣafaḥāt majhūlah min tārīkh Dimashq fī 'Aṣr al-Mamālīk, min sanat 871 H li-ghāyat 904 H*, edited by Akram Ḥasan 'Ulabī. Damascus; Beirut: Dār al-Ma'mūn lil-Turāth, 1988.

al-Ḥalabī, Ibrāhīm Ibn Muḥammad (1460–1549): *Kitāb Multaqá al-abḥur,* edited by Wahbī Sulaymān Ghāwijī al-Bānī: Dar al-Bayrouti, 2005.

al-Jawbarī, 'Abd al-Raḥmān Ibn 'Umar: Al-Ǧawbarī und sein Kašf al-asrār: ein Sittenbild des Gauners im arabisch-islamischen Mittelalter (7./13. Jahrhundert): Einführung, Edition und Kommentar, edited by Manuela Höglmeier, Islamkundliche Untersuchungen, 1939–1940;Bd. 267. Berlin: Klaus Schwarz, 2006.

al-Maqrīzī, Aḥmad Ibn 'Alī (1364–1442): *Kitāb al-sulūk li-ma'rifat duwal al-mulūk*, edited by Said A. F. 'Ashour. Le Caire: Matba'at Dar-al-Kutub, 1970–1973.

al-Maqrīzī, Aḥmad Ibn 'Alī (1364–1442): Kitāb al-Mawā'iẓ wa-al-i'tibār fī dhikr al-khiṭaṭ wa-al-āthār. Cairo: Maktabat al-Thaqafa al-Dīniyya, n.d.

al-Qalqashandī, Aḥmad Ibn 'Alī (1355 or 6–1418): *Ṣubḥ al-a'ša fī Kitābāt al-inšā.* 14 vols. Cairo: 1913–1914.

al-Qudsī, Abū Ḥāmid (d. 1483): *Badhl al-nasā'iḥ al-shar'iyya fī mā 'ala l-sulṭān wa-wulāt al-umūr wa-sā'ir al-rā'iyya*, MA thesis, Riyadh, edited by Sālim al-Shamarī. 2 vols., 1996.

al-Saḥmāwī, Shams al-Dīn Muḥammad (d. 868/1464): *al-Thaghr al-bāsim fī ṣinā'at al-kātib wa-al-kātim: al-ma'rūf bi-ism al-Maqṣad al-rafī' al-manshā al-hādī li-dīwān al-inshā lil-Khālidī*, edited by Ashraf Muḥammad Anas and Ḥusayn Naṣṣār. 2 vols. Cairo: Maṭba'at Dār al-Kutub wa-al-Wathā'iq al-Qawmīyah bi-al-Qāhirah, 2009.

al-Subkī, Taqī al-Dīn 'Alī Ibn 'Abd al-Kāfī (1284–1355): *Fatāwā al-Subkī*, edited by Ḥusām al-Dīn Qudsī. 2 vols., 1355 h.

al-Suyūṭī, Muḥammad Ibn Shihāb al-Dīn (1413 or 14–1475): *Jawāhir al-'Uqūd wa-Mu'īn al-Quḍāh wa-al-Muwaqqi'īn wa-al-Shuhūd* Edited by Muḥammad Ḥāmid al-Fiqī. 2 vols., Cairo: Maṭba'at al-Sunnah al-Muḥammadīyah, 1992.

al-Ṭarsūsī, Najm al-Dīn Ibrāhīm Ibn 'Alī (1320-ca.1356): *Tuḥfat al-Turk fī-mā Yajib an Yu'mal fī al-Mulk*, edited by Riḍwān al-Sayyid. Ibn al-Azraq Center for Political Heritage Studies. Beirut, 2012.

Ghazzī, Najm al-Dīn Muḥammad Ibn Muḥammad (1570–1651): *al-Kawākib al-sā'irah bi-a'yān al-mi'ah al-'āshirah*. 3 vols. Bayrūt: Dār al-Kutub al-'Ilmīyah, 1997.

Ibn al-Ḥanbalī, Muḥammad Ibn Ibrāhīm al-Ḥalabī (1502 or 3–1563): *Durr al-habab fi tarikh a'yan Halab*, edited by Mahmud Fakhuri and Yahya Abbarah, *Wizarat al-Thaqafah wa-al-Irshad al-Qawmi, Ihya' al-Turath al-Qadim*. Damascus: Wizārat al-Thaqāfah, 1972–1974.

Ibn al-Ḥimṣī, Aḥmad Ibn Muḥammad (1473–1527 or 8): *Ḥawādith al-zamān wa-wafayāt al-shuyūkh wa-al-aqrān*, edited by 'Abd al-'Aziz Fayyad Harfush. Vol. 1–3. Beyrouth: Dār al-Nafā'is, 1996.

Ibn al-Ḥimṣī, Aḥmad Ibn Muḥammad (1473–1527 or 8): *Ḥawādith al-zamān wa-wafayāt al-shuyūkh wa-al-aqrān*, edited by 'Umar Tadmurī. 3 vols. Ṣaydā: al-Maktabah al-'Aṣrīyah, 1999.

Ibn al-Ṣayrafī, 'Alī Ibn Munjib (1071–1147): *al-Qānūn fī dīwān al-rasā'il*, edited by Ayman Fu'ād Sayyid. al-Qāhirah: al-Dār al-Miṣrīyah al-Lubnānīyah, 1990.

Ibn Fahd, Jār Allāh Muḥammad Ibn 'Abd al-'Azīz al-Hāshimī (1486–1547): *Kitāb Nayl al-Muná bi-Dhayl Bulūgh al-Qurá li-Takmilat Itḥāf al-Wará* (*Tārīkh Makkah al-Mukarramah min Sanat 922 H ilá 946 H*), edited by Muḥammad al-Ḥabīb al-Hīlah. 2 vols. Beirut: Mu'assasat al-Furqān lil-Turāth al-Islāmī, Fir' Mawsū'at Makkah al-Mukarramah wa-al-Madīnah al-Munawwarah, 2000.

Ibn Ḥajar al-'Asqalānī, Aḥmad Ibn 'Alī (1372–1449): *al-Durar al-Kāminah fī A'yān al-Mi'ah al-Thāminah*, edited by Muḥammad Sayyid Jād al-Ḥaqq. 5 vols. Cairo: Dār al-Kutub al-Ḥadīthah, 1966–1968.

Ibn Ḥajar al-'Asqalānī, Aḥmad Ibn 'Alī: *Inbā' al-Ghumr bi-Anbā' al-'Umr*, edited by Ḥasan Ḥabashī. 3 vols., Cairo: al-Majlis al-Ā'lá lil-Shu'ūn al-Islāmīyah, 1969–1972.

Ibn Ḥijjah al-Ḥamawī, Taqī al-Dīn Abū Bakr Ibn 'Alī (d.1434): *Thamarāt al-Awrāq*, edited by Muḥammad Abū al-Faḍl Ibrāhīm. Beirut: Beirut: Dār al-Kutub al-'Ilmīyah, 1983.

Ibn Ḥijjah al-Ḥamawī, Taqī al-Dīn Abū Bakr Ibn 'Alī (d.1434): *Qahwat al-Inshā'*, edited by Rudolf Veselý, *Adab*. Beirut: Klaus Schwarz Verlag, 2005.

Ibn Iyās, Muḥammad Ibn Aḥmad (1448-ca.1524): *Badā'i' al-Zuhūr fī Waqā'i' al-Duhūr*, edited by Muḥammad Muṣṭafá. 5 vols. Wiesbaden: Franz Steiner Verlag, 1960–1975.

Ibn Jubayr, Muḥammad Ibn Aḥmad (1145–1217): *Riḥlat Ibn Jubayr*. Beirut: Dar wa Maktabat al-Hilal, n.d.

Ibn Khaldūn, 'Abd al-Raḥmān (1332–1406): Muqaddimat Ibn Khaldûn. Prolégomènes d'Ebn-Khaldoun. 3 vol, Notices et extraits des manuscrits de la Bibliothèque impériale et autres bibliothèques, edited by Etienne Quatremère, Paris: Institut impérial de France, 1858.

Ibn Nujaym, Zayn al-Dīn Ibn Ibrāhīm (1520–63): *al-Ashbāh wa-al-naẓā'ir 'alá madhhab Abī Ḥanīfah al-Nu'mān* Damascus: Dar al-Fikr, 2005.

Ibn Qayyim al-Jawzīyah, Muḥammad Ibn Abī Bakr (1292–1350): *al-Ṭuruq al-ḥukmīyah fī al-siyāsah al-shar'īyah*. Mecca: Dar al-'Alim, 1428 h.

Ibn Taghrībirdī, Abū al-Maḥāsin Yūsuf (1411–1470): *al-Nujūm al-Zāhirah fī Mulūk Miṣr wa-al-Qāhirah*, edited by Muḥammad Ḥusayn Shams al-Dīn. Beirut: Dār al-Kutub al-'Ilmīyah, 1992.

Ibn Ṭawq, Aḥmad Ibn Muḥammad (1430 or 31–1509): al-Ta'līq: yawmīyāt Shihāb al-Dīn Aḥmad Ibn Ṭawq, 834–915 H/1430–1509 M: mudhakkirāt kutibat bi-Dimashq fī awākhir al-'ahd al-Mamlūkī, 885–908 H/1480–1502 M, edited by Ja'far Muhājir. Damascus: IFEAD, 2000–2007.

Ibn Taymīyah, Aḥmad Ibn 'Abd al-Ḥalīm (1263 -1328): *al-Siyāsah al-shar'īyah fī iṣlāḥ al-rā'ī wa-al-ra'īyah*, edited by 'Alī Ibn Muḥammad 'Umrān. Mecca: Dār 'Ālam al-Fawā'id, 1429 H.

Ibn Ṭūlūn, Shams al-Dīn Muḥammad Ibn 'Alī (1485?-1546): *Qudāt Dimashq: al-Thaghr al-Bassām fī Dhikr Man Wulliya Qaḍā' al-Shām*, edited by Ṣalāḥ al-Dīn al-Munajjid. Damascus: al-Majma' al-'Ilmī al-'Arabī, 1956.

Ibn Ṭūlūn, Shams al-Dīn Muḥammad Ibn 'Alī (1485?-1546): *I'lām al-Wará bi-Man Wulliya Nā'iban min al-Atrāk bi-Dimashq al-Shām al-Kubrá*, edited by Muḥammad Aḥmad Dahmān. Damascus: Wizārat al-Thaqāfah wa-al-Irshād al-Qawmī, 1964.

Ibn Ṭūlūn, Shams al-Dīn Muḥammad Ibn 'Alī (1485?-1546): *Naqd al-talib li-zaghal al-manasib*, edited by Muhammad Ahmad Dahman and Khalid Muhammad Dahman. Beyrouth, 1992.

Ibn Ṭūlūn, Shams al-Dīn Muḥammad Ibn 'Alī (1485?-1546): *Mufākahat al-Khillān fī Ḥawādith al-Zamān*, edited by Khalīl al-Manṣūr. Beirut: Dār al-Kutub al-'Ilmīyah, 1998.

Khaṣṣāf, Aḥmad Ibn 'Umar (d. 874): *Ādāb al-Qadi = Islamic legal and judicial system*, edited by 'Umar Ibn 'Abd al-'Azīz Ṣadr al-Shahīd and Munir Ahmad Mughal, *Islamic legal and judicial system*. New Delhi, 2017.

Moritz, Bernhard: "Ein Firman des Sultans Selim für die Venzianer vom Jahre 1517" in: *Festschrift Eduard Sachau*, Edited by G. Weil, 422–443. Berlin: Georg Reiner, 1915.

Ragib, Yusuf: Actes de vente d'esclaves et d'animaux d'Egypte médiévale. 2 vols., Cahier des annales islamologiques. Le Caire: Institut français d'archéologie orientale, 2006.

Ruiz-Orsatti, Reginaldo: "Tratado de Paz entre Alfonso V de Aragón y el sultán de Egipto, al-Malik al-Ashraf Barsbay", *Al-Andalus,* 4, 1939: 333–389.

Sarakhsī, Muḥammad Ibn Aḥmad: *Kitāb al-mabsūṭ*, edited by Muḥammad Rādī, 30 vols., Cairo, Maṭba'at al-Sa'adah, 1324–31H, 1906–13.

Stern, S.M.: *Fātimid Decrees: Original Documents from the Fātimid Chancery*. London: Faber and Faber, 1964.

Turki, Abdelmagid: "Lisan al-Dīn Ibn al-Khatib (713–76/1313–74), juriste d'après son oeuvre inédite: Muthla l-Tariqa fi dhamm al-wathiqa", *Arabica* 16, 1969: 155–211 and 279–312.

c) Latin and Christian Sources

A.Beugnot: *Abrégé du Livre des Assises de la Cour des Bourgeois*, in *Recueil des historiens des croisades*, vol. 2: Lois. Paris, 1843.

Ambasciata straordinaria al sultano d'Egitto (1489–1490), edited by Franco Rossi. Venice: Comitato per la pubblicazione delle fonti relative alla storia di Venezia, 1988.

Bigaglia, Marco. A.: *Capitulare legum notariis publicis Venetiarum*. Venice: Apud Andream Poleti, 1689.

Chronique de Denys de Tell-Mahré: quatrième partie, edited by Jean Baptiste Chabot. Vol. 1. Paris: É. Bouillon, 1895.

The Codex of Justinian: a new annotated translation, with parallel Latin and Greek text based on a translation by Justice Fred H. Blume, 3 vols., Cambridge: Cambridge University Press, 2016.

Dario, Giovanni: *22 dispacci da Costantinopoli al doge Giovanni Mocenigo*, edited by Giuseppe Calò. Venezia Corbo e Fiore, 1992.

De Mas Latrie, René: *Histoire de l'île de Chypre sous le règne des princes de la maison de Lusignan*, 3 Vols. Paris: Imprimerie impériale, 1852–1861.

The Decameron, by Giovanni Boccaccio, translated by J. M. Rigg, London - A. H. Bullen, 1903.

E.Kausler: *Les Livres des Assises et des Usages dou Reaume de Jerusalem sive Leges et Instituta Regni Hierosolymitani*, Stuttgart: Adolf Krabbe, 1839.

Edbury, P. W.: *John of Ibelin: Le livre des assises*, Leiden; Boston: Brill, 2003.

Holt, P. M.: *Early Mamluk diplomacy, 1260–1290: treaties of Baybars and Qalāwūn with Christian rulers*, Leiden; New York: E.J. Brill, 1995.

Kołodziejczyk, Dariusz: *Ottoman-Polish diplomatic relations (15th-18th century): an annotated edition of 'ahdnames and other documents*, Leiden; Boston: Brill, 2000.

Las siete partidas del rey Don Alfonso el Sabio, cotejadas con varios codices antiguos por la Real academia de la historia, Real Academia de la Historia, 3 vols. Madrid: Imprenta real, 1807.

Nicola de Boateriis notaio in Famagosta e Venezia (1355–1365), edited by Antonio Lombardo. Venezia: Comitato per la pubblicazione delle fonti relative alla storia di Venezia, 1973.

Opúsculos legales del rey don Alfonso el Sabio, Tomo I, El espéculo o espejo de todos los derechos, Real Academia de la historia, Madrid, Imprenta Real, 1836.

Relazioni degli ambasciatori veneti al Senato, Serie III, Volume I, edited by E. Albèri. Florence: Società Editrice Fiorentina, 1840.

Relazioni degli ambasciatori veneti al Senato, Serie III, Volume III, edited by E. Albèri. Vol. IX. Florence: Società editrice fiorentina, 1855.

Sanudo, Marino: *I diarii di Marino Sanuto*,edited by F. Stefani, Nicolò Barozzi and G. Berchet, Venezia, 1889, Vol. XXV.

Tafel, G.L.F., and G. M. Thomas: *Urkunden zur älteren Handels- und Staatsgeschichte der Republik Venedig, mit besonderer Beziehung auf Byzanz und die Levante: Vom neunten bis zum Ausgang des fünfzehnten Jahrhunderts*: Hof- und Staatsdruckerei, 1856.

Thomas, G.M., and R. Predelli (eds.): *Diplomatarium veneto-levantinum sive acta et diplomata res venetas graecas atque levantis illustrantia*, Vol. 1. Venice: Deputazione veneta di storia patria, 1880.

Thomas, G.M., and R. Predelli (eds.): *Diplomatarium veneto-levantinum sive acta et diplomata res venetas graecas atque levantis illustrantia*, Vol. 2. Venice: Deputazione veneta di storia patria, 1880–1899.

De Villalón, Cristóbal: *Viaje de Turquía*, edited by Antonio G. Solalinde. 2 vols., *Colección Austral; 246*. Madrid: Espasa-Calpe, 1965.

d) Works Cited

Airaldi, Gabriella: "Genovesi nel mondo islamico: « carta sarracenica » e « carta in arabico »", *Critica Storica*, IX/1, 1972: 106–121.

Airaldi, Gabriella: *Studi e documenti su Genova e l'oltremare, Collana storica di fonti e studi*: 19. Genova: Università di Genova, 1974.

Akarlı, Engin Deniz: "Law in the Marketplace: Istanbul, 1730–1840" in: *Dispensing Justice in Islam: Qadis and Their Judgments*, edited by Muhammad Khalid Masud, Rudolph Peters, and David S. Powers, 245–70. Leiden: Brill, 2006.

al-Ashqar, Muḥammad 'Abd al-Ghanī: *Tujjār al-Tawābil fī Miṣr fī al-'Aṣr al-Mamlūkī*, Cairo: al-Hay'ah al-Miṣrīyah al-'Āmmah lil-Kitāb, 1999.

Al-Qattan, Najwa: "Dhimmis in the Muslim Court: Legal Autonomy and Religious Discrimination", *International Journal of Middle East Studies,* 31, 3, 1999: 429–44.

'Alī Ibrāhīm Mīlād, Salwā: "Registres judiciaires du tribunal de la Salihiyya Nagmiyya - Etude des archives", *Annales Islamologiques,* 12, 1974: 161.

Alsabagh, Munther H.: "Before Banks: Credit, Society, and Law in Sixteenth-Century Palestine and Syria". Ph.D., University of California, Santa Barbara, 2018.

Amari, Michele: *I diplomi arabi del R. Archivio Fiorentino*. Florence: Le Monnier, 1863.

Amelotti, Mario: "Fides, fides publica in età romana" in: *Hinc publica fides. Il notaio e l'amministrazione della giustizia,* edited by Vito Piergiovanni, Milan: A. Giuffrè, 2006, 9–20.

Ankum, Hans A.: "Les tabellions romains, ancetres directs des notaires modernes", *Atlas du Notariat. Le Notariat Dans Le Monde. Huit Siecles de Notariat Latin, Ars notariatus* 42, Amsterdam: Kluwer, 1989: 5–44.

Apellániz, Francisco: Pouvoir et finance en Méditerranée pré-moderne: le deuxième état mamelouk et le commerce des épices (1382–1517). Barcelona: CSIC, 2009.

Apellániz, Francisco: "Judging the Franks: Proof, Justice, and Diversity in Late Medieval Alexandria and Damascus", *Comparative Studies in Society and History,* 58, 02 2016: 350–378.

Apellániz, Francisco: "News on the Bulaq: a Mamluk-Venetian Memorandum on Asian Trade, AD 1503", *EUI Working Paper HEC 2016/01,* 2016.

Apellániz, Francisco: "The Funduqs of Damascus seen by Frankish Notaries and Merchants" in: Travel in the Middle Ages, edited by M. Sureda, Barcelona, IEMed, 2015, p. 65–71, 266–270.

Argenti, Philip P.: *The religious minorities of Chios: Jews and Roman Catholics.* Cambridge, Cambridge University Press, 1970.

Atçil, Abdurrahman: *Procedure in the Ottoman court and the duties of kadis, M. A. Thesis, Bilkent University*: the Institute of Economics and social sciences of Bilkent University, 2002.

Atiyya, 'Aziz: "An unpublished XIVth century Fatwā" in: *Studien zur Geschichte und Kultur des Nahen und Fernen Ostens,* edited by W. Heffening, P. Kahle and W. Kirfel, 55–68: E.J. Brill, 1935.

Ayalon, David: "The End of the Mamluk Sultanate (Why did the Ottomans spare the Mamluks of Egypt and wipe out the mamluks of Syria?)", *Studia Islamica,* 65, 1987: 125–148.

Aykan, Yavuz: *Rendre la justice à Amid: procedures, acteurs et doctrines dans le contexte ottoman du XVIIIème siècle.* Leiden-Boston: Brill, 2016.

Ayoub, Samy: ""The Sulṭān Says": State Authority in the Late Ḥanafī Tradition", *Islamic Law and Society,* 23, 3 2016: 239–278.

Bakhit, Muhammad A. S.: "The Ottoman Province of Damascus in the sixteenth century". Ph.D., University of London, School of Oriental and African Studies (United Kingdom), 1972.

Balard, Michel: "Note sull'amministrazione Genovese di Cipro nel Quattrocento" in: *La storia dei Genovesi; Atti del Convegno di studi sui ceti dirigenti nelle istituzioni della Repubblica di Genova*, Genova, 11–14 Giugno 1991, Tipolitografia Sorriso Francescano, XII, Genova 1994, parte 1, 83–94.

Balard, Michel: "Il notaio e l'amministrazione della giustizia nell'oltremare genovese" in: *Hinc publica fides. Il notaio e l'amministrazione della giustizia*, Edited by Vito Piergiovanni, Milan: A. Giuffrè, 2006, 355–369.

Balard, Michel: "Relations économiques entre l'Occident et le monde islamique à la fin du Moyen Age" in: Istituto Internazionale di Storia Economica "F. Datini", *Atti della xxxviii Settimana di Studi: Relazioni economiche tra Europa e mondo islamico*. Secoli XIII-XVIII, Firenze 2007, t.1, 193–218.

Balard, Michel: "La Massaria Génoise de Famagouste" in: *Diplomatics in the Eastern Mediterranean 1000–1500: Aspects of Cross-Cultural Communication*, Edited by Alexander D. Beihammer, Maria G. Parani and Christopher D. Schabel: Brill, 2008, 235–249.

Baldwin, James E.: *Islamic Law and Empire in Ottoman Cairo*: Edinburgh University Press, 2017.

Baldwin, James E.: "Petitioning the Sultan in Ottoman Egypt", *Bulletin of the School of Oriental and African Studies*, 75, 03 October 2012: 499–524.

Bartoli Langeli, Attilio: Notai: scrivere documenti nell'Italia medievale, Roma: Viella, 2006.

Bauden, F.: "Mamluk Era Documentary Studies: The State of the Art", *Mamlūk Studies Review*, 9, 2005: 15–60.

Bauden, F.: "Du destin des archives en Islam. Analyse des données et éléments de réponse" in: La correspondance entre souverains, princes et cités-États. Approches croisées entre l'Orient musulman, l'Occident latin et Byzance (XIIIe-début XVIe s.), Edited by Denise Aigle: Turnhout, 2013, 9–30.

Bechor, Guy: *God in the Courtroom: The Transformation of Courtroom Oath and Perjury Between Islamic and Franco-Egyptian Law*, Leiden; Boston: Brill, 2011.

Becker, Brian Nathaniel: *Life and Local Administration in Fifteenth-Century Genoese Chios*, Ph.D. Thesis, Western Michigan University: ProQuest Dissertations and Theses, 2010.

Behrens-Abouseif, Doris: *Practising Diplomacy in the Mamluk Sultanate: Gifts and Material Culture in the Medieval Islamic World*. London and New York: IB Tauris, 2014.

Benzoni, Gino: "Ranke's favourite source. The Venetian relazioni. Impressions with allusions to later historiography", in: *Leopold von Ranke and the Shaping of the Historical Discipline. Edited by Georg G. Iggers and James M. Powell.* Syracuse University Press, 1990: 45–57.

Bishop, Adam: *Criminal law and the development of the assizes of the crusader Kingdom of Jerusalem in the twelfth century*, PhD Dissertation, University of Toronto, 2011.

Black, Antony: *The history of Islamic political thought: from the Prophet to the present*, Edinburgh: Edinburgh University Press, 2001.

Bora, Fozia: "Did Ṣalāḥ al-Dīn Destroy the Fatimids' Books? An Historiographical Enquiry", *Journal of the Royal Asiatic Society,* 25, 1 2015: 21–39.

Bravmann, M. M.: "The State Archives In The Early Islamic Era" in: *The spiritual background of early Islam: studies in ancient Arab concepts,* Leiden; Boston: Brill, 2009, 311–314.

Brunschvig, Robert: *La Berbérie orientale sous les Hafsides*. Paris: Adrien-Maisonneuve, 1940.

Burak, Guy: *The Abū Hanīfah of his Time: Islamic Law, Jurisprudential Authority and Empire in the Ottoman Domains (16th-17th Centuries)*, PhD thesis, New York University, 2012.

Burak, Guy: "Faith, law and empire in the Ottoman 'age of confessionalization' (fifteenth–seventeenth centuries): the case of 'renewal of faith' ", *Mediterranean Historical Review,* 28:1, 2013: 1–23.

Burak, Guy: "Between the Ḳānūn of Qāytbāy and Ottoman Yasaq: A Note on the Ottomans' Dynastic Law", *Journal of Islamic Studies,* 26, 1 2015: 1–23.

Burak, Guy: *The second formation of Islamic law: the Hanafi school in the early modern Ottoman empire,* Cambridge studies in Islamic civilization: Cambridge University Press 2015.

Burak, Guy: "Evidentiary truth claims, imperial registers, and the Ottoman archive: contending legal views of archival and record-keeping practices in Ottoman Greater Syria (seventeenth–nineteenth centuries)", *Bulletin of the School of Oriental and African Studies,* 79.2, 2016: 233–254.

Burak, Guy: ""In Compliance with the Old Register": On Ottoman Documentary Depositories and Archival Consciousness", *Journal of Economic and Social History of the Orient* 62, 2019: 799–823.

Burns, Robert Ignatius: *Jews in the notarial culture: Latinate wills in Mediterranean Spain, 1250–1350*. Berkeley: University of California Press, 1996.

Buskens, Léon: "From Trash to Treasure: Ethnographic Notes on Collecting Legal Documents in Morocco" in: *Legal Documents as Sources for the History of Muslim Societies,* edited by P. Sijpesteijn, Maaike Van Berkel and Léon Buskens, Leiden - Boston: Brill, 2017, 180–208.

Buzov, Snjezana: The lawgiver and his lawmakers: the role of legal discourse in the change of Ottoman imperial culture, PhD. Thesis: University of Chicago, 2005.

Cahen, Claude: "A propos des Shuhud", *Studia Islamica,* 31 1970: 71–79.

Cahen, Claude: "Y a-t-il eu des corporations professionnelles dans le monde musulman classique? quelques notes et réflexions" in: *The Islamic City,* Edited by S. M. Stern and A. Hourani, 51–63. Oxford, 1970.

Cahen, Claude: "Du Moyen Âge aux Temps modernes" in: Les Arabes par leurs archives: XVIe-XXe siècles: [colloque international du Centre national de la recherche scientifique], Paris, 9–11 avril 1974, Colloques internationaux du Centre national de la recherche scientifique n. 555, edited by Jacques Berque and Dominique Chevallier, Paris: Centre national de la recherche scientifique, 1976, 9–15.

Cahen, Claude: "Considérations sur l'utilisation des ouvrages de Droit musulman par l'historien" in: *Les peuples musulmans dans l'histoire médiévale,* Cairo: Presses de l'Ifpo, 2014, 71–76.

Camerani, Sergio: "Contributo alla storia dei trattati commerciali fra la Toscana e i Turchi", *Archivio Storico Italiano,* 47, 1939, 83–101.

Campbell, Caroline: "The 'Reception of the Venetian Ambassadors in Damascus': Dating, Meaning and Attribution" in: *The Renaissance and the Ottoman World,* Edited by Anna Contadini, Routledge, 2016, 125–138.

Canbakal, Hülya: *Society and politics in an Ottoman town 'Ayntab in the 17th century,* Leiden; Boston: Brill, 2007.

Carosi, Carlo: "Il tradimento della fides: il falso" in: *Hinc publica fides. Il notaio e l'amministrazione della giustizia,* Edited by Vito Piergiovanni, 127–151. Milan: A. Giuffrè, 2006.

Carro, Sergio, and Amalia Zomeño: "Identifying the 'udūl in Fifteenth-Century Granada" in: *Legal Documents as Sources for the History of Muslim Societies: Studies in Honour of Rudolph Peters,* edited by Léon Buskens and Petra M. Sijpesteijn, 109–129. Leiden; Boston: Brill, 2017.

Catlos, Brian A.: *The victors and the vanquished: Christians and Muslims of Catalonia and Aragon,* 1050–1300, Cambridge studies in medieval life and thought. New York: Cambridge University Press, 2004.

Chamberlain, Michael: *Knowledge and social practice in medieval Damascus, 1190–1350,* Cambridge studies in Islamic civilization. Cambridge; New York: Cambridge University Press, 1994.

Charrière, Ernest: *Négociations de la France dans le Levant, ou Correspondances, mémoires et actes diplomatiques des ambassadeurs de France à Constantinople*. Vol. I. Paris: Imprimerie Imperiale, 1848.

Christ, Georg: Trading Conflicts: *Venetian Merchants and Mamluk Officials in Late Medieval Alexandria,* Leiden-Boston: Brill, 2012.

Çizakça, Murat: "Was Shari'ah indeed the culprit?", *MPRA Paper 22865, University Library of Munich, Germany,* 2010.

Çizakça, Murat, and Macit Kenanoğlu: "Ottoman Merchants and the Jurisprudential Shift Hypothesis" in: *Merchants in the Ottoman Empire,* edited by S. Faroqhi and Gilles Veinstein, 195–213. Paris-Louvain-Dudley, MA: Peeters, 2008.

Clayer, Nathalie: "L'Autorité religieuse dans l'islam ottoman sous le contrôle de l'État?", *Archives de sciences sociales des religions* [*En ligne*], 125, janvier - mars 2004: 45–62.

Cohen, Amnon: *Jewish life under Islam: Jerusalem in the sixteenth century*. Cambridge, Mass.: Harvard University Press, 1984.

Çolak, Hasan: *Co-existence and conflict between Muslims and non-Muslims in the 16th century Ottoman Istanbul, MA thesis, Department of History, Bilkent University,* 2008.

Contuzzi, Francesco: *La istituzione dei consolati ed il diritto internazionale europeo nella sua applicabilità in Oriente*. Naples: E. Anfossi, 1885.

Cook, Michael A.: *Commanding right and forbidding wrong in Islamic thought*. Cambridge: Cambridge University Press, 2004.

Cook, Michael A.: "Early Medieval Christian and Muslim Attitudes to Pagan Law: A Comparison" in: *Islam and Its Past: Jahiliyya, Late Antiquity, and the Qur'an*, Edited by Carol Bakhos and Michael Cook. Oxford: Oxford University Press, 2017.

Costamagna, Giorgio: *Il notaio a Genova tra prestigio e potere, Studi storici sul notariato italiano*. Roma: Consiglio nazionale del notariato, 1970.

Davis, James Cushman: Shipping and Spying in the Early Career of a Venetian Doge, 1496–1502, Florence: Leo S. Olschki, 1974 (Estr. da: *Studi veneziani*, vol. 16).

Delsalle, Paul: *Une histoire de l'archivistique*. Sainte-Foy, 1998.

Deny, J.: "Reʾīs ül-Küttāb" in: *Encyclopaedia of Islam, Second Edition*, VIII:481b: Brill.

Dumas, Juliette "Müste'min Dealing with the Ottoman Justice: Role and Strategy of the Ambassador", *Oriente Moderno*, 93 2013: 477–494.

Dursteler, Eric: *Venetians in Constantinople: nation, identity, and coexistence in the early modern Mediterranean*, Baltimore, Md.: Johns Hopkins University Press, 2006.

Edbury, P. W.: "Latins and Greeks on Crusader Cyprus" in: *Medieval Frontiers; Concepts and Practices*, Edited by David Abulafia and Nora Berend, 133–143. Aldershot: Ashgate, 2002.

Eddé, Anne-Marie: "Documents et archives d'Orient: conclusions provisoires et tendances de la recherche actuelle" in: *L'autorité de l'écrit au Moyen Âge (Orient-Occident): XXXIXe Congrès de la SHMESP* (*Le Caire, 30 avril-5 mai 2008*), Éditions de la Sorbonne, Paris, 2009, 385–400.

El-Leithy, Tamer: "Living Documents, Dying Archives: Towards a Historical Anthropology of Medieval Arabic Archives", *al-Qantara* 32, 2011: 389–434.

el-Shayyal, Gamal el-Din: "Ibn al-Sayrafi" in: *Encyclopaedia of Islam, Second Edition*, III:932a: Brill.

Emon, Anver M.: *Religious Pluralism and Islamic Law: "Dhimmīs" and Others in the Empire of Law*. Oxford: Oxford University Press, 2012.

Epstein, Steven: *Genoa & the Genoese, 958–1528*. Chapel Hill, N.C.: University of North Carolina Press, 1996.

Ergene, Boğaç: "Document Use in Ottoman Courts of Law", *Turcica*, 37, 2005: 83–111.

Ergene, Boğaç: "Evidence in Ottoman Courts: Oral and Written Documentation in Early-Modern Courts of Islamic Law", *Journal of the American Oriental Society*, 124, 3 2004: 471–491.

Escovitz, Joseph H.: *The Office of Qâdî al-Quḍât in Cairo under the Baḥrî Mamlûks*. Berlin: Klaus Schwarz, 1984.

Fadel, Mohammad: "al-Qaḍi" in: *The Oxford Handbook of Islamic Law*, Edited by Rumee Ahmed and Anver M. Emon. Oxford: Oxford University Press, 2015.

Faroqhi, Suraiya: "The Venetian presence in the Ottoman Empire (1600–1630)", *The journal of European economic history*, 15, 2 1986: 345–384.

Faroqhi, Suraiya: "Sidjill", in: *Encyclopaedia of Islam, Second Edition*, IX: 538–545, Brill, 1996.

Faroqhi, Suraiya: "Before 1600: Ottoman attitudes towards merchants from Latin Christendom", *Turcica*, 34, 2002: 69–104.

Faroqhi, Suraiya: *Artisans of Empire: Crafts and Craftspeople Under the Ottomans*. London and New York: I. B. Tauris, 2011.

Faroqhi, Suraiya: *A cultural history of the Ottomans. The imperial elite and its artefacts*, London and New York: I. B. Tauris, 2016.

Fattal, Antoine: *Le statut légal des non-musulmans en pays d'Islam*. Beyrouth: Impr. catholique, 1958.

Ferguson, Heather. L.: *The proper order of things: language, power, and law in Ottoman administrative discourses*. Stanford, California, Stanford University Press, 2018.

Fischel, Walter J.: "The Spice Trade in Mamluk Egypt: A Contribution to the Economic History of Medieval Islam", *Journal of the Economic and Social History of the Orient*, 1, 1958: 157–174.

Fischel, Walter J.: *Ibn Khaldun in Egypt: his Public Functions and his Historical Research (1382–1406); A Study in Islamic Historiography*. Berkeley and Los Angeles: University of California 1967.

Fitzgerald, Timothy Jude: *Ottoman methods of conquest: Legal imperialism and the city of Aleppo*, 1480–1570, PhD dissertation. Vol. 3365261. Ann Arbor: Harvard University, 2009.

Fleet, Kate: "Turkish-Latin Diplomatic Relations in the Fourteenth Century: The Case of the Consul", *Oriente Moderno*, 22 (83), 3 2003: 605–611.

Fleet, Kate: "Turks, Mamluks, and Latin Merchants: Commerce, Conflict, and Cooperation in the Eastern Mediterranean" in: *Byzantines, Latins, and Turks in the Eastern Mediterranean world after 1150*, Edited by Jonathan Harris, Catherine J. Holmes and Eugenia Russell, 327–344: Oxford University Press, 2012.

Fodor, P.: "How to forge documents? (A Case of Corruption within the Ottoman Bureaucracy around 1590)", *Acta Orientalia Academiae Scientiarum Hungaricae*, 48, 3 1995: 383–389.

Fossati Raiteri, Silvana: *Genova e Cipro: l'inchiesta su Pietro de Marco, capitano di Genova in Famagosta, 1448–1449*, Collana storica di fonti e studi. Genova: Università di Genova, Istituto di medievistica, 1984.

Fotic, Aleksandar: "Non-Ottoman Documents in the Kadis,' Courts (Moloviya, Medieval Charters): Examples from the Archive of the Hilandar Monastery (15th- 18th C.)" in: *Frontiers of Ottoman studies* (*Vol. 2*), Edited by Colin Imber, Keiko Kiyotaki and Rhoads Murphey, 63–73. London, 2005.

Francesca, Ersilia: "L'applicazione del diritto islamico nell'impero ottomano (1500–1800)" in: *Storia dell'Europa e del Mediterraneo: Culture, religioni, saperi, vol. XI* (*II della sezione Età moderna*), edited by Roberto Bizzocchi, 475–523. Roma: Salerno editrice, 2011.

Frantz-Murphy, Gladys: "Identity and Security in the Mediterranean World ca. AD 640–ca. 1517", in: Proceedings of the Twenty-Fifth International Congress of Papyrology, Ann Arbor 2007, (2010).

Friedmann, Yohanan: "Classification of unbelievers in Sunni Muslim Law and Tradition" in: *Tolerance and coercion in Islam: interfaith relations in the Muslim tradition*, 54–86. New York: Cambridge University Press, 2003.

Fuess, Albert: "Zulm by Mazālim? The Political Implications of the Use of Mazālim Jurisdiction by the Mamluk Sultans", *Mamluk Studies Review*, 13, 2009: 121–147.

Gallego, Isidoro González: "El Libro de los privilegios de la nación genovesa", *Historia. Instituciones. Documentos*, 1 1974: 275–358.

Garcin, Jean-Claude: "The Regime of the Circassian Mamluks" in: *The Cambridge History of Egypt. Vol.1: Islamic Egypt, 640–1517*, Edited by Carl F. Petry, 290–317: Cambridge University Press, 1998.

Gerber, Haim: *State, Society, and Law in Islam: Ottoman Law in Comparative Perspective*. Albany: State University of New York Press, 1994.

Ginio, Eyal: "Coping with the State's Agents 'from Below': Petitions, Legal Appeal and the Sultan's Justice in Ottoman Legal Practice" in: *Popular Protest and Political Participation in the Ottoman Empire—Studies in Honor of Suraiya Faroqhi*, edited by Eleni Gara, M. Erdem Kabadayı and Christoph K, İstanbul Bilgi University Press, 41–57.

Glick, Thomas F.: Islamic and Christian Spain in the Early Middle Ages, Leiden Boston: Brill, 2005 reprint.

Gökbilgin, Tayyib: "La preuve et le témoignage dans la jurisprudence des Fetva d'Ebussud et quelques exemples d'application aux tribunaux ottomans du XVI siècle", in: *Recueils de la Société Jean Bodin, La Preuve, 3, Civilisations archaïques, asiatiques et islamiques*, Bruxelles, Editions de la Librairie encyclopédique, 1964, 205–9.

Goldberg, Jessica: *Trade and Institutions in the Medieval Mediterranean: the Geniza Merchants and their Business World*, Cambridge studies in economic history. Cambridge: Cambridge University Press, 2012.

Gourdin, Philippe: "Les marchands étrangers ont-ils un statut de dhimmi?" in: *Migrations et diasporas méditerranéennes (Xe-XVIe)*, Edited by M. Balard and A. Ducellier, 435–446. Paris: Publications de la Sorbonne, 2002.

Gradeva, Rossitsa: "Orthodox Christians in the Kadi Courts: The Practice of the Sofia Sheriat Court, Seventeenth Century", *Islamic Law and Society*, 4, 1 1997: 37–69.

Gradeva, Rossitsa: "On the Judicial Functions of Kadi Courts: Glimpses from Sofia in the Seventeenth Century," *Wiener Zeitschrift zur Geschichte der Neuzeit*, 5/2, 2005: 15-43.

Graf, Tobias P.: *The Sultan's renegades: Christian-European converts to Islam and the making of the Ottoman elite, 1575-1610*. Oxford-New York, 2017.

Grignaschi, Mario: "La valeur du témoignage des sujets non musulmans (dhimmi) dans l'Empire ottoman" in: *Recueils de la Société Jean Bodin, La Preuve, III, Civilisations archaïques, asiatiques et islamiques,* 211-323. Bruxelles Editions de la Librairie encyclopédique, 1963.

Gronke, Monika: "La rédaction des actes privés dans le monde musulman médiéval: théorie et pratique", *Studia Islamica,* 59 1984: 159-174.

Groot, Alexander H. De: "The Historical Development of the Capitulatory Regime in the Ottoman Middle East from the Fifteenth to the Nineteenth Centuries", *Oriente Moderno,* 22 (83), 3 2003: 575-604.

Guo, Li: *Commerce, Culture, and Community in a Red Sea Port in the Thirteenth Century: The Arabic Documents of Quseir.* Leiden: Brill, 2004.

Hallaq, Wael B.: "The qāḍī's dīwān (sijill) before the Ottomans", *BSOAS,* 61, 1998: 415-436.

Hallaq, Wael B.: "Qadis Communicating: Legal Change and the Law of Documentary Evidence", *al-Qantara,* 20 2, 1999, 437-66.

Hallaq, Wael B.: *The origins and evolution of Islamic law, Themes in Islamic law.* Cambridge, UK; New York: Cambridge University Press, 2005.

Hallaq, Wael B.: "Islamic law: history and transformation" in: *The New Cambridge History of Islam, Vol. 4* 142-183, 2010.

Ḥamzah, ʿĀdil ʿAbd al-Ḥāfiz: *Niyābat Ḥalab fī ʿAṣr Salāṭīn al-Mamālīk (1250-1517 M/648-923 H),* 2 Vols. Tārīkh al-Miṣrīyīn: al-Hay'ah al-Miṣrīyah al-ʿĀmmah lil-Kitāb, 2000.

Head, R. C.: "Religious Boundaries and the Inquisition in Venice: Trials of Jews and Judaizers, 1548-1580," *The Journal of medieval and Renaissance studies* 20, 2 1990: 175-204.

Heck, Paul L.: "Scrolls" in: *Encyclopaedia of the Qur'ān,* edited by General Editor: Jane Dammen McAuliffe, 4:569b: Brill, 2003.

Heffening, W.: "Ta'zīr" in: *The Encyclopaedia of Islam: Second Edition,* X, 406.

Heyd, Uriel: *Studies in old Ottoman criminal law,* edited by V. L. Ménage. Oxford, 1973.

Hirschler, Konrad: "Document Reuse in Medieval Arabic Manuscripts", *COMSt Bulletin 3/1 (2017),* 2016.

Hirschler, Konrad: "From Archive to Archival Practices: Rethinking the Preservation of Mamluk Administrative Documents", *Journal of the American Oriental Society* 136.1, 2016.

Horii, Yutaka: "The Venetian Consul and Residents in Egypt under the Ottoman Conquest", *Quaderni di Studi Arabi,* 15, 1997: 121–132.

Horii, Yutaka: "Venetians in Alexandria in the First Half of the Sixteenth Century", *Annals of Japan Association for Middle East Studies,* 20, 2 2005: 131–144.

Hosainy, Hadi: "Ottoman legal practice and non-judicial actors in Seventeenth-century Istanbul" in: *Law and Legality in the Ottoman Empire and Republic of Turkey,* Edited by Kent F. Schull, 23–33. Bloomington: Indiana University Press, 2016.

Houssaye Michienzi, Ingrid Datini, Majorque et le Maghreb (14e-15e siecles): réseaux, espaces Mediterranéens et stratégies marchandes, Leiden, Brill, 2013.

Howard, D.: "The Historical Development of the Ottoman Imperial Registry "(Defter-i Hakanî)": Mid-Fifteenth to Mid-Seventeenth Centuries", *Archivum Ottomanicum*, 11, 1986: 213.

Humfress, Caroline: "Institutionalisation between Theory and Practice: Comparative Approaches to Medieval Islamic and Late Roman Law" in: *Diverging paths?: the shapes of power and institutions in medieval Christendom and Islam*, edited by John Hudson and Ana Rodríguez López, 16–29. Leiden-Boston: Brill, 2014.

Humphreys, R. Stephen: *Islamic history: a framework for inquiry*. Princeton, N.J.: Princeton University Press, 1991.

Humphreys, R. Stephen: "Ayyubids, Mamluks, and the Latin East in the thirteenth century", *Mamluk studies review*, 2, 1998: 1–17.

Hurvitz, Nimrod: "Legal Doctrines, Historical Contexts and Moral Visions: The case of Sectarians in the Courts of Law" in: *The Islamic Scholarly Tradition*, 239–264: Brill, 2011.

Hurvitz, Nimrod: "The Contribution of Early Islamic Rulers to Adjudication and Legislation: The Case of the Mazalim Tribunals". dans *Law and empire: ideas, practices, actors*, edited by Jeroen Duindam, Jill Harries, Caroline Humfress and Nimrod Hurvitz, Leiden–Boston: Brill, 2013, 133–156.

Ibrahim, Ahmed Fekry: "Al-Shaʿrānī's Response to Legal Purism: A Theory of Legal Pluralism", *Islamic Law and Society*, 20, 1–2 2013: 110–140.

Imber, Colin: *Ebu's-suʿud: the Islamic legal tradition*. Edinburgh. Edinburgh University Press, 1997.

Imber, Colin: *The Ottoman Empire, 1300–1650: The Structure of Power*: Palgrave, 2002.

İnalcık, Halil: *Suleiman the Lawgiver and Ottoman law*: Mouton, 1969.

Iorga, Nicolae: "Notes et extraits pour servir à l'histoire des croisades au XVe siècle", *Revue de l'Orient Latin*, 4 1896.

Irwin, Robert: "The Privatization of "Justice" under the Circassian Mamluks", *Mamluk Studies Review*, 5, 2002: 63–70.

Işiksel, Güneş: "Hierarchy and Friendship: Ottoman Practices of Diplomatic Culture and Communication (1290s-1600)", *The Medieval History Journal*, 2019: 1–20.

Jacoby, David: "Citoyens, sujets et protégés de Venise et de Gênes en Chypre du XIIIe au XVe siècle" in: *Recherches sur la Méditerranée orientale du XIIe au XVe siécle: peuples, sociétés, économies*, 159–188. London: Variorum Reprints, 1979.

Jacoby, David: "The fonde of Crusader Acre and its tariff: some new considerations" in: *Dei gesta per Francos: études sur les croisades dédiées à Jean Richard, crusader studies in honour of Jean Richard*, Edited by B. Z. Kędar, Jonathan Simon, Christopher Riley-Smith and Michel Balard, 277–293, Aldershot; Burlington (Vt.); Singapore: Ashgate, 2001.

Jecker, Mélanie: "Jurer selon sa religion. La figure de l'autre dans le droit médiéval castillan" in: *La culture judiciaire: Discours, représentations et usages de la justice du*

Moyen Age à nos jours, Edited by Lucien Faggion, Christophe Regina and Bernard Ribemont, 241–269. Dijon, 2014.

Jennings, R.: "Kadi, court, and legal procedure in 17th C. Ottoman Kayseri", *Studia Islamica*, 48, 1978: 133.

Jensen, De Lamar: "The Ottoman Turks in Sixteenth Century French Diplomacy", *The Sixteenth Century Journal,* 16, 4 1985: 451–470.

Johansen, Baber: "Le jugement comme preuve. preuve juridique et vérité religieuse dans le Droit Islamique Hanéfite", *Studia Islamica*, 72 1990: 5–17.

Johansen, Baber: "Commercial exchange and social order in Hanafite Law" in: *Law and the Islamic World. Past and Present*, Edited by Ch. Toll and J. Skovgaard-Petersen, 81–95. Copenhagen: Royal Danish Academy of Sciences and Letters, 1995.

Johansen, Baber: "Vérité et torture. Ius commune et droit musulman entre le Xe et le XIIIe siècle" in: *De la violence*, edited by Françoise Héritier, 123–168. Paris: Odile Jacob, 1996.

Johansen, Baber: " Formes de langage et fonctions publiques: Stéreotypes, témoins et offices dans la preuve par l'écrit en droit musulman", *Arabica,* 44, 3 1997: 333–376.

Johansen, Baber: "Entre revelation et tyrannie: le droit des non-musulmans d'apres les juristes musulmans" in: *Contingency in a Sacred Law,* 219–37. Leiden: Brill, 1999.

Johansen, Baber: "Signs as Evidence: The Doctrine of Ibn Taymiyya (1263–1328) and Ibn Qayyim al-Jawziyya (d. 1351) on Proof", *Islamic Law and Society,* 9, 2 2002: 168–193.

Johansen, Baber: "Le contrat salam: Droit et formation du capital dans I'Empire abbasside (XIe-XIIe siècle)", *Annales. Histoire, Sciences Sociales,* 61, 4 2006: 863–899.

Johansen, Baber: *A Perfect Law in an Imperfect Society. Ibn Taymiyya's Concept of 'Governance in the Name of the Sacred Law*, in: *The law Applied: contextualizing the islamic shari 'a*, edited by Peri Bearman, Bernard G. Weiss and Wolfhart Heinrichs, IB Tauris, 2008, 259–94.

Kaiser, Wolfgang: "L'économie de la rançon en Méditerranée occidentale (xvie-xviie siècle)", *Hypothèses,* 10, 1 2007: 359–368.

Kermeli, Eugenia: "Ebussuud Efendi" in: Christian Muslim relations. Volume 7, Central and Eastern Europe, Asia, Africa and South America (1500–1600): a bibliographical history, 715–23. Leiden; Boston 2015.

Köhler, Michael A.: *Alliances and treaties between Frankish and Muslim rulers in the Middle East: cross-cultural diplomacy in the period of the Crusades*, edited by P. M. Holt and Konrad Hirschler, Leiden-Boston: Brill, 2013.

Kosei, Morimoto: "What Ibn Khaldun Saw: The Judiciary of Mamluk Egypt", *Mamluk Studies Review,* 6, 2002: 109–131.

Krstić, Tijana: *Contested Conversions to Islam: Narratives of Religious Change in the Early Modern Ottoman Empire:* Stanford University Press, 2011.

Krstić, Tijana: "Contesting Subjecthood and Sovereignty in Ottoman Galata in the Age of Confessionalization: The Carazo Affair, 1613–1617", *Oriente Moderno,* 93, 2013: 422–453.

Kuehn, Thomas: "Law and Arbitration in Renaissance Florence" in: *Law, family & women: toward a legal anthropology of Renaissance Italy,* 19–74. Chicago: University of Chicago Press, 1991.

Kumakura, Wakako: "Who handed over Mamluk Land Registers to the Ottomans? A Study on the Administrators of Land Records in the Late Mamluk Period", *Mamluk Studies Review*, XVIII, 2015: 279–298.

Kuran, Timur: *The Long Divergence: How Islamic Law Held Back the Middle East.* Princeton: Princeton University Press, 2010.

Kuran, Timur, and Scott Lustig: "Structural Inefficiencies of Islamic Courts: Ottoman Justice and Its Implications for Modern Economic Life", *ERID Working Paper Number,* 52, 2010: 1–35.

Laiou-Thomadakis, Angeliki E.: "Institutional Mechanisms of Integration" in: *Studies on the Internal Diaspora of the Byzantine Empire,* edited by H. Ahrweiler and A. E. Laiou, 161–81. Washington, D.C.: Dumbarton Oaks Research Library and Collection, 1998.

Lambton, Ann K. S.: State and Government in Medieval Islam: an Introduction to the Study of Islamic Political Theory: the Jurists. Oxford; New York: Oxford University Press, 1981.

Lane, Frederic Chapin: "The Venetian Galleys to Alexandria, 1344" in: *Wirtschaftskräfte und Wirtschaftswege. I: Mittelmeer und Kontinent: Festschrift für Hermann Kellenbenz,* edited by Jürgen Schneider, 431–440. Stuttgart, 1978.

Lapidus, Ira M.: *A History of Islamic Societies, 2nd. Edition.* Cambridge: Cambridge University Press, 1988.

Lapidus, Ira M.: Review of: "Religion and Trade: Cross-Cultural Exchanges in World History, 1000–1900 ed. by Francesca Trivellato, Leor Halevi, and Cátia Antunes", *Journal of Interdisciplinary History,* 46, 2 2015: 267–269.

Lévy, Jean-Philippe: Autour de la preuve dans les droits de l'antiquité. Naples: Jovene, 1992.

Lewis, Bernard: "Siyasa" in: *In Quest of an Islamic Humanism: Arabic and Islamic studies in memory of Mohamed al Nowaihi,* Edited by A.H. Green, 3–14. Cairo: American University in Cairo Press, 1984.

Little, Donald P.: "The Significance of the Haram Documents for the Study of Medieval Islamic History", *Der Islam,* 57, 2 1980: 189–219.

Little, Donald P.: "Sidjill", Encyclopaedia of Islam, Second Edition, IX: 538–545, Brill, 1996.

Macrides, Ruth J.: "The competent court" in: *Kinship and Justice in Byzantium, 11th-15th Centuries,* Ashgate, Variorum collected studies series 642, 1999, 117–130.

Madero, Marta: "Façons de croire. Les témoins et le juge dans l'œuvre juridique d'Alphonse X le Sage, roi de Castille", *Annales. Histoire, Sciences Sociales,* 1999: 197–218.

Malka, Orit: "Disqualified Witnesses between Tannaitic Halakha and Roman Law: The Archeology of a Legal Institution", *Law and History Review*: 1–34.

Mamluk Cairo, a Crossroads for Embassies, edited by Frédéric Bauden and Malika Dekkiche. Leiden, The Netherlands: Brill, 2019.

Mandaville, Jon Elliot: *The Muslim Judiciary of Damascus in the Late Mamluk Period*, Princeton University, Ph.D., 1969.

Marglin, Jessica M.: "Written and Oral in Islamic Law: Documentary Evidence and Non-Muslims in Moroccan Shari'a Courts", *Comparative Studies in Society and History* 59, 4 2017: 884–911.

Martel-Thoumian, Bernadette: *Les civils et l'administration dans l'État militaire mamlūk (IXe/XVe siècle)*. Damas: Institut français de Damas, 1992.

Martel-Thoumian, Bernadette: "Al-Diyā' al-Mawfūr fi A'yān Banī Farfūr: un exemple de litterature prosopographique" in: *Actas XVI Congreso UEAI*, Edited by C. Vázquez and Miguel Ángel Rodriguez, Salamanca: Agencia Espanola de Cooperación Internacional, 1995, 323–335.

Masters, Bruce Alan: *Christians and Jews in the Ottoman Arab world: the roots of sectarianism*, Cambridge studies in Islamic civilization. New York: Cambridge University Press, 2001.

Masud, Muhammad Khalid: "The Doctrine of Siyāsa in Islamic Law", *Recht van de Islam*, 18, 2001: 1–29.

Mayer, Hans Eberhard: "Latins, Muslims And Greeks In The Latin Kingdom Of Jerusalem", *History*, 63, 208 1978: 175–192.

McKee, Sally: *Uncommon dominion: Venetian Crete and the Myth of Ethnic Purity*. Philadelphia: University of Pennsylvania Press, 2000.

Meloy, John L.: "Mecca Entangled" in: Reuven Amitai and Stephan Conermann, editors, *The Mamluk Sultanate from the Perspective of Regional and World History: Economic, Social and Cultural Development in an Era of Increasing International Interaction and Competition*: Vandenhoeck & Ruprecht; 2019. p. 453–79.

Meshal, Reem A, *The State, the Community and the Individual; Local Custom and the Construction of Orthodoxy in the Sijills of Ottoman-Cairo, 1558–1646, PhD dissertation*, Institute of Islamic Studies, McGill University, 2006.

Meshal, Reem A: "Antagonistic Sharī'as and the Construction of Orthodoxy in Sixteenth-Century Ottoman Cairo", *Journal of Islamic Studies*, 21 (2), 2010: 183–212.

Meshal, Reem A: Sharia and the Making of the Modern Egyptian: Islamic Law and Custom in the Courts of Ottoman Cairo: I.B.Tauris, 2014.

Messick, Brinkley Morris: *The Calligraphic State: Textual Domination and History in a Muslim Society*. Berkeley: University of California Press, 1993.

Michel, Nicolas: "Les Circassiens avaient brûlé les registres" in: *La Conquête ottomane de l'Égypte en 1517*, edited by Benjamin Lellouch, 225–268: Brill, 2012.

Miltitz, Alexandre: *Manuel des consuls*. Londres & Berlin: A. Asher, 1837–1841.

Moukarzel, Pierre: "La législation des autorités religieuses et politiques sur les marchands Européens dans le sultanat mamelouk (1250–1517)" in: *Religious cohabitation in European towns (10th-15th centuries)*121–139. Turnhout: Brepols Publishers, 2014.

Muʿammar, Muḥammad Ben: "al-maghāriba wa manṣab qadi al-quḍḍat al-mālikī fī dimashq al-mamlūkiyya", *Majalla al-ādāb wa al-ʿulūm al-insāniya—Jāmiʿa Dimashq*, 2008: 145–162.

Müller, Christian: "A Legal Instrument in the Service of People and Institutions: Endowments in Mamluk Jerusalem as Mirrored in the Ḥaram Documents", *Mamluk Studies Review*, 12, 1 2008: 173–191.

Müller, Christian: "Ecrire pour établir la preuve orale en Islam: la pratique d'un tribunal à Jérusalem au XIVe siècle" in: *Les outils de Ia pensee: Etude historique et comparative des textes*, edited by Akira Saito and Yusuke Nakamura, 63–99: Editions de la Maison des sciences de I'homme, 2010.

Müller, Christian: "The Haram al-Sharīf collection of Arabic legal documents in Jerusalem: a Mamlūk court archive?", *al-Qantara*, XXXII, 2 2011: 435–459.

Müller, Christian: "The Power of the Pen: Cadis and their Archives in Medieval Islam" in: *Manuscripts and Archives: Comparative Views on Record-Keeping*, edited by Alessandro Bausi, 361–385: Walter de Gruyter, 2018.

Müller, Christian: "Mamluk Law: a reassessment" in: *Ubi sumus? Quo vademus?* Volume 3, 263–284: V&R Unipress, 2013.

Müller, Christian: "Non-Muslims as part of Islamic law: Juridical casuistry in a fifth/eleventh-century law manual," in: The Legal Status of Dhimmis in the Islamic West, edited by Maribel Fierro and John Tolan, 2013, 21–65.

Müller, Joseph: Documenti sulle relazioni delle città toscane coll' Oriente cristiano e coi Turchi fino all'anno MDXXXI. Firenze: M. Cellini e c., 1879.

Mumcu, Serap: "1619 il trattato di pace fra l'Impero Ottomano e la Serenissima Reppubblica Ahidname di Osman II (electronic article)", *https://www.academia.edu/26041394/LAhidname_di_Osman_II, last accessed January* 2020.

Muranyi, Miklos: "Geniza or hubus: some observations on the library of the great mosque in Qayrawan", *Jerusalem Studies in Arabic and Islam*, 42, 2015: 183–201.

Murray, James M.: *Notarial instruments in Flanders between 1280 and 1452*. Bruxelles, Académie royale de Belgique, 1995.

Musso, Giangiacomo: "Gli ebrei nel Levante genovese: ricerche di archivio", *La Berio*, a. 10., n.2 maggio-agosto, 1970: 5–27.

Nader, Marwan: *Burgesses and Burgess law in the Latin Kingdoms of Jerusalem and Cyprus*, 1099–1325: Ashgate, 2006.

Nader, Marwan: "Urban Muslims, Latin Laws, and Legal Institutions in the Kingdom of Jerusalem", *Medieval Encounters*, 13, 2 2007: 243–270.

Najjar, Fauzi M.: "Siyasa in Islamic Political Philosophy" in: *Islamic theology and philosophy: studies in honor of George F. Hourani*, Edited by George F. Hourani and Michael E. Marmura, 92–110. Albany: State University of New York Press, 1984, 92–110.

Nielsen, Jørgen S.: *Secular Justice in an Islamic State: Maẓālim under the Baḥrī Mamlūks, 662/1264–789/1387*. Leiden: The Nederlands, Historisch-Archaeologisch Instituut te Istanbul, 1985.

Nussdorfer, Laurie: *Brokers of Public Trust: Notaries in Early Modern Rome*. Baltimore: Johns Hopkins University Press, 2009.

Orlando, Ermanno: "Venezia, il diritto pattizio e il commercio mediterraneo nel basso medioevo", *Reti Medievali Rivista*, 17 (1), 2016: 3–33.

Otten, Cathérine: "Quelques aspects de la justice à Famagouste pendant la période génoise" in: *Πρακτικά του Τρίτου Διεθνούς Κυπρολογικού Συνεδρίου, Λευκωσία, 16–20 Απριλίου 1996. 2, Μεσαιωνικό Τμήμα*, 333–351. Nicosia, 2001.

Otten, Cathérine: "Le registre de la Curia du capitaine Génois de Famagouste au Milieu du XV siècle: Une source pour l'étude d'une société multiculturelle" in: *Diplomatics in the Eastern Mediterranean 1000–1500: Aspects of Cross-Cultural Communication*, Edited by Alexander D. Beihammer, Maria G. Parani and Christopher D. Schabel, 251–274: Brill, 2008.

Otten, Cathérine: "Les droits du consul des Vénitiens à Famagouste au xve siècle" in: *Mélanges Cécile Morrisson*, 619–631. Paris: Collège de France—CNRS, 2010.

Oulddali, Ahmed: "Recevabilité du témoignage du ḏimmī d'après les juristes mālikites d'Afrique du Nord" in: *The legal status of ḏimmī-s in the Islamic West (second/eighth-ninth/fifteenth centuries)*, 275–292. Turnhout: Brepols, 2013.

Ozil, Ayse: *Orthodox Christians in the late Ottoman Empire: a Study of Communal Relations in Anatolia*, SOAS/Routledge studies on the Middle East. London; New York: Routledge, 2013.

Pedani, Maria: "Gli ultimi accordi tra i sultani mamelucchi d'Egitto e la repubblica di Venezia", *Quaderni di Studi Arabi*, 12, 1994: 49–64.

Pedani, Maria: *Veneta auctoritate notarius: storia del notariato veneziano: 1514–1797*, Studi storici sul notariato italiano. Milano: Giuffrè, 1996.

Peirce, Leslie P.: *Morality tales: law and gender in the Ottoman court of Aintab*. Berkeley: University Of California Press, 2003.

Piergiovanni, V.: *Hinc publica fides. Il notaio e l'amministrazione della giustizia*. Milan: A. Giuffrè, 2006.

Popper, William: *Egypt and Syria under the Circassian sultans, 1382–1468 AD; systematic notes to Ibn Taghrî Birdî's chronicles of Egypt*, University of California Publications in Semitic Philology, vols. 15, 16, 24. Berkeley: University of California Press, 1955, 1957, 1963.

Posner, Ernst: "Archives in Medieval Islam", *The American Archivist*, 35, 3–4 1972: 291.

Power, Timothy C.: "Trade Cycles and Settlement Patterns in the Red Sea Region (c. AD 1050–1250)" in: *Navigated Spaces, Connected Places. Proceedings of Red Sea Project V. Held at the University of Exeter 16–19 September 2010.* (British Foundation for the Study of Arabia Monographs No. 12), Edited by D.A. Agius, J.P. Cooper, A. Trakadas and C. Zazzaro, 137–45. Oxford: Archaeopress, 2012.

Pratesi, Alessandro: *Genesi e forme del documento medievale.* Roma: Jouvence, 1979.

Prawer, Joshua: "Social Classes in the Crusader States: The "Minorities"" in: *A History of the Crusades: The Impact of the Crusades on the Near East,* Edited by K. M. Setton, N. P. Zacour and H. W. Hazard, V, 59–115: University of Wisconsin Press, 1985.

Rapoport, Yossef: "Legal Diversity in the Age of Taqlid: the Four Chief Qadis Under the Mamluks", *Islamic Law and Society,* 10, 2 2003: 210–28.

Rapoport, Yossef: "Royal Justice and Religious Law: Siyāsa and Shariʿah under the Mamluks", *Mamluk Studies Review,* XVI 2012: 71–102.

Reimer, Michael J.: "Ottoman Alexandria: The Paradox of Decline and the Reconfiguration of Power in Eighteenth-Century Arab Provinces", *Journal of the Economic and Social History of the Orient,* 37, 2 1994: 107–146.

Richard, Jean: "La diplomatique royale dans les royaumes d'Arménie et de Chypre (XIIe-XVe siècles)", *Bibliothèque de l'école des chartes,* 1986: 69–86.

Richard, Jean: "La cour des Syriens de Famagouste d'après un texte de 1448" in: *Croisades et États latins d'Orient: points de vue et documents,* 383–398: Aldershot, 1992.

Richard, Jean: "Le plurilinguisme dans les actes de l'Orient latin" in: La langue des actes. Actes du XIe Congrès international de diplomatique (Troyes, jeudi 11-samedi 13 septembre 2003). Paris, Guyot Jeannin, Olivier, 2004.

Richard, Jean: "Aspects du Notariat public à Chypre sous les Lusignan" in: *Diplomatics in the Eastern Mediterranean 1000–1500: Aspects of Cross-Cultural Communication,* edited by Alexander D. Beihammer, Maria G. Parani and Christopher D. Schabel: Brill, 2008.

Richards, D. S.: "A Damascus Scroll Relating to a Waqf for the Yūnusiyya", *Journal of the Royal Asiatic Society of Great Britain and Ireland,* 2 1990: 267–281.

Riley-Smith, Jonathan: "Government and the indigenous in the Latin Kingdom of Jerusalem" in: *Medieval frontiers: concepts and practices,* edited by David Abulafia and Nora Berend, 121–131. Aldershot, Burlington, VT: Ashgate, 2002.

Robert, Louis: "Jean Sauvaget", *Revue Historique,* 207, 1 1952: 173–184.

Robinson, Chase F.: *Islamic historiography.* Cambridge-New York: Cambridge University Press, 2003.

Robinson, Neal: "Clay" in: *Encyclopaedia of the Qur'ān,* Edited by General Editor: Jane Dammen McAuliffe, 1:339a: Brill, 2003.

Rothman, E. Natalie: *Brokering empire: trans-imperial subjects between Venice and Istanbul.* Ithaca, N.Y.: Cornell University Press, 2012.

Rovere, Antonella: "I "publici testes" e la prassi documentale genovese (secc. XII-XIII)", *Serta antiqua et mediaevalia,* I, 1997: 291–332.

Rustow, Marina: "A petition to a woman at the Fatimid court (413–414a.h./1022–23 c.e.)", *Bulletin of the School of Oriental and African Studies,* 73 01 2010: 1–27.
Sadan, Joseph: "Genizah and genizah-like practices in Islamic and Jewish traditions. Customs concerning the disposal of worn-out Sacred Books in the Middle Ages, according to an Ottoman source", *Bibliotheca orientalis,* 43, 1–2 1986: 36–58.
Saint-Priest, François-Emmanuel Guignard, and Charles Henri Auguste Schefer: *Mémoires sur l'ambassade de France en Turquie et sur le commerce des Français dans le Levant,* Publications de l'École des langues orientales vivantes. Paris,: E. Leroux, 1877.
Salati, M.: "Un documento di epoca mamelucca sul waqf di 'Izz al-Dîn Abû l-Makârim Hamza b. Zuhra al-Husaynî al-Ishâqî al-Halabî (ca. 707/1307)", *Annali di Ca' Foscari,* XXXIII, 3, 1994: 97–137.
Santschi, Elisabeth: "Contribution à l'étude de la communauté juive en Crète vénitienne au XIVe siècle, d'après des sources administratives et juridiques", *Studi Veneziani,* XV, 1973: 177–211.
Sauvaget, Jean: *Introduction à l'histoire de l'Orient musulman: éléments de bibliographie.* Paris, 1943.
Sauvaget, Jean: "Comment étudier l'histoire du monde arabe, p. 5", *Revue Africaine,* 90, 1946: 5.
Sauvaget, Jean: Introduction to the history of the Muslim East: a bibliographical guide, Center Near Eastern. Berkeley, 1965.
Schacht, Joseph: *An introduction to Islamic law.* Oxford Oxfordshire; New York: Clarendon Press, 1982.
Schacht, Joseph: " 'Ahd" in: *The Encyclopaedia of Islam: Second Edition,* I: 255. Leiden, Brill, 1986.
Schacht, Joseph: "Amān" in: *The Encyclopaedia of Islam: Second Edition,* I: 429–430. Leiden, Brill, 1986.
Sijpesteijn, Petra M.: "The archival mind in early Islamic Egypt: Two Arabic papyri.", In: *From al-Andalus to Khurasan,* edited by P. Sijpesteijn and L. Sundelin, Brill, 2007. 163–186.
Smail, Daniel Lord: "Notaries, Courts and the legal Culture of Late Medieval Marseille" in: *Urban and Rural Communities in Medieval France, Provence and Languedoc, 1000–1500,* Edited by Kathryn Reyerson and John Victor Drendel, 23–51. Leiden–Boston: Brill, 1998.
Smail, Daniel Lord: *The consumption of justice: emotions, publicity, and legal culture in Marseille,* 1264–1423, Ithaca: Cornell University Press, 2003.
Sonbol, Amira: "Women in Shari'ah courts: a historical and methodological discussion, (Fourteenth Annual Philip D. Reed Memorial Issue Islam and the Law: Contemporary Views and Perspectives)", *Fordham International Law Journal,* 27, 1 2003: 225–253.
Speros, Vryonis, Jr.: "The Byzantine Legacy and Ottoman Forms", *Dumbarton Oaks Papers,* Vol. 23/24, 1969/1970: 251–308.

Steensgaard, Niels: "Consuls and nations in the Levant from 1570 to 1650", *Scandinavian Economic History Review*, 15, 1–2 1967: 13–55.

Steinhoff, Mark Wayne: *Origins and Development of the Notariate at Ravenna (Sixth through Thirteenth Centuries)*, Thesis (Ph. D.), New York University: Ann Arbor, Mich.: University Microfilms International, 1976.

Stilt, Kristen: *Islamic law in action: authority, discretion, and everyday experiences in Mamluk Egypt*. Oxford-New York: Oxford University Press, 2012.

Suriano, Francesco: *Il trattato di Terra Santa e dell'Oriente*, edited by Girolamo Golubovich. Milano, Tipografia editrice Artigianelli, 1900.

Theunissen, Hans: *Ottoman-Venetian Diplomatics: The 'Ahd-Names: the historical background and the development of a category of political-commercial instruments together with an annotated edition of a corpus of relevant documents*. Vol. I, EJOS, 1998.

Theunissen, Hans: "Cairo Revisited (I): Four Documents Pertinent to the Ottoman-Venetian Treaty of 1517", *Electronic Journal of Oriental Studies*, 2, 1999.

Tillier, Mathieu: "Le statut et la conservation des archives judiciaires dans l'Orient abbasside (IIe/VIIIe-IVe/Xe siècle): un réexamen" in: *L'autorité de l'écrit au moyen âge: orient-occident: XXXIXe congrès de la SHMESP, Le Caire, 30 avril-5 mai 2008*, Publications de la Sorbonne, Paris, 2009, 263–276.

Tillier, Mathieu: "Qadis and the political use of the mazalim jurisdiction under the Abbasids" in: *Public Violence in Islamic Societies: Power, Discipline, and the Construction of the Public Sphere, 7th-19th centuries C.E*, Edited by Christian Robert Lange and Maribel Fierro, 42–67. Edinburgh: Edinburgh University Press, 2009.

Tillier, Mathieu: "Califes, émirs et cadis: le droit califal et l'articulation de l'autorité judiciaire à l'époque umayyade", *Bulletin d'Etudes Orientales*, 63, 2014: 147–190.

Tillier, Mathieu: "The Mazalim in Historiography" in: *The Oxford Handbook of Islamic Law*, edited by Anver M. Emon and Rumee Ahmed, 356–380. Oxford: Oxford University Press, 2015.

Troupeau, Gérard: "Metals and Minerals" in: *Encyclopaedia of the Qur'ān*, Edited by General Editor: Jane Dammen McAuliffe, 3:383a: Brill, 2003.

Turki, Abdelmagid: "Lisan al-Dīn Ibn al-Khatib (713–76/1313–74), juriste d'après son oeuvre inédite: *Muthla l-Tariqa fi dhamm al-wathiqa*", *Arabica* 16, 1969: 155–211 and 279–312.

Tyan, Émile: *Histoire de l'organisation judiciaire en pays d'Islam*. 2 vol, *Annales de l'Université de Lyon*. Paris: Librairie du Recueil Sirey (société anonyme), 1938.

Tyan, Émile: *Le notariat et le régime de la preuve par écrit dans la pratique du droit musulman*. Harissa: Imp. St. Paul, 1945.

Udovitch, A. L.: "Les échanges de marché dans l'Islam médiéval: théorie du droit et savoir local", *Studia Islamica*, 65 1987: 5–30.

Ursinus, Michael: *Grievance Administration (şikayet) in an Ottoman Province: the Kaymakam of Rumelia's 'Record book of complaints' of 1781–1783*, Royal Asiatic Society books. New York: Routledge, Curzon, 2004.

Valérian, Dominique: "La résolution des conflits dans les communautés européennes dans les ports du Maghreb médiéval, entre métropoles et pouvoir local", *Mélanges de l'École française de Rome, Moyen Âge*, 115/1 2003: 543–564.

Valérian, Dominique: "Les marchands latins dans les ports musulmans méditerranéens: une minorité confinée dans des espaces communautaires?", *Revue des mondes musulmans et de la Méditerranée*, 107–110, 2005: 437–458.

Valérian, Dominique: "Le recours à l'écrit dans les pratiques marchandes en contexte interculturel. Les contrats de commerce entre Chrétiens et Musulmans en Méditerranée" in: *L'autorité de l'écrit au moyen âge: orient-occident: XXXIXe congrès de la SHMESP, Le Caire, 30 avril-5 mai 2008*, 59–73: Publications de la Sorbonne, 2009.

Vallet, *Eric: L'Arabie marchande: Etat et commerce sous les sultans rasûlides du Yémen (626–858/1229–1454)*. Paris: Publications de la Sorbonne, 2011.

Van Berkel, Maaike: "Reconstructing Archival Practices in Abbasid Baghdad", *Journal of Abbasid Studies*, 1, 1 2014: 7–22.

Van den Boogert, Maurits H.: *The Capitulations and the Ottoman Legal System Qadis, Consuls, and Beratlıs in the 18th Century*. Leiden; Boston: Brill, 2005.

Van den Boogert, Maurits H.: "Legal Reflections on the 'Jurisprudential Shift Hypothesis' ", *Turcica* 41, 2009: 373–382.

Veinstein, Gilles: "Les Fondements Juridiques de la Diplomatie Ottomane en Europe", *Oriente Moderno*, 88, 2 2008: 509–522.

Veselý, Rudolf: "Die Hauptprobleme der Diplomatik arabischer Privaturkurden aus dem spätmittelalterlichen Ägypten", *Archiv Orientali*, XL, 40 1972: 312–43.

Vogel, Frank E.: "Siyasa" in: *Encyclopaedia of Islam, Second Edition*, IX:693b. Leiden: Brill, 1997.

Wakin, Jeanette A.: The function of documents in Islamic law: the chapters on sales from Ṭaḥāwī's Kitāb al-shurūṭ al-kabīr. Albany: State University of New York Press, 1972.

Wansbrough, John: "A Mamluk Commercial Treaty Concluded with the Republic of Florence, 894/1489" in: *Documents from Islamic Chanceries*, Edited by S. M. Stern, 39–79. Oxford: Bruno Cassirer 1965.

Wansbrough, John: "Venice and Florence in the Mamluk Commercial Privileges", *Bulletin of the School of Oriental and African Studies*, 28, 3 1965: 483–523.

Wansbrough, John: "The Safe-Conduct in Muslim Chancery Practice", *Bulletin of the School of Oriental and African Studies*, 34, 1 1971: 20–35.

Wheeler, Joseph Russell: *The sestiere of San Polo: a cross section of Venetian society in the second half of the fifteenth century. PhD thesis, University of Warwick*, 1995.

Wilkins, Charles L.: "Witnesses and Testimony in the Courts of Seventeenth-Century Ottoman Aleppo" in: *Lire et écrire l'histoire ottomane*, edited by Vanessa Guéno and Stefan Knost, 107–129. Beirut and Damascus: IFPO–Orient-Institut Beirut, 2015.

Windler, Christian: "Diplomatic History as a Field for Cultural Analysis: Muslim-Christian Relations in Tunis, 1700–1840", *The Historical Journal,* 44, 1 2001: 79–106.

Winter, Michael: "The judiciary of late Mamluk and early Ottoman Damascus: The administrative, social and cultural transformation of the system", *History and Society During the Mamluk Period (1250–1517),* 5, V&R unipress, Göttingen 2014: 193–220.

Wittmann, Richard: *Before Qadi and Grand Vizier: Intra-communal Dispute Resolution and Legal Transactions Among Christians and Jews in the Plural Society of Seventeenth Century Istanbul,* PhD Dissertation: Harvard University, 2008.

Wray, Shona Kelly: "Instruments of Concord: Making Peace and Settling Disputes through a Notary in the City and Contado of Late Medieval Bologna", *Journal of Social History,* 42, 3 2009: 733–760.

Wright, Diana Gilliland, and Pierre A. MacKay: "When the Serenissima and the Gran Turco Made Love: The Peace Treaty of 1478", *Studi Veneziani,* 53, 2007: 261–277.

Zaʿrūr, Ibrāhīm: "al-Quḍāh al-Andalusīyūn wa-al-Maghāribah fī Bilād al-Shām fī ʿAṣr al-Mamālīk", *Dirāsāt Tārīkhīyah,* 53–54, 1995: 59–79.

Zecevic, Selma: "Translating Ottoman Justice: Ragusan Dragomans As Interpreters of Ottoman Law", *Islamic Law and Society,* 21, 4 2014: 388–418.

Zomeño, Amalia: "From Private Collections to Archives; How Christians Kept Arabic Legal Documents in Granada", *Al-Qantara: Revista de Estudios Arabes,* 32, 2 2011: 461–479.

Zomeño, Amalia: "Notaries and Their Formulas: The Legacies from the University Library of Granada". dans *From al-Andalus to Khurasan. Documents from the Medieval Islamic World,* edited by P. Sijpesteijn and L. Sundelin, Leiden–Boston: E. J. Brill, 2007, 59–80.

Index

www.ingramcontent.com/pod-product-compliance
Ingram Content Group UK Ltd.
Pitfield, Milton Keynes, MK11 3LW, UK
UKHW022224051026
15624UKWH00002B/345